Controlled Environment Engineering

This integrated book introduces broad engineering concepts and relevant physiological principles and environmental parameters and demonstrates, in a step-by-step approach, how controlled environment engineering knowledge can be put in use in designing plant, animal, and aquatic bioenvironmental systems. It covers integrated energy and waste management approaches and clean energy technologies, including real-world case studies and complete design examples with state-of-the-art technologies and updated material globally.

- Provides exclusive material on plant, animal, and aquatic systems engineering and relevant waste management systems in a single book
- Introduces broad engineering concepts, relevant physiological principles, and environmental parameters
- Incorporates a design and integrated energy and waste management approach
- Demonstrates how controlled environment engineering knowledge can be put in use in designing plant, animal, and aquatic production facilities
- Includes calculations and controlled environment design examples including homework problems

This book is aimed at researchers and graduate students in agricultural, aquacultural, and biosystems engineering; control systems; aquaculture; and agriculture.

Controlled Environment Engineering

Plant, Animal, and Aquatic Systems

İlhami Yıldız

CRC Press
Taylor & Francis Group
Boca Raton London New York

CRC Press is an imprint of the
Taylor & Francis Group, an **informa** business

Designed cover image: İlhami Yıldız

First edition published 2026
by CRC Press
2385 NW Executive Center Drive, Suite 320, Boca Raton FL 33431

and by CRC Press
4 Park Square, Milton Park, Abingdon, Oxon, OX14 4RN

CRC Press is an imprint of Taylor & Francis Group, LLC

ISBN: 978-1-032-39340-7 (hbk)
ISBN: 978-1-032-39341-4 (pbk)
ISBN: 978-1-003-34929-7 (ebk)

DOI: 10.1201/9781003349297

Typeset in Times New Roman
by Apex CoVantage, LLC

Access the Support Material at: https://www.routledge.com/9781032393407

I would like to dedicate this book to Dr. Dennis P. Stombaugh and Dr. H. Erdal Özkan.

I first met Dr. Dennis Stombaugh in the old Agricultural Engineering Building when I took his Measurements and Controls course during my first quarter at The Ohio State University (Spring Quarter of 1986). His class was my first exposure to controlled environment engineering. I soon fell in love with the field. After I completed my M.Sc. in 1987, Dr. Stombaugh became my Ph.D. advisor. He was the architect of my professional development and has been a superb mentor and father figure since then. He has tirelessly and unconditionally provided ongoing support throughout my career. I have remained in touch with him over the years, and I wish him and his family the best of health.

My second dedication goes to Dr. Özkan (Erdal Ağam), Professor, Department of Food Agricultural and Biological Engineering at The Ohio State University. I first met Erdal Ağam in the old Agricultural Engineering Building in March 1986. I believe it was his first year at Ohio State, and Dr. Stombaugh was the one who introduced Erdal Ağam to me during a class break. He has been my Erdal Ağam (a.k.a. Erdal Baba), honorary advisor, and mentor since then, providing endless and unconditional logistical, technical, and intellectual support throughout my professional and personal life. I have been in touch with Erdal Ağam regularly and wish him and his family the best of health.

I respectfully dedicate this book to Drs. Özkan and Stombaugh as a small thank you for their support and trust in me.

I will always remember them both with great respect and gratitude.

İlhami Yıldız, Ph.D., MSME, MSAE, MSc.

Contents

Author Biography

İlhami Yıldız is a professor and controlled environment systems engineer and has expertise in energy, the environment, sustainability, and the relevant innovation areas covering environmental biotechnology, bioreactors, biofuels and industrial bioproducts production using waste streams, greenhouse engineering, aquatic systems engineering, complex systems modeling, heat pumps, combined heat and power generation, flue gas recovery and greenhouse gas mitigation, and in particular, hydronic heating, drying and hot water storage systems, as well as climate change mitigation and adaptation.

He has provided extensive services at different capacities and contributed to the training of highly qualified personnel at various institutions in Canada, China, Turkiye, the United States, and Central Asia. He has also served as a member of editorial boards for several international journals, and review panels for a number of research granting agencies and institutions internationally, including the U.S. Secretary of Energy's review panel. He has published numerous books and book chapters, peer-reviewed articles in scholarly journals, and conference proceedings; organized several conferences and symposia; and delivered multiple keynote speeches and invited lectures internationally at conferences, symposia and institutions. He has also designed and built numerous environmentally friendly commercial energy and controlled environment systems, transferring extensive research findings to commercial settings in Canada, Mexico, Russia, Turkiye, and the United States. He is a recipient of a number of international honors and awards and a firm human rights defender for all, everywhere, and all the time.

Overview: President—Istinye University, Istanbul, Turkiye; Professor—Mechanical Engineering, Istinye University; Professor—International Agriculture University, Tashkent, Uzbekistan, in partnership with the Royal Agricultural University, Cirencester, Gloucestershire, United Kingdom; Adjunct (Retired) Professor—Interdisciplinary PhD Program, Dalhousie University, Halifax, Canada; Professor—Engineering, Dalhousie University; Director—Bioenvironmental Engineering Centre, Dalhousie University; Professor, Director, and Department Chair—Engineering Programs, Dalhousie University—Truro Campus; Associate Professor—BioResource & Agricultural Engineering, Cal Poly—San Luis Obispo, California, USA; Assistant Professor—Earth & Environmental Sciences, University of Windsor, Ontario, Canada; PDF in Biological & Environmental Engineering, Cornell University, Ithaca, New York, USA; Ph.D. in Food, Agricultural and Biological Engineering, M.Sc. in Mechanical Engineering, and M.Sc. in Applied Micrometeorology & Microclimatology, The Ohio State University, Columbus, Ohio, USA; B.Sc. & M.Sc. in Agricultural Engineering, University of Ankara, Turkiye. He has been honored with the CFAES 2022 International Alumni Award at The Ohio State University; the Grand Challenge Award by the U.S. Government (2008); Outstanding Faculty Award at Cal Poly (2008), San Luis Obispo, California; Presidential Letter of Appreciation at Cal Poly (2008); Gold Medal Performance Award by the University of Windsor (2004), Windsor, Ontario, Canada; Postdoctoral Fellowship (1997) by the OECD Co-operative Research Programme; Graduate Fellowship (1985), Ministry of Education, Turkiye; and Graduate Fellowship (1984) by the Scientific and Technological Research Council of Turkiye.

Preface

Establishing and sustaining civilization under cover . . .

İlhami Yıldız
Montreal, QC, Canada
Istanbul, Turkiye
iyildiz@dal.ca

Acknowledgments

The contributions of my former professors, colleagues, students, project partners, funding agencies, and producers in many different forms throughout my 40+ years of active professional service are greatly acknowledged. In addition, the following individuals and organizations kindly granted permissions and provided much practical support and assistance. Their contributions are also greatly acknowledged. In alphabetical order, they are:

ASABE, Standard EP270.5 DEC94, used with permission

ASABE, Standard EP406.2 OCT95, used with permission

ASABE, Standard EP344.3 JAN2005, used with permission

MWPS, multiple tables and figures from Structures and Environment Handbook (1987), used with permission

Taylor & Francis Group, multiple tables and figures from *Greenhouse Engineering: Integrated Energy Management*

Mr. Balaji Karuppanan, Lead—Project Management, Apex CoVantage, LLC, and his team members

Dr. Gagandeep Singh, Senior Publisher (STEM), CRC Press Taylor & Francis Group, and his team members Ms. Amrita Sood and Ms. Rishita Mondal

1 Introduction

The environment is defined as the total of all external conditions surrounding plants and animals, which comprise physical, chemical, and biological factors; no plant or animal exists or can exist independently from an environment. Parameters of an environment can be classified under two major groups: thermal and nonthermal. Thermal environmental parameters include radiation, temperature, humidity (and precipitation), and air movement, while nonthermal parameters are light, water (and soil or growth medium), air composition, altitude, and biotic environment. Modification of these environmental parameters provides tremendous engineering application opportunities in plant and animal biosystems.

Physiology, meanwhile, is defined as the science of normal function that can be investigated at the cellular level, the functioning of discrete organs, and the functioning of whole plant or animal. Green plants are photosynthetic autotrophs that store energy from sunlight and synthesize matter from inorganic raw materials. Farm animals, on the other hand, are heterotrophs and receive their energy and matter from the food they consume directly or indirectly from plants, other animals, microorganisms or combination of these sources, but their food always comes from their environment.

Engineering is defined as putting scientific knowledge into practical use. An engineer is not the same as a scientist: An engineer has a background in mathematics and science, as would a chemist or physicist, but applies these basic principles to practical problems; the keyword for engineering is synthesis, which literally means putting things together.

Within the context of controlled environment engineering, probably one of the best questions one can ask is how environment, physiology, and engineering get along; the simplest answer for this question is that they get along quite well. From the engineering point of view, the environment provides the initial and boundary conditions in both plant and animal biosystems, and an engineer is concerned with i) how much food, feed, or fiber plants or animals produce in specific environments; ii) if the environment is not ideal, what needs to be changed and how much; iii) whether the cost of changing the environment will be offset by increased output; and iv) the social and environmental impacts of the changes if any.

To answer these questions, engineers need to know plants' and animals' physiological responses to their environments and changes in those environments, although it is only important for them to understand the basic physiology of animals and plants. Having a basic understanding of physiological principles and thorough knowledge of the environmental factors, engineers then turn to scientists for collaboration in areas closely related to plant or animal physiology, such as a plant scientist, an animal scientist, or a veterinarian in order that best possible engineering approach can be taken to modify (design) an environment for economically, environmentally, and socially justified profitable plant or animal welfare, growth, production, and reproduction.

After setting the stage for controlled environment engineering, we can now turn to a more important challenge we are facing today on our increasingly populated planet, that is, how to live within such that its resources (i.e., the four sacred elements: earth, air, water, and fire) can sustain human civilization and the natural environment in a healthy and functional state for many generations in the foreseeable future. How can we continue to improve human welfare and ensure that food, energy, and environmental justice are reasonably realized within the limits of our planet's resources?

At base, the question is "What is the maximum number of people our planet can sustain?" while sustaining the natural environment, its species, and its systems in a healthy and functional state. We are using our renewable and nonrenewable environmental resources faster than they can be replenished. We are extracting minerals, oil, and ground water without sufficient concern for their limits or the need to recycle them. As a result, there are already shortages of some resources, and there will be more shortages of other resources in the future. Therefore, we must learn how to sustain our resources so that we can continue to fuel human civilization and other living things locally and globally on our planet within natural limits for many generations.

Sustainability means leaving our children and future generations a planet at least as rich as we live in today, and requires maintaining a fully functional and healthy natural environment, which forms the foundation for human civilization. Due to the population increase, however, sustainability is a moving target rather than a fixed one; more resources will be required to maintain a healthy and functional status. Our choices, up to a certain extent, determine the local and global sustainable carrying capacity for the human population of our planet. For instance, more people mean more food, energy, and freshwater demands, and people living in less developed parts of the world today will be consuming more food, energy, and fresh water per capita when they reach higher standards of living or if and when we can actualize food, energy, and environmental justice.

The projections already provide warnings about declining freshwater availability and increased drought risk due to climate-driven impacts and water pollution in many regions around the globe. Approximately 70% of freshwater resources are used for agri-food production (Pimentel et al., 2004), and available freshwater resources are becoming increasingly stressed as a result of increasing and competing demands, exacerbated by pollution and climate change-driven impacts (Goonetilleke and Vithanage, 2017). Even worse, according to the United Nations, it will be necessary to produce 60% more food globally and 100% more in developing countries by 2050 (UN, 2015). Similarly, about 9 billion people are projected to require affordable energy services by 2050 (IEA, 2012).

DOI: 10.1201/9781003349297-1

Consequently, reducing the vulnerability of agri-food production systems primarily to climate change and strengthening the adaptive capacity of such systems are extremely important for sustaining the moving target of improved agri-food production and food justice and, thereby, the livelihoods of billions of people around the globe (Yildiz, 2019). Moreover, reducing emissions due to fossil fuel consumption and production is vital for the well-being of the global population, and the agri-food industry has tremendous potential for reducing inefficiencies and emissions while playing its crucial roles in food and energy security and poverty reduction.

Climate change and other environmental stress functions exert site-specific stresses over agri-food industry; therefore, agri-food adaptations, innovations, and mitigation efforts need to be locally based responses to these forcing functions that will ultimately speed up agri-food innovations, generating significant technical, methodological, managerial, and institutional innovations particularly in the domains of AI, biotechnology, controlled environments, efficiency, safety, insurance, and policy (Yildiz, 2020). In return, and as part of a reciprocal relationship, agri-food innovations—complementing other forms of mitigation and adaptation to climate change and other forcing functions—will become one of the prominent societal responses to climate change and other stressors, accelerating overall mitigation and adaptation processes as well as preventing or facilitating population migrations (Yildiz, 2020).

Having the human capital and capability to innovate and apply technology, however, does not guarantee that a sustainable future for our plant will automatically be created; the biggest challenge for a sustainable future is still the human aspect, the *will* to comprehensively change our ways of living and thinking. The overall goal of sustainable agri-food and energy production in a sustainable environment requires and maintains multiple linkages among agri-food production and justice, energy production and justice, and environmental protection and justice. At this junction, food justice for over a billion people living in cold and hot food deserts and other extreme environments becomes vitally important, and for this, local sustainability is the key.

And exactly at this point, the author would like to strongly emphasize the concept of controlled environments. Adopting controlled environment technologies for agri-food production, whenever and wherever it is possible, would allow us to truly care for the aforementioned sacred elements within fully enriching cycles and provide more diverse, reliable, evenly distributed, and local agri-food and energy productions, encompassing efficiency and conservation measures and protecting our local and hence global environments. We can only then ensure that our future generations live in a healthy and functional environment with sufficient rich resources and eventually sustain our civilization.

A well-established and well-designed bioeconomy that is full of controlled environment innovations and commercialization opportunities would bring food, energy, and the environment together for economically, environmentally and socially sustainable operations and rural communities while the industry is adapting to a warmer and more variable climate in many geographic regions and numerous other forcing functions such as extreme environments, pandemics, security threats, trade wars, and tariffs. Every day, engineers and scholars are developing novel AI-based cutting-edge autonomous controlled environment technologies for site-specific agri-food production systems, such as agrochemicals, heating, cooling, ventilation, lighting, aeration, biofiltration, irrigation, harvesting, other plant/animal management technologies, and monitoring systems.

In summary, controlled environment technologies have tremendous potential to reduce the amounts of agrochemicals, energy, water, labor and other resources required for agri-food production, leading to extremely resource-efficient, cost-effective production systems with reduced or no environmental footprints and thereby sustaining and improving our agri-food operations and communities on our planet globally. Furthermore, fully autonomous, confined-space, controlled environment technologies that we are developing for the extreme environments of our planet today will soon help establish joint human, plant, and animal communities in space and sustain our novel societies and civilization beyond.

BIBLIOGRAPHY

Goonetilleke, A and Vithanage, M (2017) Water resources management: Innovation and challenges in a changing world. *Water*, 9: 281.

IEA (2012) *International Energy Agency, World Energy Outlook 2012*. OECD, Paris.

IPCC (2013) Climate change 2013: The physical science basis. Contribution of Working Group I to the Fifth Assessment Report of the Intergovernmental Panel on Climate Change, Cambridge University Press, Cambridge, New York.

Pimentel, D, Berger, B, Filiberto, D, Newton, M, Wolfe, B, Karabinakis, E, Clark, S, Poon, E, Abbett, E and Nandagopal, S (2004) Water resources: Agricultural and environmental issues. *BioScience*, 54: 909–918.

UN (2015) U.N. world water assessment program. Water for a Sustainable World: The United Nations World Water Development Report, UNESCO, Paris.

World Commission on Environment and Development (1987) *Our Common Future*. Oxford University Press, Oxford. ISBN 019282080X.

Yildiz, I (2018) Energy fundamentals volume (ed.). In: Dincer, I (ed.), *Comprehensive Energy Systems*. Elsevier, Oxford.

Yildiz, I (2019) Review of climate change issues: A forcing function perspective in agricultural and energy innovation. *International Journal of Energy Research*: 1–16.

Yildiz, I (2020) Innovation and commercialization dynamics in agri-food industry under contemporary forcing functions. *International Journal of Research, Innovation and Commercialization*, 3(1): 1–21.

Yildiz, I (2021) *Greenhouse Engineering: Integrated Energy Management*. CRC Press, Taylor and Francis Group, FL USA. ISBN: 978-1-4987-4313-6.

2 Fundamentals

2.1 INTRODUCTION

This chapter establishes the essential groundwork for this book. Therefore, it starts with covering the energy units and conversions. The base and derived quantities, relationships between quantities, and quantity equations will be discussed. The multiples and submultiples in SI units will be presented. Furthermore, conversions for work and energy measurements between SI and inch-pound (I-P) units will also be provided.

The chapter will then move into and briefly review thermodynamics, introducing the basic concepts of energy and mass and the laws of thermodynamics before getting into different modes of heat transfer for different geometries. The chapter will then be relating heat transfer analyses to different energy applications, for instance, heat transfer through walls or windows, which is essential for determining heating or cooling loads of buildings. Furthermore, analysis of heat exchangers for numerous applications (also common in controlled environment systems) will be presented as well.

The chapter then introduces the study of psychrometrics, which deals with thermodynamic properties of moist air. Using these properties, we will analyze engineering processes involving moist air, covering perfect gas relations and describing their use in thermal comfort problems in controlled environments. In addition to the heat transfer analyses, further refinements in heat transfer mechanisms will be covered to account for variations in surface and air space heat transfer resistances along with in-depth analyses of water vapor transfer, surface condensation process, and thermal insulation. Another subsection will cover fuels, combustion of fuels, theoretical and excess air requirements for combustion, flue gas and carbon emissions, and the costs of different heating fuels.

After that, as the piping systems for water, natural gas, oil, etc. are all designed using the principles of fluid mechanics, a dedicated subsection will cover fundamental fluid mechanics, which is essential for handling general heating, ventilating, and air-conditioning processes. In this subsection, chemical and physical properties of water, the affinity or similarity laws for pumps/fans, and essentials of gas transfer between aqueous and gaseous phases will briefly be reviewed as well.

Finally, as solar energy powers virtually everything on Earth and in the atmosphere and is vitally important for controlled environment operations, this chapter will have a fully dedicated section on light, radiation, characteristics of solar radiation, and relevant data. The solar spectrum and direct and diffuse components of solar radiation, as well as the temporal and spatial distributions of solar energy, will be studied in depth before the chapter is concluded. Many numerical examples are provided in every subsection of the chapter to demonstrate the relevant concepts for the readers to easily grasp them.

2.2 UNITS AND CONVERSIONS

2.2.1 Introduction

When dealing with engineering and scientific relationships, to appreciate the magnitudes of physical quantities, it is essential to have a solid grasp of units and to recognize two types of equations: quantity and numerical. Both types are found in texts and reference books, and the concept of units and quantities is useful in understanding their respective features. This chapter covers the main features of quantities and quantity equations and provides the most important units and conversions relating to energy. Quantity equations are also called equations between quantities, or physical equations, and numerical equations are alternatively called measure equations.

2.2.2 Quantities

In 1954, the 10th General Conference on Weights and Measures (CGPM) decided that an international system should be derived from six base units to provide for the measurement of temperature and optical radiation in addition to mechanical and electromagnetic quantities. The six units recommended at this conference were meter, kilogram, second, ampere, degree Kelvin (later renamed Kelvin), and candela. In 1960, the 11th CGPM named the system the International System of Units, abbreviated as SI from the French name, Le Système International d'Unités (CGMP, 1961). A seventh base unit, the mole, was added in 1971 by the 14th CGPM (CGMP, 1972). SI is the modern form of the metric system and is the most widely used measurement system today.

Therefore, the International System of Quantities (ISQ) is based on seven base quantities, length, mass, time, thermodynamic temperature, electric current, luminous intensity, and amount of substance. Other quantities such as area, pressure, and electrical resistance are all derived from these base quantities. The ISQ defines quantity as any physical property that can be measured with in SI (BIPM, 2012). A quantity may also be a physical constant, such as the gas constant or Planck's constant. Several hundred quantities are employed to describe and measure the physical world, including the following (Wildi, 1995):

length	viscosity	area	electromotive force
time	energy	luminance	entropy
mass	speed	angle	pressure
force	power	temperature	momentum

2.2.2.1 Relationships between Quantities

Physics is defined as the study of mathematical relationships among various physical properties. Physical quantities are defined, as noted, when these properties allow a reasonable mathematical description. The relationship of

DOI: 10.1201/9781003349297-2

all other quantities can be established in terms of a few base quantities properly selected either by definition, by geometry, by physical law, or by a combination of the base quantities. For instance, pressure is a quantity that is related by definition to a quantity force divided by a quantity area; area is a quantity related by geometry to the product of two quantities of length; and force is a quantity related (by Newton's Second Law) to the quantity mass times the quantity acceleration.

The relationships between quantities are expressed in the form of quantity equations. We can relate even an isolated quantity such as temperature to the quantities pressure, volume, and mass. We can further relate the quantities length and time by using the universal constant and the speed of light. Therefore, if we define our concepts correctly, we can relate any quantity to any other quantity. For instance, the equation area = length × width is a quantity equation that states that the quantity (area of a rectangle) is equal to the quantity (length) times the quantity (width).

2.2.2.2 Base Quantities

To reduce a set of quantity equations, we have to first establish a number of base quantities, the building blocks upon which we develop the entire structure and relationships of the physical world. As mentioned earlier, the international system of units, or SI, makes use of seven based quantities: mass (kg), length (m), time (s), temperature (K), electric current (A), luminous intensity (cd), and amount of substance (mol). The number of base quantities, as well as their choice, is quite arbitrary, but generally, we select quantities that are easy to understand and frequently used and for which accurate and measurable standards can be established.

2.2.2.3 Derived Quantities

As mentioned in the relationship section earlier, using the selected base quantities as building blocks, derived quantities are those which can be deducted by definition, geometry, or physical law. Some derived quantity examples are area (the product of two lengths), velocity (length/time), force (mass × acceleration), pressure, and power. There are also what are called supplementary units (as a class of derived units), namely the plane angle (radian = rad = m/m) and solid angle (steradian = sr = m^2/m^2).

2.2.2.4 Multiples and Submultiples of Quantities

Note that the magnitude of a quantity can have an extremely large range. In an effort to handle such a large range, the SI unit system generated 20 prefixes shown in Table A.1.

2.2.2.5 Types of Quantity Equations

The energy of wind, the pressure at the bottom of an air or water column, the weight of an object, and the viscosity of a liquid are all physical quantities of nature, and whether they are measured or not, these quantities are always there interacting with each other according to fundamental laws. Physicists often express these laws in terms of quantity equations because quantities conform to these laws. Quantity equations possess two important features; first, they show the relationship between quantities, and second, they can be used with any system of units.

There are three basic types of quantity equations:

1. Quantity equations developed from the laws of nature, for instance, Newton's Second Law of motion:

$$F = m \times a$$

where F is the force, m is the mass, and a is acceleration.

2. Quantity equations developed from geometry, for instance, area of a circle:

$$A = \pi r^2$$

where A is area, π is a mathematical coefficient based on the geometry of a circle, and r is the radius.

3. Quantity equations developed from a definition, for instance, definition of pressure:

$$p = F / A$$

where p is the pressure, F is the force, and A is the area.

Many quantity equations can be developed as a combination of these basic equations, and in all cases, we can use any units we want to describe the magnitudes of the relevant physical quantities.

2.2.3 Units and Conversions

This section, as modified after ASHRAE (1984) and (1989), references ASTM Standard E 380–84 Standard for Metric Practice (ASTM, n.d.) as one of the basic standards for SI usage (National Bureau of Standards, n.d; CSA, n.d.; ASME, n.d.; IOS, n.d.). Table A.2 provides conversion factors rounded to three or four significant figures for conversion between SI and I-P, and Table A.3 provides conversion factors for different physical quantities related to energy.

2.2.3.1 Useful Units in Electricity

2.2.3.1.1 Coulomb

In an electric circuit, the unit of electric charge in SI is coulomb (C). An ampere (A) is one coulomb of electric charge transported through any cross-section of a conductor per second and is equivalent to the amount of charge on about 6,241,510,000,000,000,000 electrons.

2.2.3.1.2 Volt

In an electric circuit, the unit of electric potential, potential difference, and electromotive force in SI is volt (V). If we consider our house wiring as plumbing, volts can then be considered a measure of the water pressure. One volt is the potential difference between two points on a conductor when the current flowing is one ampere and the power dissipated between the points is one watt. The volt is a derived unit, and in terms of base units, it can be expressed as follows:

$$\text{Volt} = \text{Watt} / \text{Ampere} = \left(\text{m}^2.\text{kg}\right) / \left(\text{s}^3.\text{A}\right)$$

2.2.3.1.3 *Watt*

In an electric circuit, one watt (joules per second) is a current of one ampere at a pressure of one volt. In terms of base units,

$$\text{Watt} = \text{J} / \text{s} = \left(\text{m}^2.\text{kg}\right) / \text{s}^3$$

2.2.3.1.4 *Ohm*

In an electric circuit, the unit of electrical resistance (a derived unit) in SI is the ohm (Ω). One ohm is defined as the electrical resistance between two points on a conductor when a constant potential difference of one volt applied to these points produces in the conductor a current of one ampere. The ohm is a derived unit, and in terms of base units it can be expressed as follows:

$$\text{Ohm} = \text{Volt} / \text{Ampere} = \left(\text{m}^2.\text{kg}\right) / \left(\text{s}^3\text{A}^2\right)$$

2.2.4 Rules for Using SI Units

2.2.4.1 Capitalization

The names of units start with a lowercase letter when writing the units out except for in a title or the beginning of a sentence. The only exception is degree Celsius. Unless they come from an individual's name (in which case the first letter of the symbol is capitalized), lowercase is used in writing symbols for units. The only exception is L for liter. Symbols for numerical prefixes (multiples and submultiples) are also lowercase, except for those representing multipliers of 10^6 or more, for instance, mega (M), giga (G), tera (T), peta (P), exa (E), zetta (Z), and yotta (Y). It means that all prefixes are written in lowercase when spelled out.

Lowercase units: m, kg, s, mol, etc.
Uppercase units: A, K, Hz, Pa, C, etc.

Symbols rather than self-styled abbreviations should always be used to represent units.

Correct usage: A, s **Incorrect usage:** amp sec

2.2.4.2 Use of Plurals

Remember that symbols are never expressed as plural; that is, they never take an "s". However, when the names of units are spelled out, they can take a plural inflection; fractions, on the other hand, are always written as singular. Plurals are used as required when writing unit names. For example, henries is plural for henry. The following exceptions are noted:

Singular: lux, Hertz, Siemens
Plural: lux, Hertz, Siemens

Example 2.1: Correct and Incorrect Usages of Units

Correct Usage	Incorrect Usage
5 kg	5 kgs
5 kilograms	5 kilogram
5.57 kg	–
5.57 kilograms	5.57 kilogram
0.57 kilogram	0.57 kilograms

2.2.4.3 Use of Hyphens and Spaces

Also remember that a hyphen or a space is not used to separate a prefix from the name of the unit. A space, however, is left between a symbol and the number to which it refers, with the exception of the symbols for degree, minute, and second of angles and for degree Celsius. In three cases, the final vowel in the prefix is omitted: megohm, kilohm, and hectare.

Example 2.2: Correct and Incorrect Usages of Units

Correct Usage	Incorrect Usage
5 kg	5kg
40° 45' 30"	40° 45' 30"
30°C	30 °C
5 km	5km
MJ	M J
5 s	5s
5 milliseconds	5 milli-seconds

2.2.4.4 Use of Numerals and Periods

Remember that scientific and technical writing is different from newspaper, magazine, and other writings. In scientific and technical writing, numerals are used for all numbers expressing physical quantities; however, it is a common practice to write out the numbers from one to nine and use numerals for other numbers in newspapers. In ordinary books and magazines, for instance, whole numbers from one through ninety-nine, and any of these followed by "hundred," "thousand," "million," "billion," etc., are spelled out. Also, keep in mind that the associated number is written as numerals when the unit is represented by an abbreviation or symbol.

Periods are never used after SI symbols unless the symbol is at the end of a sentence.

Example 2.3: Correct and Incorrect Usages of Units

Correct Usage	Incorrect Usage
J/kg	
J kg^{-1}	joules.kg^{-1}
joules per kilogram	joules/kilogram
N m	newton.meter
newton meter	
newton-meter	

2.2.4.5 Use of Symbols for Mathematical Operations

Units are represented by symbols, not by their spelled-out names when the units (SI) are used with symbols for mathematical operations.

Notes to remember:

1. When writing unit names as a product, always use a space (preferred) or a hyphen.
   Correct usage: newton meter or newton-meter
2. When expressing a quotient using unit names, always use the word per and not a solidus (/). The solidus or slush mark is reserved for use with symbols.

   Correct usage: meter per second
   Incorrect usage: meter/second
3. When writing a unit name that requires a power, use a modifier, such as squared or cubed, after the unit name. For area or volume, the modifier can be placed before the unit name.
   Correct usage: millimeter squared or square millimeter
4. When denoting a quotient by unit symbols, any of the following are accepted form:

   Correct usage: m/s or m.s–1

In more complicated cases, consider using negative powers or parentheses. For acceleration, use m/s^2 or $m.s^{-2}$ but not m/s/s. For electrical potential, use $kg.m^2/(s^3.A)$ or $kg.m^2.s^{-3}.A^{-1}$ but not $kg.m^2/s^3/A$.

2.2.5 Overall Examples

Example 2.4: Area

Show the unit of area in SI.

Solution

Area equations are quantity equations arising from geometry; for example, the area equation for a pipe is expressed as follows:

$$\text{Area}(A) = \pi r^2 = \pi(\text{m})^2$$

$$\text{Area}(A) = \text{m}^2$$

where A is the area in m^2, π is 3.14 (dimensionless), and r is the radius in m.

In another example, the area for a rectangle is expressed as follows:

$$\text{Area}(A) = w \times l = \text{m} \times \text{m}$$

$$\text{Area}(A) = \text{m}^2$$

where A is the area in m^2, w is the width in m, and l is the length in m.

Example 2.5: Volume

Show the unit of volume in SI.

Solution

Volume equations are quantity equations arising from geometry; for example, the volume equation for a pipe is expressed as follows:

$$\text{Volume}(V) = \pi r^2 l$$

$$\text{Volume}(V) = \pi\ (\text{m}^2)(\text{m})$$

$$\text{Volume}(V) = \text{m}^3$$

where V is the volume in m^3, π is 3.14 (dimensionless), and r is the radius in m.

The volume for a rectangular cross-section is expressed as follows:

$$\text{Volume}(V) = w \times l \times h = \text{m} \times \text{m} \times \text{m}$$

$$\text{Volume}(V) = \text{m}^3$$

where V is the volume in m^3, w is the cross-sectional width in m, l is the cross-sectional length in m, and h is the height.

Example 2.6: Volume

Determine the unit of volume (m^3) in SI for a given volume in I-P.

Solution

The volume unit in I-P is ft^3 and remember that 1 ft = 0.3048 m:

$$\text{Volume}(V) = \text{ft}^3\left(0.3048\frac{\text{m}}{\text{ft}}\right)^3$$

$$\text{Volume}(V) = 0.028317\text{m}^3$$

Example 2.7: Mass

Determine the unit of mass (kg) in SI for a given mass in I-P.

Solution

The mass unit in I-P is lb_m and remember that 1 lb_m = 0.45359 kg:

$$\text{Mass}(m) = \text{lb}_\text{m}\left(0.45359\frac{\text{kg}}{\text{lb}_\text{m}}\right)$$

$$\text{Mass}(m) = 0.45359\text{kg}$$

Example 2.8: Force

Show that the unit of force is Newton (N) in SI.

Solution

The unit of force in SI is the force that when applied to a mass of one kilogram produces an acceleration of one meter per second per second. Newton's second law of motion, a quantity equation established from the laws of nature, is expressed as

$$\text{Force}(F) = \text{mass} \times \text{acceleration} = m \times a$$

$$\text{Force}(F) = (\text{kg})(\text{m/s}^2) = \text{kg.m/s}^2$$

$$\text{Force}(F) = \text{N}$$

where m is the mass in kg, a is the acceleration in m/s², and F is the force in N and dependent on the mass and the acceleration.

Example 2.9: Force

Show that the unit of force is N in SI for a given force in I-P.

Solution

The force unit in I-P is lb_f, and remember that 1 lb_m = 0.45359 kg, 1 ft = 0.3048 m, and gravitational acceleration g is 32.174 lb_m/s^2:

$$\text{Force}(F) = 1\,\text{lb}_f = (1\,\text{lb}_m)(32.174\,\text{ft/s}^2) = 32.174\,\text{lb}_m.\text{ft/s}^2$$

$$\text{Force}(F) = (32.174\,\text{lb}_m.\text{ft/s}^2)(0.45359\,\text{kg/lb}_m)(0.3048\,\text{m/ft})$$

$$\text{Force}(F) = 4.45\,\text{kg.m/s}^2 = 4.45\,\text{N}$$

where g is the gravitational acceleration and F is the force.

Note that in I-P, acceleration of 9.80665 m/s² corresponds exactly to 32.174048 ft/s² as shown:

$$g = 9.80665\,\text{m/s}^2 \frac{\left(\dfrac{100\,\text{cm}}{1\,\text{m}}\right)}{\left(\dfrac{12\,\text{in}}{1\,\text{ft}}\right)\left(\dfrac{2.54\,\text{cm}}{1\,\text{in}}\right)} = 32.174084\,\text{ft/s}^2$$

Example 2.10: Pressure

Show the unit of pressure (Pascal, Pa) in SI.

Solution

In solids, we deal with stresses; in liquids and gasses, however, we deal with pressure, which is defined as the normal component of force per unit area. Therefore, the unit for pressure, Pa, is a derived unit. One Pascal is the pressure resulting from a force of one Newton acting uniformly over an area of one square meter. Therefore, the pressure equation is a quantity equation established from a definition and is expressed as follows:

$$\text{Pressure}(p) = \text{force / area} = F/A = mg/A$$

$$\text{Pressure}(p) = (\text{kg})(\text{m/s}^2)/\text{m}^2 = (\text{kg.m/s}^2)/\text{m}^2 = \text{N/m}^2$$

$$\text{Pressure}(p) = \text{Pa}$$

where F is the force (the mass times the acceleration) in N and A is the area in m². The pressure is dependent on the mass, the gravitational acceleration, and the area.

Example 2.11: Pressure

Show that the unit of pressure is Pa in SI for a given force in I-P.

Solution

The pressure unit in I-P is psi (lb_f/in^2); 1 lb_f = 4.45 kg m/s² = 4.45 N and 1 in = 0.0254 m:

$$\text{Pressure}(p) = \text{force / area} = F/A$$

$$= \frac{\left[\dfrac{(1\,\text{lb}_f)\left(4.45\,\text{kg}\dfrac{\text{m}}{\text{s}^2}\right)}{1\,\text{lb}_f}\right]}{\{[\text{in}^2(6.452\times10^{-4}\,\text{m}^2/\text{in}^2]\}}$$

$$= 6{,}894.8(\text{kg})(\text{m/s}^2)/\text{m}^2$$

$$= 6{,}894.8\,\text{Pa} = 6.89\,\text{kPa}$$

where F is the force and A is the area.

Example 2.12: Work

Show that the unit of work is joule (J) in SI.

Solution

The unit of work or energy in SI is joule, which has the symbol J. It represents the work done when the point of application of a force of one Newton is displaced one meter in the direction of the force. One watt-second is equal to 1 J. The work equation is then a quantity equation established from a definition and is expressed as follows:

$$\text{Work} = \text{force} \times \text{distance}$$

$$= (\text{mass} \times \text{acceleration}) \times \text{distance}$$

$$W = ma\Delta x$$

$$= (\text{kg})(\text{m/s}^2)(\text{m})$$

$$= \left(\text{kg}\frac{\text{m}}{\text{s}^2}\right)(\text{m})$$

$$\text{Work} = \text{N.m} = \text{J}$$

where W is the work in J, m is the mass in kg, a is the acceleration in m/s², and Δx is the distance traveled in m. The work is dependent on the mass, the acceleration, and the distance traveled, for instance.

Example 2.13: Energy

Show that the unit of energy is J in SI.

Solution

Energy is defined as the ability to perform work, and as expressed earlier, the unit of energy in SI is joule (J). The energy equation is a quantity equation established from a definition and expressed as follows:

$$\text{Energy} = \text{work} = \text{force} \times \text{distance}$$

$$= (\text{mass} \times \text{acceleration}) \times \text{distance}$$

$$E = m\,a\,\Delta x$$

$$(\text{kg})(\text{m/s}^2)(\text{m}) = (\text{kg.m/s}^2)(\text{m})$$

$$\text{N.m} = \text{J}$$

where m is the mass in kg, a is the acceleration in m/s², and Δx is the distance traveled in m. In this case, the energy is dependent on the mass, the acceleration, and the distance traveled.

Example 2.14: Power

Show that the unit of power is watt (W) in SI.

Solution

The unit of power in SI is the watt and has the symbol W. Power is the rate at which energy is expended or work done. The watt in thermodynamics is defined as "the power which in one second gives rise to energy of one joule." In mechanical terms, however, a power of one watt can move a mass of one kilogram in one second, through a distance of one meter with such force that the kilogram mass' velocity at the end of the meter will be one meter per second greater than it was at the beginning. In an electric circuit, on the other hand, one watt is a current of one ampere at a pressure of one volt. Therefore, the power equation is a quantity equation established from a definition and is expressed as follows:

$$\text{Power} = \frac{\text{work}}{\text{time}} = \frac{\text{energy generation}}{\text{time}} = \frac{\text{energy consumption}}{\text{time}}$$

Since the unit of work or energy is J, then

$$\text{Power}(P) = \frac{\text{J}}{\text{s}} = \text{W} = \frac{E}{t} = \frac{\text{energy}}{\text{time}} = ma\Delta x/t$$

where t is the time in seconds. In this case, the power is dependent on the mass, the acceleration, the distance traveled, and the time taken to travel.

Example 2.15: Volt

Express the unit of volt in terms of base units.

Solution

As mentioned earlier, in an electric circuit, the unit of electric potential, potential difference, and electromotive force in SI is the volt and has the symbol V. One volt is the potential difference between two points on a conducting wire carrying a constant current of one ampere, and the power dissipated between the points is one watt.

The volt is a derived unit, and in terms of base units, it can be expressed as follows:

$$E = \frac{P}{I} = \text{electromotive force} = \frac{\text{power}}{\text{current}}$$

$$\text{Volt} = \frac{\text{Watt}}{\text{Ampere}} = \text{V} = \frac{\text{W}}{\text{A}}$$

$$= \frac{\text{J/s}}{\text{A}}$$

$$= \frac{\text{J}}{\text{C}}$$

$$= \frac{\text{Nm}}{\text{C}}$$

$$= \text{A}\Omega$$

$$= \text{A}\left[\left(\text{kg.m}^2\right)/\left(\text{s}^3.\text{A}^2\right)\right]$$

$$\text{Volt} = \left(\text{kg.m}^2\right)/\left(\text{A.s}^3\right)$$

Example 2.16: Radiant Flux Density

Determine how many units of radiant flux density I (W/m²) in SI are in a given amount of Btu/h.ft².

Solution

Radiant flux density is the amount of energy received on a unit surface in unit time.

$$\text{Radiant flux density}\left(\frac{\text{W}}{\text{m}^2}\right) = \left(\frac{\text{Btu}}{\text{h.ft}^2}\right)\left(1{,}054.35\frac{\text{J}}{\text{Btu}}\right)\left(\frac{\text{h}}{3{,}600\ \text{s}}\right)\left(\frac{\text{ft}}{0.3048\ \text{m}}\right)^2$$

$$\text{Radiant flux density}\left(\frac{\text{W}}{\text{m}^2}\right) = 3.15\frac{\text{W}}{\text{m}^2}$$

Example 2.17: Boiler Horsepower

Determine how many kilowatts of power in SI are in one boiler horsepower (bhp) in I-P.

Solution

Remember that 1 bhp is the energy rate needed to evaporate 34.5 lb_m of water at 212°F (100°C) in one hour; therefore, it is equal to 33,475 Btu/h. A boiler horsepower is approximately 13 times larger than mechanical horsepower (engine output).

Also remember that 1 Btu = 1,054.35 J and 1 h = 3,600 s; therefore,

$$\text{Boiler horsepower} = 1\text{bhp}\left[\frac{33{,}475\frac{\text{Btu}}{\text{h}}}{\text{bhp}}\right]\left(1{,}054.35\frac{\text{J}}{\text{Btu}}\right)\left(\frac{\text{h}}{3{,}600\ \text{s}}\right)$$

$$\text{Boiler horsepower} = 9{,}804\ \text{W} = 9.8\ \text{kW}$$

Example 2.18: Horsepower

Determine how many kilowatts of power in SI are in one horsepower (hp) in I-P.

Solution

Horsepower is a unit in I-P that is sometimes used to express the rate at which mechanical energy is expended. It was originally defined as 550 foot-pounds per second (ft-lb_f/s).

Remember that 1 lb_f = 4.45 kg.m/s² = 4.45 N and 1 ft = 0.3048 m:

$$\text{Horsepower} = \left(550\text{ft.lb}_\text{f}/\text{s}\right)\left[\left(\frac{0.3048\ \text{m}}{\text{ft}}\right)\left(4.45\text{kg}\frac{\text{m}}{\text{s}^2}\right)/\text{lb}_\text{f}\right]$$

$$\text{Horsepower} = 746\,\text{W} = 0.746\text{kW}$$

Example 2.19: Refrigeration Ton

Determine how many Btu/h and kWh of refrigeration is provided by one ton of refrigeration.

Solution

Remember that 1 Btu = 1,054.35 J, 1 h = 3,600 s, 1 lb_m ice = 1 lb_m of liquid water, latent heat of fusion = 144 Btu/lb_m, and 1 short ton = 2,000 lb_m.

Refrigeration is commonly expressed in tons. One ton of refrigeration is the latent heat of fusion needed to melt one short ton (2,000 lb_m) of ice in 24 hours. Therefore,

$$1 \text{ refrigeration ton} = \frac{\left(2{,}000 \text{ lb}_m\text{ice} \times 144 \dfrac{\text{Btu}}{\text{lb}_m}\right)}{24 \text{ h}}$$

$$1 \text{ refrigeration ton} = 12{,}000 \text{ Btu/h}$$

$$1 \text{ refrigeration ton} = \left(12{,}000 \frac{\text{Btu}}{\text{h}}\right)\left(1{,}0.54.35 \frac{\text{J}}{\text{Btu}}\right)\left(\frac{\text{h}}{3{,}600 \text{ s}}\right)\left(\frac{\text{kJ}}{1{,}000 \text{ J}}\right)$$

$$1 \text{ refrigeration ton} = 3.52 \text{ kW}$$

Example 2.20: Energy

Show how many MJ of energy 1 kWh is equal to.

Solution

Remember that 1 MJ is equal to 10^6 J of energy, and 1 kWh is equal to 1,000 J/s. Therefore,

$$1 \text{ kWh} = \left(1{,}000 \frac{\text{J}}{\text{s}}\right)(1 \text{ h})\left(60 \frac{\text{min}}{\text{h}}\right)\left(60 \frac{\text{s}}{\text{min}}\right) = 3{,}600{,}000 \text{ J}$$

$$1 \text{ kWh} = (3{,}600{,}000 \text{ J}) / \left(1{,}000{,}000 \frac{\text{J}}{\text{MJ}}\right)$$

$$1 \text{ kWh} = 3.6 \text{ MJ}$$

Example 2.21: Energy

Show how many MJ and kWh of energy 1 therm (US) has.

Solution

The therm is a unit of heat energy equal to 100,000 Btu units. It is the energy equivalent of burning approximately 100 ft^3 (2.83 m^3) of natural gas. Natural gas meters measure volume rather than energy content; therefore, natural gas companies use a therm factor to convert the volume of gas used to its heat equivalent and thence calculate the actual energy use. Please remember that natural gas with a higher-than-average concentration of butane, ethane, or propane has a higher therm factor, while impurities lower the therm factor.

$$1 \text{ therm} = (100{,}000 \text{ Btu})\left(1{,}054.35 \frac{\text{J}}{\text{Btu}}\right)\left(\frac{\text{MJ}}{10^6 \text{ J}}\right) = 105.5 \text{ MJ}$$

$$1 \text{ therm} = (105.5 \text{ MJ})\left(\frac{\text{kWh}}{3.6 \text{ MJ}}\right) = 29.3 \text{ kWh}$$

Example 2.22: Potential Energy

Show that the unit of potential energy is J in SI.

Solution

$$\text{Potential energy} = \text{force} \times \text{elevation}$$

$$\text{Potential Energy} = (\text{mass} \times \text{acceleration}) \times \text{elevation}$$

$$E_{pe} = mgh$$

$$E_{pe} = (\text{kg})\left(\frac{\text{m}}{\text{s}^2}\right)(\text{m}) = \left(\text{kg}\frac{\text{m}}{\text{s}^2}\right)\text{m}$$

$$E_{pe} = \text{N.m} = \text{J}$$

where m is the mass in kg, g is the gravitational acceleration in m/s^2, and h is the elevation from a datum in m.

Example 2.23: Kinetic Energy

Show that the unit of kinetic energy is J in SI.

Solution

$$\text{Kinetic energy} = 0.5 \times \text{mass} \times (\text{speed})^2$$

$$E_{ke} = 0.5 \times m \times u^2$$

$$E_{ke} = \text{kg}\left(\frac{\text{m}}{\text{s}}\right)^2 = \left(\text{kg}\frac{\text{m}}{\text{s}^2}\right).\text{m}$$

$$E_{ke} = \text{N.m} = \text{J}$$

where m is the mass in kg and u is the speed in m/s.

Example 2.24: Pressure Energy

Show that the unit of pressure energy is J in SI.

Solution

$$\text{Pressure energy} = \frac{\text{mass} \times \text{pressure difference}}{\text{density}} = \frac{m\Delta P}{\rho}$$

$$= \frac{(\text{kg})(\text{Pa})}{\left(\dfrac{\text{kg}}{\text{m}^3}\right)} = (\text{kg})(\text{kg}/\text{m.s}^2) / \left(\frac{\text{kg}}{\text{m}^3}\right)$$

$$= \left(\text{kg}\frac{\text{m}}{\text{s}^2}\right)(\text{m})$$

$$\text{Pressure energy} = \text{N.m} = \text{J}$$

where m is the mass in kg, ΔP is the pressure difference between two points in Pa, and ρ is the density in kg/m^3.

Example 2.25: Kinematic Viscosity

Show the units of kinematic viscosity v in SI.

Solution

The quantity equation for v is given as follows:

$$v = \frac{\text{dynamic viscosity}}{\text{density}} = \frac{\mu}{\rho} = Ft/\rho$$

$$v = \frac{(\text{Pa})(\text{s})}{\text{kg}/\text{m}^3} = \left(\frac{\text{kg}}{\text{m.s}^2}\right)\text{s} / \left(\frac{\text{kg}}{\text{m}^3}\right)$$

$$v = \frac{\text{m}^2}{\text{s}}$$

where Pa is equal to kg/(m.s²), μ is the dynamic viscosity in Pa.s, and ρ is the density in kg/m³. Kinematic viscosity is dependent on force F, time t, and density ρ.

Example 2.26: Dimensionless Number

Show that the Reynolds number (Re) is a dimensionless number.

Solution

The quantity equation for the Reynolds number is given as follows:

$$Re = \rho u \frac{D}{\mu} = \frac{\left[\left(\frac{kg}{m^3}\right)\left(\frac{m}{s}\right)(m)\right]}{\left[\left(\frac{kg}{m.s^2}\right)s\right]} = \text{Dimensionless}$$

where ρ is the density of the fluid in kg/m³, u is the average velocity of the fluid in m/s, D is the diameter of the pipe in which the fluid flows in m, and μ is the dynamic viscosity in Pa.s.

Example 2.27: Pump Head

Show that the pump head h_p unit is m in SI.

Solution

$$h_p = \frac{\text{pumping energy}}{\text{gravitational acceleration}} = \frac{E_p}{g}$$

$$= \frac{J/kg}{m/s^2} = \frac{N.m/kg}{m/s^2} = \left[\left((kg.m/s^2)m/kg\right)\left(s^2/m\right)\right]$$

$$h_p = m$$

where J is the unit of energy in SI (= force × distance = N.m), N is the unit of force (= kg.m/s²), and g is the gravitational acceleration, 9.81 m/s². Pump head is dependent on pumping energy E_p and g.

Example 2.28: Pump Power

Show that the pump power θ unit is W in SI.

Solution

$$\text{Pump Power}(\theta) = \dot{m}E_p = \left(\frac{kg}{s}\right)\left(\frac{J}{kg}\right)$$

$$\text{Pump Power}(\theta) = \frac{J}{s} = W$$

where $\dot{m}$ is the mass flow rate in kg/s and E_p is the magnitude of pumping energy in J/kg.

If the pump has a conversion efficiency of η, then the pump input power θ_{bk} is calculated as follows:

Pump Input Power = pump break power (θ_{bk})
= pump fluid power/efficiency = θ_{fl}/η

where θ_{fl} is the pump output (fluid) power and efficiency η is in decimals (i.e., 0.6 is used for 60% conversion efficiency).

Example 2.29: Pump Fluid Power

Show that θ_{fl} is W in SI.

Solution

$$\text{Pump Fluid Power}(\theta_{fl}) = \dot{m}g\,h_p = \left(\frac{kg}{s}\right)\left(\frac{m}{s^2}\right)m$$

$$= \left(\frac{kg.m}{s^2}\right)\left(\frac{m}{s}\right) = N\frac{m}{s}$$

$$\text{Pump Fluid Power}(\theta_{fl}) = \frac{J}{s} = W$$

where $\dot{m}$ is the mass flow rate in kg/s, g is the gravitational acceleration in m/s², and h_p is the pump head in m.

Example 2.30: Pump Break Power

Show that θ_{bk} unit is W in SI.

Solution

$$\text{Pump Break Power}(\theta_{bk}) = \Omega\omega = \left(\frac{kg.m}{s^2}\right)\left(\frac{m}{s}\right) = N\frac{m}{s}$$

$$\text{Pump Break Power}(\theta_{bk}) = \frac{J}{s} = W$$

where Ω is the torque in kg.m/s² and ω is the angular velocity of the shaft in m/s.

Example 2.31: Temperature

Show the temperature in SI for a thermodynamic temperature of 350 K.

Solution

By definition, the temperature in °C is equal to the thermodynamic temperature in K minus 273.15. Note that by definition, a temperature interval of 1°C is equal to a temperature interval of 1 K, and 0°C (also known as the ice point) corresponds to 273.15 K (Figure 2.1). Then, the temperature in SI (°C) can be expressed as follows:

$$\text{Temperature}(°C) = \text{Thermodynamic temperature (K)} - 273.15$$

$$= 350\text{ K} - 273.15$$

$$\text{Temperature}(°C) = 76.85°C$$

Example 2.32: Temperature

Determine the thermodynamic temperature equivalent of 550°C.

Solution

Thermodynamic temperature is expressed as follows:

$$\text{Thermodynamic temperature}(K) = \text{Temperature}(°C) + 273.15\text{ K}$$

$$= 550°C + 273.15\text{ K}$$

$$\text{Thermodynamic temperature (K)} = 823.15\text{ K}$$

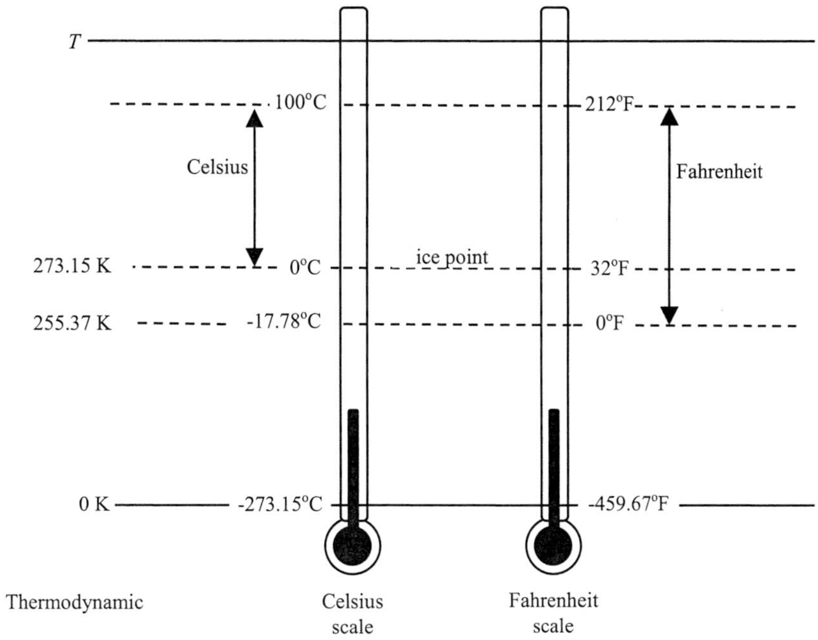

FIGURE 2.1 Relationships between temperature scales.

Source: *Greenhouse Engineering*, Edn. 1 by İlhami Yıldız, Copyright © 2021 by Imprint. Reproduced by permission of Taylor & Francis Group (*Greenhouse Engineering*).

Note that K was substituted for °C because the units are identical (Figure 2.1).

Example 2.33: Temperature

Determine the temperature in Fahrenheit for 21°C.

Solution

Note that the freezing and boiling temperatures on the Celsius scale, 0°C and 100°C, respectively, correspond to 32°F and 212°F, respectively. Therefore, the Celsius temperature range of (100°C – 0°C) = 100°C corresponds exactly to the Fahrenheit temperature range of (212°F – 32°F) = 180°F (Figure 2.1):

$$(212°F - 32°F) = (100°C - 0°C) = (180°F = 100°C)$$

Therefore,

$$1.0°C = 1.8°F,$$

and,

$$K = 1.8°F$$

Then, the relationship between °C and °F can be defined as follows:

$$\text{Temperature}(°F) = 1.8\frac{°F}{°C} + 32°F$$

where 32°F again is the freezing temperature for water in Fahrenheit and is equal to 0°C in Celsius (Figure 2.1). Then, °F in this example can be determined as

$$\text{Temperature}(°F) = \frac{1.8°F}{°C}(21°C) + 32°F$$

$$\text{Temperature}(°F) = 69.8°F \approx 70°F$$

The absolute temperature scale related to the Fahrenheit temperature scale is known as the Rankine scale and is designated R. The relationship between the Rankine and Fahrenheit scales is expressed as follows:

$$\text{Rankine} = °F - 459.67$$

Example 2.34: Temperature

Determine the temperature in SI for 70°F.

Solution

The relationship between Celsius and Fahrenheit is

$$\text{Temperature}(°C) = (°F - 32°F)\left(\frac{5}{9}\right)\left(\frac{°C}{°F}\right)$$

where 32°F is the freezing temperature for water in Fahrenheit, which is equal to 0°C in Celsius. Then, the temperature in SI (°C) can be determined as

$$\text{Temperature}(°C) = (70°F - 32°F)\left(\frac{5}{9}\right)\left(\frac{°C}{°F}\right)$$

$$\text{Temperature}(°C) = 21.1°C \approx 21°C$$

Example 2.35: Density

Determine the unit (kg/m^3) of density in SI if the unit is given in I-P.

Solution

Remember that the density is defined as the mass m per unit volume V. Therefore, the density equation is a quantity equation established from a definition and is expressed in I-P as follows:

$$\rho = m / V = \text{lb}_m / \text{ft}^3$$

Remember that 1 ft = 0.3048 m and 1 lb_m = 0.45359 kg:

$$\rho = \left[lb_m\left(0.45359\frac{kg}{lb_m}\right)\right] / \left[ft^3\left(0.3048\frac{m}{ft}\right)^3\right]$$

$$\rho = 16.0184\ kg/m^3$$

Density is dependent on the mass *m* and the volume *V*.

Example 2.36: Specific Volume

Determine the unit (m^3/kg) of specific volume in SI if the unit is given in I-P.

Solution

Remember that the specific volume is defined as the volume *v* per unit mass *m* and is therefore a reciprocal of density. Thus, the specific volume equation is a quantity equation established from a definition and is expressed in I-P as follows:

$$v = 1/\rho = ft^3/lb_m$$

Remember that 1 ft = 0.3048 m and 1 lb_m = 0.45359 kg:

$$v = \left[ft^3\left(0.3048\frac{m}{ft}\right)^3\right] / \left[lb_m\left(0.45359\frac{kg}{lb_m}\right)\right]$$

$$v = 0.062428 m^3/kg$$

Example 2.37: Specific Heat or Entropy

Determine the unit (J/(kg.K)) of specific heat in SI if the unit is given in I-P.

Solution

The constant pressure specific heat (c_p) and constant volume specific heat (c_v) are useful functions, particularly for gases. Remember that specific heat is defined as the heat energy required for raising the temperature of a unit mass of substance one degree. Therefore, the specific heat equation is a quantity equation established from a definition and is expressed in I-P as follows:

$$c = Btu/lb_m{}^\circ F$$

Remember that 1 Btu = 1,054.35 J; 1 lb_m = 0.45359 kg; and K = 1.8°F.

$$c = \left[Btu\left(1{,}054.35\frac{J}{Btu}\right)\right] / \left[lb_m\left(0.45359\frac{kg}{lb_m}\right)^\circ F\left(\frac{K}{1.8^\circ F}\right)\right]$$

$$c = 4{,}184\ J/(kg.\ K)$$

Therefore, in SI, specific heat is defined as the heat energy (J) required for raising the temperature of 1 kg of a substance by 1 K (or 1°C).

Example 2.38: Enthalpy

Determine the unit (J/kg) of enthalpy in SI if the unit is given in I-P.

Solution

Remember that enthalpy is defined as the heat energy available per unit mass of a substance. So, the enthalpy equation is a quantity equation established from a definition and is expressed in I-P as follows:

$$h = Btu/lb_m$$

Remember that 1 Btu = 1,054.35 J and 1 lb_m = 0.45359 kg.

$$h = \left[Btu\left(1{,}054.35\frac{J}{Btu}\right)\right] / \left[lb_m\left(0.45359\frac{kg}{lb_m}\right)\right]$$

$$h = 2{,}324.5\ J/kg$$

Example 2.39: Entropy

Determine the unit (J/g) of entropy in SI if the unit is given in I-P.

Solution

Remember that entropy is a measure of molecular disorder: Higher disorder in any system correlates with higher entropy; conversely, high order gives low entropy. Boltzmann showed that there exists a simple relationship between the entropy of a given system of molecules and its probability (thermodynamic) of occurrence. So, the entropy equation is a quantity equation established from a definition, has the symbol *s*, and is expressed in I-P as follows:

$$s = Btu/lb_m$$

Remember that 1 Btu = 1,054.35 J and 1 lb_m = 0.45359 kg.

$$s = \left[Btu\left(1{,}054.35\frac{J}{Btu}\right)\right] / \left[lb_m\left(1{,}000\frac{g}{kg}\right)\left(0.45359\frac{kg}{lb_m}\right)\right]$$

$$s = 2.3245\ J/kg$$

Example 2.40: Thermal Conductivity

Determine the unit (W/(m.K)) of thermal conductivity in SI if the unit is given in I-P.

Solution

Remember that thermal conductivity is the amount of heat that passes through a unit area when the temperature difference between the two sides is one degree per unit distance. So, the thermal conductivity equation is a quantity equation established from a definition and is expressed in I-P as follows:

$$k = Btu/(h\ ft\ ^\circ F)$$

Keep in mind that 1 Btu = 1,054.35 J; 1 ft = 0.3048 m; K = 1.8°F; and 1 h = 3,600 s.

$$k = \left[Btu\left(1{,}054.35\frac{J}{Btu}\right)\right] / \left[h\left(3{,}600\frac{s}{h}\right)ft\left(0.3048\frac{m}{ft}\right)^\circ F\left(\frac{K}{1.8^\circ F}\right)\right]$$

$$k = \left(1.7296\frac{J}{s}\right)/(m.K) = 1.7296\ W/(m.K)$$

Example 2.41: Heat Transfer Coefficient

Determine the unit (W/(m^2.K)) of heat transfer coefficient in SI if the unit is given in I-P.

Solution

Remember that the heat transfer coefficient is defined as the amount of heat that is transferred from/to a unit area per unit time when the temperature difference between the surface and the ambient is one degree. So, the heat transfer coefficient equation is a quantity equation established from a definition and is expressed in I-P as follows:

$$h = \text{Btu}/\left(\text{h.ft}^2.°\text{F}\right)$$

Keep in mind that 1 Btu = 1,054.35 J; 1 ft^2 = 0.0929 m^2; K = 1.8°F; and 1 h = 3,600 s.

$$h = \left[\text{Btu}\left(1{,}054.35\frac{\text{J}}{\text{Btu}}\right)\right] / \left[\text{h}\left(3{,}600\frac{\text{s}}{\text{h}}\right)\text{ft}^2\left(0.0929\frac{\text{m}^2}{\text{ft}^2}\right) °\text{F}\left(\frac{\text{K}}{1.8}\right)\right]$$

$$h = 5.68\ \text{W}/\left(\text{m}^2.\text{K}\right)$$

Example 2.42: Thermal Transmittance

Determine the unit (W/(m^2.K)) of thermal transmittance in SI if the unit is given in I-P.

Solution

Remember that thermal transmittance is the amount of heat that passes through an entire wall, ceiling, etc. section of a unit area per unit time when the temperature difference between the air on the warm side and the air on the cold side is one degree. So, the thermal transmittance equation is a quantity equation established from a definition and is expressed in I-P as follows:

$$U = \text{Btu}/\left(\text{h.ft}^2.°\text{F}\right)$$

Remember that 1 Btu = 1,054.35 J; 1 ft^2 = 0.0929 m^2; K = 1.8°F; and 1 h = 3,600 s.

$$U = \left[\text{Btu}\left(1{,}054.35\frac{\text{J}}{\text{Btu}}\right)\right] / \left[\text{h}\left(3{,}600\frac{\text{s}}{\text{h}}\right)\text{ft}^2\left(0.0929\frac{\text{m}^2}{\text{ft}^2}\right) °\text{F}\left(\frac{\text{K}}{1.8°\text{F}}\right)\right]$$

$$U = 5.68\ \text{W}/\left(\text{m}^2.\text{K}\right)$$

Example 2.43: Thermal Resistance

Determine the unit (m^2.K/W or m^2.°C/W) of thermal resistance in SI if the unit is given in I-P.

Solution

Remember that thermal resistance is the resistance of one unit area to heat flow through a substance per unit time when the temperature difference between the two sides is one degree and has the symbol *R*. Thermal resistance is an additive quantity; that is, 2 in. material has twice the R of 1 in. material. It also does not include the boundary layer resistances. So, the heat transfer coefficient equation is a quantity equation established from a definition and is expressed in I-P as follows:

$$R = \text{h.ft}^2.°\text{F}/\text{Btu}$$

Remember that 1 Btu = 1,054.35 J; 1 ft^2 = 0.0929 m^2; K = 1.8°C = 1.8°F; and 1 h = 3,600 s.

$$R = \left[\text{h}\left(3{,}600\frac{\text{s}}{\text{h}}\right)\text{ft}^2\left(0.0929\frac{\text{m}^2}{\text{ft}^2}\right)°\text{F}\left(\frac{\text{K}}{1.8}\right)\right] / \left[\text{Btu}\left(1{,}054.35\frac{\text{J}}{\text{Btu}}\right)\right]$$

$$R = 0.176\ \text{m}^2.\text{K}/\text{W}$$

Example 2.44: Wavelength

Calculate the maximum wavelength λ_{max} for the Sun's surface radiation.

Solution

Wien's displacement law states that the wavelength at which a blackbody emits its maximum amount of radiation is inversely proportional to its absolute temperature in Kelvin, $\lambda_{max} = a/T$, where λ is in μm, *a* is 2,897 μm.K, and *T* is temperature in K.

Assuming an average surface temperature of approximately 6,000 K, its maximum emission is

$$\lambda_{max} = 2{,}897\ \mu\text{m.K}/6{,}000\ \text{K}$$

$$\lambda_{max} = 0.48\ \mu\text{m} \approx 0.5\ \mu\text{m}$$

A wavelength of approximately 0.5 μm lies within the visible spectrum.

Example 2.45: Frequency

Determine the frequency of a radiational wavelength of 0.5 μm.

Solution

Wavelength (λ) and frequency (v) are related, and this relationship can formally be expressed through the speed of light *c* as $c = \lambda v$, where *c* is the speed of light, 3 × 10^8 m/s, λ is the wavelength (μm), and *v* is the frequency (cycles/s, also known as Hz). Then, if one is known, then the other can easily be determined using the speed of light, which is a constant.

$$c = \lambda v$$

$$v = c/\lambda$$

$$v = \left(3\times10^8\text{m}/\text{s}\right)/\left(0.5\ \mu\text{m}\right)$$

$$= \left(3\times10^{14}\mu\text{m}/\text{s}\right)/\left(0.5\ \mu\text{m}\right)$$

$$= \left(6\times10^{14}\text{cycles}/\text{s}\right)$$

$$v = 6\times10^{14}\ \text{Hz}$$

Example 2.46: Energy Content of a Single Wavelength of Solar Radiation

Determine the energy content E of a solar radiation frequency of 6×10^{14} Hz (or 0.5 μm wavelength).

Solution

Solar radiation consists of photons, which can be considered packets of energy, which is related to frequency v as $E = hv$, where E is the energy content (J) and h is the Planck constant (6.626×10^{-34} J.s).

$$E = hv$$

$$E = \left(6.626 \times 10^{-34} \text{ J.s}\right)\left(\frac{6 \times 10^{14} \text{ cycles}}{\text{s}}\right)$$

$$E = 3.98 \times 10^{-19} \text{ J}$$

2.3 ESSENTIALS OF THERMAL ENERGY

2.3.1 Introduction

In this section, first, the laws of thermodynamics, which form the foundation of heat transfer analyses, were covered. Along the way, temperature was defined, and then energy and mass balance concepts were introduced before covering different modes of heat transfer using different objects and geometries. Toward the end, a subsection was also provided on fossil fuel combustion and carbon sequestration. A number of numerical examples were included in the section to demonstrate and grasp the relevant concepts easily.

A system can have various forms of energy including chemical, electrical, kinetic, magnetic, nuclear, potential, or thermal. Potential and kinetic energy are examples of what is called macroscopic forms, which are possessed by a system as a lump sum in reference to a particular outside frame. The forms of energy related to the molecular structure and level of activity are called microscopic forms, and the internal energy of a system is defined as the sum of different microscopic energy forms, which include sensible, latent, chemical, and nuclear. Sensible energy is associated with the kinetic energy of molecules (i.e., with a temperature change), whereas latent energy is associated with a change in phase in a system, and chemical and nuclear energy are related with the atomic bonds and the bonds within the atomic nucleus, respectively. Some of these concepts are covered further in depth in the following sections; however, we must first understand the laws of thermodynamics and energy and mass balance concepts before moving any further.

2.3.2 Laws of Thermodynamics

Thermodynamics is defined as the science of energy that covers the entire characteristics of energy transformations. When two bodies at different temperatures are brought together, a transfer of energy occurs between these bodies until a thermal equilibrium is reached. The zeroth law of thermodynamics indicates that when two bodies have equal temperatures, they are in thermal equilibrium with each other. This process of energy flow due to the temperature difference is called heat transfer, and the rate of heat transfer is function of this existing temperature gradient.

We use the conservation of energy principle in dealing with heat transfer analyses, which basically states that energy can neither be created nor destroyed; rather, it can only change its form. This principle is called the first law of thermodynamics, the study of which deals with quantifying the amount of energy transferred from one body or system to another; however, it does not specify the direction of energy transfer, which is actually defined by the second law of thermodynamics. The second law states that energy has not only a quantity but also a quality, and the direction of energy transfer is determined by the direction in which the quality of energy decreases.

Furthermore, the second law dictates theoretical performance and efficiency limits of energy systems as well. For instance, when heat (energy in transition) is added to a system, the system's thermal energy content rises, and both molecular activity in the system and therefore the temperature increase as a result. So the temperature, a property of a given system, mirrors the thermal energy content of the system and is related to hotness or coldness of an object; however, it is difficult to give an exact definition of temperature. As explained, the zeroth law explains the temperature relationships of two bodies with respect to a third body, while the third law, which defines thermodynamic temperature, states that a body or a system has its minimum possible energy content at the theoretically lowest temperature (i.e., absolute temperature). At this point (absolute zero), the particle constituents of matter have minimal motion and can become no colder.

2.3.2.1 First Law of Thermodynamics

The first law, as introduced briefly, states that energy can neither be created nor destroyed; rather, it can only change its form from one to another. Using the first law, we can analyze the process of energy transfer between a system or a body and its surroundings by simply employing the energy balance approach.

Overall, energy is transferred to or is extracted from a system in the form of heat, work, or mass flow. Employing a control volume and a system boundary over which the energy gains or losses occur, we can easily analyze the process of energy transfer. As mentioned earlier, when heat is transferred to a system, it ends up with a rise in molecular energy and therefore in temperature. Hence, this process increases the internal energy of the system. If heat is transferred from a system to the surrounding, the internal energy of the system decreases due to the reduced molecular energy content.

Energy transfer can also take place by performing work on a system or by a system performing work itself. If work is performed by the system itself, the energy of the system decreases, while the energy of the system increases when the work is done on the system. Furthermore, mass flow is another mechanism through which energy transfer can take place over the boundary of a system. Similarly, when mass flows into a given control volume, the energy content of this volume increases as a result of the amount of energy introduced by the mass flowing into the system, whereas it decreases when the mass leaves the system.

2.3.2.1.1 Energy Balance

In any process, the difference between the total amount of energy entering and leaving a system is equal to the change in total energy of the system. This is called the energy balance and can be expressed as follows:

$$E_i - E_o = \Delta E_{sys} \tag{2.1}$$

where E_i and E_o are the amounts of energy entering and leaving the system, respectively, and ΔE_{sys} is the change in total energy of the system.

If we apply the energy balance concept per unit time basis, then we can write the energy balance equation as

$$\dot{E}_i - \dot{E}_o = \frac{dE_{sys}}{dt} \tag{2.2}$$

where $\dot{E}_i$ and $\dot{E}_o$ are the amounts of energy entering and leaving the system per unit time, respectively.

We can apply the energy balance approach to any system for any process; however, we need to make sure that all forms of energy entering or leaving the system are properly identified and quantified. In heat transfer analyses, since we only consider the energy in the form of heat, the energy (heat) balance equation takes the following form:

$$Q_i - Q_o + Q_{gen} = \Delta E_{sys,\,thermal} \tag{2.3}$$

where Q_i and Q_o are the amounts of heat transfer into and out of the system, respectively, Q_{gen} is the amount of heat generation, and $\Delta E_{sys,\,thermal}$ is the change in thermal energy content of the system.

2.3.2.1.2 Energy Balance for Closed Systems

A system is called closed when there is no mass exchange with surroundings, that is, no inflow or outflow across its boundary. Since no mass enters or leaves closed systems, we can only talk about work and heat transfer in such systems; when applied to closed systems, Equation (2.1) can be written as

$$(Q_i + W_i) - (Q_o + W_o) = \Delta E_{sys} \tag{2.4}$$

$$(Q_i - Q_o) + (W_i - W_o) = \Delta E_{sys} \tag{2.5}$$

In stationary systems that do not involve a change in kinetic or potential energies, the system's total energy is internal, and the energy balance for such closed systems can be written as follows:

$$E_i - E_o = \Delta U \tag{2.6}$$

where ΔU is the change in internal energy of the system, which can be expressed in terms of the mass m, constant volume specific heat c_v, and the change in temperature ΔT as

$$\Delta U = m c_v \Delta T \tag{2.7}$$

Thence, the heat balance equation for closed stationary systems can be written as

$$Q_{net} = m c_v \Delta T \tag{2.8}$$

where Q_{net} represents the net transfer of heat to or from the system.

2.3.2.1.3 Energy and Mass Balance for Open Systems

An open system (or a control volume) is a system where both mass and energy are allowed to cross the control boundary. In control volumes, energy can be transferred in the form of heat, work, and mass transfer, and compressors, heat exchangers, pumps, and turbines are just a few examples of such systems. For such systems, the principle of conservation of mass is particularly important and is expressed as the difference between the total mass entering m_i and leaving m_o the system (control volume) m_{cv} in a specific time interval:

$$m_i - m_o = \Delta m_{cv} \tag{2.9}$$

If it is considered on a per unit time basis, then the mass balance equation becomes

$$\dot{m}_i - \dot{m}_o = \frac{dm_{cv}}{dt} \tag{2.10}$$

where $\dot{m}_i$ and $\dot{m}_o$ represent the rates of mass inflow and outflow, respectively.

In heat transfer applications, quite often we encounter fluid flow within pipes and ducts. In such cases, we can determine the mass flow rate from the cross-sectional area of the flow passage, density, and velocity of the fluid. In one-dimensional approximations, the fluid properties are assumed to only change in the flow direction. The mass flow rate can then be expressed as

$$\dot{m} = \rho u A \tag{2.11}$$

where ρ is the density of the flowing fluid, u is the average velocity of the fluid in the direction of flow, and A is the cross-sectional area of the flow passage.

We can also express the mass flow rate $\dot{m}$ in terms of the volumetric flow Q rate of the fluid as follows:

$$Q = u A \tag{2.12}$$

$$\dot{m} = \rho Q \tag{2.13}$$

In steady-flow processes, the amount of mass within the control volume does not change over time. Thence, the total mass inflow rates are equal to the total mass outflow rates:

$$\sum \dot{m}_i = \sum \dot{m}_o \tag{2.14}$$

In many heat transfer applications, only one inlet and outlet are encountered. Under such conditions, for steady-flow processes, the inlet and outlet mass flow rates are equal:

$$\dot{m}_i = \dot{m}_o \tag{2.15}$$

In open systems (control volumes), energy transfer takes place by mass transfer as well; therefore, it is necessary to determine the total energy of the fluid that enters or leaves the system. The total energy of a flow per unit mass entering or leaving a system can be expressed as follows:

$$E_{total} = \text{Flow Energy} + \text{Internal Energy} + \text{Kinetic Energy} + \text{Potential Energy}$$

$$E_{total} = \left(h + \frac{u^2}{2} + gz\right) \quad (2.16)$$

where Q is the volumetric flow rate of the fluid The middle term on the right-hand side of the equation represents the kinetic energy while the far right term represents the potential energy.

The energy balance equation for steady-flow processes can thus be written as

$$\dot{E}_i - \dot{E}_o = \frac{dE_{sys}}{dt} \quad (2.17)$$

For steady-flow processes, $\frac{dE_{sys}}{dt} = 0$; hence, $\dot{E}_i = \dot{E}_o$, then

$$\dot{Q}_i + \dot{W}_i + \sum \dot{m}_i \left(h + \frac{u^2}{2} + gz\right) = \dot{Q}_o + \dot{W}_o + \sum \dot{m}_o \left(h + \frac{u^2}{2} + gz\right) \quad (2.18)$$

where $\dot{Q}_i$ and $\dot{Q}_o$ are the rates of heat added to and removed from the system, respectively, and $\dot{W}_i$ and $\dot{W}_o$ are the rates of work done on and by the system; the terms in the parentheses were explained earlier.

In processes that do not include energy transfer by work, and also when the kinetic and potential energy changes are negligible, the energy balance equation for steady-flow conditions can be reduced to

$$\dot{Q}_{net} = \dot{m}(h_o - h_i) = \dot{m} c_p (T_o - T_i) \quad (2.19)$$

where $\dot{Q}_{net}$ is the net rate of heat transfer, c_p is the constant pressure specific heat, and h and T are the enthalpy and temperature of the flowing fluid, respectively.

2.3.2.2 Second Law of Thermodynamics

When energy transfer happens between a system and its surroundings, the transfer occurs only in a certain direction. The first law of thermodynamics does not say anything about the direction of heat flow as it only deals with the transformation and conservation of energy. The second law of thermodynamics, on the other hand, employs entropy, which is a thermodynamic property that measures degree of randomization or disorder at the microscopic level and defines the direction of an energy transfer or conversion process. Processes can be reversible or irreversible.

The second law also states that any process that reduces the entropy of the isolated system is impossible. If the entropy is unchanged by the forward process, then it will also be unchanged by the reverse process. Since no entropy is produced, such a process can go in either direction without violating the second law and is called reversible. If the entropy is produced by the forward process, however, then the entropy will decrease for the reverse process, and hence, the reverse process is impossible. That is, the forward process is irreversible.

Entropy is produced in irreversible processes, and all real processes are subjected to some measure of irreversibility. Recognition of the irreversibilities in real processes is quite important in engineering as they reflect an increase in the amount of disorganized energy (requires further work before it can be used effectively) at the expense of organized energy (can easily be put to practical use). In short, the second law of thermodynamics states that entropy can be produced but never destroyed. For the first law, we presented the energy balance approach; however, the second law deals with the entropy and exergy balance instead, and the entropy balance for a system can be expressed as follows:

$$S_i - S_o + S_{gen} = \Delta S_{sys} \quad (2.20)$$

where S represents the entropy.

As discussed earlier, the first law deals with the quantity of energy, whereas the second law is concerned about the associated quality. At this point, we must introduce the concept of exergy, a measure of the maximum useful work potential of a system in a particular reference environment. A system that is in equilibrium with its surroundings has zero exergy and is therefore said to be at the dead state. In irreversible processes, exergy destruction takes place while entropy is being generated. Thence, the amount of destroyed exergy can be expressed as a function of entropy generation as follows:

$$Ex_{dest} = T_o S_{gen} \quad (2.21)$$

where Ex_{dest} is the amount of exergy destroyed and T_o is the temperature of the dead state (the system is in equilibrium with the surroundings, where no useful work can be obtained).

The exergy balance equation for a given system can then be developed as follows:

$$Ex_i - Ex_o - Ex_{dest} = \Delta Ex_{sys} \quad (2.22)$$

If we express it on a rate basis (per unit time), Equation (2.22) takes the following form:

$$\dot{Ex}_i - \dot{Ex}_o - \dot{Ex}_{dest} = \frac{dEx_{sys}}{dt} \quad (2.23)$$

Just like we employ the energy balance analysis for control volumes, exergy balance can also be analyzed for control volumes. However, heat transfer analyses are based on the first law of thermodynamics and the energy balance concept. The following sections cover the types and mechanisms of heat transfer.

2.3.3 Heat Transfer and Its Mechanisms

Heat is the form of energy in transition caused by a temperature difference. The process and study of energy flow due

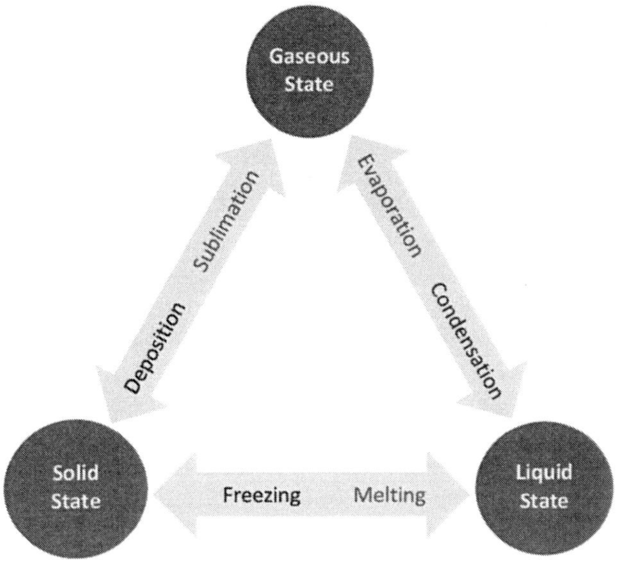

FIGURE 2.2 Phase changes and relevant processes.
Source: ***Greenhouse Engineering*** **(2021).**

to the presence of a temperature difference is called heat transfer, and the rate of heat transfer is a function of this existing temperature gradient. As mentioned earlier, thermodynamic analysis deals with determining the amount of heat transferred between systems using the conservation of energy principle, whereas the study of heat transfer focuses on determining the rate of heat transfer.

All substances in principle can hold a certain amount of heat, and the property that signifies this amount of heat is known as heat (thermal) capacity. When heat is added to a system, the system's thermal energy increases and molecular activity in the system and therefore the temperature rise as a result. When a solid substance is heated, for instance, its temperature keeps rising until it reaches the melting point of that substance, which is the highest temperature that can be achieved before the phase change from solid to liquid starts taking place. The heat that is absorbed by the solid until melting point is called sensible heat Q_S; however, the heat that is used to convert the solid to a liquid is called latent heat of fusion (melting) Q_L (Figure 2.2).

Similarly, when a liquid is heated, the temperature of the liquid keeps rising until it reaches the boiling point, which is the highest temperature that can be reached at the liquid phase at the measured pressure. The heat that is absorbed by the liquid while increasing the temperature up until the boiling point is again sensible heat, while the heat that is used to convert the liquid to the vapor phase is called latent heat of vaporization.

The internal energy related to the kinetic energy of molecules, as a result of energy absorbed or released by a system (as discussed earlier), that leads to an increase or decrease in temperature is known as sensible heat. The quantity of sensible heat associated with a temperature change, Q_S, can be expressed as follows:

$$Q_S = mc_p\left(T_2 - T_1\right) \tag{2.24}$$

where Q_S is the total sensible heat of the system in J (Btu), m is the mass of the system in kg (lb), c_p is the specific

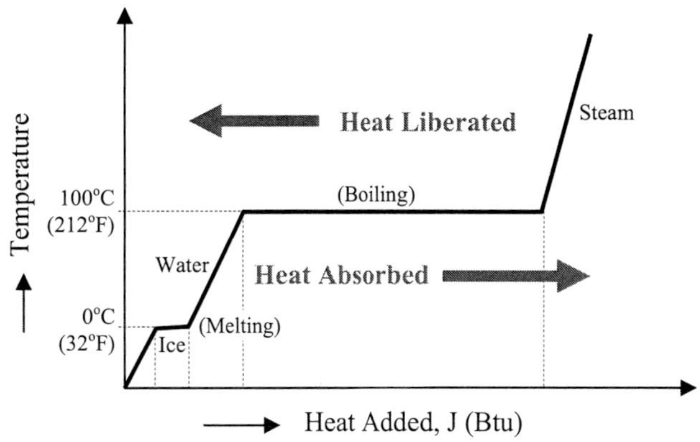

FIGURE 2.3 Phase changes of water.
Source: ***Greenhouse Engineering*** **(2021).**

heat in J/kg.°C (Btu/lb.°F), and $\left(T_2 - T_1\right)$ is the temperature change in °C (°F). Specific heat is the heat to raise 1 kg (1 lb) of a substance 1°C (1°F); the specific heats of some common substances are provided in Table A.4. For example, the specific heat of air is 1004 J/kg.°C (0.24 Btu/lb°F), and that for water is 4186 J/kg°C (1.00 Btu/lb.°F).

Example 2.47: Sensible Heat

Determine the total amount of sensible heat required to increase the temperature of 1,000 kg of water from 4 to 40°C.

Solution

From Table A.1, specific heat c_p of water is 4,186 J/kg°C.

Using Equation (2.24),

$$Q_S = mc_p\left(T_2 - T_1\right)$$

$$Q_S = 1{,}000\,\text{kg}\left(4{,}186\ \text{J/kg.°C}\right)\left(40 - 4\right)\text{°C}$$

$$Q_S = 150{,}696{,}000\ \text{J}\left(\approx 142{,}840\ \text{Btu}\right)$$

Observable properties such as temperature and pressure characterize the state or phase of a system. Every specific property of a substance at a given state has a certain magnitude that is independent of the path of the substance arriving at the given state. For instance, heat applied to a substance at a given condition changes its state or phase. The substance temperature remains constant during this phase change until the process is completed. This is true for a solid to liquid phase change, liquid to vapor phase change, or the reverse. Figure 2.3 is a typical diagram for water to change phase from solid (ice) to liquid to vapor as heat is added. As shown in the figure, the temperature during these phase changes remains constant.

As discussed, sensible energy is associated with the kinetic energy of molecules (i.e., with a temperature change), whereas latent energy is associated with a change in phase in a system. The energy required to convert a unit mass of a particular substance from solid to liquid is known as latent heat of fusion, whereas the energy required to convert a unit mass of liquid of a particular substance from liquid to gaseous state is known as latent heat of vaporization.

The temperature of a system remains constant during a phase change until the process is complete. Quantity of latent heat associated with a phase change in a system, Q_L, can be expressed as follows:

$$Q_L = mh \tag{2.25}$$

where Q_L is the total latent heat of the system in J (Btu), m is the mass of the system in kg (lb), and h is the latent heat (enthalpy) of fusion or vaporization in J/kg (Btu/lb). Latent heat (enthalpy) of fusion and vaporization of some common substances are provided in Table A.5.

When ice melts, for instance, latent heat of fusion is absorbed, and liquid water is formed while the temperature remains at 0°C (32°F). The latent heat of fusion of ice, Q_L, that is, the amount of energy that has to be supplied to change 1 kg of ice (at its melting point) to liquid water at the same temperature, is 3.39×10^5 J/kg. In imperial units, 1 lb of ice requires 144 Btu of energy to melt. The quantity of heat associated with a change in phase from liquid water to water vapor—latent heat of vaporization Q_L—that is, the energy required to evaporate 1 kg liquid water without a change in temperature at 100°C—is about 2.26×10^6 J. In imperial units, 1 lb of water requires 1,044 Btu of energy to evaporate.

Example 2.48: Latent Heat of Fusion

Determine the total amount of latent heat of fusion of one ton (short) of ice melted in 24 hours.

Solution

From Table A.5, enthalpy of fusion h of water is 144 Btu/lb:

1 ton (short) of ice = 2,000 lb

Using Equation (2.25),

$$Q_L = mh$$

$$Q_L = 2{,}000 \text{ lb ice} \times 144 \frac{\text{Btu}}{\text{lb}} = 288{,}000 \text{ Btu} (\approx 303{,}840 \text{ J})$$

This means that 288,000 Btu of energy (latent heat of fusion) is needed for melting 2,000 lb (1 ton) of ice in 24 hours. Then,

$$288{,}000 \text{ Btu} / 24 \text{ h} = 12{,}000 \text{ Btu/h} \left(12{,}660 \frac{\text{J}}{\text{h}} = 3.516 \text{ W}\right)$$

That is, one ton of refrigeration equals to 12,000 Btu/h (3.516 W).

Example 2.49: Latent Heat of Vaporization

Determine the total latent heat of vaporization of 1,000 kg of water evaporated at 100°C.

Solution

From Table A.5, enthalpy of fusion h of water is 2.26×10^6 J/kg.

Using Equation (2.25),

$$Q_L = (1{,}000 \text{ kg water}) \times \left(2.26 \times 10^6 \frac{\text{J}}{\text{kg}}\right)$$

$$= 2.26 \times 10^9 \text{ J} (\approx 2.14 \times 10^6 \text{ Btu})$$

This section also covers modes of heat transfer. Thermal energy is transferred from one region (or a body) to another by three modes, namely conduction, convection, and radiation. Heat transfer is one of those transport phenomena, which include mass transfer (discussed earlier), momentum transfer or fluid friction (discussed later in this chapter), and electrical conduction. Transport phenomena have similar rate equations, and the flux in each case is proportional to a potential difference. In heat transfer by conduction and convection, as we will see shortly, the potential difference is the temperature difference, and heat transfer takes place through matter due to this temperature gradient. In radiation heat transfer, we have a change in energy from internal to electromagnetic energy at the source for transmission and then back to internal energy where it is received. The amount of thermal energy transferred by radiation increases quickly as the temperature increases.

In order to be able to handle heat transfer problems, some heat transfer equations have been developed from fundamentals; however, they are most of the time obtained from correlations of observed data. Typically, these correlations utilize certain dimensionless numbers developed from dimensional analysis or analogy, and Table A.6 lists some dimensionless numbers that are important for heat transfer.

2.3.3.1 Conduction Heat Transfer

Thermal conduction is the mode of heat transfer where energy is transferred from more energetic to less energetic neighboring particles or group of particles at the atomic level. Conduction occurs in solids, liquids, and gases. In liquids and electrically non-conducting solids, it occurs by longitudinal lattice molecular oscillations as well as movement of free electrons; in gases, the molecules are under constant random motion, and the energy transfer occurs due to elastic collision and diffusion of molecules. In solid opaque bodies, thermal conduction is the mode of energy transfer because there is no material flow within the medium, and in flowing fluids, thermal conduction also dominates in the laminar flow region very close to a solid boundary where the flow is parallel to the surface and there is no eddy motion. As a consequence of the second law, thermal energy transfer occurs in the direction of decreasing temperature (Figure 2.4), and conduction heat transfer can be classified as one-, two-, or three-dimensional. The classification obviously depends on the variation of temperature within the medium, and depending on the geometry, the Cartesian, cylindrical or spherical coordinate system is utilized.

2.3.3.1.1 Steady-State Heat Conduction

During the heat transfer process, if the surface temperature of the body in question does not change with time, it is termed a steady-state condition, and the transfer process is analyzed by utilizing simple thermal resistance without employing any differential equations.

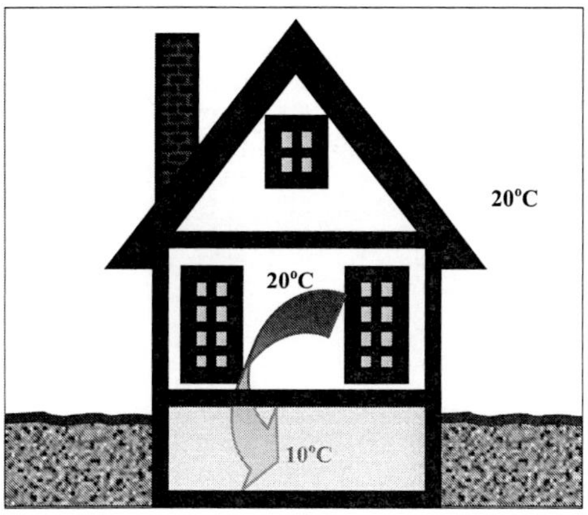

FIGURE 2.4 Heat flows when there is a temperature gradient.
Source: ***Greenhouse Engineering*** **(2021).**

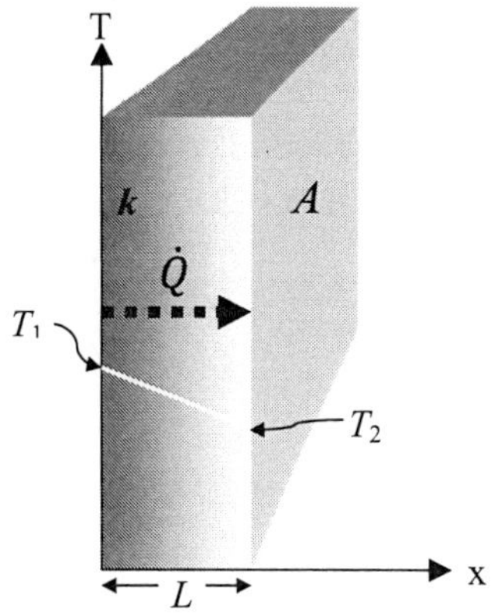

FIGURE 2.5 Heat conduction through a plain wall.
Source: ***Greenhouse Engineering*** **(2021).**

2.3.3.1.1.1 Steady-State Heat Conduction in Walls If energy balance per unit time developed in Equation (2.2) is applied to a plane wall, then we can write the energy balance equation as follows:

$$\dot{Q}_i - \dot{Q}_o = \frac{dE_w}{dt} \tag{2.26}$$

Since energy content of the wall does not change with time in steady-state conditions—that is, $\frac{dE_w}{dt} = 0$—then the energy balance equation simply reduces to

$$\dot{Q}_i = \dot{Q}_o \tag{2.27}$$

For steady-state heat conduction in one dimension, the Fourier equation applies:

$$\dot{Q} = -kA\frac{dT}{dx} \tag{2.28}$$

where $\dot{Q}$ is the heat transfer rate (W or Btu/h), k is the thermal conductivity (W/m.°C or Btu/h.ft.°F), A is the cross-sectional area normal to the flow (m²), and $\frac{dT}{dx}$ is the temperature gradient (°C/m).

Since thermal conductivity can be given in different units, one must be very careful in using consistent units. Under steady-state conditions, since the conduction heat transfer rate through the plane wall is constant and k and A do not vary, $\frac{dT}{dx}$ is a constant as well. With a linear temperature distribution across the wall (Figure 2.5), the conduction heat transfer rate can be expressed as follows:

$$\dot{Q} = -kA\frac{(T_1 \quad T_2)}{L} \tag{2.29}$$

The Fourier equation presented earlier states that the heat flow rate $\dot{Q}$ in the direction of x is directly proportional to the temperature gradient $\frac{dT}{dx}$ and to A normal to the heat flow; additionally, k is the proportionality constant in between (Tables A.7 and A.8). The minus sign on the right-hand side of the equation states that the heat transfer is positive in the direction of decreasing temperature.

Example 2.50: Heat Conduction through a Plain Wall

Consider a 20 cm thick concrete wall of 3 m × 6 m with uniform thermal conductivity. Determine the steady-state conduction heat transfer rate when the inside surface temperature of the wall is 20°C and the outside surface temperature of the wall is −10°C. Assume that steady-state conditions prevail and the heat transfer is one-dimensional.

Solution

T_1 = 20°C and T_2 = −10°C; A = 3 m × 6 m = 18 m²; L = 0.2 m; k = 1.385 W/m°C (Table A.4).

For steady-state heat conduction in one dimension, Fourier Equation (2.29) applies:

$$\dot{Q} = -kA\frac{(T_1 - T_2)}{L}$$

Hence, the rate of conduction heat transfer through the concrete wall is

$$\dot{Q} = -\left(1.385\frac{\text{W}}{\text{m.}^\circ\text{C}}\right)\left(18\text{m}^2\right)\frac{\left(20-(-10)\right)^\circ\text{C}}{0.2\text{ m}}$$

$$= -3{,}740\text{ W}\left(\approx -12{,}763\text{ Btu/h}\right)$$

2.3.3.1.1.2 Steady-State Heat Conduction in Cylinders In steady-state one-dimensional conduction heat transfer in a hollow cylinder (Figure 2.6) with a uniform wall thickness, inner radius r_i, outer radius r_o, inner and outer surface temperatures (T_1 and T_2, respectively) and with no heat generation within the cylinder, according to Fourier's equation, the rate of heat conduction through the cylinder's wall and A normal to the heat flow in the direction of the temperature gradient takes the following form:

$$\dot{Q} = -kA\frac{dT}{dr} \tag{2.30}$$

where the surface area A is equal to $2\pi rL$ at distance r from the center of the cylinder.

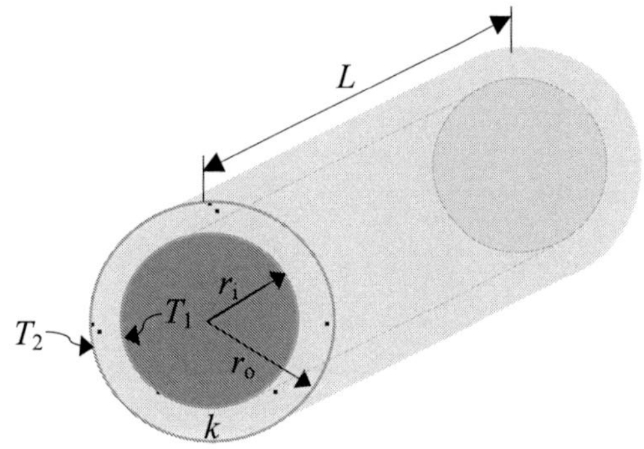

FIGURE 2.6 Heat conduction through a hollow cylinder.
Source: *Greenhouse Engineering* (2021).

Thermal conductivity being the proportionality constant in between, Equation (2.30) provides the following:

$$\dot{Q} = -2\pi L k \frac{(T_1 - T_2)}{ln\left(\frac{r_o}{r_i}\right)} \quad (2.31)$$

Example 2.51: Heat Conduction through a Cylinder

Now consider a 51 mm steel tube with a 2.1 mm wall thickness and inside diameter of 46.7 mm with a uniform thermal conductivity of 17 W/(m°C). Determine the steady-state conduction heat transfer rate when the inside surface temperature of the cylinder is 93°C and the outside temperature is 90°C. Assume the pipe length is 30 m, steady-state conditions prevail, and the heat transfer is one-dimensional.

Solution

T_i = 93°C and T_o = 90°C; r_i = 0.0467 m and r_{wall} = 0.0021 m; L = 30 m; k = 17 W/m°C

The rate of heat transfer through the pipe can then be determined using Equation (2.31) as follows:

$$\dot{Q} = -2\pi L k \frac{(T_1 - T_2)}{ln\left(\frac{r_o}{r_i}\right)}$$

The outer radius of the pipe is $r_o = r_i + r_{wall} = 0.0467$ m $+ 0.0021$ m $= 0.0488$ m.

$$\dot{Q} = -2\pi(30\text{m})\left(17\frac{\text{W}}{\text{m.°C}}\right)\frac{(93-90)\text{°C}}{\ln\left(\frac{0.0488}{0.0467}\right)}$$

$$= -218,442 \text{ W}(\approx -745,534 \text{ Btu/h})$$

2.3.3.1.1.3 Steady-State Heat Conduction in Spheres In steady-state one-dimensional conduction heat transfer in a hollow sphere (Figure 2.7) with uniform wall thickness, inner radius r_i, outer radius r_o, and inner and outer surface temperatures of T_1 and T_2, respectively, as well as constant thermal conductivity and no heat generation within the sphere, the rate of heat conduction through the sphere's wall and A normal to the heat flow in the direction of temperature gradient, Fourier's equation takes the following form:

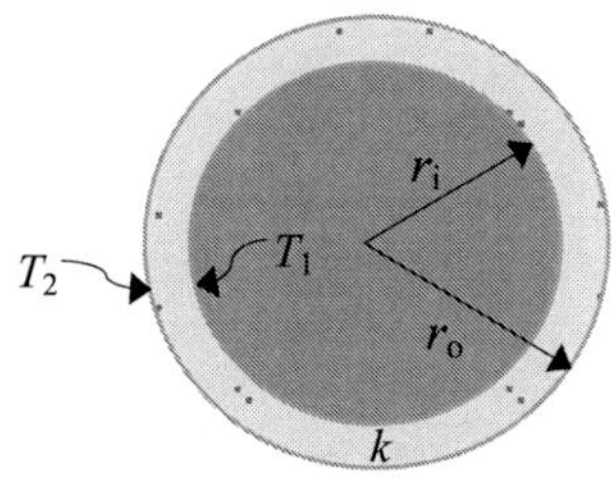

FIGURE 2.7 Heat conduction through a hollow sphere.
Source: *Greenhouse Engineering* (2021).

$$\dot{Q} = -k A \frac{dT}{dr} \quad (2.32)$$

where the surface area A is equal to $4\pi r^2$.

After integrating Equation (2.32), the following equation is obtained:

$$\dot{Q} = -\frac{4\pi k(r_i r_o)(T_1 - T_2)}{r_o - r_i} \quad (2.33)$$

Example 2.52: Heat Conduction through a Spherical Tank

Let's now consider an aluminum spherical tank with a 20 mm wall thickness and an inside diameter of 1.50 m with uniform thermal conductivity. Determine the steady-state conduction heat transfer rate when the inside surface temperature of the sphere is 30°C and the outside temperature is 10°C. Assume that the steady-state conditions prevail and the heat transfer is one-dimensional.

Solution

T_i = 30°C and T_o = 10°C; r_i = 1.50 m and r_{wall} = 0.02 m; k = 204.2 W/m°C (Table A.7). The outer radius of the tank is $r_o = r_i + r_{wall}$ = 1.50 + 0.02 m = 1.52 m.

Then, the rate of heat transfer through the sphere can be determined using Equation (2.33) as follows:

$$\dot{Q} = -\frac{4\pi k(r_i r_o)(T_1 - T_2)}{r_o - r_i}$$

$$\dot{Q} = -\frac{(4\pi)\left(204.2\frac{\text{W}}{\text{m.°C}}\right)(1.50\text{ m})(1.52\text{ m})(30-10)\text{°C}}{(1.52-1.50)\text{ m}}$$

$$= -5,847,635 \text{ W}(\approx -19,957,797 \text{ Btu/h})$$

2.3.3.1.2 Resistance Concept and Overall Heat Transfer

In heat transfer applications, the concept of thermal resistance is useful for analyzing steady-state problems. The steady-state heat conduction equation for a wall can be rearranged as

$$\dot{Q} = -k.A_m \frac{(T_1 - T_2)}{L_m} = -\frac{\Delta T}{R_t} \quad (2.34)$$

where A_m is the mean cross-sectional area normal to flow (m^2 or ft^2), L_m is the mean length of heat flow path (m or ft), ΔT is the overall temperature difference (°C or °F), and

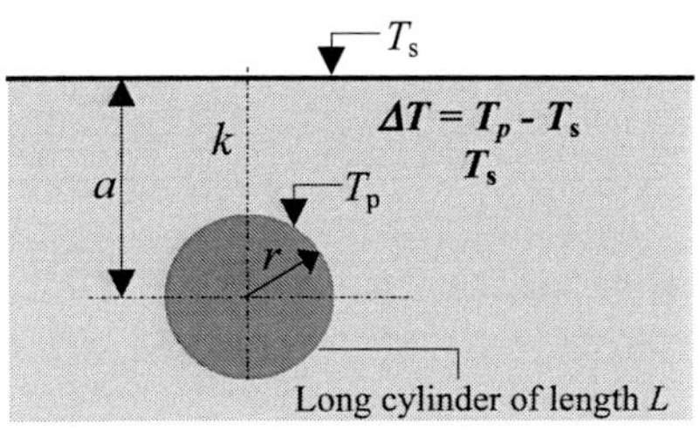

FIGURE 2.8 Heat conduction through a buried hollow cylinder.
Source: ***Greenhouse Engineering*** **(2021).**

$R_t = L_m / k.A_m$ is the thermal resistance (°C/W or °F.h/Btu), which is directly proportional to the mean length of the heat flow path, L_m, and inversely proportional to k and to the mean cross-sectional area normal to the flow, A_m. In essence, it is the resistance to heat flow of 1 m^2 (1 ft^2) of a material when the temperature difference between the two sides is 1°C (1°F).

Thermal resistance R_t is an additive quantity and does not include the boundary layer resistances; for instance, R for 10 cm thick material is twice as high as R for the same material but 5 cm thick. Equation (2.34) is similar to Ohm's law for electrical circuits, and this analogy can be utilized for heat conduction in complex shapes that cannot be easily solved by exact analytical means.

If the thermal resistance through the wall is expressed as R_w, then employing the resistance concept, Equation (2.29) takes the following form for the rate of heat conduction through the wall:

$$\dot{Q} = -\frac{(T_1 - T_2)}{R_w} \tag{2.35}$$

Other thermal resistance equations for a few common shapes are provided next. For instance, resistance for radial flow through a cylinder (Figure 2.6) is given by

$$R = \frac{\ln(r_o - r_i)}{2\pi\, k\, L} \tag{2.36}$$

and thermal resistance for a buried cylinder (Figure 2.8) is

$$R = \frac{\ln\left[\left(a + \sqrt{a^2 - r^2}\right)/r\right]}{2\pi\, k\, L} = \frac{Cosh^{-1}(a/r)}{2\pi\, k\, L}(L >> 2r) \tag{2.37}$$

Example 2.53: Heat Conduction through a Buried Polyethylene Tube

Now consider a polyethylene tube with a diameter of 25 mm buried in a 15 cm deep dense concrete floor with a uniform thermal conductivity of 1.4 W/(m.K). Determine the conduction heat transfer rate when the floor surface temperature is 17.5°C and the surface temperature of the buried tube is 32.5°C. Assume the tube length is 7.2 m, the steady-state conditions prevail, and the heat transfer is one-dimensional.

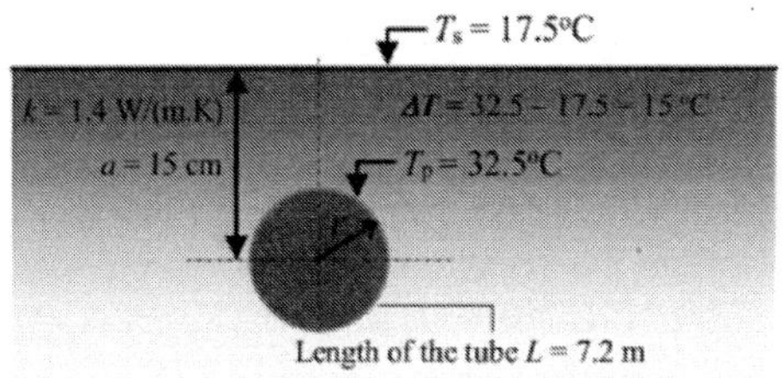

Solution

The tube radius is r = 0.025/2 = 0.0125 m; the depth of the tube is a = 0.15 m; and

$$L = 7.2 \text{ m} >> 2r = 0.025 \text{ m}.$$

The steady state one-dimensional rate of heat transfer through the tube can then be determined using Equation (2.38) as follows:

$$R = \frac{\ln\left[\left(a + \sqrt{a^2 - r^2}\right)/r\right]}{2\pi k L} = \frac{\ln\left[\left(0.15\text{m} + \sqrt{(0.15\text{m}^2) - (0.0125\text{m})^2}\right)/0.0125\text{m}\right]}{2\pi\left(1.4\frac{\text{W}}{\text{m.K}}\right)7.2\text{m}}$$

$$R = 0.0502 \text{ K/W}$$

Now using Equation (2.34), we can calculate the rate of heat transfer as follows:

$$\dot{Q} = -\frac{\Delta T}{R} = -\frac{15\text{K}}{0.0502\text{K/W}} = 299 \text{ W} (\approx 1{,}021 \text{ Btu/h})$$

Furthermore, resistance for radial flow through a sphere (Figure 2.6) can be expressed as

$$R = \frac{(1/r_i - 1/r_o)}{2\pi\, k} \tag{2.38}$$

The thermal resistance concept is quite useful for heat transfer problems involving combinations of conduction, convection, and radiation because in most steady-state heat transfer problems, more than one mode of heat transfer is involved. In these cases, different heat transfer coefficients can be combined into an overall coefficient so that we can calculate the total heat transfer from the terminal temperatures.

Next, consider a three-step steady-state heat transfer from a warmer fluid to a solid wall, through the wall, and then to a colder fluid. An overall coefficient of heat transfer U, based on the difference between the bulk temperatures of the two fluids, $T_i - T_o$, is calculated as follows:

$$\dot{Q} = -\frac{(T_i - T_o)}{R_t} = -UA(T_i - T_o) \tag{2.39}$$

where A is the surface area m^2 (ft^2) normal to the heat flow and U is the overall heat transfer coefficient W/m^2.°C (Btu/h.ft^2.°F). This gives us the heat transfer per unit time through an entire wall, ceiling, etc. section of 1 m^2 (1 ft^2) when the difference between the warmer and cool fluids is 1°C (1°F).

Here, U and the total resistance $R_t = 1/UA$ include the inside and outside boundary layer resistances, R_i and R_o, respectively, which are provided in Table A.9 for different flow directions and surface emissivities. The surface resistances combine the effects of conduction, convection, and radiation. The resistance for inside surfaces is usually for still air, and the outside surface resistances for winter and summer are based on 6.6 m/s (15 mph) and 3.3 m/s (7.5 mph) wind speeds, respectively.

Heat transfer resistances of some common materials are provided in Table A.11; however, the readers of this book are referred to *2013 ASHRAE Handbook of Fundamentals* (ASHRAE, 2013) for a long list of typical thermal properties of common building and insulating materials. Remember that the units of resistances in both Tables A.4 and A.5 are given per unit area. The surface resistances combine the effects of conduction, convection, and radiation. Surface resistances increase with decreasing emissivity. In buildings, the resistance for inside surfaces is usually taken for still air; for moving outside air, forced convection is predominant. Hence, R is independent of the slope and heat flow direction.

2.3.3.1.2.1 Heat Flow in Series Now consider the overall heat transfer through a wall composed of layers of different materials with different thicknesses (Figure 2.9).

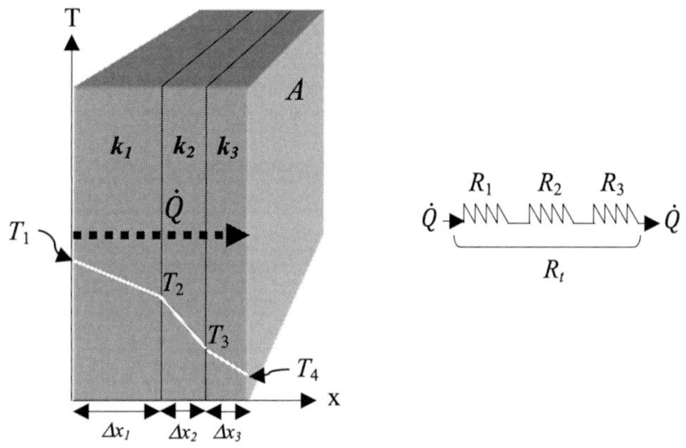

FIGURE 2.9 Thermal resistances in series.

Source: ***Greenhouse Engineering*** **(2021).**

The overall resistance of the thermal circuit becomes

$$R_t = R_1 + R_2 + R_3 = \frac{\Delta x_1}{k_1\,A} + \frac{\Delta x_2}{k_2\,A} + \frac{\Delta x_3}{k_3\,A} \tag{2.40}$$

Additionally, since the heat transfer $\dot{Q}_x$ through the circuit is constant, $\dot{Q}_x$ becomes

$$\dot{Q}_x = -\frac{(T_1 - T_2)}{\dfrac{\Delta x_1}{k_1\,A}} = -\frac{(T_2 - T_3)}{\dfrac{\Delta x_2}{k_2\,A}} = -\frac{(T_3 - T_4)}{\dfrac{\Delta x_3}{k_3\,A}}$$

Then, using the overall thermal resistance and terminal temperatures, the overall heat transfer rate can be determined as follows:

$$\dot{Q}_x = -\frac{(T_1 - T_4)}{R_t} \tag{2.41}$$

Thermal resistance also exists for the heat flow from the surface of an object to the surrounding environment. The surface resistances combine the effects of conduction, convection, and radiation. We can determine surface resistance from Newton's law of cooling:

$$\dot{Q} = -h\,A\left(T_s - T_\infty\right) \tag{2.42}$$

$$R_i = \frac{1}{h_i\,A} = \frac{\left(T_s - T_\infty\right)}{\dot{Q}} \tag{2.43}$$

where R_i is the surface thermal resistance, h_i is the convective heat transfer coefficient, A is the surface area normal to the direction of heat flow, T_s is the surface temperature, and T_∞ is the ambient temperature. The resistance for inside surfaces is usually for still air, and the outside surface resistances for winter and summer are based on 6.6 m/s (15 mph) and 3.3 m/s (7.5 mph) wind speeds, respectively (Table A.9).

Now, let's consider the composite wall configuration given in Figure 2.9, with interior surface temperature T_i and external surface temperature of T_o (Figure 2.10). Then, since $\dot{Q}_x$ through the circuit is constant, $\dot{Q}_x$ becomes

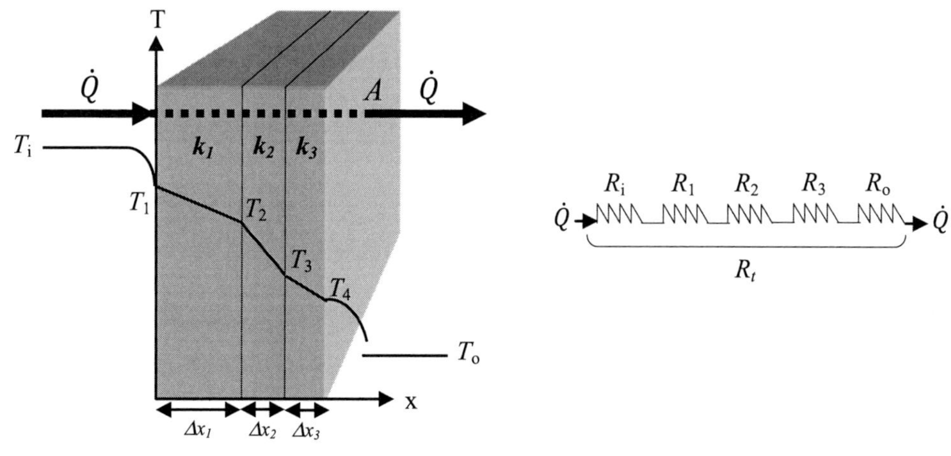

FIGURE 2.10 Thermal resistances in series including the surface resistances.

Source: ***Greenhouse Engineering*** **(2021).**

$$\dot{Q}_x = -\frac{(T_i - T_1)}{\frac{1}{h_i\ A}} = -\frac{(T_1 - T_2)}{\frac{\Delta x_1}{k_1\ A}} = -\frac{(T_2 - T_3)}{\frac{\Delta x_2}{k_2\ A}}$$
$$= -\frac{(T_3 - T_4)}{\frac{\Delta x_3}{k_3\ A}} = -\frac{(T_4 - T_o)}{\frac{1}{h_o\ A}} \tag{2.44}$$

where the overall resistance of the thermal circuit is

$$R_t = R_i + R_1 + R_2 + R_3 + R_o = \frac{1}{h_i\ A}$$
$$= \frac{\Delta x_1}{k_1\ A} = \frac{\Delta x_2}{k_2\ A} = \frac{\Delta x_3}{k_3\ A} = \frac{1}{h_o\ A} \tag{2.45}$$

When dealing with composite walls, it is easier to work with the overall heat transfer coefficient U, whose relationship with the total thermal resistance is given as

$$U = \frac{1}{R_t\ A} = \frac{1}{h_i} + \frac{\Delta x_1}{k_1} + \frac{\Delta x_2}{k_2} + \frac{\Delta x_3}{k_3} + \frac{1}{h_o} \tag{2.46}$$

Then, using R_t or U and the inside and outside temperatures, the heat transfer rate can be determined using Equation (2.39).

Example 2.54: Heat Transfer through Single-Pane Glazing

Consider a 2.2 m × 4.0 m × 5 mm thick single-pane glass sidewall. Determine the rate of overall heat transfer from a glasshouse with an indoor temperature of 20°C through the glass sidewall if the outside temperature is −10°C.

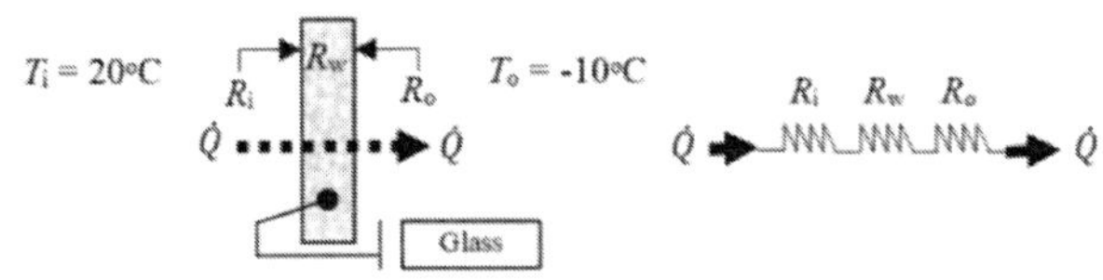

Solution

T_i = 20°C and T_o = − 10°C; A = 2.2 m × 4.0 m = 8.8 m²; R_i = 0.120 m² °C/W and R_o = 0.030 m²°C/W (Table A.9); R_w = 1.3×10⁻² m² °C/W (Table A.11).

Since the glass is 5 mm thick, R_w = (1.3 × 10⁻² m² °C/W)/2 = 0.65 × 10⁻² m² °C/W. Then, total R_t is

$$R_t = R_i + R_w + R_o$$
$$= 0.120\text{m}^2\frac{°\text{C}}{\text{W}} + 0.65\times10^{-2}\text{m}^2\frac{°\text{C}}{\text{W}} + 0.030\ \text{m}^2\frac{°\text{C}}{\text{W}}$$
$$R_t = 0.1565\ \text{m}^2\frac{°\text{C}}{\text{W}}$$

and the rate of heat transfer is

$$\dot{Q} = -\frac{A}{R_t}(T_1 - T_2) = -\frac{8.8\ \text{m}^2}{0.1565\ \text{m}^2.°\text{C/W}}\left[20°\text{C} - (-10°\text{C})\right]$$
$$Q = -1{,}687\dot{\text{W}}\left(\approx 5{,}758\frac{\text{Btu}}{\text{h}}\right)$$

Or, using the overall heat transfer coefficient,

$$U = \frac{1}{R_t A} = \frac{1}{0.1565\ \text{m}^2.°\text{C/W}} = 6.39\,\text{W/m}^2°\text{C}$$
$$\dot{Q} = -UA\,(T_1 - T_2)$$
$$= -\left(6.39\frac{\text{W}}{\text{m}^2}°\text{C}\right)(2.2\,\text{m}\times4.0\,\text{m})\left[20°\text{C} - (-10°\text{C})\right]$$
$$\dot{Q} = -1{,}687\ \text{W}\left(\approx 5{,}758\frac{\text{Btu}}{\text{h}}\right)$$

2.3.3.1.2.2 Heat Flow in Parallel Now consider the overall heat transfer through the parallel wall shown in Figure 2.11, which has three layers of the same thickness but with different materials, properties, and areas. The wall face temperatures are kept at T_1 and T_2.

One-dimensional heat flow in this configuration in essence has three different pathways: through the first section, through the second, and through the third, and the total heat flow rate is equal to the sum of the three. The individual wall areas normal to heat flow are A_1, A_2, and A_3. Here we can simply calculate the total resistance by employing the parallel circuit approach. We can first add the individual resistances to obtain R_t.

$$R_t = \left(\frac{1}{\frac{1}{\Delta x/k_1A_1} + \frac{1}{\Delta x/k_2A_2} + \frac{1}{\Delta x/k_3A_3}}\right) \tag{2.47}$$

Then, the rate of heat transfer through the composite wall can be determined using Equation (2.39).

Example 2.55: Heat Transfer through a Parallel Wall

Consider the 2.2 m high, 6.0 m wide, and 10 cm thick parallel wall provided in Figure 2.10 and the thermal conductivities given in Table A.7. The first 2 m of the wall is made of fir with thermal conductivity of 0.156 W/m°C, the next 2 m is sandstone (2.90 W/m°C), and the last 2 m is common brick (0.692 W/m°C). Determine the

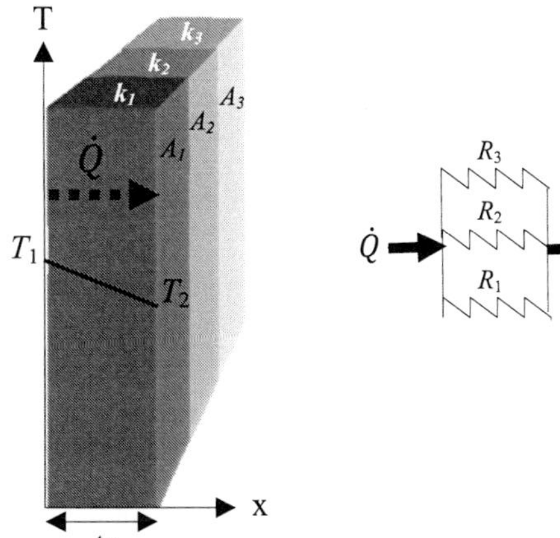

FIGURE 2.11 Thermal resistances in parallel circuits.
Source: *Greenhouse Engineering* (2021).

rate of heat transfer from the wall if the left side is kept at 20°C while the right side is kept at –20°C.

Solution

$$A_1 = A_2 = A_3 = 2.2 \text{ m} \times 2.0 \text{ m} = 4.4 \text{ m}$$

First calculate the individual resistances for different wall sections:

$$R_{\text{fir}} = \frac{0.1\text{m}}{\left(0.156 \text{ W/m.°C} \times 4.4 \text{ m}^2\right)} = 0.1457 \text{ °C/W}; \frac{1}{R_{\text{fir}}} = 6.683$$

$$R_{\text{sandstone}} = \frac{0.1\text{m}}{\left(2.90 \text{ W/m.°C} \times 4.4 \text{ m}^2\right)} = 0.00784 \text{ °C/W};$$

$$\frac{1}{R_{\text{sandstone}}} = 127.551$$

$$R_{\text{brick}} = \frac{0.1\text{m}}{\left(0.692 \text{ W/m.°C} \times 4.4 \text{ m}^2\right)} = 0.0328 \text{ °C/W};$$

$$\frac{1}{R_{\text{brick}}} = 30.488$$

Then, R_t and $\dot{Q}$ are calculated as

$$R_t = \left(\frac{1}{\frac{1}{R_{\text{fir}}} + \frac{1}{R_{\text{sandstone}}} + \frac{1}{R_{\text{brick}}}}\right)$$

$$= \left(\frac{1}{6.663 + 127.551 + 30.488}\right)$$

$$= \left(\frac{1}{164.902}\right) = 0.006064 \text{ °C/W}$$

$$\dot{Q} = -\frac{\Delta T}{R_t} = -\frac{[20-(-20)]\text{°C}}{0.006064\text{°C/W}}$$

$$= -6,596 \text{ W}\left(\approx -22,512\frac{\text{Btu}}{\text{h}}\right)$$

As a check,

$$\dot{Q}_{\text{fir}} = -\frac{\Delta T}{R_{\text{fir}}} = -\frac{[20-(-20)]\text{°C}}{0.1457 \text{ °C/W}} = -274.5 \text{ W}$$

$$\dot{Q}_{\text{sandstone}} = -\frac{\Delta T}{R_{\text{sandstone}}} = -\frac{[20-(-20)]\text{°C}}{0.00784 \text{ °C/W}} = -5,102 \text{ W}$$

$$\dot{Q}_{\text{brick}} = -\frac{\Delta T}{R_{\text{brick}}} = -\frac{[20-(-20)]\text{°C}}{0.0328\text{°C/W}} = -1,219.5 \text{ W}$$

$$\dot{Q}_{\text{total}} = -\left(\dot{Q}_{\text{fir}} + \dot{Q}_{\text{sandstone}} + \dot{Q}_{\text{brick}}\right)$$

$$= -(274.5 + 5,102 + 1,219.5)$$

$$= 6,596 \text{ W}(\approx 22,512 \text{ Btu/h})$$

The result of Example 2.51 shows quite a large heat loss from a bare pipe. To decrease the loss from a bare pipe, we usually insulate the pipe with a material with low thermal conductivity. This produces two resistances in a series thermal circuit, each having variable heat transfer areas (Figure 2.12).

Applying Equation (2.31) to each of the cylinders in Figure 2.12 yields the following. For the inner cylinder,

$$\dot{Q} = -2\pi k_1 L \frac{(T_1 - T_2)}{\ln\left(\frac{r_2}{r_1}\right)} \text{ and } R_{ti} = -\frac{\ln\left(\frac{r_2}{r_1}\right)}{2\pi k_1 L} \tag{2.48}$$

For the outer cylinder,

$$\dot{Q} = -2\pi k_2 L \frac{(T_2 - T_3)}{\ln\left(\frac{r_3}{r_2}\right)} \text{ and } R_{to} = -\frac{\ln\left(\frac{r_3}{r_2}\right)}{2\pi k_2 L} \tag{2.49}$$

We can now directly write that the overall resistance is the sum of R_{ti} and R_{to}, and using Equations (2.48) and (2.49), we can obtain the following:

$$\dot{Q} = -2\pi L \frac{(T_1 - T_3)}{\left(\frac{1}{k_1}\right) ln\left(\frac{r_2}{r_1}\right) + \left(\frac{1}{k_2}\right) \ln\left(\frac{r_3}{r_2}\right)} \tag{2.50}$$

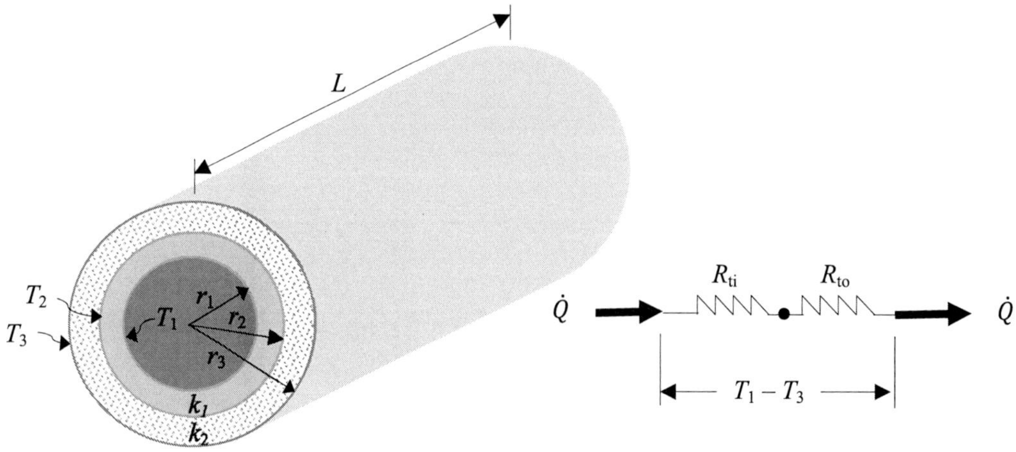

FIGURE 2.12 Heat transfer through a compound cylinder.
Source: *Greenhouse Engineering* (2021).

Example 2.56: Heat Conduction through an Insulated Cylinder

Now consider the same 51 mm steel tube with a 2.1 mm wall thickness and inside diameter of 46.7 mm with uniform thermal conductivity of 17 W/(m.°C) insulated with 2.5 cm mineral wool (k = 0.045 W/(m.°C)). Determine the steady-state conduction heat transfer rate when the inside surface temperature of the cylinder is 93°C and the outside temperature of the insulation is 25°C. Assume the pipe length is 30 m, the steady-state conditions prevail, and the heat transfer is one-dimensional.

Solution

T_1 = 93°C and T_3 = 90°C; r_i = 0.0467 m; r_{wall} = 0.0021 m; k_1 = 17 W/m.°C; k_2 = 0.045 W/m.°C; L = 30 m.

The outer radius of the pipe is $r_2 = r_1 + r_{wall} = 0.0467 +$ 0.0021 m = 0.0509 m, and the outer radius of the insulation is $r_3 = r_2 + r_{ins} = 0.0509 + 0.025$ m = 0.0759 m.

$$\left(\frac{1}{k_1}\right)\ln\left(\frac{r_2}{r_1}\right) = \left(\frac{1}{17\dfrac{\text{W}}{\text{m.°C}}}\right)\ln\left(\frac{0.0509\text{ m}}{0.0467\text{ m}}\right) = 0.005$$

$$\left(\frac{1}{k_2}\right)\ln\left(\frac{r_3}{r_2}\right) = \left(\frac{1}{0.045\dfrac{\text{W}}{\text{m.°C}}}\right)\ln\left(\frac{0.0759\text{ m}}{0.0509\text{ m}}\right) = 8.879$$

The rate of heat transfer through the compound pipe can then be determined using Equation (2.50) as follows:

$$\dot{Q} = -2\pi(30)\frac{(93-25)}{0.005+8.879}$$
$$= -1,442\text{ W } (\approx -4,922\text{ Btu/h})$$

2.3.3.1.3 Mean Temperature Difference (Heat Exchangers)

In industrial processes, unmixed fluids are separated with devices known as heat exchangers. The heat flow in a heat exchanger typically involves conduction heat transfer within the wall that separates the two heat-exchanging fluids and convection within each fluid in the exchanger.

Heat exchangers have various shapes depending on the function, such as condensers, evaporators, economizers, air heaters, water heaters, or re-heaters. Typically, heat exchangers are designated by the geometric shape and the relative directions of the flow of the heat exchanging fluids. For example, Figure 2.13a shows a concentric-tube (or double-pipe) heat exchanger, the simplest exchangers, which contain two pipes of different diameters with fluids flowing in the same (parallel flow) or opposite (counterflow) directions at different temperatures in each pipe.

Other common heat exchangers shown in the figure include shell-and-tube, plate-fin, compact plate, and frame type, which are designed to achieve greater heat transfer area per unit volume. Moreover, when a hot fluid flows through a porous storage mass, for instance, the porous storage mass stores the thermal energy temporarily until it is transferred back to a cold fluid. These types of heat exchangers are called regenerative.

When heat is exchanged between two fluids flowing through a heat exchanger, the local temperature difference, ΔT, varies along the flow path. Heat transfer is calculated using the following equation, which was developed for heat exchangers using Newton's law of cooling:

$$\dot{Q} = UA\Delta T_m \tag{2.51}$$

where U is the overall heat transfer coefficient with the same units as the convection heat transfer coefficient h, W/m^2 °C (Btu/h.ft^2.°F); A is the heat transfer surface area associated with U in m^2, and ΔT_m is an appropriate mean temperature difference in °C (°F).

U in Equation (2.51) is not usually constant for all locations in the heat exchanger; its local value is a function of the local fluid temperatures. However, it is a common practice to evaluate the individual heat transfer coefficients based on the arithmetic average of fluid temperatures. Similar to convection, we have thermal resistance equal to $1/UA$, the use of which will be best illustrated in a few example problems.

Figure 2.14 shows that the temperatures of the fluids vary continuously as they proceed through the heat exchanger. That is why the true mean temperature difference of fluids exchanging heat cannot be determined by simply subtracting temperatures. The true mean temperature difference can be obtained by assuming that overall U and the specific heat of each fluid are constant over the entire heat exchanger, the fluid flows are in steady state, and the exchange process is adiabatic (i.e., the heat losses are negligible). In Figure 2.14, h is for the hot fluid, and c is for the cold fluid. Furthermore, i denotes the temperature at the entry of a fluid of a heat exchanger, and o is the temperature of the fluid at the exchanger exit.

The arrowheads on the temperature curves show the direction of the flow of each flow. The largest temperature difference between the fluids in the heat exchanger at either inlet or outlet is expressed as θ_A, and the smallest is θ_B. Now, we can develop ΔT_m, also known as the logarithmic mean temperature difference, in Equation (2.51) for parallel-flow, counterflow, and constant-temperature heat exchangers as follows:

$$\Delta T_m = \frac{\theta_A - \theta_B}{ln\left(\dfrac{\theta_A}{\theta_B}\right)} \tag{2.52}$$

Since we assumed that the heat exchange is a steady-state, adiabatic process, applying the energy balance approach, the rate of heat transfer $\dot{Q}$ from the hot fluid must be equal to that received by the cold fluid as shown:

$$\dot{Q} = \dot{m}_h c_{p,h}\left(T_{h,o} - T_{h,i}\right) = m_c c_{p,c}\left(T_{c,o} - T_{c,i}\right) \tag{2.53}$$

where $\dot{m}_h$ and $\dot{m}_c$ denote the mass flow rates of the hot and cold fluids, respectively; $c_{p,h}$ and $c_{p,c}$ represent the specific heats; and $T_{h,i}$, $T_{h,o}$, $T_{c,i}$, and $T_{c,o}$ are the inlet and outlet temperatures of the hot and cold fluids. In the analysis of heat exchangers, the product of mass flow rate and specific heat, $C_h = (\dot{m}c_p)_h$ or $C_c = (\dot{m}c_p)_c$ is expressed as a single quantity known as the heat capacity rate.

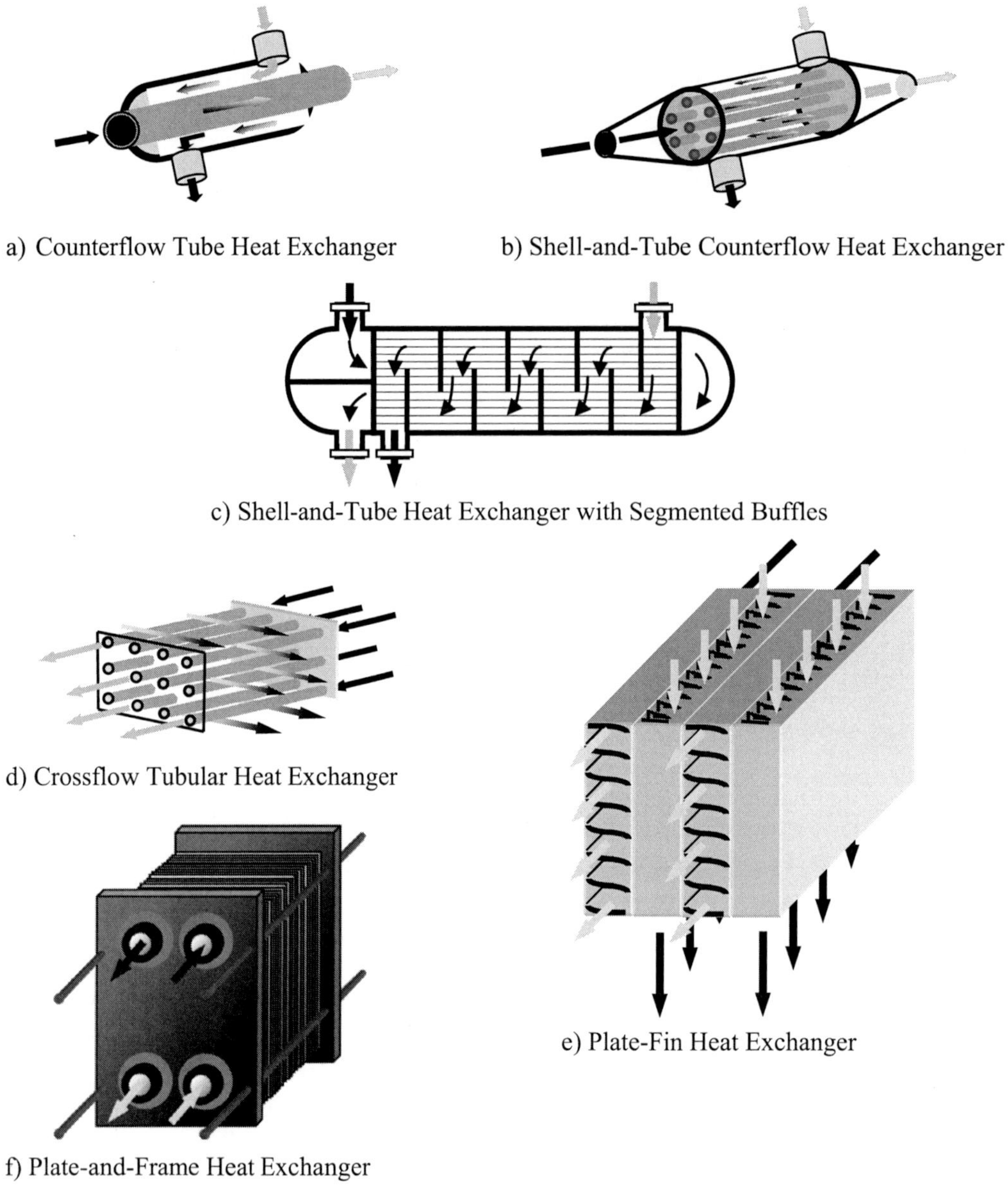

FIGURE 2.13 Schematic views of some heat exchanger types.
Source: ***Greenhouse Engineering*** **(2021).**

The relationship developed for the log mean temperature difference is specifically for parallel flow, counterflow, and constant-temperature heat exchangers. In crossflow or shell-and-tube heat exchangers with multiple passes, correction factors have been developed to account for the deviation from parallel or counterflow. Correction factor F is applied to the log mean temperature difference to obtain the true temperature difference. The log mean temperature difference approach is quite useful when the inlet and outlet temperatures of the fluids and the overall heat transfer coefficient are known and the required heat transfer area needs to be determined; this method can be employed because ΔT_m can easily be calculated in heat transfer design.

Generally speaking, along with the inlet temperatures and mass flow rates of the hot and cold fluids, the desired temperature of the cold or hot fluid is known as well. After picking the type of heat exchanger, the heat transfer surface area that will provide the desired outlet temperature is determined to select the appropriate heat exchanger size. Using Equation (2.51), the heat transfer area A can easily be determined, although for the same task, we can also use an alternate method, the number of heat transfer units (NTU), which uses three nondimensional parameters and is discussed in more depth later in the chapter. For the problem just given, we first determine the effectiveness ε and the capacity ratio Z and then calculate NTU.

In many cases, however, the temperatures of the fluids leaving the exchanger are not known. For instance, if the inlet temperatures, heat exchanger type, and size are known but the heat transfer rate and the fluid outlet temperatures need to be determined (a performance type problem), even though the log mean temperature difference method can be employed to solve the problem, it may take some time as it requires trial and error. In such cases, the NTU method becomes quite handy, and the problem can be solved easily because it requires no iteration.

Further, if the heat exchanger type and dimensions as well as the mass flow rates are known, NTU and Z can be

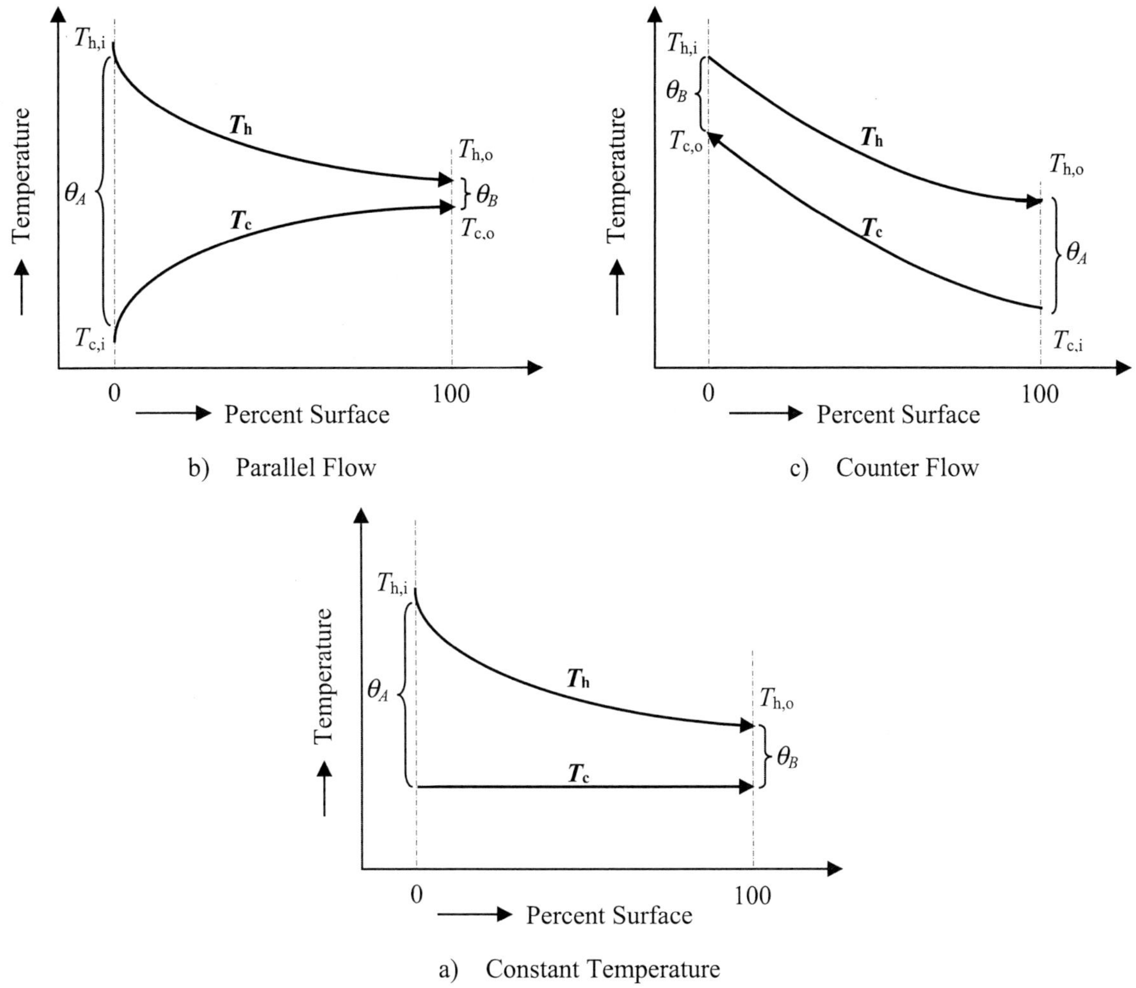

FIGURE 2.14 Fluid temperatures in different flow heat exchangers.
Source: ***Greenhouse Engineering*** **(2021).**

calculated first, and then ε can be determined using an appropriate diagram or an equation. As we will see shortly, since $\dot{Q}_{max}$ can be calculated, we can easily determine the actual heat transfer rate using Equation (2.54). Then, using Equation (2.53), we can determine the outlet temperatures of both fluids.

One of the nondimensional parameters used with *NTU* is known as the exchanger heat transfer effectiveness, ε; this is the ratio of the actual rate of heat transfer to the maximum possible rate of heat transfer and can be expressed as

$$\varepsilon = \frac{\dot{Q}}{\dot{Q}_{max}} \tag{2.54}$$

where $\dot{Q}$ is the actual rate of heat transfer and $\dot{Q}_{max}$ represents the maximum possible rate of heat transfer. Employing the energy balance approach on either the hot or cold fluid, $\dot{Q}_{max}$ is determined from the maximum temperature difference between the hot and cold fluids and the minimum heat capacity rate of the fluids as follows:

$$\Delta T_{max} = (T_{h,i} - T_{c,i}) \tag{2.55}$$

$$C_h = (\dot{m}_h c_{p,h}) \text{ and } C_c = (\dot{m}_c\, c_{p,c}) \tag{2.56}$$

If we consider $C_c < C_h$, then cold fluid will experience higher temperature difference, and the temperature of the cold fluid will approach the inlet temperature of the hot fluid ($T_{c,o} = T_{h,i}$) as the heat exchanger length approaches infinity. Therefore, based on Equation (2.53),

$$\dot{Q}_{max} = C_c\, \Delta T_{max} \tag{2.57}$$

Similarly, if $C_h < C_c$, then hot fluid will experience the higher temperature difference, and the temperature of the hot fluid will approach the inlet temperature of the cold fluid ($T_{h,o} = T_{c,i}$). Then, again based on Equation (2.53),

$$\dot{Q}_{max} = C_h\, \Delta T_{max} \tag{2.58}$$

Based on these discussions, we can write the following general equation:

$$\dot{Q}_{max} = C_{min}\, \Delta T_{max} \tag{2.59}$$

where C_{min} is either C_c or C_h, whichever is smaller.

Then, with the help of Equations (2.53) and (2.59), we can develop

$$\varepsilon = \frac{C_h\ (T_{h,i} - T_{h,o})}{C_{min}(T_{h,i} - T_{c,i})} \tag{2.60}$$

or

$$\varepsilon = \frac{C_h\,(T_{c,o} - T_{c,i})}{C_{min}(T_{h,i} - T_{c,i})} \tag{2.61}$$

As mentioned earlier, ε is nondimensional and has a value between zero and one ($0 \le \varepsilon \le 1$). If the effectiveness and the inlet and outlet temperatures of the hot and cold fluids are known, then the actual heat transfer rate can be determined from the following equation:

$$\dot{Q} = \varepsilon C_{min}\,(T_{h,i} - T_{c,i}) \tag{2.62}$$

Generally, for any heat exchanger, the heat transfer effectiveness can be expressed as a function of NTU and $Z = C_{min}/C_{max}$, the second nondimensional quantity. Then, we can write the following relationship:

$$\varepsilon = f\left(NTU, \frac{C_{min}}{C_{max}}, flow\,arrangement\right) \tag{2.63}$$

In this relationship, Z, depending on the heat capacity rates of the hot and cold fluids, can take the form of C_c/C_h or C_h/Cc. NTU, the third nondimensional quantity, is used extensively in heat exchanger analyses:

$$NTU = \frac{UA}{C_{min}} \tag{2.64}$$

Although ε is independent of the temperatures in the heat exchanger, it is dependent on the exchanger flow arrangement and geometry. Hence, the effectiveness relations are different for different types of heat exchangers.

For any heat exchanger in which Z is zero (for instance, if one fluid undergoes a phase change in a condenser or evaporator), the effectiveness becomes

$$\varepsilon = 1 - \exp(-NTU) \tag{2.65}$$

and heat transfer rate can be determined from

$$\dot{Q} = c_h\left(T_{h,o} - T_{h,i}\right) = c_c\left(T_{c,o} - T_{c,i}\right) \tag{2.66}$$

The heat transfer rate in terms of entering fluid temperatures was developed and presented earlier in Equation (2.62). The proper mean temperature difference for the rate equation (Equation (2.51)) is then provided by

$$\Delta T_m = \varepsilon\frac{(T_{h,i} - T_{c,i})}{NTU} \tag{2.67}$$

The effectiveness for parallel-flow heat exchangers is

$$\varepsilon = \frac{1 - e^{[-NTU\,(1+Z)]}}{1+Z}$$

If $Z = 1$, then the effectiveness equation becomes

$$\varepsilon = \frac{1 - e^{(-2\,NTU)}}{2} \tag{2.68}$$

The effectiveness of a counterflow heat exchanger is expressed as

$$\varepsilon = \frac{1 - e^{[-NTU\,(1-Z)]}}{1 - Z e^{[-NTU(1-Z)]}} \tag{2.69}$$

and if $Z = 1$, then the effectiveness equation becomes

$$\varepsilon = \frac{NTU}{1 + NTU} \tag{2.70}$$

Incropera and DeWitt (1996) and Kays and London (1964) show the relations of ε, NTU, and Z for other flow arrangements. Afgan and Schlunder (1974) and these authors also present graphical representations for further convenience.

Example 2.57: Water Preheating with a Plate-and-Frame Counterflow Heat Exchanger

Water is to be preheated in a plate-and-frame counterflow heat exchanger from 4 to 21°C at a mass flow rate of 12.5 kg/s. A hot stream of water from the boiler mixing group on manifold is used for this purpose. The temperature of the hot water at the heat exchanger inlet is 95°C, and it flows with a mass flow rate of 8.5 kg/s. The overall heat transfer coefficient of the exchanger is determined to be 5,750 W/m².K. Determine the total heat exchanger surface area required if the surface area per plate is 0.126 m². Assume that steady-state operation prevails and that heat losses and gains to/from the surroundings as well as the fouling effect are negligible.

Solution

The log mean temperature difference approach is quite useful when the inlet and outlet temperatures of the fluids and the overall heat transfer coefficient are known and the required heat transfer area needs to be determined. Since the inlet and outlet temperatures of the hot and cold fluids are known in this problem, we will employ the log mean temperature difference method because ΔT_m can easily be calculated. As mentioned earlier, this type of problem is a heat transfer design problem.

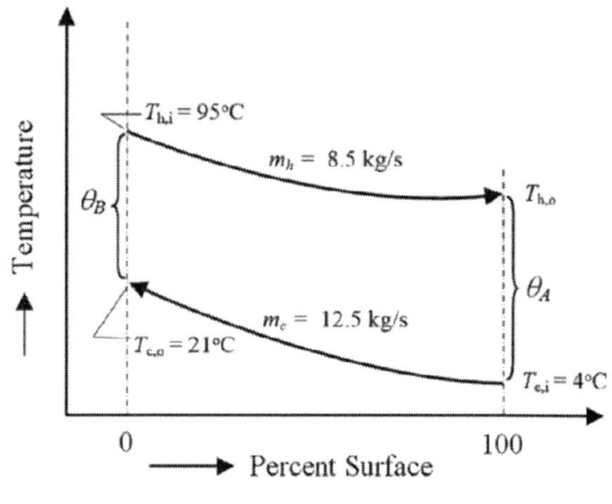

First, we can calculate the rate of heat transfer using Equation (2.53) as follows:

$$\dot{Q} = \dot{m}_c c_{p,c}\left(T_{c,o} - T_{c,i}\right) = \left(12.5\frac{\text{kg}}{\text{s}}\right)\left(4{,}180\frac{\text{J}}{\text{kg.°C}}\right)(21 - 4)\text{°C}$$

$$= 888{,}250\,\text{W} = 888.3\,\text{kW}$$

After the rate of heat transfer is calculated, we can use the energy balance approach to determine the outlet temperature of the hot water side:

$$\dot{Q} = 888.3\text{ kW} = \dot{m}_h c_{p,h}\left(T_{h,i} - T_{h,o}\right)$$

$$T_{h,o} = T_{h,i} - \frac{\dot{Q}}{\dot{m}_h\, c_{p,h}} = [95°\text{C}] - \frac{888.3\text{ kW}}{\left(8.5\frac{\text{kg}}{\text{s}}\right)\left(4.18\frac{\text{kJ}}{\text{kg.°C}}\right)} = 70°\text{C}$$

Now that the inlet and outlet temperatures are known, the log mean temperature difference can be determined using Equation (2.52) as follows:

$$\Delta T_m = \frac{\theta_A - \theta_B}{ln\left(\frac{\theta_A}{\theta_B}\right)}$$

For a counterflow heat exchanger,

$$\theta_A = \left(T_{h,i} - T_{c,o}\right) = (95 - 21)°\text{C} = 74°\text{C}$$

$$\theta_B = \left(T_{h,o} - T_{c,i}\right) = (70 - 4)°\text{C} = 66°\text{C}$$

Thence,

$$\Delta T_m = \frac{\theta_A - \theta_B}{ln\left(\frac{\theta_A}{\theta_B}\right)} = \frac{[74 - 66]°\text{C}}{\ln\left(\frac{74}{66}\right)} = 69.9°\text{C}$$

and the required heat transfer area can be calculated as

$$A = \frac{\dot{Q}}{U\Delta T_m} = \frac{888{,}250\text{ W}}{\left(5{,}750\frac{\text{W}}{\text{m}^2.°\text{C}}\right)(69.9°\text{C})} = 2.21\text{m}^2$$

The required number of heat exchanger plates can then be calculated as

$$\text{Number of plates} = \frac{A}{\text{Area per plate}} = \frac{\left(2.21\text{m}^2\right)}{\left(0.126\text{ m}^2\right)}$$
$$= 17.5 \approx 18$$

Example 2.58: Water Preheating with a Double-Pipe Counterflow Heat Exchanger

A double-pipe counterflow heat exchanger is used to heat water for a greenhouse or aquaculture operation. The inlet temperatures of the hot and cold streams are 35°C and 4°C, respectively. The cold stream of irrigation/culture water flows with a mass flow rate of 1.0 kg/s, and the hot stream flows at 1.5 kg/s. If the overall heat transfer coefficient is 800 W/m².°C and the heat transfer area is 4 m², determine the outlet temperatures of both the hot and cold streams.

Solution

Here, we can use the *NTU* method to determine the outlet temperatures easily. Based on the givens in this problem (performance type), we will first calculate *NTU* and *Z*; then, using Equation (2.69) (or a diagram) for counterflow heat exchanger, ε will be determined. Additionally, we will use Equation (2.59) to determine $\dot{Q}_{max}$, and after that, we can easily determine the actual heat transfer rate using Equation (2.54); then, using Equation (2.53), we will determine the outlet temperatures of both fluids.

The heat capacity rates for the cold and hot water can be calculated as follows:

$$C_c = \left(\dot{m}_c c_{p,c}\right) = \left(1.0\frac{\text{kg}}{\text{s}}\right)\left(4{,}180\frac{\text{J}}{\text{kg.°C}}\right) = 4{,}180\frac{\text{W}}{°\text{C}}$$

$$C_h = \left(\dot{m}_h c_{p,h}\right) = \left(1.5\frac{\text{kg}}{\text{s}}\right)\left(4{,}180\frac{\text{J}}{\text{kg.°C}}\right) = 6{,}270\frac{\text{W}}{°\text{C}}$$

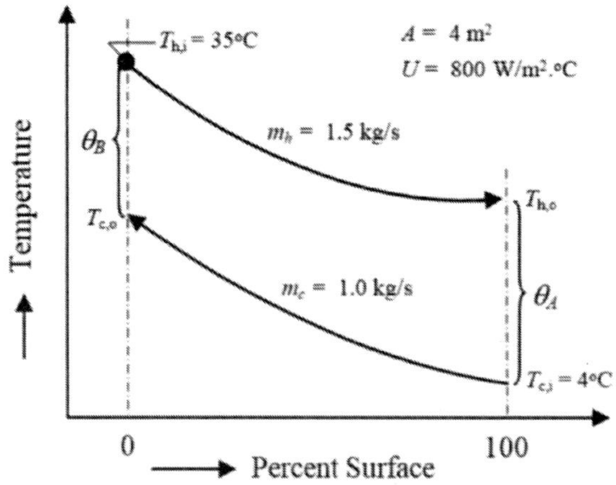

This shows that the minimum heat capacity rate C_{min} is for the cold fluid. Now, we can first calculate *NTU* using Equation (2.64),

$$NTU = \frac{UA}{C_{min}} = \frac{\left(800\frac{\text{W}}{\text{m}^2.°\text{C}}\right)\left(4\text{ m}^2\right)}{\left(4{,}180\frac{\text{W}}{°\text{C}}\right)} = 0.77$$

and then the capacity ratio $Z = \frac{C_{min}}{C_{max}}$ as follows:

$$Z = \frac{C_{min}}{C_{max}} = \frac{\left(4{,}180\frac{\text{W}}{°\text{C}}\right)}{\left(6{,}270\frac{\text{W}}{°\text{C}}\right)} = 0.67$$

Knowing *NTU* and *Z*, and using Equation (2.69) for counterflow heat exchanger, ε can be determined as

$$\varepsilon = \frac{1 - e^{[-NTU(1-Z)]}}{1 - Ze^{[-NTU(1-Z)]}} = \frac{1 - e^{[-0.77(1-0.67)]}}{1 - 0.67\ e^{[-0.77(1-0.67)]}} = 0.47$$

Then, using Equation (2.59), the maximum rate of heat transfer can be determined as follows:

$$\dot{Q}_{max} = C_{min}\Delta T_{max} = C_c\ (T_{h,i} - T_{c,i}) = 4{,}180\frac{\text{W}}{°\text{C}}(35 - 4)°\text{C}$$

$$\dot{Q}_{max} = 129{,}580\text{ W} = 129.6\text{ kW}$$

After that, we can easily determine the actual heat transfer rate using Equation (2.54) and ε:

$$\dot{Q} = \varepsilon\dot{Q}_{max} = 0.47(129.6\text{kW}) = 60.9\text{kW}$$

Now, using Equation (2.53), we can determine the outlet temperatures of both fluids. First, the temperature of the irrigation/culture water is

$$T_{c,o} = \frac{\dot{Q}}{\dot{m}_c\ c_{p,c}} + T_{c,i} = \frac{60.9\text{ kW}}{4.18\frac{\text{kW}}{°\text{C}}} + 4°\text{C} = 18.6°\text{C}$$

Then, the outlet temperature of the hot water stream is calculated as

$$T_{h,o} = T_{h,i} - \frac{\dot{Q}}{\dot{m}_h\, c_{p,h}} = (35°C) - \frac{60.9\ \text{kW}}{6.27\ \frac{\text{kW}}{°C}} = 25.3°C$$

Example 2.59: Finned-Pipe Crossflow Heat Exchanger for Recovering Waste Heat from Flue Gas

Let's now consider a crossflow finned-pipe compact heat exchanger used to recover waste heat from boiler flue gas in a greenhouse operation. The heat-exchanger body is made of aluminum, the pipe inside diameter is 13.8 mm, and all the other dimensions and overall configuration are shown in the figure provided in the solution.

The mass flow rate on the flue gas (hot) side is 1.25 kg/s normal to the heat exchanging pipes at 1 atm pressure and 550°C and the cross-sectional area of the heat exchanger is 0.20 m². The overall heat transfer coefficient based on the flue gas (hot) side is 95 W/m².°C. If the mass flow rate of water is 5 kg/s and the inlet and outlet temperatures are 75°C and 95°C, respectively, determine the heat exchanger volume, the number of pipelines in the flue gas flow direction, and the outlet temperature of the flue gas. Assume that the flue gas has the thermal properties of air at the same temperature and pressure, and the ratio of flue gas (hot) side heat transfer surface area to heat exchanger volume, α, is 269 m²/m³.

Solution

Flue gas (air) at$\bar{T}$ = 430°C: $c_p = 1{,}075\ \frac{\text{J}}{\text{kg.°C}}$

Water at $\bar{T}$ = 85°C: $c_p = 4{,}201\ \frac{\text{J}}{\text{kg.°C}}$

We first calculate the heat capacity rates for the cold and hot fluids:

$$C_c = (\dot{m}_c\, c_{p,c}) = \left(5.0\ \frac{\text{kg}}{\text{s}}\right)\left(4{,}201\ \frac{\text{J}}{\text{kg.°C}}\right) = 21{,}005\ \frac{\text{W}}{°C}$$

$$C_h = (\dot{m}_h c_{p,h}) = \left(1.25\ \frac{\text{kg}}{\text{s}}\right)\left(1{,}075\ \frac{\text{J}}{\text{kg.°C}}\right) = 1{,}344\ \frac{\text{W}}{°C}$$

Then, we can calculate the required rate of heat transfer using Equation (2.53) as follows:

$$\dot{Q} = C_c\left(T_{c,o} - T_{c,i}\right) = \left(21{,}005\ \frac{\text{W}}{°C}\right)[95-75]°C = 420{,}100\ \text{W}$$

C_{min} is for the hot fluid (flue gas), so using Equation (2.59), the maximum rate of heat transfer can then be determined as follows:

$$\dot{Q}_{max} = C_{min}\,\Delta T_{max} = C_h\left(T_{h,i} - T_{c,i}\right) = 1{,}344\ \frac{W}{°C}\ (550-75)°C$$

$$\dot{Q}_{max} = 638{,}400\ \text{W}$$

Then, using Equation (2.54), the heat exchanger effectiveness can be calculated as

$$\varepsilon = \frac{\dot{Q}}{\dot{Q}_{max}} = \frac{420{,}100\ \text{W}}{638{,}400\ \text{W}} = 0.658$$

and we can calculate the capacity ratio $Z = \frac{C_{min}}{C_{max}}$ as follows:

$$Z = \frac{C_{min}}{C_{max}} = \frac{\left(1{,}344\ \frac{\text{W}}{°C}\right)}{\left(21{,}005\ \frac{\text{W}}{°C}\right)} = 0.06$$

Using ε as calculated in the next equation or ε, the capacity ratio Z and the effectiveness graph for a non-mixing single-pass crossflow heat exchanger provided by Incropera and DeWitt (1996), the NTU can be determined as 1.00.

$$\varepsilon = 1 - \exp\left[\left(\frac{1}{Z}\right)(NTU)^{0.22}\left\{\exp\left[-Z(NTU)^{0.78}\right]-1\right\}\right]$$

$$NTU = \frac{U_h\, A_h}{C_{min}} = 1.00$$

Then, using this NTU, the hot side heat transfer surface area A_h is found:

$$A_h = \frac{1.00 \times 1{,}344\ \text{W/°C}}{95\ \text{W/m}^2.\text{°C}} = 14.15\ \text{m}^2$$

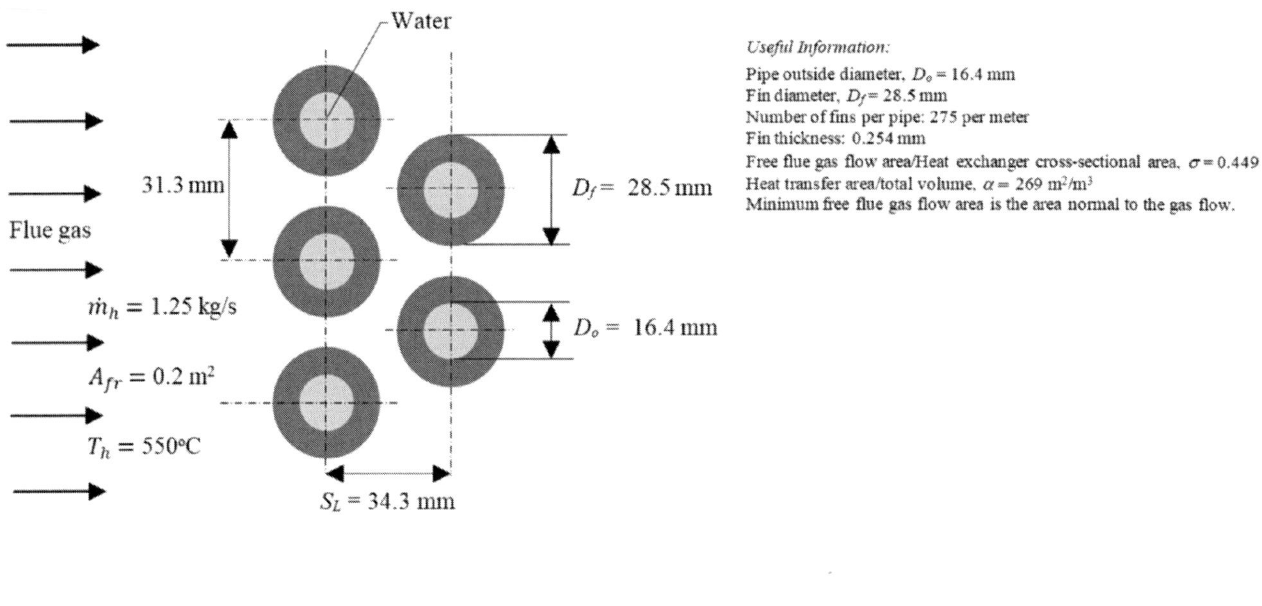

Since the hot (flue gas) side heat transfer surface area per heat exchanger volume $\alpha = \frac{A_h}{V}$ is already given as 269 m^2/m^3, we can easily determine the required heat exchanger volume V:

$$V = \frac{A_h}{\alpha} = \frac{14.15\ m^2}{269\ m^2/m^3} = 0.0526\ m^3$$

Since we know V, we can now determine the heat exchanger dimensions in the flue gas flow direction using the following expression for N_L, which gives the number of pipe rows in the direction of flue gas flow:

$$N_L \approx \frac{L - D_f}{S_L} + 1$$

where $L = \frac{V}{A_{fr}} = \frac{0.0526\ m^3}{0.20\ m^2} = 0.263\ m$, D_f is the fin diameter, S_L is the distance between the front and back row pipe centers.

$$N_L \approx \frac{(263 - 28.5)\ mm}{34.3\ mm} + 1 = 7.84 \approx 8$$

The outlet temperature of the flue gas is,

$$T_{h,o} = T_{h,i} - \frac{\dot{Q}}{C_h} = (550°C) - \frac{420{,}100\ W}{1{,}344\ \frac{W}{°C}} = 237°C$$

The result of this equation for calculating outside temperature shows that our previously assumed average flue gas (air) temperature of 430°C was reasonable.

2.3.3.1.4 Transient Heat Conduction

So far we have focused on steady-state conduction heat transfer, where the temperatures do not change over time. However, quite often the heat transfer and temperature distribution under unsteady or transient conditions ought to be known. In transient conditions, the temperature of the body of interest at any point as well as its heat content varies temporally and spatially. Time-dependent transient heat transfer problems are encountered in many energy engineering applications, for instance differing external air temperatures and varying solar irradiation that affects the heating and cooling loads of a controlled environment such as a greenhouse or a livestock or aquatic building, or differing temperatures of the building materials.

The fundamental equation for three-dimensional transient conduction in solids or fluids is

$$\frac{\partial T}{\partial t} = \alpha\left(\frac{\partial^2 T}{\partial x^2} + \frac{\partial^2 T}{\partial y^2} + \frac{\partial^2 T}{\partial z^2}\right) \tag{2.71}$$

where $\alpha = \frac{k}{\rho c_p}$ is the thermal diffusivity, k is the thermal conductivity, ρ is the density, and c_p is the specific heat. This equation tells us that in the case of high thermal conductivity and low density and/or specific heat, thermal diffusivity is high, that is, heat diffusion is fast.

To predict the rate of temperature change of a material being held at constant volume with uniform temperature, such as a well-stirred volume of fluid that changes temperature due to a net rate of heat gain or loss, the proper equation can be written as

$$m c_v \frac{\partial T}{\partial t} = -Q_{net} \tag{2.72}$$

where m is the mass of the material and c_v is the specific heat at constant volume. If the heating or cooling takes place at constant pressure, then c_v is replaced by c_p.

We should also keep in mind that for liquids and solids, c_v and c_p are almost equal; therefore, we can use c_p, and the small error caused can be ignored. In Equation (2.72), Q_{net} can include heat transfer due to conduction, convection, or radiation. When convective boundary conditions are needed in the solution for slower transient conditions, then we often use the heat transfer coefficients (h) based on steady-state correlations.

2.3.3.2 Convection Heat Transfer

Convection heat transfer can involve energy transfer by conduction and eddy mixing and diffusion due to bulk fluid motion. The transferred thermal energy is usually sensible heat, and in cases where the phase changes from liquid to vapor, the convection heat transfer also includes latent heat exchange.

Let's now consider heat transfer to a flowing fluid inside a conduit. If the Reynolds number (Re) is large enough, then there exist three flow regions: i) a laminar sublayer, the layer immediately adjacent to the wall where the heat transfer mode is thermal conduction; ii) a buffer layer, the transition region outside the laminar sublayer, where both conduction and eddy mixing (convection) are significant; and iii) a turbulent region beyond the buffer layer extending to the center of the conduit, where the dominant mode of transfer is convection.

The main body of fluid in most equipment is in turbulent mode, and a laminar layer is always present at the solid walls. In low-velocity or viscous flows where low Reynolds numbers prevail, the entire flow can be laminar with no transition or eddy region. Heat transfer by convection can be further categorized based on the nature of flow. When the flow is induced by buoyancy forces arising from density variations caused by temperature variations in the fluid, it is termed free or natural convection. On the other hand, if fluid currents are caused by external sources, for instance, a fan, pump, or atmospheric wind, then the mode of heat transfer is termed forced convection.

The fundamental equation for convection heat transfer is known as Newton's law of cooling and is expressed as follows:

$$\dot{Q} = -h A_s (T_s - T_\infty) \tag{2.73}$$

where h represents the convection heat transfer coefficient (W/m².°C or Btu/h.ft².°F), which includes all effects regarding the convection heat transfer and dependent upon the boundary layer conditions; that is, the nature of the fluid flow, thermal properties, surface geometry, and physical properties all impact the magnitude of convection heat transfer. Typical values of h are provided in Table A.12. In Equation (2.73), A_s is the surface area (m^2 or ft^2), T_s is the surface temperature (°C or °F), and T_∞ is the temperature of the fluid flowing over the surface (°C or °F). Newton's law of cooling does not consider radiation heat transfer, which in many cases is smaller than conduction or convection involving a surface and a fluid; however, in the case of high surface temperatures and natural convection, the rate of heat transfer due to radiation is comparable with that of natural convection.

When compared with Equation (2.34), we can write the convective resistance term (boundary layer resistance or film resistance) R_c in Equation (2.73) as follows:

$$R_c = \frac{1}{h A_s} \tag{2.74}$$

The thermal resistance for R_c is treated in the same manner as the resistance in conduction discussed earlier.

Example 2.60: Heat Convection from a Steel Tube

Now assume that the 51 mm steel tube in Example 2.51 has a heat transfer coefficient on the outside due to natural convection to the surrounding air at 20°C of 5.1 W/m².°C. Determine the heat transfer from the given tube to the greenhouse air.

Solution

T_i = 93°C; r_i = 0.0467 m and r_{wall} = 0.0021 m; L = 30 m; k = 17 W/m.°C

The outer radius of the pipe $r_o = r_i + r_{wall}$ = 0.0467 m + 0.0021 m = 0.0488 m

The resistance of the pipe is

$$R_{pipe} = \frac{\ln\left(\frac{r_o}{r_i}\right)}{2\pi k L} = \frac{\ln\left(\frac{0.0488}{0.0467}\right)}{2\pi(17\,\text{W/m.°C})\times 30\,\text{m}}$$

$$= 1.3734\times 10^{-5}\,°\text{C/W}$$

$$R_{convection} = \frac{1}{hA} = \frac{1}{(5.1\,\text{W/m}^2.°\text{C})(\pi\times 0.0976\,\text{m}\times 30\,\text{m})}$$

$$= 0.02133\,°\text{C/W}$$

$$R_{total} = 6.025\times 10^{-4} + 0.02133 = 0.02134\,°\text{C/W}$$

$$\dot{Q} = -\frac{(T_1 - T_2)}{R} - \frac{(93-20)°\text{C}}{0.02193\,°\text{C/W}}$$

$$= -3,420\,\text{W}\left(\approx -11,673\frac{\text{Btu}}{\text{h}}\right)$$

Here, we have observed that the outside film resistance acted as an insulator, and the resistance of the pipe is negligible (see Example 2.51).

Furthermore, with the help of the Nusselt number (Nu), also known as the dimensionless convection heat transfer coefficient, h is nondimensional, and the relationship is defined in the following equation:

$$\text{Nu} = \frac{hD}{k} \text{ or } \frac{hL}{k} \tag{2.75}$$

where D and L are the characteristic length and diameter, respectively, and k is the fluid thermal conductivity. Nu in essence provides the ratio of the convective heat transfer through a fluid layer—from temperature T_1 on one side to T_2 on the other in the presence of motion—to the conductive heat transfer through the same fluid when stagnant:

$$\text{Nu} = \frac{\dot{Q}_{conv}}{\dot{Q}_{cond}} = \frac{h(T_1 - T_2)}{\frac{k(T_1 - T_2)}{L}} = \frac{hL}{k} \tag{2.76}$$

As will be discussed later in the sections on natural and forced convection, using dimensional analysis, the nondimensionalized expression of Nu reduces the total number of variables for natural and forced convection heat transfers. Once Nu is determined using a proper correlation, then using the relationship expressed in Equation (2.76), we can easily determine the convective heat transfer coefficient.

Let's now recall that earlier in this chapter, we considered a composite wall configuration that faced indoors at temperature T_i and faced outdoors at T_o (Figure 2.10) and determined the overall resistance of the thermal circuit considering the inside and outside boundary layer resistances as provided in Equation (2.45):

$$R_t = \frac{1}{h_i A} + \frac{\Delta x_1}{k_1 A} + \frac{\Delta x_2}{k_2 A} + \frac{\Delta x_3}{k_3 A} + \frac{1}{h_o A} \tag{2.77}$$

We also stated that when dealing with composite walls, it is much easier to work with the overall heat transfer coefficient U, whose relationship with the total thermal resistance is provided earlier in Equation (2.46) as

$$U = \frac{1}{R_t} = \frac{1}{h_i} + \frac{\Delta x_1}{k_1} + \frac{\Delta x_2}{k_2} + \frac{\Delta x_3}{k_3} + \frac{1}{h_o} \tag{2.78}$$

Then, using the overall R_t or overall U and the inside and outside temperatures, the heat transfer rate can be easily determined using Equation (2.39).

2.3.3.2.1 *Natural Convection*

As mentioned earlier, when the flow is induced by buoyancy forces arising from density variations caused by temperature variations in the fluid, natural convection occurs. Using dimensional analysis, the following nondimensionalized expression that reduces the total number of variables was developed for natural convection:

$$\frac{hD}{k} = \text{Nu} = A(\text{Gr})^a (\text{Pr})^b \tag{2.79}$$

where Nu is the Nusselt number; Gr is the Grashof number; Pr is the Prandtl number, the details of which are provided in Table A.6; and A, a, and b are constants that change depending on the system under consideration. We should remember that Nu, Gr, and Pr are dimensionless numbers and that consistent units must be employed throughout. The properties of fluid such as density, viscosity, specific heat and thermal conductivity are usually evaluated at the film temperature of fluid. The characteristic length is the diameter of a tube or the length of the plane plate if a plate is involved.

The boundary layer of the fluid is either laminar or turbulent, and this situation will affect the constants in Equation (2.79). Experimentally observed values indicate that h increases (moving from laminar to turbulent) when the product of (Gr) × (Pr) exceeds 10^8. Heat transfer literature provides natural and forced heat transfer coefficients for different geometries and orientations. As mentioned earlier, natural convection is a consequence of density differences in the fluid; since the warmer fluid molecules rise, the orientation of the hot surface impacts the rate of heat transfer. As Table A.9 shows, a hot horizontal surface has lower thermal resistance than a vertical surface because the hot fluid molecules from the horizontal surface rise more easily and the flow becomes more turbulent. However, the thermal resistance increases if the heat transfer from the horizontal surface is downward because the hot fluid molecules have the most difficult time rising from this orientation. Furthermore, fluid flow over hot surfaces experiences frictional resistance that can slow down the fluid motion especially in forced convection, which is discussed in the following section.

2.3.3.2.2 Forced Convection

Forced convection flow takes place between a solid surface and a moving fluid and can be either laminar or turbulent, inside or outside tubes; it is also involved in phase change processes such as boiling and condensation. For instance, heat exchangers and forced water and air heaters, coolers, condensers, and evaporators are all forced convection devices.

Before applying Newton's law of cooling provided in Equation (2.73), as with natural convection as well, we first need to determine whether the boundary layer is laminar or turbulent and then calculate h accordingly. Nu-Re correlations can be employed to calculate forced convection heat transfers; these are provided in heat transfer literature for different fluids and flow regimes. The expressions for Nu and Re are provided in Table A.6.

As mentioned earlier, in the turbulent region, the dominant mode of transfer is eddy mixing, that is, turbulent mixing (convection). In the region where we have the transition from laminar to turbulent, Re is called the critical Reynolds number and is a function of the surface geometry and fluid flow conditions. Because of the number of cases involved in forced convection, h can be determined from the following generalized equation whose specific constants and shapes are provided in the heat transfer literature:

$$\frac{hD}{k} = \text{Nu} = C(\text{Re})^a(\text{Pr})^b \tag{2.80}$$

where Nu is the Nusselt number; Re is the Reynolds number; Pr is the Prandtl number; and a, b, and C are constants that vary. It must be remembered that Nu, Re, and Pr are dimensionless numbers, and consistent units must be employed throughout. The density, viscosity, specific heat, and thermal conductivity are usually calculated at the film temperature of fluid. D is the diameter of a tube, inside or outside, when a tube is involved; however, the relevant parameter is the length of the plane plate if a plate is involved; for other shapes, the hydraulic diameter is used, and u is the average velocity of the flowing fluid.

2.3.3.2.2.1 External Flow Forced Convection If a fluid flows over an external surface (flat surface, pipe, cylinder, sphere, etc.), the flow is external. In such flows, to determine the heat transfer rates, we employ experimentally developed correlations between Nu, Re, and Pr. The fluid properties used in determining these dimensionless parameters are typically evaluated at the film temperatures T_f, which is the average of the surface temperature and the bulk temperature of the fluid. Forced convection correlations for different geometries and different external flow regimes are covered in depth in the heat transfer literature and therefore will not be repeated here.

Example 2.61: External Flow Convective Heat Loss from a Flat Plate Collector

Let's now consider a horizontal flat plate solar collector. As shown, the top cover (3 m × 1 m) of the collector has a surface temperature of 50°C. If the wind speed is 6.5 m/s and the ambient air temperature is 4°C, assuming steady-state conditions, determine the rate of convective heat loss from the cover.

Solution:

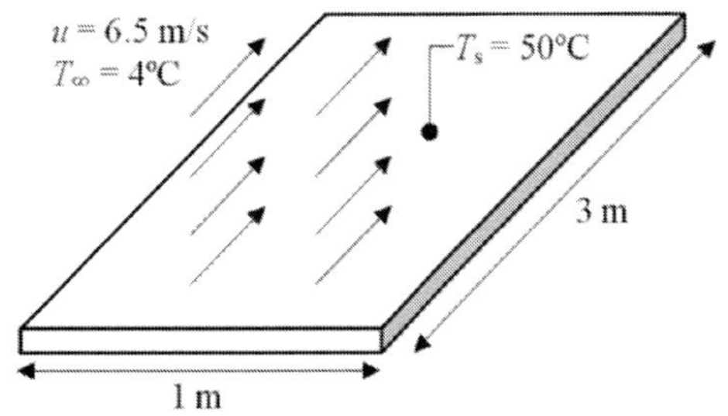

We first need to evaluate the properties of air at T_f needed for the dimensionless Re and Pr, and T_f is given by

$$T_f = \frac{T_s + T_\infty}{2} = \frac{(50+4)°\text{C}}{2} = 27\,°\text{C}$$

Using T_f = 27°C and Table A.13, we can now determine air density ρ as 1.1765 kg/m³, kinematic viscosity v as 15.76 × 10^{-6} m²/s, thermal conductivity k as 0.0261 W/m.K, and Pr as 0.714.

Re at the cover end can be determined as

$$\text{Re} = \frac{uL}{v} = \frac{\left(6.5\frac{\text{m}}{\text{s}}\right)(3\,\text{m})}{\left(15.76\times10^{-6}\right)\text{m}^2/\text{s}} = 1{,}237{,}310$$

For Re ≤ 10^8 and 0.6 < Pr < 60, the proper equation from the literature is provided as

$$\text{Nu} = \left(0.037\,\text{Re}^{0.8} - 871\right)\text{Pr}^{1/3}$$

Then,

$$\text{Nu} = \left(0.037\,(1{,}237{,}310)^{0.8} - 871\right)0.714^{\frac{1}{3}} = 1{,}696$$

After determining Nu, we can then calculate h using the following:

$$h = \text{Nu}\frac{k}{L} = 1{,}696\left(\frac{0.0261\,\text{W/m.K}}{3\,\text{m}}\right) = 14.75\frac{\text{W}}{\text{m}^2.\text{K}}$$

Now that we have h, using Newton's law of cooling (Equation (2.73)), the rate of convective heat loss from the cover can finally be determined as follows:

$$\dot{Q} = -hA(T_s - T_\infty) = -\left(14.75\frac{\text{W}}{\text{m}^2.\text{K}}\right)(3\text{ m}\times 1\text{ m})(50-4)°\text{C}$$

$$= -2{,}036\text{ W}(\approx -6{,}948\text{ Btu/h})$$

Example 2.62: External Flow Convective Heat Loss from a Hot Water Storage Tank

The thermal energy generated during CO_2 dosing in a greenhouse operation is stored in a hot water storage tank for later use. Let's now consider that a cylindrical thermal energy storage tank with a diameter of 4 m and a height of 6 m is used to store the heat generated during the dosing. Assuming that the storage tank has a uniform and constant outside surface temperature of 50°C, and the steady-state conditions prevail, determine the rate of convective heat loss from the hot water storage tank if the average wind speed is 1.25 m/s and the ambient air temperature is −10°C. The efficiency of a thermal energy storage decreases as the thermal losses from the tank increase. If the overall efficiency of a hot water storage tank is defined as the ratio of the heat recovered to the heat input, determine the overall thermal efficiency of the tank for a day for the conditions just provided if the total energy input was 33 GJ.

Solution

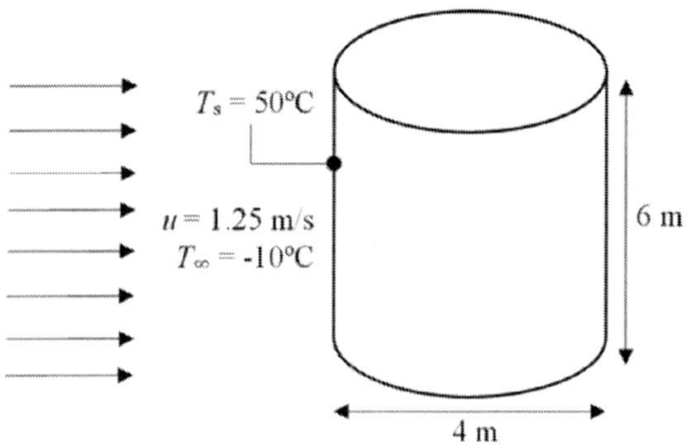

The outside surface temperature of the tank is given as 50°C. The velocity of crossflow wind and temperature are known as well.

We need to determine the properties of air at the film temperature T_f:

$$T_f = \frac{(T_s + T_\infty)}{2} = \frac{[50-10]°\text{C}}{2} = 20°\text{C}$$

At 20°C, from Table A.13, ρ = 1.2047 kg/m³, ν = 18.21 × 10^{-6} m²/s, Pr = 0.716, and k = 0.0256 W/m.K. We can now determine Re using the tank diameter as the characteristic length:

$$\text{Re} = \frac{uD}{\nu} = \frac{\left(1.25\frac{\text{m}}{\text{s}}\right)(4\text{m})}{\left(18.21\times 10^{-6}\frac{\text{m}^2}{\text{s}}\right)} = 274{,}574$$

For 40,000 < Re < 400,000 and Pr ≥ 0.7, the proper equation in the literature is

$$\text{Nu} = 0.027\text{Re}^{0.805}\text{Pr}^{\frac{1}{3}} = 0.027\left(274{,}574^{0.805}\right)(0.716)^{\frac{1}{3}}$$

$$= 653.9$$

Using this Nu, we can now determine h as

$$h = \text{Nu}\frac{k}{L} = (653.9)\frac{0.0261\text{W/m.K}}{4\text{m}} = 4.19\frac{\text{W}}{\text{m}^2.\text{K}}$$

Now, using Newton's law of cooling, the rate of convective heat loss from the storage tank is

$$\dot{Q} = -h\dot{A}(T_s - T_\infty) = -4.19\frac{\text{W}}{\text{m}^2\text{K}}[\pi(4\text{ m})(6\text{ m})]$$

$$[50-(-10)]^\circ\text{C}$$

$$= -18{,}923\text{W}(-64{,}582\text{Btu/h}),$$

and the overall thermal efficiency of the storage tank η is expressed as

$$\eta_{\text{overall}} = \frac{\text{Heat recovered}}{\text{Heat input}} = \frac{Q_{recovered}}{Q_{in}} = 1 - \frac{Q_{lost}}{Q_{input}}$$

where Q_{lost} is the amount of heat lost during 24 hours, and Q_{input} is the total heat input to the storage tank during this time. The overall energy efficiency of the storage tank can then be calculated as:

$$\eta_{\text{overall}} = 1 - \frac{(18{,}923\text{ J/s}\times 24\text{h}\times 3{,}600\text{s/h})}{(33\times 10^9\text{J})} = 0.95$$

Thence, assuming that the given conditions remain constant, the overall efficiency of the tank is 95%.

2.3.3.2.2.2 Internal Flow Forced Convection The fluid flow in a conduit such as a pipe or duct is classified as internal flow. For determining internal flow forced convection heat transfers as well, we employ the experimental correlations, which are provided in heat transfer literature for different flow regimes. In determining the Reynolds and Nusselt numbers, we use the characteristic length, and when a tube is involved, the characteristic length is the tube's D; however, for other shapes, it is the hydraulic diameter. After determining Nu, the heat transfer coefficient is calculated for a given situation.

Example 2.63: Internal Flow Convective Heat Transfer from a Boiler Tube to Water

Hot air is being used for generating hot water for hydronic heating in a boiler. Approximately 550 K of hot air enters a 6 m long × 5 cm diameter tube with an average velocity of 8 m/s and boils water at 373 K. Assuming that steady-state conditions prevail and that the outer surface temperature of the tube is uniform, constant, and equal to the boiling temperature with negligible thermal resistance, determine the convective heat transfer coefficient of the air flowing in the tubes.

Solution

We know that the air inlet temperature is 550 K, but we do not know the exit temperature of the air, so we will assume a mean air temperature of 500 K for determining the properties of air. At the end of our calculations, the air exit temperature will be calculated to see if the mean air temperature is in agreement with the assumed mean air temperature. If they are not in agreement, we will repeat the calculations using the new mean temperature. At an assumed mean temperature T_{avg} = 500 K, Table A.13 gives air density of 0.6964 kg/m³, specific heat of 1,030 J/kg.K, kinematic viscosity of 3.88 × 10⁻⁵ m²/s, thermal conductivity of 0.0407 W/m.K, and Prandtl number of 0.684.

The average velocity of air through the tube is known to be 5 m/s, then Re can be calculated:

$$\mathrm{Re} = \frac{uD}{\nu} = \frac{\left(8\frac{\mathrm{m}}{\mathrm{s}}\right)(0.05\,\mathrm{m})}{\left(3.88\times10^{-5}\frac{\mathrm{m}^2}{\mathrm{s}}\right)} = 10{,}309$$

Therefore, the flow is turbulent. Assuming that the flow is fully developed as it flows through the tube, the literature provides the proper Nu calculation as

$$\mathrm{Nu} = 0.023\mathrm{Re}^{0.8}\mathrm{Pr}^{0.3} = 0.023\left(10{,}309^{0.8}\right)(0.684)^{0.3} = 33.3$$

Using this Nu, we can now calculate h:

$$h = \mathrm{Nu}\frac{k}{D} = (33.3)\frac{0.0407\frac{\mathrm{W}}{\mathrm{m}}.\mathrm{K}}{0.05\,\mathrm{m}} = 27.1\frac{\mathrm{W}}{\mathrm{m}^2.\mathrm{K}}$$

Now using h, we can determine the exit temperature of the air as follows:

$$T_o = T_s - (T_s - T_i)e^{\left(-\frac{hA}{\dot{m}c_p}\right)}$$

$$T_o = (373\ \mathrm{K}) - (373\ \mathrm{K} - 550\ \mathrm{K})e^{\left[-\frac{\left(27.1\frac{\mathrm{W}}{\mathrm{m}^2.\mathrm{K}}\right)\pi(0.05\ \mathrm{m}\times 6\ \mathrm{m})}{0.011\frac{\mathrm{kg}}{\mathrm{s}}\times 1{,}030\frac{\mathrm{J}}{\mathrm{kg.K}}}\right]}$$
$$= 391.6\ \mathrm{K}$$

We assumed that the surface temperature of the tube is uniform and constant; therefore, the log mean temperature difference can be calculated as

$$\Delta T_m = \frac{(T_o - T_i)}{ln\left(\frac{T_s - T_o}{T_s - T_i}\right)} = \frac{(391.6 - 550)\mathrm{K}}{\ln\left(\frac{373 - 391.6}{373 - 550}\right)} = 70.3\ \mathrm{K}$$

Thence, the rate of convective heat transfer can be determined as follows:

$$\dot{Q} = -hA\Delta T_m = -\left(27.1\frac{\mathrm{W}}{\mathrm{m}^2.\mathrm{K}}\right)\pi(0.05\ \mathrm{m}\times 6\ \mathrm{m})(70.3\mathrm{K})$$
$$= -1{,}795\,\mathrm{W}\,(\approx -6{,}126\ \mathrm{Btu/h})$$

In these calculations, we used the air properties determined at the initially assumed mean temperature of 500 K. The new mean temperature can now be calculated as $T_{avg} = \frac{391.6 + 550}{2} = 470.8\mathrm{K}$, and we can see that it is not in agreement with 500 K. Therefore, the calculations need to be repeated, this time evaluating the air properties at the new mean temperature of 470.8 K.

Example 2.64: Internal Flow Convective Heat Loss from an Air Duct

Waste heat is being used to heat a greenhouse, and the 50°C air at the source is sent directly to the greenhouse with an air blower at a flow rate of 5 m/s through a 40 m rectangular duct with the dimensions shown in the next. The duct temperature stays almost constant at 40°C. Assuming that the steady-state conditions prevail and the duct has a smooth inside surface, determine the rate of heat loss from the duct as the air flows through before reaching the greenhouse.

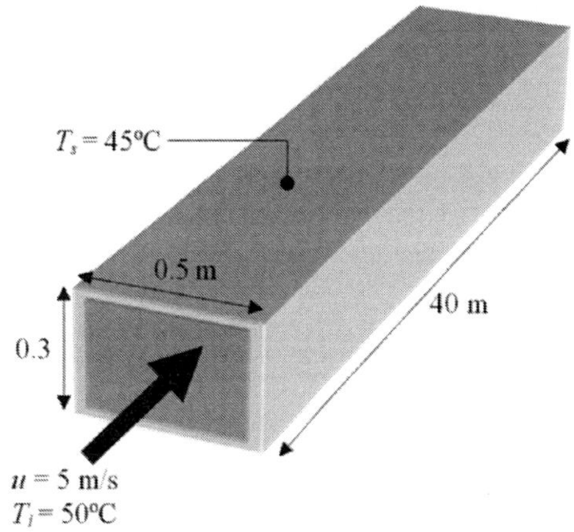

Solution

We know that the air inlet temperature is 50°C, but we do not know the outlet temperature; so, we will assume a mean air temperature of 40°C for determining the properties of air. At the end of our calculations, the air outlet temperature will be calculated to see if the mean air temperature is in agreement with the assumed mean air temperature. If they are not in agreement, we will repeat the calculations using the new mean temperature. At the assumed temperature of 40°C, the air has a density of 1.1275 kg/m³, specific heat of 1,007 J/kg.K, thermal conductivity of 0.0271 W/m.K, kinematic viscosity of 1.915 × 10⁻⁵ m²/s, and Pr of 0.712 (Table A.13).

Since the duct is noncircular, we have to calculate the hydraulic diameter D_e to calculate Re using either hydraulic radius r_h or the cross-sectional area for flow A_c and the total wetted perimeter p:

$$D_e = 2r_h = \frac{4A_c}{p} = \frac{4\,(0.3\ \mathrm{m}\times 0.5\ \mathrm{m})}{2\,(0.3\ \mathrm{m} + 0.5\ \mathrm{m})} = 0.375\ \mathrm{m}$$

Then, Re can be calculated as

$$\mathrm{Re} = \frac{uD_e}{\nu} = \frac{\left(5\frac{\mathrm{m}}{\mathrm{s}}\right)(0.375\,\mathrm{m})}{\left(1.915\times10^{-5}\frac{\mathrm{m}^2}{\mathrm{s}}\right)} = 94{,}937$$

Here, the flow is turbulent, and the entry lengths are approximately $10D_e = 3.75\mathrm{m}$. Since the duct is 40 m long, which is much longer than 3.75 m, the flow through the duct is fully developed. For a turbulent, fully developed flow with 0.7 ≤ Pr ≤ 160 and Re ≥ 10,000, the literature properly calculates Nu as

$$\mathrm{Nu} = 0.023\mathrm{Re}^{0.8}\mathrm{Pr}^{0.3} = 0.023(94{,}937)^{0.8}(0.712)^{0.3}$$
$$= 199.3$$

Using this Nu and D_e, we can now calculate h:

$$h = \mathrm{Nu}\frac{k}{D_e} = (199.3)\frac{0.0271\frac{\mathrm{W}}{\mathrm{m}}.\mathrm{K}}{0.375\,\mathrm{m}} = 14.4\frac{\mathrm{W}}{\mathrm{m}^2.\mathrm{K}}$$

And using h, we can determine the exit temperature of the air as follows:

$$T_o = T_s - (T_s - T_i)e^{\left(-\frac{hA}{\dot{m}c_p}\right)}$$

$$T_o = (45°C) - (45°C - 50°C)e^{\left[-\frac{\left(14.4\frac{W}{m^2.K}\right)(2\times(0.3m+0.5m)\times(40m))}{0.846\frac{kg}{s}\times 1{,}007\frac{J}{kg.K}}\right]}$$

$$T_o = 46.7°C$$

We assumed that the surface temperature of the duct is uniform and constant; therefore, the log mean temperature difference can be calculated as

$$\Delta T_m = \frac{(T_o - T_i)}{ln\left(\frac{T_s - T_o}{T_s - T_i}\right)} = \frac{(46.7 - 50)}{\ln\left(\frac{45 - 46.7}{45 - 50}\right)} = 3.1°C$$

Thence, the rate of convective heat loss from the duct can be determined as follows:

$$\dot{Q} = -hA\Delta T_m = -\left(16.3\frac{W}{m^2.K}\right)$$
$$(2\times(0.3\ m + 0.5\ m)\times 40\ m)(3.1°C)$$
$$= -3{,}191W(\approx -10{,}891\ Btu/h)$$

In these calculations, we used the air properties evaluated at the initially assumed mean temperature of 40°C. The new mean temperature can now be calculated as $T_{avg} = \frac{46.7 + 50}{2} = 48.4°C$, and we can see that it is not in agreement with 40°C. Therefore, the calculations need to be repeated for the air properties at the new mean temperature of 48.4°C.

2.3.3.2.3 Boiling and Condensation Heat Transfer

Many engineering problems involve boiling and condensation including problems in controlled environment engineering. This section briefly and generally covers boiling and condensation phenomena and their types.

2.3.3.2.3.1 Boiling Heat Transfer If vaporization takes place at a solid–liquid interface, it is called boiling; at a temperature beyond the saturation temperature and at a certain pressure, a liquid will start boiling. Since a fluid flow is involved in the process, boiling is also a form of convective heat transfer, and it depends on the latent heat of vaporization of the liquid and the surface tension at the liquid–vapor interface. The heat transfer coefficient during boiling is typically higher than the coefficients involved in convection with no phase changes. Using Newton's law of cooling, the rate of heat transfer during a boiling process from a given surface to a fluid can be expressed as

$$\dot{Q} = -hA_s(T_s - T_{sat}) \tag{2.81}$$

where T_s and T_{sat} are the surface and fluid saturation temperatures, respectively; A_s is the heat transfer area; and h is the boiling heat transfer coefficient.

Boiling that takes place without any fluid flow is known as pool boiling, whereas flow boiling occurs during a bulk fluid motion generally caused by external means. Heat transfer literature provides the equations for different boiling types.

2.3.3.2.3.2 Condensation Heat Transfer Condensation is observed when the temperature of a given vapor decreases below its saturation temperature. If vapor comes in contact with a solid, free liquid, or gas surface that has a temperature lower than its saturation temperature, then it starts condensing. There are two types of condensation, film and dropwise. In film condensation, the vapor condenses as a film of liquid on the surface; in dropwise condensation, the vapor condenses as liquid droplets on the surface. Typically, the rates of heat transfer in dropwise condensation are much higher than those of film condensation because the condensate droplets slide down the surface as they increase in size, allowing more vapor to be in contact with the condensing surface. In film condensation, however, the condensing surface is covered by the liquid film, which resists further exchange of heat between the vapor and the surface. The rate of condensation heat transfer is expressed in the following equation:

$$\dot{Q} = h_L A_s(T_{sat} - T_s) = \dot{m}h'_{fg} \tag{2.82}$$

where h_L is the heat transfer coefficient; A_s is the heat transfer area; T_{sat} and T_s are the saturated temperature of the vapor and the condensing surface temperature, respectively; $\dot{m}$ is the condensate mass flow rate; and h'_{fg} is the modified latent heat of vaporization, which considers whether the vapor is superheated and therefore requires cooling to its saturation temperature before condensation begins and whether the liquid is further cooled below its saturation temperature. The expression for the modified latent heat of vaporization h'_{fg} is

$$h'_{fg} = h_{fg} + 0.68c_{p,l}(T_{sat} - T_s) + c_{p,v}(T_v - T_{sat}) \tag{2.83}$$

where h_{fg} is the latent heat of vaporization, $c_{p,l}$ is the specific heat of the liquid at the film temperature, and $c_{p,v}$ is the specific heat of the vapor at the average vapor temperature T_v and saturation temperature T_{sat}. The heat transfer coefficient h_L depends on variables including the properties of the fluids, nature of flow, and the surface geometry. Heat transfer literature provides the correlation equations for heat transfer coefficients for different geometries and condensation types.

2.3.3.3 Radiation Heat Transfer

Radiation is a form of energy that is emitted by all objects that have a temperature above absolute zero. It is an electromagnetic phenomenon that differs from both conduction and convection as it does not require a medium to transfer heat and it can travel through a vacuum.

A body that absorbs all incident radiation at all wavelengths is called a blackbody, and the heat radiated by a body is composed of electromagnetic waves of many different wavelengths or frequencies. Radiation laws govern the relationships between a radiating surface and the heat radiated into the space in all directions, and the first of these laws is

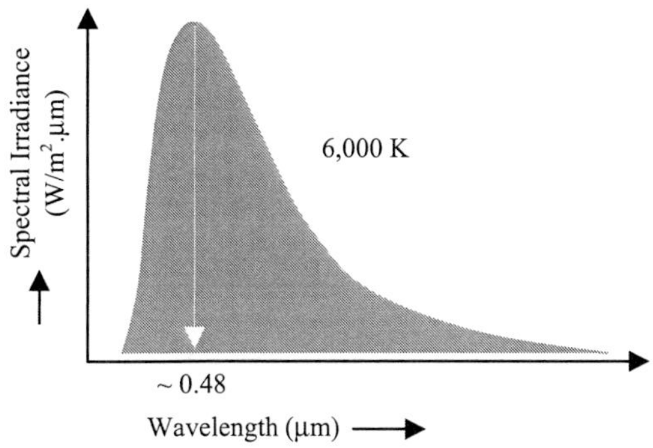

FIGURE 2.15 Schematic view of standard blackbody curves for solar radiation assuming a surface temperature of 6,000 K. The total radiant energy (E^*), given by the Stefan–Boltzmann equation, corresponds to the area beneath the curve.

Source: ***Greenhouse Engineering*** **(2021).**

Planck's law, which states that the wavelength of emission from a perfectly black object—a blackbody, that is a perfect emitter and absorber of radiation (Kirchhoff's law)—depends on the absolute temperature of the emitting body:

$$E_\lambda^* = \frac{C_1}{\left\{\lambda^5\left[\exp\left(\frac{C_2}{\lambda T}\right) - 1\right]\right\}} \quad (2.84)$$

where E_λ^* is the amount of energy in W/m².µm (Btu/h.ft².µm) emitted at a single wavelength λ (µm) (blackbody monochromatic radiation) at temperature T in K (°R). Constants C_1 and C_2 have magnitudes of 3.74 × 10⁸ W.µm⁴/m² (1.187 × 10⁸ Btu.µm⁴/h.ft²) and 1.44 × 10⁴ µm.K (25,898 µm °R), respectively. Solar radiation, a very good approximation to that of a blackbody that follows Planck's curve, obtained using Equation (2.84) for an average outer surface temperature (6,000 K) of the Sun is presented in Figure 2.15. Actual (nonblack) bodies emit less than the theoretical blackbody emission, $E_\lambda = \varepsilon_\lambda E_\lambda^*$ where ε_λ is the emissivity, which is a measure of emission efficiency.

Absorption of radiation or photons is the exact reverse of photon emission. A photon is absorbed as soon as it interacts with a molecule, an atom or one of its electrons, and this interaction destroys the photon and brings the molecule or an atom to an excited state. The excited molecule or atom subsequently starts reemitting one or more photons and continues the exchange of energy from radiant energy to atomic excitation and back again.

Two other particularly useful radiation laws, which can be derived from Equation (2.84) by differentiation and integration, respectively, are Wien's law for the wavelength of a maximum emission for a blackbody at a given temperature, and the Stefan–Boltzmann law for the total energy emitted by a blackbody. Wien's displacement law states that the wavelength at which a blackbody emits its maximum amount of radiation is inversely proportional to its absolute temperature in Kelvin (or in Rankine):

$$\lambda_{max} = \frac{a}{T} \quad (2.85)$$

where λ is the wavelength in µm, a is 2897 µm.K (5216 µm.°R), and T in K (or in °R).

Example 2.65: Blackbody Monochromatic Radiation

Determine the blackbody monochromatic radiation at a single wavelength of 0.5 µm emitted by the Sun's surface, E_λ^*.

Solution

Using Equation (2.84) and assuming a surface temperature of 6,000 K, the blackbody monochromatic radiation at 0.5 µm is

$$E_\lambda^* = \frac{3.74 \times 10^8 \text{ W.µm}^4/\text{m}^2}{\left\{(0.5 \text{ µm})^5\left[\exp\left(\frac{1.44 \times 10^4 \text{ µm.K}}{(0.5 \text{ µm})(6{,}000 \text{ K})}\right) - 1\right]\right\}}$$

$$E_\lambda^* = 99.3 \times 10^6 \text{ W/m}^2.\text{µm}$$

That is, it is simply 99.3 MJ per square meter solar surface area every second.

Example 2.66: Wavelength of Maximum Radiation (Solar) Emission

Determine the wavelength of maximum radiation emission λ_{max} for the Sun's surface.

Solution

Using Equation (2.85) and assuming an average surface temperature of approximately 6,000 K, its maximum emission is

$$\lambda_{max} = \frac{a}{T}$$

$$\lambda_{max} = \frac{2897 \text{ µm.K}}{6{,}000 \text{ K}}$$

$$\lambda_{max} = 0.48 \text{µm}$$

A wavelength of approximately 0.5 µm lies within the visible spectrum.

Example 2.67: Wavelength of Maximum Radiation (Terrestrial) Emission

Determine the wavelength of maximum radiation emission λ_{max} for the Earth's surface.

Solution

Assuming an average surface temperature of 15°C (i.e., 288.15 K), we can determine the maximum emission:

$$\lambda_{max} = \frac{2897 \text{ µm.K}}{288.15 \text{ K}}$$

$$\lambda_{max} = 10.0 \text{µm}$$

A wavelength of 10.0 µm lies within the infrared spectrum, way beyond the visible spectrum. This shows that a hotter object emits a greater portion of its energy at shorter wavelengths than a cooler object.

To determine how much radiation a given object emits over all wavelengths combined, we need to sum up all the energy contributions of the photons at all wavelengths. This total energy is represented by the area between the object's blackbody curve and the x-axis as shown in Figure 2.15. That area, and therefore the radiant energy emission, is given by the Stefan–Boltzmann law, given in the following equation:

$$E^* = -\sigma T^4 \tag{2.86}$$

where E^* is the total energy emitted in J (Btu) per unit time per unit area of black surface, σ is known as the Stefan–Boltzmann constant and equals 5.679 × 10^{-8} J/s.m^2.K^4 (0.1714 × 10^{-8} Btu/h.ft^2.R^4), and T is the absolute temperature of the surface in K (R).

Example 2.68: Sun's Total Radiation Emission

Estimate the Sun's total radiation emission per unit time per unit surface area.

Solution

Using Equation (2.86) and assuming an average surface temperature of the Sun as 6,000 K, we can estimate that the Sun emits

$$E^* = \left(5.679 \times 10^{-8} \frac{\text{J}}{\text{s.m}^2.\text{K}^4}\right)(6{,}000\,\text{K})^4$$

$$E^* = 73.6 \frac{\text{MJ}}{\text{s.m}^2}$$

In all directions, with an approx. radius of 6.96 × 10^{11} m, we can easily calculate the Sun's emission as approximately 4.5 × 10^{20} MJ of energy every second.

When radiant energy falls on a surface, it can be absorbed, reflected, or transmitted through the material. Hence, from the first law of thermodynamics,

$$a + t + r = 1 \tag{2.87}$$

where a is the fraction of incident radiation absorbed, or absorbance; t is the fraction of incident radiation transmitted, or transmittance; and r is the fraction of incident radiation reflected, or reflectance. Most solids behave as opaque materials in the infrared region, and for opaque materials, $t = 0$ and $a + r = 1$. For black surfaces, $a = 1$, $t = 0$, and $r = 0$.

The net interchange of heat by radiation between two bodies at different temperatures can be written as

$$\dot{Q} = -F_E F_A A \sigma\left(T_1^4 - T_2^4\right) \tag{2.88}$$

where F_E is the emissivity factor, F_A is the geometric (or view) factor (Table A.14), A is the area in m^2 (ft^2), σ is the Stefan–Boltzmann constant of 5.679 × 10^{-8} J/s.m^2.K^4 (0.1714 × 10^{-8} Btu/h.ft^2.°R^4), and T_1 and T_2 are the absolute temperatures of the surfaces in K (or °R).

Example 2.69: Radiation Heat Loss from a Steel Tube

Now consider the 51 mm steel tube in Examples 2.51 and 2.60 with an outside diameter of 0.0488 m that runs through a greenhouse whose walls are at 20°C. Determine the heat loss by radiation from 30 m of the tube if its outside surface temperature is 9°C. Assume that there is no crop canopy in the greenhouse yet.

Solution

From Table A.14, case 2, F_A = 1, A is the area of tube, and F_e is the emissivity of the steel tube. From Table A.15, for oxidized steel, ε = 0.79. Then, using Equation (2.88),

$$\dot{Q} = -F_e F_A A \sigma\left(T_1^4 - T_2^4\right)$$

$$\dot{Q} = -(0.79)(1)(\pi)(0.0488\,\text{m} \times 30\,\text{m})\left(5.679 \times 10^{-8} \frac{\text{J}}{\text{s.m}^2.\text{K}^4}\right) \times\left[(90°\text{C} + 273.15\text{ K})^4 - (20°\text{C} + 273.15\text{ K})^4\right]$$

$$\dot{Q} = -2{,}064\,\text{W}\left(\approx -7{,}044 \frac{\text{Btu}}{\text{h}}\right)$$

2.3.4 Air–Water Vapor Mixtures: Psychrometrics

The study of psychrometrics deals with thermodynamic properties of moist air, and using these properties, we analyze engineering and nonengineering situations and processes involving moist air. This section covers perfect gas relations and describes their use in thermal comfort problems.

Atmospheric air normally contains many gaseous components, some water vapor, and contaminants such as smoke, pollen, etc.; dry air contains no water or contaminants, and therefore, its composition is relatively constant. Even when there is little moisture in the air, it plays a significant role in thermal comfort for humans, plants, and animals as well as in air-conditioning applications. Within the range of thermal comfort and air-conditioning applications, we can treat both dry and moist (atmospheric) air as perfect gases, with negligible error. Therefore, the perfect gas equation, $pV = nRT$, still holds, and the total pressure of atmospheric air (mixture) can be expressed in terms of the partial pressure as follows:

$$p = p_a + p_w \tag{2.89}$$

where p_a is the partial pressure of dry air and p_w is the partial pressure of water vapor. Then, we can write the following expressions:

$$p_a = \frac{n_a RT}{V} \text{ (dry air)} \tag{2.90}$$

$$p_w = \frac{n_w RT}{V} \text{ (water vapor)} \tag{2.91}$$

where V is the total volume of the mixture, n_a is the number of moles of dry air, n_w is the number of moles of water vapor, R is the universal gas constant (for SI: 8314.41 J/kg$_{mol}$·K and for I-P: 1545.32 ft.lb$_f$/lb$_{mol}$·°R), and T is the absolute temperature in Kelvin or Rankine.

At low pressures, the interaction between the air and gas molecules is modest, the mixture obeys the perfect gas equation, and thence, we can write the following equation:

$$(p_a + p_w)V = (n_a + n_w)RT \tag{2.92}$$

where (p_a + pw) is the total mixture pressure (the sum of partial pressures) and (n_a + nw) is the total number of moles in the mixture.

The enthalpy of a mixture of perfect gases equals the sum of the individual partial enthalpies of the components, in this case, of the air and water vapor. Then,

$$H = H_a + H_g = m_a h_a + m_g h_g \tag{2.93}$$

Dividing the total enthalpy of the mixture by the mass of the dry air, we can obtain the enthalpy of the mixture per unit mass of dry air, which will be expressed as h without any subscript:

$$\left(\frac{H}{m_a}\right) = h = h_a + \left(\frac{m_g}{m_a}\right) h_g = h_a + Wh_g \tag{2.94}$$

It should be emphasized that the new enthalpy h is the sum of the specific enthalpy of the dry air h_a and the product of the specific enthalpy of the saturated water vapor h_g and the ratio of the mass of water vapor m_g to the mass of air m_a (a.k.a. the humidity ratio or specific humidity W) in the mixture.

Within the range of interest, if we take 0°C as the reference temperature and the constant-pressure specific heat for dry-air is c_p = 1.006 kJ/kg °C, we obtain

$$h_a\left(\text{kJ / kg}\right) = c_p T \approx \left(1.006 \text{ kJ / kg.°C}\right) T \tag{2.95}$$

$$\Delta h_a\left(\text{kJ / kg}\right) = c_p \Delta T \approx \left(1.006 \text{ kJ / kg.°C}\right) \Delta T \tag{2.96}$$

where T is the dry-bulb temperature in °C. The enthalpy of water vapor at 0°C is 2,501 kJ/kg. The average c_p of water vapor within the temperature range of interest is 1.86 kJ/kg°C. Then, the enthalpy of water vapor can be determined from the following equation:

$$h_g\left(\text{kJ / kg}\right) \approx 2501 + 1.86T \tag{2.97}$$

The enthalpy of the mixture (moist air) per unit mass of dry air is

$$h\left(\text{kJ / kg}\right) \approx 1.006T + W\left(2501 + 1.86T\right) \tag{2.98}$$

In I-P units, it is

$$h_a\left(\text{Btu / lb}\right) = c_p T \approx \left(0.240 \text{ Btu / lb.°F}\right) T \tag{2.99}$$

$$h_g\left(\text{Btu / lb}\right) \approx 1061 + 0.444\ T \tag{2.100}$$

where T is the dry-bulb temperature in °F. The moist air enthalpy then becomes

$$h\left(\text{Btu / lb}_{\text{da}}\right) \approx 0.240\ T + W\left(1061 + 0.444T\right) \tag{2.101}$$

2.3.4.1 Some Definitions and Relationships

The humidity ratio, moisture content, or mixing ratio, W, is the ratio of mass of water vapor to mass of air in a mixture. The advantage with humidity ratio is that if the amount of water vapor in the air remains unchanged, the ratio remains constant. The humidity ratio is expressed in grams or kg of moisture (water vapor) per kilogram of dry air. Alternatively, in I-P units, it is given as $\text{lb}_{\text{water}}/\text{lb}_{\text{da}}$ or grains of water vapor per pound of dry air where 7,000 grains is equal to 1 lb of mass.

$$W\left(\frac{\text{kg}_{\text{water}}}{\text{kg}_{\text{da}}}\right) = \frac{m_w}{m_a} = 0.622\left(\frac{x_w}{x_a}\right) = 0.622\left(\frac{p_w}{p_a}\right)$$
$$= 0.622\left(\frac{p_w}{p - p_w}\right) \tag{2.102}$$

where m_w is the mass of water vapor, m_a is the mass of dry air, 0.622 is the ratio of molecular masses (18.01528/28.9645 = 0.62198), x_w is the mole fraction for water vapor, x_a is the mole fraction for dry air, $\frac{x_w}{x_a}$ is the mole fraction ratio (by definition $x_a + x_w = 1$), p_w is the partial pressure of water vapor, and p is the total pressure ($p = p_a + p_w$ = 101,325 Pa or 14.696 psia at sea level).

Specific humidity, q, is the ratio of mass of water vapor to the total mass of the moist air as shown:

$$q\left(\frac{\text{kg}_{\text{water}}}{\text{kg}_{\text{air}}}\right) = \frac{m_w}{m_w + m_a} = \frac{W}{1 + W} \tag{2.103}$$

The degree of saturation, W, is expressed as the air humidity ratio W to the humidity ratio W_s of saturated air at the same temperature and pressure as in the following equation:

$$\mu = \frac{W}{W_s} \tag{2.104}$$

where

$$W_s = 0.622\frac{p_{ws}}{p - p_{ws}} \tag{2.105}$$

and p_{ws} is the saturation pressure of water vapor at the given temperature T. This is the amount of moisture the air can hold in vapor form (a.k.a. the limit saturation) before condensation happens. It is required in determining the saturation pressure of water vapor, is only a function of temperature, and can be found in Tables A.56 and A.57 as tabulated values or calculated using the following equations:

The saturation pressure p_{ws} over ice for −100°C (−148°F) < T < 0°C (32°F):

$$\ln(p_{ws}) = C_1/T + C_2 + C_3T + C_4T^2 + C_5T^3$$
$$+ C_6T^4 + C_7 ln(T) \tag{2.106}$$

where the constants in SI are

C_1 = −5.647 535 9 E + 03
C_2 = 6.392 524 7 E + 00
C_3 = −9.677 843 0 E − 03
C_4 = 6.221 570 1 E − 07
C_5 = 2.074 782 5 E − 09
C_6 = −9.484 024 0 E − 13
C_7 = 4.163 501 9 E + 00

In I-P, the constants are

$C_1 = -1.021\,416\,462\text{ E} + 04$
$C_2 = -4.893\,503\,01$
$C_3 = -5.376\,579\,44\text{ E} - 03$
$C_4 = 1.920\,237\,69\text{ E} - 07$
$C_5 = 3.557\,583\,16\text{ E} - 10$
$C_6 = -9.034\,468\,83\text{ E} - 14$
$C_7 = 4.163\,501\,9$

The saturation pressure p_{ws} over liquid water for 0°C (32°F) < T < 200°C (392°F) is

$$\ln(p_{ws}) = C_8 T + C_9 + C_{10}T + C_{11}T^2 + C_{12}T^3 + C_{13}\,ln(T) \quad (2.107)$$

where

$C_8 = -5.800\,220\,6\text{ E} + 03$
$C_9 = 1.391\,499\,3\text{ E} + 00$
$C_{10} = -4.864\,023\,9\text{ E} - 02$
$C_{11} = 4.176\,476\,8\text{ E} - 05$
$C_{12} = -1.445\,209\,3\text{ E} - 08$
$C_{13} = 6.454\,967\,3\text{ E} + 00$

In I-P, the constants are

$C_8 = -1.044\,039\,708\text{ E} + 04$
$C_9 = -1.129\,479\,96\text{ E} + 01$
$C_{10} = -2.702\,235\,5\text{ E} - 02$
$C_{11} = 1.289\,036\,0\text{ E} - 05$
$C_{12} = -2.478\,068\text{ E} - 09$
$C_{13} = 6.5459673$

In both Equations (2.106) and (2.107), *ln* is the natural logarithm, p_{ws} is the saturation pressure in Pa for SI (and psia for I-P), and T is the absolute temperature in Kelvin (K = °C + 273.15) for SI (and °R = °F + 459.67 for I-P).

Water vapor density or absolute humidity, d_v, on the other hand, is the ratio of the mass of water vapor to the total volume and can be expressed as follows:

$$d_v = \frac{m_w}{V} \quad (2.108)$$

Density ρ of the moist air mixture is the ratio of the total mass or air to the total volume of air as shown:

$$\rho = \frac{m_a + m_w}{V} = \frac{1}{v}(1+W) \quad (2.109)$$

where v is the moist air volume in m^3/kg_{da} (or ft^3/lb_{da}), m_a is the total mass of dry air, m_w is the total mass of water vapor, V is the total volume of the mixed air, and W is the humidity ratio.

Moist air volume v can be expressed as follows:

$$v = \frac{V}{m_a} = \frac{V}{28.9645(n_a)} \quad (2.110)$$

$$v = \frac{RT}{28.9645(p-p_w)} = \frac{R_a T}{(p-p_w)} = \frac{R_a T}{p}(1+1.6078W) \quad (2.111)$$

where 28.964 is the apparent molecular mass or weighted average molecular weight of all components for dry air, n_a is the number of moles of dry air, R is the universal gas constant, R_a is the gas constant for dry air (for SI: 0.287055 kJ/kg.K; for I-P: 53.352 $ft.lb_f/lb_m.°R$), T is the absolute temperature, p is the total pressure, p_w is the partial pressure of water vapor, and W is the humidity ratio.

The relative humidity, ϕ, is the ratio of the mole fraction of water vapor, x_w, in moist air to the mole fraction x_{ws} in the air, saturated at the same temperature and pressure and expressed as:

$$\phi = \left.\frac{x_w}{x_{ws}}\right|T,p = \left.\frac{p_w}{p_{ws}}\right|T,p = \frac{\mu}{1-(1-\mu)\left(\frac{p_{ws}}{p}\right)} \quad (2.112)$$

Please note that both ϕ and degree of saturation μ are zero for dry air and unity for saturated mixed air.

Also note that the standard temperature we are familiar with, a measure of internal energy, is called the dry-bulb temperature, T. We can also define dew-point temperature, T_d, which is the temperature of moist air at which water vapor condenses into liquid. Air is maximally saturated at the same p with the same W as that of the given moist air, and it is defined as the solution $T_d\ (p,\ W)$ of the following equation:

$$W_s(p, T_d) = W \quad (2.113)$$

When the air temperature is lowered to the dew point at constant pressure, the relative humidity rises to 100% and condensation take place. T_d is a direct measure of W because $W = W_s$ at $T = T_d$. T_d for dew point temperatures below 0°C (32 °F) is calculated as

$$T_d = 6.09 + 12.608\alpha + 0.4959\alpha^2 \quad (2.114)$$

where T_d is the dew point temperature in °C, $\alpha = ln\ (p_w)$, and p_w is the water vapor partial pressure in kPa.

In I-P, the dew point temperature equation is provided as follows:

$$T_d = 90.12 + 26.142\alpha + 0.8927\alpha^2 \quad (2.115)$$

where T_d is the dew point temperature in °F, $\alpha = ln\ (p_w)$ and p_w is the water vapor partial pressure in psia.

T_d for the dew point temperatures between 0°C (32°F) and 93°C (200°F) is calculated as

$$T_d = \text{a} + \text{b}\alpha + \text{c}\alpha^2 + \text{d}\alpha^3 + e(p_w)^{0.1984} \quad (2.116)$$

where

T_d = dew point temperature in °C
a = 6.54
b = 14.526
c = 0.7389
d = 0.09486
e = 0.4569
$\alpha = ln\ (p_w)$
p_w = water vapor partial pressure in kPa.

In I-P, T_d is calculated as

$$T_d = a + b\alpha + c\alpha^2 + d\alpha^3 + e(p_w)^{0.1984} \qquad (2.117)$$

where

T_d = dew point temperature in °F
a = 100.45
b = 33.193
c = 2.319
d = 0.17074
e = 1.2063
$\alpha = ln\ (p_w)$
p_w = water vapor partial pressure in psia.

Thermodynamic wet-bulb temperature, T_w, is the temperature at which water, by evaporation into moist air at a given T and humidity ratio W, can bring air to saturation (100% relative humidity) adiabatically at the same T_w while p is held constant.

To measure the wet-bulb temperature, we use a psychrometer, which consists of two thermometers; one of the thermometer's bulbs is covered by a thoroughly wetted wick. When the wet-bulb is left in an air stream, water evaporates from the wick, and the bulb reaches the equilibrium wet-bulb temperature. This process is not exactly adiabatic saturation, which defines the thermodynamic wet-bulb temperature just discussed; rather, it is a simultaneous heat and mass transfer from the wet-bulb. The energy needed to evaporate liquid water (heat of vaporization) is taken from the air in the form of sensible heat and converted to latent heat, which lowers the temperature at constant enthalpy.

Equation (2.118) is used to calculate the wet-bulb temperature in SI, which requires iterative solving until we find the T_w that satisfies the equation:

$$W = \frac{(2501 - 2.381T_w)W_{s@T_w} - 1.006(T - T_w)}{2501 + 1.805T - 4.186T_w} \qquad (2.118)$$

where both T_w and T are in degree Celsius.

For I-P units, the following equation is used, solving the equation iteratively until we find the T_w that satisfies the equation:

$$W = \frac{(1093 - 0.556T_w)W_{s@T_w} - 0.240(T - T_w)}{1093 + 0.444T - T_w} \qquad (2.119)$$

where T_w and T are in degree Fahrenheit.

To specify the state of moist air, we need three moist air properties, so that the others can be calculated either using the equations provided above for greater accuracy and automation or determined using the psychrometric charts (Figures 2.16). Total atmospheric pressure, p, is typically

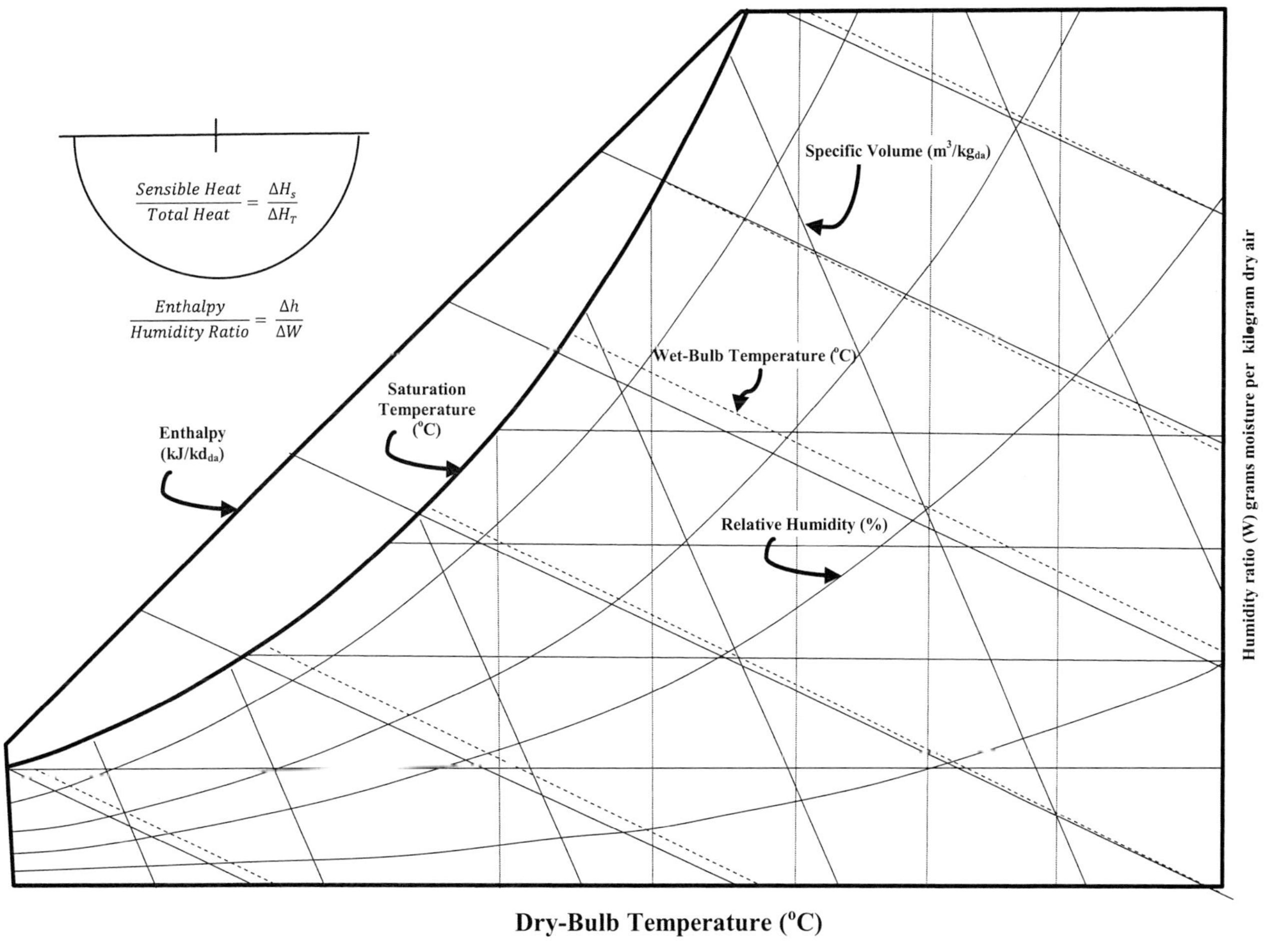

FIGURE 2.16 Components of a psychrometric chart—normal temperature sea level (SI).
Source: *Greenhouse Engineering* (2021).

known; therefore, we only need to know two other properties. Thermodynamic properties of moist air at standard atmospheric pressure and thermodynamic properties of water at saturation were adopted from the *ASHRAE* (2013) and presented here in Tables A.56 and A.57.

Example 2.70: Psychrometric Calculations

Consider a moist air parcel with T = 25°C and 10°C dew point temperature. Calculate the following: a) W; b) μ; c) ϕ; d) h; e) v; f) ρ; and g) T_w.

Solution

a) For W, assuming p = 101.325 kPa (sea level) and using Equation (2.102),

$$W\left(\frac{\text{kg}_{\text{water}}}{\text{kg}_{\text{da}}}\right) = 0.622\left(\frac{p_w}{p-p_w}\right)\Bigg|_{@T=25°\text{C}}$$

For T_d = 10°C, the actual W is by definition the same as W_s at air temperature 10°C; therefore,

$$W_{@T=25°\text{C}} = W_{s@T=10°\text{C}} = 0.622\left(\frac{p_{ws}}{p-p_{ws}}\right)\Bigg|_{@T=10°\text{C}}$$

From Table A.57, we find $p_{ws@10°\text{C}}$ = 1.2280 kPa; then,

$$W_{@T=25°\text{C}} = W_{s@T=10°\text{C}}$$
$$= 0.622\left(\frac{1.2280\text{ kPa}}{101.325\text{ kPa} - 1.2280\text{ kPa}}\right)\Bigg|_{@T=10°\text{C}}$$
$$= 0.00763\ \text{kg}_{\text{water}}/\text{kg}_{\text{da}}$$

b) Degree of saturation μ can be calculated using Equation (2.105); for this, we need the saturation humidity ratio W_s at 25°C as we just determined the humidity ratio W at 25°C:

$$W_{s@T=25°\text{C}} = 0.622\left(\frac{p_{ws}}{p-p_{ws}}\right)\Bigg|_{@T=25°\text{C}}$$

From Table A.57, we find $p_{ws@25°\text{C}}$ = 3.1692 kPa; then,

$$W_{s@T=25°\text{C}} = 0.622\left(\frac{3.1692\text{ kPa}}{101.325\text{ kPa} - 3.1692\text{ kPa}}\right)\Bigg|_{@T=25°\text{C}}$$
$$= 0.02001\ \text{kg}_{\text{water}}/\text{kg}_{\text{da}}$$

Then, using Equation (2.104), we can now calculate the degree of saturation as

$$\mu = \frac{W}{W_s} = \frac{0.00763}{0.02001} = 0.38$$

c) Relative humidity ϕ can be calculated from Equation (2.112) as we now know both p_w and p_{ws}:

$$p_{w@25°\text{C}} = p_{ws@10°\text{C}} = 1.2280\text{kPa}$$

$$\phi = \frac{p_w}{p_{ws}}\Bigg|T,p = \frac{1.2280\text{ kPa}}{3.1692\text{ kPa}} = 0.387 = 38.7\%$$

d) Enthalpy h can now be calculated using Equation (2.98) as follows:

$$h\left(\frac{\text{kJ}}{\text{kg}_{\text{da}}}\right) \approx 1.006T + W(2501 + 1.86T)$$
$$\approx 1.006(25°\text{C}) + 0.00763(2501 + 1.86(25°\text{C}))$$
$$h \approx 44.6\text{kJ}/\text{kg}_{\text{da}}$$

e & f) Specific volume v and density ρ can be calculated using Equations (2.111) and (2.109), respectively, with the gas constant for dry air R_a = 0.287055 kJ/kg·K and temperature in Kelvin:

$$v = \frac{R_aT}{p}(1+1.6078W) = \frac{0.287055(25+273.15)}{101.325}$$
$$= (1+1.6078(0.00763)) = 0.855\text{ m}^3/\text{kg}_{\text{da}}$$

$$\rho = \frac{1}{v}(1+W) = \frac{1}{0.855}(1+0.00763) = 1.179\text{kg}/\text{m}^3$$

g) Wet-bulb temperature T_w fits Equation (2.118), which requires iteration:

$$W = \frac{(2501 - 2.381\boldsymbol{T_w})W_{s@T_w} - 1.006(T - \boldsymbol{T_w})}{2501 + 1.805T - 4.186\boldsymbol{T_w}} = 0.00763$$

where T=25°C, T_w is unknown (°C), and $W_{s@T_w=?} = 0.622\left(\frac{p_{ws}}{p-p_{ws}}\right)\Bigg|_{@T_w=?}$

We just need to guess T_w first, calculate p_{ws}, calculate W_s for that temperature, and then repeat the iteration until the calculated W based on those values (and original temperature) in Equation (2.118) is equal to actual W (0.00763 in our case). After satisfying this, the iteration process stops, and we determine T_w as follows:

$$T_w = 16.0°\text{C}$$

Example 2.71: Using the Psychrometric Chart

Now consider a moist air parcel with T = 24°C and 50% relative humidity. Assuming p = 101.325 kPa (sea level) and this time using the psychrometric chart, determine W, W_s, T_w, h, v, ρ, and T_d.

Solution

Using the psychrometric chart (ASHRAE, 2013) (see Figure 2.16 for the components), we can determine all the required quantities as shown in the following figure.

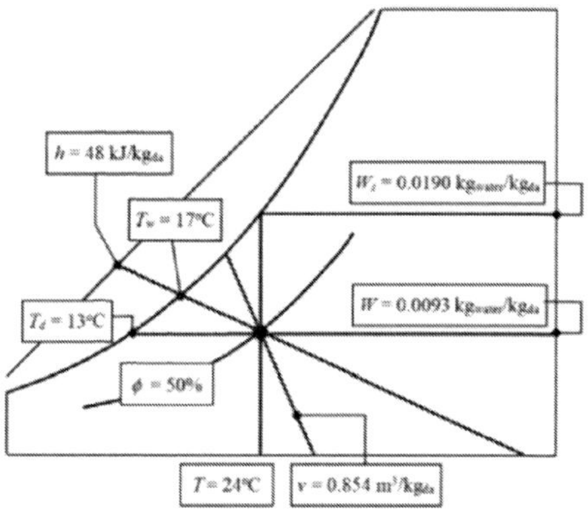

2.3.4.2 Psychrometric Processes

2.3.4.2.1 *Sensible Heating*

Sensible heating does not change the moisture content; that is, the humidity ratios at the initial (State 1) and final (State 2) conditions are the same as no moisture is added to or removed from the air, as shown in the psychrometric sketch in Figure 2.17.

Note that the relative humidity of air decreases during sensible heating. This is because the moisture-holding capacity of the air increases with increasing air temperature.

The conservation of mass reduces to

$$\dot{m}_{a1} = \dot{m}_{a2} = \dot{m}_a \tag{2.120}$$

$$W_1 = W_2 \tag{2.121}$$

Neglecting any fan work and losses, the conservation of energy yields

$$\dot{Q}_S = \dot{m}_a \left(h_2 - h_1\right) \tag{2.122}$$

where $\dot{Q}_S$ is the rate of sensible heating provided; $\dot{m}_a$ is the mass flow rate of dry air; W_1 and W_2 are the humidity ratios at the inlet and exit of the heating section, respectively; and h_1 and h_2 are enthalpies per unit mass of dry air at the inlet and exit of the heating section, respectively.

Example 2.72: Sensible Heating

Now consider moist air saturated at 5°C entering a unit heater at a flow rate of 10 m³/s. Determine the rate of heating if the moist air must leave the unit heater at T = 30°C.

Solution

Figure 2.17 schematically presents the solution. However, State 1 is located on the saturation curve, meaning that T, T_d, and T_w all equal 5°C. From the psychrometric chart, h_1 = 18.6 kJ/kg$_{da}$, W_1 = 0.0054 kg$_{water}$/kg$_{da}$, and v_1 = 0.795 m³/kg$_{da}$. State 2 is located at the intersection of 30°C and $W_2 = W_1$ = 0.0054 kg$_{water}$/kg$_{da}$. Hence, h_2 = 45.0 kJ/kg$_{da}$.

Then, the mass flow rate is:

$$\dot{m}_a = \left(10\text{m}^3/\text{s}\right)/(0.795\ \text{m}^3/\text{kg}_{da}\right) = 12.6\text{kg}_{da}/\text{s},$$

and from Equation (2.122), the rate of heating Q_s is

$$\dot{Q}_S = \dot{m}_a\left(h_2 - h_1\right) = 12.6(45.0 - 18.6)$$

$$= 332\text{kW}\left(\approx 1{,}133{,}106\frac{\text{Btu}}{\text{h}}\right).$$

2.3.4.2.2 *Sensible Heating and Humidification*

Sometimes, humidification is applied in sensible heating, as shown in Figure 2.18, to adjust the relative humidity. Humidification takes place by passing the air through a humidifying section. If steam is used in the humidifier, there will be additional heating; therefore, the temperature at Point 3 will be greater than the temperature at Point 2. However, if water is sprayed instead, part of the latent heat of vaporization will be absorbed from the air, which will result in some cooling of the air; therefore, the temperature at Point 3 will be lower than the temperature at Point 2.

In sensible heating and humidification, the conservation of mass reduces to

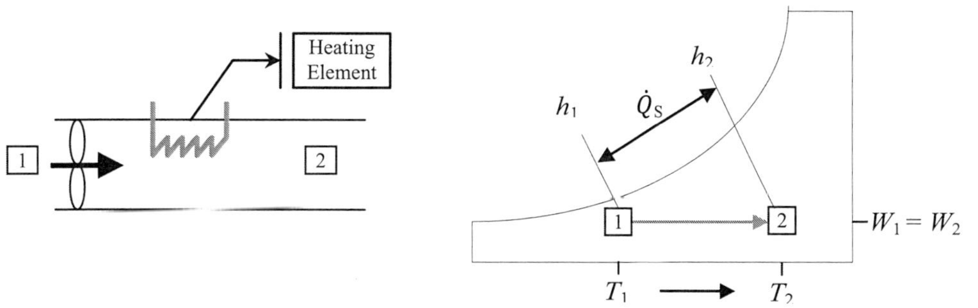

FIGURE 2.17 Sensible heating.

Source: *Greenhouse Engineering* (2021).

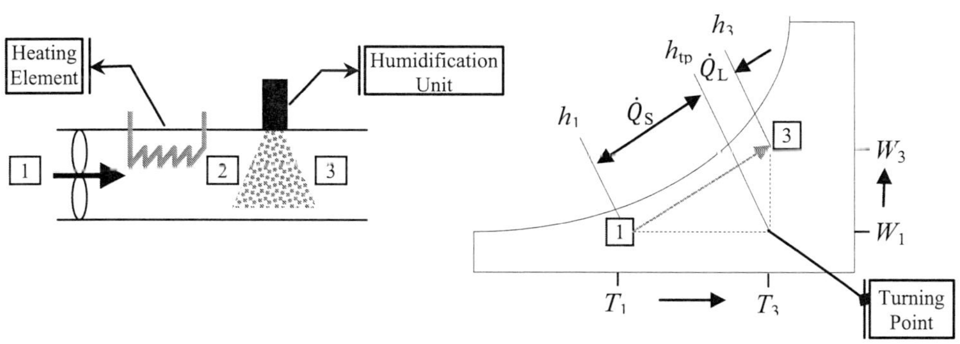

FIGURE 2.18 Sensible heating and humidification.

Source: *Greenhouse Engineering* (2021).

$$\dot{m}_{a1} = \dot{m}_{a2} = \dot{m}_{a3} = \dot{m}_a \quad (2.123)$$

$$W_1 = W_2 \text{ and } W_1 < W_3 \quad (2.124)$$

$$\dot{m}_w = \dot{m}_a \, \Delta W = \dot{m}_a (W_3 < W_1) \quad (2.125)$$

where $\dot{m}_a$ is the mass flow rate of dry air; $\dot{m}_w$ is the rate of moisture addition to the moist air; and W_1 and W_3 are the humidity ratios at Point 1 and 3, respectively.

Neglecting any fan work and losses, the conservation of energy yields

$$\dot{Q}_T = \dot{Q}_S + \dot{Q}_L \quad (2.126)$$

$$\dot{Q}_T = \dot{m}_a \left(h_{tp} - h_1\right) + \dot{m}_a \left(h_3 - h_{tp}\right) \quad (2.127)$$

where $\dot{Q}_T$ is the rate of heating provided; $\dot{Q}_S$ is the rate of sensible heating provided; $\dot{Q}_L$ is the rate of latent heating provided; $\dot{m}_a$ is the mass flow rate of dry air; h_1 and h_3 are the enthalpies per unit mass of dry air at the inlet and exit of the heating and humidification section, respectively; and h_{tp} is the enthalpy per unit mass of dry air at the turning point.

2.3.4.2.3 Sensible Cooling

This is cooling without changing the moisture content of the air; that is, the humidity ratios at the initial (State 1) and final (State 2) conditions are the same as no moisture is added to or removed from the air, as shown in Figure 2.19.

Note that during sensible cooling, as opposed to during sensible heating, the relative humidity of air increases. This is because the moisture-holding capacity of the air decreases with decreasing air temperature.

Just as in sensible heating, the conservation of mass reduces to

$$\dot{m}_{a1} = \dot{m}_{a2} = \dot{m}_a \quad (2.128)$$

$$W_1 = W_2 \quad (2.129)$$

where $\dot{m}_a$ is the mass flow rate of dry air, and W_1 and W_2 are the humidity ratios at Point 1 and 2, respectively.

Neglecting any fan work and losses, the conservation of energy yields

$$\dot{Q}_S = \dot{m}_a \left(h_2 - h_1\right) \quad (2.130)$$

where $\dot{Q}_S$ is the rate of sensible cooling provided and h_1 and h_2 are the enthalpies per unit mass of dry air at the inlet and exit of the cooling section, respectively.

2.3.4.2.4 Sensible Cooling and Dehumidification

This cooling involves removing some moisture from the air; to do this, we need to cool the air below its dew-point temperature. First, as the air passes through a cooling coil, the air temperature decreases, and its relative humidity increases until the air temperature reaches its dew-point temperature. Then, any further cooling beyond the dew-point temperature condenses part of the moisture in the air while the air remains saturated and moves down the saturation curve during the entire condensation process as shown in Figure 2.20.

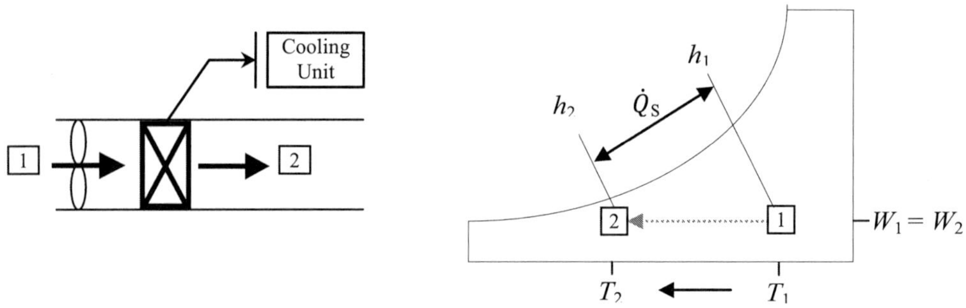

FIGURE 2.19 Sensible cooling process.
Source: ***Greenhouse Engineering*** **(2021).**

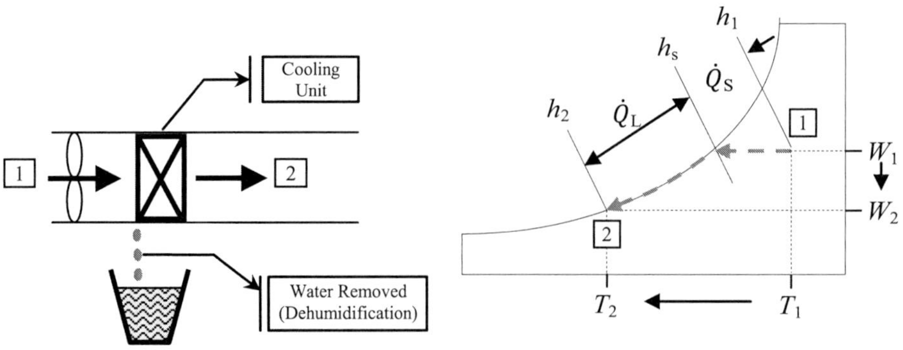

FIGURE 2.20 Sensible cooling and dehumidification.
Source: ***Greenhouse Engineering*** **(2021).**

In sensible cooling and dehumidification, the conservation of mass reduces to

$$\dot{m}_{a1} = \dot{m}_{a2} = \dot{m}_a \text{ and } W_1 \neq W_2 \quad (2.131)$$

$$\dot{m}_w = \dot{m}_a \Delta W = \dot{m}_a \left(W_2 - W_1\right) \quad (2.132)$$

where $\dot{m}_a$ is the mass flow rate of dry air, $\dot{m}_w$ is the rate of moisture removal from the moist air, and W_1 and W_2 are the humidity ratios at States 1 and 2, respectively.

Neglecting any fan work and losses, the conservation of energy yields

$$\dot{Q}_T = \dot{Q}_S + \dot{Q}_L \quad (2.133)$$

$$\dot{Q}_T = \dot{m}_a \left(h_s - h_1\right) + \dot{m}_a \left(h_2 - h_s\right) = \dot{m}_a \left(h_2 - h_1\right) \quad (2.134)$$

where $\dot{Q}_T$ is the total cooling provided per unit time; $\dot{Q}_S$ is the rate of sensible energy removed; $\dot{Q}_L$ is the rate of latent energy removed; h_1 and h_2 are the enthalpies per unit mass of dry air at the inlet and exit of the heating and humidification section, respectively; and h_s is the enthalpy per unit mass of dry air at the saturation point.

Example 2.73: Sensible Cooling and Dehumidification

Now consider moist air entering a chiller at T = 35°C and 60% relative humidity with a flow rate of 0.425 m³/s and leaving the chiller and entering the conditioned space at required T = 5°C and 100% relative humidity. Determine the total refrigeration capacity required and the total water removal rate from the moist air. Note: One ton of refrigeration equals a heat removal rate of 12,000 Btu/h.

Solution

Figure 2.20 schematically shows the solution. However, State 1 is located at the intersection of T = 35°C and φ = 60%. Therefore, h_1 = 91.0 kJ/kg$_{da}$, W_1 = 0.0214 kg$_{water}$/kg$_{da}$, and v_1 = 0.903 m³/kg$_{da}$. State 2 is located on the saturation curve at 5°C. Thus, h_2 = 19.0 kJ/kg$_{da}$, and W_2 = 0.0054 kg$_{water}$/kg$_{da}$.

The mass flow rate of dry air is

$$\dot{m}_a = \left(0.425\,\text{m}^3/\text{s}\right) / \left(0.903 \frac{\text{m}^3}{\text{kg}_{da}}\right) = 0.471\,\text{kg}_{da}/\text{s}$$

Using Equation (2.134):

$$\dot{Q}_T = \dot{m}_a \left(h_2 - h_1\right)$$
$$= 0.471(19.0 - 91.0)$$
$$= -33.9\,\text{kW}\left(\approx -115{,}741 \frac{\text{Btu}}{\text{h}} \approx -9.65\,\text{ton}\right)$$

Now, using Equation (2.132), we can determine the water removal rate as follows:

$$\dot{m}_w = \dot{m}_a \Delta W = \dot{m}_a \left(W_2 - W_1\right)$$
$$= 0.425(0.0214 - 0.0054)$$
$$= 0.0068\,\text{kg}_{water}/\text{s} \approx 24.5\,\text{kg}_{water}/\text{h}$$

2.3.4.2.5 Adiabatic Water Addition to Moist Air Stream

Sometimes, steam or liquid water can be injected into a moist airstream to raise its humidity. Figure 2.21 shows a diagram of this common air-conditioning process.

We can now write conservation of mass for the steady-state process as

$$\dot{m}_{a1} = \dot{m}_{a2} = \dot{m}_a \text{ and } W_1 \neq W_2 \quad (2.135)$$

$$\dot{m}_a W_1 + \dot{m}_w = \dot{m}_a W_2 \xrightarrow{yields} \dot{m}_w = \dot{m}_a \Delta W = \dot{m}_a \left(W_2 - W_1\right) \quad (2.136)$$

Assuming an adiabatic process, we can then write the following energy balance equation:

$$\dot{m}_a h_1 + \dot{m}_w h_w = \dot{m}_a h_2 \quad (2.137)$$

Substituting Equation (2.136) into Equation (2.137) and dividing by $\dot{m}_a$, we obtain the following expression:

$$h_2 = h_1 + h_w (W_2 - W_1) \quad (2.138)$$

$$h_w = \frac{h_2 - h_1}{W_2 - W_1} = \frac{\Delta h}{\Delta W} \quad (2.139)$$

where $\dot{m}_a$ is the mass flow rate of dry air; $\dot{m}_w$ is the mass flow rate of water (any phase); h_1 and h_2 are the enthalpies of the moist air at States 1 and 2, respectively; h_w is

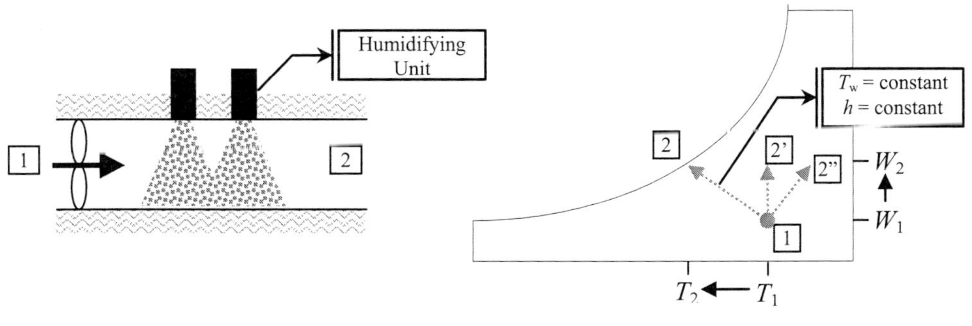

FIGURE 2.21 Adiabatic water addition to moist air stream.
Source: *Greenhouse Engineering* **(2021).**

the specific enthalpy of water (any state); and W_1 and W_2 are the humidity ratios at States 1 and 2, respectively. As shown in Figure 2.21, the direction of the final state (State 2) of the moist air is fixed by the specific enthalpy of the injected water and drawn through State 1 of the moist air.

If we spray cold water, the enthalpy of the water is lower than that of the moist air. As a result, the enthalpy of the moist air leaving the process (State 2) is equal to the enthalpy of the mixture entering (State 1). As water evaporates, the latent heat of vaporization is absorbed from the surrounding air and water body itself, causing a lower temperature. As a result, both air and water are cooled.

A simple energy balance gives the amount of cooling provided:

$$\dot{Q} = \dot{m}_a\, c_p \left(T_2 - T_1\right) = \dot{m}_a\, L_v \left(W_1 - W_2\right) \quad (2.140)$$

where c_p is the specific heat of air at constant pressure (Table A.13); T_1 and W_1 are the entering air T and W at State 1, respectively; L_v is the latent heat of vaporization or enthalpy of vaporization of water (Table A.57); and T_2 and W_2 are the leaving T and W at State 2, respectively.

This ancient method of air-conditioning is highly effective in dry climates. On the psychrometric chart, the process line 1 to 2 moves on a line of constant T_w and constant h, which is termed evaporative cooling—essentially an adiabatic saturation process—from an initial state at low W and high T to a final state at higher humidity and lower T. Since the sprayed water enthalpy is lower than that of the moist air, lines of constant enthalpy are approximately parallel to lines of constant adiabatic T_w.

If hot water or steam with progressively higher enthalpy is sprayed into the moist air (the process line 1 to 2), as Equation (2.138) indicates and Figure 2.21 shows, the enthalpy of the moist air at the end of the spraying process (State 2) will increase. If, on the other hand, the enthalpy of the steam in the spray is the same as the enthalpy of the water vapor in the moist air entering the process (the process line 1 to 2), the leaving moist air at State 2 is the same temperature as the entering mixture (Figure 2.21).

Example 2.74: Evaporative Cooling

Let's now consider a fan-and-pad evaporative cooling system used for air conditioning an institutional greenhouse. Assume that the outside air enters the evaporative cooling pad at T = 35°C and φ = 25% with a flow rate of 4 m³/s. Assuming that the cooler has a cooling efficiency of 80%, determine the supply T, ϕ, total cooling capacity provided, and the rate of water used for this evaporative cooling. Note: One ton of refrigeration equals a heat removal rate of 12,000 Btu/h.

Solution

Figure 2.21 (process line 1 to 2) schematically shows the solution. However, State 1 does not in this example lie on the saturation curve as the cooler is not 100% efficient. At 100% efficiency, the temperature at State 2 is the T_w of State 1; that is, State 1 is located at the intersection of T = 35°C and φ = 25%, and T_w is the temperature at State 2, approximately 20.2°C. With 80% efficiency, however, T_2 becomes T_1 minus the 80% of ($T_1 - T_w$), that is, T_2 = 35°C – 0.8 (35°C – 20.2°C) = 23.2°C, the supply temperature to the greenhouse.

Using the intersection of T_2 = 23.2°C and the constant T_w of 20.2°C on the psychrometric chart, we can determine the humidity ratio at State 2 as W_2 = 0.0137 kg_{water}/kg_{da}, the specific volume v_2 = 0.858 m³/kg, and the relative humidity at State 2 as ϕ = 77%. Now, we can calculate the mass flow rate of dry air using v_2 at State 2 as the fan is positioned on the opposite wall (negative pressure or suction ventilation). The mass flow rate of dry air is then

$$\dot{m}_a = (4\ \text{m}^3/\text{s})/(0.858\ \text{m}^3/\text{kg}_{da}) = 3.43\ \text{kg}_{da}/\text{s}$$

Now, using Equation (2.140)

$$\dot{Q} = \dot{m}_a c_p (T_2 - T_1) = (3.43)(1.006)(23.2 - 35)$$
$$= -40.7\,\text{kW}\left(\approx -13{,}891\frac{\text{Btu}}{\text{h}} \approx -1.16\,\text{ton}\right),$$

we can calculate the cooling capacity using the humidity ratios as well:

$$\dot{Q} = \dot{m}_a L_v (W_1 - W_2)$$
$$= (3.43)(2{,}430)(0.0088 - 0.0137)$$
$$= -40.8\,\text{kW}\,(\approx -1.16\,\text{ton})$$

where we used a latent heat of vaporization L_v of approx. 2,430 kJ/kg (at ~30°C) (Table A.57) and a humidity ratio at State 1, W_1 = 0.0088 kg_{water}/kg_{da} at the intersection of T = 35°C and φ = 25%.

Now, using Equation (2.136), the theoretical rate of water use for evaporative cooling can be calculated as

$$\dot{m}_w = \dot{m}_a (W_2 - W_1) = (3.43)(0.0137 - 0.0088)$$
$$= 0.0168\,\text{kg}_{water}/\text{s} \approx 60.5\,\text{kg}_{water}/\text{h}$$

2.3.4.2.6 Adiabatic Mixing of Two Streams of Moist Air

The adiabatic mixing of two moist airstreams, shown in Figure 2.22, is a very common practice in air-conditioning systems and is governed by three equations: the mass balance equation for the air, the mass balance equation for the water vapor, and the energy balance equation.

The mass balance equation for the entering and leaving air is

$$\dot{m}_{a1} + \dot{m}_{a2} = \dot{m}_{a3} \quad (2.141)$$

The mass balance equation for the water vapor can be written as

$$\dot{m}_{a1} W_1 + \dot{m}_{a2} W_2 = \dot{m}_{a3} W_3 \quad (2.142)$$

And assuming an adiabatic process, the energy balance equation using the mixture of enthalpies becomes

$$\dot{m}_{a1} h_1 + \dot{m}_{a2} h_2 = \dot{m}_{a3} h_3 \quad (2.143)$$

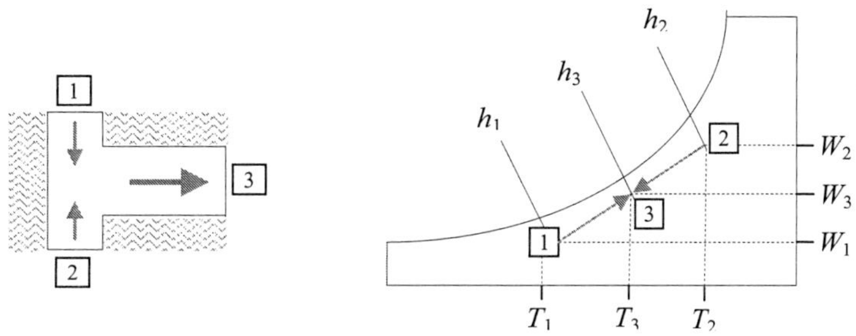

FIGURE 2.22 Adiabatic mixing of two streams of moist air.
Source: *Greenhouse Engineering* (2021).

After eliminating $\dot{m}_{a3}$ and substitution, we can develop the following:

$$\frac{\dot{m}_{a1}}{\dot{m}_{a2}} = \frac{h_2 - h_3}{h_3 - h_1} = \frac{W_2 - W_3}{W_3 - W_1} \tag{2.144}$$

Based on Equation (2.144), the state point of the resulting mixture (State 3) is always on the straight line connecting the state points of the streams 1 and 2 being mixed and divides the line into two segments, with its location set by the same ratio as the masses of dry air in streams 1 and 2.

2.3.4.2.7 Absorption of Space Heat and Moisture Gains by Moist Air

In air conditioning a space, we encounter moisture and energy gains almost all the time, and it is required to remove given amounts of energy and water from the space at a specified condition. The energy gains consist of energy alone independent of the energy contributions due to the moisture gains, and we call this the sensible heat gain $\dot{Q}_s$. The net moisture gains $\sum \dot{m}_w$ can be through the system boundaries as well as from water sources within the space. Each unit of water mass injected into the space adds an amount of energy equal to its specific enthalpy. The system is basically governed by the mass and energy balance equations.

Assuming steady-state conditions, the mass balance equation can be written as

$$\dot{m}_a W_1 + \sum \dot{m}_w = \dot{m}_a W_2 \text{ or}$$
$$\sum \dot{m}_w = \dot{m}_a (W_2 - W_1) \tag{2.145}$$

which represents the total rate of net water addition to the space. Similarly, the energy balance equation can be written as

$$\dot{m}_a h_1 + \dot{Q}_s + \sum \dot{m}_w h_w = \dot{m}_a h_2 \tag{2.146a}$$

or

$$\dot{Q}_s + \sum \dot{m}_w h_w = \dot{m}_a (h_2 - h_1) \tag{2.146b}$$

and the left side of this equation represents the total rate of energy addition to the space from all sources.

After eliminating $\dot{m}_a$ and substitution, from Equations (2.142) and (2.143), we can develop the following:

$$\frac{\Delta h}{\Delta W} = \frac{h_2 - h_1}{W_2 - W_1} = \frac{\dot{Q}_s + \sum (\dot{m}_w h_w)}{\sum (\dot{m}_w)} \tag{2.147}$$

Based on Equation (2.147), on the psychrometric chart and for a given state of the removed air, all possible states (conditions) for the supply air must lie on a straight line drawn through the state point of the removed air, which has a direction specified by $\left[\dot{Q}_s + \sum (\dot{m}_w h_w)\right] / \sum (\dot{m}_w)$. This line is the condition line for the given problem.

Example 2.75: Absorption of Space Heat and Moisture Gains

Let's now consider a research greenhouse from which moist air is removed at T = 30°C and T_w = 22°C. The sensible heat gain for the greenhouse is 20 kW. A rate of moisture gain of 10.0 kg/h occurs from crop transpiration that is assumed as saturated water vapor at 28°C. Moist air is introduced to the greenhouse at a dry-bulb temperature of 20°C. Determine the supply air's thermodynamic wet-bulb temperature and the volumetric flow rate.

Solution

The psychrometric sketch schematically shows the solution. State 2 is located at the intersection of T = 30°C and T_w = 22°C. Table A.57 provides the specific enthalpy of added water vapor at 28°C, h_g as 2434.55 kJ/kg$_{water}$.

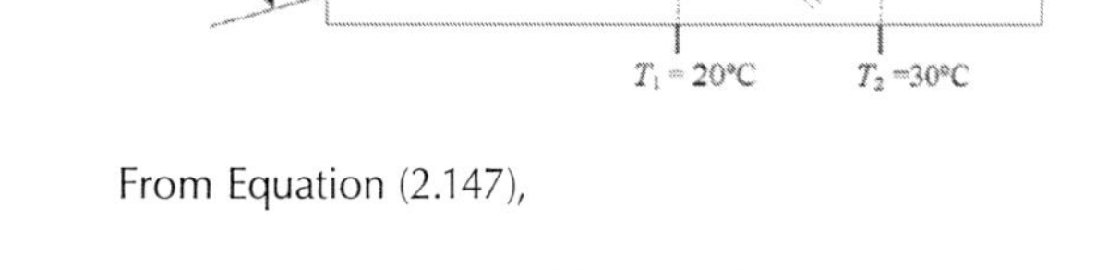

From Equation (2.147),

$$\frac{\Delta h}{\Delta W} = \frac{20 + (10/3{,}600)(2434.55)}{\left(\frac{10}{3{,}600}\right)} = 9{,}635 \text{kJ/kg}_{water}$$

With the $\Delta h/\Delta W$ protractor, we can now establish a reference line of direction: $\Delta h/\Delta W$ = 9.635 kJ/g$_{water}$. Parallel to this reference line, a straight line on the psychrometric chart can be drawn through State 2. The intersection of this line with the T = 20°C line fixes

State 1. Thus, we can determine the wet-bulb temperature at State 1, T_{w1} = 17.7°C.

Alternatively, we can establish the condition line using the protractor's sensible-total heat ratio scale instead of the $\Delta h / \Delta W$ scale. On the protractor, $\Delta H_S / \Delta H_T$ is the ratio of the rate of sensible heat gain for the space to the rate of total energy gain for the space. Therefore,

$$\frac{\Delta H_S}{\Delta H_T} = \frac{\dot{Q}_s}{\dot{Q}_s + \sum(\dot{m}_w h_w)} = \frac{20}{20 + (10/3{,}600)(2434.55)}$$
$$= 0.747$$

Observe that $\Delta H_S / \Delta H_T = 0.747$ on the protractor, which coincides closely with $\Delta h / \Delta W = 9.635$ kJ/g_{water}. The flow of dry air can now be calculated using either Equation (2.145) or Equation (2.146). Using (2.146),

$$\dot{m}_a = \frac{\left[\dot{Q}_s + \sum \dot{m}_w h_w\right]}{h_2 - h_1}$$

$$= \left[20 + (10/3{,}600)(2434.55)\right] / (64.2 - 49.9)$$

$$\dot{m}_a = 1.872 \text{kg}_{da} / \text{s}$$

At State 1, v_1 = 0.846 m³/kg$_{da}$; therefore, the volumetric air flow rate $\dot{m}_a v_1$ = (1.872)(0.846) = 1.584 m³/s.

2.3.5 Water Vapor Transfer, Condensation, and Thermal Insulation

Section 2.3.3 presented heat transfer and its mechanisms through different structural members and shapes. This section covers further refinements in these heat transfer mechanisms to account for variations in surface and air space heat transfer resistances. Further, the section covers the effects of ceilings, framing members, gable, and roof combinations on heat loss from different buildings. Finally, water vapor transfer, the temperature gradient through a wall, surface condensation, and problems and prevention in buildings as well as thermal insulation are discussed in depth.

2.3.5.1 Surface Resistance

As discussed earlier, heat transfer from a surface is affected by the surface temperature and emissivity, air velocity, and temperature difference between the surface and air. That is, the surface resistances combine the effects of conduction, convection, and radiation. The total resistance $R_t = 1/UA$ and the overall thermal transmittance, that is, the overall heat transfer coefficient U, include the inside and outside boundary layer resistances, R_i and R_o, respectively, the values of which are provided in Table A.9 for different flow directions and surface emissivities. The table shows that surface resistance increases as surface emissivity decreases. Usually, the resistance for inside surfaces is for still air, and the outside surface resistances for winter and summer are based on 6.6 m/s (15 mph) and 3.3 m/s (7.5 mph) wind speeds, respectively. For moving air, on the other hand, R is independent of surface slope and heat flow direction because heat transfer is predominantly by forced convection under such conditions. Typical thermal properties (density, thermal conductivity, thermal resistance and specific heat) of building and insulating materials can be found in Table 1 of Chapter 26 of the *ASHRAE Handbook* (ASHRAE, 2021).

2.3.5.1.1 Air Space Resistance

Heat transfer in an air space mainly takes place due to convection and radiation and depends on the mean surface temperature; the temperature difference between the two surfaces facing the air space; the surface emissivities; and the width, surface roughness, and orientation of the air space.

If the effective emissivity ε of the two surfaces facing the air space is known, then we can select effective thermal resistances of plane air spaces from *ASHRAE Handbook* Chapter 23, Table 3 (ASHRAE, 2021). Then, we can calculate the effective emissivity as follows:

$$\varepsilon = \frac{1}{\left(\frac{1}{\varepsilon_1} + \frac{1}{\varepsilon_2} - 1\right)} \quad (2.148)$$

where ε_1 and ε_2 are the emissivities of the two surfaces facing the air space. Emissivities for typical building surfaces are given in Table A.10.

2.3.5.1.2 Parallel Heat Flow Paths

Earlier, this chapter covered parallel heat flow for dissimilar components arranged side by side, which results in parallel heat flow paths with different thermal conductances. It was emphasized that when dealing with composite walls, it is easier to work with the overall thermal transmittance, that is, the overall U, whose relationship with the total thermal resistance is given in Equation (2.46). If there is no lateral heat flow between the paths, then each path is considered to extend straight from inside to outside, and U is calculated for each path.

The average transmittance or the heat transfer coefficient is calculated as follows:

$$U_{avg} = (U_a)A + (U_b)B + \ldots + (U_n)N \quad (2.149)$$

where A, B, . . ., and N are the respective areas and U_a, U_b, . . ., and U_n are the respective coefficients of thermal transmittance of the different structural components.

2.3.5.1.3 Adjustment for Framing Members

To adjust for parallel heat flow through wall sections with framing and insulated areas, we can use the following equation instead:

$$U_{avg} = U_s\left(\frac{S}{100}\right) + U_i\left(1 - \frac{S}{100}\right) \quad (2.150)$$

where U_{avg} is the average transmittance for the building section, U_s is the transmittance for area backed by framing members (studs, joists, plates, furring, headers, sills, etc.), U_i is the transmittance for area between framing members, and S is the percentage of area backed by framing members that varies widely with construction type.

2.3.5.1.4 Combined Ceiling, Gable, and Roof Resistances

When estimating heat loss through the ceiling of structures with unheated, closed attic spaces, the heat transfer characteristics of the roof and gable construction are included. Convection dominates such an attic space where the thermal resistance is very low. Therefore, the effect of the attic as an air space is simply disregarded, and the situation is treated as analogous to a simple electrical network, where the total resistance can be expressed as:

$$R_t/A = R_c/A_c + \left(R_g \times R_r\right) \div \left(R_g \times A_r + R_r \times A_g\right) \quad (2.151)$$

where the subscripts c, g, and r refer to the ceiling, gable, and roof, respectively and A_c, A_g, and A_r are the areas of the ceiling, gable, and roof, respectively. A can be any of the three areas. Letting $A = A_c$ and multiplying, we obtain the following:

$$R_t = R_c + \left(R_r \times R_g\right) \div \left(\mathrm{m} \times R_r + \mathrm{n} \times R_g\right) \quad (2.152)$$

where $\mathrm{m} = A_g / A_c$ and $\mathrm{n} = A_r / A_c$, and R_t is the combined overall heat transmission coefficient for the ceiling, roof, and gable per unit ceiling area.

2.3.5.2 Water Vapor Properties and Transfer

Water vapor is a gas that occupies the same space as the other gases in the atmosphere, that is, in the air. It migrates by a variety of transport mechanisms, including diffusion of water by partial pressure gradients as an independent gas and displacement by air flow as a constituent part of the atmosphere. Water vapor in the air, for any given temperature and degree of saturation, exerts pressure and diffuses from points of higher vapor pressure toward points of lower pressure both in the air and within other materials. Hence, it can migrate with or against air flow.

Vapor diffusion in free air is a rapid process that results in minimal vapor pressure differentials between connected spaces. As a result, vapor pressure differentials in buildings and across the building envelope are not as strong as the driving force for air convection. Therefore, when present, air movement can displace greater amounts of moisture than diffusion does. Hence, air movement has a higher priority in moisture control in buildings.

Water vapor activates sorption in the building envelope materials, and therefore, water vapor flow in and through the envelope can cause condensation on both nonporous and porous surfaces. The properties of air and water vapor mixtures are well known and have already been covered in Section 2.3.4. Changes in psychrometric properties in heating, cooling, and air mixing can easily be followed using the psychrometric charts (Figure 2.16), where the saturation line represents the limiting water vapor concentrations at different temperatures.

2.3.5.2.1 Water Vapor Flow through Materials

As mentioned earlier, diffusion transfers water vapor through air and building materials in small quantities. However, it can still be important in industrial applications, such as cold-storage facilities or in buildings where a high indoor partial water vapor pressure is present due to special activities in the space (e.g., greenhouses and livestock and aquatic buildings).

The water vapor diffusion rate through materials is calculated based on a form of Fick's law for diffusion, which closely parallels Fourier's equation for heat flux by conduction (Equation 2.28), as follows:

$$\dot{m} = -\mu \times A \times \left(\frac{dp}{dx}\right) \quad (2.153)$$

where $\dot{m}$ is the vapor transfer rate in kg/s (grains/h; 7000 grains = 1 lb), μ is the water vapor permeability in kg/s.m.Pa (grains/h.ft.in Hg). This value in essence is the steady-state water vapor flux through a unit thickness of homogeneous material, moving perpendicular to isobaric planes and induced by a unit partial water vapor pressure difference, under specified conditions of temperature and relative humidity. A is the cross-sectional area normal to the direction of vapor flow m^2 (ft^2), dp is the vapor pressure gradient in the direction of flow Pa (in Hg), and dx is the distance of flow m (ft). Typical water vapor permeance and permeability values for common building and insulation materials can be found in Table 5 of Chapter 26 of the *ASHRAE Handbook* (ASHRAE, 2021).

Water vapor transfer through materials is a complex process and not well understood. The permeability, μ, can vary and is dependent primarily on relative humidity, and moderately on temperature, which is often neglected. In the case of steady-state water vapor transfer through a plane wall of homogeneous material subjected to constant vapor pressure, the water vapor transfer rate can be obtained by separating variables and integrating Equation (2.153):

$$\dot{m} = \mu \times A \times \left(p_1 - p_2\right) \div L \quad (2.154)$$

where p_1 and p_2 are the vapor pressures at each end of the flow path, L is the length of the flow path (material thickness), and the mean permeability μ' is:

$$\mu' = 1/\left(p_2 - p_1\right) \int_{P_1}^{P_2} dp \quad (2.155)$$

Average water vapor permeabilities of common building materials can be found in any editions of the *ASHRAE Handbook of Fundamentals* as well as in ASHRAE (2021).

2.3.5.2.2 Water Vapor Flow through Walls

The rate of water vapor flow is a function of the pressure difference and the transmission characteristics of the wall. When temperatures on opposite sides of a wall are different, then the warm side usually has the higher vapor pressure. If the wall is vapor permeable, vapor then diffuses from the higher vapor pressure to the direction of the lower pressure, and if the temperature along the diffusion direction is lower than the dew-point temperature at any point, then the vapor starts condensing on that point of the wall.

The transfer rate of water vapor, $\dot{m}$, through a material is analogous to the transfer of heat and can be obtained by rewriting Equation (2.154):

$$\dot{m} = K \times A \times \left(p_1 - p_2\right) \quad (2.156)$$

or

$$\dot{m} = A \times (p_1 - p_2) \div D$$

where $\dot{m}$ is the rate of water vapor transmitted in kg/s (grains/h) and A is the area normal to water vapor flow path in m^2 (ft^2). $K = \mu' / L$ is the permeance in kg/s.m^2.Pa (grains/h.ft^2.in Hg), which indicates how quickly, in kg/s (grains/h), one square meter (or square foot) of a vapor barrier or diffusion retarder permits water vapor passage when there is a one Pascal (one inch Hg) difference in partial pressure on both sides of the barrier. p_1 is the vapor pressure at the start of the flow path in Pa (in Hg), and p_2 is the pressure at the end of the path. $D = 1/K$ is the water vapor resistance—the reciprocal of water vapor permeance—in s.m^2.Pa/kg (h.ft^2.in Hg/grains).

The overall vapor resistance, which is similar to the overall heat transfer resistance, in a nonhomogeneous wall can be expressed as

$$D_T = D_1 + D_2 + ... + D_n \quad (2.157)$$

where D_T is the overall wall water vapor transmission resistance, and D_1, D_2, . . . D_n are the vapor resistances for each wall component.

The vapor transfer rate, $\dot{m}$, through the whole wall is then

$$\dot{m} = A \times (p_i - p_o) \div D_T \quad (2.158)$$

where p_i and p_o are the vapor pressures on the inside and outside of the wall, respectively.

2.3.5.2.2.1 Vapor Pressure Gradient and Flow without Condensation We can estimate vapor pressure at any point in a wall using an equation similar to that for wall temperatures, which will be discussed in Section 2.3.5.2.3. The procedure assumes an equal vapor flow rate at all points in the wall, and vapor flow through the full thickness of the wall equals the vapor flow through any point X within the wall. Then, by analogy to Equation (2.161),

$$p_x = p_i - (D_x - D_t) \times (p_i - p_o) \quad (2.159)$$

where p_x is the water vapor pressure at the point X, p_i is the inside water vapor pressure, p_o is the outside pressure, D_t is the overall wall water vapor resistance, $D_t = 1/K_t$, and D_x is the resistance from p_i to point X, $D_x = 1/K_x$.

2.3.5.3 Condensation

Condensation is the phase change from water vapor to liquid on glass, metal, synthetic foils, etc. Technically speaking, condensation is the condition where the partial water vapor pressure is greater than saturation. Water vapor condenses in a film on surfaces that are at temperatures below the dew-point temperature of the indoor air. Frost forms if the surface temperature is below freezing, and sometimes, condensation occurs first, and then ice forms from the condensed water after temperatures drop below freezing.

Condensation frequently occurs on windows or single glazing and on metal frames without a thermal break. Condensation within or on wall or ceiling surfaces or in attic spaces can cause severe structural damage over time and reduce the value of some insulation materials. In greenhouses, condensation typically forms on indoor glazing surfaces, but in livestock or aquatic buildings, condensation can form not only on the wall surface but also within a wall.

2.3.5.3.1 Estimating Temperature within Walls

The flow of heat through a nonhomogeneous wall of unit area is already expressed in Equation (2.34) as $\dot{Q} = (T_i - T_o) \div R_t$. Let us now consider a wall in Figure 2.23 that has a total resistance R_t. Then, the overall resistance from the indoor air to a point X in the wall is R_x, and the temperature at that point becomes T_x.

Since the rate of heat flow through the entire thickness of the wall is the same as that through the wall to any point X along the direction of heat transfer, the equations for heat flow rate through a unit wall area can be set equal to each other as follows:

$$\dot{Q} = (T_i - T_o) \div R_t = \dot{Q}_x = (T_i - T_x) \div R_x \quad (2.160)$$

After solving for T_x, the temperature at any point X becomes

$$T_x = T_i - (R_x \div R_t) \times (T_i - T_o) \quad (2.161)$$

2.3.5.3.2 Surface Condensation and Prevention

Surface condensation occurs when water vapor contacts a nonporous surface that has a temperature lower than the dew point of the surrounding air; often, condensation is a problem in winter and occurs on the warm side surfaces such as inside window and wall surfaces. We therefore need thick enough insulation so that the surface temperature on the warm side of an insulated building assembly always exceeds the dew point in the space.

However, we also need to remember that even without reaching the dew-point temperature, relative humidity at the surface may reach such high levels that mold grows over

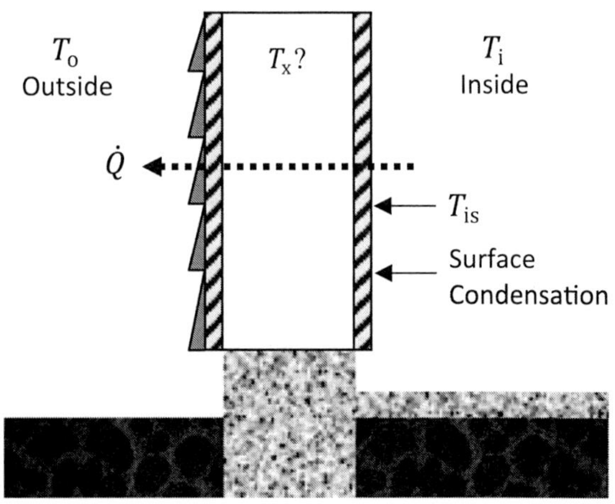

FIGURE 2.23 Heat flow through an uninsulated wall, wall temperature, and condensation.

Source: ***Greenhouse Engineering*** **(2021).**

time. To prevent condensation on interior surfaces, if it is applicable, we can reduce the moisture production by lowering the indoor relative humidity, either by increasing natural or mechanical ventilation or somehow increasing the surface temperature. The indoor surface temperature can be increased by increasing R_t or decreasing R_i in Equation (2.161); sometimes, it is possible to reduce the difference between the indoor and outdoor temperatures $(T_i - T_o)$. To prevent condensation within a wall, on the other hand, we can reduce permeability (K) of the wall by adding a vapor barrier (always on the warmer side of the wall), reducing the temperature difference between inside and outside temperatures, increasing R_t, or increasing the permeability of the cold side of the wall to let the moisture.

We can easily estimate the indoor surface temperature, t_{is}, using Equation (2.161) by substituting R_i (indoor surface resistance) for R_x, and analyze the situations accordingly, as in the following examples.

Example 2.76: Surface Temperature and Condensation

Let's now consider a building where the inside air temperature is 15.5°C (60°F), outside air temperature is −11°C (12°F). The building wall is uninsulated and composed of 0.5 in. gypsum (R = 0.079 m²°C/W or 0.45 h.ft²°F/Btu) and 3/8 in. plywood (R = 0.083 m²°C/W or 0.47 h.ft²°F/Btu) with 3.5 in. air space (R = 0.158 m² °C/W or 0.90 h.ft² °F/Btu) in between. Determine the wall surface temperature and the indoor relative humidity at which condensation starts. Note: The conversion between SI and I-P of R in Table A.3 is given as 1 h.ft²°F/Btu = 0.176 m²°C/W.

Solution

From Table A.9, the indoor and outdoor surface resistances are 0.120 m²°C/W (0.68 h.ft²°F/Btu) and 0.030 m²°C/W (0.17 h.ft²°F/Btu), respectively. We can then determine the total thermal resistance of the building wall as

$$R_t = R_i + R_{gypsum} + R_{air\,space} + R_{plywood} + R_o$$

$$R_t = 0.120 + 0.079 + 0.158 + 0.083 + 0.030$$

$$R_t = 0.470 \text{ m}^2\ ^\circ\text{C/W} \ (\approx 2.67 \text{ h.ft}^2\ ^\circ\text{F/Btu})$$

And the surface temperature as

$$T_{is} = T_i - (R_i \div R_t) \times (T_i - T_o)$$

$$T_{is} = 15.5^\circ\text{C} - (0.120 \div 0.470) \times (15.5 - (-11))^\circ\text{C}$$
$$= 8.7^\circ\text{C} \ (\approx 47.7^\circ\text{F})$$

From the psychrometric chart, we can easily determine that under these conditions, condensation on the wall surface occurs if the indoor relative humidity is greater than 64%.

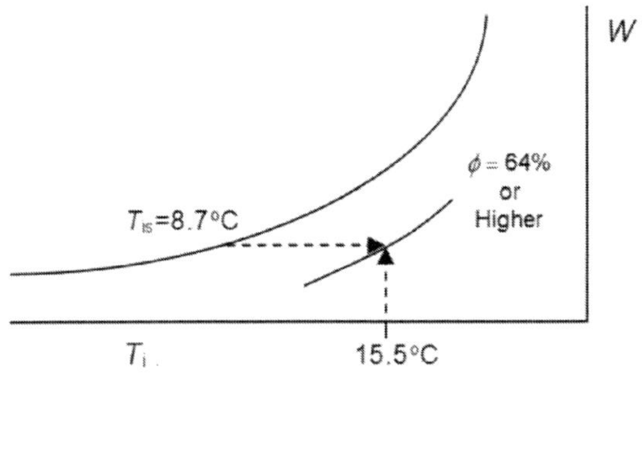

Example 2.77: Insulation, Surface Temperature, and Condensation

Now let's consider the same building with 3.5 in. of R-11 insulation added to the same wall section and the effect of this insulation on the wall surface temperature and surface condensation. That is, determine the new wall surface temperature and indoor relative humidity at which condensation starts taking place.

Solution

R_t after adding R-11 is now higher than in the previous example, 2.248 m².°C/W (12.77 h.ft².°F/Btu), and now we can determine the wall surface temperature as follows:

$$T_{is} = 15.5^\circ\text{C} - (0.120 \div 2.248) \times (15.5 - (-11))^\circ\text{C}$$
$$= 14.1^\circ\text{C} \ (\approx 57.4^\circ\text{F})$$

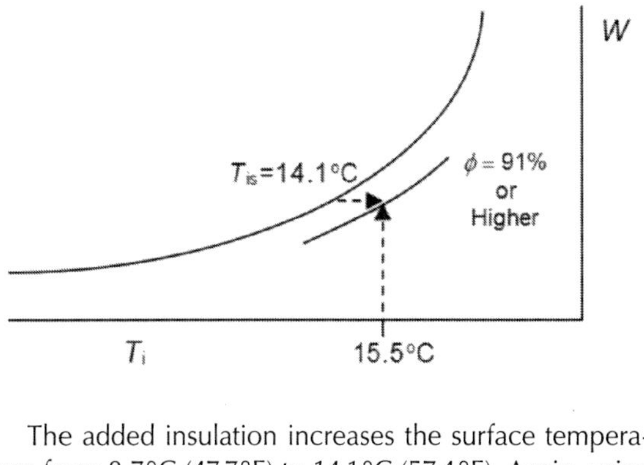

The added insulation increases the surface temperature from 8.7°C (47.7°F) to 14.1°C (57.4°F). Again, using the psychrometric chart, we can easily determine that this added insulation now prevents surface condensation as long as the indoor relative humidity stays below 91%.

Example 2.78: Surface Temperature and Condensation

Let's now consider increasing the inside surface temperature given in the example. We can increase wall surface temperature by decreasing the inside boundary layer resistance. By increasing air velocity over the surface to about 6.6 m/s (15 mph) with a fan, the inside boundary layer resistance, R_i, can be reduced from 0.120 m².°C/W (0.68 h.ft².°F/Btu) to 0.030 m².°C/W (0.17 h.ft²°F/Btu) (Table A.9). Now determine the new wall surface temperature and indoor relative humidity at which condensation starts taking place.

Solution

Under the given conditions, R_t reduces to 0.030 m².°C/W (0.17 h.ft².°F/Btu), which reduces R_t as well to 0.380 m².°C/W (2.16 h.ft².°F/Btu). Then, the resulting surface temperature, t_{is}, becomes

$$T_{is} = 15.5^\circ\text{C} - (0.030 \div 0.380) \times (15.5 - (-11))^\circ\text{C}$$
$$= 13.4^\circ\text{C} \ (\approx 56.1^\circ\text{F})$$

By increasing the boundary layer resistance, we have now increased the surface temperature from 8.7°C (47.7°F) to 13.4°C (56.1°F), and under these conditions, condensation on the inside wall surface does not occur as long as the indoor relative humidity remains below 87%.

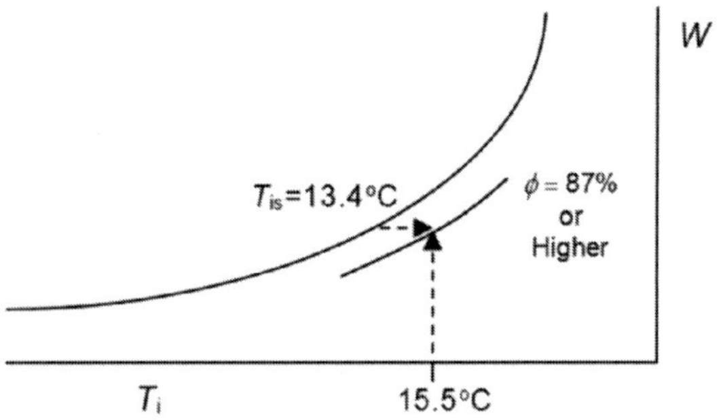

2.3.5.4 Thermal Insulation

The purpose of thermal insulation is to reduce conductive, convective, and radiant heat flows; that is, it is primarily used to limit heat gain or loss from surfaces at temperatures above or below ambient temperature. Thermal insulation applications include walls, floors, ceilings, roof, pipes, tanks, vessels and equipment, and ducts.

When properly applied in building envelopes or on mechanical systems, insulation satisfies at least one of the following objectives: i) conserving energy by reducing heat loss or gains, as well as preserving natural and financial resources, ii) controlling or minimizing condensation on or within a building assembly by keeping surface temperature above the dew point of surrounding air, iii) modulating temperature fluctuations in unconditioned or partly conditioned spaces, iv) reducing damage to building assemblies and mechanical systems from exposure to fire and freezing in the event of a system failure or even when the system is idle, v) minimizing temperature change in processing fluids where tight control is needed, and vi) controlling surface temperatures for personnel or occupant comfort to avoid hot or cold contact burns. The primary property of a thermal insulation material is low apparent thermal conductivity. However, depending on the capabilities of the materials, additional benefits may be gained, such as providing support for a surface finish, reducing air leakage into or out of controlled spaces, and providing better control of noise and to a certain extent vibration.

Thermal insulation materials have porosities typically exceeding 90% or so. Given this high porosity, heat transmission involves conduction in the solid matrix but mainly gas conduction and radiation in the pores. This is why the term apparent thermal conductivity is used rather than thermal conductivity, which is a property of a homogeneous and nonporous material. Apparent thermal conductivity is affected by structural parameters such as thickness, density, and fibrous or cellular matrix. Therefore, each piece of a given insulation material has a unique apparent thermal conductivity value that is specific to the conditions it was measured under and cannot be generalized to other combinations of temperature, temperature difference, moisture content, or age.

Thermal insulation is normally typically composed of i) inorganic fibers and cellular materials such as glass or rock wool, capillary-active insulation materials, transparent glass fibers, perlite, and vermiculite; ii) organic fibers and cellular materials such as cotton, animal hair, wood, pulp, cork, transparent plastic tubes, and plastic foams like polystyrene, polyurethane, and other polymers; and iii) metallic or metallized organic reflective materials. The physical forms of insulation include loose fill, insulating cement, flexible and semi-rigid insulation, rigid materials, reflective materials and formed-in-place insulations. Typical thermal properties of insulating materials can be found in Table 1 of Chapter 26 of the *ASHRAE Handbook* (ASHRAE, 2021). Also see Examples 2.56 and 2.77 for the effect of thermal insulation in numerical calculations.

2.3.6 Fuels and Combustion Reactions

Combustion is the chemical process of an oxidant's rapid reaction with a fuel to release the stored energy as thermal energy. The oxidant is typically oxygen in the air, and along with the thermal energy in the form of high-temperature gases, usually small amounts of electromagnetic energy (as light), electric energy (as free electrons), and mechanical energy (as noise) are also released during the process.

Hydrocarbon fuels are primarily composed of carbon and hydrogen, and most conventional fuels contain small amounts of sulfur as well. Flue gas is the product of complete or incomplete combustion including, if present, excess air. During complete combustion of hydrocarbon fuels, all the carbon is oxidized to carbon dioxide (CO_2), all the hydrogen to water (H_2O), and all the sulfur to sulfur dioxide (SO_2) or sulfur trioxide (SO_3). Excess air, usually expressed as a percentage, is supplied beyond what is required theoretically for combustion to completely oxidize the fuel. Stoichiometric or theoretical combustion is the ideal, where fuel is burned completely: In stoichiometric combustion, fuel reacts with the exact amount of oxygen needed to completely oxidize all carbon, hydrogen, and sulfur in the fuel to CO_2, H_2O, and SO_2 or SO_3, respectively. Hence, the flue gas from stoichiometric combustion contains no excess or unreacted oxygen and no carbon monoxide (CO).

The amounts of CO_2, H_2O, and SO_2 or SO_3 released during stoichiometric combustion are the maximum achievable and are called the maximum theoretical percentages or ultimate amounts. In reality, stoichiometric combustion is rarely achieved due to imperfect mixing (of air and fuel) and finite reaction rates (fixed by fuel selection); therefore, most combustion equipment needs to operate with some excess air. To determine the excess air for a combustion system, we first determine the stoichiometric—ideal—air–fuel ratio. When fuel is burned completely, then all air is depleted, but heating equipment rarely runs on perfect mixing ratios. Boilers and high-temperature furnaces typically incorporate a modest amount of excess air, approximately 10% to 20% more than what is required for burning the fuel completely. Natural gas-fired boilers can run on as little as 5% excess air, and pulverized coal-fired boilers can run with 20% excess air.

If insufficient air is provided to a burner, combustion is incomplete, and the nonoxidized components in the flue gas remain, such as C, H_2, CO, etc. Incomplete combustion does not use fuel efficiently, which produces air pollution and even more dangerously can lead to explosions due to the CO production; CO is an odorless, tasteless, and colorless toxic gas. Every combustion process has the potential to produce this toxic gas as a result of incomplete combustion. CO inhibits the blood's ability to carry oxygen to vital organs because it combines with the oxygen-carrying hemoglobin in the blood.

Health effects of CO, its concentrations, and its exposure hours are provided in Table A.16. Typical sickness symptoms are mild headache, fatigue, nausea, and dizziness; one to three minutes of exposure to 12–13,000 ppm CO concentration is deadly, and a concentration of 1600 ppm is fatal after one hour. Combustion equipment is designed and operated to ensure complete rather than stoichiometric combustion to reduce or eliminate the dangers of CO production. It should be noted that air at higher altitudes is less dense, and therefore, there is less oxygen per unit volume. Excess air must be adjusted accordingly or combustion will be incomplete and can eventually cease.

2.3.6.1 Fuel Classification and Heating Values

Hydrocarbon fuels are classified based on their physical state—gas, liquid, solid—and this section analyzes the fuels according to this classification.

Fossil fuels have always been the primary contributor to global energy consumption. Overall fossil fuel consumption increased approximately 51% in the period 1995–2015, and it is predicted that the consumption will increase approximately 18% in the period 2015–35. In 1995, oil, natural gas, and coal had total primary energy consumption shares of 38%, 22%, and 26% (87% in total), respectively, and in 2015, they had total shares of 32%, 24%, and 29% (85% in total), respectively; meanwhile, non-fossil energy consumption in 2015 included hydroelectric (6.8%), nuclear (4.4%), wind (1.4%), solar (0.45%), and other renewables (0.89%) (Yildiz, 2018). It is predicted that by 2035, the shares will be oil, 29%; natural gas, 25%; and coal, 24% (78% in total) and that total global fossil fuel consumption will decrease by 7%.

However, even though the share of fossil fuel consumption has decreased and is predicted to decrease further, these energy sources will continue to be the main contributors to the primary energy mix in the foreseeable future as more unconventional fossil fuels are explored. The highest change in fossil fuel consumption has been in the consumption of natural gas, a cleaner fossil fuel than oil and coal, and it is predicted that natural gas consumption will continue to decrease; however, fossil fuel extraction and burning in all forms has serious environmental consequences that will require further and closer attention as more unconventional fossil fuel resources are explored and their overall consumption expands further.

As discussed earlier, combustion releases thermal energy or heat; the amount of heat generated by complete combustion of a specific fuel is fixed and called the heating value, heat of combustion, or calorific value of that fuel. The combustion process typically produces water vapor as well, and we can recover the heat contained in this water vapor by condensing the water. When this water vapor is entirely condensed and the latent heat contained in the water vapor is recovered, the fuel's heating value is referred to as the higher heating value (HHV), higher calorific value (HCV), gross calorific value (GCV), or total heating value (THV). Conversely, when the latent heat of vaporization is not recovered, then the fuel's heating value is termed lower heating value (LHV), lower calorific value (LCV), net calorific value (NCV), or net heating value (NHV). Tables A.17 to A.21 provide the heating values of different fuels. If the heating values of fuels are specified without designating higher or lower, they generally mean the higher heating values, including those provided in this book as well.

2.3.6.1.1 Gaseous Fuels

In the past, many different gases were used as energy sources; however, usage is currently limited to natural gas and liquefied petroleum gases. Natural gas is an odorless and colorless hydrocarbon gas mixture consisting of primarily methane (CH_4) with varying amounts of other higher alkenes such as ethane, propane, butane, pentane, and hexane and sometimes small portions of carbon dioxide, oxygen, nitrogen, hydrogen sulfide, or helium. It is gaseous rather than liquid like oil or solid like coal.

Natural gas is a nonrenewable fossil fuel source of energy for heating, cooking, electricity generation, etc. It is most commonly transported by pipeline; however, natural gas is increasingly being transported by ship in a liquefied form to meet greater global demand for the fuel. Natural gas (primarily CH_4) is also generated by the decomposition of municipal waste in landfills and manure from livestock production. CH_4 is a greenhouse gas that is approximately 22 times more potent than CO_2. Capturing and burning the gas to produce usable heat and power prevents the CH_4 from being released directly from the landfill or feedlot into the atmosphere. Before natural gas can be used as a fuel, most of the time, it is processed to remove impurities such as water to meet market specifications.

Heating values of natural gases differ considerably between markets from 33.6 to 44.8 MJ/m^3 (900 to 1200 Btu/ft^3); however, the typical range at sea level is 37.3 to 39.2 MJ/m^3 (1000 to 1050 Btu/ft^3). For safety purposes, odorants are added to natural gas and liquid petroleum gas (LPG) so that they can be noticed in the case of leaks. LPGs are usually obtained as a byproduct of oil refineries and typically consist of propane and butane, which are gaseous under normal atmospheric conditions; however, they are liquefied by applying moderate pressures at normal temperatures so that they can be transported conveniently. We basically have three commercially available LPGs: butane, propane, and the mixture of the two.

At atmospheric pressure, commercial propane has a boiling point of approx. −40°C (−40°F), quite convenient for northern climates. It mainly consists of propane but usually contains 5% to 10% propylene as well and has heating values of approximately 93.3 MJ/m^3 (2,500 Btu/ft^3) or 50.5 MJ/kg (21,669 Btu/lb). It is available in tank trucks, cylinders, and bottles.

Commercial butane, on the other hand, has a relatively high boiling point of approx. 0°C (32°F) at atmospheric pressure, not convenient for northern climates unless it is maintained above 0°C. It mainly consists of butane but can contain up to 5% butylene as well, and it has heating values of approximately 119.4 MJ/m^3 (3,200 Btu/ft^3) or 49.7 MJ/kg (21,321 Btu/lb). It is available in tank trucks and bottles but not in cylinders. Commercial butane–propane mixtures are available with varying ratios of butane and propane. Heating values of different gases and substances occurring in common fuels are provided in Table A.17.

2.3.6.1.2 Liquid Fuels

Liquid fuels are typically mixtures of hydrocarbons derived from crude petroleum. Refining crude petroleum produces a variety of fuels and other useful products. Almost all lighter hydrocarbons are refined into gasoline, diesel fuels, jet fuels, kerosene, light heating oils, and LPGs. Heavy hydrocarbons, on the other hand, are refined into residual oils and products such as asphalt, petroleum coke, and lubricating oils. Biofuels, in contrast with both, contain over 80% renewable materials such as biomass, which is derived from photosynthesis. Biofuels are classified as first-, second-, or third-generation or waste.

The first-generation biofuels are based on food biomass such as corn, sugar cane, and plant and vegetable oils; the second-generation biofuels are based on non-food biomass such as cellulose and hemicelluloses, and the third-generation biofuels primarily refer to algae and cyanobacteria. All of the above-mentioned biomasses can be converted to alcohols, biodiesel, biogas, bio-oil, and other bioproducts using biochemical, thermo-chemical, and hydrothermal processes. Biofuels are promising alternatives to fossil fuels and have the capacity to control greenhouse gas emissions; therefore, they can have positive impacts on climate change. Bioethanol and biodiesel are currently the most widely used biofuels; however, the share of biofuels in the primary energy supply is still very small because the current costs of biofuels are not competitive with those of fossil fuels.

Fuel selection for specific applications is usually a function of availability and economic factors, including the costs of fuel, preheating, handling, and equipment, as well as the environmental requirements. Heating values of a number of different liquid fuels are presented in Table A.18.

2.3.6.1.3 Solid Fuels

Solid fuels consist of coal; coke; crop residues; wood and waste products of agricultural, forestry, and industrial operations. Out of these products, coal is the most widely used fuel for heating and cooling. Following oil (32%) is coal with a share of 29%. Currently, about 40% of global electricity production is contributed by coal, but this number is expected to decline due to increased shares from renewable energy sources and natural gas. Global coal production in 2015 increased by 25.1% compared to that in 1995 and was projected to increase further in the next 20 years (8.4% based on the 2015 production).

Importantly, coal is primarily made up of carbon; therefore, it is a carbon-intensive energy source, and burning coal is the biggest single source of CO_2 emissions from human activity. Coal combustion produces almost double the greenhouse gas emissions of burning gas for the same amount of energy. For instance, although coal generated less than 30% of the world's energy supply in 2013, it produced 46% of the global CO_2 emissions. It must be noted that coal mining also releases the potent greenhouse gas methane into the environment.

Even though there is no distinct demarcation between coal types, under proper conditions, as ancient plant material is exposed to high temperatures and pressures over geological time, the metamorphic grade of the dead material increases successively into

- *Peat*, which is considered to be a precursor of coal, which has industrial importance as a fuel in some regions.
- *Lignite*, also known as brown coal, is the lowest rank of coal and is almost solely used for electric power generation.
- *Sub-bituminous coal*, whose properties range between those of lignite and bituminous coal and which is used primarily to generate steam–electric power.
- *Bituminous coal*, which is a dense and usually black (sometimes dark brown) sedimentary rock, is mainly used in steam–electric power generation and for heat in manufacturing.
- *Steam coal* is a grade between bituminous coal and anthracite. It was once commonly used as a fuel for steam locomotives and also for domestic water heating.
- *Anthracite,* the highest rank of coal, is a harder, glossy black coal mainly used for residential and commercial space heating. It is further divided into metamorphically altered bituminous coal and petrified oil.
- *Graphite*, one of the more difficult types of coal to ignite, is not generally used as fuel.

Wood, in contrast with already dry coal, contains a considerable amount of water when it is fresh; after it is cut and stacked for a year or two, the average moisture content of wood typically decreases to about 20%. During combustion, the temperature rises to flue gas temperature, and water evaporates. Table A.22 shows the average energy content of wood based on its moisture content when specific data is not available. Please note that by volume, wet wood has about 85% of the energy of oven-dry wood, whereas by weight wet, wood has less than half (42%) of the energy of oven-dry wood.

2.3.6.1.4 Comparing Costs of Heating Fuels

For convenience, average cost comparisons for electricity, natural gas, LPG, fuel oil, coal, and wood are given in the following equations:

Electricity

$$C_E = P_E\left(\frac{100,000\,\text{Btu}}{3,412\,\text{Btu/kWh}}\right) \tag{2.162}$$

where C_E is the cost of electricity per 100,000 Btu, P_E is the price of electricity per kWh, and 1 kWh is equal to 3,412 Btu.

Natural Gas

$$C_{NG} = P_{NG}\left(\frac{100,000\,\text{Btu}}{1,036\,\text{Btu/ft}^3}\right) \tag{2.163}$$

where C_{NG} is the cost of natural gas per 100,000 Btu, P_{NG} is the price of natural gas per cubic foot, and the average energy content of natural gas is assumed to be 1,036 Btu/ft^3.

Liquid Propane Gas

$$C_{LPG} = P_{LPG}\left(\frac{100,000\,\text{Btu}}{91,500\,\text{Btu / gal}}\right) \tag{2.164}$$

where C_{LPG} is the cost of LPG per 100,000 Btu, P_{LPG} is the price of LPG per gallon, and the average energy content of LPG is assumed to be 91,500 Btu/gal.

Fuel Oil

$$C_{Oil} = P_{Oil}\left(\frac{100,000\,\text{Btu}}{139,000\,\text{Btu / gal}}\right) \tag{2.165}$$

where C_{Oil} is the cost of fuel oil per 100,000 Btu, P_{Oil} is the price of fuel oil per gallon, and the average energy content of fuel oil is assumed to be 139,000 Btu/gal.

Coal

$$C_{Coal} = P_{Coal}\left(\frac{100,000\,\text{Btu}}{20,000,000\,\text{Btu / ton}}\right) \tag{2.166}$$

where C_{Coal} is the cost of coal per 100,000 Btu, P_{Coal} is the price of coal per short ton, and the average energy content of Grade No. 2 coal used in a medium-capacity burner is assumed to be 20,000,000 Btu/ton.

Wood

$$C_{Wood} = P_{Wood}\left(\frac{100,000\,\text{Btu}}{20,000,000\,\text{Btu / cord}}\right) \tag{2.167}$$

where C_{Wood} is the cost of wood per 100,000 Btu, and P_{Wood} is the price of wood per cord; the average energy content of dry wood varies considerably but is assumed to be 20,000,000 Btu/cord. The cord is the most commonly used measurement unit for purchasing firewood, which includes bark and air space, and can be calculated as follows:

In I-P, 1 cord = (4 ft) (4 ft) (8 ft) = 128 ft^3; in SI, 1 cord = (1.22 m) (1.22 m) (2.44 m) ≈ 3.63 m^3.

It should be kept in mind that the air space in a cord can be as high as 40% due to the irregular shape of wood. In general, the net cord volume ranges between 80 and 100 ft^3, but for firewood, it averages about 85 ft^3.

Example 2.79: Calculating the Number of Cords

Determine the number of cords in the pile.

Solution

A pile of firewood is 10 ft long, 5 ft wide, and 6 ft high:

$$\text{No.of Cords} = \frac{(10\,\text{ft})(5\,\text{ft})(6\,\text{ft})}{128\,\text{ft}^3} = 2.34$$

Dimensions for another pile of firewood are given as 3 m long, 1.5 m wide, and 1.8 m high:

$$\text{No.of Cords} = \frac{(3\,\text{m})(1.5\,\text{m})(1.8\,\text{m})}{3.63\,\text{m}^3} \approx 2.23$$

Example 2.80: Comparing Costs of Fuels

A controlled-environment operation is trying to compare the costs of a number of locally available fuels ignoring the handling costs, credits, and additional fees. Local fuel prices are listed as follows: $440 per cord of firewood; $0.0029 per ft^3 of natural gas; $3.59 per gallon of fuel oil; $0.125 per kWh of electricity; and $3.40 per gallon of propane. Determine the most cost-effective fuel at this particular location.

Solution

Using the equations provided in the previous section, we can determine the fuel costs as follows:

$$C_E = 0.125\frac{\$}{\text{kWh}\left(\dfrac{100,000\,\text{Btu}}{3,412\dfrac{\text{Btu}}{\text{kWh}}}\right)} = \$3.66\,\text{per}\,100,000\,\text{Btu}$$

$$C_{NG} = 0.005\,\$/\text{ft}^3\left(\frac{100,000\,\text{Btu}}{1,036\,\text{Btu}/\text{ft}^3}\right) = \$0.48\,\text{per}\,100,000\,\text{Btu}$$

$$C_{LPG} = 2.50\,\$/\text{gal}\left(\frac{100,000\,\text{Btu}}{91,500\,\text{Btu}/\text{gal}}\right) = \$2.73\,\text{per}\,100,000\,\text{Btu}$$

$$C_{Oil} = 2.45\,\$/\text{gal}\left(\frac{100,000\,\text{Btu}}{139,000\,\text{Btu}/\text{gal}}\right)$$
$$= \$1.76\text{ per }100,000\,\text{Btu of fuel oil}$$

$$C_{Wood} = 325\,\$/\text{cord}\left(\frac{100,000\,\text{Btu}}{20,000,000\,\text{Btu}/\text{cord}}\right)$$
$$= \$1.63\,\text{per}\,100,000\,\text{Btu}$$

Based on the local fuel prices provided, the most cost-effective fuel is natural gas.

2.3.6.2 Combustion Reactions and Calculations

The exact amount of air required for complete combustion is termed theoretical or stoichiometric air. Table A.23 provides the combustion reactions for complete combustion. For instance, 12 kg (or 12 lb) carbon is oxidized by 32 kg (or 32 lb) oxygen, 2 kg (or 2 lb) hydrogen is oxidized by 16 kg (or 16 lb) oxygen.

Oxygen for combustion is usually provided by air, a mixture of nitrogen, oxygen, water vapor, carbon dioxide, and some inert gases. For combustion calculations, we can use dry air, which is composed of 20.95% oxygen and 79.05% inert gases (nitrogen, argon, etc.) by volume or 23.15% oxygen and 76.85% inert gases by weight. In the calculations, it is assumed that nitrogen does not change through the combustion process, even though small amounts of nitrogen oxides form.

The combustion process can in general terms be expressed as follows:

$$[C + H]\,(\text{fuel}) + [O_2 + N_2]\,(\text{air}) \rightarrow [CO_2 + H_2O + N_2]\,(\text{heat}) \tag{2.168}$$

where C is carbon, H is hydrogen, O is oxygen, and N is nitrogen.

From this relationship by weight, we can easily determine that the mass of dry air providing the required oxygen for complete combustion is 100/23.15 = 4.32 times the mass of oxygen. The masses of air provided in Table A.23 were calculated based on this relationship. In determining the oxygen required for complete combustion, the oxygen contained in the fuel should be deducted from the required amount as this oxygen is already part of the fuel. Allowance should also be made for water vapor as it is always present in atmospheric air. We sometimes calculate combustion based on volume; in such cases, the volume of dry air providing the required oxygen for complete combustion is 100/20.95 = 4.78 times the mass of oxygen. The volumes of required oxygen and air for specific fuel constituents are provided in the last two columns of Table A.23, respectively.

Using these relationships, we can develop the following equation for the theoretical mass of dry air required (Wa) for stoichiometric combustion of a unit mass of any hydrocarbon fuel:

$$Wa = 0.0144\,(8\,C + 24\,H + 3\,S - 3\,O) \qquad (2.169)$$

where C, H, S, and O are the mass percentages of carbon, hydrogen, sulfur, and oxygen in the fuel respectively.

Sometimes, analyses of hydrocarbon fuels, as the case for gaseous fuels, are based on hydrocarbon components rather than the elemental components. In such cases, we can use the following relationship to determine the theoretical Wa required for the stoichiometric combustion of a unit mass of gaseous fuel:

$$\begin{aligned} Wa = {} & 2.47\,CO + 34.28\,H_2 + 17.24\,CH_4 + 16.09\,C_2H_6 \\ & + 15.68\,C_3H_8 + 15.47\,C_4H_{10} + 13.27\,C_2H_2 \\ & + 14.78\,C_2H_4 + 6.08\,H_2S - 4.32\,O_2 \end{aligned} \qquad (2.170)$$

where CO, H_2, and others are the mass fractions of each constituent in the gaseous fuel.

However, if the fuel analysis is provided on volumetric basis, then it is better to calculate air requirements on volumetric bases and then convert this volumetric quantity to mass quantity if so desired. The theoretical volume of air required (Va) for the stoichiometric combustion of a unit volume of gaseous fuel is determined as follows:

$$\begin{aligned} Va = {} & 2.39\,CO + 2.39\,H_2 + 9.57\,CH_4 + 16.75\,C_2H_6 \\ & + 23.95\,C_3H_8 + 31.14\,C_4H_{10} + 11.96\,C_2H_2 \\ & + 14.38\,C_2H_4 + 7.18\,H_2S - 4.78\,O_2 \\ & + 30.47\text{ illuminants} \end{aligned} \qquad (2.171)$$

where CO, H_2, and others are the volumetric fractions of each constituent in the gaseous fuel. Illuminants, unlike typical fuel compounds, include a number of compounds not separated by common gas analysis. The main illuminant components are ethylene (C_2H_4) and acetylene (C_2H_2)—both included in Equation (2.171)—but include propylene, butylenes, pentene, benzene, toluene, and xylene; the equation also calculates the dry air required for combustion per unit volume of each of these gases. Because the illuminant percentages are small, the values can be lumped together, and we can use an approximate value of 30 unit volume of dry air per unit volume of gas.

Usually, in engineering applications, only approximate amounts of air requirements are necessary; if the approximate amounts are enough, or if fuel-specific information is not available, then we can conveniently use the figures provided in Tables A.24 and A.25. Another commonly used approximation is that 0.9 ft^3 (0.0255 m^3) air is required for 100 Btu (105.5 kJ) of fuel. In addition to the theoretical amount of air required for combustion, excess air must be provided to make sure the combustion is complete. Percent excess air can be calculated using the following equation:

$$\text{Excess Air}(\%) = \left(\frac{\text{Supplied Air} - \text{Theoretical Air}}{\text{Theoretical Air}}\right) \times 100 \qquad (2.172)$$

The unit mass amount of dry air supplied per unit mass of fuel burned can be determined from the following equation:

$$\text{Dry Air Supplied} = \left(\frac{C(3.04\,N_2)}{(CO_2 + CO)}\right) \qquad (2.173)$$

The value obtained from Equation (2.173) and the theoretical air value can be used in Equation (2.172) together to determine the amount of excess air.

The following equation, on the other hand, can be used to calculate excess air on a percentage basis for gaseous fuels from the unit volumes of stoichiometric combustion products and air and volumetric analysis of the flue gas:

$$\text{Excess Air}(\%) = \left(\frac{U - CO_2}{CO_2}\right) \times 100 \times \left(\frac{P}{A}\right) \qquad (2.174)$$

where *U* is the ultimate carbon dioxide of flue gases resulting from stoichiometric combustion (%), CO_2 is the carbon dioxide content of flue gases (%), *P* is the dry products resulting from stoichiometric combustion (unit volume per unit volume of gas burned), and *A* is the air required for stoichiometric combustion (volume per unit volume of gas burned). *P/A* is approximately 0.9 for most natural gases; therefore, we can substitute 90 for the product of 100 × (*P/A*) in Equation (2.174).

Excess air is most often calculated using the results of flue gas analyses, and the amounts of air required for stoichiometric combustion (theoretical air) are not always known. In such cases, the following relationship can conveniently substitute for Equation (2.173):

$$\text{Excess Air}(\%) = \left(\frac{100\left[O_2 - \dfrac{CO}{2}\right]}{0.264\,N_2 - \left[O_2 - \dfrac{CO}{2}\right]}\right) \qquad (2.175)$$

where O_2, CO, and N_2 are the volumetric percentages obtained from the flue gas analysis.

The general practice is to provide 5% to 50% excess air; however, as just discussed, the exact amount depends on the fuel type. Typical amounts of excess air for various fuels are provided in Table A.26.

2.3.6.2.1 Boiler Room Air Supply

As mentioned earlier, incomplete combustion generates CO and poses dangers from re-ignition for both personnel and property safety. Therefore, for complete combustion and safety, boiler rooms are provided with openings that supply sufficient fresh air. Temporary air-take openings such as doors and windows should be avoided because they can be closed and cut off the required air supply. ASME recommends providing 1 square inch free area per 2,000 Btu/h total burner input in the boiler room with a minimum of two air supply openings kept clear at all times.

2.3.6.2.2 Theoretical CO_2 Production

The theoretical (or ultimate) CO_2 concentration obtained in the combustion of a hydrocarbon fuel with air is the amount of CO_2 released when the fuel is completely burned with the theoretically required air, that is, zero excess air. Using the following equation, theoretical CO_2 amounts released during combustion with excess air present can be calculated from flue gas analyses as follows:

$$\text{Theoretical } CO_2\,(\%) = U = \frac{\%\ CO_2 \text{ in flue gas sample}}{1-\left(\dfrac{\%\ O_2 \text{ in flue gas sample}}{20.95}\right)} \tag{2.176}$$

Approximate theoretical CO_2 values and CO_2 values observed with different amounts of excess air for stoichiometric combustion of various common fuel types are provided in Table A.27.

The theoretical CO_2 emissions from combustion can also be calculated from the following relationship. To calculate the CO_2 emission from a fuel, the carbon content of the fuel must be known and multiplied by the ratio of molecular weight of CO_2 (44) to the molecular weight of C (12), which is 44/12 = 3.67. Then,

$$q_{CO_2} = \left(\frac{c_f}{h_f}\right)\left(\frac{M_{CO_2}}{M_C}\right) \tag{2.177}$$

where q_{CO_2} is the specific CO_2 emission (kg_{CO_2}/kWh), c_f is the specific carbon content of fuel (kg_{carbon}/kg_{fuel}), h_f is the specific energy content of fuel (kWh/kg_{fuel}), M_C is the molecular weight of C ($kg_{carbon}/kmol_{carbon}$), and M_{CO_2} is the molecular weight of CO_2 (kg_{CO_2} / $kmol_{CO_2}$). Specific carbon contents, energy contents, and CO_2 emissions from combustion of some common fuels are presented in Table A.32.

Example 2.81: Calculating Theoretical CO_2 and Excess Air

A controlled-environment operation is burning natural gas for heating, and the flue gas analysis provides 12.0% CO_2, 3.0% O_2, and 85.0% N_2. Determine the theoretical percent of CO_2 and the percent of excess air.

Solution

Using Equation (2.162), we can first determine the theoretical CO_2 as

$$\text{Theoretical } CO_2\,(\%) = U = \frac{\%\ CO_2 \text{ in flue gas sample}}{1-\left(\dfrac{\%\ O_2 \text{ in flue gas sample}}{20.95}\right)}$$

$$= \frac{12.0}{1-\left(\dfrac{3.0}{20.95}\right)} = 14.01\%$$

Then, using Equation (2.160), we can calculate the excess air:

$$\text{Excess Air}\,(\%) = \left(\frac{U - CO_2}{CO_2}\right)\times 100 \times \left(\frac{P}{A}\right)$$

$$= \left(\frac{14.01-12.0}{12.0}\right)\times 90 = 15.08\%$$

Example 2.82: Calculating and Comparing Theoretical CO_2 Emissions of Various Fuels

Using Equation (2.105), calculate the theoretical CO_2 emissions from natural gas, bunker oil, bituminous coal, and wood and compare the findings with the values provided in Table A.28.

Solution

Using Equation (2.162), first determine the theoretical CO_2 emissions:

$$\text{Natural gas: } q_{CO_2} = \left(\frac{0.75}{15.4}\right)\left(\frac{44}{12}\right) = 0.18\, kg_{CO_2}/kWh$$

$$\text{Bunker oil: } q_{CO_2} = \left(\frac{0.85}{11.6}\right)\left(\frac{44}{12}\right) = 0.27\, kg_{CO_2}/kWh$$

$$\text{Bituminous coal: } q_{CO_2} = \left(\frac{0.65}{8.4}\right)\left(\frac{44}{12}\right) = 0.28\, kg_{CO_2}/kWh$$

$$\text{Wood: } q_{CO_2} = \left(\frac{0.50}{4.6}\right)\left(\frac{44}{12}\right) = 0.40\, kg_{CO_2}/kWh$$

From the calculations, we can see that wood has the highest CO_2 emission; however, wood is a biofuel and is considered a carbon-neutral fuel. Therefore, from the environmental point of view, it is still an environmentally friendly fuel. If, on the other hand, we compare the other three fossil fuels with respect to their CO_2 emissions, we can easily observe that natural gas is a relatively clean burning fuel. Considering the emissions of other flue gas constituents, natural gas is truly a clean-burning fuel, and as a consequence, we use natural gas for CO_2 enrichment in commercial greenhouse operations, which will be covered in depth later in this book.

2.3.6.2.3 Quantity of Flue Gas Produced

We need the mass of dry flue gas produced per mass of fuel for heat loss and efficiency calculations, and this mass consists of the fuel (minus the ash remaining in the furnace), the theoretical air required for combustion, and excess air. This mass is calculated from the flue gas analysis for solid fuels as follows:

$$\text{Mass of Flue Gas Produced} = \frac{11CO_2 + 8O_2 + 7(CO + N_2)}{3(CO_2 + CO)}C \tag{2.178}$$

where CO_2, O_2, CO, and N_2 are the volumetric percentages from the flue gas analysis and C is the mass of carbon burned per mass of fuel, corrected for carbon in the ash.

For gaseous fuels, the volume of dry flue gas per unit volume of gaseous fuel can be determined from the following equation:

$$\frac{\text{Volume of Dry Flue Gas}}{\text{Volume of Fuel Gas}} = \frac{\begin{pmatrix}\text{Unit volume of } CO_2 \text{ produced per unit} \\ \text{volume of gas burned} \times 100\end{pmatrix}}{\text{Percent } CO_2 \text{ by analysis}} \quad (2.179)$$

The excess air quantity can be determined by subtracting the quantity of dry flue gases resulting from stoichiometric combustion from the total volume of flue gas. Theoretical flue gas values for combustion of various fuels are provided in Table A.25.

Example 2.83: Calculating the Amount of Required Dry Air and Required Flue Gas Composition

For the same flue gas analysis provided in Example 2.81, determine the m^3 of dry air required for combustion of m^3 of gaseous fuel, m^3 of each constituent in the flue gas, and the total volume of dry and wet flue gases. Fuel analysis also provided 85% CH_4, 10% C_2H_6, and 5% N_2 by volume.

Solution

Using Equation (2.173), we can first determine the volume of dry air required for combustion:

$$V_a = 9.57\ CH_4 + 16.75\ C_2H_6 = (9.57 \times 0.85) + (16.75 \times 0.10) = 9.81 m^3_{\text{dry air}} / m^3_{\text{natural gas}}$$

Note that this number is in agreement with the figure ($9.5 m^3_{\text{dry air}} / m^3_{\text{natural gas}}$) provided in Table A.25.

From Table A.23, the constituents per m^3 of flue gas are as follows:

Carbon dioxide, CO_2:

generated from methane : 0.85 (CH_4)(1.0) = 0.85 m^3
generated from ethane : 0.10 (C_2H_6)(2.0) = 0.20 m^3
Total CO_2 : 1.05 m^3

Oxygen, O_2, in excess air:

$$0.209 \times 0.151 \times 9.81\ m^3_{\text{dry air}} / m^3_{\text{natural gas}} = 0.31\ m^3$$

Nitrogen, N_2:

generated from methane : 0.85 (CH_4)(9.57–2.0) = 6.43 m^3
generated from ethane : 0.10 (C_2H_6)(16.75–3.50) = 1.33 m^3
from fuel : 0.05 m^3
in excess air : 0.791 × 0.151 × 9.81 $m^3_{\text{dry air}} / m^3_{\text{natural gas}}$ = 1.17 m^3
Total N_2 : 8.98 m^3

Water Vapor, H_2O:

generated from methane : 0.85 (CH_4)(2.0) = 1.70 m^3
generated from ethane : 0.10 (C_2H_6)(3.0) = 0.30 m^3
Total H_2O : 2.00 m^3

Total volume of dry gases per m^3 of natural gas:

$$1.05 + 0.31 + 8.98 = 10.34\ m^3_{\text{dry fuel air}} / m^3_{\text{natural gas}}$$

Note that this is in agreement with the figure (10.5 m^{3}dry flue gas/m^{3}natural gas) provided in Table A.25. The volume of dry flue gas per unit volume of gaseous fuel can also be computed using Equation (2.179).

Finally, total volume of wet gases per m^3 of natural gas:

$$10.34 + 2.00 = 12.34\ m^3_{\text{dry fuel air}} / m^3_{\text{natural gas}}$$

2.4 ESSENTIALS OF FLUID MECHANICS

2.4.1 Introduction

Fluid mechanics is the science that deals with the behavior of fluids at rest or as flowing (or deforming) bodies. The piping systems for water, natural gas, oil, etc. are all primarily designed using the principles of fluid mechanics. Flowing fluids transfer mass or heat, and hence fluid flow and its mechanics are important for heating, ventilating, air-conditioning, and refrigeration systems engineers and designers as well. This section covers fluid mechanics fundamentals needed to handle general HVAC processes for controlled environment systems.

2.4.2 Fluids and Their Properties

A substance can primarily exist in three phases: solid, liquid, or gas; it can also exist in plasma phase at very high temperatures. A substance in the liquid or gaseous phase is referred to as a fluid, although liquids and gases behave differently with respect to the nature of molecular actions, particularly the degree of compressibility and liquid formation of an interface.

When reaction of fluids to shearing is considered, fluids also separate themselves from solids. If a fluid is placed under a shear stress condition, it keeps deforming as long as the shear is present, whereas a solid can resist applied shear stress and deform just a finite amount. In fluid mechanics, a stress is a force per unit area, and the normal component of force acting on a surface (per unit area) is called normal stress (called pressure in a fluid at rest); the tangential component per unit area is shear stress (remember that a fluid at rest has zero shear stress).

Normal stress can be either tensile or compressive depending on whether it stretches or compresses the fluid, and it equals *F/A*, where *F* is the normal force and *A* is the area of the surface on which it acts. On the other hand, the shear stress τ equals *F/A*, where *F* is the tangential force and *A* is the area on which it acts. As the shear stress is directly proportional to the viscosity of the fluid, the importance of shear stress becomes obvious when the flow of viscous Newtonian fluids is considered. When fluid layers move relative to each other at different velocities—say u and $u + du$ at distance dy (Figure 2.24)—viscosity together with relative velocity cause shear stress acting between the fluid layers, and as a result, the slower layer resists and slows down the faster-moving layer.

The fluid property viscosity is a measure of internal flow resistance. All fluid flows have some degree of viscosity; there is no zero-viscosity flow. The generated shear stress τ (that is, F/A) is proportional to the rate of change of velocity with respect to y (Newton's law of viscosity), $\tau \propto (du/dy)$ or $\tau = \mu\ (du/dy)$, where μ is the proportionality constant and is known as the coefficient of dynamic viscosity or just viscosity and du/dy represents the rate of shear strain or rate of shear deformation or velocity gradient. Then, we have $\mu = \tau/(du/dy)$. Dynamic viscosity changes with changing temperature and pressure. As temperature rises, viscosity of liquids decreases whereas viscosity of gases increases. In SI, the common unit of dynamic viscosity is Pa.s, Pascal second, which is also expressed as N.s/m^2 or kg/m.s. In fluid dynamics, μ often appears in ratio to fluid density ρ as $\nu = \mu/\rho$, termed kinematic viscosity, for which the unit in SI is m^2/s. Fluid density ρ, or the mass per unit volume, is another important fluid property that is involved in most fluid models.

Fluid motion can be analysed based on one of several models (Figure 2.25), the simplest being the ideal fluid model, in which the fluid is incompressible and has no viscosity (resistance to shearing). In reality, no ideal fluid exists; rather, it is an imaginary fluid. While the ideal fluid model can be used and is valid for a wide range of applications, it is frequently necessary to include the effects of shearing (viscous action). A fluid that possesses viscosity is called a real fluid. All fluids in actual practice are in fact real fluids and possess some viscosity.

If a real fluid obeys Newton's law of viscosity—that is, the shear stress is directly proportional to the shear strain—the fluid is called a Newtonian fluid. In HVAC or controlled-environment applications, most fluids can be treated as Newtonian fluids, for which the deformation is directly proportional to the shearing stress. Examples include water, air, and emulsions. Fluids that do not obey Newton's law of viscosity are classified as non-Newtonian fluids. Moreover, if a fluid has a shear stress that is more than the yield value, and shear stress is proportional to the velocity gradient (shear strain), then this fluid is called ideal plastic fluid.

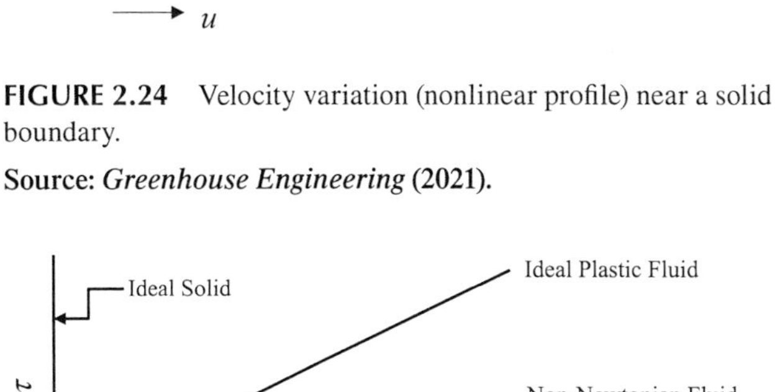

FIGURE 2.24 Velocity variation (nonlinear profile) near a solid boundary.

Source: ***Greenhouse Engineering*** **(2021).**

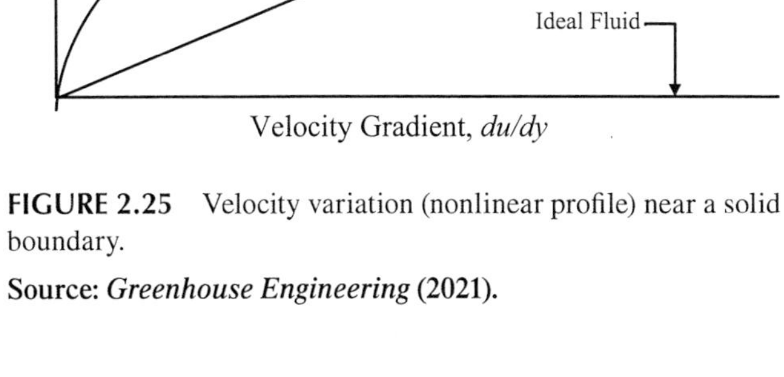

FIGURE 2.25 Velocity variation (nonlinear profile) near a solid boundary.

Source: ***Greenhouse Engineering*** **(2021).**

2.4.3 Properties of Water

2.4.3.1 Introduction

Our planet is dominated by water, a special substance consisting of two hydrogen atoms and one oxygen atom (H_20). Water covers more than 70% of Earth's surface, mostly in the liquid form; the second most common form of water on our planet is ice.

All substances including water exist in three distinct forms, solid, liquid, and gas, and we know that the form of a substance at any given time is a function of the temperature and pressure. Solids have crystalline structures, rigidity, and definite shapes; as a result, they tend to resist external forces. At temperatures below freezing (0°C), at one atmosphere, pure water exists as solid (ice). When heated, solids can be converted to liquids.

Liquids lacks rigidity, and while having definite volumes do not have definite shapes; liquids flow when an external force is applied. Between the freezing point temperature (0°C) and the boiling point temperature (100°C), at one atmosphere pressure, water exists in its liquid form. When heated beyond their boiling points, every liquid including water converts to its gaseous form. A gas, in contrast, has neither a rigid structure nor a definite volume, conforming to the shape and volume of its containing vessel. A gas expands and contracts greatly with changes in temperature and pressure and has the ability to diffuse.

Water is the major constituent of almost all life forms; therefore, it is essential for life. Without water and its unique and unusual properties, the life forms we know on Earth today would never have developed as most animals and plants contain more than 60% water by volume. Water shapes our natural environment and facilitates life by not only mediating chemical and biochemical processes but also our climate and weather. This section briefly covers the chemical and physical properties of this wonderful substance and how they are operationalized in the natural environment.

2.4.3.2 Chemical Properties of Water

To understand the causes of the unique and unusual properties of water, a basic knowledge of chemical bonding and the structure of the water molecule is necessary. Before anything else, the ability of elements to form compounds depends on the ability of their atoms to acquire or give up electrons. Elements acquiring electrons become negatively charged ions (anions), and elements giving up electrons become negatively charged ions (cations). The power of an element to form compounds is called the valence of that element. Valence is a positive or a negative whole number that is based on the number of electrons gained or lost, and for inorganic compounds, the algebraic sum of the valence numbers of the combining elements becomes zero. For instance, sodium gives up one of its electrons to become a cation with a valence of +1, while chloride attracts one electron to form an anion with a valence of −1. These oppositely charged ions that attract each other then form a molecule called sodium chloride (the algebraic sum of the valence numbers: (+1) + (−1) = 0), and this electrostatic attraction of oppositely charged ions is called ionic bonding.

The elements (hydrogen and oxygen) that combine to make water exist separately in molecules containing two atoms each (i.e., H_2 and O_2). The two atoms are held together by sharing an electron pair in a chemical bond called covalent bond that is much stronger than the ionic bond, and much more stable than the individual atoms. And it is this nature of the atomic structure of water that causes its molecules to have unique electrochemical properties. The chemical bonds in the water molecule are covalent bonds because the hydrogen atoms combine with the oxygen atom in shared electron pairs. It is the unique distribution of the electrons in the resulting chemical bond that causes the hydrogen atoms to bond with the oxygen atom at an angle of 104.5° (Figure 2.26).

The oxygen atom exerts a relatively strong pull on the shared electron pair, causing the hydrogen atoms to become excess electropositive regions and the oxygen atom to become an excess electronegative region. The positive and negative charges are not evenly distributed around a center point, which makes the water molecule polar (Figure 2.26 left). Moreover, when the electropositive hydrogen atoms on the water molecule are attracted to the electronegative oxygen atoms of neighboring water molecules, hydrogen bonding takes place (Figure 2.26 right). Water molecules form groups due to hydrogen bonding, and as a result, strong surface tension is generated within the associated groups of water molecules. It must also be noted that these hydrogen bonds are much weaker than covalent bonds even though they are responsible for most of the other unusual properties of water such as high freezing and boiling points, high heat capacity, high heats of fusion (melting) and vaporization, and excellent solvency (Figure 2.3).

Hydrogen bonds are most stable in ice, somewhat stable in liquid water, and broken in water vapor, and these loose connections between water molecules give water several properties important in supporting life and stabilizing the Earth's climate. As mentioned, water molecules in liquid form exhibit strong cohesion (we can think of this as how water molecules stick together). This cohesion facilitates the transport of chemicals, such as nutrients and waste in plants and animals. Hydrogen bonding also gives water molecules the significant capacity to resist temperature change. Initial heating of water weakens hydrogen bonds but does not speed molecular motion; as a result, water can absorb a large amount of heat with only small changes in temperature (Figure 2.3). This property has huge implications for the freezing and melting of water as well as the evaporative cooling of plants and animals.

Moreover, water molecules in ice are farther apart than the molecules in liquid water, resulting in a solid phase that is less dense than the liquid phase. At freezing, hydrogen bonding leads to a regular crystalline structure and greater voids between water molecules and therefore is the reason for the lower density of ice relative to liquid water. Liquid water, on the other hand, has a more random arrangement and more unstable hydrogen bonding between molecules, and therefore greater density. This is why ice floats on liquid water, providing an insulating effect that prevents entire water bodies from freezing solid in winter and thereby facilitating aquatic life underneath in lakes and rivers.

Furthermore, hydrogen bonding is responsible for maintaining the integrity of the water molecules during chemical reactions. While other compounds undergo ionization, the water molecules maintain their chemical integrity. In pure water, a relatively small number of molecules ionizes into hydrogen and hydroxyl ions.

The polar nature of the water molecule is why it is electrostatically attractive to other water molecules and other ions available in solution. This molecular polarity causes water to be a powerful solvent, particularly for crystalline salts. It should be remembered that as the dielectric constant (one of the properties related to dipole structure of water molecule) increases, the force between the charges decreases. By aligning in an electric field, water molecules neutralize the electric field and create resistance to the transmission of an electrostatic charge between charged bodies. As a result, the high dielectric constant of water reduces the attractive

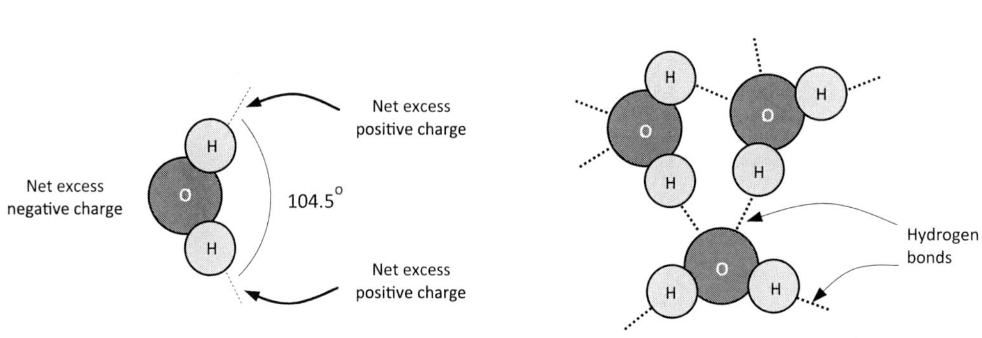

FIGURE 2.26 Polarity of water molecule (left) and hydrogen bonding between water molecules (right).

force of ionized substances in water, helping to account for the excellent ability of water to dissolve a huge variety of substances that are necessary for life.

Hydrophilic is the term for substances that dissolve readily in water. These substances are composed of ions or polar molecules that attract water molecules through electrical charge effects. Ionic substances such as sodium chloride dissolve easily in water because water molecules are attracted to the positive [Na^+] or negative [Cl^-] charge of each ion. Separately, polar substances such as urea dissolve in water because their molecules form hydrogen bonds with the surrounding water molecules. In contrast, molecules that contain predominately nonpolar bonds, usually insoluble in water, are hydrophobic. Water molecules are less attracted to these hydrophobic molecules (e.g., hydrocarbons, which contain many CH bonds) than to other molecules.

2.4.3.2.1 Hydrogen Ion Concentration

Positively charged hydrogen ions [H^+] can spontaneously move from one water molecule to another, thereby creating two ionic species. Since the process is quickly reversible, hydrogen ions continually shuttle between water molecules. A small fraction of pure water molecule dissociates into [H^+] and hydroxyl ions [OH–] and contains a steady state concentration of these ions according to the following chemical equation:

$$H_2O \leftrightarrow [H^+] + [OH^-] \tag{2.180}$$

For water, the product of [H^+] concentration and [OH^-] concentration is a constant called the dissociation constant, K_{eq}, which is expressed as follows:

$$K_{eq} = [H^+][OH^-] = 10^{-14} \tag{2.181}$$

where K_{eq} is dissociation constant, [H^+] is hydrogen ion concentration (mol/litre), and [OH^-] is hydroxide ion concentration (mol/litre).

The dissociation of pure water is balanced so that each H^+ that dissociates produces an ion of OH^-. In pure water,

$$[H^+] = [OH^-] = 10^{-7} \text{ mol / litre} \tag{2.182}$$

Separately, pH is a convenient means of expressing the concentration of H^+ ions, and it is defined as the negative logarithm to the base 10 of the hydrogen ion concentration (pH = $-\log_{10}$ [H^+]). In the case of pure water, where $[H^+] = 10^{-7}$ mol / litre, then pH = 7, making water neutral; that is, the concentration of hydrogen ions balances the concentration of hydroxide ions. Adding a source of hydrogen ions to water increases the concentration of H^+ and lowers pH below 7 to become acidic. Conversely, adding a source of hydroxide ions to water decreases the actual concentration of H^+ and increases pH above 7, making the basic.

We should remember that the pH of water has a large influence on aquatic organisms in particular and affects other chemicals that may be present in the aquatic environment. For instance, when pH rises into the basic range, ammonium ions (NH_4^-) dissolved in water combines directly with H^+ and convert to the ammonia form (NH_3), which exists in water as a dissolved gas. The pH of natural lakes varies between 6 and 9, and the pH of seawater typically varies between 7.8 and 8.3 because of the buffering capacity (to be covered later in this section) of salt water (Wheaton, 1977).

2.4.3.2.2 Salinity

Natural waters contain compounds that are either dissolved or suspended as particles. Salinity is a term generally used in reference to seawater. Salinity is defined as the concentration of total dissolved solid material in water when all the carbonate (CO_3^{2-}) has been converted to CO_2, the bromine and iodine replaced by chlorine, and all the organic material completely oxidized. Salinity is expressed as grams of dissolved solid material per kilogram of seawater; g/kg can also be stated as parts per thousand (ppt) of dissolved solids.

Typical seawater in the open ocean has a salinity of 35 ppt. Coastal seawater is often diluted with freshwater, and salinity is therefore usually in the range of 28 to 33 ppt. Na^+ and Cl^- contribute about 86% of the salinity in seawater. The ions sulphate (SO_4^{2-}), magnesium (Mg^{2+}), calcium (Ca^{2+}), and potassium (K^+) contribute 13%; and, a large number of other components contribute the remaining 1% (Landau, 1992).

Most of the constituents of seawater are said to be conservative; that is, their relative concentrations in seawater are constant even if the total salinity changes. Other constituents of seawater that often serve as nutrients for aquatic life forms are said to be non-conservative since their relative concentrations change with biological activity. Constituents of seawater such as N, P, and CO_2 are non-conservative; they are rapidly taken up by phytoplankton and other aquatic organisms. Different species of aquatic organisms including fish experience optimal growth at different levels of salinity. Generally, the higher the salinity in water, the less the risk of pathogens and parasites infecting aquaculture species.

Seawater in coastal estuaries and river mouths is mixed with freshwater so that a range of concentrations of salts or total dissolved solids (TDS) can be found. Freshwater has less than 1.5 ppt of TDS; brackish water has less than 5 ppt of TDS; saline water has TDS greater than 5 ppt; and seawater has 30 to 35 ppt of TDS (U.S. Geological Survey, 1985). Water with greater than 40 ppt salinity is called brine or hypersaline. Salt water can become hypersaline under conditions of high evaporation and low precipitation and/or low rates of freshwater runoff. Tropical saltwater estuaries are examples of hypersaline waters.

A number of techniques can be used to measure salinity. A traditional method of measuring salinity has been to measure the concentration of chlorine, or the chlorinity, the total amount of the halogens chlorine, bromine, and iodine, expressed as an equivalent concentration of chlorine (ppt) assuming that all the bromine and iodine have been hypothetically replaced by chlorine (Stickney, 1979). Since chlorine in a conservative constituent of seawater, salinity will always be directly proportional to the concentration of chlorine.

Chlorinity can be determined by titration of a measured volume of seawater with a standard silver nitrate solution in the presence of an indicator that changes color when all the halides have been precipitated by the silver nitrate. Salinity can be calculated from chlorinity by what is called Knudsen's formula as follows:

$$S = 0.30 + 1.8050(Cl) \tag{2.183}$$

where S is the salinity (ppt), and Cl is the chlorinity (ppt).

More modern techniques of measuring salinity include measurements based on *conductivity*, *density*, and *refractive index*. The least expensive method of measuring salinity is by use of a *hydrometer* to measure density. Salinity can be read directly from a hydrometer calibrated to measure salinity or by using conversion tables to determine salinity from the actual density measurement. Measurement of salinity by measuring *electrical conductivity* is the most accurate method. *Conductivity meters* to measure salinity are relatively expensive and require a large volume of sample water. A *refractometer* can be used to measure salinity simply and rapidly and requires only a drop of water to make the measurement. The refractive index of water, the water's ability to bend light rays striking the water surface at an angle, is a function of the concentration of salinity in the water sample.

Increasing salinity in water, on the other hand, affects many of the other physical properties of water. Water properties such as boiling point, osmotic pressure, viscosity, surface tension, refractive index, and density all tend to rise with increasing salinity. The properties such as thermal conductivity, specific heat capacity, vapor pressure, freezing point, and the temperature of maximum density tend to decrease with increasing salinity.

2.4.3.2.3 Hardness

Hardness is closely related to salinity. The term is generally applied to freshwater, while salinity is used when discussing saltwater. Hardness is defined as the sum concentration of all polyvalent cations in a water sample. The concentration of polyvalent cations is conventionally expressed as an equivalent concentration of $CaCO_3$ in units of mg/L. The major cations that cause hardness are calcium (Ca^{2+}) and magnesium (Mg^{2+}), but other polyvalent cations such as iron (Fe^{2+}), manganese (Mn^{2+}), strontium (Sr^{2+}), and aluminum (Al^{3+}) can be present in small quantities in freshwater sources.

Ground waters are often exposed to limestone rock, which is mainly composed of $CaCO_3$ and $MgCO_3$. Calcium carbonate and magnesium carbonate are very soluble in water and thereby produce the majority of hardness in most water sources. Hence, a reasonable estimate of hardness can be made by measuring the total concentration of Ca^{2+} and Mg^{2+} ions in water and then converting these to total equivalent concentrations of $CaCO_3$ expressed in terms of mg/L or as parts per million (ppm). Here we assume that the specific gravity of water is 1.000; then the concentration unit ppm is equivalent to mg/L.

The total hardness of water is the sum of the hardness caused by each species of polyvalent cations. The concentration of each polyvalent cation is converted to mg/L as $CaCO_3$ by the following calculation:

$$\text{mg / L of cation}_n \text{ as CaCO}_3$$

$$\text{concentration of cation}_n\left(\frac{\text{mg}}{\text{L}}\right)\times$$

$$= \frac{\left(50\dfrac{\text{mg CaCO}_3}{\text{meq of CaCO}_3}\right)}{\text{equivalent weight of cation}_n\left(\dfrac{\text{mg}}{\text{meq}}\right)} \tag{2.184}$$

where "meq" stands for milliequivalent weight.

The total of all equivalent weights of $CaCO_3$ calculated accordingly yields the total hardness as follows:

$$\text{Total Hardness} = \sum_{i=1}^{n}\left(\frac{\text{mg}}{\text{L}}\right)\text{of cation}_i \text{ as CaCO}_3 \tag{2.185}$$

Table 2.1 presents an approximate qualitative scale for hardness for water classification. We must remember that in aquatic environments, calcium is a constituent in the shells of mollusks and crustaceans and in the bones of fish; hence, aquatic animals require calcium in their environment. Calcium in water is also important for the hatching of eggs for some species of fish (Landau, 1992).

The following example illustrates the conversion of concentrations of different polyvalent cations into a single number expressed as hardness.

Example 2.84: Water Hardness

Let's now consider a ground water sample with a measured concentration of 150 mg/L of Ca^{2+} and 22 mg/L of Mg^{2+}. Determine the total hardness in mg/L as $CaCO_3$.

Solution

$CaCO_3$ has a molecular weight of 100.09, and since the typical valence of the dissociated ions of $CaCO_3$ is 2, it has an equivalent weight of 50.05 grams per equivalent weight. If we divide both the numerator and denominator of this unit by 1000, the equivalent weight can be expressed as 50.05 mg/meq.

Ca^{2+} has an atomic weight of 40.1 g/mole; therefore, it has an equivalent weight of 20.05 mg/meq. Then, using Equation (2.184), the concentration of Ca^{2+} (mg/L) expressed as an equivalent concentration of $CaCO_3$ is calculated as

TABLE 2.1
Qualitative Classification of Water According to Hardness

Description	Hardness	
	meq/L	mg/L as $CaCO_3$
Soft	<1	<50
Moderately Hard	1–3	50–150
Hard	3–6	150–300
Very Hard	>60	>300

Source: Greenhouse Engineering (2021).

$$= \frac{150\left(\frac{\text{mg Ca}^{2+}}{\text{L}}\right)\times\left(50\frac{\text{mg CaCO}_3}{\text{meq}}\right)}{20.05\left(\frac{\text{mg Ca}^{2+}}{\text{meq}}\right)} = 374\frac{\text{mg}}{\text{L}} \text{ as CaCO}_3$$

Mg^{2+} has an atomic weight of 24.3 g/mole; therefore, it has an equivalent weight of 12.15 mg/meq. Then, using Equation (2.184), the concentration of Mg^{2+} (mg/L) expressed as an equivalent concentration of $CaCO_3$ is

$$= \frac{22\left(\frac{\text{mg Mg}^{2+}}{\text{L}}\right)\times\left(50\frac{\text{mg CaCO}_3}{\text{meq}}\right)}{12.15\left(\frac{\text{mg Mg}^{2+}}{\text{meq}}\right)} = 90.5\frac{\text{mg}}{\text{L}} \text{ as CaCO}_3$$

Then, using Equation (2.185), the total hardness of the water sample is calculated as follows:

$$\text{Total Hardness} = (374 + 90.5) = 464.5\frac{\text{mg}}{\text{L}} \text{ as CaCO}_3$$

2.4.3.2.4 *Alkalinity and Buffering Capacity*

A buffered solution is one that can resist changes in pH that would normally result from the addition of acids (H^+) or bases (OH^-). An important source of buffering for natural surface waters is the carbonate buffering system produced by the absorption of atmospheric CO_2 in water. The absorption of dissolved CO_2 in water produces carbonic acid according to the following reaction:

$$CO_2 + H_2O \leftrightarrow H_2CO_3 \text{ (carbonic acid)} \quad (2.186)$$

At pH less than 4.5, most of the carbonate is present in the form of carbonic acid (H_2CO_3). When the water pH is between 6.5 and 10.5, carbonic acid is highly dissociated to the bicarbonate form:

$$H_2CO_3 \leftrightarrow H^+ + HCO_3^- \text{ (bicarbonate)} \quad (2.187)$$

At pH greater than 11.5, all of the carbonate is present as CO_3^{2-}:

$$HCO_3^- \leftrightarrow H^+ + CO_3^{2-} \text{ (carbonate)} \quad (2.188)$$

Carbonate acts as a buffer in water systems. If acid is added to water buffered with bicarbonate ions, Equation (2.187) shifts to the left to form more carbonic acid thus decreasing the hydrogen ion concentration and maintaining the pH near neutral. If a base is added to water buffered with bicarbonate ions, the bicarbonate further dissociates according to Equation (2.188) to produce additional H^+ in solution that will neutralize the added OH^- and resist a rise in pH.

Alkalinity is the opposite of hardness; it is a measure of the total anion concentration in water that will react to neutralize added H^+. Alkalinity is obviously different from hardness, but it is reported in the same units as mg/L of basic anions converted to equivalent mg/L of $CaCO_3$. The majority of alkalinity in natural waters is produced by the presence of carbonate and bicarbonate ions; hydroxide ions also contribute to alkalinity, and H^+ reduces it. The total H^+ that can be neutralized by water containing primarily carbonate ion species is determined by the following equation (Davis and Cornwell, 1991):

$$\text{Alkalinity} = [HCO_3^-] + 2[CO_3^{2-}] + [OH^-] - [H^+] \quad (2.189)$$

In other words, the equivalent concentration of H^+ used to titrate the ions in Equation (2.188) are expressed in terms of equivalent mg of $CaCO_3$/L. In practice, all carbonate species in solution are converted to H_2CO_3 below pH 4.5. Alkalinity can be measured by titrating a water sample to pH 4.5 with a standard H^+ solution to convert all of the bicarbonate and carbonate ions to carbonic acid. If one is given the concentrations of the ionic species $[HCO_3^-]$, $[CO_3^{2-}]$, $[OH^-]$, and $[H^+]$, the alkalinity of each of these ions expressed as equivalents mg/L as $CaCO_3$ can be calculated by the following equation:

$$\text{Alkalinity (mg/L as CaCO}_3) = \left(\frac{\text{mg}}{\text{L}} \text{ of species}\right)\frac{(\text{Equivalent weight of CaCO}_3)}{(\text{Equivalent weight of species})} \quad (2.190)$$

Example 2.85: Water Alkalinity

Let's now consider a water sample with pH 11.0, 40 mg/L of CO_3^{2-}, and 70 mg/L HCO_3^-. What is the alkalinity expressed as mg/L of $CaCO_3$?

Solution

The molecular weight of (CO_3^{2-}) is 60; therefore, the equivalent weight of (CO_3^{2-}) is 30. We can convert carbonate (CO_3^{2-}) to equivalent concentration of $CaCO_3$ and calculate alkalinity due to (CO_3^{2-}) as follows:

$$\text{Alkalinity}(CO_3^{2-}) = \left(40\frac{\text{mg}}{\text{L}}\right)\frac{\left(50\frac{\text{mg CaCO}_3}{\text{meq}}\right)}{\left(30\frac{\text{mg}}{\text{meq}}\right)} = 66.7\frac{\text{mg}}{\text{L}} \text{ as CaCO}_3$$

The molecular weight of (HCO_3^-) is 61; therefore, the equivalent weight of (HCO_3^-) is 61. We can convert (HCO_3^-) to equivalent concentration of $CaCO_3$ and calculate alkalinity due to (HCO_3^-) as follows:

$$\text{Alkalinity}(HCO_3^-) = \left(70\frac{\text{mg}}{\text{L}}\right)\frac{\left(50\frac{\text{mg CaCO}_3}{\text{meq}}\right)}{\left(61\frac{\text{mg}}{\text{meq}}\right)} = 57.4\frac{\text{mg}}{\text{L}} \text{ as CaCO}_3$$

And because pH = 11, OH^- = 10^{-3} mole/L, and the equivalent weight of (OH^-) is 17mg / meq. Then,

$$[OH^-] = 10^{-3}\frac{\text{mole}}{\text{L}} \times 17\frac{\text{g}}{\text{mole}} \times 10^3\frac{\text{mg}}{\text{g}} = 17\frac{\text{mg}}{\text{L}}$$

We can now convert [OH⁻] to the equivalent concentration of $CaCO_3$ and calculate alkalinity due to [OH⁻] as follows:

$$\text{Alkalinity}(OH^-) = \left(17\frac{mg}{L}\right)\frac{\left(50\frac{mg\ CaCO_3}{meq}\right)}{\left(17\frac{mg}{meq}\right)}$$

$$= 50\frac{mg}{L}\text{ as }CaCO_3$$

Again, because pH = 11, [H⁺] = 10^{-11} mole/L. The molecular weight of H^+ = 1 g/mole, and the equivalent weight of (H^+) is 1mg / meq. Then,

$$\left[H^+\right] = 10^{-11}\frac{mole}{L}\times 1\frac{g}{mole}\times 10^3\frac{mg}{g} = 10^{-8}\frac{mg}{L}$$

We can now convert [H⁺] to equivalent concentration of $CaCO_3$ and calculate alkalinity due to [H⁺] as follows:

$$\text{Alkalinity}(H^+) = \left(10^{-8}\frac{mg}{L}\right)\frac{\left(50\frac{mg\ CaCO_3}{meq}\right)}{\left(1\frac{mg}{meq}\right)}$$

$$= 5\times 10^{-7}\frac{mg}{L}\text{ as }CaCO_3$$

Finally, we can calculate

$$\text{Total Alkalinity} = 66.7 + 57.4 + 50 - \left(5\times 10^{-7}\right)$$

$$\text{Total Alkalinity} = 174.1\frac{mg}{L}\text{ as }CaCO_3$$

Now that both alkalinity and hardness have been discussed, the concepts of carbonate hardness, non-carbonate hardness, and total hardness can be explained. Total hardness is the concentration of all multivalent metallic cations in solution. The concentration of each multivalent cation needs to be converted to equivalent mg/L as $CaCO_3$. Carbonate hardness, on the other hand, is the hardness associated with the anions HCO_3^- and CO_3^{2-}. Carbonate hardness is sometimes referred to as temporary hardness because it can be removed from water by heating the water, since heating drives the following precipitation reaction, which removes Ca^{2+} from solution:

$$Ca^{2+} + 2HCO_3^- \rightarrow CaCO_3\downarrow + CO_2 + H_2O \qquad (2.191)$$

Non-carbonate hardness is associated with anions other than HCO_3^- and HCO_3^{2-}. If carbonate hardness exceeds the total hardness, then carbonate hardness is given the same value as total hardness.

In aquatic systems, dissolved CO_2 can be acquired not only by absorption from the atmosphere but also from the respiration of aquatic plants and animals; in contrast, photosynthesis in phytoplankton or algae utilizes CO_2. Significant growths of algae can affect the pH and buffering capacity of water. At dusk in natural ponds when photosynthesis ceases but respiration continues, the level of CO_2 and thus bicarbonate ion increases and hence the pH drops. During the daylight hours, photosynthesis utilizes CO_2, the buffering capacity decreases, and the pH rises. Salinity also affects carbonate alkalinity. For example, at a pH of 8.0 and a temperature of 24°C, less than 0.5% of alkalinity is in the form of carbonate ions, while seawater at the same pH and temperature has approximately 8% of the alkalinity in the form of carbonate ions (Stickney, 1979).

2.4.3.3 Physical Properties of Water

Physical property values of water at normal atmospheric pressure on the sea level at 0°C are provided in Table A.58. In this section, these and some other physical properties, and their relevance to controlled-environment systems, are discussed in depth.

2.4.3.3.1 Density and Thermal Stratification

As discussed earlier, fluid density ρ and mass (kg) per unit volume (m^3) are important fluid properties, and water is the only common substance that expands when it freezes. Water reaches maximum density at 4°C and expands as it is warmed beyond 4°C. Water also expands as its temperature drops below 4°C and is converted to ice; ice forms on open water surfaces when temperatures reach 0°C or lower. As ice forms, the liquid water surrounding the ice will be a little warmer than 0°C. In view of the density properties, the ice will be less dense than liquid water surrounding it and will float on the surface. If freezing water could sink, it would be possible for lakes to freeze solid all the way down to the bottom in winter. This would obviously kill all fish life and much of the plant life. Water at 4°C is at its maximum density and tends to sink to the lower depths of the lake, while water that is warmer or colder than 4°C is more buoyant tending to float creating seasonal thermal stratification of water in lakes.

In temperate climates, thermal stratification of lake water into stable temperature layers inhibits vertical mixing. The deeper cold-water layers in a lake are not replenished with dissolved oxygen until fall and spring when the thermal layers overturn. As lake water warms in spring and cools in fall, the surface waters approach 4°C, and the density differences that maintain the stable thermal layers temporarily disappear, allowing the wind and currents to mix the lower water layers.

Typical thermal layers in a temperate climate lake exhibit three different stratification layers: the epilimnion, the thermocline, and the hypolimnion. In summer, the surface water (epilimnion) is warmed by the Sun and is less dense than the colder water beneath. The epilimnion is well mixed by wind and wave action and is well supplied with dissolved oxygen from the atmosphere. Beneath the epilimnion, the temperature drops quickly in a transition zone called the thermocline. The bottom layer of the lake, on the other hand—the hypolimnion—retains water at 4°C. Since water at 4°C is at its maximum density, this water stays at the bottom.

As fall approaches and the surface waters cool, the entire water column approaches a uniform 4°C, and all the water in the lake becomes mixed by wind and wave action, especially during storms. In fall, the steep temperature transition of the thermocline disappears as the surface waters cool below 4°C. Ice forms and floats on the surface, acting as insulation that protects deeper water from freezing.

The formation of stable thermal layers in lakes inhibits the replenishment of dissolved oxygen and the distribution of nutrients in lake water. Thus, bottom layers of lakes can become anaerobic in summer and winter when the hypolimnion is stable and does not mix with the upper layers of water. Dead algae and plants in freshwater ponds or fish waste and unused feed in aquaculture operations can sink and accumulate on the bottom. Microbial degradation of this detrital organic matter rapidly utilizes existing dissolved oxygen in the hypolimnion, and anaerobic microbial degradation processes begins. Anaerobic degradation of organic matter produces CH_4 and hydrogen sulfide (H_2S). H_2S gas in water is toxic to fish and is a hazard in poorly managed aquaculture ponds that have low water turnover.

Additionally, as saltwater freezes, the salts present in liquid water are separated so that the ice becomes nearly salt free and hence undergoes the same type of expansion as freshwater ice (Wheaton, 1977). Seawater in the immediate vicinity of ice formation can be higher in salinity than normal. Adding salts or silt to liquid water increases its density, and seawater is therefore denser than freshwater.

2.4.3.3.2 Latent Heats of Vaporization and Fusion

The latent heat of vaporization of a solid or a liquid was defined earlier in Section 2.3 as the amount of heat energy that must be added at a constant temperature to vaporize a unit quantity of that substance. The latent heat of vaporization for water at 100°C is 2.26×10^6 J, among the highest values for any liquid. Therefore, water vapor stores a great deal of heat energy, which is released to the surroundings upon condensation. In imperial units, 1 lb of water requires 1,044 Btu of energy to evaporate. The latent heat of vaporization for seawater is the same as that for freshwater (Wheaton, 1977). Latent heat can be a major source of heat energy loss in controlled environments such as intensive greenhouse, livestock, and aquatic operations.

The latent heat of fusion (melting) was also defined earlier in Section 2.3 as the amount of heat energy that must be added to convert a unit quantity of solid substance (e.g, ice) to liquid (e.g., water). The latent heat of fusion for pure water ice is 3.39×10^5 J/kg at 0°C, and in imperial units, 1 lb of ice requires 144 Btu to melt. Salinity affects the latent heat of fusion of water. The heat of fusion of ice made from seawater at a freezing point of −8.7°C is 226×10^5 J/kg (West, 1968).

2.4.3.3.3 Specific Heat

Specific heat was defined in Section 2.3 as the amount of heat energy required to raise the temperature of a unit mass of a substance by one degree of temperature. Specific heats of common substances are provided in Table A.4. The specific heat of water is 4186 J/kg°C (1.00 Btu/lb°F), while it is 1004 J/kg°C (0.24 Btu/lb°F) for air. The specific heat of water is about five times greater than the specific heats of most common heavy solids. This means that water is slower to heat and cool than most other materials; therefore, water acts as a heat sink to moderate rapid environmental temperature changes, which can be lethal to living organisms. Furthermore, ground water from wells tends to be more consistent in temperature on an annual basis than surface waters. If we follow the seasons closely, we can clearly observe that coastal areas are warmer in winter and cooler in summer than corresponding inland areas.

In cases where water must be heated or cooled for controlled environments such as greenhouse, livestock, and aquatic operations, the high specific heat of water means that relatively large amounts of energy will have to be input or removed to artificially control water temperature; artificial temperature control in such operations should be avoided or minimized as much as possible. If it is necessary to artificially control water temperature in controlled environments by adding or removing heat energy, the systems should be designed by an engineer with proper professional expertise to ensure that an optimally economic system is achieved. A number of numerical examples related to water heating are provided in Sections 2.3 and 2.5.

2.4.3.3.4 Thermal Conductivity

The thermal conductivity of water is quite low. In fact, water is a good insulator as long as circulation is prevented in the water body so that there are no convection currents. The majority of heat transfer that occurs in water is convection. Convection in water mixes the liquid, which causes rapid temperature change. Water circulating against a solid surface can transfer heat at a high rate because circulation continually renews the temperature difference between the water and the solid. The circulation of water keeps transporting heat to or away from the solid.

2.4.3.3.5 Melting and Boiling Points

As discussed earlier, under ambient atmospheric pressure, water melts at 0°C and boils at 100°C; similar compounds such as H_2S, H_2Se, and H_2Te boil at normal ambient temperatures. Water has an unusually large difference between freezing and boiling points. Because of this, water exists as a liquid over much of the Earth's surface. Also, increasing salinity raises the boiling point of water.

2.4.3.3.6 Vapor Pressure

Vapor pressure refers to the pressure exerted by the gaseous phase of a material that is in equilibrium with its liquid or solid phase. The vapor pressure of pure water is a function of temperature and at 100°C is 101,325 Pa, which is equivalent to standard atmospheric air pressure (Table A.58). When water reaches 100°C, vapor pressure becomes equal to the ambient atmospheric pressure and boiling commences. Liquid water at the freezing point, 0°C, has a vapor pressure of 610.4 Pa. Tables for the vapor pressure of water as a function of temperature between 0°C and 100°C are commonly used; alternatively, graphs of temperature versus water vapor pressure can be used, but these are less accurate than tabulated values. The vapor pressure increases 166-fold between 0°C and 100°C.

Salinity, on the other hand, tends to lower the vapor pressure of water. Wheaton (1977) gave the following equation to calculate the vapor pressure of water as a function of salinity given the vapor pressure for pure water:

$$p_{ws} = p_w \left[1.000016 - 0.000537 \times S\right] \qquad (2.192)$$

where p_{ws} is the vapor pressure of salt water, p_w is the vapor pressure of pure water, and S is the salinity of water (ppt).

Evaporation from open water surfaces increases with increasing temperature if other environmental parameters remain constant. Increasing water temperatures could cause poorly selected pumps in controlled environment systems to experience cavitation, which occurs in pumping systems when the local static pressure inside the pump or piping system is reduced to equal to or less than the vapor pressure of water. Vapor bubbles rapidly form in a boiling action at these low-pressure points and then suddenly collapse when the pressure again increases downstream. Cavitation results in vibration, knocking noises, and excessive wear of the components in the pump or piping system and also reduces the capacity and efficiency of a pumping system. A pumping system that appears to operate well when water temperatures are low can experience cavitation when the water temperature rises a few degrees if the water vapor pressure becomes greater than the low-pressure points in the pumping system.

2.4.3.3.7 Osmotic Pressure

Osmotic pressure is defined as the pressure required to prevent movement of water across a semipermeable membrane. A semi-permeable membrane allows water molecules to pass through but does not allow the passage of other molecules or ions dissolved in water. The osmotic pressure of pure water is obviously zero. The osmotic pressure of seawater at 35 ppt salinity is approximately 21 atm (Wilson and Arons, 1955). The blood of freshwater fish has an osmotic pressure of approximately 6 atm, the same as a sodium chloride solution of about 7 ppt (Boyd, 1982). Freshwater is defined as water with dissolved solids content of less than 1.5 ppt. The cells of organisms are surrounded by semipermeable membranes. Aquatic animals must expend energy to maintain the salt balance in their cells if the concentration of internal cell fluids is different from the salinity of the aquatic environment. Intracellular concentration against an external concentration difference is called osmoregulation. In general, fish species in freshwater must expend energy to maintain the higher salt concentration inside their cells against the osmotic pressure of freshwater surrounding their bodies. On the other hand, fish in seawater must expend energy to maintain the lower salt concentration in their cells against higher salinity seawater. Animals in freshwater expend energy for osmoregulation to keep their cells from diluting and swelling with water; however, marine organisms must expend energy for osmoregulation to keep their cells from shrinking due to water loss to the more concentrated surrounding seawater. A rapid change in water salinity can produce too high osmoregulation energy requirements for aquatic organisms, causing undue stress, disease susceptibility, or even death. Furthermore, increasing water temperatures also increase osmotic pressure, although the difference is smaller than the change in osmotic pressure due to increasing salinity.

2.4.3.3.8 Viscosity

As discussed earlier in this section, internal flow resistance is quantified by the fluid property called viscosity (also known as dynamic viscosity), μ, which is a measure of a fluid's resistance to flow shear (Figure 2.24). The shear stress, τ (i.e., *F/A*), in a liquid like water is proportional to the rate of change of velocity between layers of liquid. This differential velocity is measured perpendicular to the direction of liquid flow, and τ is the frictional force per unit area between layers of fluid that are sliding over one another. Then, we have the relationship between shear stress and velocity gradients in a fluid as $\mu = \tau/(du/dy)$, where μ is the viscosity of liquid (Pa.s, which can also be expressed as N.s/m^2 or kg/m.s), τ is the shear stress or the shear force per unit area between moving liquid layers (N/m^2), and du/dy represents the rate of shear strain or velocity gradient (s^{-1}). All fluid flows have some degree of viscosity; there is no zero-viscosity flow.

Viscosity changes with the change of temperature and pressure. As temperature rises, viscosity of liquids decreases. Therefore, cold water requires a little more energy to pump. Also, fish utilize more energy to swim in cold water than in warm water. There is very little difference in the viscosity of freshwater and seawater. The viscosity of pure water at 20°C is 1.002×10^{-3} Pa.s (West, 1968). We should also remember that in fluid dynamics, μ often appears in a ratio as $\nu = \mu/\rho$, termed kinematic viscosity, for which the unit in SI is [m^2/s]. Density, or mass (kg) per unit volume (m^3), is another important fluid property that has already been discussed and is involved in most fluid models.

2.4.3.4 Turbidity and Color

2.4.3.4.1 Turbidity

Turbidity is defined as the haziness of a fluid caused by large numbers of individual particles. The differently sized suspended solid particles cause the turbidity. Suspended particulate matter in water bodies is measured by measuring the amount of light that is scattered and absorbed by the particles (Henry and Heinke, 1989). Turbidity is an extremely important parameter in aquatic systems, and it can be due to a variety of reasons such as high populations of plankton, algae, or humic substances transported in runoff water, suspended particles of clay, etc. (Boyd, 1982).

Turbidity caused by plankton is usually desirable in pond aquaculture. Plankton growth indicates a fertile system that is producing high amounts of organisms on the lower levels of the food chain that serve as food for the cultured species. However, turbidity reduces the penetration of light (including natural sunlight) into water. This limits the growth of aquatic plants, which can be beneficial to aquaculture operations in ponds. However, high turbidity due to excessive plankton growth is not desirable because the oxygen demand of the plankton bloom results in serious dissolved oxygen depletion. Turbidity due to humic substances, on the other hand, is not directly harmful to fish (Boyd, 1982). These waters are usually acidic with low nutrient levels and limited photosynthesis due to poor light penetration. Turbidity from humic substances is undesirable in systems that depend on aquatic fertility to achieve good growth rates of cultured species.

Furthermore, turbidity due to clay particles is undesirable in aquatic systems. Clay or silt turbidity restricts light penetration and settles to the bottom where it is harmful to fish eggs and bottom-dwelling organisms. It is difficult to avoid turbidity in aquaculture ponds with clay bottoms because clay soils provide a good seal against leakage; constructed ponds therefore require lining with clay if the natural soils are sandy or porous. It should be noted, however, that behavioral changes have been observed in fish exposed to 20,000 mg/L of clay turbidity, and significant fish mortality has been observed at turbidity above 175,000 mg/L (Boyd, 1982).

Other studies have shown that growth rates of sunfish, trout, and largemouth bass were generally best in clear ponds with less than 25 mg/L of turbidity; ponds with turbidity of 25 to 100 mg/L or muddy ponds with turbidities greater than 100 mg/L had reduced fish growth rates (Buch, 1956). It was also observed that clear ponds with low turbidity had more phytoplankton.

Moreover, particulate matter suspended in water provides a large surface area for the growth of fungi and bacteria. High turbidity can increase the risk of disease in aquaponics and aquaculture systems. Suspended particles can also adsorb nutrients such as phosphates, making them unavailable for plant growth. Continuous use of turbid waters in aquaponics or aquaculture ponds leads to rapid sedimentation that can fill ponds with silt in a few years. High levels of suspended solids also clog piping and filtration systems.

Turbidity can be measured by directly measuring light absorption with a spectrophotometer. A standard unit for spectrophotometric turbidity measurement is the nephelometric turbidity units (NTUs), which are calibrated using a suspension of formazin polymers at specified concentrations. NTUs are typically used in wastewater treatment systems and are less relevant to the assessment of turbidity in aquaponics or aquaculture systems. Different types of suspended solids exhibit different levels of absorption for different wavelengths of light. Therefore, spectrophotometric measurements of turbidity for aquaponics and aquaculture waters should be compared against known standards containing suspended solids similar to the suspended solids causing turbidity in the systems.

2.4.3.4.2 Color

Suspended particulate matter in water causes turbidity. The color of water is a result of scattering of light of different wavelengths by the impurities in water. Color in water does not reduce its transparency in the way that turbidity does. Pure water appears blue because blue light penetrates farther into water bodies and is scattered more than other wavelengths in the visible band. Waters from boggy soils, on the other hand, tend to be tea colored due to the high concentrations of tannic and humic acids. Dissolved soil often gives a reddish hue to water. The color of water in aquaponics or aquaculture ponds is often a function of algae populations. Experienced farmers can often detect the onset of oxygen shortages in water by observing water color.

2.4.4 Mass Continuity

Let's now consider incompressible fluid flow—similar to water—and apply the principle of conservation of mass to fluid flow through a conduit (Figure 2.27).

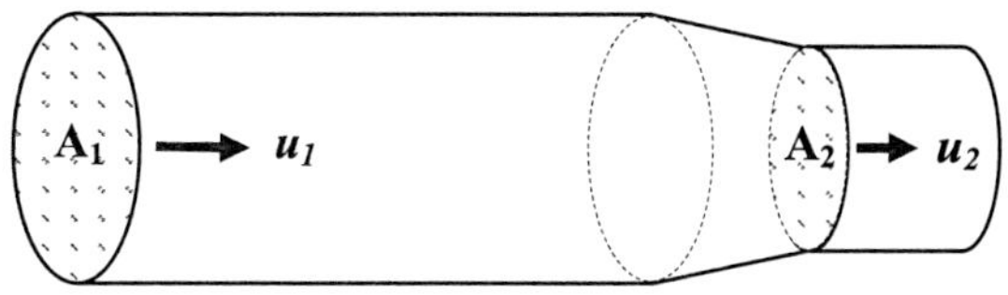

FIGURE 2.27 Fluid flow in a conduit with varying cross-sectional areas.

Source: ***Greenhouse Engineering* (2021).**

The flow equation is given by $Q = Au$, where Q is the volumetric flow rate, A is the cross-sectional area of the conduit, and u is the average velocity of the fluid. If we have a conduit having varying cross-sections as given in Figure 2.27, then the continuity relationship states that except when branches occur, Q is the same at all sections along the conduit as no fluid can be gained or lost. Therefore,

$$Q = A_1u_1 = A_2u_2 = \text{constant} \tag{2.193}$$

This is called the continuity equation, and it is true for any number of changes along the conduit.

Velocity of fluid in a conduit is not uniform across the cross-sectional area; thence, an average velocity that is calculated by the continuity equation for a steady flow is used:

$$u = \frac{Q}{A} = \frac{4Q}{\pi D^2} \tag{2.194}$$

The conduit (pipe) diameter can be determined easily when volumetric flow rate and velocity are known:

$$D = \sqrt{\left(\frac{4Q}{\pi u}\right)} \tag{2.195}$$

where D is the internal pipe diameter, Q is the volumetric flow rate, u is the velocity, and A is the pipe cross-sectional area.

If mass flow rate is known instead, then the diameter can be calculated using the following equation:

$$D = \sqrt{\left(\frac{4\dot{m}}{\pi \rho u}\right)} \tag{2.196}$$

where D is the internal pipe diameter, D is the mass flow rate, ρ is the fluid density, and u is the velocity, and A is the pipe cross-sectional area.

2.4.5 Laminar and Turbulent Fluid Flows

In a conduit (pipe) flow, streamlines are in straight parallel lines if the fluid velocity is small, and the streamlines will keep being straight and parallel with the pipe wall up to a velocity—known as critical velocity—where the streamlines break into diffused patterns and start wavering, after which the fluid flow involves random perturbations without any order or periodicity; the velocity of such flows varies temporally and spatially. When the velocity is lower than the critical velocity, the flow is called laminar, for which the flow velocity is highest on the pipe axis, and equal to zero on the pipe wall. The regime of flow is turbulent if the flow velocity is greater than the critical velocity. In turbulent flow, there is always a thin layer of fluid moving in laminar flow at the pipe wall, and this layer is known as the boundary layer.

The Reynolds number, a dimensionless quantity, giving the ratio of inertia to viscous forces, is used to determine flow regime. For flows through round pipes and tubes, the following form of Re is used:

$$\mathrm{Re} = \frac{uD}{\nu} = \frac{\rho u D}{\mu} \tag{2.197}$$

where u is the fluid velocity, D is the internal pipe diameter, ρ is the fluid density, ν is the kinematic viscosity, and μ is the dynamic viscosity. Laminar flow is generally expected if Re is less than 2,000; for 2,000 < Re < 4,000, the flow is said to be in transition state, and predictions are not reliable, and fully developed turbulent flow exists when Re > 4,000.

When Re needs to be calculated for noncircular cross-sections such as square, rectangular or oval conduits, equivalent diameter ($D_{eq} = 4\ A/P_w$) is used, where A is the flow area and P_w is the wetted perimeter of the cross section. This approach is what is used for air ducts, for instance, which are often rectangular.

Example 2.86: Reynolds Number Calculation

Water at 30°C is flowing through a 30 meter 2 in. nominal steel pipe (Schedule 40) with a mass flow rate of 2 kg/s. Determine the Reynolds number for this fluid flow.

Solution

Internal diameter of 2 in. nominal standard steel pipe (Sch. 40) (Table A.31) is 0.0525 m.

Water density ρ at 30°C (Table A.58) = 995.7 kg/m³

Dynamic viscosity μ at 30°C (Table A.58) = 798 × 10^{-6} kg/m s = 798×10^{-6} Pa.s

From the continuity equation,

$$u\left(\frac{\mathrm{m}}{\mathrm{s}}\right) = \frac{Q(\mathrm{m}^3/\mathrm{s})}{A(\mathrm{m}^2)} \text{ and } Q\left(\frac{\mathrm{m}^3}{\mathrm{s}}\right) = \frac{\dot{m}(\mathrm{kg/s})}{\rho(\mathrm{kg/m}^3)}$$

$$Q\left(\frac{\mathrm{m}^3}{\mathrm{s}}\right) = \frac{2(\mathrm{kg/s})}{995.7(\mathrm{kg/m}^3)} = 0.002\frac{\mathrm{m}^3}{\mathrm{s}}$$

Then,

$$u = \frac{4\left(0.002\dfrac{\mathrm{m}^3}{\mathrm{s}}\right)}{\pi(0.0525\ \mathrm{m})^2} = 0.93\mathrm{m/s}$$

And then,

$$\mathrm{Re} = \frac{\rho u D}{\mu} = \frac{995.7\dfrac{\mathrm{kg}}{\mathrm{m}^3}0.93\dfrac{\mathrm{m}}{\mathrm{s}}0.0525\,\mathrm{m}}{798\times10^{-6}\dfrac{\mathrm{kg}}{\mathrm{m.s}}} = 60{,}813$$

(the flow is turbulent)

2.4.6 Bernoulli's Equation: Energy Conservation in an Ideal Fluid

Bernoulli's principle states that an increase in the speed of a fluid occurs simultaneously with a decrease in static pressure or a decrease in the fluid's potential energy. Although Daniel Bernoulli deduced that pressure decreases when the flow speed increases, it was Leonhard Euler who derived Bernoulli's equation in its usual form in 1752 (Darrigol and Frisch, 2008; Anderson, 2016). The principle can be applied to different types of fluid flow, and therefore, there are different forms of the equation. The simple form of Bernoulli's equation is used for incompressible flows (i.e., most liquid flows and gas flows moving at low Mach number).

Fluid energy is composed of kinetic, potential (due to elevation), and internal energy. Therefore, Bernoulli's equation can be derived from the principle of conservation of energy, which states that in a steady flow, ignoring viscosity, compressibility, and thermal effects, the sum of all forms of energy in a fluid along a streamline is the same at all points on that streamline. This requires that the sum of kinetic, potential, and internal energy remains constant (Batchelor, 2000). Bernoulli's principle can also be derived directly from Newton's second law of motion. If a volume of fluid is flowing horizontally from a region of high pressure to a region of low pressure, then there must be more pressure behind than in front. This gives a net force on the mentioned volume, moving it along the streamline.

Bernoulli performed his experiments on liquids, so as mentioned, earlier his original equation is valid only for incompressible flow. A common form of Bernoulli's equation, valid at any arbitrary point along a streamline, can then be presented as

$$\frac{u^2}{2} + gz + \frac{p}{\rho} = \text{constant} \tag{2.198}$$

where u is the fluid flow speed at a point on a streamline, g is the acceleration due to gravity, z is the elevation (in the direction opposite g) of the point above a reference plane, p is the pressure at the chosen point, and ρ is the density of the fluid at all points in the fluid. The constant on the right-hand side of the equation depends only on the streamline chosen, whereas v, z, and p depend on the particular point on that streamline. For this form of Bernoulli equation, the flow must be steady (i.e., the flow parameters velocity, density, etc. at any point cannot change with time), the flow must be incompressible (even though pressure varies), and friction by viscous forces must be negligible.

Dividing Equation (2.198) by g produces the head equation derived from Bernoulli's principle, noting that each term can be described in the length dimension such as meters or feet:

$$\frac{u^2}{2g} + z + \frac{p}{\rho g} = \text{constant}$$

And for two points:

$$\frac{u_1^2}{2g} + z_1 + \frac{p_1}{\rho_1 g} = \frac{u_2^2}{2g} + z_2 + \frac{p_2}{\rho_2 g} \quad (2.199)$$

where the far-left term $\frac{u^2}{2g}$ is called the velocity head, which represents the internal energy of the fluid due to its motion; the middle term z is called the elevation head, which represents the potential energy of the fluid due to its elevation (with respect to a reference plain); and the far-right term $\frac{p}{\rho g}$ is called the pressure head, which represents the internal energy of the fluid due to the exerted pressure on the container (pipe wall). As mentioned earlier, if friction losses are neglected and no energy is added to or taken from a piping system, the sum of these three heads give the total head.

Multiplying Equation (2.198) by ρ gives us an equation with three pressure terms:

$$\frac{\rho u^2}{2} + \rho g z + p = \text{constant} \quad (2.200)$$

It must be noted that in this form of Bernoulli's equation, the total pressure is constant. If the middle term (the pressure due to elevation) is constant and the far-right term (the static pressure) increases, then the far-left term (the dynamic pressure) must have decreased. That is, an increase in static pressure due to flow resistance results in a decrease in the flow speed if it is not due to elevation difference.

All three versions of Bernoulli's equation just given are just simplified versions of energy balance on a system, and from the equation, many other equations are derived for different energy losses and gains that happen in real-world systems. In the real world, energy is added to or extracted from the fluid (e.g., via pumps and turbines), and when the viscosity of the fluid is taken into account, energy is lost due to friction due to the fluid rubbing against itself and the pipe and fitting walls, and by the flow turbulence; therefore, the total energy head presented in Equation (2.199) is no longer constant along the pipe. This lost energy is the energy converted to the sound and heat generated by fluids as they flow through the pipes. The amount of frictional loss is affected by the length, roughness, diameter, and changes in the shape or section of the pipe as well as the velocity and type of flow of the fluid. Fittings, valves, bends, etc. all increase frictional losses. We can then take Equation (2.170), the head equation for two points of one streamline in a fluid flow, modify it accordingly, and restore the equality taking all the above-mentioned terms into account as follows:

$$\frac{u_1^2}{2g} + z_1 + \frac{p_1}{\rho_1 g} + h_p = \frac{u_2^2}{2g} + z_2 + \frac{p_2}{\rho_2 g} + h_{f,major} + h_{f,minor} + h_t \quad (2.201)$$

where u is the fluid flow speed at the chosen point on a streamline; g is the acceleration due to gravity; z is the elevation of the chosen point above the reference plane; p is the pressure at the chosen point; ρ is the density of the fluid at the chosen point of the fluid; h_p is the pump head; $h_{f,\,major}$ is the head loss due to friction in the pipe; $h_{f,\,minor}$ is the frictional head losses due to fittings, valves, bends, etc.; and h_t is the turbine head.

Head losses due to friction and fittings and turbine head are included as terms in Equation (2.201) on the right-hand side because they all represent losses or extractions (in the presence of a turbine) of energy in the system, whereas the pump head is included in the left-hand side because the pump adds mechanical energy to the system. Flow in a pipe always creates energy loss due to friction. Energy loss can be measured using two gauges just like static pressure in the direction of fluid flow. In practice, $h_{f,\,major}$ and $h_{f,\,minor}$ are calculated using tabulated values (typically provided by manufacturers) that require pipe type, diameter, length and flow rate, and type of fitting.

2.4.6.1 Pressure Loss due to Friction

The general equation for pressure loss due to friction, also known as the Darcy–Weisbach equation—expressed in Pascals—can also be used as follows:

$$\frac{\Delta P}{L} = f\frac{\rho}{2}\frac{u^2}{D} \text{ or } \Delta P = \rho f \frac{L}{D}\frac{u^2}{2} \quad (2.202)$$

Head loss expressed in meters of fluid is expressed as

$$\frac{\Delta h_{f,major}}{L} = f\frac{1}{2g}\frac{u^2}{D} \text{ or } h_{f,major} = f\frac{L}{D}\frac{u^2}{2g} \quad (2.203)$$

where ΔP is the pressure loss due to friction (Pa), $\Delta h_{f,major}$ or $h_{f,major}$ is the head loss due to friction (m), f is the friction factor (unitless), L is the pipe length (m), D is the hydraulic diameter of the pipe (m), g is the acceleration due to gravity (9.81 m/s^2), and u is the mean flow velocity (m/s). Frictional head loss (Equation (2.203)) can be converted into frictional loss as pressure (Equation (2.202)) by the relationship $\Delta P = h_{f,\,major}\,\rho\, g$.

The Darcy–Weisbach equation can be used for both laminar and turbulent flows as well as for any liquid in a pipe; it can also be used for gases and vapors. Friction factor f is a function of the Reynolds number and the roughness of the pipe and is almost constant for turbulent flow, varying from only about 0.01 to 0.05.

For fully developed laminar flow (also known as Hagen–Poiseuille flow) in a pipe, Re < 2,000, f is independent of relative roughness and depends only on Re, giving

$$f = \frac{64}{\text{Re}} \quad (2.204)$$

where f is the friction factor (dimensionless) and Re is the Reynolds number (dimensionless). In short, with laminar flow, f varies inversely with Re.

If Re > 4,000, then the flow is turbulent, f is a function of the relative pipe roughness and Re, and it can be determined using Moody's chart (Figure 2.28), one of the most famous and important charts. It can also be determined using the Colebrook equation or its approximations as shown in the next example. On Moody's diagram, f can be determined based on Re and relative pipe roughness, which is equal to ε/D, where ε is the surface roughness factor (also called absolute roughness, the height or measure of the roughness element) and D is the internal diameter of the pipe. Surface roughness factors have been measured and tabulated for different materials used in piping systems (Table A.32).

Example 2.87: Determining Friction Factor Using Moody Charts

Determine $\frac{\varepsilon}{D}$ and f using a Moody chart for Example 2.86.

Solution

Surface roughness factor ε for new steel pipe (Table A.32) = 4.5×10^{-5} m. Then,

$$\text{Relative Roughness} = \frac{\varepsilon}{D} = \frac{4.5\times10^{-5}\text{ m}}{0.0525\text{ m}}$$
$$= 0.000857 \approx 0.0009$$

Now using the Moody chart (Figure 2.28) with Re = 60,813 and relative pipe roughness of 0.0009, we can find f as 0.023.

For very large Re, f is independent of the number. A very thin laminar sublayer exists for this completely turbulent flow along the pipe wall where the surface roughness completely dominates the character of the flow.

It should be remembered that there is a head loss in any pipe no matter how smooth the pipe surface, and even for smooth pipes, the friction factor is not zero. This is because of the existing no-slip boundary condition that requires fluid to stick to the solid surface over which it flows. There is always some surface roughness that produces the no-slip behavior (therefore f is not zero) on the molecular level. Such pipes are called hydraulically smooth. For hydraulically smooth pipe, f is approximated by Blasius's (1911) equation:

$$f = (100\,\text{Re})^{-1/4} \quad \text{for Re up to } 10^5 \tag{2.205}$$

$$f = 0.0032 + \frac{0.221}{\text{Re}^{0.237}}$$
$$\text{for } 10^5 < \text{Re} < 3\times10^6 \text{ at least} \tag{2.206}$$

Colebrook combined all data for transition state and turbulent flow in smooth and rough pipes into a relationship known as the Colebrook equation, the most commonly used equation for calculating the friction coefficient for turbulent flow. The Colebrook equation is implicit in f, and determination of f requires trial and error (iteration) as shown:

$$\frac{1}{\sqrt{f}} = -2\log\left[\frac{\varepsilon}{3.7D} + \frac{2.51}{\text{Re}\sqrt{f}}\right] \tag{2.207}$$

We can also use an approximate explicit relationship (Swamee and Jain, 1976) for $5{,}000 < \text{Re} < 10^8$ and $0.000001 < \varepsilon < 0.01$ as follows:

$$\frac{1}{\sqrt{f}} = -2\log\left[\frac{\varepsilon}{3.7D} + \frac{5.74}{\text{Re}^{0.9}}\right] \tag{2.208}$$

Another approximate explicit relationship for f (Haaland, 1983) can also be used for $4{,}000 < \text{Re} < 10^8$ and $0.000001 < \varepsilon < 0.05$ as follows:

$$\frac{1}{\sqrt{f}} = -1.8\log\left[\left(\frac{\varepsilon/D}{3.7}\right)^{1.11} + \frac{6.9}{\text{Re}}\right] \tag{2.209}$$

A transition region appears in Figure 2.28 for Re between 2000 and 4000, where the flow might be laminar, turbulent,

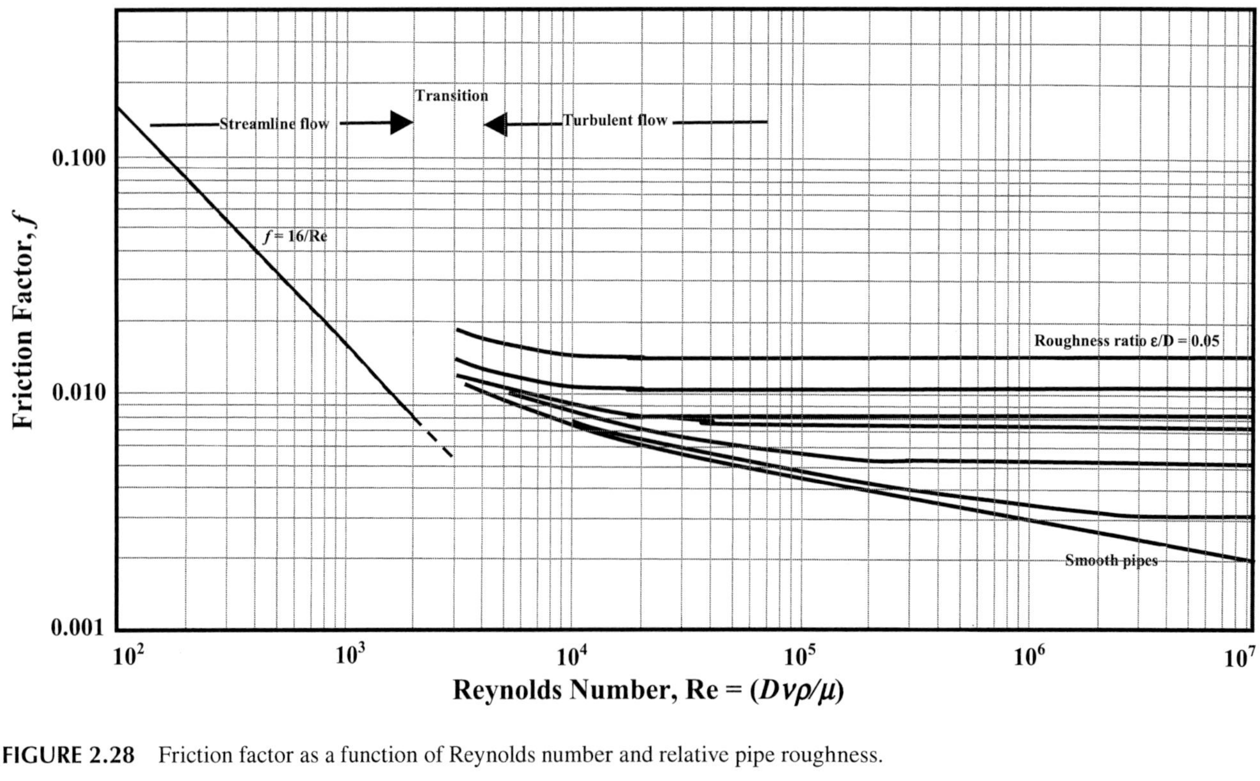

FIGURE 2.28 Friction factor as a function of Reynolds number and relative pipe roughness.
Source: *Greenhouse Engineering* (2021).

or a mixture of both, and an interpolation approach is needed in this region for estimating *f*.

Example 2.88: Determining Friction Factor Using Colebrook Equation Approximations

Find *f* using Equations (2.208) and (2.209) for Example 2.81.

Solution

Step 1: Find the Reynolds number.
Reynolds number has already been determined as Re = 60,813.

Step 2: Classify flow
Since Re = 60,813 > 4,000, then the flow is turbulent.

Step 3: Determine *f*.
Using Equation (2.208) (Swamee equation),

$$\frac{1}{\sqrt{f}} = -2\log\left[\frac{\varepsilon}{3.7D} + \frac{5.74}{\text{Re}^{0.9}}\right]$$

$$\frac{1}{\sqrt{f}} = -2\log\left[\frac{4.5\times10\text{-}5}{3.7\times0.0525} + \frac{5.74}{60{,}813^{0.9}}\right] \Rightarrow f = 0.023$$

Similarly, Equation (2.209) (Haaland equation) gives

$$\frac{1}{\sqrt{f}} = -1.8\log\left[\left(\frac{\varepsilon/D}{3.7}\right)^{1.11} + \frac{6.9}{\text{Re}}\right]$$

$$\frac{1}{\sqrt{f}} = -1.8\log\left[\left(\frac{4.5\times10^{-5}/0.0525}{3.7}\right)^{1.11} + \frac{6.9}{60{,}813}\right]$$

$$\Rightarrow f = 0.023$$

Notably, the two calculated approximations and the friction factor determined earlier using the Moody chart are in agreement with each other.

2.4.6.2 Valve and Fitting Head Losses, Resistance Coefficients, and Equivalent Lengths

Most pipe systems consist of many more components than just straight pipes; therefore, the term $h_{f,minor}$ in Equation (2.201) needs to be calculated. As stated earlier, these losses are minor and can be due to the pipe entrance or exit; sudden or gradual expansions or contractions; bends, elbows, tees, and other fittings; or open or partially closed valves. In some cases, the minor losses are greater than the major losses; for instance, a partially closed valve can have higher pressure drop or head loss than that of a long pipe. For most fluid dynamic designs, it is assumed that the flow through valves and fittings is turbulent; therefore, almost all pressure drop or head loss data provided by manufacturers is reported for turbulent flow conditions and is proportional to the velocity in valves and fittings.

The pressure or head loss data for valves and fittings is experimentally measured, tabulated, and provided by manufacturers. However, due to the time-consuming and costly nature of experimental testing, it is impossible to obtain test data for every size and type of valve and fitting. It is therefore desirable to provide a reliable means of extrapolating available test information to cover items that have not been tested. Commonly used concepts for accomplishing this task are equivalent length (*L/D*), resistance coefficient (*K*), and flow coefficient (C_v).

Since velocity in a pipe is obtained at the expense of static head, the decrease in static head due to velocity is $\frac{u^2}{2g}$ (Equation (2.199)), the velocity head. *K* used for determining minor head losses or pressure drops in Equation (2.210) therefore is defined as the number of velocity heads lost due to a valve or fitting:

$$h_{f,minor} = \frac{\Delta P}{\rho g} = K\frac{u^2}{2g} \tag{2.210}$$

where *K* is a dimensionless resistance coefficient (also called the loss coefficient). Equation (2.210) is valid for computing head loss due to valves and fittings for all conditions of flow, including laminar flow, using *K* as given in Table A.33 or other relevant literature.

The pressure or head loss due to valve and fitting resistance is always associated with the *D* where the velocity in $\frac{u^2}{2g}$ occurs. *K* is independent of *f* and Re and is a constant for any given valve or fitting in a piping system under all flow conditions, including laminar flow. If *K* is needed for a different *D* from the diameter for which we can find *K* from tables or diagrams, then we can use the relationship $K_a = K_b\left[D_a/D_b\right]^4$ where *a* defines *K* and *D* with reference to the internal diameter of the pipe and *b* defines *K* and *D* with reference to the internal diameter of the pipe for which the values of *K* were established and can be found in tables and diagrams.

The pressure or head loss due to friction in a straight pipe, on the other hand, is expressed by the Darcy–Weisbach equation (Equation (2.203)) as $h_{f,major} = \left(f\frac{L}{D}\right)\frac{u^2}{2g}$. It then follows that $K = \left(f\frac{L}{D}\right)$, where $\frac{L}{D}$ is the equivalent length in diameters of straight pipe that will cause the same pressure drop as the valve or fitting under the same flow conditions. Since K is a constant under all flow conditions, $\frac{L}{D}$ for any given valve or fitting must vary inversely with the change in *f* for different flow conditions. When the equivalent length approach is used for determining the minor losses due to valves and fittings, then tabulated $\left(\frac{L}{D}\right)$ values (Table A.34) need to be employed instead of resistance coefficients.

Using *K*, the minor head losses and therefore the total head loss due to friction can be calculated using the following equation:

$$h_{f,minor} = \sum_{i=1}^{n} K_i\left(\frac{u^2}{2g}\right)$$

Then,

$$h_{f,major} + h_{f,minor} = \left[f\frac{L}{D} + \sum_{i=1}^{n} K_i\right]\frac{u^2}{2g} \tag{2.211}$$

where n is the total number of valves and fittings and K needs to be determined for each valve and fitting from tables (Table A.33).

K due to sudden enlargements is calculated using the following equation:

$$K = \left[1 - \frac{D_1^2}{D_2^2}\right]^2 \tag{2.212}$$

and K due to sudden contractions is calculated using

$$K = 0.5\left[1 - \frac{D_1^2}{D_2^2}\right] \tag{2.213}$$

where D_1 is the smaller internal diameter and D_2 is the larger internal diameter.

Example 2.89: Determining Major and Minor Head Losses

Determine $h_{f,major}$ and $h_{f,minor}$ for the previous example assuming that the pipe is horizontal with two fully open gate valves and two regular flanged 90° elbows.

Solution

Using Equation (2.203), we can first calculate $h_{f,major}$ as follows:

$$h_{f,major} = f\frac{L}{D}\frac{u^2}{2g} = 0.023\frac{30\text{ m}}{0.0525\text{ m}}\frac{\left(0.93\frac{\text{m}}{\text{s}}\right)^2}{2\times 9.81\text{ m/s}^2}$$

$$h_{f,major} = 0.57\text{m}$$

Then, using Equation (2.211), we can calculate $h_{f,minor}$:

$$h_{f,minor} = \Sigma_{i=1}^{n} K_i\left(\frac{u^2}{2g}\right)$$

where we can substitute $K_1 = 0.15$ (gate valve), $K_2 = 0.15$ (gate valve), and $K_3 = 0.30$ (90° regular flanged elbow).

$$h_{f,minor} = \Sigma_{i=1}^{n} K_i\left(\frac{u^2}{2g}\right) = (K_1 + K_2 + 2\times K_3)\left(\frac{u^2}{2g}\right)$$

$$h_{f,minor} = (0.15 + 0.15 + 2\times 0.30)\left(\frac{\left(0.93\frac{\text{m}}{\text{s}}\right)^2}{2\times 9.81\text{ m/s}^2}\right)$$

$$h_{f,minor} = 0.04\text{m}$$

Similarly, using Equation (2.211), we can calculate both the major and minor head losses as follows:

$$h_{f,major} + h_{f,minor} = \left[f\frac{L}{D} + \Sigma_{i=1}^{n} K_i\right]\times\left(\frac{u^2}{2g}\right)$$

$$= \left[0.023\frac{30\text{ m}}{0.0525\text{ m}} + (0.15 + 0.15 + 0.30 + 0.30)\right]$$

$$\times\left(\frac{\left(0.93\frac{\text{m}}{\text{s}}\right)^2}{2\times 9.81\text{m/s}^2}\right)$$

$$h_{f,major} + h_{f,minor} = 0.61\text{m}$$

2.4.6.3 Flow Coefficients, Pressure Drops, and Control Valve Flow Rates

The control valve industry uses a standard number called the flow coefficient to express the valve capacity and valve flow characteristics in terms of this coefficient. The valve flow coefficient represents a valve's ability to pass flow. In the UK and the US, the symbol C_v is used and is defined as the flow of water at the reference temperature of 60°F in gallons per minute at a pressure drop of one pound per square inch across the valve. The larger the C_v, the more flow a valve can pass with a given pressure drop. For liquids, flow for a given C_v is typically calculated from the following equation:

$$Q = C_v\sqrt{\left(\frac{\Delta p 62.4}{\rho}\right)} \tag{2.214}$$

where Q is the valve flow rate (USGPM), Δp is the difference between upstream and downstream pressure (psi), 62.4 is the conversion factor for fluids computed in relation to water, and ρ is the density of fluid (lb/ft^3).

Equation (2.214) uses a conversion factor of 62.4, which is the density of water in lb/ft^3. The following equation is frequently utilized using a fluid's specific gravity G where the relationship to water density is already considered.

$$Q = C_v\sqrt{\left(\frac{\Delta p}{G}\right)} \tag{2.215}$$

Likewise, C_v for particular flow characteristics can be determined from the following equation:

$$C_v = Q\sqrt{\left(\frac{G}{\Delta p}\right)} \tag{2.216}$$

Similar equations are also available using metric units:

$$Q = K_v\sqrt{\left(\frac{\Delta p 1{,}000}{\rho}\right)} \tag{2.217}$$

$$K_v = Q\sqrt{\left(\frac{\rho}{\Delta p 1{,}000}\right)} \tag{2.218}$$

$$Q = K_v\sqrt{\left(\frac{\Delta p}{G}\right)} \tag{2.219}$$

$$K_v = Q\sqrt{\left(\frac{G}{\Delta p}\right)} \text{ or } K_v = 0.8646\, C_v \tag{2.220}$$

where Q is the valve flow rate (m^3/h); K_v is the flow coefficient (primarily used in Europe), which describes the amount of water at 15.5°C that flows through a valve with a pressure loss of 1 bar at a given valve position (K_{vs} is used instead of K_v when the flow that generates a pressure drop of 1 bar when the valve is fully open); Δp is the difference between upstream and downstream pressure (bar); 1,000 is the conversion factor for fluids computed in relation to water; ρ is the density of fluid (kg/m^3); and G is the specific gravity.

Liquid flow capacity of a valve in metric units can also be directly converted to C_v using the following equation:

$$C_v = 0.0694 Q \sqrt{\left(\frac{\rho}{\Delta p 999}\right)} \text{ or } C_v = 1.157\, K_v \quad (2.221)$$

where Q is the flow rate (l/m), ρ is the density (kg/m^3), and Δp is the pressure drop (bar).

The mixing valve can be sized by setting the maximum and minimum pressure drops over the valve. A mixing valve should have maximum and minimum pressure drops of 10 kPa = 0.1 bar (= 1.45 psi) and 3.5 kPa = 0.035 bar (0.5 psi), respectively. Then, using Equation (2.89) or Equation (2.220) and this pressure difference, C_v can be calculated and the correct mixing valve can be chosen easily.

Example 2.90: Calculating Flow Coefficient and Choosing a Mixing Valve

For a given volumetric flow rate of 300 USGPM of water and a maximum pressure drop of 1.45 psi, calculate the flow coefficient C_v (and K_v).

Solution

Using Equation (2.216), we can calculate C_v as follows:

$$C_v = Q \sqrt{\left(\frac{G}{\Delta p}\right)} = 300 \text{ gpm} \sqrt{\left(\frac{0.999}{1.45 \text{psi}}\right)} = 249$$

Then, using this flow coefficient, we can choose the closest mixing valve from the manufacturer's list, which has a minimum C_v of 249. Then, using the chosen mixing valve's C_v from the manufacturer's tables, we can go ahead and calculate the actual pressure drop Δp to see if $\Delta p < 1.45$ psi.

Using Equation (2.220), we can do the same calculation in metric units for the water flow rate of 68.2 m^3/h and a maximum pressure drop of 0.1 bar and find K_v as follows:

$$K_v = 68.2 \text{ m}^3/\text{h} \sqrt{\left(\frac{0.999}{0.1 \text{bar}}\right)} = 216$$

Similarly, we can choose the closest mixing valve from the manufacturer's list that has a minimum K_v of 216, and then, using the chosen mixing valve's K_v from the manufacturer's tables, we can now go ahead and calculate the actual pressure drop Δp to see if $\Delta p < 0.1$ bar.

2.4.6.4 Proper Sizing of Check Valves

Many difficulties are encountered with check valves, both lift and swing types. These difficulties are due to over-sizing, which ends up in noisy operation and premature wearing of moving parts. The minimum velocity required to lift the disc to the fully open and stable position has been determined by testing various types of check valves and provided by check valve manufacturers. Sizing a check valve based on the specified minimum velocity for full disc lift often results in valves that are smaller than the pipes in which they are installed, but the pressure drop is comparable with that of a full-size valve with the disc not fully open. The losses due to sudden and gradual contraction and expansion, which will occur in such installations, can be easily calculated using the formulations provided earlier or tabulated data for fittings.

2.4.6.5 Gas Pressure Regulator Capacity, and Flow Coefficients

For gas pressure regulators operating in the range of inlet pressures up to 100 bar, the following equations for subcritical and critical flows are used:

Subcritical Flow:

$$Q = \frac{13.57}{\sqrt{S(T_u + 273)}} C_g \frac{p_u + p_b}{2} \sin\left[K_l \sqrt{\frac{p_u - p_d}{p_u + p_b}}\right] \quad (2.222)$$

Critical Flow:

$$Q = \frac{13.57}{\sqrt{S(T_u + 273)}} C_g \frac{p_u + p_b}{2} \quad (2.223)$$

where Q is the gas flow rate (m^3/h), S is the gas relative density (for air $S = 1$), Tu is the gas upstream temperature (°C), C_g is the gas flow coefficient, pu is the gas upstream pressure (bar), p_d is the gas downstream pressure (bar), p_b is the ambient atmospheric pressure (bar), and K_l is the valve body shape factor.

If $K_l < 130$ and $(p_u - p_d) > 0.1\,(p_u + p_b)$, then we can use the simplified equation for subcritical flow rate where $(p_u - p_d) < 0.5\,(p_u + p_b)$:

$$Q = \frac{13.57}{\sqrt{S(T_u + 273)}} C_g \sqrt{(p_d + p_b)(p_u - p_d)} \quad (2.224)$$

Subcritical flow rate where $(p_u - p_d) > 0.5\,(p_u + p_b)$:

$$Q = \frac{13.57}{\sqrt{S(T_u + 273)}} C_g \sqrt{\frac{p_u + p_b}{2}} \quad (2.225)$$

If desired, Equation (2.225) can also be expressed with the flow coefficient K_g rather than C_g as follows:

$$Q = K_g \sqrt{\frac{p_u + p_b}{2}} \quad (2.226)$$

Using these equations provided, if C_g or K_g provided by manufacturers is known, Q can be easily calculated for a given gas regulator's upstream and downstream pressures. All the equations provided are for regulators with fully open control element, as flow coefficients are measured for fully open regulators. Capacities for different positions of the control element in a regulator are calculated based on the curve of relevant regulator.

2.4.6.6 Natural Gas Control Valve Flow Rates for Pressure Ratios at Normal Conditions

Natural gas control valves use flow coefficient KG, which is defined for under the critical pressure ratio ($p_a/p_e > 0.54$) as follows:

$$KG = \frac{Q_n}{\sqrt{p_a\left(p_e - p_a\right)}} \tag{2.227}$$

For overcritical pressure ratio ($p_a/p_e < 0.54$):

$$KG = \frac{2Q_n}{p_e} \tag{2.228}$$

where KG is the flow coefficient (m³/h), Q_n is the flow rate (m³/h) at normal conditions (p = 101,325 Pa and T = 273.15 K), p_a is the pressure after the control valve (bar), and p_e is the pressure in front of the control valve (bar).

For other gases, the following correction factor should be used:

$$Q_{n\,\text{Gas}} = f\,Q_{n\,\text{Natural Gas}} \tag{2.229}$$

$$f = \sqrt{\frac{0.78}{\rho\text{Gas}}} \tag{2.230}$$

where $Q_{n\,\text{Natural Gas}}$ is the flow rate at normal conditions of natural gas (m³/h), $Q_{n\,\text{Gas}}$ is the flow rate of the other gas at normal conditions (m³/h), and f is the correction factor.

2.4.7 Lift, Circulation, and Pumping Energy Requirements

For a system containing a pump and a turbine, Bernoulli's equation was developed as Equation (2.201). Multiplying each side of Equation (2.201) ρg gives

$$\frac{\rho_{u1}^{\;2}}{2} + \rho g z_1 + p_1 + p_p = \frac{\rho_{u2}^{\;2}}{2} + \rho g z_2 + p_2 + \rho g(h_{f,major} + h_{f,minor}) + p_t \tag{2.231}$$

For the application of Bernoulli's equation in certain situations, we can make the following simplifications and cancellations where applicable:

a. *Tanks, reservoirs and nozzles exposed to the atmosphere*
First, when the fluid at a reference point is exposed to the atmosphere, the pressure at that reference point is zero, and therefore, the pressure head term can be canceled. Second, the tank the fluid is drawn from is much larger than the flow area inside the pipe. Since $u = Q/A$, the velocity at the surface of a reservoir or tank is very small, and because with velocity head, velocity u is squared, its value. In conclusion, the velocity head at the surface of a reservoir or tank can be considered zero; therefore, it is canceled from the Bernoulli equation.

b. *When both reference points are in the same pipe*
Under the conditions of steady flow assumption, the velocity will be the same throughout the pipe due to an existing uniform flow area. Therefore, under steady-flow conditions, when the two reference points for the Bernoulli equation are inside the same size pipe, the velocity head terms on both sides of the equation are equal, and hence are canceled.

c. *When elevations are equal at both reference points*
When both reference points for the Bernoulli equation are at the same elevation, the elevation head terms are equal as well, and therefore, they are cancelled.

d. *Closed circuit conduit flow*
When both reference points for the Bernoulli equation are in the same closed-circuit flow, the pressure head terms are equal at both reference points, and therefore, they are cancelled as well. Additionally, the fluid lifting (elevation head) is compensated by the same amount of fluid falling; therefore, the elevation head terms are cancelled.

The relationship between the pump head and the pump pressure p_p, on the other hand, is

$$h_p = \frac{p_p}{\rho g} \tag{2.232}$$

After calculating p_p, as explained next, we can determine the power delivered by the pump (pump fluid power) θ_{fl}, the pump input power (pump break power) θ_{bk}, the pump break horsepower (Bhp), and the pump's current requirement provided the voltage available.

The relationship between θ_{fl} and p_p is given by

$$\theta_{fl} = p_p Q \tag{2.233}$$

Pump efficiency (η) represents the losses in the pump due to friction and electrical efficiency and is usually expressed in decimals (e.g., 0.6 is used for 60% conversion efficiency), and it is the ratio of the pump output power θ_{fl} to the pump input power (pump break power) θ_{bk}. Therefore, we can write the following relationship:

$$\theta_{fl} = \eta\theta_{bk} \tag{2.234}$$

Then, combining Equations (2.232) and (2.233) gives the following:

$$p_p = \eta\theta_{fl} / Q \tag{2.235}$$

This expression can be substituted in the Bernoulli equation, Equation (2.231), and the required pump power or alternatively the system flow rate for a given pump power rate can be determined. Further, θ_{bk} can then be determined from the following equation:

$$\theta_{bk} = p_p Q / \eta \tag{2.236}$$

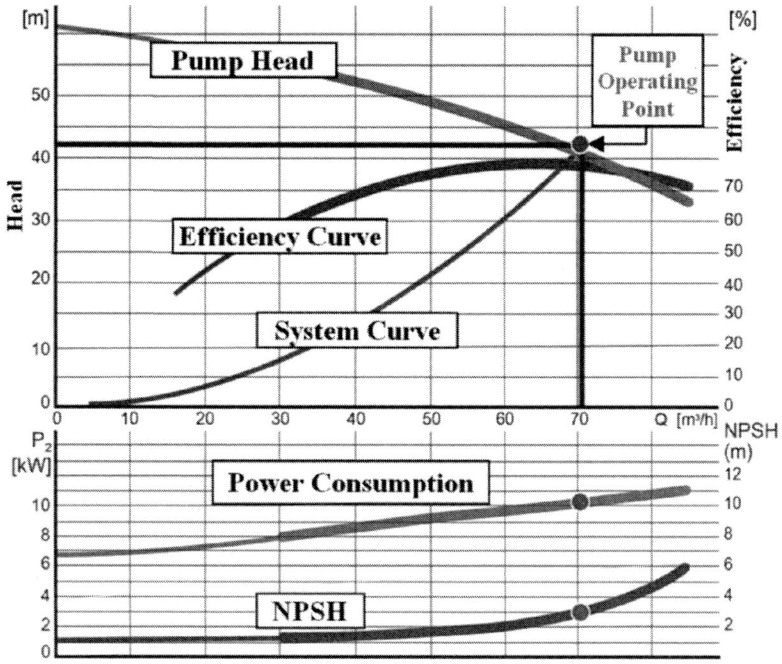

FIGURE 2.29 Characteristic curves of a centrifugal pump and a system curve. NPSH: net positive suction head.
Source: *Greenhouse Engineering* (2021).

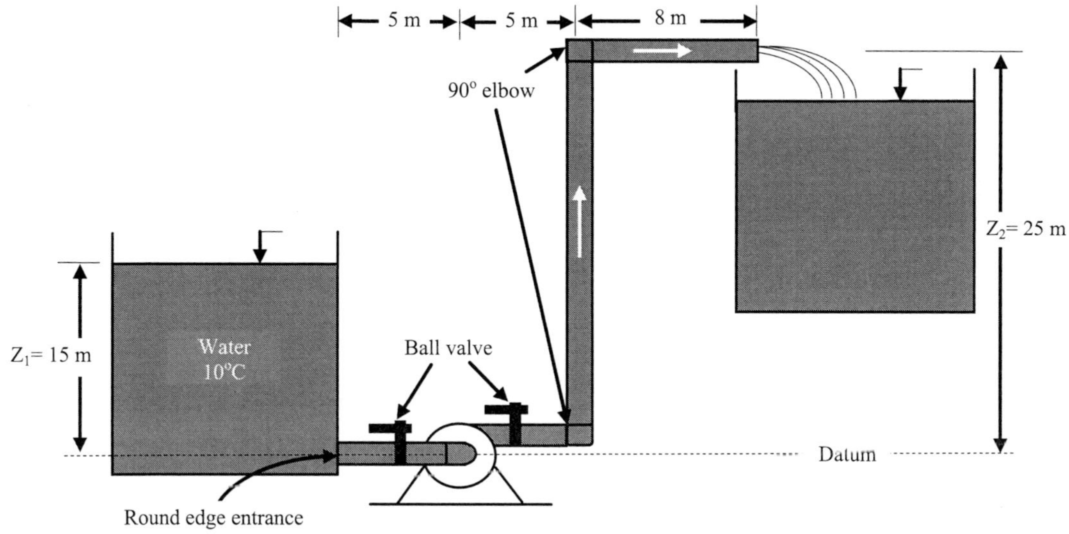

FIGURE 2.30 Water lifting between two reservoirs.
Source: *Greenhouse Engineering* (2021).

In other words, θ_{bk} is equal to the electrical power supplied P_e, which is given by the following relationship:

$$\theta_{bk} = P_e = IV \quad (2.237)$$

where P_e is the electrical power (in watt), I is the current (in ampere), and V is the voltage (in volt). Typically, the voltage is known (e.g., 110 V, 220 V), so the current needs to be determined for cables, fuses, and controllers.

After determining the pump head for the system's required flow, we can develop the system curve (i.e., the flow capacity versus the system head), plot it on the pump characteristic curves provided by pump manufacturers, and determine the operating point of the chosen point. Figure 2.29 depicts the typical efficiency curve, power curve, and rate of discharge curve for a centrifugal pump at constant operational speed and the system curve plotted against and the pump's operating point accordingly.

Example 2.91: Water Lifting between Reservoirs

Water at 10°C from a large reservoir is lifted with a pump to another reservoir at 1 kg/s through 1 ½ in. and 2 in. standard (Schedule 40) steel piping systems shown in Figure 2.30. Determine the total head loss and size of the required pump.

Solution

Internal diameter of 1 ½ in. standard (Schedule 40) steel pipe (Table A.31) = 0.04089 m. Internal diameter of 2 in. standard (Schedule 40) steel pipe (Table A.31) = 0.05250 m. Surface roughness factor ε for new steel pipe

(Table A.32) = 4.5 × 10^{-5} m. Water density ρ at 10°C (Table A.58) = 999.7 kg/m³. Dynamic viscosity μ at 10°C (Table A.58) = 1306 × 10^{-6} kg/m.s = 1306 × 10^{-6} Pa.s.

Using Equation (2.231) form of the Bernoulli equation, we can easily solve this problem as follows. Remember that the reservoir on the left side and the discharge on the right side are exposed to the atmospheric pressure, so $p_1 = p_2$. And since the reservoir has a very large hydraulic diameter compared with the diameter of the discharge pipe, average water velocity at the top of the reservoir u_1 can be treated as zero ($u_1 = 0$). Since there is no turbine in the system, the energy extracted by the turbine, p_t, is zero as well.

$$\cancelto{0}{\frac{\rho u_1^2}{2}} + \rho g z_1 + \cancelto{0}{p_1} + p_p = \frac{\rho u_2^2}{2} + \rho g z_2 + \cancelto{0}{p_2} + \rho g(h_{f,major} + h_{f,minor}) + \cancelto{0}{p_t}$$

$$p_p = \rho\left[\frac{u_2^2}{2} + g(z_2 - z_1) + g(h_{f,major} + h_{f,minor})\right]$$

Step 1: Find the Reynolds number.

From the continuity equation,

$$u\left(\frac{m}{s}\right) = \frac{Q(m^3/s)}{A(m^2)} \text{ and } Q\left(\frac{m^3}{s}\right) = \frac{\dot{m}(kg/s)}{\rho(kg/m^3)}$$

and then,

$$Q\left(\frac{m^3}{s}\right) = \frac{1(kg/s)}{999.7(kg/m^3)} = 0.0010\,m^3/s$$

Then, for 1 ½ in. pipe,

$$u = \frac{4\left(0.001\frac{m^3}{s}\right)}{\pi(0.04089\ m)^2} = 0.76m$$

and for 2 in. pipe u = 0.46 m/s

For 1 ½ in. pipe,

$$Re = \frac{\rho u D}{\mu} = \frac{999.7\frac{kg}{m^3}0.76\frac{m}{s}0.04089m}{1306\times10^{-6}\frac{kg}{m.s}} = 23{,}818$$

and for 2 in. pipe,

$$Re = 18{,}579$$

Step 2: Classify flow.

Re = 23,818 > 4,000 and Re = 18,551 > 4000, so both flows are turbulent.

Step 3: Determine *f*.

Using Equation (2.235) for a 1 ½ in. pipe,

$$\frac{1}{\sqrt{f}} = -2\log\left[\frac{\varepsilon}{3.7D} + \frac{5.74}{Re^{0.9}}\right]$$

$$\frac{1}{\sqrt{f}} = -2\log\left[\frac{4.5\times10^{-5}}{3.7\times0.04089} + \frac{5.74}{23{,}818^{0.9}}\right]$$

f = 0.027 and for 2 in. pipe f = 0.028

Step 4: Determine the pump pressure.

Using Equation (2.211), we first can calculate the major and minor head losses together as follows:

$$h_{f,major} + h_{f,minor} = \left[f\frac{L}{D} + \sum_{i=1}^{n} K_i\right]\frac{u^2}{2g}$$

where K_1 = 0.05 (ball valve fully open), K_2 = 1.50 (90° threaded regular elbow), and the resistance coefficient due to sudden contractions is

$$K_5 = 0.5\left[1 - \frac{D_1^2}{D_2^2}\right] = 0.5$$

For a 1 ½ in. pipe,

$$h_{f,major} + h_{f,minor} =$$

$$\left[0.027\frac{(5+5+25+8)\ m}{0.04089\ m} + (2\times0.05 + 2\times1.50 + 0.5)\right]$$

$$\times\left(\frac{\left(0.76\frac{m}{s}\right)^2}{2\times9.81m/s^2}\right)$$

$$h_{f,major} + h_{f,minor} = 0.838\ m + 0.107\ m = 0.945\ m$$

For a 2 in. pipe,

$$h_{f,major} + h_{f,minor} = 0.287\ m$$

Then, using the velocity head, elevation head, and the head due to friction, p_p can be calculated as follows:

$$p_p = \rho\left[\frac{u_2^2}{2} + g(z_2 - z_1) + g\,(h_{f,major} + h_{f,minor})\right]$$

For a 1 ½ in. pipe,

$$p_p = 999.7\frac{kg}{m^3}\left[\frac{\left(0.762\frac{m}{s}\right)^2}{2} + 9.81\frac{m}{s^2}(25-15)m + 9.81\frac{m}{s^2}(0.945\ m)\right]$$

$$p_p = 107{,}627Pa$$

For a 2 in. pipe,

$$p_p = 100{,}993Pa$$

Step 5: Determine the total pump head.

Dividing p_p calculated in Step 4 by ρ g, we can easily determine h_p pump head as follows:

For a 1 ½ in. pipe,

$$h_p = \frac{p_p}{\rho\, g} = \frac{107{,}627\ Pa}{999.7\ kg/m^3\ 9.81m/s^2} = 11.0m$$

For a 2 in. pipe,

$$h_p = 10.3m$$

Step 6: Determine the required pump power θ_{fl}

Using Equation (2.233) $p_p = \frac{\theta_{fl}}{Q}$ or $\theta_{fl} = p_p Q$, the required pump (fluid) power (the pump delivering) can be calculated as follows:

$$\theta_{fl} = p_p Q$$

For a 1 ½ in. pipe,

$$\theta_{fl} = (107{,}627\ Pa)(0.0010\ m^3/s) = 107.7\ W$$

For a 2 in. pipe:

$$\theta_{fl} = 101.0 \text{ W}$$

Step 7: Determine the required pump input (break) power θ_{bk}.

Assuming a pump conversion efficiency of 60% and using Equation (2.234), we can calculate θ_{bk} as

$$\theta_{fl} = \eta \theta_{bk}$$

For a 1 ½ in. pipe,

$$\theta_{bk} = \frac{\theta_{fl}}{\eta} = \frac{107.7 \text{ W}}{0.6} = 179.4 \text{ W}$$

For a 2 in. pipe,

$$\theta_{bk} = 168.4 \text{ W}$$

Since 1 hp = 746 W, the pump break horsepower (Bhp) for 1 ½ in. pipe is $179.4\,\text{W}\frac{1\text{hp}}{746\,\text{W}}$ = 0.24 hp (and 0.23 hp for 2 in. pipe). Assuming no electrical losses, θ_{bk} is then equal to P_e, which has the following relationship:

$$\theta_{bk} = P_g = IV$$

Then, the pump current requirement can be determined as follows:

For a 1 ½ in. pipe,

$$I = \frac{P_e}{V} = \frac{179.4 \text{ W}}{120 \text{ V}} = 1.50 \text{ A}$$

For a 2 in pipe,

$$I = 1.40 \text{ A}$$

Using commercially available pump characteristic curves and the required pump capacity of 60 LPM with an 11.0 m system head (15.9 USGPM with a 36.0 ft system head) for the 1 ½ in. pipe system and 60 LPM with a 10.3 m system head (15.9 USGPM with a 33.8 ft system head) for the 2 in. pipe system, and we can easily choose pumps for this task.

Example 2.92: Water Circulating in a Closed-Loop Piping System

Water at an average temperature of 60°C is circulated between two large reservoirs in a closed-circuit piping system with a pump at 5 kg/s through a 2 in. standard (Sch. 40) steel piping system shown in Figure 2.31. Determine the total head loss and size of the required pump.

Solution

Internal diameter of 2 in. standard (Sch. 40) steel pipe (Table A.31) = 0.05250 m. Surface roughness factor ε for new steel pipe (Table A.32) = 4.5 × 10⁻⁵ m. Water density ρ at 60°C (Table A.58) = 983.13 kg/m³. Dynamic viscosity μ at 60°C (Table A.58) = 4.67 × 10⁻⁴ kg/m.s = 4.67 × 10⁻⁴ Pa.s.

Using Equation (2.231), we can easily solve this problem as follows. First, remember that under steady flow conditions, the velocity head terms on both sides of the equation are equal, and therefore, they cancel each other. When both reference points for the Bernoulli equation are in the same closed-circuit flow, the pressure head terms are equal at both points, and therefore, they are cancelled as well. Also, when both reference points for the Bernoulli equation are in the same closed-circuit flow, the fluid lifting is compensated by the same amount of fluid falling; therefore, the elevation head terms are cancelled.

$$\cancelto{0}{\frac{\rho u_1^2}{2}} + \cancelto{0}{\rho g z_1} + \cancelto{0}{p_1} + p_p = \cancelto{0}{\frac{\rho u_2^2}{2}} + \cancelto{0}{\rho g z_2} + \cancelto{0}{p_2} + \rho g(h_{f,major} + h_{f,minor}) + \cancelto{0}{p_t}$$

$$p_p = \rho g(h_{f,major} + h_{f,minor})$$

Step 1: Find the Reynolds number.

From the continuity equation,

$$u\left(\frac{\text{m}}{\text{s}}\right) = \frac{Q(\text{m}^3/\text{s})}{A(\text{m}^3)} \text{ and } Q\left(\frac{\text{m}^3}{\text{s}}\right) = \frac{\dot{m}(\text{kg/s})}{\rho(\text{kg/m}^3)}$$

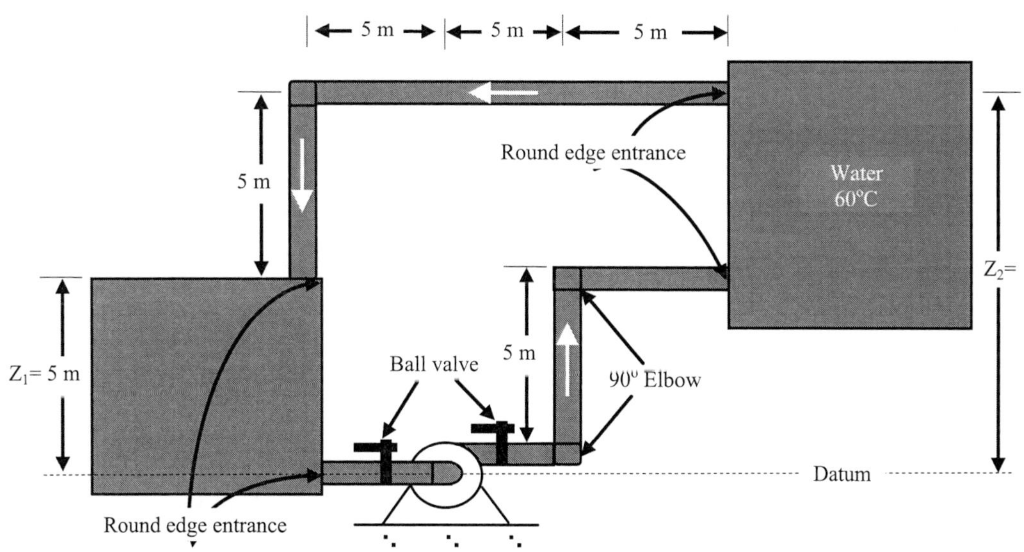

FIGURE 2.31 Closed-loop water circulation.
Source: *Greenhouse Engineering* (2021).

and

$$Q\left(\frac{m^3}{s}\right)=\frac{5(\text{kg}/\text{s})}{983.13(\text{kg}/\text{m}^3)}=0.0051\text{m}^3/\text{s}$$

Then,

$$u=\frac{4\left(0.001\dfrac{\text{m}^3}{\text{s}}\right)}{\pi(0.05250\text{ m})^2}=2.35\text{m}/\text{s}$$

and

$$\text{Re}=\frac{\rho uD}{\mu}=\frac{983.13\dfrac{\text{kg}}{\text{m}^3}1.07\dfrac{\text{m}}{\text{s}}0.05250\text{ m}}{4.67\times10^{-4}\dfrac{\text{kg}}{\text{m.s}}}=259{,}791$$

Step 2: Classify flow.
Re = 259,791 > 4000; then, the flows is turbulent.

Step 3: Determine *f*.
Using Equation (2.235),

$$\frac{1}{\sqrt{f}}=-2\log\left[\frac{\varepsilon}{3.7D}+\frac{5.74}{\text{Re}^{0.9}}\right]$$

$$\frac{1}{\sqrt{f}}=-2\log\left[\frac{4.5\times10^{-5}}{3.7\times0.05250}+\frac{5.74}{259{,}791^{0.9}}\right]$$

$f = 0.020$

Step 4: Determine the pump pressure.
Using Equation (2.211), we can calculate the major and minor head losses together as follows:

$$h_{f,major}+h_{f,minor}=\left[f\frac{L}{D}+\Sigma_{i=1}^{n}K_i\right]\frac{u^2}{2g}$$

where $K_1 = 0.05$ (ball valve fully open), $K_3 = 0.3$ (90° regular elbow flanged), and the resistance coefficients due to sudden contraction and sudden expansion are

$$K_5=0.5\left[1-\frac{D_1^2}{D_2^2}\right]=0.5\quad K=\left[1-\frac{D_1^2}{D_2^2}\right]^2=1.0$$

$$h_{f,major}+h_{f,minor}$$

$$=\left[\begin{array}{l}0.027\dfrac{(5+5+25+8)\text{m}}{0.04089\text{m}}+(2\times0.05+3\times0.3\\+2\times(0.5+1.0))\end{array}\right]$$

$$\times\left(\frac{\left(0.76\dfrac{\text{m}}{\text{s}}\right)^2}{2\times9.81\text{m}/\text{s}^2}\right)$$

$$h_{f,major}+h_{f,minor}=4.298\text{m}+1.126\text{m}=5.425\text{m}$$

Then, using the velocity head, elevation head, and the head due to friction, p_p can be calculated as follows:

$$p_p=\rho g\,(h_{f,major}+h_{f,minor})$$

$$p_p=983.13\text{kg}/\text{m}^3\left(9.81\frac{\text{m}}{\text{s}^2}\right)(5.425\text{m})$$

$$p_p=52{,}318\text{ Pa}$$

Step 5: Determine the total pump head.
Dividing p_p calculated in Step 4 by ρ g, we can easily determine h_p as follows:

$$h_p=\frac{p_p}{\rho\,g}=\frac{52{,}318\text{ Pa}}{983.13\text{ kg/m}^3\ 9.81\text{ m/s}^2}=5.425\text{ m}$$

Step 6: Determine the required pump power.
Using Equation (2.233) $p_p=\dfrac{\theta_{fl}}{Q}$ or $\theta_{fl}=p_pQ$, θ_{fl} can be calculated as follows:

$$\theta_{fl}=p_pQ$$

$$\theta_{fl}=(52{,}318\text{ Pa})(0.0051\text{m}^3/\text{s})=266.1\text{W}$$

Step 7: Determine the required pump input (break) power.
Assuming pump conversion efficiency of 60% and using Equation (2.234), we can calculate θ_{bk} as

$$\theta_{fl}=\eta\theta_{bk}$$

$$\theta_{bk}=\frac{\theta_{fl}}{\eta}=\frac{224.6\text{W}}{0.6}=443.5\text{W}$$

Since 1 hp = 746 W, Bhp = $374.4W\dfrac{1\text{hp}}{746\text{ W}}$ = 0.59 hp. Assuming no electrical losses, θ_{bk} is then equal to P_e, which has the following relationship:

$$\theta_{bk}=P_e=I\times V$$

Then, the pump current requirement can be determined as

$$I=\frac{P_e}{V}=\frac{443.5\text{ W}}{120\text{ V}}=3.70\text{ A}$$

Using commercially available pump characteristic curves and the required pump capacity of 305.1 LPM with a 5.4 m system head (80.6 USGPM with a 17.8 ft system head), we can easily choose a pump for this task.

2.4.8 Pump/Fan Affinity Laws

2.4.8.1 Introduction

Pumps and fans are mainly used for the transmission of liquids and gases, respectively, and they are extensively used in controlled-environment systems. Centrifugal pumps transfer liquid by the centrifugal force generated by the rotation of an impeller. Impellers are essentially discs mounted on the pump shaft and within the pump's casing; when rotated, they impart the motive centrifugal force to the fluid being pumped. Centrifugal pumps are the most widely used pumps in controlled-environment systems mainly due to their low cost, simple construction, silent operation, and low maintenance. They are designed based on both mechanical and hydraulic considerations. The mechanical variables used in their designs are speed, impeller diameter, and power, while the hydraulic variables are head, capacity (volumetric flow rate), and efficiency.

Throughout the years, a wonderful database and charts and equations have been developed to allow accurate predictions of pump and fan performances from the results of pump and fan testings. The performance charts are plotted for constant shaft rotation with the speed expressed in rotation per minute (rpm). The basic independent variable considered is the discharge or volumetric flow rate Q, with head h, power θ, and efficiency η taken as the dependent variables.

From the typical performance curves for centrifugal pumps, we know that the head is approximately constant at low discharge and then drops to zero at Q_{max}. This indicates that at the given speed and impeller size, the pump's delivery rate is maximized to Q_{max} when head is zero. Test results show that η is always zero at zero flow and at Q_{max}, and it reaches a maximum, approximately 85%, at about $0.6 \times Q_{max}$. This is the Q, or the best efficiency point, where $\eta = \eta_{max}$, with the head and brake power expressed as h_p and θ_{bk}, respectively. Here, we should note that η is not independent of h_p and θ_{bk} but rather is calculated using the relationship given in Equation (2.234), $\eta = \theta_{fl} / \theta_{bk} = p_p.Q / \theta_{bk} = (\rho.g).Q.h_p / \theta_{bk} = \gamma.\ Q.h_p / \theta_{bk}$, where γ is the specific weight of the discharged fluid.

Moreover, centrifugal pump casings are fabricated to accept a given maximum impeller diameter. The impeller diameter is important because the peripheral velocity is directly proportional to the diameter and hence is directly relatable to the pump head developed. The maximum impeller dimension can be trimmed to produce a smaller diameter and thereby results in a smaller centrifugal force delivered to the fluid. On the other hand, a pump operating with an impeller with a smaller diameter can be retrofitted with a larger-diameter impeller up to the maximum as long as it can be accommodated by the pump's casing. The affinity laws for pumps and fans play an important role in determining performance. They are also used for scaling up when performance parameters exceed those of existing pumps or fans.

2.4.8.2 Affinity Laws

As just mentioned, affinity or similarity laws for pumps and fans are used in hydraulics and HVAC to predict the performance of a given machine under different conditions of operation (such as head, volumetric flow rate, shaft rotational speed, or impeller diameter) from those under which it has been tested. Once a pump or a fan has been selected and the impeller diameter has been determined to deliver a defined flow rate for a required level of head (energy), the affinity laws can be used to determine what new speed or impeller diameter is required to satisfy the alternative operating conditions. A revised flow versus head (energy) versus horsepower curve can be developed and plotted from these relationships.

The affinity laws apply both to centrifugal and axial flows. The only requirement is that the two pumps/fans are dynamically similar, i.e., the ratios of the fluid forced are the same. We must note that there are two sets of affinity laws: the laws for known pumps and fans and laws for geometrically similar or homologous centrifugal pumps and fans (meaning the same velocity diagrams at entrance and exit); both sets are used to approximate head, volumetric flow rate, and power curves for different motor speeds and impeller diameters.

These relations do not hold exactly if the ratio of speed change is greater than 1.5–2.0 or if suction conditions become limiting, such as net positive suction head.

2.4.8.2.1 *Pump Affinity Laws for a Specific Centrifugal Pump*

a) Volumetric Flow Rate: The volumetric flow rate of a centrifugal pump can be expressed as

$$Q_1 / Q_2 = (N_1 / N_2) + (D_1 / D_2) \tag{2.238}$$

where Q is the volumetric flow rate (m³/s, L/s, gpm, cfm, etc.), N is the shaft rotational speed (rpm), and D is the impeller diameter (mm, in, etc.).

b) Head or Pressure: The head or pressure of a centrifugal pump can be expressed as

$$dp_1 / dp_2 = (N_1 / N_2)^2 + (D_1 / D_2)^2 \tag{2.239}$$

where dp is the head or pressure (m, ft, Pa, psi, etc.).

c) Power: The power consumption (pump break horsepower) of a centrifugal pump can be expressed as

$$P_1 / P_2 = (N_1 / N_2)^3 + (D_1 / D_2)^3 \tag{2.240}$$

where P is the power (W, bhp, etc.)

A) When Impeller Diameter Held Constant: If D is constant, then N can simplify the affinity laws to the following:

a) Volumetric Flow Rate: In this relationship, Q is proportional to N, and the volume capacity of a centrifugal pump can be expressed as:

$$Q_1 / Q_2 = (N_1 / N_2) \tag{2.241}$$

b) Head or Pressure: In this relationship, centrifugal pump pressure or head is proportional to the square of shaft speed and can be expressed as

$$dp_1 / dp_2 = (N_1 / N_2)^2 \tag{2.242}$$

c) Power: In this relationship, centrifugal pump power is proportional to the cube of shaft speed and can be expressed as

$$P_1 / P_2 = (N_1 / N_2)^3 \tag{2.243}$$

Note that if N increases by 10%, Q increases by 10%, dp increases by 21%, and P increases by 33%. Therefore, if we want to increase Q for an existing system by 10%, then we have to increase P by 33%.

Example 2.93: Same Impeller Diameter

An axial flow pump delivers 300 L/s of water at a head of 6 m when rotating at 2000 rpm. If its efficiency is 80%, determine the power delivered by the shaft to the pump (in kW). Additionally, if this same pump were operated at 2400 rpm, the efficiency would increase to 83%. Using the applicable affinity law equations, determine Q_2, dp_2, and P_2. Remember that $\eta_1 \neq \eta_2$.

Solution

Using the specific weight of water, $\gamma = \rho.g = 9.807\,\text{kN/m}^3$, and the following relationships that were covered earlier,

$$\eta = \gamma.Q.h_p / \theta_{bk} => \eta.\theta_{bk} = \gamma.Q.h_p$$

And letting $\eta.\theta_{bk} = P_1 = \gamma Q_1 h_{p1}$, we can calculate P_1 as follows:

$$P_1 = \left(9.807\frac{\text{kN}}{\text{m}^3}\right)\left(0.3\frac{\text{m}^3}{\text{s}}\right)(6\,\text{m})\left[\frac{\text{kJ}}{\text{kN.m}}\right]\left[\frac{\text{kW.s}}{\text{kJ}}\right]$$

Then, $P_1 = 17.65\,\text{kW}$. Considering 80% pump efficiency, we can determine the power the shaft delivers to the pump (P_1') as follows:

$$P_1' = \frac{P_1}{\eta_p} = \frac{17.65\,\text{kW}}{0.8} => P_1' \approx 22.06\text{ kW}$$

Since the same pump is used, $D_1 = D_2$, and the final volumetric flow rate (Q_2) can be calculated using Equation (2.241) as follows:

$$Q_2 = Q_1(N_2/N_1)$$
$$Q_2 = 300\,\text{L/s}(2400\,\text{rpm}/2000\,\text{rpm})$$
$$Q_2 = 360\,\text{L/s}$$

Again, since the same pump is used, $D_1 = D_2$, and dp_2 can be calculated using Equation (2.242) as follows:

$$dp_2 = dp_1(N_2/N_1)^2$$
$$dp_2 = 6\,\text{m}\left((2400\,\text{rpm})/(2000\,\text{rpm})\right)^2$$
$$dp_2 = 8.64\,\text{m}$$

Assuming $\rho = 1000\,\text{kg/m}^3$ and $g = 9.807\,\text{kg/m}^3$, we can calculate P_2 as

$$P_2 = \rho.g.Q_2.h_{p_2}$$
$$P_2 = \left(1000\frac{\text{kg}}{\text{m}^3}\right)\left(9.807\frac{\text{m}^3}{\text{s}}\right)(8.64\,\text{m})\left(0.36\frac{\text{m}^3}{\text{s}}\right)\left(\frac{1}{1000}\right)$$
$$P_2 = 30.5\,\text{kW}$$

Now using the pump's new efficiency (83%), we can determine P_2' as follows:

$$P_2' = \frac{P_2}{\eta_P} = \frac{30.5\text{ kW}}{0.83} \approx 36.75\text{ kW}$$

Note that in this example, when N increases by 20%, Q increases 20% as well, dp increases 44%, and P increases approximately 66%. This means that if we want to increase Q for an existing system 20%, then we have to increase P supply 66%.

Example 2.94: Changing Shaft Speed

In this example, the shaft speed is changed while the impeller size is kept constant. The initial volumetric flow rate is 5 L/s, the initial head is 30 m, the initial pump power is 3.75 kW, the initial shaft speed is 1750 rpm, and the final shaft speed is 3500 rpm. Using the applicable affinity law equations, find the final Q_2, dp_2, and P_2.

Solution

Q_2 can be calculated using Equation (2.241):

$$Q_2 = Q_1(N_2/N_1)$$
$$Q_2 = 5\,\text{L/s}(3500\,\text{rpm}/1750\,\text{rpm})$$
$$Q_2 = 10\,\text{L/s}$$

Similarly, dp_2 can be calculated using Equation (2.242):

$$dp_2 = dp_1(N_2/N_1)^2$$
$$dp_2 = 30\,\text{m}\left((3500\,\text{rpm})/(1750\,\text{rpm})\right)^2$$
$$dp_2 = 120\,\text{m}$$

And P_2 can be calculated using Equation (2.243):

$$P_2 = P_1(N_2/N_1)^3$$
$$P_2 = 3.75\,\text{kW}\left((3500\,\text{rpm})/(1750\,\text{rpm})\right)^3$$
$$P_2 = 30\,\text{kW}$$

In this example, we see that if we double the shaft speed of an existing system, then we double the volumetric flow rate and increase the pressure four times, and as a result, we have to increase the power supply eight times. The cubic nature of this relationship between power and the rotational shaft speed shows how even for small performance gains, large amounts of additional power are needed.

B) When Shaft Speed Held Constant: If N is constant, then a change in D can simplify the affinity laws to the following:

a) Volumetric Flow Rate: In this case, flow is proportional to the impeller diameter, and the volumetric flow rate can be expressed as

$$Q_1 / Q_2 = D_1 / D_2 \tag{2.244}$$

b) Head or Pressure: In this case, pressure or head is proportional to the square of the impeller diameter, and the head or pressure can be expressed as

$$dp_1 / dp_2 = (D_1 / D_2)^2 \tag{2.245}$$

c) Power: In this case, power is proportional to the cube of impeller diameter, and the power can be expressed as

$$P_1 / P_2 = (D_1 / D_2)^3 \quad (2.246)$$

These laws assume that the pump or fan efficiency remains constant—$\eta_1 = \eta_2$—which is rarely true; however, it can be a good approximation when used over appropriate frequency or diameter ranges. The exact relationship between speed, diameter, and efficiency depends on the particulars of an individual fan or pump design. Product testing becomes necessary if the acceptability range is not known, or if a high level of accuracy is required. Additionally, interpolation from accurate data sometimes provides more reliable predictions than do the affinity laws.

Example 2.95: Changing Impeller Diameter

In this example, the diameter of the pump impeller is reduced while the shaft speed is constant. The diameter is changed from 200 to 150 mm. Using the applicable affinity law equations, find the final Q_2, dp_2, and P_2.

Solution

The final Q_2 can be calculated using Equation (2.244):

$$Q_2 = Q_1 (D_2 / D_1)$$

$$Q_2 = 5\text{L/s}(150\text{mm} / 200\text{mm})$$

$$Q_2 = 3.75\text{L/s}$$

The final dp_2 can be calculated using Equation (2.245):

$$dp_2 = dp_1 (D_2 / D_1)^2$$

$$dp_2 = 30\text{m}(150\text{mm} / 200\text{mm})^2$$

$$dp_2 = 16.9\text{m}$$

The final P_2 can be calculated using Equation (2.246):

$$P_2 = P_1 (D_2 / D_1)^3$$

$$P_2 = 3.75\text{kW}(150\text{mm} / 200\text{mm})^3$$

$$P_2 = 1.58\text{kW}$$

Note that in this example, when D is reduced by 25%, Q is reduced 25% as well, dp is reduced 44%, and P is reduced approximately 58%. This means that if we use a 25% smaller diameter, then the power needed for the pump reduces by 58% while the volumetric flow rate is reduced by only 25%.

Example 2.96: Changing Shaft Speed

In this example, a 5 m³/s fan operates at 500 rpm using a 3.5 kW. This fan generates 25 mm of static pressure. The unit is installed, and the actual static pressure at the job site is found to be 37.5 mm. Using the applicable affinity law equations, find the new brake horsepower as well as the volumetric flow rate.

Solution

First, the new shaft speed (rpm) can be calculated by rearranging Equation (2.242):

$$N_2 = N_1 \sqrt{(dp_2 / dp_1)}$$

$$N_2 = 500\text{rpm}\sqrt{(37.5\text{mm} / 25\text{mm})}$$

$$N_2 = 612.4\text{rpm}$$

Now, the new P_2 can be calculated using Equation (2.243):

$$P_2 = P_1 (N_2 / N_1)^3$$

$$P_2 = 3.5\text{kW}(612.4\text{rpm} / 500\text{rpm})^3$$

$$P_2 = 6.4\text{kW}$$

In this example, we see that the increased static pressure after installing the fan at the job site increased N for the fan by 22.5% and increased the needed P by 83%. Again, the cubic nature of the relationship between power and rotational shaft speed shows that for small performance gains, large amounts of additional power are needed.

2.4.8.2.2 *Geometrically Similar Pumps*

a) Volume Capacity: The volumetric flow rate of a centrifugal pump can be expressed as

$$Q_1 / Q_2 = (N_1 / N_2).(D_1 / D_2)^3 \quad (2.247)$$

where Q is the volumetric flow rate (e.g., m³/s, L/s, gpm, cfm, etc.), N is the shaft speed (rpm), and D is the impeller diameter (e.g., mm, in, etc.).

b) Head or Pressure: The head or pressure of a centrifugal pump (m, ft, Pa, psi, etc.) can be expressed as

$$dp_1 / dp_2 = (N_1 / N_2)^2 .(D_1 / D_2)^2 \quad (2.248)$$

c) Power: The power of a centrifugal pump (W, bhp, etc.) can be expressed as

$$P_1 / P_2 = (N_1 / N_2)^3 .(D_1 / D_2)^5 \quad (2.249)$$

Overall, the affinity laws are used to estimate the effect of speed and impeller diameter on pump or fan performance. Therefore, it should be kept in mind that these relationships are solely to be used on pump or fan curve predictions or friction-only system curves. Using them to directly determine the effects of impeller or speed changes for system curves with static head can result in prediction errors.

For friction-controlled curves, the performance and system curves intersect at matching points at both speeds. This confirms that the affinity laws (for systems with no static head) can be used to forecast where the pump or fan and system will operate. The affinity laws, however, cannot predict how the pump or fan and system interact under new conditions if static head is present in the mix.

2.4.9 Essentials of Gas Transfer

2.4.9.1 Gas Transfer

Gas transfer, that is, exchange of gases between aqueous and gaseous phases, is an essential element for many environmental processes. Aquatic systems and wastewater treatment plants, for instance, require enhanced transfer of oxygen to sustain aquatic life and maintain aerobic degradation. Exchange of a dissolved compound with the atmosphere such as oxygen is controlled by the degree of mixing in the aqueous and gaseous phases, the compound concentration in the two phases, the compound's equilibrium distribution, and the interface surface area. Throughout the years, we have developed technologies to enhance gas transfer, which include aeration diffusers, packed-tower air stripping, and membrane stripping. Each of these technologies basically creates a high interface surface area to enhance gas transfer.

2.4.9.2 Dissolved Gases and Their Solubilities in Water

Gases dissolve in liquids to form solutions. To understand why gases dissolve in water, we first need to remember Dalton's law, which tells us that the pressure a gas exerts in a fixed volume is dependent on the total pressure of the gas mixture and the individual concentration of the gas in the mixture. This is an important relationship because it is used to create the driving force to remove or dissolve gases in water and can be expressed as follows:

$$P_t = P_a + P_b + .. + P_n \tag{2.250}$$

where P_t is the total pressure of the gas mixture, P_a is the partial pressure of the gas a, P_b is the partial pressure of gas b, and P_n is the partial pressure of gas n. The partial pressure of a is expressed as $P_a = P_t.C_a$, where C_a is the concentration of a in the mixture.

Gas dissolution in liquids is a process for which we can write an equilibrium constant. For instance, the equilibrium between oxygen gas and dissolved oxygen in water is $O_2(\text{aq.}) \leftrightarrow O_2(\text{gas})$. The constant for this equilibrium is $K = P_{O_2} / C_{O_2}$. The form of the equilibrium constant shows that the concentration of a solute gas in solution is directly proportional to the partial pressure of that gas above the solution provided no chemical reaction takes place between the liquid and the gas. This is the next relationship we need, and it is known as Henry's law.

This relationship is generally valid at low dissolved gas concentrations. As concentrations and partial pressures increase, we observe deviations from Henry's law, just as gases also deviate from the ideal gas law as pressures increase and temperatures decrease. Stating the pressure–concentration ratio as an equation and use of the usual modern symbol (H_a) for the constant on a concentration basis, Henry's law is expressed as follows:

$$C_a = P_a / H_a \tag{2.251}$$

where C_a is the mole fraction (mol/L) of the dissolved gas, P_a is the partial pressure of the gas (atm) in the total gas mixture (e.g., atmosphere), and H_a is the Henry's law constant (atm/(mol/L)) for the given gas at the system temperature. In the technical literature, there are various other forms of Henry's law; therefore, with respect to the use of Henry's law constants, readers of the technical literature must be very careful to note which version of the equation is being used.

Henry's law constants for many gases in many different solvents at different temperatures have been measured and tabulated. For the Equation (2.251) version of Henry's law, at 25°C, H_{O_2} is 769.23 atm/(mol/L), H_{CO_2} is 29.41 atm/(mol/L), H_{H_2} is 1282.05 atm/(mol/L), and H_{N_2} is 1639.34 atm/(mol/L). The inverse of the constant, multiplied by the partial pressure of the gas above the solution, is called the molar solubility of the gas. Thus, oxygen at one atmosphere and at 25°C has a molar solubility of [1 atm/769.23 atm/(mol/L) = 1.3 × 10^{-3} mol/L] or 1.3 mmol/L.

The Henry's law constant generally increases with increasing temperature. Hence, the solubility of gases generally decreases with increasing temperature. For instance, when water is heated on a stove, the gas bubbles appear on the sides of the pan well below the boiling point of water, and as a consequence, dissolved gases leave water because it can no longer contain the amount of air it did before at lower temperatures. Adding boiled or distilled water to a fish tank, therefore, will kill the fish due to suffocation unless the water is re-aerated before adding the water to the tank.

Another example of Henry's law is carbonated drinks. The drink itself contains dissolved carbon dioxide. The gas above the drink before opening the bottle is almost pure carbon dioxide at a pressure slightly higher than atmospheric pressure. Since the pressure above the liquid is lower after opening the bottle, part of the dissolved carbon dioxide comes out of liquid as bubbles and escapes. If the drink is left in the open for a while, the concentration of carbon dioxide in the drink will eventually come into equilibrium with the carbon dioxide in the air.

Example 2.97: Henry's Law—Calculating Dissolved O_2 Concentration

Using Henry's law, find the amounts of dissolved oxygen and dissolved carbon dioxide in air-saturated water under normal atmospheric conditions at 25°C.

Solution

Normal atmospheric conditions are 20.946 mol% oxygen (Table A.45), which makes the partial pressure of oxygen (P_{O_2}) 0.20946 atm, and Henry's law constant for oxygen (H_{O_2}) at 25°C is 769.23 atm/(mol/L).

Using Henry's law (Equation (2.251)), the concentration of oxygen (C_{O_2}) can be calculated as

$$C_{O_2} = P_{O_2} / H_{O_2} = 0.20946 \text{ atm} / \left[769.23 \text{ atm} / (\text{mol} / \text{L})\right]$$

$$C_{O_2} = 2.723 \times 10^{-4} \text{ mol} / \text{L} = 0.2723 \text{ mmol} / \text{L}$$

Again, normal atmospheric conditions are 0.033 mol% carbon dioxide (Table A.45), which makes P_{CO_2} 0.00033 atm, and H_{CO_2} at 25°C is 29.41 atm/(mol/L).

Using Henry's law, the concentration of C_{O_2} can be calculated as

$$C_{CO_2} = P_{CO_2} / H_{CO_2} = 0.00033\,\text{atm} / [29.41\,\text{atm}/(\text{mol}/\text{L})]$$

$$C_{CO_2} = 1.1\times10^{-5}\,\text{mol}/\text{L} = 0.011\,\text{mmol}/\text{L}$$

Henry's law tells us that the amount of gas that will dissolve in water is proportional to its partial pressure in the total gas mixture. By increasing or decreasing the partial pressure of a gas in contact with water, we can increase or decrease the amount of gas that will dissolve in the water. It is evident from Dalton's law that the partial pressure of a specific gas can be lowered by lowering the total pressure of the gas mixture, the concentration of the specific gas in that mixture, or a combination of these two.

The solubility in water of a single component of a mixture of gases such as air depends not only on the properties of the particular gas being considered but also on the temperature, salinity, mole fraction of the given gas in the mixture, and the total pressure. As discussed, the solubility of gases in water is predicted quite well by Henry's law as long as the partial pressure of the gas is less than one atmosphere, and the Henry's law constant depends on the gas and on system temperature.

The standard gas composition of air exclusive of water vapor is provided in Table A.45. Air is approximately 78% nitrogen, 20.9% oxygen, 0.033% carbon dioxide, and 0.934% argon along with a number of other gases at very low concentrations. Nitrogen and oxygen have relatively low solubilities in water; carbon dioxide is about 41 times more soluble than oxygen on a weight per volume basis, but since CO_2 is a small component of air, it is not usually present in high concentrations in water. Table A.46 lists the solubility of the major atmospheric gases in water at 15°C. As the table shows, dissolved gases in water are typically reported as mass per volume (mg/L); however, dissolved gas concentrations can also be given as percent of saturation, or gas tension, which is the partial pressure of the dissolved gas in the liquid phase.

As will be discussed shortly, the solubility of a gas on a mass per volume basis (mg/L) decreases with increasing temperature and salinity. The solubility of a gas increases proportionally as the total pressure increases; for instance, the solubility of oxygen in water can be increased by increasing the pressure within an aerator (e.g., by introducing air in a submerged diffuser whereby the total pressure is increased by the hydrostatic pressure of the water). The solubility of a gas under increased hydrostatic pressure can be determined using the following equation:

$$C_h = C\,(P_{ambient} + \rho gh - P_w)/(760 - P_w) \qquad (2.252)$$

where C_h is the solubility of gas at depth h (mg/L), C is the solubility of gas at 1 standard atmosphere (mg/L), $P_{ambient}$ is the ambient atmospheric pressure (mm Hg), ρg is the specific pressure of water per meter depth (73.22 mm Hg/m), h is the depth of water (m), and P_w is the vapor pressure of water (mm Hg).

2.4.9.2.1 Dissolved Oxygen

2.4.9.2.1.1 Solubility of Oxygen in Water

There are two major sources of dissolved oxygen (DO) in water: i) photosynthesis of aquatic plants and plankton and ii) oxygen transfer from the atmosphere. As mentioned earlier, oxygen forms approximately 21% by volume of the atmosphere; however, it is not quite soluble in water (Table A.45).

The saturation oxygen content of water is a function of temperature and salinity. As mentioned earlier, the oxygen-holding capacity of water decreases as temperature increases above 0°C. Increasing salt concentration in water decreases the dissolved saturation oxygen capacity of water, as well. Table A.47 provides the solubility of oxygen in water exposed to air at 760 mm Hg (1 atm) pressure for temperatures from 0 to 30°C and for increasing salt concentration. Water containing less oxygen than its equilibrium capacity for a given temperature and salinity is called undersaturated, while water containing more oxygen than its equilibrium capacity is called supersaturated. We can also express oxygen concentration in terms of percentage saturation.

Example 2.98: Percent Saturation

Using the data from Table A.47, calculate the percent saturation of three levels of DO in freshwater at 10 °C and 1 atm pressure. The three DO concentrations to be expressed as percent of saturation are a) 6.0 mg/L, b) 11.3 mg/L, and c) 12.7 mg/L.

Solution

From Table A.47, the DO concentration for pure water at 10°C and 1 atm is 11.3 mg/L. We can calculate percent saturation of DO using the following equation:

$$\%\ \text{Saturation of DO} = \left(\frac{\text{Actual DO}(\text{mg}/\text{L})}{\text{Saturated DO}(\text{mg}/\text{L})}\right)\times100 \qquad (2.253)$$

a) Using Equation (2.253), let's convert 6.0 mg/L DO to percent saturation:

$$\%\ \text{Saturation of DO} = \left(\frac{6\,\text{mg}/\text{L}}{11.3\,\text{mg}/\text{L}}\right)\times100 \cong 53\%$$

(undersaturation)

b) Now let's convert 11.3 mg/L DO:

$$\%\ \text{Saturation of DO} = \left(\frac{11.3\,\text{mg}/\text{L}}{11.3\,\text{mg}/\text{L}}\right)\times100 = 100\%$$

(saturation)

c) And now let's convert 12.7 mg/L DO.

$$\%\ \text{Saturation of DO} = \left(\frac{12.7\,\text{mg}/\text{L}}{11.3\,\text{mg}/\text{L}}\right)\times100 \cong 112\%$$

(supersaturation)

Example 2.99: Oxygen Partial Pressure or Oxygen Tension

Dissolved oxygen content in water can also be expressed in terms of oxygen partial pressure or oxygen tension.

Oxygen tension is the partial pressure of atmospheric oxygen that is required to hold a given concentration of dissolved oxygen in water. Table A.45 gives the standard volumetric fraction of oxygen in the atmosphere as 0.2095. The oxygen tension in air at an atmospheric pressure of 760 mm Hg is 159.23 mm Hg (760 mm Hg × 0.2095). Determine the oxygen tension of pure water at 10°C and 1 atm for the DO values in Example 2.98.

Solution

The DO values can be calculated using the following equation:

$$O_2\text{Tension} = \left(\frac{\text{Actual DO (mg/L)}}{\text{Saturated DO (mg/L)}}\right) \times \left[\text{Atm. } O_2\text{Tension (mm Hg)}\right] \quad (2.254)$$

a) Using Equation (2.254), let's convert 6.0 mg/L DO in water to O_2 tension:

$$O_2\text{Tension} = \left(\frac{6\text{ mg/L}}{11.3\text{ mg/L}}\right) \times [159.2\text{ mm Hg}]$$
$$\cong 84.5\text{ mm Hg}$$

b) Now let's convert 11.3 mg/L DO:

$$O_2\text{Tension} = \left(\frac{11.3\text{mg/L}}{11.3\text{ mg/L}}\right) \times [159.2\text{ mm Hg}]$$
$$= 159.2\text{ mm Hg}$$

c) And now let's convert 12.7 mg/L DO:

$$O_2\text{Tension} = \left(\frac{12.7\text{ mg/L}}{11.3\text{ mg/L}}\right) \times [159.2\text{ mm Hg}]$$
$$\cong 178.9\text{ mm Hg}$$

Gas transfer rate can be modeled as the product of a driving force (the difference between the equilibrium concentration and the actual concentration) and an overall volumetric gas transfer coefficient (a function of the geometry, mixing levels of the system, and the solubility of the compound). In equation form, it can be expressed as follows:

$$\frac{dC}{dt} = K_L.a.(C_s - C) \quad (2.255)$$

where dC/dt is the rate of change in gas concentration (mg/L.h), $K_L.a$ is the overall volumetric gas transfer coefficient (1/h), C_s is the equilibrium dissolved gas concentration at saturation (mg/L), and C is the existing dissolved gas concentration in the liquid at any point in time (mg/L).

$K_L.a$ is a function of the interface surface area, the liquid volume, the oxygen diffusion coefficient in water, and the thickness of the laminar boundary layer through which the gas must diffuse before the much faster turbulent mixing can disperse the dissolved gas; it is equal to $K_L(A/V)$, where K_L is the gas transfer coefficient (m/h), A is the area of gas/liquid interface (m^2), and V is the liquid volume (m^3). $K_L.a$ is system specific and therefore must be evaluated separately for each system of interest (Weber and Digiano, 1996). It increases with the interface area and the diffusion coefficient and decreases with the liquid volume and the thickness of the boundary layer. The functional form of this relationship is given by $K_La = AD/V\delta$, where D is the oxygen diffusion coefficient in water and δ is the thickness of the laminar boundary layer.

2.4.9.2.1.2 Oxygen Transfer to Water The O_2 transfer rate equation can be written as Equation (2.255) and governed by the same criteria as expressed. Equation (2.255) can then be integrated with appropriate initial conditions to obtain the O_2 concentration as a function of time. However, care must be taken to ensure that K_La is not a function of the dissolved oxygen concentration. This dependency can occur where air is pumped through diffusers on the bottom of, for instance, activated sludge tanks; rising air bubbles are significantly depleted of O_2 as they rise through the activated sludge tank, and the extent of O_2 depletion is a function of the O_2 concentration in the activated sludge. Integrating Equation (2.255) with initial conditions of $C = C_0$ at $t = t_0$:

$$\int_{C_0}^{C} \frac{dC}{C_s - C} = \int_{t_0}^{t} K_L a\, dt \quad (2.256)$$

$$\ln\left[\frac{C_s - C}{C_s - C_0}\right] = -K_La(t - t_0) \quad (2.257)$$

The simple gas transfer model given in Equation (2.257) is acceptable when the gas transfer coefficient is independent of the dissolved gas concentration, which is met in systems where the gas bubbles do not change concentration significantly as they rise through the water column. This condition is met when the water column is shallow, the bubbles have large diameters, or the difference between the concentration of dissolved gas and the equilibrium concentration is small. Equation (2.257) can be evaluated using linear regression so that K_La is determined from the slope of the straight line as follows:

$$K_La = \frac{\ln(C_s - C_0) - \ln(C_s - C)}{(t - t_0)} \quad (2.258)$$

In actual aeration systems, it can be difficult to directly measure K_L and A (the surface area of air/water interface area). In practice, K_La can be measured in a test apparatus under standard conditions for a given aerator system. The measured K_La can then be adjusted to apply to nonstandard conditions in actual operating aeration systems. If the change in DO concentration is measured over time in a standard test system, Equation (2.258) plotted as a graph will produce a straight line with slope of ($-K_La$). The graph will have a *y*-axis showing the natural logarithm of O_2 deficit [$\ln(C_s - C)$] as a function of time (*x*-axis). The slope of the straight line will be the K_La.

In the equations we have been using, for oxygen, C_s is equal to the concentration in gas–liquid interface, which is equal to the saturation O_2 concentration. C_s is determined by the partial pressure of oxygen in the gas phase (P_{O_2}) and the liquid's Henry's constant for oxygen (H_{O_2}): $C_s = P_{O_2}/H_{O_2}$. A liquid's H_{O_2} is dependent on temperature and the concentration of impurities.

2.4.9.2.1.3 Factors Influencing O_2 Transfer The main factors that influence the rate of O_2 transfer into water are i) the differential gas deficit gradient, ii) water temperature, iii) impurities in water, and iv) the area of gas–liquid interface.

The differential gradient between the partial pressure of O_2 in the gas phase and that in the liquid phase is the driving force for diffusion across the gas/liquid interface. In other words, if DO in water is near saturation level, there is little driving force to absorb more O_2, and if the DO tension of water exceeds the partial pressure of O_2 in the gas phase (supersaturation), then DO leaves the water and is transported into the gas phase.

Temperature affects the oxygen-holding capacity of water; as water temperature increases, the solubility of O_2 decreases. Raising water temperature at a given level of DO reduces the differential gradient. Impurities in water can affect the surface tension at the gas/liquid interface and hence change the rate of diffusion and change the saturation holding capacity for DO in water.

The surface area of the gas/liquid interface is the most important parameter affecting the rate of O_2 uptake by water. The rate of O_2 transport is increased by increasing the surface area of the air/water boundary. Thus, introducing air into water as small bubbles instead of large, breaking water into small droplets, or creating turbulence increases the area of gas/liquid interface. A large interfacial area between air and water promotes rapid exchange of DO between the air and water. This explains why turbulence, mixing, or introducing air into water as small bubbles increases the rate of oxygen transfer.

For instance, if the same volume of air is introduced into water as one thousand small bubbles instead of one large bubble, the air/water interfacial area created by the small bubbles will be much larger, promoting quicker uptake of DO. In fact, the total surface area of the small bubbles is ten times larger than the surface area created by the single large bubble of equivalent total air volume. It is important to have large surface areas of contact between air and water and have good circulation of water to rapidly increase DO in water bodies by transfer from atmospheric air. The diffusion rate of oxygen into water beyond the surface air/water boundary layer is very slow. For instance, to increase the DO from 0 to 0.4 ppm at a depth of 10 m by diffusion alone would require about 600 years (Wheaton, 1977).

2.4.9.2.2 Carbon Dioxide

2.4.9.2.2.1 Sources and Sinks of CO_2 in Aquatic Systems CO_2 comprises approximately 0.033% by volume of the atmosphere (Table A.45). Therefore, it is a relatively minor component of the atmosphere; however, it is the major medium for transport of carbon in Earth's carbon cycle. Ocean waters contain more than 50 times the amount of inorganic carbon as the atmosphere. Carbonate rock and organic carbon in Earth's crust contain almost 40,000 times the amount of carbon as the atmosphere.

CO_2 is the major carbon source for photosynthesis and also a major byproduct of respiration of living organisms. CO_2 and water together contain three of the most important elements forming living matter: carbon, hydrogen, and oxygen. As mentioned earlier, CO_2 provides a natural bicarbonate buffering system for natural bodies of water. The major sources of CO_2 in natural waters include: i) direct absorption from the atmosphere; ii) absorption into raindrops falling in the atmosphere; iii) the respiration of aquatic organisms; iv) aerobic and anaerobic decomposition of organic matter, and iv) chemical reactions in water between acids and carbonate compounds. The major sinks for CO_2 in aquatic systems i) the demand for CO_2 by the photosynthetic processes of plants and plankton and ii) return of CO_2 to the atmosphere from water that is supersaturated in dissolved CO_2.

2.4.9.2.2.2 Solubility of CO_2 in Aquatic Systems CO_2 is more soluble in water than oxygen. As discussed earlier, the solubility of a gas in water is a direct function of the partial pressure of that gas in equilibrium with the water, that is, Henry's Law.

Boyd (1982) provided the solubility of carbon dioxide in pure water at different temperatures. If the values provided by Boyd are compared with the levels of dissolved O_2 at saturation in water that is in equilibrium with air at 1 atm, then we will see that the absolute amounts of saturated DO are greater, but the partial pressure of O_2 at 1 standard atm is (760 mm Hg × 0.2095% O_2) 159.2 mm Hg; in contrast, that of CO_2 (760 mm Hg × 0.033% CO_2) is 0.251 mm Hg. That is, the normal atmospheric partial pressure of oxygen is 634 times that of CO_2.

Comparing Boyd's (1982) values of dissolved CO_2 with those of dissolved O_2 in Table A.47 shows that the solubility of CO_2 is much greater than that of O_2. For instance, let's now consider the equilibrium concentrations of O_2 and CO_2 in pure water at 20°C. From Table A.47, DO is 9.2 mg/L, which is equal to 2.88×10^{-4} mol/L, and dissolved CO_2 is 0.56 mg/L, which is equal to 1.22×10^{-5} mol/L. The mole ratio of DO to CO_2 is 23.6, which is much less than the difference in atmospheric partial pressures. This shows that the solubility of CO_2 in water decreases as temperature and salinity increase.

2.4.9.2.2.3 CO_2 and Bicarbonate Buffering Carbonate buffering was discussed earlier; as shown in Equation (2.286), when CO_2 dissolves in water, a fraction of the dissolved CO_2 gas reacts with water to form carbonic acid (H_2CO_3), and H_2CO_3 in water strongly dissociates as hydrogen $\left[H^+\right]$ and bicarbonate [HCO_3^-] ions (Equation (2.187)). The equilibrium constant for H_2CO_3 dissociation in water at 25°C is $K_{eq1} = 10^{-6.35}$:

$$K_{eq1} = \frac{\left[H^+\right] + \left[HCO_3^-\right]}{\left[H_2CO_3\right]} = 10^{-6.35} \quad (2.259)$$

where K_{eq1} is the equilibrium constant for dissociation of carbonic acid (mol/L), $\left[H^+\right]$ is the hydrogen ion concentration (mol/L), $\left[HCO_3^-\right]$ is the bicarbonate ion concentration (mol/L), and $[H_2CO_3]$ is the carbonic acid concentration (mol/L).

HCO_3^- are more weakly dissociated in water (Equation (2.188)) to form $\left[H^+\right]$ and carbonate ions $\left[CO_3^{2-}\right]$. The equilibrium constant for (2.188) is expressed as follows at 25°C:

$$K_{eq2} = \frac{[H^+]+[CO_3^{2-}]}{[HCO_3^-]} = 10^{-10.33} \quad (2.260)$$

Equations (2.187) and (2.188) indicate that the equilibrium concentrations of H_2CO_3 (or CO_2), HCO_3^-, and CO_3^{2-} in water are all influenced by pH (the hydrogen ion concentration). Given an initial concentration of dissolved CO_2, Equations (2.259) and (2.260) can be used to determine the relative concentrations of H_2CO_3, HCO_3^-, and CO_3^{2-} in water as a function of pH as shown in the curves in Figure 2.32.

Using Equation (2.259), the percentage concentrations in water on a molar basis of carbonic acid and bicarbonate ions can be calculated as presented in Table A.48. The percentage concentrations were calculated assuming that all of the dissolved carbon dioxide was present either as bicarbonate ion or as carbonic acid at pH less than 8.5. Furthermore, using Equation (2.260), the percentage concentrations in water on a molar basis of bicarbonate ions and carbonate ions can be calculated as presented in Table A.49. The percentage concentrations were calculated assuming that all of the dissolved carbon dioxide was present either as bicarbonate ion or as carbonate ion at pH greater than 8.5.

Again, as discussed earlier, the equilibrium between carbonic acid and bicarbonate and carbonate ions forms an important buffering system in natural bodies of water. If an acid is added to water, bicarbonate ions react with the added hydrogen ions to produce more dissolved carbonic acid and carbon dioxide, which is in equilibrium with atmospheric CO_2; this effectively prevents increases in hydrogen ion concentration and keeps pH from dropping. Buffering will continue until the bicarbonate ions are all used. If base is added to water, hydrogen ions will be removed by reaction with the added OH ions, and more of the bicarbonate ions in Equation (2.188) will dissociate, increasing the concentration of carbonate ions and thus preventing the increase of pH. Buffering will continue until most of the bicarbonate is dissociated to carbonate ions.

2.4.9.2.2.4 CO_2 Levels in Freshwater and Saltwater Lakes and streams in areas of predominantly igneous bedrock tend to have low carbon dioxide levels. Igneous rocks are insoluble, and there tends to be low levels of nutrients and relatively low levels of biological activity in waters associated with igneous rock.

Low levels of biological growth in these waters means that CO_2 as a byproduct of respiration is not in abundant supply. On the other hand, waters in areas that have predominantly sedimentary bedrock tend to have high levels of carbonate and bicarbonate. Waters with low pH have most of their soluble carbon in the form of dissolved CO_2, and waters in areas of igneous rock can have pH values less than 6. Waters with higher pH between 6 and 9 can contain a high level of bicarbonate ions in solution in equilibrium with dissolved CO_2. These waters are better buffered against pH fluctuations and tend to be more biologically active.

Flowing waters do not tend to have annual CO_2 concentration cycles. The chemical composition of the watershed area is the main determinant of pH in inland lakes and rivers. If pH is low, most of the inorganic carbon in water is present as CO_2; if pH is near neutral or above, bicarbonate concentrations are higher, and good buffering capacity exists.

Moreover, CO_2 concentrations in water decrease with increasing salinity or temperature. The presence of excess base in saltwater causes it to be more buffered than freshwater; the pH for ocean water ranges from 7.5 to 8.4. Cold polar surface waters have a greater capacity for CO_2 than warm waters. In ocean circulation, sinking cold polar water circulates to replace warmer tropical waters. The overall effect is that dissolved CO_2 is transported from polar regions to tropical regions. The saltwater buffering system increases the ability of salt water to absorb CO_2.

Brackish waters occur where river waters meet and mix with seawater. The CO_2 levels in freshwater are variable and mainly dominated by biological activity, temperature, and

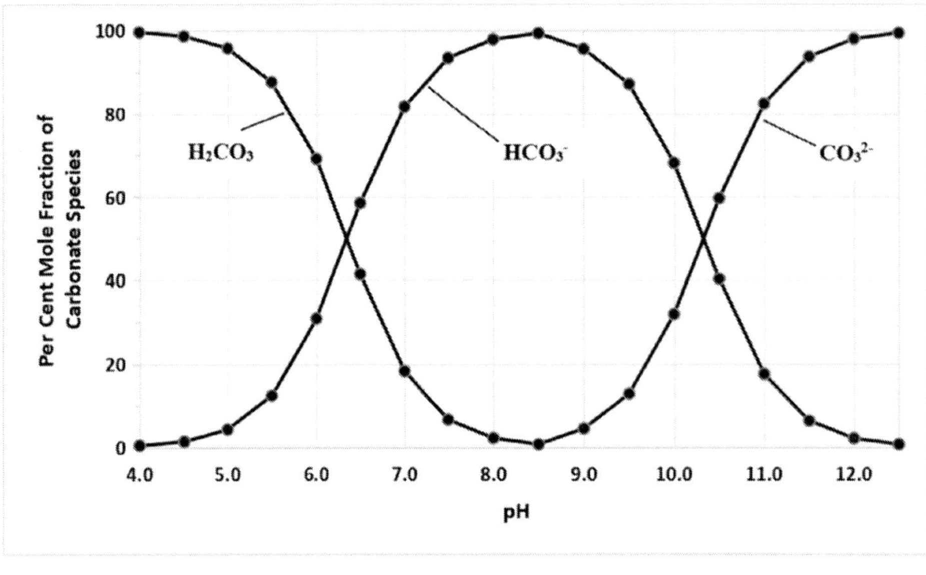

FIGURE 2.32 Relationship between mole fractions of H_2CO_3, HCO_3^-, and CO_3^{2-} as a function of pH.

pH. Seawater has lower solubility of CO_2 but has high holding capacity for bicarbonate and carbonate ions.

2.4.9.2.3 Nitrogen

Nitrogen (N_2) forms 78% by volume of the gas mixture in air (Table A.45). Nitrogen is relatively inert and does not directly take part as a nutrient in aquatic plant or animal growth. A few microorganisms can use N_2 directly for incorporation into organic cellular components, and a small group of anaerobic bacteria can convert nitrate to N_2. The main source of dissolved N_2 in water is the atmosphere, and the main mechanism for removal of dissolved N_2 from water is by transfer back to the atmosphere or by exposure to air/water interfaces under conditions of supersaturation. Natural waters contain more dissolved N_2 than O_2 because the partial pressure of N_2 in the atmosphere is 3.73 times greater than that of O_2.

2.4.9.2.4 Ammonia

The main sources of ammonia (NH_3) in natural waters are the bacterial decomposition of proteins and the deamination of amino acids, which are constituents of organic detritus. Fish also excrete NH_3 directly. Agricultural fertilizer runoff is another source of NH_3 in water bodies. Since NH_3 is a product of decomposition of organic matter, concentrations will vary over depth in natural lakes and ponds. The well-aerated surface layers of water with high levels of photosynthetic activity will take up NH_3 as a nutrient; there will be more NH_3 in the lower layers of the lake where decomposition processes outweigh photosynthesis. These concentrations of NH_3 are evenly distributed in lakes during the fall and spring turnovers.

In water, dissolved NH_3 gas exists in equilibrium with ammonium ions (NH_4^-), according to the following equation:

$$K_{eq} = \frac{[NH_4^+]+[OH^-]}{[NH_3]} = 10^{-4.74} \quad (2.261)$$

The equation shows that as pH increases, the concentration of OH^- increases, so that the concentration of the unionized form of NH_3 increases. Table A.50 presents the percentage of unionized ammonia in aqueous solutions at different temperatures as a function of pH. Thus, for instance, water containing 3.0 mg/L of total ammonia nitrogen at pH of 7 and temperature 20°C would contain (3.0 mg/L × 0.004) = 0.012 mg/L of NH_3, and if the pH is raised to 9.4, then the NH_3 concentration becomes (3.0 mg/L × 0.50) = 1.5 mg/L of NH_3.

2.4.9.2.5 Methane and Hydrogen Sulfide

Aerobic decomposition of organic matter by microorganisms produces CO_2 as the main form of mineralized carbon, according to the following overall biological growth equation:

$$\text{Organic Matter} + O_2 \rightarrow CO_2 + H_2O + \text{New Cells} \quad (2.262)$$

The anaerobic decomposition of organic matter by microorganisms produces both CH_4 and CO_2 as carbon-based byproducts as follows:

Anaerobic Acidogenesis

$$\text{Organic Matter} \xrightarrow{\text{Acid-forming bacteria}} \text{Volatile Fatty Acids} + CO_2\uparrow + H_2S\uparrow + H_2O \quad (2.263)$$

Anaerobic Methanogenesis

$$\text{Organic Acids} \xrightarrow{\text{Methane bacteria}} CH_4\uparrow + CO_2\uparrow \quad (2.264)$$

The gases produced by anaerobic degradation of organic matter typically consist of 50% to 60% CH_4, 30% to 40% CO_2, and small percentages of H_2S and other trace gases. CH_4 is only produced under anaerobic conditions. Anaerobic decomposition can take place at a reduced rate down to about 5°C in freshwater lakes and rivers, and therefore, CH_4 is more likely to be produced in deeper unaerated water.

Hydrogen sulfide (H_2S) is another byproduct gas formed during anaerobic decomposition of organic matter. H_2S in water dissociates into HS^- ions, which can further dissociate into S^{2-} ions according to the following reactions:

$$H_2S \leftrightarrow HS^- + H^+ \quad (2.265)$$

$$HS^- \leftrightarrow S^{2-} + H^+ \quad (2.266)$$

The equilibrium constant for Equation (2.265) is $K_{eq1} = 10^{-7.01}$, so that

$$K_{eq1} = \frac{[HS^-]+[H^+]}{[H_2S]} = 10^{-7.01} \quad (2.267)$$

The equilibrium constant for Equation (2.266) is $K_{eq2} = 10^{-13.89}$, so that

$$K_{eq2} = \frac{[S^{2-}]+[H^+]}{[HS^-]} = 10^{-13.89} \quad (2.268)$$

Obviously, pH regulates the distribution of unionized and ionized hydrogen sulfide species. The fraction of H_2S existing as unionized H_2S decreases with increasing pH. This behavior is opposite to the dissociation of NH_3, which becomes less ionized with increasing pH. Boyd (1982) reported data on the percentage of H_2S that exists as the unionized form in 25°C aqueous solution at different pH values.

2.5 LIGHT, RADIATION, AND CHARACTERISTICS OF SOLAR RADIATION

2.5.1 Introduction

Solar energy powers virtually everything on Earth and in the atmosphere. This energy warms the air and Earth's surfaces and drives the winds, currents, evaporation, clouds, rain, etc. Essentially, the process starts when the Sun's energy in the form of electromagnetic radiation (radiation, in short) enters the atmosphere. The seasonal distribution of this energy obviously depends on the orbital characteristics of the Earth revolving around the Sun. Solar radiation is significantly impacted by the inclination of the Earth's axis. This axis (the line joining the two poles through the center of the globe) is tilted 23.5° from the perpendicular. The axis maintains the same orientation with respect to the galaxy; therefore, the amount of incoming solar radiation (or "insolation") at the top of the atmosphere, hence at the surface of the Earth, varies considerably creating seasons (by impacting the duration of daylight and elevation of the Sun in the sky with respect to time).

The solar spectrum covers the range of radiation from very short to very long wavelengths. The spectrum belonging to a blackbody at 6,000 K is pretty close to the incident radiation from the Sun at the top of the Earth's atmosphere. When the Sun's radiation passes through the atmosphere, it is reflected, scattered, and absorbed by dust particles, gas molecules, ozone, and water vapor. The magnitude of solar radiation's attenuation at a given time and location is determined by atmospheric composition and the length of the atmospheric pathway the solar radiation travels. Compared with scattering, atmospheric absorption of solar radiation is relatively small. Ozone's absorption of ultraviolet radiation is a significant component that is vital for sustaining life on Earth. In addition to the ultraviolet absorption, at other wavelengths, absorption by nitrous oxide, carbon dioxide, oxygen, ozone, and water vapor takes place as well.

The solar radiation reaching a surface on Earth has both direct and diffuse components. Energetically they act in the same way. Since diffuse radiation comes from the entire sky, it is difficult to predict its intensity as it varies as moisture and pollutant contents of the atmosphere change throughout any given day at any location. This radiant energy entering Earth's atmosphere is eventually transformed into a variety of other energy forms. As the radiation laws are already covered in the thermal energy section, this section begins with the introduction of light and solar physics and moves to the characteristics of solar radiation, covering it in depth with relevant numerical examples.

2.5.2 Light and Solar Physics

Newton was among the first to realize that a sunbeam is the composite of beams of different colors that comprise what is termed the visible spectrum. We have spent the last three centuries understanding the nature of light, and now we have a fairly complete model of the nature of light, or more generally speaking, of radiation.

Light is considered to behave as the composite waves of different wavelengths. Wavelength is the distance between two given points (let's say the peak) on one wave and the same point (the next peak) on the next wave in the same wavelength. The wavelengths of visible light are extremely short; therefore, we use a small unit to measure wavelength: the micrometer. A micrometer is one-millionth of a meter and is abbreviated as μm where μ stands for one-millionth, and m indicates "meter". In the visible spectrum, the red light has the longest wavelengths, while violet has the shortest. As will be discussed later in depth, the atmosphere scatters light differently based on wavelengths. In the visible spectrum, for instance, shorter-wavelength blue and violet light are more readily scattered than longer wavelengths. This is in fact why we see the sky as blue (the human eye is less sensitive to violet wavelengths).

Performed experiments also indicated that light energy behaves like a stream of particles. For instance, it is difficult to understand how sunlight travels through an empty space, as waves require a medium to travel in or propagate their energy. Ocean and sound waves for instance travel through water and air using the water and air molecules to propagate their energy, respectively; without the mentioned media, it would not be possible for them to exist as there would be no molecules to vibrate. Light waves, however, require no medium to propagate. Light is a stream of particles of radiant energy known as photons. Photons have no mass, occupy no space, and travel at the speed of light; in this sense, they are quite different from particles of matter. As will be presented further in this section, the amount of energy carried by an individual photon is extremely small and inversely related to wavelength: The shorter the light's wavelength, the higher the energy content of its photons. In the visible spectrum, for instance, violet light photons have more energy than red light photons.

As just presented, each of the wave and particle model describes certain behavior of light that the other model cannot. So, light is only explained by this wave-particle duality model. Another important aspect is that the photons are different from particles of matter. Countless photons are created and destroyed at any time; however, these creations and destructions do not violate the law of conservation of energy as they simply change forms through emission and absorption. Energized particles of matter (molecules, electrons, etc.), before discharging their excess energy fully or partially and returning to their lower energy levels, remain at energized or excited stage for a very short time. This discharging or shedding of excess energy is done by releasing a photon. Obviously, more energy interactions result in higher energy photons, therefore emissions at shorter wavelengths. The energy source leading to photon production can be in any form, such as nuclear, combustion, or electricity. In summary, a photon can be described as a bundle of energy emitted by an excited molecule, atom, or electron as it returns from a high energy level to a lower energy level. Molecules, atoms, and electrons (that is, the basic particles of matter) create and destroy photons, which form the basic particles of radiant energy.

Solar energy comes from the thermonuclear reactions in the core of the Sun, where most of the solar mass is concentrated (Figure 2.33). In the core, the temperature at a pressure of about 10×10^6 bar is estimated to be 15×10^6 K. This energy moves outwards by radiative diffusion in the interior section, and from there; it is transferred by convection to the photosphere. Then from this outer surface zone, the energy, mostly shortwave radiation, is emitted into space in all directions. The outer surface or photosphere has an average temperature of 5,780 K; however, it fluctuates due to the sunspots, which are relatively cooler regions with temperatures closer to 4,000 K. When it strikes a material, the radiation can be absorbed, transmitted, or reflected. The full discussion on the laws of radiation can be found in the thermal energy section of this book.

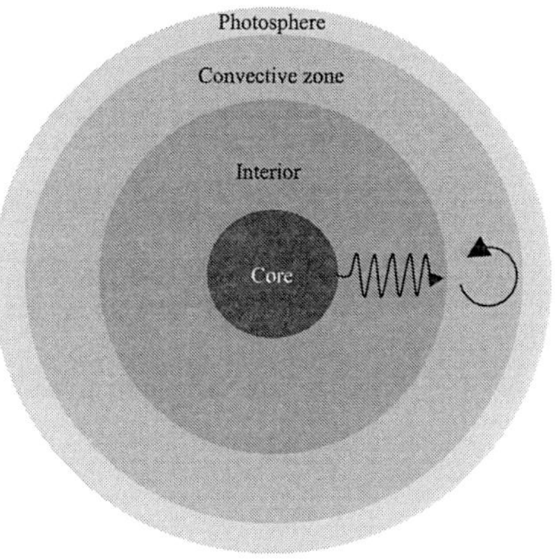

FIGURE 2.33 Main radial zones of the Sun (Taylor, 2005).

The Sun has a radius (696,000 km) approximately 100 times larger than that of the Earth, and its mass is estimated as 2 × 10^{30} kg, which is 300 million times bigger than that of the Earth. Just like a planet, it is believed that the Sun is 4.5 billion years old and is likely to last for 5.5 billion more years.

As noted earlier, solar energy comes from the thermonuclear reactions in the core of the Sun, where most of the solar mass is concentrated. Spectroscopic measurements of sunlight reaching the Earth from the photosphere (Figure 2.33) dictates that the solar mass is mainly composed of two elements, hydrogen, which makes up about 70% of solar mass, and helium (He), about 27%; the remaining 3% of solar mass is made up of all the other elements (McAlester, 1983). The reaction is:

$$4(^1H) \rightarrow 4He + \text{energy} + 2 \text{ neutrinos}$$

The conversion of H into He through solar fusion serves as the source of solar radiation (electromagnetic radiation) received on Earth. Every second, approximately 630 tons of H are estimated to be converted to 625 tons of He, with the remaining five tons released as mass equivalent energy ($E = m \times c^2$), which gives an approximate energy release of 4 × 10^{20} MJ every second.

2.5.3 Characteristics of Solar Radiation

This section starts with the solar radiation reaching the outer space of the atmosphere and analyzes the spectral distribution of this radiation. Then, variations in the amount of this radiation are studied, covering the impacts of different variables such as time, solar geometry, and atmospheric effects.

2.5.3.1 Solar Constant

It is useful to know a few solar energy terms. First, the amount of solar radiation reaching a horizontal surface is called insolation. The second term is power, the rate of energy flow. As presented earlier, the unit of power in SI is the watt. One watt is an energy flux of one joule per second, and therefore, the unit of time is built into the definition of watt. The third term is solar constant, the insolation at the top of Earth's atmosphere, that is, the amount of solar energy passing thorough unit area every second at the mean distance of the Earth (perpendicular to the Earth–Sun line). The term solar constant provides a convenient way of referring to the flux of energy radiated by the Sun.

Measurements over approximately the past 100 years show that the Sun's energy output is nearly constant. The measured flux is virtually the same. The solar constant varies only slightly with time and expressed in watts has a mean measured value of 1,368 W/m². The watt is the unit in which people buy electric energy for their households, greenhouses, and livestock and aquaculture buildings, which is typically given in the unit of kilowatt-hour (kWh). Ignoring the losses from the solar beam as it goes through the atmosphere, one square meter area of Earth's surface oriented toward the solar beam at about a 45° angle receives 1.368 kilowatt-hours of solar energy every hour. Considering that a kilowatt-hour is 1,000 watts of energy flow lasting for an hour, we can now imagine the amount of solar energy reaching a controlled environment roof or wall surface exposed to the Sun during a day.

Example 2.100: Calculating the Solar Constant

Using the measured solar surface temperature of 5,780 K, calculate the solar constant.

Solution

$$E^* = \left(5.679 \times 10^{-8} \text{ W/m}^2\text{K}^4\right)(5,780 \text{ K})^4$$

$$E^* = 63,384,522 \text{ W/m}^2$$

Using the surface area of the Sun (= $4\pi R_S^2$ = 6.09×10^{12} km²), we can calculate E^* as follows:

$$E^* = \left(63,384,522 \frac{\text{W}}{\text{m}^2}\right)\left(6.09 \times 1,018 \text{ m}^2\right)$$

$$E^* = 3.86 \times 10^{26} \text{ W}$$

Dividing this by the surface area of a sphere at the mean distance of Earth to the Sun ($4\pi R_{SE}^2$ = 2.83 × 10^{17} km² = 2.83 × 10^{23} m²), we can calculate the solar constant as

$$\text{Solar Constant} = \frac{E^*}{4\pi R_{SE}^2}$$

$$= \frac{3.86 \times 10^{26} \text{W}}{2.83 \times 10^{23} \text{ m}^2}$$

$$\text{Solar Constant} = 1,364 \frac{\text{W}}{\text{m}^2}$$

This calculated solar constant is pretty close to the mean measured value (by satellites) of 1,368 W/m² at an average distance (Earth's orbit radius), R_{SE}, of 1.495 × 10^8 km. Note that we employ the inverse-square law here (that is, radiation emitted from a spherical source decreases with the square of the distance from the center of the sphere, $E_2 = E_1 (R_1/R_2)^2$, where R is the radius from the center of the sphere).

If we multiply, let's say, the measured solar constant 1,368 W/m² by the projected area of the Earth (πR_E^2, where R_E is the mean radius of the Earth, 6,371 km), then using the area of a disk rather than the surface area of a sphere, we can find the intercepted solar radiation by the Earth as a whole at the top of the atmosphere every second approximately 1.74×10^{11} MJ. In reality, incoming radiation is the solar constant minus reflected sunlight times the area of the intercepting surface; that is, $[(1-A)(\text{Solar Constant})(\pi R_E^2)]$, where A is the global albedo (0.3), which will be discussed further in depth later. Overall, the top of the atmosphere receives about 4.5×10^{-10} of the total energy output of the Sun.

Example 2.101: Calculating the Intercepted Solar Radiation by the Earth

Determine the intercepted solar radiation by Earth as a whole at the top of the atmosphere every second.

Solution

As discussed,

$$\text{Intercepted Solar Radiation} = (1-A)(\text{Solar Constant})(\pi R_E^2)$$

where albedo A is 0.3, the measured solar constant is 1,368 W/m², and the mean radius of the Earth is 6.371×10^6 m.

$$\text{Intercepted Solar Radiation} = (1-0.3)(1368\,\text{W/m}^2)(\pi[6.371 \times 10^6\,\text{m}]^2)$$

$$\text{Intercepted Solar Radiation} = 1.22 \times 10^{11}\,\text{MJ}$$

2.5.3.2 Solar Spectrum

Figure 2.34 shows the complete set of possible wavelengths, also known as the electromagnetic spectrum. The solar spectrum covers the range of radiation from very short wavelengths (high frequency) to very long wavelengths (low frequency), as shown in Figure 2.35. The smooth spectrum (dashed line) in the figure belongs to a blackbody at 6,000 K, and this spectral shape is pretty close to the incident radiation from the Sun at the top of the atmosphere. The observed deviations are mostly in the high-energy (ultraviolet) region ($\lambda < 0.4$ μm), where the molecular bands responsible for ozone production are found.

The wavelength of peak radiation emission at approx. 0.5 μm actually suggests the Sun to have a blue-green color; however, we see the Sun as yellow-white due to the radiation's interaction with the Earth's atmosphere and the sensitivity of the human eye (maximum sensitivity is about 0.55 μm). And as shown in Figure 2.35, the effect of the Earth's atmosphere is to reduce the total insolation at the atmosphere to the magnitudes reaching sea level. The solar radiation emission mainly falls in the wavelength region of 0.15 to 4.0 μm, with about 9% in the form of ultraviolet radiation ($\lambda < 0.40$ μm), 38% in the visible (0.40–0.70 μm), and 53% in the near infrared (0.7–4.0 μm).

Figure 2.35 shows that at very short wavelengths, the Sun emits a relatively small amount of radiation and that as we move to the right towards the visible spectrum, the emission increases sharply, reaching a maximum at about the middle of the visible range. However, the visible spectrum is still just a segment in a much broader range of solar radiation. Earlier we called this electromagnetic radiation because it has both electric and magnetic properties. The properties of light presented earlier in this section apply to radiation throughout the electromagnetic spectrum and solar spectrum. Therefore, moving to the left in Figures 2.34 and 2.35 also means moving toward radiation of shorter wavelengths and hence more energetic photons. Moving to the right, on the other hand, means that we move towards longer wavelengths and therefore less energetic photons.

As will be discussed further later, the photons of ultraviolet radiation, for instance, are highly energetic, and stratospheric ozone plays an important role in shielding Earth's surface from these photons. The Sun also emits shorter-wavelength x-rays and gamma rays and longer-wavelength infrared (4.0–100 μm), microwave (0.1–10 μm) and radio (> 1 cm) photons; however, these do not make any significant contribution to the total solar energy reaching Earth's atmosphere.

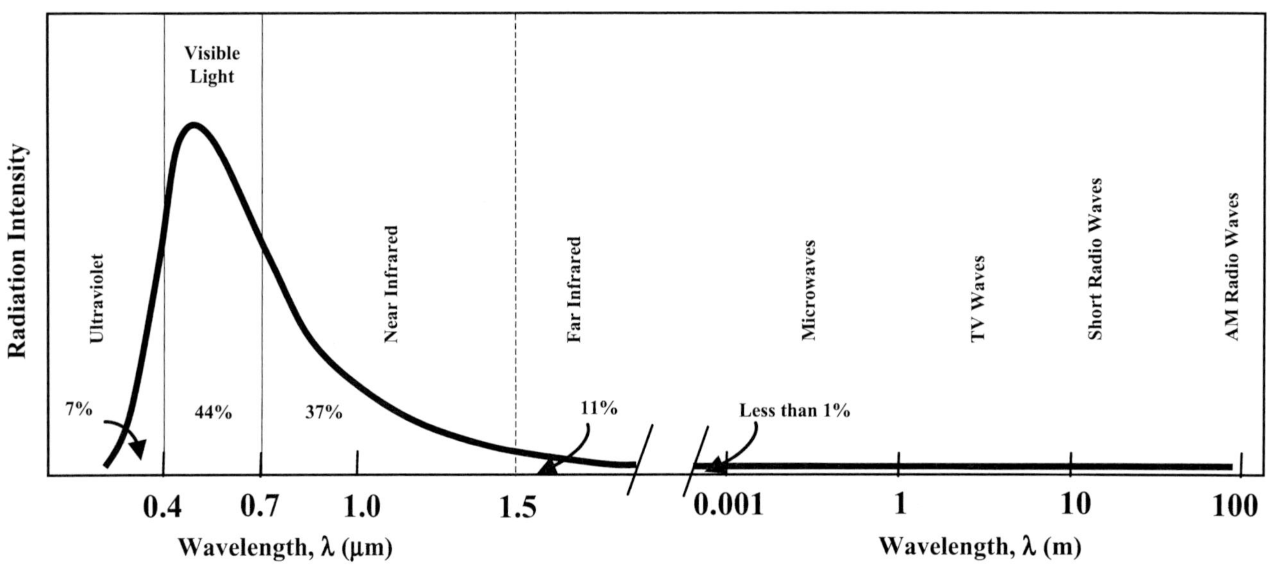

FIGURE 2.34 Electromagnetic spectrum: The complete set of possible wavelengths.
Source: *Greenhouse Engineering* (2021).

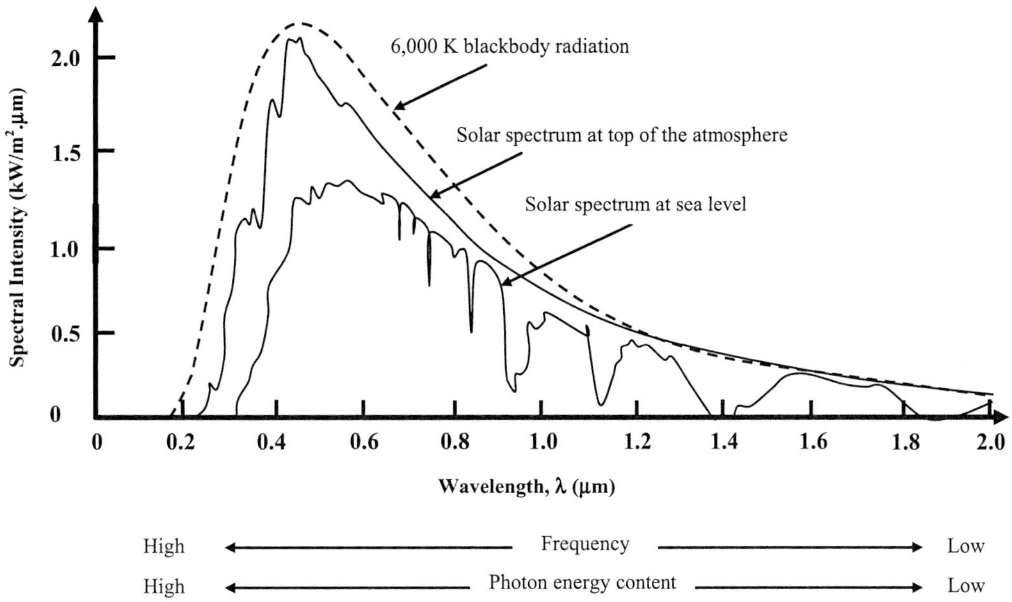

FIGURE 2.35 Solar spectrum. The smooth dashed line is the spectrum the Sun would have if it were a blackbody at the "best fit" temperature of 6,000 K.

Source: ***Greenhouse Engineering*** **(2021).**

As mentioned earlier, the smooth spectrum (dashed line) in Figure 2.35 belongs to a blackbody at 6,000 K.

Blackbody is the term employed to describe a perfectly efficient radiator (not referring to its actual color). Therefore, the blackbody radiation given in Figure 2.35 represents the maximum radiation the Sun at a given average temperature of 6,000 K can emit. We should also remember that most solids, liquids and dense gases radiate very close to the blackbody radiation and have, what we call, continuous spectra of radiation.

Air molecules, as in the atmosphere, however, are quite separated and have much less interaction with each other than molecules of denser substances such as liquids and solids. Each air molecule has distinct mass and structure, vibrates, rotates, and goes through different processes more or less independent from the surrounding molecules. Substances such as CH_4, N_2O, O_2 and O_3, CO_2, and H_2O that interact with only certain photons with discrete energy levels and wavelengths are called selective absorbers and emitters. These substances are not considered blackbodies because they emit (or absorb) radiation only at specific wavelengths, producing line spectra of radiation rather than wavelengths. For this reason, the Stefan–Boltzmann equation (which gives the total amount of radiation emitted or absorbed) is not applicable for selective emitters or absorbers like the atmosphere; therefore, the solar spectrum at sea level has a different shape from the other two in Figure 2.35.

As Figures 2.34 and 2.35 show, wavelength (λ) and frequency (v, measured in cycles/s, also known as Hz) are related, and this relationship can formally be expressed through the speed of light, c, as follows:

$$c = \lambda\, v \tag{2.269}$$

where c is the speed of light, which has a value of 3×10^8 m/s. If one is known, then the other can easily be determined using the speed of light, which is a constant.

Solar radiation consists of photons, which can be considered packets of energy and are related to frequency v as:

$$E = hv \tag{2.270}$$

where E is the energy content (J), and h is the Planck constant (6.626×10^{-34} J.s). This shows that shorter wavelengths have higher frequencies and more energy than those of longer wavelengths (lower frequencies), as shown in Figures 2.34 and 2.35.

Example 2.102: Calculating the Frequency of Radiation

Determine the frequency for a radiational wavelength of 0.5 μm.

Solution

Using Equation (2.269),

$$c = \lambda v$$
$$v = c / \lambda$$
$$v = (3\times10^8 \text{ m/s})/(0.5\ \mu\text{m}) = (3\times10^{14}\ \mu\text{m/s})/(0.5\ \mu\text{m})$$
$$= 6\times10^{14} \text{ cycles/s} = 6\times10^{14} \text{ Hz}$$

Similarly, a radiational wavelength of 10 μm has a frequency of 3×10^{13} cycles/s (or 3×10^{13} Hz).

Example 2.103: Calculating the Energy Content of Light at a Certain Frequency

Determine E of light at a frequency of 6×10^{14} Hz (or 0.5 μm).

Solution

Using Equation (2.270),

$$E = hv$$

$$E = (6.626\times10^{-34}\ \text{J.s})(6\times10^{14}\ \text{cycles/s})$$
$$E = 3.98\times10^{-19}\ \text{J}$$

Similarly, a radiation frequency of 3 × 10^{13} Hz (or 10 μm) has energy of approximately 1.99 × 10^{-20} J. This clearly shows the abovementioned difference between the energy contents of two different wavelengths.

2.5.3.3 Solar Time

Earth's orbital velocity varies throughout the year, as does apparent solar time (AST), which is determined by a sundial. AST changes by a small amount from mean time, which is kept by a clock running at a constant rate. This change is called the equation of time. The position of the Sun in the sky is dictated by local solar time (LST), which is determined by adding the equation of time to the local civil time. Local civil time, on the other hand, is derived from LST by adjusting for longitude differences based on the correction of the 4 min/° difference between the local longitude and the standard time meridian for a given location (e.g., 75° for Eastern standard time in Canada and the United States).

The relationship between Apparent Solar Time and Local Standard Time is provided in Equation (2.271):

$$\text{AST} = \text{LST} + \text{ET} + 4(\text{LSTM} + \text{LON}) \quad (2.271)$$

where ET is the equation of time in minutes (Table A.35), LSTM is the local standard time meridian in ° of arc (in Canada and the U.S., the LSTMs are 60° for Atlantic Standard Time (ST); 75° for Eastern ST; 90° for Central ST; 105° for Mountain ST; 120° for Pacific ST; 135° for Yukon ST; 150° for Alaska–Hawaii ST), LON is the local longitude in ° of arc, and 4 minutes is how long it takes for the Earth to rotate 1°.

The next section discusses that Earth's equatorial plane is tilted 23.5° to the orbital plane (Figure 2.36). Therefore, the solar declination δ (the angle between the Earth–Sun line and the equatorial plane) varies throughout the year (Figure 2.36 and Table A.35). It is these nonuniform periods of daylight and darkness that give us seasons.

2.5.3.4 Solar Geometry

We know that the planets in the solar system have elliptical orbits around the Sun. The main features of the Earth's orbital geometry are shown in Figure 2.36. The ellipticity in the solar system is relatively small: The Earth's motion around the Sun follows a nearly circular elliptical orbit with a period of 365–1/4 days. Kepler found that for circular orbits, the time period (Y) of the orbit is related to the distance (R) of the planet from the Sun; that is, $Y = aR^{3/2}$. Parameter $a = 0.1996\,d/(\text{Gm})^{3/2}$, where d is Earth days and Gm is gigameters (10^9 m).

Figure 2.36 shows that the Earth is closest to the sun in January (winter) in the northern hemisphere and farthest away in July.

Example 2.104: Calculating the Earth's Orbital Period

Calculate the Earth's orbital period, *Y*.

Solution

Using an average Sun to Earth distance of R = 149.6 Gm,

$$Y = aR^{3/2}$$
$$Y = 0.1996\,d/(\text{Gm})^{3/2}\,(149.6\,\text{Gm})^{3/2}$$
$$Y = 365.2\ \text{days}$$

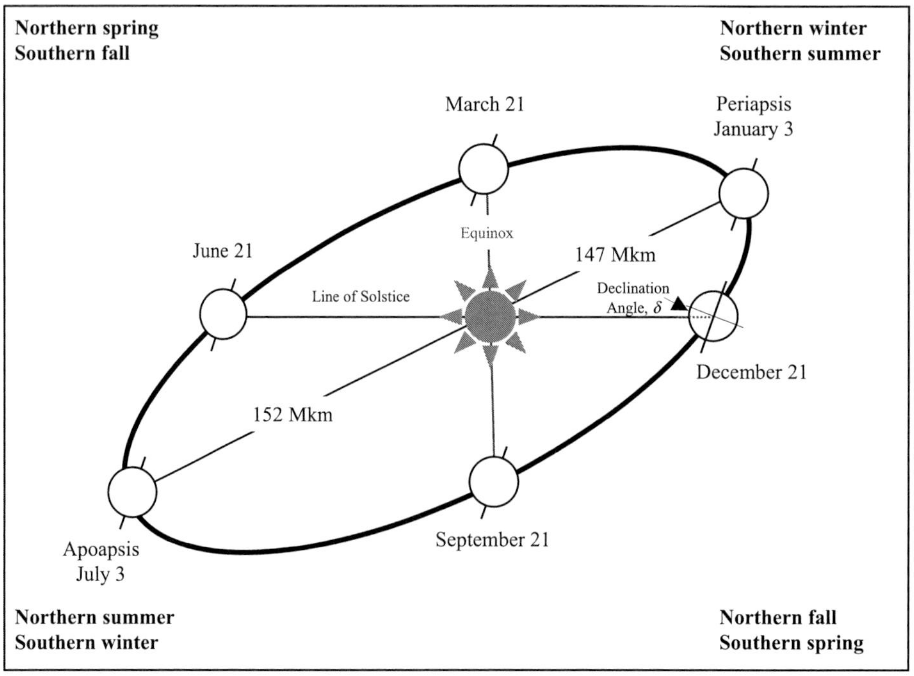

FIGURE 2.36 Earth's motion around the Sun (time of orbit is one revolution of 365.242 days).
Source: *Greenhouse Engineering* (2021).

By using a leap year every four years, the shortfall of the 365-day calendar we use is corrected accordingly.

As mentioned earlier, solar radiation that reaches the Earth is significantly impacted by the inclination of Earth's axis. This axis (the line joining the two poles through the center of the planet) is tilted 23.5° from the perpendicular, and the axis maintains the same orientation with respect to the galaxy; therefore, the insolation at the top of the atmosphere—at Earth's surface—varies considerably. This process creates seasons by impacting the duration of daylight and elevation of the Sun in the sky with respect to time. Although the Sun's output is nearly constant, the insolation reaching a unit horizontal area at any specific location on Earth's surface varies between 0 and 1,050 W/m^2, depending on latitude, season, time of day, cloudiness, and air pollution. This section focuses on variations due to latitude, season, and time.

Figure 2.37 shows the Sun's apparent paths across the sky on different dates. It is called the apparent path because it is not the actual motion of the Sun; rather, it is the Earth's rotation about its own axis one per day. As Earth rotates, different regions start receiving the sunlight and end again. On September 21 (and on March 21), anywhere on Earth, the Sun seems to rise from the eastern side of the sky and set 12 hours later in the west. However, the actual apparent path varies considerably based on the latitude. In fact, solar beams are very rarely perpendicular to the Earth's surface at any given location. Figure 2.37 shows that the smaller the angle between a solar beam and Earth's surface, the larger the area the solar energy spreads; as a result, less solar energy is received per unit area. In late September, for instance, at the top of the atmosphere over equatorial regions, more solar energy is received per unit area than over other regions, but a location directly on the equator will receive the solar beam directly downward for a very short time around noon.

The daily path of the Sun is not constant throughout the year; rather it shifts slightly and steadily every day, and this has a huge impact on insolation. The latitude and the

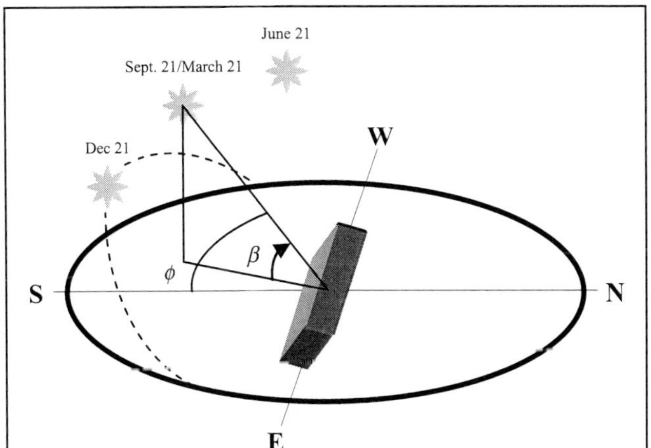

FIGURE 2.37 Annual and hourly changes in the Sun's position for 40°N. β is the solar altitude—angle above the horizon, and φ is the solar azimuth—angle from the true south.

Source: ***Greenhouse Engineering*** **(2021).**

season are a result of the solar geometry. For instance, the North Pole tilts toward the sun in summer in the Northern Hemisphere and away from the Sun in winter. Consequently, the Northern Hemisphere has more daylight hours in the summer, reaching a maximum value on the summer solstice, June 21 (the position on the left side of Figure 2.36) when the tilt toward the Sun is greatest. In the Northern Hemisphere, the Sun does not directly shine down on the equator but rather on latitude 23.5° N. On the summer solstice, the daily path is highest in the sky, and the day is the longest of the year.

In the Southern Hemisphere, on the other hand, the situation is the reverse: The hemisphere is tilted farthest from the Sun on or about June 21 and receives the least sunlight, marking the onset of winter. In the northern hemisphere's winter, however, because the tilt angle (and the North Pole) is the farthest from the Sun, solar radiation needs to go through a longer path, resulting in more absorption and scattering in the atmosphere; therefore, less radiation reaches Earth's surface when the need for heat is the greatest. At this time of the year, the Sun is directly overhead at latitude 23.5°S on the winter solstice, when the Southern Hemisphere starts enjoying the beginning of summer with long hours of the Sun high in the sky.

Figure 2.37 schematically shows the annual and hourly changes in the Sun's position and relevant solar angles for latitude. At the extreme, the noontime sun is directly overhead at either the Tropic of Cancer (summer solstice) or the Tropic of Capricorn (winter solstice). On the winter solstice, the areas north of the Arctic Circle have 24 hours of darkness, while the areas south of the Antarctic Circle have 24 hours of daylight. When the Sun is directly overhead at the equator (on the spring and autumnal equinoxes, on March 21 and September 21, respectively), everywhere on Earth, there are 12 hours of daylight; that is, all latitudes have equal amounts of day and night (Figure 2.36).

Figure 2.37 shows that as summer moves into fall and then into winter, the sunrise and sunset times gradually move southward. As a result, the days get shorter, and the solar path gets lower in the sky. In December, the Sun rises and sets quite a bit south of east and south of west, respectively. The figure depicts the Sun's annual and hourly position only for 40°N. Note that the Sun's summertime daily path never goes below the horizon; therefore, these regions experience continuous sunlight (24 hours of daylight every day) during summer months, which gives them the highest insolation at the top of the atmosphere at that time of the year. Also note that there is a lack of significant variation in insolation at the top of the atmosphere throughout the year along the equator, in contrast with seasonal changes at higher latitudes, and that January insolation values decrease drastically as we go from the equator toward the North Pole but remain constant south of the equator. In June, however, this pattern is reversed, with northern latitudes experiencing little change.

2.5.3.5 Solar Radiation

As will be discussed in depth later in this section, some solar radiation scattered by air molecules and dust reaches the Earth in the form of diffuse radiation, I_d. I_d intensity is difficult to measure because it comes from all directions in the sky, and it varies as moisture and dust content of the atmosphere change throughout any given day. For instance, on an overcast day, the diffuse component is all the solar radiation reaching Earth.

The total solar radiation reaching a terrestrial surface, I_t, is composed of direct solar radiation I_D, diffuse sky radiation I_d, and the solar radiation reflected from the surrounding surfaces I_r (Robinson and Sellers, 1999). I_D is the product of the direct normal solar radiation I_{DN} and the cosine of the angle of incidence θ between the incoming solar rays and a line normal (perpendicular) to the surface (ASHRAE, 1989):

$$I_t = I_{DN} \cos\theta + I_d + I_r \tag{2.272}$$

2.5.3.6 Solar Angle

The position of the Sun in the sky is expressed in terms of the solar altitude and the solar azimuth. The height of the Sun is usually given in terms of solar altitude β (Figure 2.37). This is the angular distance between the Sun's rays and the horizon and is given by Equation (2.273), and φ is the angle measured from true south (Figure 2.37), given by Equation (2.274) (ASHRAE, 1989).

$$\sin\beta = \cos L \cos\delta \cos H + \sin L \sin\delta \tag{2.273}$$

$$\cos\phi = \left(\sin\beta \, \sin L - \sin\delta\right) / \left(\cos\beta \cos L\right) \tag{2.274}$$

where these angles depend on the local latitude L; δ (the angle between the Earth–Sun line and the equatorial plane (Figure 2.36 and Table A.35), which is a function of the time of year, and therefore varies from +23.5° on June 21 to −23.5° on December 21; and AST, expressed as the *hour angle H*, where $H = 0.25 \times$ number of minutes from local solar noon), in degrees. H is zero at local noon and increases in magnitude by π/12, (15°), for every hour before or after noon.

Further, assuming that Earth's orbit is circular, the solar declination angle for any day of the year can be approximated using $\delta = \Phi \cos\left[C\left(d - d_r\right)/d_y\right]$, where Φ is the tilt angle of 23.5°, C is a full 360°, d is the Julian day, d_r is the Julian day for summer solstice on June 21 (172 for non-leap years), and d_y is the number of days per year, that is, 365 or 366 in a leap year.

Example 2.105: Calculating the Solar Declination Angle

Given Φ = 23.5°; C = 360°; d_r = 172 days; and d_y = 365 days, calculate the solar declination angle on March 21.

Solution

Assume: Not a leap year.

Using $\delta = \Phi \cos\left[C\left(d - d_r\right)/d_y\right]$,

$$\delta = \Phi \cos\left[360^\circ \left(80 \text{ days} - 172 \text{ days}\right)/365 \text{ days}\right]$$

$$\delta = -0.3^\circ$$

On the spring equinox (March 21), the declination angle should be zero, so our calculation is a good approximation. It is still winter before March 21, and the declination angle should be negative, and the values in spring and summer are positive.

Example 2.106: Calculating the Solar Azimuth and Altitude

Determine the solar azimuth and altitude at 0830 Atlantic ST on March 21 at 45° north latitude and 63° west longitude.

Solution

Using Equation (2.271), local time is 0830 + 4 (60° − 63°) = 0818. Table A.35 gives the Equation of Time as −7 min, so AST = 0818 −7 = 0811 or 229 minutes from local solar noon; therefore, H = 0.25 × 229 min = 57.3°. As calculated in Example 2.93 and also provided in Table A.35, the solar declination on March 21 is 0°. Therefore, using Equation (2.273), β can be calculated as follows:

$$\sin\beta = \cos 45^\circ \cos 0^\circ cos57.3^\circ + \sin 45^\circ \sin 0^\circ$$

$\sin\beta = 0.382$; therefore, $\beta = 22.5°$

Then, ϕ can be calculated using Equation (2.274) as

$$\cos\phi = \left(\sin 22.5^\circ \sin 45^\circ - \sin 0^\circ\right) / \left(\cos 22.5^\circ \cos 45^\circ\right)$$

$\cos\phi = 0.414$; therefore, $\phi = 65.5°$

Figure 2.38 shows the solar angles and incident angles for horizontal and vertical surfaces. Line OV is perpendicular to the horizontal plane in which the solar azimuth, angle HOS, and the surface azimuth, angle POS (Ψ), are located. Angle HOP is the surface solar azimuth (γ) and is given by

$$\gamma = \phi - \Psi \tag{2.275}$$

The solar azimuth angle ϕ is negative for morning hours and positive for afternoon hours. The absolute γ is used in Equation (2.276). The surface is considered in the shade if γ is greater than 90° or less than 270°.

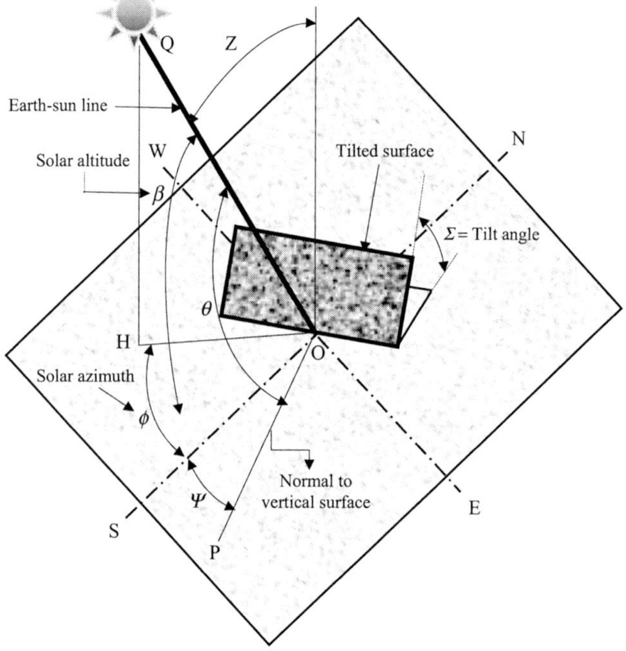

FIGURE 2.38 Solar angles. β is the solar altitude—angle above the horizon, and φ is the solar azimuth—angle from true south.

Source: ***Greenhouse Engineering*** **(2021).**

For any surface, the angle between the incoming solar rays and a line normal to the surface is called the angle of incidence θ. For the horizontal surface given in Figure 2.38, the angle of incidence, θ_H, is QOV; the incident angle for the vertical surface, θ_V, is QOP. For any surface, θ is related to β, γ, and the tilt angle of the surface from the horizontal Σ as shown:

$$\cos\theta = \cos\beta\cos\gamma\sin\Sigma + \sin\beta\cos\Sigma \qquad (2.276)$$

When the surface is horizontal, $\Sigma = 0°$ and $\cos\theta_H = \sin\beta$. When the surface is vertical, however, $\Sigma = 90°$ and $\cos\theta_v = \cos\beta\cos\gamma$.

The surface azimuth angle, Ψ, measured from the south, on the other hand, was provided by ASHRAE (1989) as follows: 180° (for N), −135° (for NE), −90° (for E), −45° (for SE), 0° (for S), 45° (for SW), 90° (for W), and 135° (for NW).

Example 2.107: Calculating the Incident Angle

For the conditions of Example 2.106, calculate θ at a vertical surface facing southeast.

Solution

As we have determined that Ψ is to the east (AST < 1200), and the surface azimuth is also to the east, they both must be negative: that is, $\varphi = -65.5°$ and $\Psi = -45°$. Therefore, γ can be calculated using Equation (2.275) as

$$\gamma = -65.5° - (-45°) = -20.5°$$

The negative γ that we calculated indicates that the sun at the given time is east of the normal to the surface.

As mentioned, when the surface is vertical, $\Sigma = 90°$, and θ_v can then be determined using $\cos\theta_v = \cos\beta\cos\gamma$. Therefore,

$$\cos\theta_v = \cos 22.5° \cos 20.5° = 0.865$$

Then, θ_v, the angle between the incoming solar rays and a line normal to the surface, can be calculated:

$$\theta_v = 30.1°$$

2.5.3.7 Atmospheric Effects

The previous section mainly focused on the distribution of insolation at the top of the atmosphere at different times and locations. This section traces a beam of solar energy and explores its interactions with the atmosphere. It is assumed that the beam has 100 units of energy as it enters the atmosphere. The fate of this 100 units of energy can be tracked by referring to Figure 2.39 as we read. From our daily observations, it is clear that if the air is free of clouds and pollution, solar beams reach the ground with little interference from the atmosphere. The solar disk, on the other hand, becomes completely obscured under cloudy or dirty sky conditions.

Clear air is transparent to sunlight. Normally, when the Sun's radiation passes through Earth's atmosphere, it is reflected, scattered, and absorbed by dust particles, gas molecules, ozone, and water vapor. The magnitude of solar radiation's attenuation at a given time and location is determined by atmospheric composition and length of the atmospheric pathway the solar radiation travels. The length of atmospheric path is given in terms of the air mass m, which is the ratio of the atmospheric air mass in the actual Earth–Sun path to the mass that would exist if the Sun were overhead at sea level ($m = 1.0$). Obviously, above the atmosphere,

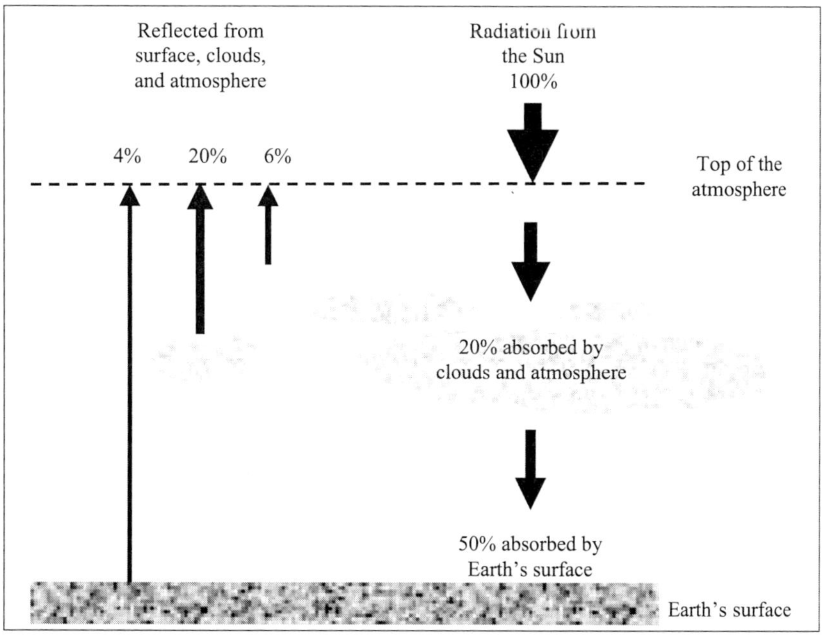

FIGURE 2.39 Schematic distribution of what happens to solar radiation after it falls on Earth.
Source: *Greenhouse Engineering* (2021).

$m = 0$. For most purposes, m at any given time and location equals the cosecant of β (Figure 2.38) multiplied by the ratio of the existing barometric pressure to standard pressure. Interactions of solar radiation within the atmosphere take place almost simultaneously; however, these are separate processes with different consequences, and therefore, they must be treated separately.

When a photon hits a small object without being absorbed, radiation scatters, that is, interacts with small particles. However, when radiation hits a larger object, then there is a complete change in the photon's travel direction, and this special condition is called reflection. This change in travel direction can be in any or multiple directions.

Air molecules are between 0.0001 and 0.001 μm in size, much smaller than the wavelength of the visible band (0.4–0.7 μm). Particles smaller than the wavelength of solar radiation, such as atmospheric gas molecules, generate something called Rayleigh scattering. How much scattering generated is directly proportional to the fourth power of the wavelength; therefore, the scattering of blue light ($\lambda \approx 0.4$ μm) within a cloudless atmosphere without pollution is about ten times greater than that for red light ($\lambda \approx 0.7$ μm). As a result of this Rayleigh scattering process, we see the daytime sky as blue. In the evening, however, we observe a much redder sun and reddish sky. This is because of the very long pathway the radiation travels through the atmosphere, which causes most of the visible wavelengths to be scattered many more times, leaving only the red wavelength.

When larger particles—such as water droplets and pollution particles—exist in the atmosphere, and the scattering particles and the wavelength of the radiation have similar sizes, a different type of scattering process takes place. This process is called Mie scattering. In this case, all wavelengths are scattered almost uniformly because of the particle size and the wavelength of the radiation. When the atmosphere is polluted or overcast, then no wavelength is preferentially scattered, creating a light blue/grayish sky. Under clear sky conditions, 6% of the original 100 units are scattered back to space.

There is comparatively less atmospheric absorption of solar radiation than there is scattering. Ozone's absorption of ultraviolet radiation is a significant component that is vital for sustaining life on Earth. In addition to the ultraviolet absorption, at other wavelengths, absorption by nitrous oxide, carbon dioxide, oxygen, ozone, and water vapor takes place as well. Overall, as shown in Figure 2.39, approximately 20% of the energy reaching the top of the atmosphere is absorbed by the clouds (2%, i.e., 2 units of our sample 100 units) and the atmosphere (18%, i.e., 18 units of our sample 100 units).

As mentioned, some of the atmospheric absorption takes place due to the molecules of ozone and monatomic oxygen in the upper atmosphere; this absorbed energy causes the high temperatures observed in the stratosphere and thermosphere. As a result of the scattering, and the absorption by the clouds and the atmosphere, only 74 units of the sample 100 units of energy can reach Earth's surface as direct beam or diffuse solar radiation (which will be discussed further in the following section). Figure 2.39 shows that clouds alone significantly reduce the amount of solar energy reaching Earth's surface.

As mentioned, clouds absorb 2% and reflect 20% back to space. All these atmospheric processes (absorption, reflection, and scattering) leave only 54% of the original energy to eventually reach Earth's surface. It is obvious that the areas that receive the most radiation are deserts, and the polar regions receive the least radiation. It should also be remembered that Figure 2.39 represents the long-term, global averages; local values can differ drastically. At high latitudes, where the solar angle is low, for instance, solar beams must travel a much longer atmospheric path than at low latitudes; therefore, the beams are more likely to be scattered or absorbed further.

As briefly mentioned, the solar radiation that reaches Earth's surfaces is both direct (the radiation that casts a shadow) and diffuse (radiation scattered from clouds, particles, and air molecules, coming from the entire sky) components. Energetically though, they act in the same way. Since diffuse radiation originates from the entire sky, its intensity is difficult to predict due to fluctuations in atmospheric moisture and pollution throughout the day and across locations. Because cloud cover and atmospheric pollution vary considerably and are hard to predict, the most practical approach is to average the solar radiation received at a location over several recent years and assume that on average, similar levels will be received in the future. At the Earth's surface on clear days, solar radiation is approximately 85% direct and 15% diffuse. On a completely overcast day, all solar radiation reaching Earth's surface is diffuse.

2.5.3.7.1 Direct Normal Solar Radiation

As mentioned earlier, most of the solar radiation reaching Earth's surface on clear days is direct normal radiation or solar intensity I_{DN}, which is determined by the following equation:

$$I_{DN} = \frac{A}{\left[e^{\left(\frac{B}{\sin\beta}\right)}\right]} \tag{2.277}$$

where A is the apparent insolation at $m = 0$ (Table A.35), and B is the atmospheric extinction coefficient, a dimensionless ratio (Table A.35). Both values vary during the year due to the seasonal changes in atmospheric pollution and water vapor content and the Earth–Sun distance. Using Equation (2.277) and Table A.35, I_{DN} at Earth's surface on a clear day for any given latitude can be calculated and tabulated for the daylight hours for the 21st day of each month.

Example 2.108: Calculating the Direct Component of Insolation on a Horizontal Surface

For a clear day, determine the direct component of insolation on a horizontal surface, I_{DH}, for the conditions given in our original example.

Solution

We have already determined that $\beta = 22.5°$ and that $\sin\beta = 0.382$. Using Equation (2.277) and the A, B, and C values provided in Table A.35, I_{DN} can be calculated as

$$I_{DN} = \frac{A}{\left[e^{\left(\frac{B}{\sin\beta}\right)}\right]} = \frac{1184.4\ \text{W/m}^2}{\left[e^{\left(\frac{0.156}{0.382}\right)}\right]}$$

$$I_{DN} = 787.3\ \text{W/m}^2$$

Then,

$$I_{DH} = I_{DN}\sin\beta = \left(787.3\frac{\text{W}}{\text{m}^2}\right)\sin 22.5^{\circ}$$

$$I_{DH} = 301.3\ \frac{\text{W}}{\text{m}^2}$$

Example 2.109: Calculating the Direct Component of Insolation on an Inclined Surface

For a clear day, determine the direct component of insolation on an inclined surface, I_D, with a tilt angle of 45° and facing southeast under the aforementioned conditions. The solar azimuth and altitude are at 0830 Atlantic ST on March 21 at 45° north latitude and 63° west longitude.

Solution

Earlier, the local time was established to be 0818. Table A.35 provided the equation of time as –7 min, so AST was determined as 0811 or 229 minutes from local solar noon; the hour angle was then calculated as 57.3°. As also determined previously and provided in Table A.35, the solar declination on March 21 is 0°. Then, $\beta = 22.5°$ and $\varphi = 65.5°$. We also determined that Ψ is to the east and must be negative, that is, $\Psi = -45°$. Therefore, γ was calculated using Equation (2.275) as –20.5°. The negative γ obviously indicates that the Sun at the given time is east of the normal to the surface. Finally, $\theta_v = 30.1°$.

In Example 2.96, I_{DN} was calculated as 787.3 W/m². As was expressed earlier, I_t is composed of I_D, I_d, and I_r. I_D is the product of I_{DN}, the cosine of θ—which as shown earlier in Equation (2.276) is related to β, γ, and Σ:

$$I_D = I_{DN}\cos\theta = I_{DN}\left(\cos\beta\cos\gamma\sin\Sigma + \sin\beta\cos\Sigma\right)$$

$$I_D = \left(787.3\text{W/m}^2\right)\left(\begin{matrix}\cos 22.5\cos 20.5\sin 45.0 + \\ \sin 22.5\cos 45.0\end{matrix}\right)$$

$$I_D = 694.8\ \text{W/m}^2 \approx 695\ \text{W/m}^2$$

2.5.3.7.2 Diffuse Solar Radiation

The diffuse solar radiation reaching a surface on Earth can come from the sky or from solar radiation reflected off adjacent surfaces. On a clear sky, a simplified equation for the diffuse solar radiation reaching any surface from the sky is

$$I_{ds} = C\, I_{DN}\, F_{ss} \tag{2.278}$$

where C is the diffuse radiation factor (dimensionless), I_{DN} is the sky radiation falling on a horizontal surface, and F_{ss} is the angle factor between the surface and the sky (dimensionless)— 0.5 and 1.0 for vertical and horizontal surfaces, respectively—determined by Equation (2.279) for other surfaces.

$$F_{ss} = (1.0 + \cos\Sigma)/2 \tag{2.279}$$

where Σ is the tilt angle measured upward from the horizontal plane (Figure 2.38).

Solar radiation reflected by ground has the components of diffuse sky and direct solar radiation falling on a horizontal surface. The total solar radiation reaching the ground is determined by Equation (2.280):

$$I_{tH} = I_{DN}(C + \sin\beta) \tag{2.280}$$

where I_{DN} times sin β gives the direct radiation falling on a horizontal surface. Then, the ground-reflected diffuse solar radiation on any surface can be estimated by Equation (2.281):

$$I_{dg} = I_{tH}\,\varsigma_g F_{sg} \tag{2.281}$$

where ς_g is the reflectance of the foreground (dimensionless) and F_{sg} is the angle factor between the surface and the ground (dimensionless). Obviously, the sum of the angle factors equals 1.0, and the angle factor for surface to ground is determined by Equation (2.282):

$$F_{sg} = (1.0 - \cos\Sigma)/2 \tag{2.282}$$

If the surface is exposed to only the ground and the sky, then F_{ss} = (1.0 – F_{sg}). The reflectance values for different ground surfaces and foreground surfaces are provided by Threlkeld (1962) and reported by ASHRAE (1989).

Example 2.110: Calculating the Overall Total Solar Radiation Reaching a Solar Collector's Surface

Determine the diffuse solar radiation incident on a solar collector and then determine the overall total solar radiation reaching the collector's surface with a 45° slope that faces southeast at 0830 Atlantic time on March 21 at 45° north latitude.

Solution

Equation (2.273) for l = 45°, δ = 0° (Table A.35), and H = 57.3° gave β as 22.5°. Then, we calculated I_{DN} as 787.3 W m^{-2} for 45° north latitude at 0830 Atlantic time (i.e., AST of 0811), and Table A.35 gives C = 0.071 for March 21. Using Equation (2.282), we can calculate F_{sg}, from the collector to the ground as follows:

$$F_{sg} = \left(1.0 - \cos 45^{\circ}\right)/2 = 0.146$$

Remember that if the surface is exposed to only the ground and the sky, then F_{ss} = (1.0 – F_{sg}). Therefore, F_{ss} is calculated as

$$F_{ss} = \left(1.0 - 0.146\right) = 0.854$$

As mentioned earlier, solar radiation reflected by ground has the components of diffuse sky and direct solar radiation falling on a horizontal surface. Then, the diffuse solar radiation reaching the collector from the sky can be determined using Equation (2.278) as

$$I_{ds} = 0.071\,(787.3\text{W/m}^2)\ 0.854$$

$$I_{ds} = 47.7\ \text{W/m}^2$$

The amount of total solar radiation reaching the ground is calculated using Equation (2.280):

$$I_{tH} = \left(787.3\,\text{W/m}^2\right)(0.071 + \sin 22.5°)$$

$$I_{tH} = 357.2\ \text{W/m}^2$$

If the ground is crushed rock with solar reflectance of 20% (Table A.36), I_{dg} falling on the collector can then be easily determined using Equation (2.281) as

$$I_{dg} = \left(357.2\,\text{W/m}^2\right)(0.20)(0.146)$$

$$I_{dg} = 10.4\,\text{W / m}^2$$

The total diffuse solar radiation falling on the collector is then determined as the sum of diffuse sky radiation and the ground-reflected diffuse radiation, that is, 47.7 + 10.4 = 58.1 W/m^2. Then, the total solar radiation reaching a collector surface with a 45° slope that faces southeast at 0830 Atlantic on March 21 45° north latitude can be determined by Equation (2.272) as

$$I_t = I_D + I_d + I_r$$

$$I_t = 694.8 + 47.7 + 10.4$$

$$I_t = 752.9\ \text{W/m}^2 \approx 753\ \text{W/m}^2$$

2.5.4 Insolation at Earth's Surface

As explored in the previous section, on a long-term, global average, only 54% of the solar energy passing through the top of the atmosphere reaches Earth's surface (Figure 2.39). Just a portion of this energy is absorbed at the surface; the rest is reflected. The ratio of the reflected portion to the incoming radiation is defined as the albedo:

$$\text{Albedo}, A = \frac{\text{Reflected Solar Radiation}}{\text{Incoming Solar Radiation}} \tag{2.283}$$

Albedo is normally expressed as a percentage, and satellite measurements indicate that Earth's surface has an average albedo of 8%; that is, on average, 8% of insolation is reflected back. Albedos for typical surfaces are presented in Table A.37. When both sides of Equation (2.283) are multiplied by the incoming solar radiation, the expression becomes

$$\text{Reflected Solar Radiation} = A \times \begin{pmatrix}\text{Incoming Solar}\\ \text{Radiation}\end{pmatrix} \tag{2.284}$$

Example 2.111: Calculating the Reflected Solar Radiation from a Surface

Using Equation (2.284) and an average surface albedo of 8% (that is, 0.08), determine the reflected solar radiation.

Solution

$$\text{Reflected Solar Radiation} = A \times (\text{Incoming Solar Radiation})$$

$$= 0.08 \times 54\ \text{units}$$

$$\text{Reflected Solar Radiation} = 4.32\,\text{units}\,(\approx 4\,\text{units})\,\text{reflected}$$

If the albedo concept is applied to all of Earth and its atmosphere, it is called planetary albedo, and it has a long-term average value of 30%, that is, 30 units (20 from clouds, 6 from atmosphere, and 4 from Earth's surface) of the sample 100 units of solar energy is reflected back to space (Figure 2.39). The remaining 70% of the solar energy reaching Earth's atmosphere is absorbed and transformed into heat.

Earlier, we defined insolation, how much sunlight is shining down on us. By knowing the insolation of a particular region, we can determine the size of a solar collector required for a particular application and how much energy it can produce. Obviously, an area with poor insolation will need a larger collector area than those with high levels.

For application purposes, insolation is generally expressed in kWh/m^2.day. Btu or MJ are also used, in which case the conversion is 1 kWh/m^2.day = 317.1 Btu/ft^2.day = 3.6 MJ/m^2.day. The raw energy conversions are 1 kWh = 3,412 Btu = 3.6 MJ. As discussed earlier in depth, insolation changes throughout the year, at its lowest in winter and highest in summer; close to the equator, the difference throughout the year is minimal; and at high latitudes, winter levels can be just a fraction of summer levels. A very high summer value, as in a hot desert area, is 7 kWh/m^2.day. For comparison purposes, the readers of this book are referred to the average annual insolation of 2.27 kWh/m^2.day for Oslo, Norway (considered very low), and 5.26 kWh/m^2.day for Miami, Florida (considered very high).

Tables A.38–A.43 list the average insolation values for main cities in each region of the world, which can be used in solar energy project designs (Taylor, 2005). The NASA Surface Meteorology and Solar Energy (SSE) data set consisting of resource parameters was developed for assessing and designing renewable energy systems. The monthly total solar radiation incident on a horizontal surface at Earth's surface for a given month was averaged for that month every month for ten years; each monthly average was evaluated as the numerical average of values at three-hour intervals.

Renewable energy technologies range in complexity from solar ovens and simple photovoltaic panels installed in rural communities to commercial buildings with integrated photovoltaics and large thermal and wind-generating power plants. The availability of accurate, global solar radiation and meteorology data is extremely important for successful controlled environment and other renewable energy projects. NASA's Earth Science Enterprise makes the SSE data set available free of charge online (NASA, 2016).

Historically, climatological profiles of insolation and meteorology parameters calculated from ground measurements have been used for determining the viability of controlled environments and other renewable energy projects. Although ground measurement data has been used successfully, ground measurement stations are situated mainly in populated regions. In remote areas, where many controlled-environment and other renewable energy projects are being implemented, measurement stations are quite limited; additionally, available data can be sporadic, inconsistent, or incomplete. In contrast to ground measurements, the SSE data is continuous, consistent, long term, and global, although the SSE data within a particular grid cell is not necessarily representative of a particular microclimate or point within the cell; rather, the data is considered the average over the entire area of the cell. SSE data, however, is not intended to replace ground measurement data.

NASA emphasizes that it is prepared to fill the gap where ground measurements do not exist and to augment areas where ground measurements are available. Utilizing the SSE data, the renewable energy resource potential can be determined for any location on the globe and is considered to be accurate for preliminary feasibility studies for controlled-environment and other renewable energy projects. Detailed insolation information (including direct solar normal and diffuse components) for any particular location or time can be directly obtained from NASA. Table A.44 lists annual estimated solar resource availabilities and ranks around the world and may prove to be useful for potential controlled-environment and other renewable energy projects.

Example 2.112: Determining a Flat Plate Collector Area Required for Water Heating

Assuming that the flat plate collector (FPC) efficiency is 50% and ignoring the ground-reflected radiation falling on the collector, determine the collector size to be tilted from the horizontal surface facing south (for both Los Angeles, California, USA [34° north latitude], and Halifax, Nova Scotia, Canada [45°N]) required to heat a total of 150 L of hot water per day (needed for four-people in a family-operated greenhouse) from 5 to 45°C in March.

Solution

We know that the amount of heat required for the given temperature difference can be calculated using Equation (2.24); that is, $Q_S = mc_p(T_2 - T_1)$, where m is the mass of water, c_p is the specific heat of water (Table A.4), and Q_S is the heat required.

$$Q_S = 150\ \text{kg}(4{,}186\ \text{J/kg °C})(45-5)\text{°C}$$

$$Q_S = 25{,}116{,}000\ \text{J/day}\ (\approx 25.1\ \text{MJ/day})$$

The heat available from the FPC is $Q_S = (\text{Insolation})\times(\text{FPC Area})\times(\text{Efficiency})$. Table A.43 provides the average insolation for March in Los Angeles as 5.09 kWh/m^2.day. Then, the heat available from the FPC is

$$25.1\frac{\text{MJ}}{\text{day}} = \left(5.09\frac{\text{kWh}}{\text{m}^2\text{day}}\right)\left(3.6\frac{\text{MJ}}{\text{kWh}}\right)\text{FPC Area}\left(\text{m}^2\right)(0.5)$$

$$\text{FPC Area} = 2.7\ \text{m}^2$$

As a useful rule of thumb, the collector should be tilted 44–45° from the horizontal surface facing south for Los Angeles (34°N plus 10°).

Table A.40 provides the average insolation for March in Halifax as 3.46 kWh/m^2.day. Then, the heat available from the FPC is

$$25.1\frac{\text{MJ}}{\text{day}} = \left(3.46\frac{\text{kWh}}{\text{m}^2\text{day}}\right)\left(3.6\frac{\text{MJ}}{\text{kWh}}\right)\text{FPC Area}\left(\text{m}^2\right)(0.5)$$

$$\text{FPC Area} = 4.0\ \text{m}^2$$

Again, the collector should be tilted 55° from the horizontal surface facing south for Halifax (45°N plus 10°). As insolation is higher in Los Angeles than in Halifax, ~32% less collector area is required to heat the same amount of water in Los Angeles.

This example gives us rough approximations for the required collector areas. For a detailed analysis, we would use the average monthly direct and diffuse components of solar radiation per square meter per day, or daily or hourly direct and diffuse components of solar radiation per square meter when available, and perform detailed calculations for the average direct, diffuse, and ground-reflected solar radiation reaching the collector surface for a given collector orientation and tilt angle. Then, we would decide what the design collector area is for this application.

Example 2.113: Determining the FPC Area Required for Heating a Small Greenhouse

Assuming that the FPC efficiency is 50%, determine the collector size required to heat a small greenhouse for one day when the heating load is 5.86 kW. Use mean daily insolation of 2.25 kWh/m^2.day (or 8.1 MJ/m^2.day) and local latitude 40°N.

Solution

Remember that $Q_S = (\text{Insolation})\times(\text{FPC Area})\times(\text{Efficiency})$. The thermal energy needed for a day is 5.86 kW × (24 h/day) (3600 s/h) = 506,304 kJ/day = 506.3 MJ/day.

The solar energy collected in one day will be $\left(8.1\frac{\text{MJ}}{\text{m}^2.\text{day}}\right)\times(\text{FPC Area})\times(\text{Eff.})$. Therefore,

$$\text{FPC Area} = \left(506.3\frac{\text{MJ}}{\text{day}}\right) / \left[\left(8.1\frac{\text{MJ}}{\text{m}^2.\text{day}}\right)(0.5)\right]$$

$$\text{FPC Area} = 125\,\text{m}^2$$

A useful rule of thumb is that the collector for space heating should be south facing and inclined at a collector tilt angle from the horizontal equal to the local latitude plus 10°. Therefore, the collector in the example is facing south at 40° plus 10° = 50°.

This example gives us a rough approximation for the required collector area. For a detailed analysis, we would use the average monthly direct and diffuse components of solar radiation per square meter per day, or daily or hourly direct and diffuse components of solar radiation per square meter when available, and perform detailed calculations for the average direct, diffuse, and ground-reflected solar radiation reaching the collector surface for a given collector orientation and tilt angle. Then, considering a thermal storage device for nighttime heating, we would decide the design collector area for this application.

Example 2.114: Determining the Size of Hot Water Thermal Storage

Now, using the FPC (125 m^2 surface area with efficiency of 50%) and the greenhouse in the previous example (located at 40°N latitude with a heating load is 5.86 kW), calculate how many tons of water would be necessary to store enough thermal energy for three days of space heating. Assume the water in the storage tank begins providing heat at 65°C and has a lower temperature limit of 30°C (i.e., ΔT = 35°C).

Solution

The three-day capacity of the thermal storage system would be

$$Q = (506.3 \text{ MJ/day}) \times (3 \text{ days}) = 1{,}518.9 \text{ MJ}$$

For a given temperature difference, using Equation (2.24), $Q_S = m\, c_p\, \Delta T$, where Q_S is the heat taken from the system, ΔT is the temperature difference after lowering, c_p is the specific heat of water (4,186 J/kg °C; Table A.4), and m is the mass of water required for thermal storage. Then,

$$Q_S = mc_p(\Delta T)$$

$$1{,}518.9 \text{ MJ} = \text{m (kg)}\ (4{,}186 \text{ J/kg.°C})(35\text{°C})$$

$$m = 10{,}367 \text{ kg} = 10.4 \text{ ton}$$

Example 2.115: Determining Solar Collector Area for Power Generation

Assuming collector conversion efficiency of 8% and ignoring the ground-reflected radiation falling on the collector, determine the collector size to be tilted from the horizontal surface facing south (for both Los Angeles and Halifax) needed to provide power for a greenhouse community that has an average annual power requirement of 5,000,000 kWh.

Solution

The required daily power output for the community can be calculated as

$$\left(5{,}000{,}000\frac{\text{kWh}}{\text{year}}\right)\left(3.6\frac{\text{MJ}}{\text{kWh}}\right)/\left(\frac{1\text{year}}{365\text{day}}\right)$$

$$\text{Required Daily Power Output} = 49{,}315\frac{\text{MJ}}{\text{day}}$$

The power available from the collector is (insolation) × (collector area) × (eff.). Table A.43 provides the annual average daily insolation in Los Angeles as 5.40 kWh/m².day (=19.44 MJ/m².day). Then, the collector area required for the daily power output of 49,315 MJ/day in Los Angeles is

$$\text{Collector Area}\left(\text{m}^2\right) = \left(49{,}315\ \frac{\text{MJ}}{\text{day}}\right)/\left[\left(19.44\frac{\text{MJ}}{\text{m}^2.\text{day}}\right)(0.08)\right]$$

$$\text{Collector Area} = 31{,}710 \text{ m}^2 = 3.1 \text{ ha}$$

As a useful rule of thumb, the collectors should be tilted 44–45° from the horizontal surface facing south for Los Angeles.

Table A.40 provides the annual average daily insolation in Halifax as 3.40 kWh/m².day (= 12.24 MJ/m².day). Then, the collector area required for the daily power output of 49,315 MJ/day in Halifax is

$$\text{Collector Area}\left(\text{m}^2\right) = \left(49{,}315\frac{\text{MJ}}{\text{day}}\right)/\left[\left(12.24\frac{\text{MJ}}{\text{m}^2.\text{day}}\right)(0.08)\right]$$

$$\text{Collector Area} = 50{,}363 \text{ m}^2 = 5.0 \text{ ha}$$

Again, the collector should be tilted 55° from the horizontal surface facing south for Halifax. As insolation is higher in Los Angeles than in Halifax, much less (~38%) collector area is required to generate the same amount of power required. For a detailed analysis, we would use the average monthly direct and diffuse components of solar radiation per square meter per day, or daily or hourly direct and diffuse components of solar radiation per square meter when available.

NOMENCLATURE

a	atto
a	absorbance; acceleration; constant 2897 μm.K (5,216 μm.°R); or (0.1996 d/(Gm)$^{3/2}$)
a to ***e***	constants in Equations (2.116) and (2.117)
A	ampere, a unit of electric current
A	area, cross-sectional area
AST	apparent solar time (minutes)
B	dimensionless ratio
BEP	best efficiency point
Btu	British thermal unit
c	centi
c	speed of light, 3 × 10^8 m/s; specific heat; specific carbon content; constant; dimensionless ratio
cd	luminous intensity unit
C	Coulomb, a unit of electric charge
C	diffuse radiation factor (dimensionless); heat capacity rate; cost; mass of carbon; flow coefficient; concentration
C_1, C_2	constants
C_1 to C_{13}	constants in Equations (2.106) and (2.107)
°C	degree Celsius, a unit of temperature
d	deci
d	day; Julian day (number); number of days per year (365)
D	diameter; water vapor resistance ($1/K$); impeller diameter
da	deka
d_v	absolute humidity of moist air; mass of water per unit volume of mixture
E	exa; amount of energy; east
E	energy
ET	equation of time (minutes)
Ex	amount of exergy
E^*	total radiation emission
E_λ	real body (nonblack) radiation
E_λ^*	blackbody monochromatic radiation
f	femto
f	friction factor; correction factor
F	Fahrenheit
F	force; angle factor (dimensionless)
F_A	geometric (or view) factor

F_E	emissivity factor
Fo	Fourier number
FPC	flat plate collector
g	gravitational acceleration
G	giga
G	specific gravity
Gm	gigameter
Gr	Grashof number
Gz	Graetz number
GCV	gross calorific value
h	hector
h	height; head; head loss; enthalpy; heat transfer coefficient; specific energy content; Planck's constant (6.626×10^{-34} J.s)
H	hour angle (° of arc); rate of heat gain for space; Henry's law constant
hp	horsepower, a unit of power in I-P
Hz	Hertz (cycle/s)
h'_{fg}	modified latent heat of vaporization
$h_s@Tw$	specific enthalpy of saturated moist air at thermodynamic wet-bulb temperature
$h_w@Tw$	specific enthalpy of condensed water (liquid or solid) at thermodynamic wet-bulb temperature and pressure of 101.325 kPa
HCV	higher calorific value
HHV	higher heating value
I	current; radiant flux density; amount of solar radiation per unit area per unit time
J	Joule, a unit of energy and work
k	kilo
k	dimensionless coefficient; thermal conductivity
K	Kelvin, the unit of thermodynamic temperature
K	resistance coefficient; permeance; dissociation constant; equilibrium constant
kg	kilogram, a unit of mass
KG	flow coefficient for natural gas control valves
K_l	valve body shape factor
K_v	flow coefficient
kWh	kilowatt-hour (= 3.6 MJ), a unit of energy
l	length
L	local latitude
L	length
LCV	lower calorific value
LHV	lower heating value
LON	local longitude
LPG	liquid petroleum gas
LST	local solar time
LSTM	longitude of the standard time meridian
m	meter, a unit of length in SI; *milli* – one thousandth
m	mass
$\dot{m}$	mass flow rate; vapor transfer rate
M	mega
M	mass; molecular weight
meq	milliequivalent
mol	mole, a unit of amount of substance
n	nano
n	number of moles in a moist air sample
N	Newton; north
N	shaft rotational speed
NE	northeast
Nu	Nusselt number
NW	northwest
N_L	number of pipe rows
NCV	net calorific value
NHV	net heating value
NTU	number of heat transfer units; nephelometric turbidity unit
p	pico
p	pressure; vapor pressure; wetted perimeter
P	peta; unit price
P	power; perimeter; partial pressure
Pa	Pascal
Pe	Peclet number
Pr	Prandtl number
ppm	parts per million
ppt	parts per thousand
q	specific emission
Q	volumetric flow rate; amount of heat energy
$\dot{Q}$	rate of heat
QOP	incident angle for the vertical surface
QOV	angle of incidence
r	radius, reflectance
R	universal gas constant, 8,314.41 J/(kg_{mole}·K); gas constant for dry air or water vapor; radius
R	thermal resistance
Re	Reynolds number
rad	radian, a unit of plane angle
s	entropy
sr	steradian, a unit of solid angle
S	south
S	entropy; gas relative density; salinity
SE	southeast
ST	standard time
SW	southwest
S_L	distance between front and back row pipe centers
St	Stanton number
t	time, transmittance
T	tera
T	time; temperature; dry-bulb temperature of moist air
T_d	dew-point temperature of moist air
T_w	thermodynamic wet-bulb temperature of moist air
THV	total heating value
ton	refrigeration ton, 12,000 Btu/h or 3.52 kW
therm	a unit of heat energy, 105.5 MJ or 100,000 Btu
u	average velocity
U	internal energy; overall heat transfer coefficient; thermal transmittance; ultimate (theoretical) carbon dioxide
v	specific volume
V	volt
V	volume; heat exchanger volume, theoretical volume; total volume of moist air sample; electric potential; potential difference; electromotive force
w	width
W	watt; west
W	work; humidity ratio of moist air; mass of water per unit mass of dry air

$\dot{W}$	rate of work
W_a	theoretical mass of dry air
W_s	humidity ratio of moist air at saturation at thermodynamic wet-bulb temperature
x	distance; mole fraction
y	yocto
Y	yotta; time period (days)
z	elevation of the point above a reference plane
z	zepto
Z	zetta
Z	capacity rate ratio

Greek Letters

α	thermal diffusivity; transfer surface area per heat exchanger volume; ln (p_w), parameter used in Equations (2.114) through (2.117)
β	solar altitude
γ	specific humidity of moist air; mass of water per unit mass of mixture; specific weight, surface solar azimuth
δ	solar declination
Δ	difference
ε	emissivity; effectiveness; surface roughness factor
ς	reflectance
η	conversion efficiency; pump efficiency; thermal efficiency
θ	angle of incidence
θ	pump power; temperature difference
λ	wavelength
µ	micro
μ	dynamic viscosity; degree of saturation W/W_s; water vapor permeability
μ'	mean value of permeability
v	frequency; kinematic viscosity; specific volume; frequency, cycles/s = Hz
ρ	density
σ	Stefan–Boltzmann constant, 5.679×10^{-8} W/m^2.K^4 (0.1714×10^{-8} Btu/h.ft^2.R^4)
Σ	tilt angle of the surface
τ	shear stress
φ	relative humidity; solar azimuth angle
Φ	tilt angle of the Earth
Ψ	surface azimuth angle
Ω	ohm, a unit of electric resistance

Subscripts

a	air; dry air; after the control valve; constituent a
A	cross-sectional area
b	ambient
bk	break power
c	cold; convective
cv	control volume
C	carbon
CO_2	carbon dioxide
d	dew point; diffuse
D	direct
dg	ground reflected diffuse
ds	diffuse solar
DH	direct component of insolation on a horizontal surface
DN	direct normal
e	electrical; in front of the control valve
eq	equivalent
E	electricity; Earth
f	film; fuel; saturated liquid water; frictional
fl	fluid power
g	water vapor; gas; ground
h	hot; hydraulic
H	horizontal
i	in; inside
is	inside surface
ke	kinetic energy
L	latent
LPG	liquid propane gas
m	mean
n	normal conditions
max	maximum
NG	natural gas
o	out; outside
p	constant pressure; pipe; pump
pe	potential energy
r	ground-reflected; summer solstice
s	surface; saturated moist air
S	Sun
S	sensible
sg	between the surface and the ground
ss	between the surface and the sky
SE	Earth to the Sun
sat	saturation
t	total; turbine; direct solar radiation
T	total
u	upstream
v	volume, constant volume, valve
V	vertical
w	wall; window; water in any phase; wetted
ws	saturated water vapor
X	point X
λ	certain wavelength (monochromatic)
a, b, . . ., n	different surfaces
1, 2, 3	numbers; state points
∞	ambient

Superscripts

*	blackbody
o	degree
'	minute (angle)
''	second (angle)

BIBLIOGRAPHY

Introduction: Units and Conversions

Andrews, J and Jelley, N (2007) *Energy Science: Principles, Technologies, and Impacts*. Oxford University Press Inc., New York, pp. 13–14.

ASHRAE (1984) *SI Guide for HVAC & R*, 6th ed., American Society of Heating, Refrigeration and Air Conditioning Engineers, Inc., Atlanta, GA 30329.

ASHRAE (1989) *ASHRAE Handbook of Fundamentals*. American Society of Heating, Refrigeration and Air Conditioning Engineers, Inc., Atlanta, GA 30329.

ASME (n.d.) *ASME Orientation and Guide for Use of Metric Units*. American Society of Mechanical Engineers, 345 East 47th Street, New York, NY 10017.

ASTM (n.d.) Standards for Metric Practice, ASTM Standard E 380 (American Society for Testing and Materials), Philadelphia.

BIPM (2012) *International Vocabulary of Metrology—Basic and General Concepts and Associated Terms (VIM)*. 3rd Edition. The Bureau International des Poids et Mesures.

CGMP (1961) Comptes Rendus de la 11e CGPM (1960), p. 87.

CGMP (1972) Comptes Rendus de la 11e CGPM (1971), p. 78.

CSA (n.d.) *Metric Practice Guide*. CSA Standard CAN 3-Z234-1 Canadian Standards Association, 178 Rexdale Blvd., Rexdale, Ontario M9W 1R3.

IOS (n.d.) *SI Units and Recommendations for the Use of Their Multiples and of Certain Other Units*. ISO Standard 1000 International Organization for Standardization, Geneva, Switzerland; available from American National Standards for Metric Institute, 1430 Broadway, New York, NY 10018.

National Bureau of Standards (n.d.) *The International System of Units (SI)*. National Bureau of Standards Special Publication 330. Superintendent of Documents, U.S. Government Printing Office, Washington, DC 20402.

Wildi, T (1995) *Metric Units and Conversion Charts: A Metrication Handbook for Engineers, Technologists, and Scientists*. IEEE Press, New York, pp. 65–66.

Yildiz, I (2021) *Greenhouse Engineering: Integrated Energy Management*. 1st Edition. CRC Press, Taylor and Francis Group, FL, USA. ISBN: 978-1-4987-4313-6.

Yildiz, I and Liu, Y (2018) Energy units, conversions, and dimensional analysis. In: Dincer, I (ed.), *Comprehensive Energy Systems*. Vol. 1, Elsevier, Oxford, UK, pp. 1–23.

Thermal Energy

Afgan, NH and Schlunder, EU (1974) *Heat Exchangers: Design and Theory Sourcebook*. McGraw-Hill Book Co., Inc., New York.

ASHRAE (1989) *Handbook of Fundamentals*. American Society of Heating, Refrigeration and Air Conditioning Engineers, Inc., Atlanta, GA, pp. 15.1–15.14.

ASHRAE (2013) *Handbook of Fundamentals*. American Society of Heating, Refrigeration and Air Conditioning Engineers, Inc., Atlanta, GA, pp. 6.1–6.17.

ASHRAE (2021) *Handbook of Fundamentals*. American Society of Heating, Refrigeration and Air Conditioning Engineers, Inc., Atlanta, GA, pp. 26.1–26.23.

Bassett, MR and Trethowen, HA (1984) Effect of condensation on emittance of reflective insulation. *Journal of Thermal Insulation*, 8(October): 127.

Brown, AI and Marco, SM (1958) *Introduction to Heat Transfer*. 3rd Edition. McFraw-Hill Book Co., Inc., New York.

Çengel, YA and Boles, MA (2002) *Thermodynamics: An Engineering Approach*. McGraw-Hill, Boston.

Çengel, YA and Ghajar, AJ (2015) *Heat and Mass Transfer: Fundamentals and Applications*. McGraw-Hill Education, New York.

Chaddock, JB (1965) Moist air properties from tabulated virial coefficients. *Humidity and Moisture Measurement and Control in Science and Industry*, 3: 273.

Demirel, Y (2018) Biofuels. In: Dincer, I (ed.), *Comprehensive Energy Systems*, Vol. 1, Elsevier, Oxford, UK, pp. 875–908.

Duffie, JA and Beckman, WA (2013) *Solar Engineering of Thermal Processes*. Wiley, New York.

Engineering ToolBox (2003) Available at: https://www.engineeringtoolbox.com (Retrieved on 23.03.2020).

Goff, JA (1949) Standardization of thermodynamic properties of moist air. *Heating, Piping, and Air Conditioning*, 21(11): 118.

Goff, JA, Anderson, JR and Gratch, S (1943) Final values of the interaction constant for moist air. *ASHVE Transactions*, 49: 269.

Goff, JA and Gratch, S (1945) Thermodynamic properties of moist air. *ASHVE Transactions*, 51: 125.

Granet, I and Bluestein, M (2004) *Thermodynamics and Heat Power*. Pearson Prentice Hall, NJ.

Haines, RW (1961) How to construct high altitude psychrometric charts. *Heating, Piping, and Air Conditioning*, 33(10): 144.

Hardy, HC, Telfair, D and Pielemeier, WH (1942) The velocity of sound in air. *Journal of the Acoustical Society of America*, 13: 226.

Harrison, LP (1965) Fundamental concepts and definitions relating to humidity. In: Wexler, A and Wildhack, WA (eds.), *Humidity and Moisture Measurement and Control in Science and Industry* 3:3. Reinhold Publishing, New York.

Hyland, RW and Wexler, A (1983a) Formulations for the thermodynamic properties of dry air from 173.15 K to 473.15 K, and of saturated moist air from 173.15 K to 372.15 K, at pressures to 5 MPa. *ASHRAE Transactions*, 89(2A): 520–535.

Hyland, RW and Wexler, A (1983b) Formulations for the thermodynamic properties of the saturated phases of H2O from 173.15 K to 473.15 K. *ASHRAE Transactions*, 89(2A): 500–519.

Incropera, FP and DeWitt, DP (1996) *Fundamentals of Heat Transfer*. John Wiley and Sons, Inc., New York.

Karig, HE (1946) Psychrometric charts for high altitude calculations. *Refrigerating Engineering*, 52(11): 433.

Kays, WM and London, AL (1964) *Compact Heat Exchangers*. 2nd Edition. McGraw-Hill Book Co., Inc., New York.

Keenan, JH and Kaye, J (1945) *Gas Tables*. John Wiley and Sons, New York.

Keenan, JH, Keyes, FG, Hill, PG and Moore, JG (1969) *Steam Tables*. John Wiley and Sons, New York.

Khademi, F and Yildiz, I (2018) Energy and solid wastes. In: Dincer, I (ed.), *Comprehensive Energy Systems*. Vol. 1, Elsevier, Oxford, UK, pp. 980–1020.

Kuehn, TH, Ramsey, JW and Threlkeld, JL (1998) *Thermal Environmental Engineering*. 3rd Edition. Prentice-Hall, Upper Saddle River, NJ, p. 188.

Kusuda, T (1970) Algorithms for psychrometric calculations. NBS *Publication* BSS21 (January) for sale by Superintendent of Documents, U.S. Government Printing Office, Washington, DC.

Mason, EA and Monchick, L (1965) Survey of the equation of state and transport properties of moist gases. In: *Humidity and Moisture Measurement and Control in Science and Industry*, 3:257. Reinhold Publishing, New York.

Midwest Plan Service (1987) *Structures and Environment Handbook*. 11th Edition. Ames, IA.

NASA (1976) U.S. Standard atmosphere, 1976. National Oceanic and Atmospheric Administration, National Aeronautics and Space Administration, and the United States Air Force. Available from National Geophysical Data Center, Boulder, CO.

NIST (1990) Guidelines for realizing the international temperature scale of 1990 (ITS-90). NIST *Technical Note* 1265. National Institute of Technology and Standards, Gaithersburg, MD.

Olivieri, J (1996) *Psychrometrics—Theory and Practice*. ASHRAE, Atlanta.

Palmatier, EP (1963) Construction of the normal temperature: ASHRAE psychrometric chart. *ASHRAE Journal*, 5: 55.

Peppers, VW (1988) Unpublished paper. Available from ASHRAE. Preston-Thomas, H. 1990. The international temperature scale of 1990 (ITS90). *Metrologia*, 27(1): 3–10.

Rohsenow, WM (1946) Psychrometric determination of absolute humidity at elevated pressures. *Refrigerating Engineering*, 51(5): 423.

Smithsonian Institution (1954) *Smithsonian Physical Tables*. 9th rev. Edition. Available from the Smithsonian Institution, Washington, DC.

VDI (1999) Environmental meteorology-Interactions between atmosphere and surfaces-Calculation of short-wave and long-wave radiation. *Standard* 3789 Part 2. Verein Deutscher Ingenieure (Association of German Engineers), Dusseldorf.

Yildiz, I (2018) Fossil fuels. In: Dincer, I (ed.), *Comprehensive Energy Systems*. Vol. 1, Elsevier, Oxford, UK, pp. 521–567.

Yildiz, I (2021) *Greenhouse Engineering: Integrated Energy Management*. 1st Edition. CRC Press, Taylor and Francis Group, FL, USA. ISBN: 978-1-4987-4313-6.

Fluid Mechanics, Water Properties and Gas Transfer

American Public Health Association (1976) *Standard Methods for the Examination of Water and Wastewater*. 14th Edition. Rand, MC et al. (eds.). American Public Health Association, Washington, DC.

Anderson, JD (2016) Some reflections on the history of fluid dynamics. In: Johnson, RW (ed.), *Handbook of Fluid Dynamics* (2nd ed.). CRC Press. ISBN 9781439849576.

Batchelor, GK (2000) *An Introduction to Fluid Dynamics*. Cambridge University Press, Cambridge, UK. ISBN 978-0-521-66396-0.

Beck, S and Collins, R (2008) Moody Diagram. Available on: Wikipedia Commons https://commons.wikimedia.org/wiki/File:Moody_diagram.jpg (Retrieved on 21.03.2020).

Blakely, DR and Hrusa, CT (1989) *Inland Aquaculture Development Handbook. Fishing News Books*. Oxford, Great Britain, Chap. 4.

Blasius, H (1911) Das Ähnlichkeitsgesetz bei Reibungsvorgängen. *Physikalische Zeischrift*, 12: 1175–1177.

Boyd, CE (1982) *Water Quality Management for Pond Fish Culture*. Elsevier Scientific Publishing Company, New York, NY, Chap. 2.

Buch, DH (1956) Effects of turbidity on fish and fishing. *Transactions of the North American Wildlife Confederation*, 21: 249–261.

Çengel, YA and Cimbala, JM (2017) *Fluid Mechanics Fundamentals and Applications*. 4th Edition. McGraw-Hill Education, New York.

Chemical Engineers' Handbook (1973) *Chemical Engineers' Handbook*. 5th Edition. Perry, RH and Chilton, CH (eds.). McGraw-Hill Book Company, New York, NY, pp. 3–45.

The Chemical Rubber Co. (1968) *Handbook of Chemistry and Physics*. Weast, RC (ed.), The Chemical Rubber Co., Cleveland, OH, p. F–81.

Darrigol, O and Frisch, U (2008) From Newton's mechanics to Euler's equations. *Physica D: Nonlinear Phenomena*, 237(14–17): 1855–1869. https://doi.org/10.1016/j.physd.2007.08.003

Davis, ML and Cornwell, DA (1991) *Introduction to Environmental Engineering*. 2nd Edition. McGraw-Hill, Inc., New York, NY, Chap. 3.

European Inland Fisheries Advisory Commission (1973) Water quality criteria for European freshwater fish. *Report on Ammonia and Inland Fisheries*, 7: 1011–1022.

Frankland, TM (1984) *The Pipe Fitter's and Pipe Welder's Handbook*. Revised Edition. Glencoe Publishing Co., Mission Hills, CA. ISBN 0-02-802500-8.

Geankoplis, CJ (1983) *Transport Processes and Unit Operations*. 2nd Edition. Allyn and Bacon, Inc., Boston, p. 800.

Haaland, SE (1983) Simple and Explicit Formulas for the Friction Factor in Turbulent Pipe Flow. *Journal of Fluids Engineering*, 105: 89–90. https://doi.org/10.1115/1.3240948

Henry, JG and Heinke, GW (1989) *Environmental Science and Engineering*. Prentice Hall, Englewood Cliffs, NJ, Chap. 6.

Landau, M (1992) *Introduction to Aquaculture*. John Wiley & Sons, Inc., New York, NY, Chap. 2.

Masters, GW (1991) *Introduction to Environmental Engineering and Science*. Prentice Hall, Englewood Cliffs, NJ, Chap. 4.

McLarney, W (1984) *The Freshwater Aquaculture Book: A Handbook for Small Scale Fish Culture in North America*. Hartley & Marks, Publishers, Vancouver, BC, Chap. X–1.

Piper, RC, McElwain, IB, Orme, LE, McCraren, JP, Fowler, LG and Leonard, JR (1982) *Fish Hatchery Management*. United States Department of the Interior, Washington, DC, Chap. 1.

Russo, RC and Thurston, RV (1991) *Aquaculture and Water Quality—Advances in World Aquaculture*. Vol. 3, The World Aquaculture Society, Baton Rouge, Louisiana.

Stickney, RR (1979) *Principles of Warm Water Aquaculture*. John Wiley & Sons, New York, Chap. 3 and 4.

Swamee, DK and Jain, AK (1976) Explicit equations for pipe flow problems. *Journal of the Hydraulics Division*, 102: 657–664.

U.S. Geological Survey (1985) *Study and Interpretation of the Chemical Characteristics of Natural Water*. Water—Supply paper 2254. Department of the Interior, Washington, DC.

Weber, WJJ and Digiano, FA (1996) *Process Dynamics in Environmental Systems*. John Wiley & Sons, Inc., New York.

West, RC (ed.) (1968) *Handbook of Chemistry and Physics*. The Chemical Rubber Co., Cleveland, OH, p. A–10.

Wheaton, FW (1977) *Aquacultural Engineering*. John Wiley & Sons, New York, NY, Chap. 2 and 4.

Wilson, KG and Arons, AB (1955) Osmotic pressures of seawater solutions computed from experimental vapor pressure lowering. *Journal of Marine Science*, 14(2): 195–198.

Yildiz, I (2021) *Greenhouse Engineering: Integrated Energy Management*. 1st Edition, CRC Press, Taylor and Francis Group, FL, USA. ISBN: 978-1-4987-4313-6.

Light, Radiation, and Characteristics of Solar Radiation

Ahrens, CD (2004) *Essentials of Meteorology: An Invitation to the Atmosphere*. 4th Edition. Thomson Learning, Inc.

ASHRAE (1989) *Handbook of Fundamentals*. American Society of Heating, Refrigeration and Air Conditioning Engineers, Inc., Atlanta, GA 30329, pp. 27.1–27.38.

Hinrichs, RA and Kleinbach, M (2006) *Energy: Its Use and the Environment*. 4th edition. Thomson Brooks/Cole, Belmont, CA, pp. 160–204.

McAlester, AL (1983) *The Earth: An Introduction to the Geological and Geophysical Sciences*. Prentice-Hall, Englewood Cliffs.

NASA (2016) Surface Meteorology and Solar Energy Data Set. http://eosweb.larc.nasa.gov/sse/ (Retrieved on 11.11.2016).

Robinson, PJ and Sellers, AH (1999) *Contemporary Climatology*. 2nd Edition. Pearson Education Ltd, Essex, England, pp. 17–38.

Sen, Z (2008) *Solar Energy Fundamentals and Modeling Techniques*. Springer-Verlag London Limited, pp. 47–98.

Stull, RB (2000) *Meteorology for Scientists and Engineers*. 2nd Edition. Thomson Brooks/Cole, Pacific Grove, CA, pp. 23–38.

Taylor, FW (2005) *Elementary Climate Physics*. Oxford University Press Inc., New York, pp. 29–39.

Threlkeld, JL (1962) *Thermal Environmental Engineering*. Prentice Hall, New York, p. 321.

Yildiz, I (2018) Solar energy. In: Dincer, I (ed.), *Comprehensive Energy Systems*. Vol. 1, Elsevier, Oxford, pp. 638–664.

Yildiz, I (2021) *Greenhouse Engineering: Integrated Energy Management*. 1st Edition. CRC Press, Taylor and Francis Group, FL, USA. ISBN: 978-1-4987-4313-6.

3 Plant, Animal, and Aquatic Structures

3.1 INTRODUCTION

Controlled-environment buildings serve a variety of purposes, but they primarily provide suitable environments for intensive production of a variety of crops, livestocks and fish. They conserve both land and water, enabling sustainable waste management and maximizing sustainable production, which is particularly important in regions where land and water are expensive or not readily available; they offer viable solutions for water use conflicts, water quality, and waste disposal.

Controlled-environment operations, particularly greenhouses and recirculating aquaculture systems (RAS) facilitate profitable year-round agricultural and aquacultural production to competitively respond to market demand, especially in areas of geographic extremes such as high aridity, cold, or heat and in food deserts, considerably reducing transportation distances and costs. Old warehouse spaces are wonderful properties for both controlled-environment aquaculture and greenhouse/factory operations in urban environments. Overall, controlled environments serve agrifood production and justice, energy production and justice, and environmental protection and justice extremely well, which after all is the goal.

Controlled environments are provided for optimal plant, livestock, and fish growth under safe and efficient work conditions. Therefore, controlled-environment buildings and all their parts should be designed and built by first choosing a proper site and safely supporting all loads, including dead load, without exceeding the allowable stresses for the construction materials. However, load analyses and requirements for designing controlled-environment buildings and their components is not the focus of this chapter or of the book; therefore, they will not be dealt with in depth here. Rather, the readers of this book are referred to the existing literature and design manuals such as NGMA (2004) for design loads and structural designs. These manuals provide guidelines for engineers in the form of building code provisions for production and commercial greenhouses and livestock and aquatic buildings. The codes include the provisions for roof live load, snow load, wind loads, and seismic and collateral loads and often provide structural design examples. Overall, this chapter will cover site selection and layout and design loads for controlled-environment buildings and then provide typical foundations, floors, framing materials and structural forms, covering materials, curtain systems, and insect screens. The chapter will conclude with aquatic culture systems including round tanks, and cross-flow and mixed-cell raceways.

3.2 SITE SELECTION AND LAYOUT

Choosing a good building site and a careful plan is the first and foremost consideration before a controlled environment facility is designed and built as it affects a number of vital requirements of the operations. Careful planning includes reviewing the present, assessing the future, and providing flexibility for higher capacity and expansion. Obviously, it is much less costly to correct an error on paper at the planning stage than after construction begins or is completed. We should keep in mind that a controlled-environment facility in the wrong place is at least a 15- to 20-year mistake that has multiple long-term effects on operation.

3.2.1 PLANNING FACTORS

Sites for controlled-environment facilities are typically designed using a zoning approach. A zoning approach divides a controlled-environment operation into separate activity zones, which can include a family living area decorated with guest parking, recreation, and flower and vegetable gardens; a zone for shop, machinery, chemical, pesticide, fertilizer, and fuel storage; a zone for greenhouse, small livestock, grain, and feed storage; and yet another zone for aquaculture, large livestock units, or expansion for other controlled-environment operations.

For an operation with a family living area, the family house is the center of the planning zones surrounded by easily landscaped facilities and farther away from less attractive facilities. This area is protected from dust, noise, and odors with space and trees and needs wind protection via plantings or fences that block summer breezes and cause snow drifts over driveways and walkways. Heavy and large equipment, and fire hazards are also kept away from this zone. For operations without a family house, the activity zones are centered around the business operation court. In this approach, even though wider zones are preferred, depending on the land availability, typically 30 m apart circles are drawn using the house or the business operation court as the center.

Fertilizers and pesticides create serious storage problems. For instance, ammonium nitrate is an explosive fertilizer if contaminated with fuel, oil, or certain other materials. Moreover, some herbicides and insecticides can contaminate each other and fertilizers, and some combinations are flammable. Chemicals must therefore be stored in their original containers so that they can remain identified with instructions for their use and special care. Local laws and regulations must be followed closely for proper storage, use, and disposal of these products.

Again, it must be emphasized that the zone planning has to accommodate expansion and new facilities; we therefore typically assume that our operation will at least double in size in the future. For livestock operations, we have to plan for not only animals themselves but also grain storage. It should be noted that large livestock and aquaculture operations need adequate space, drainage, waste management, vehicle access, loading facilities, feed distribution, and other services; hence these facilities are located within the last zone. Dust, noise, odor, micro-organisms, bio-aerosol emissions, and waste disposal create serious problems in

DOI: 10.1201/9781003349297-3

intensive operations; therefore, such units and high-labor facilities are located as far from the family living and neighboring residential areas as possible.

Animals in closed and fan-ventilated controlled-environment buildings typically create less nuisance. The milkhouse of a dairy center as well has little effect on family living depending on the visits of the bulk milk truck. Most of the time, nuisances are either air- or water-borne; hence waste handling facilities should be located downwind from living centers, and any runoff must be controlled before it leaves the operation.

Clearance between buildings as well is an important consideration for vehicle access, snow storage (where applicable), and fire protection. Furthermore, naturally ventilated buildings require certain (approx. 15 m or 50 ft) clearance. Further considerations affecting controlled environment planning and essential factors for different operations are discussed next.

3.2.2 Topography, Climate, and Utilities

Topography, local climate, and availability of utilities all affect operations with respect to drainage, building location, access routes, prevailing wind directions, and overall feasibility of the operation in the first place. The following sections will review each of these specific items under individual subsections.

3.2.2.1 Topography and Drainage

First, topography determines overall drainage patterns, prevailing wind directions, and therefore, the building location. It can restrict the space available for a large center and can also affect access routes and convenience to utilities. Operations should not be located on a flood plain, on low ground, in peat or very rocky soil, etc. Surface and subsurface drainage as well as ground slope are all important features for effective operations. We should avoid sites near high water tables (rivers, springs, etc.) and rough topography or any other features limiting expansion. A well-drained site will make the construction much easier and help prevent the environmental stress on the controlled-environment facility in the long run.

Proper surface drainage is naturally provided if the site has a slope somewhere between 2% and 6%. While natural drainage is preferred, sometimes improvements may be needed, and that is typically done by grading. In such cases, surface water is intercepted and diverted by constructing terraces. In brief, well-drained soil is desirable, and therefore, the highest part of the site is typically selected for buildings. Furthermore, subsurface drainage is needed around building foundations and below-ground storage when available. At the planning stage, special attention must be given to ensuring that subsurface drainage does not pollute nearby water resources.

3.2.2.2 Wind and Snow Control, and Orientation

In addition to taking advantage of topography, planners also need to consider windbreaks to help reduce winter winds and control snow for a comfortable living area and operation space. High winds drain the interior heat and increase the operational costs. If a windbreak is provided naturally by choosing a sloping landscape in the planning stage, that is perfect; if not, however, a site that allows for planting trees as a windbreak that will not cause too much shadowing is still effective.

Windbreaks slow down wind speed, causing blowing snow and dust to accumulate along the breaks. When locating windbreaks, special consideration should be given to allow proper drainage and air movement in in particular in warmer months. Multi-row windbreaks of tall and short trees, evergreens, and shrubs reduce wind velocities 5 to 10 tree heights upwind and 10 to 20 tree heights downwind. Solid fences provide short-range wind protection but tend to cause snow accumulation on both sides of the fence. In contrast, fences with about 80% solidity reduce wind speed over longer distances and help distribute snow more evenly, promoting faster melting.

Moreover, we should keep in mind that summer ground vegetation cover such as bushes and grass considerably reduces the infrared and shortwave radiation reflection from ground and surroundings; therefore, the buildings should be located where they can benefit most, particularly in warmer months, from the surrounding vegetation cover, and the building surroundings should be properly vegetated. Depending on the type of controlled-environment operation, buildings can be further protected from excess solar radiation via awnings, roof overhangs, trees, and reflective roofing materials.

When possible, a south-facing slope in the Northern Hemisphere (a north-facing slope in the Southern Hemisphere) should be preferred for all controlled environment buildings as the exposure promotes solar radiation, drying, crop growth, and all sorts of other good things during winter months and protect buildings from strong northerly winds in the Northern Hemisphere (southerly winds in the Southern Hemisphere) providing a natural windbreak as well as adequate drainage. Overall, a good site for a controlled-environment facility meets the requirements of easy access to a good all-weather road, well-drained ground with enough family guest and business visitor parking, and sufficient distance from living areas to avoid creating nuisance from dust, flies, noise and odor, particularly from livestock and labor-intensive operations.

3.2.2.3 Utilities

In addition to the above-mentioned requirements, adequate quality and quantity water, electricity, gas, reliable telephone and internet interception, fire prevention, safety, and security are other essential services, which are required for a successful controlled-environment operation. In some cases, water storage tanks for peak needs or liquid propane tanks are needed.

Buried utility lines, typically 35–40 cm deep, are preferred to avoid damage from heavy machinery, reduce hazards, and improve aesthetic of the operation. It must be ensured that water and sewer lines are buried below average frost penetration depth. When possible, buildings should be spaced at least 15 m apart for reducing fire spread and permitting access for firefighting. Just as any other business, controlled-environment operations are also vulnerable to arson, theft, vandalism, etc.; therefore, security for these investments is a problem and must be taken seriously.

3.2.3 Essential Factors

So far, this chapter has discussed the overall planning factors for controlled-environment facilities. This subsection will cover the essential factors specifically for greenhouses and livestock and aquaculture facilities.

3.2.3.1 Greenhouses

Greenhouses are controlled environments that are built in many different variations to meet grower and other site specific conditions and functional needs. NGMA (2004) defines a greenhouse as a production greenhouse if it is occupied for growing plants on a production or research basis and is not normally occupied (no public access). Incidental use such as tending or moving products is considered production activity and is therefore not considered occupied. Any greenhouse building used for the display and sale of horticultural crops and supplies is termed a commercial greenhouse.

These two groups of greenhouses function in different ways and hence their functional and orientation needs are different and sometimes dictated by the existing buildings and other structures. Hence, the best location, layout, and orientation (the direction the greenhouse is facing) strongly depends on how the greenhouse is planned to be used and the position of the property with respect to a nearby public road, and as mentioned, other nearby buildings and structures. For instance, gutters oriented north–south give a constantly changing shadow while the Sun moves from east to west throughout the day. However, gutters oriented east–west cast a relatively permanent shadow as the Sun moves throughout the day and hence produce poorer light uniformity. In any case, the layout of a greenhouse facility should consider future expansion, locating the headhouse (the work and indoor storage area where boiler rooms are typically located) to the north of the crop production area to prevent shading (Figure 3.1), separating the commercial area from the production area, providing accessible customer and supplier traffic and parking for disabled persons, and permitting windbreaks and privacy to a residence as discussed earlier, when possible.

Two of the most essential requirements for controlled-environment greenhouse operations are dependable supply of energy and quality water. Dependable supply of energy means electricity, and depending on the location, fuel for heating if needed. Electricity can be obtained from the grid or from a cogeneration system built within the operation complex if the cost is justified. If the operation relies on the grid, we need to make sure that the cost of bringing power will be reasonable, that is that access to the grid will be easy, so that it will not put excessive burden on the operation.

Another consideration for providing electricity is the environment. This entails checking the utility company's energy mix to see whether it is using clean, renewable, and local energy sources for power generation. If greenhouse heating will be required in the region, then we also need to know what kinds of energy sources are available for heating at a reasonable cost. If the heating fuel will be outsourced,

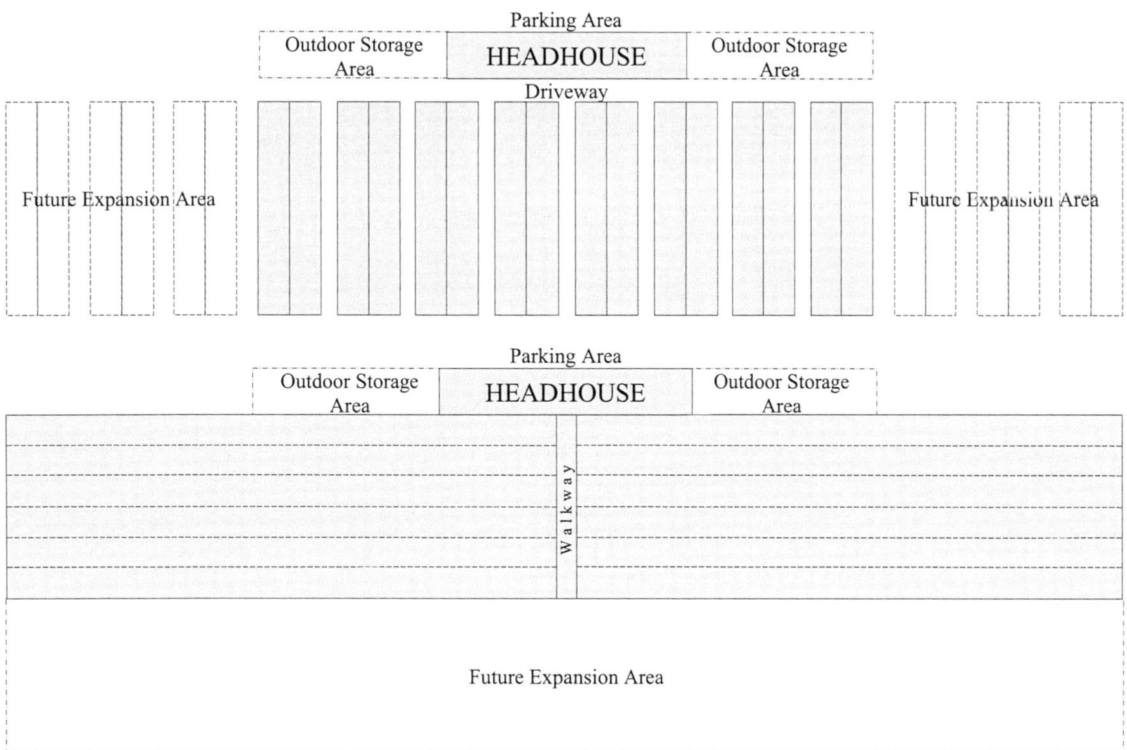

FIGURE 3.1 Layout examples for individual (or twin) greenhouses with separate headhouse (top) and gutter connected greenhouse with attached headhouse (bottom).

Source: ***Greenhouse Engineering*, Edn. 1 by İlhami Yıldız, Copyright © 2021 by Imprint. Reproduced by permission of Taylor & Francis Group (*Greenhouse Engineering*).**

we also need to make sure that the operation has easy access to, for instance, a natural gas line or an all-season public road so that the operation has an uninterruptable supply of heating fuel. As discussed earlier, having easy access to a public road is also important for delivering other supplies to the operation and delivering the operation output to market with little or no problems.

Dependable supply of quality water is a vitally important requirement. There will not be any controlled-environment plant, livestock, or aquacultural production unless there is enough quality water; therefore, availability of municipal water, surface water, or groundwater potential, relevant regulations, and their qualities must be known before an intensive production under controlled environment is attempted.

Municipal water supplies are of high quality for humans, plants, and livestock; however, it is not the case for aquaculture while it is available in sufficient quantities. Municipal water is designed for human consumption, and therefore, treated to destroy pathogenic organisms; hence it often contains chemicals such as chlorine and fluorine, which are very toxic to fish. These chemicals must be neutralized before using municipal water in aquaculture. When municipal water is too expensive or not available, then surface water and groundwater can definitely be used for plant and livestock production, again depending on the availability and quality. Each of these sources has different characteristics, and advantages and disadvantages for specific controlled environments.

Surface waters coming from turbulent, well-mixed sources such as small rivers and streams are fine; however, while they are well supplied with dissolved oxygen, they should not be the sole source of water due to the extensive damage they can cause if they are contaminated with poisons, predators, or competitor organisms. Less-turbulent, low-circulation surface water sources include large rivers, lakes, ponds, and reservoirs. Turbidity and suspended solids can be problems in surface waters. Surface waters typically have low dissolved minerals and low hardness, and their pH and mineral contents depend on the watershed area.

Groundwater from springs, wells, and depressions is typically free from bacteria, viruses, and turbidity, but it has low or no dissolved oxygen and can contain high levels of toxic dissolved gases such as nitrogen, carbon dioxide (CO_2), methane (CH_4), and hydrogen sulfide (H_2S) that are not desirable for aquaculture. Therefore, before using, the groundwater must be conditioned properly through aeration and degassing in order to add oxygen and remove the excess gasses.

High concentrations of iron can also be a problem with groundwater and have to be removed via aeration. When the water is aerated, iron precipitates as iron oxide, which can be removed by granular filtration or settling basin. In limestone areas, groundwater tends to have high hardness and high alkalinity. In areas of granite bedrock, groundwater tends to be soft and low in dissolved minerals, but it can have high dissolved CO_2 content. Groundwater also tends to have more consistent year-round temperatures, whereas surface waters can have wide seasonal temperature fluctuations, which can be undesirable thing for a specific species under the controlled environment in mind. To see the importance of available water resources for controlled-environment operations, we can easily look at very successful greenhouse operations around the world and see that some of these successful operations are having extremely difficult times with operational expansion due to limited water resources available in the region.

TABLE 3.1
Classification of Water Quality

Class	Electrical Conductivity (mS/cm)	Sodium (mg/L)	Chloride (mg/L)	Sulphate (mg/L)
1	0.5	30	50	100
2	0.5–1.0	30–60	50–100	100–200
3	1.0–1.5	60–90	100–150	200–300

Source: OMAFRA (2001).

As discussed, the available water quantity can be less of an issue than water quality. Therefore, the quality of available water must also be tested before planning a controlled environment facility. Table 3.1, for instance, provides water quality classification; it categorizes water quality for greenhouse use into three classes: Class 1 water is good for all purposes; Class 2 water can be used in soil culture and substrate where leaching can be performed, but it should not be used for nutrient film technique plants or when nutrient solution recirculation is considered; Class 3 water should not be used for crops that are sensitive to salt.

Additionally, ponds and cisterns for rainwater collection and storage might be considered; however, this as a sole source will not supply enough water for year-round crop production. It should be kept in mind that the irrigation system must be able to provide an average of 8 L/m^2.day of water for a fully grown crop. Some crops such as cucumbers require up to 5 L/plant.day on sunny days (see Figure 3.2).

3.2.3.2 Animal Facilities

In addition to the planning requirements, just like for greenhouses, dependable supply and quality water is the most essential requirement for controlled-environment animal (i.e., livestock and poultry) operations. Water is followed by drainage and waste disposal.

Optimum livestock and poultry production requires cool, clean, and adequate quantity water from any of the water sources just discussed. Livestock operations also require water for cleaning (warm water for milking and milking equipment), sanitation spraying, waste removal, and hot weather cooling of animals. Therefore, operations must make sure that sufficient supply of quality water is available for these purposes. An adequately designed water supply provides enough water needed in the operation every single day all year long. The supply system must also meet the demands for peak use periods continuously, which are typically cleanup and feeding times, and may last up to two hours. Livestock water consumption is affected by species, size, age, milk or egg production, air temperature, type of ration, dry matter consumed, and other variables. Typical water requirements for livestock production are provided by MWPS (1987).

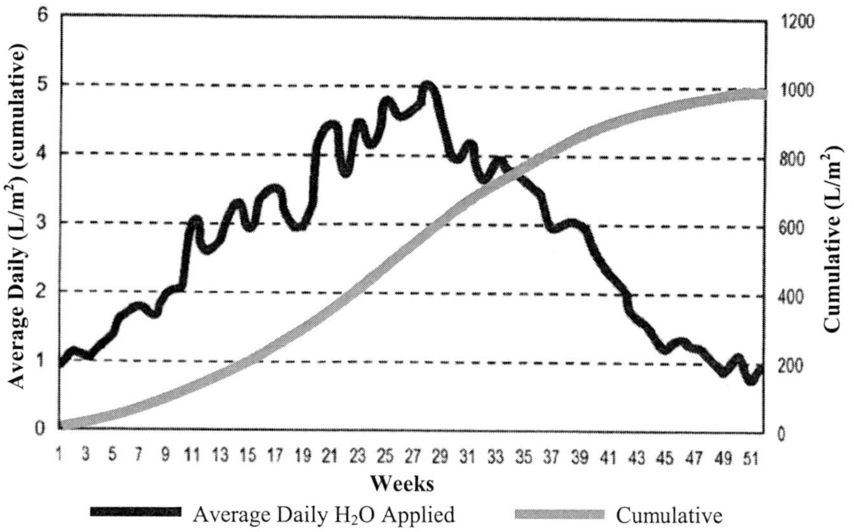

FIGURE 3.2 Average daily water application to greenhouse vegetables in Harrow, ON, Canada (OMAFRA, 2001).

Water from natural water sources typically contains impurities (e.g., silt, clay, algae, colored matter, insects, worms) that impact the water usability and suitability. While drinking water must be safe to drink and have acceptable flavor, appearance, and odor, water for livestock does not have to be as free of impurities as much as long as it does not contain biological and chemical impurities that adversely impact animal health. Maximum chemical concentrations for safe drinking water (free from disease-causing organisms) are also provided by MWPS (1987).

Drainage is another essential element for a livestock operation; however, it is usually possible to fix potential issues. Poor drainage can cause serious headaches and even destroy potentially successful operations due to waste management problems. Therefore, livestock facilities should be sited on relatively high ground to support gravity-based waste handling and ensure effective drainage of rain, snowmelt, and subsurface water around building foundations and pits.

The issue of waste handling and disposal in large modern livestock facilities has become a critical factor, and as will be discussed later, a separate cash-generating activity center by itself. As a result, good livestock management practices and justified public demands have placed new regulations and requirements on modern livestock operation centers. Local regulations must be followed closely because they can prohibit certain developments in the first place and require or prohibit some management practices and hence limit planning decisions.

Furthermore, an east–west orientation is usually preferable for livestock buildings to minimize exposure to the Sun, etc. In hot weather, light breezes or natural ventilation across the buildings are highly desirable, and therefore, a prevailing wind can justify slight deviations from the preferred east–west orientation. In summary, keeping in mind that we have to feed an ever-growing world population, we must continuously work on the moving target of sustainable livestock production that will ensure profitable operations for safe and secure food for all while providing high standards of welfare for animals and a clean environment for all to enjoy for many generations to come.

3.2.3.3 Aquatic Facilities

Aquaculture is the farming of aquatic organisms, and it supplies more than half of all seafood consumed today. Aquaculture systems are classified as extensive (pond culture), semi-intensive (cage culture), or intensive (recirculating aquaculture systems), depending on the water source and the number of aquatic organisms grown per water volume. Extensive and semi-intensive systems are difficult to control as open air systems for which we have almost no control on their environments; these are always susceptible to airborne and waterborne contaminants, and they have long-term sustainability issues. This book focuses on fully sustainable, infinitely scalable controlled-environment recirculating aquaculture systems (RAS), although some fundamental information is provided for other systems.

As mentioned, we have total control of the environment in RAS, where fish are grown in secure tanks in enclosed environments where the aerial environment of the building is also controlled, just like in greenhouses and dairy and poultry buildings. In RAS, all the environmental variables, such as temperature, oxygen, pH, salinity, alkalinity, and ammonia, nitrite, CO_2, and suspended solids, are continuously monitored and fully controlled. Water recirculates through the system components, and due to their low water makeup requirements (e.g., 10% per day or less), they capture 96% to 100% of the waste produced (Heinen et al., 1996). This means that the solid wastes are filtered and removed, adequate dissolved oxygen (DO) levels are maintained through oxygen addition, and effluent water is sent through a biofilter for biological conversion of ammonia nitrogen to nitrate nitrogen.

As discussed earlier, one of the most critical factors to consider when choosing a proper site for a controlled-environment operation is a source of quality water with continuous sufficient supply. Since surface waters are usually not used for aquaculture due to the risks mentioned earlier, the two most common water sources left are municipal water and groundwater, assuming that they are economically available at required quantities, and their regulations and qualities permit their use for intensive aquaculture production under controlled environments.

The amount of water needed for an aquaculture operation is a function of different factors such as species, density, production technology, and management practices. At minimum, enough water should be available for filling production tanks within 24 to 48 hours and for routine and emergency flushing of tanks, filter backwashing, facility clean up, and other domestic needs (Timmons and Ebeling, 2010). As a good rule of thumb, Timmons and Ebeling (2010) further suggest having sufficient water for 100% exchange of total system volume per day. Beyond the minimum, the amount of additional new water for a given system is obviously a function of the degree of reuse and recycling of the water already in the system.

While a number of water quality concerns have been discussed earlier, the effect of certain organic compounds and high fine solid contents is still not known; there is no single list of water quality standards for aquaculture as the species, temperature regimes, and production techniques all affect the specific needs. What might be considered good water for one type of organism may not be good for another.

Water quality typically has the most important impact on fish growth, and a list of recommended water quality characteristics for aquaculture is provided in Table 3.2. Typically, DO is the primary water quality factor that limits fish growth and population density. The other two most important factors are unionized ammonia and dissolved CO_2 concentrations. Full RAS reuse much more of their total water volume; as a result, the depletion or accumulation rates of water quality parameters are affected and therefore must be treated accordingly. All these and other environmental factors, as well as their interrelationships and effects on fish growth will be discussed further in Chapter 4.

TABLE 3.2

Recommended Water Quality Characteristics for Aquaculture[a]

Parameter	Concentration (mg/L)	Parameter	Concentration (mg/L)
Alkalinity (as $CaCO_3$)	50–300	Nitrate (NO_3)	0–400 or higher
Aluminum (Al)	<0.01	Nickel (Ni)	<0.1
Ammonia (NH_3-N unionized)	<0.0125 (Salmonids)	Dissolved Oxygen (DO)	>5
Ammonia (TAN) Cool-water	<1.0		>90 mm Hg partial pressure
Ammonia (TAN) Warm-water	<3.0	Ozone (O_3)	<0.005
Arsenic (As)	<0.05	PCBs	<0.002
Barium (Ba)	<5	pH	6.5 – 8.5
Calcium (Ca)	4–160	Phosphorous (P)	0.01 – 3.0
Carbon dioxide—Tolerant sp.	<60	Potassium (K)	<5
Carbon dioxide—Sensitive sp.	<20	Salinity	Depends on salt or fresh sp.
Chlorine (Cl)	<0.003	Selenium (Se)	<0.01
Total hardness (as $CaCO_3$)	>100	Silver (Ag)	<0.003
Hydrogen cyanide (HCN)	<0.005	Sodium (Na)	<75
Hydrogen sulfide (H_2S)	<0.002	Sulfate (SO_3)	<50
Iron (Fe)	<0.15	TGP (total gas pressure)	<105% (species dependent)
Lead (Pb)	<0.02	Sulfur (S)	<1
Magnesium (Mg)	<15	TDS (total dissolved solids)	<400 (site and sp. specific)
Manganese (Mn)	<0.01	TSS (total suspended solids)	10–80 (species dependent)
Mercury (Hg)	<0.02	Uranium (U)	<0.1
Nitrogen (N_2)	<110% total gas pressure	Vanadium (V)	<0.1
	<103% as nitrogen gas	Water temperature	4–18°C[b]
Nitrite (NO_2)	<1, 0.1 in soft water		

Heavy Metal Water Quality Criteria

Metal (µg/L)	Freshwater				Seawater
	500*	100*	10*	1*	
Copper	35	9	1.3	0.18	3.1
Zinc	460	120	17	2.4	81
Cadmium	0.75	0.25	0.049	0.01	8.8

*Hardness (mg/L as $CaCO_3$)

Sources: [a]Meade (1985), Piper et al. (1982), Lawson (1995); [b]New Brunswick Department of Fisheries and Aquaculture (1995).

3.3 DESIGN LOADS

Controlled-environment greenhouse and livestock and aquaculture buildings and all structural parts are designed and built to support four general types of loads: dead loads, the weight of the structure; live loads, loads associated with building use; and snow and wind loads that do not exceed the allowable stresses for the materials. Sometimes, the building is a lean-to or an attached even-span structure that is added to an existing building. In these cases, we need to make sure that the existing structure is properly strengthened to withstand existing loads and the additional loads imposed by the lean-to or attached structure. It should be kept in mind that the most severe effect from any load combination governs the design of the structure, foundation, or structural member under consideration.

3.3.1 Dead Loads

These loads are the weight of all permanent construction material including but not limited to floor, walls, roofs, framing and glazing materials, and fixed service equipment. The actual weights of relevant building materials are used in estimating dead loads. These weights are typically tabulated and provided in handbooks for building designs.

3.3.2 Live Loads

These are the temporary loads generated by the use and occupancy of the controlled-environment building and do not include dead, wind, or snow loads. Live loads both static and dynamic can be exterior or interior based on building use and occupancy. Static loads result from the weights of equipment, plants, animals, stored products, and maintenance materials. Dynamic loads, on the other hand, result from the dynamic effect of material handling equipment: forklifts, tractors, etc.

Exterior live loads can include the temporary loads of workers on building roofs and temporary equipment such as scaffolds. Interior live loads, on the other hand, are temporary loads imposed on the structure by objects hanging for less than 30 days. If a live load is imposed on the structure for a continuous period of 30 days or more, then such loads are handled as permanent and hence included as part of the dead load. Live dynamic loads due to materials handling, irrigation equipment, etc. are also treated as live loads.

3.3.3 Snow Loads

Snow load is a vertical load based on expected ground snow accumulation at the building site, roof slope, covering material, and type of building structure, that is, whether the building is a livestock building or a single-span or gutter-connected greenhouse, for instance, and whether it is heated or unheated.

Snow ground loads are based on the 50-year mean recurrence interval (with an annual probability of 0.02), and the readers of this book are referred to local sources for pertinent data. The nearby buildings should also be taken into account as they influence snow sliding from their roofs and snow drifting. Building gutter slopes should be designed carefully so that snow melt and rain can smoothly run off without accumulating. Enough space (at least 3 m or 10 ft) between buildings should also be left for snow accumulation, vehicle access, and fire protection. This space also prevents the risk of sidewall crushing due to snow sliding off greenhouse roof, for instance. As mentioned earlier, naturally ventilated buildings also require certain (approx. 15 m or 50 ft) clearance.

3.3.4 Wind Loads

The design wind loads for controlled-environment structures as a whole or for individual components and glazing are based on wind speed data with a 50-year mean recurrence interval (with an annual probability of 0.02) and determined by an analytical procedure that covers the velocity pressure, gust response, and appropriate pressure or force coefficients. The readers of this book are referred to local sources for pertinent wind speed data.

Wind design pressure, that is, the equivalent static pressure, is used for determining wind loads on buildings. The pressure due to wind load is assumed to act in a direction normal to the surface under consideration, either as positive pressure directed toward the surface or as negative pressure (suction) directed away from the surface. In calculating the wind loads for components and covering materials, the pressure difference between opposite faces of the surface are taken into consideration. Exposure is also determined for the building site and reflects variations in surrounding ground surface roughness due to natural topography and vegetation as well as existing construction. Greenhouses particularly have very small deadweights, that is, there is not much to resist the uplift force created by the wind; therefore, we need to make sure that enough anchorage to the ground is provided against the uplift. It must be noted that no cookie cutter approach will work for determining controlled-environment building design loads as every project is unique, and each structural element must be designed with great care. The readers of this book are referred to MWPS (1987), NGMA (1996), and ASABE Standards (2018) for further information on loads.

3.4 FOUNDATIONS AND FLOORS

As for any other building, a proper foundation is essential for the usefulness and life of a controlled-environment building. A building foundation resists external and internal forces, including the weight of the structure itself—such as a greenhouse or livestock or aquaculture building—as well as snow loads, wind pressure, and the weight of stored contents. A good foundation is able to distribute these forces so that movement of the building is small and uniform.

Greenhouse foundations can be flagpole type placed directly in the ground, spread footings, or continuous concrete footings. Figure 3.3 shows some typical greenhouse foundations. Flagpole footings are designed for vertical and horizontal loads. The footing must be able to handle the vertical loads and distribute the lateral bearing to the soil produced by a lateral load applied to the column. Sometimes the flagpole type footings are backfilled with concrete or other filling material.

The footings are typically the unrestrained type; however, a restrained flagpole footing design can also be employed when there is a slab on grade surrounding the footing. Spread footings as well are designed to distribute vertical and horizontal loads to the soil. The size of the footing is determined by the vertical load acting on the footing and the bearing pressure of the soil. Continuous footings are also frequently used and often constructed as a base for the glazing, for instance. The footings rest on level, undisturbed soil and are set below the frost depth, typically at a minimum depth of 60 cm (24 in.) beneath the ground surface as shown in Figure 3.3.

Concrete masonry, reinforced concrete, galvanized steel, and treated wood can be used for piers, and the walls between the piers can be built using poured concrete, precast concrete, masonry, or any decay-resistant material. A continuous concrete or masonry foundation wall is built on poured concrete footings. When for instance, 6 m (24 ft) span greenhouses are in consideration, then a 15 cm (6 in.) wall is considered sufficient, whereas a 20 cm (8 in.) wall should be used for wider-span greenhouses (Aldrich and Bartok, 1994). Pipe or tube foundations driven into the ground without footing should not be used for permanent controlled-environment buildings as they do not provide much resistance to uplift in high-wind locations. When they are used for temporary (less than a year) buildings such as pipe frame (hoop) greenhouses, then a minimum of 75 cm (30 in.) ground insert from grade should be used.

The best greenhouse flooring materials are gravel, pea stone, and trap rock, which are designed specifically for greenhouses. A thickness of 15 to 20 cm (6 to 8 in.) allows for good drainage while blocking weeds. When a smoother and harder flooring surface is desired, then a 5 to 7.5 cm (2 to 3 in) thick porous concrete made from uniform-sized aggregate and a cement–water paste can be used. Porous

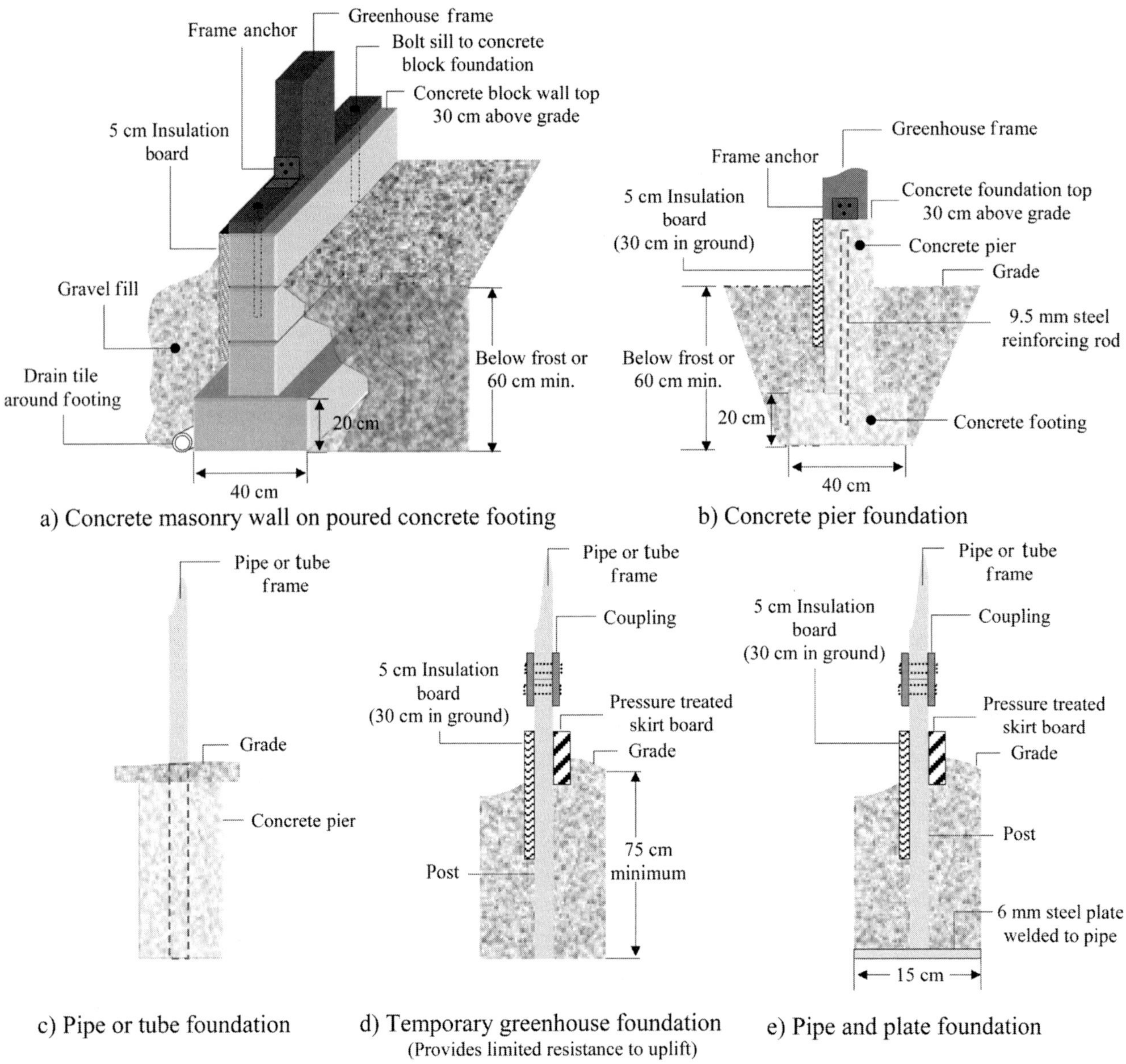

FIGURE 3.3 Typical greenhouse foundations.
Source: *Greenhouse Engineering* (2021).

concrete flooring should only be used in areas where personnel and light vehicles such as garden tractors operate; they are not recommended for soil mixing areas where small particles have the potential to clog the pores of concrete and therefore prevent downward movement of water. The flooring in heavy traffic areas and aisles should be asphalt or concrete whose thickness depends on the traffic load, but 7.5 to 10 cm (3 to 4 in.) is recommended in such areas (ASABE Standards, 2018). Floor surfaces should have a proper slope for drainage and a final rough broom finish for safety purposes and adequate traction for pedestrians and material-handling machinery.

Livestock building foundations can also be of the type floating slab with or without raised sill, curtain wall (concrete, concrete masonry, or treated wood), manure pit and basement walls (reinforced concrete, concrete masonry, or treated wood), and post and grade beam (concrete or wood). Figure 3.4 shows some typical livestock building foundations. Footings are included under posts in post-frame buildings as well because snow loads and wind uplift can

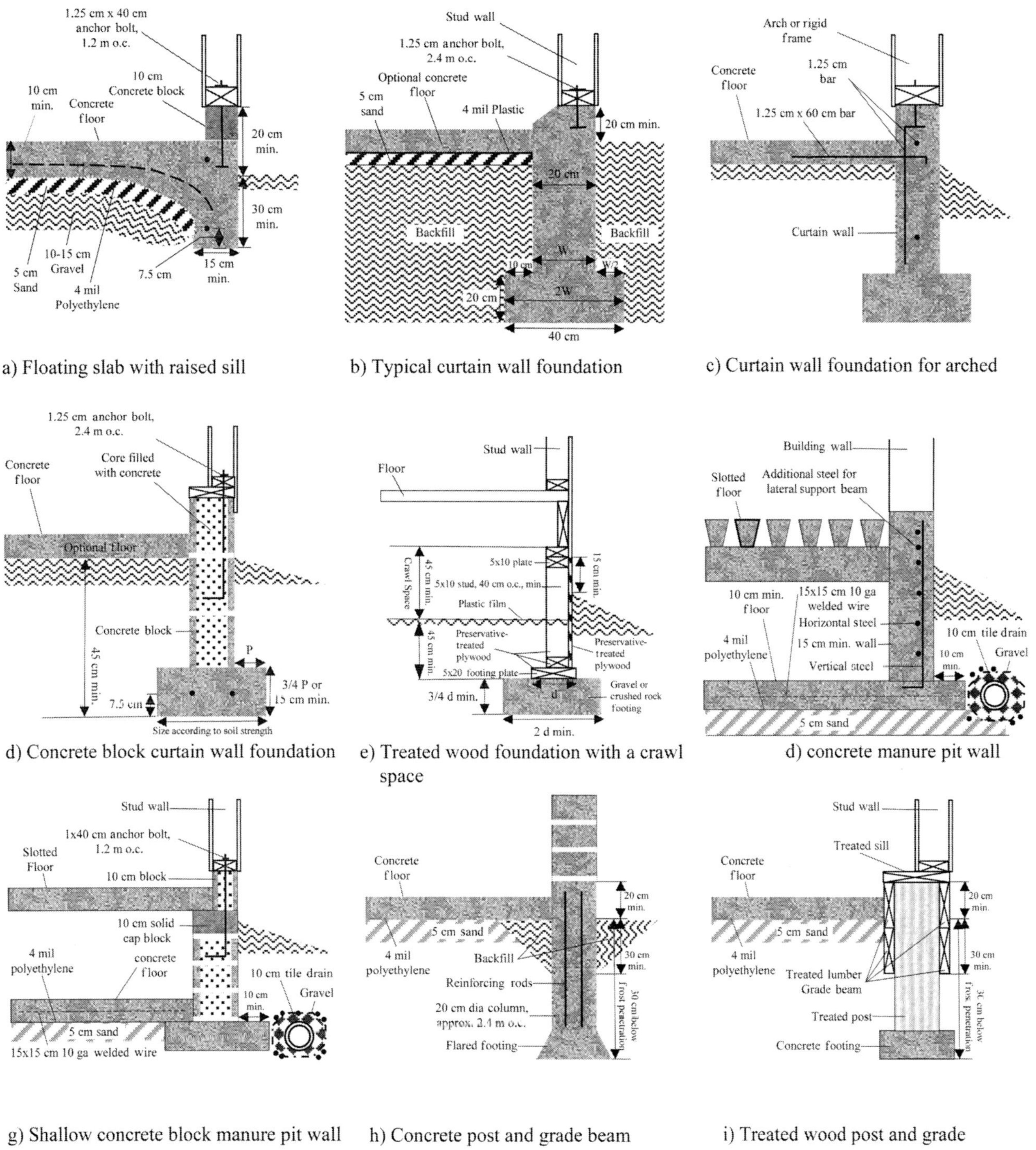

FIGURE 3.4 Typical livestock building foundations.
Source: ***Structures and Environment Handbook*** **(1987), 11th Edition, Copyright © 1987 by Midwest Plan Service, Iowa State University, Ames, Iowa, USA (*Structures & Environment*).**

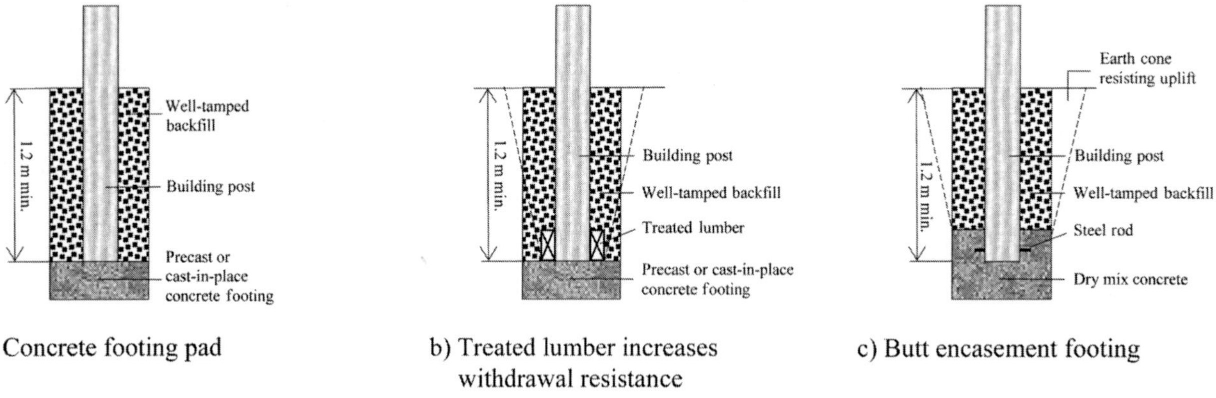

FIGURE 3.5 Post footings.
Source: *Structures & Environment* (1987).

both cause building failures if adequate footings are not provided. Concrete footing with or without treated lumber or butt encasement footings are common (Figure 3.5). Aquaculture buildings can use any of the existing greenhouse or livestock building foundations and usually do not require separate foundations or footings.

3.5 FRAMING, STRUCTURAL FORMS, AND ROOFING SYSTEMS

3.5.1 Frame Materials

Aluminum, reinforced concrete, steel, and wood are the materials typically used for building greenhouse and livestock and aquaculture building frames either individually or in combination. Reinforced concrete is used for foundations and low walls only. Aluminum, galvanized steel, and wood are used for building almost all structural members, and aluminum and galvanized steel usually require no maintenance; however, they should be kept from direct contact with the ground so that they can be protected from corrosion.

Stud frame (e.g., platform framing for single-story buildings and balloon framing for two-story buildings) and pole construction are the most common types for livestock buildings. In pole construction, treated round timber posts are set in the ground as the primary structural members. Painting the indoor surfaces of all frame materials naturally improves the light reflection in greenhouses. Wood, on the other hand, needs to be preserved against decay, especially when in contact with soil, and the chemical preservatives used should pose no danger to plants, animals, or humans. Further details of the construction types are outside the scope of this book.

3.5.2 Structural Forms and Roofing Systems

The most common structural form of greenhouses is a straight side wall and a gable roof. Post-and-beam and post-and-truss arches are employed for forming the greenhouse gable structure. Any of these forms can easily be used to build a single span, a twin span, or a large gutter-connected greenhouse complex. Figure 3.6 shows some typical greenhouse shapes.

Further, Figure 3.7 shows some common livestock housing roofing types. A shed roof, for instance, is very simple to build, particularly on narrow buildings. Gable and hip roofs can be used on many buildings, and hip roofs require no gable-end maintenance. When high head room is required, then gambrel, gothic, and gable rigid roofs are used. Monitors and half-monitors are, on the other hand, mainly used for better lighting.

3.6 BUILDING ENVELOPE, COVERING MATERIALS, AND INSECT SCREENS

3.6.1 Building Envelope and Covering Materials

This section focuses on the building envelope and covering materials for greenhouse type buildings because they have significant impacts on plant production under controlled environments; many of these building covering and shading materials are used with livestock and aquaculture buildings as well.

Covering materials (also called glazing) have considerable impacts on plant production in greenhouses. Different glazing materials have different radiative properties that influence the amount and spectral distribution of solar radiation that reaches the plant canopy, which directly affects plant growth. Furthermore, the type of covering material used directly influences condensation on the surface of covering material itself, creating unfavorable environments for plants, and indirectly affects moisture and CO_2 concentrations in a greenhouse.

Giacomelli (1999) states that if the goal is to produce high-quality plant products within efficient and cost-effective controlled-environment systems, then the greenhouse design must consider (i) the type of crop and its associated cultural demands and (ii) the production system that will support the growth of the crop; all features and components including the greenhouse covering should properly support these two essential needs. This is quite a fair statement, and there are many commercially available transparent covering materials including rigid materials (glass derivatives), thin films (normally double layer, air inflated polyethylene (PE), ethylene vinyl acetate (EVA), polyvinyl chloride (PVC), polyvinyl fluoride (PVF), and polyester), and

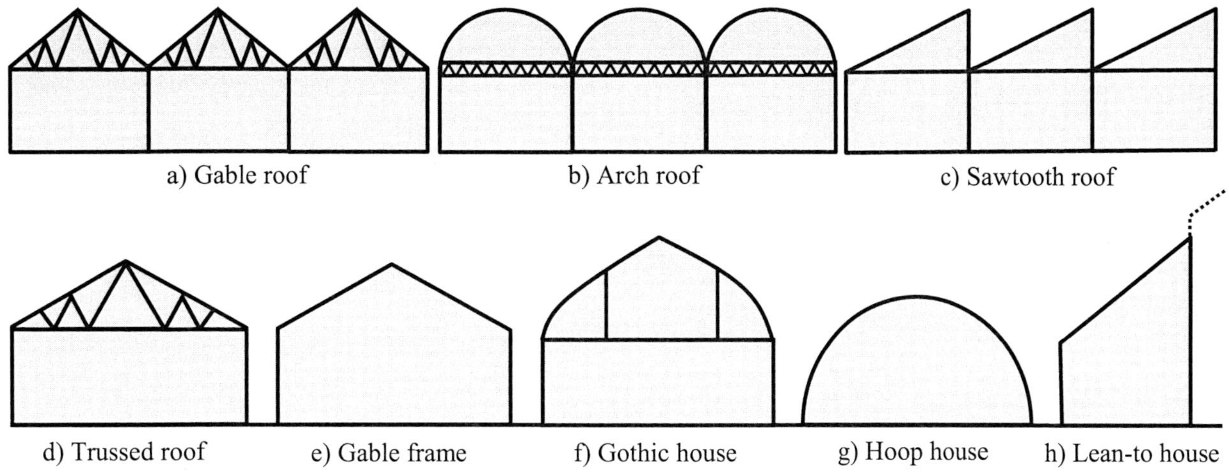

FIGURE 3.6 Common greenhouse roof types.
Source: *Greenhouse Engineering* (2021).

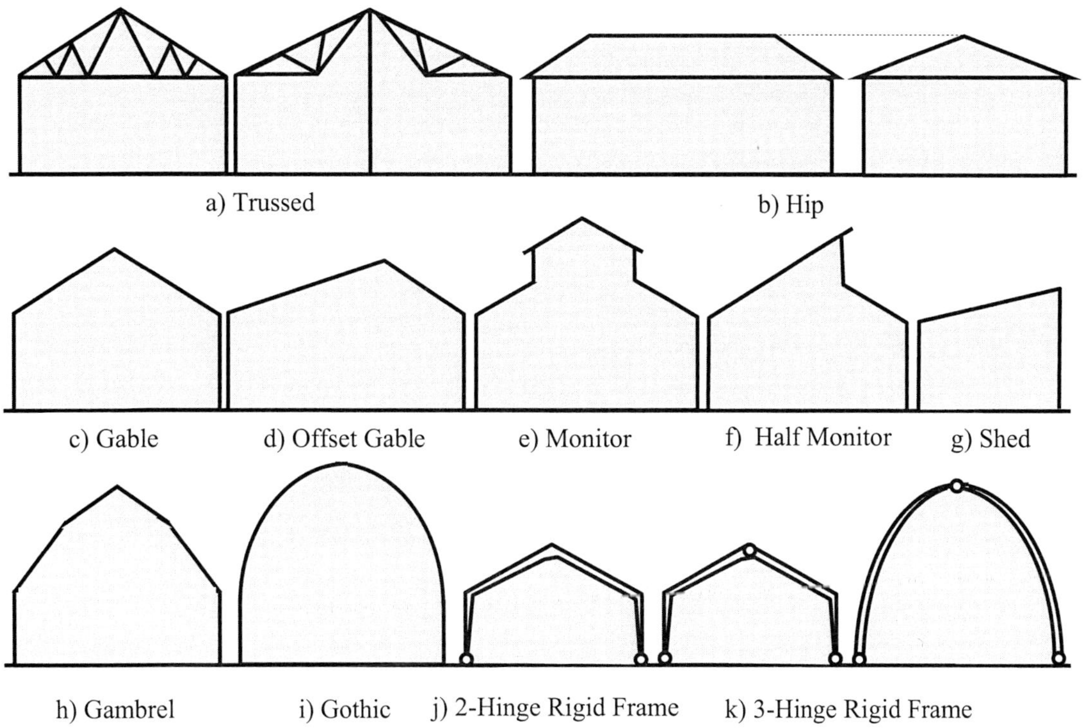

FIGURE 3.7 Common livestock building roof types.
Source: *Structures & Environment* (1987).

flexible plastics (such as fiberglass-reinforced plastic (FPR), acrylic (PMMA), polycarbonate (PC), polypropylene (PP), polyethylene terephthalate glycol (PETG), and laminar composites).

Statistics Canada (2019), for instance, reported that greenhouse production in Canada in 2018 was a $3 billion industry with a total greenhouse area of approx. 2,337 ha, 57% of which were plastic greenhouses while the remaining 43% were glass. This distribution was 62% and 38%, respectively, in the province of Ontario, where 65% of Canada's greenhouses are located. Labor and heating are the major costs of greenhouse operations, particularly in mid- and high-latitude countries including Canada. Annual labor costs typically range from 30% to 35% of the total cost of production, while heating costs typically ranging from 20% to 35% of the total production cost depending on the fuel prices. The covering material directly influences not only plant growth but also the costs of heating, cooling, and ventilation in greenhouses due to their varying radiative properties and insulation effects.

As covered earlier in Chapter 2, when radiant energy falls on a surface such as a greenhouse covering, it is absorbed, reflected, or transmitted through the material; from the first law of thermodynamics, this relationship was expressed in Equation (2.87) as $a + t + r = 1$. As mentioned, there exists a range of different greenhouse covering materials with different radiative properties, and Table A.59 lists the transmissivities of different glazing materials. However, there is no universal covering material that would provide the ideal environment for all plants in all climatic conditions. What is critical for plant growth and development is that the visible portion of the spectrum, that is, the photosynthetically active radiation (PAR) range, covers the wavelengths from 0.4 to 0.7 μm. The right covering material for a greenhouse is determined not only by the plant light requirements but also by the climate, longevity, handling requirements, and budget, and sometimes by personal preferences. Plant requirements dictate the amount of PAR that the covering material would provide; however, the climate and budget determine the number of panels involved for any type of glazing material used.

As Table A.59 shows, the shortwave (solar) and longwave (thermal) transmissivities of covering materials are different. For instance, PE film is quite transparent to thermal radiation, which is more important on clear rather than overcast nights because under such conditions, greenhouses have direct radiation exchange with the sky and transmit more thermal energy. In double PE greenhouses, infrared (IR) film that blocks thermal radiation is typically installed as the inner layer to block the inside IR radiation and reduce heat losses throughout the night. In heated greenhouses, the savings have been reported to total from 10% to 20% depending on the sky conditions (Bartok, 2010). IR film costs only a few more cents per square meter than ordinary PE film and pays for itself after only a few months.

As also discussed earlier in Chapter 2, the overall rate of heat transfer coefficient (U) (Table A.60) of greenhouse glazing is important for heat loss, and the lower the tabulated U, the higher the resistance to heat loss. The commercially used R ($1/U$) expresses the resistance to heat transfer: A higher R in glazing material means higher resistance to heat transfer. Air infiltration is another important component of heat loss that is not encompassed under U but must also be dealt with. Greenhouse covering systems vary quite a bit in air infiltration performance, which is included in separate heat loss calculations. Air exchange values for greenhouses with different glazings are provided in Table A.61.

Single-layer covering usually provides higher total solar (shortwave) radiation—that is, higher PAR—than multilayer coverings (used in many high-latitude greenhouses); however, the latter provides more insulation against heat loss, which is a significant consideration in colder climates. Inflated double-PE films and acrylic or PC panels are popular in high-latitude regions. For normal operation conditions, a single-layer PE film is quite enough, but the double layer reduces heat loss at night by about 40% (Bartok, 2010).

Normally, covering materials influence the spectral distribution of solar radiation that reaches the plant canopy, and high light transmission is essential to plant growth. Light passing through a greenhouse glazing is split into direct and diffused radiation. Direct light creates shadows, while diffusion eliminates shadows and ensures more uniform light distribution in the greenhouse so that it can even reach the lower parts of the plants, prevent burning, and offer a moderate cooling effect. Therefore, manufacturers have added light diffusion to their glazing materials to enhance the amount of diffused solar radiation reaching the lower parts of the plant canopy while keeping the PAR levels the same, which is particularly important for tall vegetable crops. Increased diffused solar radiation in the lower parts is important for reducing fungal development and insect propagation as well.

Where plenty of solar radiation is available, photoselective films that reflect or absorb specific wavelengths of the solar spectrum that are not required for plant growth can significantly decrease greenhouse temperatures, lowering the need for greenhouse cooling and irrigation and improving the thermal comfort in work environments. Ultraviolet (UV)-blocking films, which absorb UV-radiation up to 390 nm, for instance, control some insects and fungal diseases as they block out a significant part of the visible range of insects. However, greenhouses that use bumblebees to pollinate plants should use a film that transmits at least some UV light because bees need UV light for navigational purposes.

Moreover, PE films tend to attract dust particles that can accumulate on the PE film over long periods of exposure and reduce light transmission, which has a negative impact on plant growth and yield. Manufacturers now use special anti-dust agents in the film that prevent dust particles from sticking on the film. Furthermore, condensation is a common problem in greenhouses, where droplets reduce light transmission and commonly lead to disease problems when they drip onto plants. Anti-dripping (AD) films, also referred to as anti-condensation (AC) films, prevent early morning and late afternoon fog formation in the greenhouse. The AC agent reduces the surface tension and allows condensation to flow rather than forming droplets. The additives are slowly washed out by water; indeed, the effect lasts for up to two years.

Mechanical strength (e.g., tear and debris impact resistance), longevity (the usable life of covering materials to function effectively internally and externally in a commercial greenhouse operation), flammability, and chemical and material compatibility are also important considerations in choosing an adequate glazing material. Sometimes, the reduced light transmission has more impact on the useful life of glazing material than reduced physical strength.

Generally, while there is no magic glazing material, all of the discussed materials perform very well in greenhouses. The specific choice for any given situation depends on the crop light requirement, cost, climate, and sometimes personal preferences. Similar arguments are valid for greenhouse-type aquaculture and livestock buildings.

3.6.2 Animal Housing

Animal housing type is based on the building environment desired, and the insulation capacity of these buildings depends on a number of factors such as the local climate (i.e., the outside design temperature), inside design temperature, number of animals, and costs. Operations usually provide separate housing for different animal groups, but groups of different animals can be housed in the same building as long as they are managed separately. Figure 3.8 shows some examples of different livestock buildings.

Cold buildings are the least expensive buildings to build, and designed for winter time indoor temperatures within a few degrees of the outdoor temperatures (see Figure 3.8a). These buildings are usually uninsulated with unregulated natural ventilation, and remain cold in winter; hence, they are normally used for machinery storage, and only for large animals, such as cattle or finishing and gestating swine. Since they are not insulated buildings, winter time freezing and condensation as well as summer time solar heat gain can be a problem; hence, a minimum roof insulation is typically recommended.

Modified-environment or modified open-front buildings are insulated buildings designed for controlled natural and fan ventilation (Figure 3.8b). They have higher indoor air temperatures (typically above 0°C or 32°F in temperate climates) than outdoor temperatures and are typically used for warm free-stall livestock buildings, poultry production buildings, and swine finishing and gestating buildings. Desired indoor temperature and moisture levels are maintained through animal heat and controlled natural ventilation. Obviously, the added insulation increases the cost of livestock housing, but it helps keep the building cooler during summer months and reduces the risk of condensation in cold winter months as long as natural ventilation is managed properly.

Supplementally heated buildings, on the other hand, have insulation and mechanical ventilation for better environmental control and more comfortable working conditions (Figure 3.8c). They are relatively expensive to build, have winter inside temperatures above 4°C (40°F) in temperate climates, and are typically used for farrowing buildings, calf buildings, farm shops, and office buildings.

3.6.3 Curtain Systems

See Chapter 5 for curtain systems and relevant discussions.

3.6.4 Insect Screens

Integrated pest management (IPM) of greenhouse crops does not depend upon pesticide application but includes screening to physically control flying insect pests moving into the greenhouse from outside. Common greenhouse pests such as aphids, thrips, and whiteflies as well as less common pests like tarnished plant bug are all included in this control.

Benefits of insect screens have been demonstrated: When they are properly chosen, installed, and used, they reduce pest levels as well as pesticide use and exposure and improve pest control effectiveness. Disadvantages of using insect screens include difficulties in sizing and fastening to structure, reduced ventilation, and screen maintenance. For instance, smaller-mesh screens exclude more pests than larger mesh, but they are much more resistant to airflow, causing higher static pressure drops, insufficient air exchange, higher greenhouse temperatures, higher energy consumption by the fans, and excessive wear on the fan motors. The insufficient air exchange—reduced ventilation—can however be addressed by enlarging the vent area and adding a screened framework to ensure sufficient air exchange for greenhouse cooling.

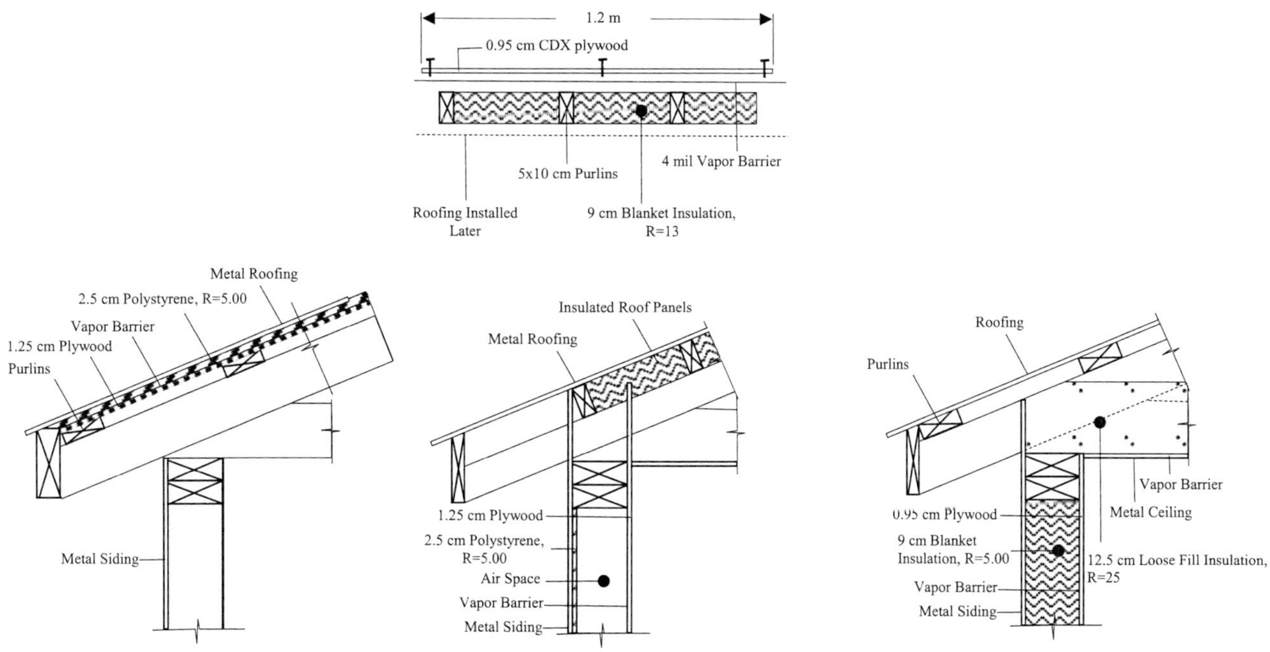

a) Cold building. b) Modified environment building insulation. c) Supplementary heated building insulation.

FIGURE 3.8 Examples of livestock buildings.
Source: ***Structures & Environment*** **(1987).**

There exist many insect screen types the market today with their own merits depending on their applications. The essential challenge is to match the screen mesh to the need, that is to the size of the target pests, and this is true for controlled-environment buildings other than greenhouses. Large-mesh household screens are not adequate for excluding most major greenhouse pests; common greenhouse pests require smaller mesh sizes. Typically, growers have multiple pests of concern and need to choose a mesh size that will exclude the smallest of the concerned pests, usually thrips or whiteflies. Table A.62 shows the screen mesh sizes needed to exclude major greenhouse pest species.

The installation and type of screening structure to be installed depends upon a number of considerations including the ventilation, duration of use, cost, durability, maintenance, etc. Figure 3.9 shows some common insect screen installations. Generally speaking, passively ventilated greenhouses are more difficult for screened operations, and the ventilation in such greenhouses is less effective than forced ventilation. Some buildings only require screens during certain times of year depending on when the concerned pests appear. Sometimes, heavy duty long-term use considerations are required, and in such cases stronger screens should be used.

There are mainly four types of screens available on the market based on the materials used: stainless steel and brass, polyethylene, acrylic, and nylon. The first of these lasts the longest but cost the most. PE screens come in different shapes such as monofilament (each thread is a solid strand), multifilament (strands made of many fibers make a strong screen), and low-cost film (with punched micro holes restricting airflow). Acrylic yarns are multifilament screens as well, while nylon is a low-cost, less-durable, shorter-term, light-duty option. Most commercial screens have life spans of three to five years. Small mesh sizes are susceptible to blockage with dust and other deposits that reduce air flow through the screens and can cause excessively high temperatures in greenhouses if they are not maintained properly. Hence, it is extremely important to keep the screens clean.

3.7 AQUATIC CULTURE SYSTEMS

Needless to say, water is required to raise fish. In aquaculture operations, fish are reared in ponds, raceways, or tanks.

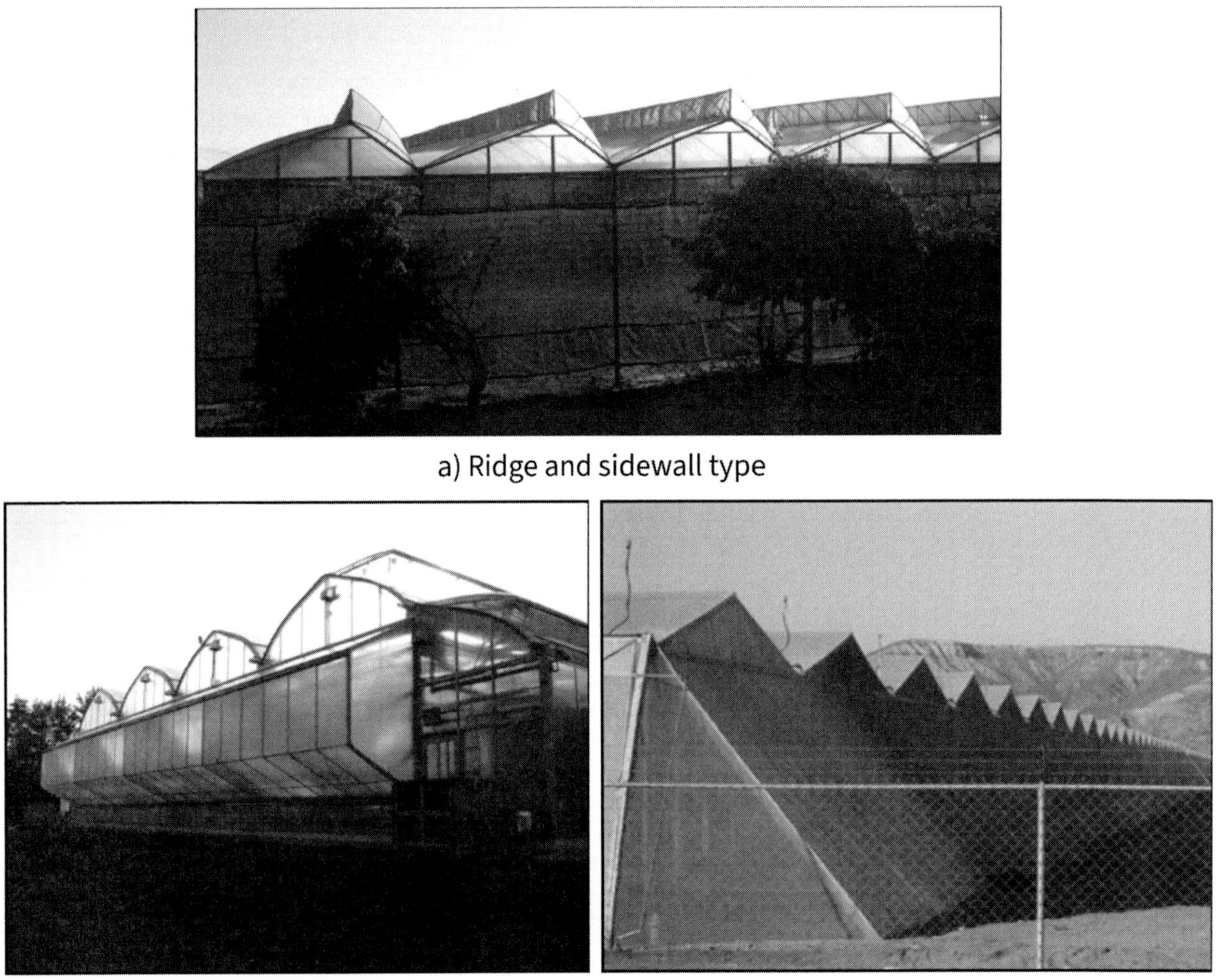

a) Ridge and sidewall type

b) Box type

c) Lean-to type

FIGURE 3.9 Common insect screen installations.

Source: *Greenhouse Engineering* (2021).

Many raceway systems are called open or flow-through systems because the water makes only one pass through the system and is then discarded. In extensive open pond culture, fish are reared free in large ponds and therefore are subject to losses from diseases, parasites, predation, pollutants, stress, and seasonally changing growing conditions. In contrast, intensive RAS facilities raise fish in indoor tanks at high stocking densities in relatively small spaces in fully controlled environments. Unlike flow-through systems, rather than discarding the water, recirculating systems filter and recondition the water for recycling back through fish culture tanks. As stated earlier, this book is restricted to controlled-environment recirculating systems and focuses on tank culture only.

If that is the case, we then need to talk about the main issue in all tank cultures, which is the self-cleaning issue. Inability to self-clean the growth medium is the predominant reason for the failure of tank culture design. Unless these systems perform effective and quick self-cleaning—removing fish fecal matter, uneaten feed, and other solids from the culture water—they are guaranteed to fail.

Therefore, the first legitimate question to ask is whether the aquatic culture tank employed will effectively and quickly remove solids. Other questions include the shape and depth of tanks. This section will address these questions relevant to culture tanks, offer a short review of raceway and hybrid designs, and discuss numerous other issues.

3.7.1 Round Tank Systems

Again, water is required to raise fish, and the water needs to be contained. The water-containing structures—culture tanks—are designed considering parameters such as cost of production, effective utilization of space, maintenance of good water quality, and overall management of fish health and growth. Different tank designs for intensive fish culture have been reviewed by a number of scholars in the literature (Wheaton, 1977; Klapsis and Burley, 1984; Piper et al., 1982; Cripps and Poxton, 1992; Timmons et al., 1998; Summerfelt et al., 2000a; Timmons and Ebeling, 2010).

The literature indicates that large round tanks (Figure 3.10) cost significantly less than more but smaller tanks for growing the same amount of fish (Timmons et al., 1998; Summerfelt et al., 2000a). It should be remembered that the capital cost of tank size is not directly proportional to tank volume, and there seems to be a definite trend toward larger round tanks for fish production. Timmons and Ebeling (2010) emphasize that substantial savings in both capital and labor costs can be realized by using fewer but larger culture tanks and that the time it takes to service a tank is similar whether the tank is small or large. While many intensive aquaculture operations use tanks with tank diameter to depth ratios as low as 3:1, the recommended ratios usually vary from 5:1 to 10:1 (Burrows and Chenoweth, 1970; Chenoweth et al., 1973; Larmoyeux et al., 1973; Timmons and Ebeling, 2010).

Here, we should also keep in mind that not all fish (except tilapia) effectively utilize the entire water column; therefore, deeper water columns make no sense and many times are useless. Round tanks are reported to make good aquatic culture systems simply because they are easy to maintain, provide a uniform culture environment, allow a wide range of rotational velocities to optimize fish health and growth, and permit the rapid concentration and removal of settled solids through the center drain (Timmons and Ebeling, 2010). Having said that, one must also remember that when larger tanks fail, the economic loss is much greater as they have more fish and typically the entire tank is lost.

It is very important to note that round tanks can achieve almost instant complete mixing reactor status (Davidson and Summerfelt, 2005). This means that any dissolved constituent in the water flowing into the tank instantly changes to the existing condition throughout the tank. Therefore, when adequate mixing is achieved, this literally means that all fish within the tank experience the same water quality. This also means that as Timmons et al. (1998) stated, adequate water quality can be maintained throughout the round culture tank by optimizing the water inlet structure and by selecting an adequate water exchange rate that does not decrease fish production when the system reaches its carrying capacity, which is the maximum fish biomass that can be supported at a given feeding rate.

The rapid hydraulic exchange in small tanks results in reasonably good water quality due to the high turnover rate, that is, hydraulic retention time (HRT, usually 10 minutes per tank volume exchange) that carries more oxygen into the tank and rapidly flushes waste from the tank. The HRT in larger tanks, however, is typically longer (45 minutes or more), and therefore the water injection, flow rate, and removal become dominant factors in larger tanks, affecting the quality and uniformity of water conditions in the

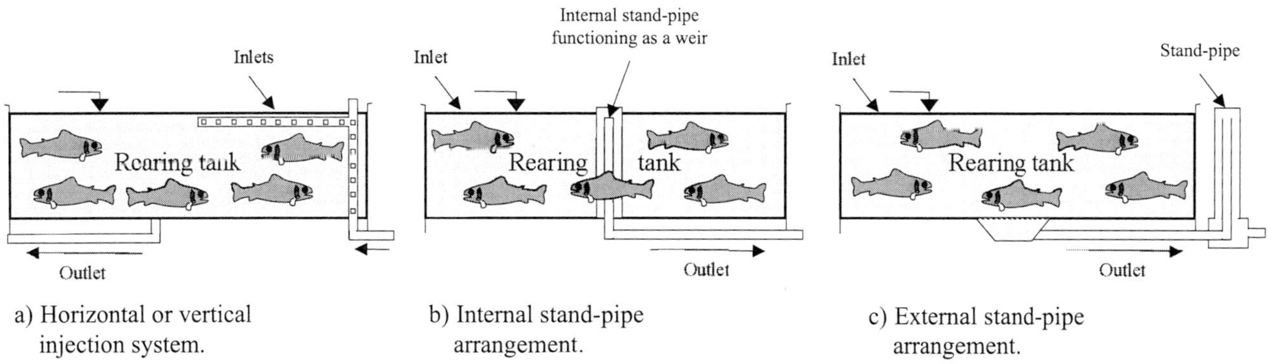

FIGURE 3.10 Round culture tank configurations.

tank. Moreover, it should be emphasized that the carrying capacity of a culture tank is influenced by water exchange, feeding rate, oxygen consumption, and waste production. Quality engineering is required to adequately design the water inlet and outlet structures and fish grading and/or removal mechanisms so that the labor required for fish handling can be reduced as well and achieve uniform water quality, rotational velocities, and solid removal from the round tank (see Figure 3.10).

3.7.2 Raceway Systems

Raceways have a rectangular or trapezoidal cross-sectional area, through which water enters the raceway at one end and flows through the raceway in a plug-flow manner, basically with minimal back mixing. Hence, the best water quality is observed where the water enters the raceway tank and then deteriorates steadily through the raceway toward the outlet. As a result, it is common to see fish clustering toward the raceway inlet to reach oxygen easily or toward the outlet to avoid accumulating concentrations of contaminants.

Raceways are the most common fish-rearing tank designs and are typically associated with salmonid culture. The length-to-width ratio is important in raceway design to prevent circular movement among the fish that promote the accumulation of waste in the center. The typical length-to-width ratio is 10, but a ratio under 6 is recommended. The width should not be great enough to cause a large cross-section and hence weaker water velocity. Widths of 2–5 m and lengths of 20–50 m have been reported to give good results with average depths of around 1 m. The slope of the raceway floor is generally 1%. Raceways are usually constructed side by side, with common walls for maximizing the space use efficiency and reducing construction costs; this is why they are favored for their better utilization of floor space and easier handling.

However, raceways also require high water exchange rates in order to maintain sufficient water quality for fish growth, and the typical exchange rate is one tank volume exchange every 10 to 15 minutes (Summerfelt et al., 2000a). Their inability to rapidly self-clean under typical operating conditions is reported as one of their main disadvantages. Velocity through the raceway is usually 2–4 cm/s, which is not fast enough to effectively remove settled solids. Hence, frequent cleaning is required.

When there is sufficient fish load, however, the fish themselves clean the tank by continuously preventing the solid waste from settling (e.g., 20 kg/m^3 for trout). Other mechanisms can also be utilized to remove more settled solids, such as placing series of baffles across the raceway to artificially increase water velocity (20 to 40 cm/s) and its sweeping action or simply lowering the water level in the tank, which increases the water velocity, clusters the fish, and moves the solid waste to the water surface to eventually evacuate it.

The flow rates requires for cleaning velocity versus for oxygen supply are quite different; raceways are primarily designed based on minimizing cross-sectional area to promote maximum water velocity. As a consequence, many raceway systems are operated in series, with the discharge of the upstream raceway serving as the inflow water for the next downstream raceway. In the meantime, water can be treated between different raceway sections to restore a specific water quality parameter for the next downstream utilization or removal (e.g., oxygen addition, CO_2 removal).

3.7.3 Round Tank Systems versus Raceway Systems

The ideal round tank is described as a continuous-flow stirred tank reactor in which the dissolved gas concentrations are well mixed and equal to the concentration in the effluent (Colt and Watten, 1988). Because of this, round tanks do not have a distinct water quality gradient, and therefore, the rearing environment is frequently somewhat homogeneous. As a consequence, ammonia and CO_2 are fully mixed into the culture medium, and this is not the case in raceways. High DO concentrations entering a round tank, unlike raceways, are quickly diluted with lower DO water already present in the culture tank. Moreover, in contrast with raceways, water velocities in round tanks can be controlled independent of the water exchange rate (Davidson and Summerfelt, 2005).

Additionally, while raceways have uniform water velocities along the channel, the round tank water velocities are somewhat lower farther away from the tank walls, which facilitates fish in selecting more comfortable zones in the rearing environment. The existing literature states that the most critical factors for round tanks for optimal fish growth are the inlet and outlet design, the flow through the tank, and the flow through the center drain (Tvinnereim and Skybakmoen, 1989; Summerfelt et al., 2000b; Davidson and Summerfelt, 2005). Properly designed systems contribute to self-cleaning.

Overall, round tanks have a number of other advantages over raceways. First, when constructed without common walls, raceways typically require 1.5 to 2.0 times as much wall length as round tanks require due to raceways' large length-to-width ratios. Furthermore, round tanks can handle the weight of the confined water better structurally with thinner walls than raceways. As a result, round tanks provide much cheaper growth environments and make lower water exchange rate operations possible while not only creating desirable hydraulic patterns but also continuous self-cleaning characteristics.

Moreover, fish naturally crowd to their feeders during feeding activities, and it is more difficult to distribute feed throughout raceways. In contrast, round tanks can easily be equipped with feeders, typically requiring only one or just a few feeding stations at most. Round tanks often make it possible to accomplish the maximum cumulative oxygen consumption within a single culture tank; this means that the water is used up before passing through only one tank instead of a series of them. This also means that since there is no serial going through of water, all culture units can be operated on the same elevation.

3.7.4 Crossflow and Mixed-Cell Raceway Systems

As discussed, the relevant experience and literature show that round tanks have uniform water quality properties, increased water velocities, and hence good solid removal characteristics. Raceways, on the other hand, make more efficient utilization of floor space and facilitate easier sorting, grading, and handling fish than traditional round tanks. The main disadvantages of raceways are the large water volume requirements, high water discharge rates, slow water velocities, and hence highly limited self-cleaning ability (Timmons et al., 1998).

Crossflow tanks, a hybrid approach that incorporates the desirable characteristics of both round tanks and linear raceways, are offered as a solution to the shortcomings of raceways. In this hybrid design, water is uniformly distributed along one side of the crossflow tank through a submerged manifold and collected back in a submerged perforated drain line running along the opposite side (Watten and Johnson, 1990).

Another solution is converting a single raceway into a series of counter-rotating mixed-cell raceways in an effort to eliminate metabolite concentration gradients, increase water velocities, and improve solid removal at low water exchange rates (Figure 3.11) (Watten et al., 2000). Acting as hydraulically separated round tanks, the approach provides the advantages of raceways with the improved solid removal capability of round tanks. The study findings showed excellent bottom water velocities for moving solids toward the center drains in each cell of a mixed-cell raceway at different exchange rates. They demonstrate excellent potential for new production system designs or retrofitting existing raceways.

3.7.5 Tank Materials

In essence, anything that holds water can be used as an aquaculture production unit. Earthen raceways can be easily built by excavation, compaction, and leveling, without any particular problem. In these systems, the tank ends are usually concrete, and the floor is impervious with a slope, ensuring adequate stability to the whole system. Due to the need for future access by mechanized maintenance equipment, the ratio of water surface area to total surface area in earthen raceway tanks is typically lower than in concrete raceways. As a result, land use efficiency is low. Additionally, cleaning and disinfecting earthen raceway tanks are difficult.

3.7.5.1 Wood

Wood is used for building inexpensive fish-culture tanks that are sealed with nontoxic materials such as epoxy or fiberglass resin paint or lined with a high-density polyethylene (HDPE) or butyl rubber liner. Once these materials are cured properly, they form smooth semirigid surfaces that are hard yet with some flexibility that are easy to clean. Construction is easy, fast, and inexpensive.

Where needed, covering the tank bottom with an insulation board is a good idea. Normally marine plywood is used, but other grades can also be used. The sides of a wooden tank should be braced to resist the forces exerted by the water, and all the exposed wood surfaces should be painted against rotting with a paint containing no lead or heavy metals. Since many treated woods contain substances that are toxic to fish, they should never be in contact with the culture water. Long and shallow raceways can be constructed using plywood panels and a liner.

3.7.5.2 Steel and Aluminum

For larger and more permanent tanks, we can build galvanized or epoxy-coated steel tanks that can be bolted together to form large, deep tanks (32 m [105 ft] × 4.3 m [14 ft]). The bottoms of these tanks are often partially buried

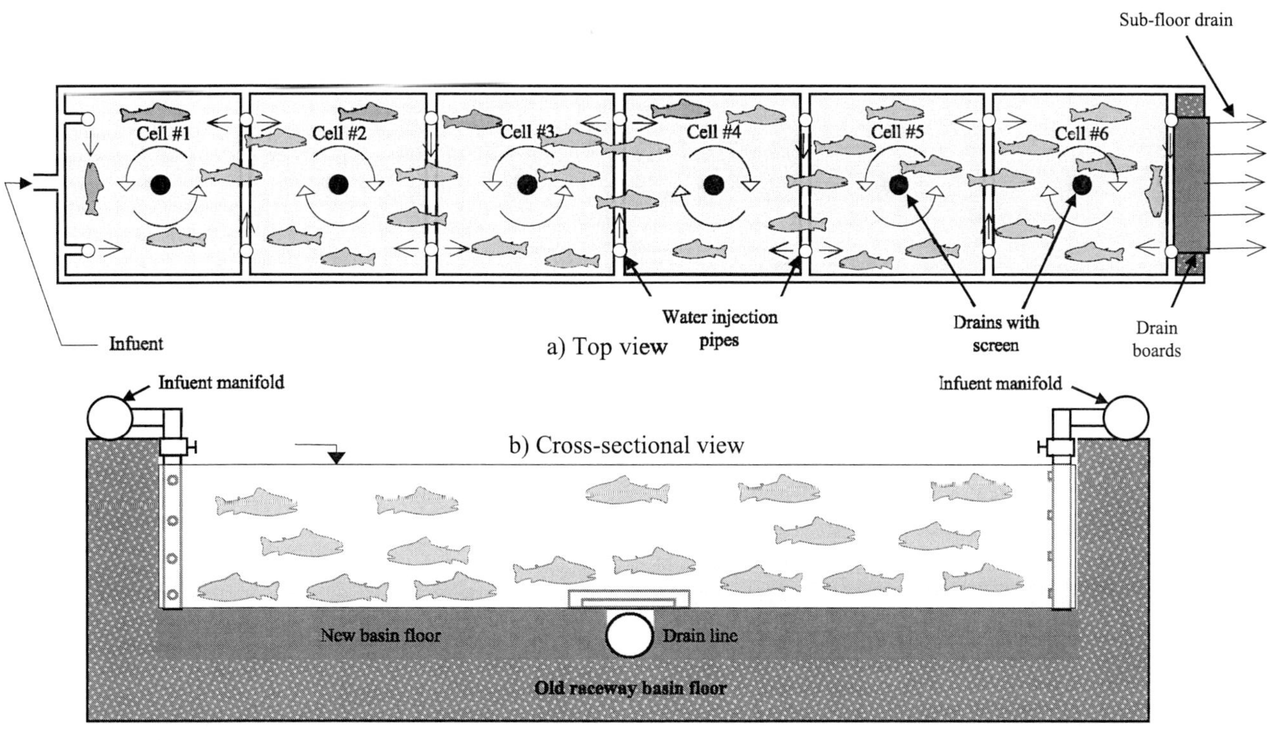

FIGURE 3.11 A mixed-cell raceway water tank.

poured concrete that can contain heating coils. Partially burying tanks helps conserve heat and eases fish observation because these tanks are usually deep.

Carbon steel is probably the cheapest good structural support for tanks; with proper coating, steel resists corrosion in both freshwater and saltwater systems. However, the protective coatings required to prevent steel corrosion in saltwater are expensive. Stainless steel is excellent but expensive for freshwater in particular, but no matter how resistant, stainless steel will eventually corrode in saltwater. Because it is an expensive material, stainless steel is not a suitable material for large tanks, although it is suitable for use in small tanks with either freshwater or saltwater.

Aluminum also serves well in freshwater, and some aluminum alloys provide wonderful service in saltwater systems. Aluminum and steel are, indeed, the two commonly used metals for fabricating small tanks, which are quite popular in hatcheries for rearing small fry and fingerlings. In principle, since even a little erosion over long periods can cause ion buildup in the culture water, all metals in recirculating systems should be used with great care. Aluminum becomes soluble in acidic water and leaches into the water (Lawson, 1995). Furthermore, unless they are properly coated or lined, galvanized metal tanks should not be used for aquaculture purpose at all. It must be noted that zinc eventually leaches from the galvanized coating and can cause heavy-metal poisoning in certain fish species.

3.7.5.3 Plastics

As discussed earlier, plastic materials cover a large spectrum of polymers including PE, vinyl, PMMA, PVC, fiberglass, PP, and similar materials. Each of these plastics has positive and negative properties. We can use most plastics in both freshwater and saltwater systems. However, they can be expensive and vulnerable to UV radiation, which is important in aquacultural applications. Generally speaking, plastics are lightweight (and therefore, easily portable), durable, and inert and can easily be formed into any desired shape. Today, plastic tanks come in various shapes and sizes.

3.7.5.3.1 Polyethylene

PE tanks are popular because of their low cost. PE is also commonly used for making impermeable liners for low-cost rigid structures. It is typically available in rolls of up to 6 m wide and 75 m long, and it comes in many thicknesses depending on the need. Since these tanks are lightweight, setup, shipment, and relocation are quick and easy. They have smooth surfaces that make cleaning relatively easy. However, they sometimes need to be supported on the bottom and sides, and watertight connections must be ensured as well. Also, we have to be careful with PE before using because it is initially toxic due to the hardeners and die lubricants used during manufacturing; therefore, they should be conditioned with clean running water for a minimum of two weeks prior to use for marine culture. As is well known, PE does not have a long life, and it can be easily punctured.

3.7.5.3.2 Vinyl

Similar to PE, vinyl is a flexible plastic material widely used as liners in aquacultural facilities. Just like PE, vinyl is relatively low cost, smooth, and easily cleaned, but it also does not have a long life, and it can be easily punctured. Provided a good sand base, vinyl can provide relatively long service.

3.7.5.3.3 Acrylic

Acrylics come in different colors, and transparent acrylic is widely used in fish culture. Acrylic plastic is essentially a rigid and inert material in both freshwater and saltwater; however, it has limited flexibility for smaller thicknesses less than 0.35 cm or so. Acrylic does not shatter like glass; hence it can easily be drilled, taped, and worked with machine tools into desired shapes and dimensions. It should be kept in mind that the humidity difference between the sides of acrylic sheets causes the sheets to swell, warp, and eventually break. Immersion of one side of the acrylic sheet in water, however, does not cause any similar problems. Acrylics have many applications in aquacultural facilities including small tanks; however, they are not used for large tank construction due to the cost.

3.7.5.3.4 PVC

PVC is a durable and inert material in both freshwater and saltwater and is widely used in aquaculture applications. Since it is relatively inexpensive and available in a wide range of sizes and wall thicknesses, PVC is mainly used for piping and fittings in aquacultural systems. PVC can also be used for tank construction, but the sheets are expensive and require structural support for large tanks.

3.7.5.3.5 Fiberglass

Fiberglass is the most common material used for constructing aquaculture tanks. These tanks can be manufactured in any shape and size because fiberglass is relatively inexpensive, flexible, lightweight, and easy to cut and drill, which facilitates easy repairs and modifications. Fiberglass is a popular material for building circular tanks because of the strong tension loading, the stress experienced in circular tank walls. It is UV resistant and inert in both freshwater and saltwater. Up to several hundred cubic-meter tank sizes are commercially available, while smaller tanks less than 2.4 m in diameter are usually available as a single-molded unit.

3.7.5.3.6 Polypropylene

PP is also an inert material in both freshwater and saltwater, and available PP tanks can hold at least 1500 liters. Larger PP tanks are more expensive than similar-size fiberglass tanks. The main problem with PP other than the cost is that it is very difficult to bond anything else to it besides PP.

3.7.5.4 Concrete and Masonry

Reinforced concrete or masonry has been widely used to build raceways and large tanks for many years around the world due to their cost and strength advantages over other materials for handling large quantities of water. They are easy to work with, can have all shapes, and last indefinitely for both fresh- and saltwater as long as properly reinforced with steel bars.

Before building and investing in such an option, however, one must make sure that operational water can be sustained for many years to come. Concrete tanks are quite watertight and have the longest life, but they are expensive. Concrete slabs and blocks are more cost-effective than single-unit concrete structures. The slab is cast in one piece and set in properly compacted soil. A layer of gravel is spread before casting the slab, particularly on sandy soil. The interior surfaces of reinforced concrete or masonry tanks should be smoothed with proper curing periods so that fish skin can be protected from abrasion, and the cleaning and disinfection operations become relatively easy. When needed, they can even be coated with a commercially available sealer.

NOMENCLATURE

a	absorptivity
AC	anti-condensation
AD	anti-dripping
DO	dissolved oxygen
EVA	ethylene vinyl acetate
FPR	fiberglass-reinforced plastic
HDPE	high-density polyethylene
HRT	hydraulic retention time
IPM	integrated pest management
IR	infrared
NFT	nutrient film technique
PAR	photosynthetically active radiation
PC	polycarbonate
PE	polyethylene
PETG	polyethylene terephthalate glycol
PMMA	acrylic
PP	polypropylene
PVC	polyvinyl chloride
PVF	polyvinyl fluoride
RAS	recirculating aquaculture system
r	reflectivity
t	transmissivity
TDS	total dissolve solids
TGP	total gas pressure
TSS	total suspended solids
U	overall heat transfer coefficient
UV	ultraviolet

BIBLIOGRAPHY

Aland, A and Banhazi, T (eds.) (2013) *Livestock Housing: Modern Management to Ensure Optimal Health and Welfare of Farm Animals*. Wageningen Academic Publishers, The Netherlands. ISBN: 978-90-8686-217-7.

Aldrich, RA and Bartok, JW (1994) *Greenhouse Engineering*. NRAES-33. Northeast Regional Agricultural Engineering Service. Cornell University, Ithaca, NY, USA.

ASABE (2018) *Standards, Engineering Practices, and Data*. V21(1). American Society of Agricultural and Biological Engineers, St. Joseph, MI, USA.

Baker, JR, Bell, ML and Shearin, EA (1997) *Insect Screening. Information Note 104: Ornamentals and Turf, Integrated Pest Management*. North Carolina Cooperative Extension Service. NCSU, p. 4.

Bartok, JW (2010) Choose the right plastic film for your needs. In: *Greenhouse Management*, May, Valley View, Ohio, USA.

Bell, ML and Baker, JR (1997) Choose a greenhouse screen based on its pest exclusion efficiency. *North Carolina Flower Growers' Bulletin*, 42(2): 7–13.

Bethke, JA (1994) Considering Installing Screening? This is what you need to know. *Greenhouse Manager*, April: 34–36.

BP (2017) Available on: http://bp.com/statisticalreview (Retrieved on 21.10.2017).

Burrows, R and Chenoweth, H (1970) The rectangular circulating rearing pond. *Progressive Fish-Culturist*, 32: 67–80.

Chenoweth, HH, Larmoyeux, JD and Piper, RG (1973) Evaluation of circular tanks for salmonid production. *Progressive Fish-Culturist*, 35: 122–131.

CIGR (1999) *Handbook of Agricultural Engineering*, Volume II: *Animal Production & Aquacultural Engineering*. CIGR—The International Commission of Agricultural Engineering (ed.). ASAE, St. Joseph, MI. ISBN 0-929355-98-9.

Colt, J and Watten, B (1988) Application of pure oxygen in fish culture. *Aquacultural Engineering*, 7: 397–441.

Cripps, SJ and Poxton, MG (1992) A review of the design and performance of tanks relevant to flatfish culture. *Aquacultural Engineering*, 11: 71–91.

Davidson, J and Summerfelt, ST (2005) Solids removal from a coldwater recirculating system-comparison of a swirl separator and a radial-flow settler. *Aquacultural Engineering*, 33: 47–61.

Ebeling, JM, Timmons, MB, Joiner, JA and Labatut, RA (2005) Mixed-cell raceway: Engineering design criteria, construction, hydraulic characterization. *Journal of North American Aquaculture*, 67(3): 193–201.

EIA (2016) *United States Energy Information Administration, International Energy Outlook 2016*. Washington, DC.

Friend, HH and Decoteau, DR (1990) Spectral transmission properties of selected row cover materials and implications in early plant development. Proceedings of the 22nd National Agricultural Plastics Congress, Montreal, Quebec, Canada, pp. 1–6.

Giacomelli, GA (1999) *Greenhouse Glazings: Alternatives Under the Sun*. Department of Bioresource Engineering, Cook College, Rutgers University. Available on: https://ceac.arizona.edu/sites/default/files/G%20Systems.pdf (Retrieved on 10.05.2020).

Godbey, LC, Bond, TE and Zornig, HF (1979) Transmission of solar and long-wavelength energy by materials used as covers for solar collectors and greenhouses. *Transactions of the ASAE*, 22(5): 1137–1144.

Goldammer, T (2019) *Greenhouse Management: A Guide to Operations and Technology*. Apex Publishers, Centreville, Virginia, USA.

Gray, HE (1992) Greenhouse glazing uncovered. *Greenhouse Grower*, 10(8): 5457.

Hahn, RH and Rosentreter, EE (1989) Heating, ventilating and cooling greenhouses. In: Hahn, RH and Rosentreter, EE (eds.), *ASAE Standards*. American Society of Agricultural and Biological Engineers, St. Joseph, MI, pp. 452–455.

Heinen, JM, Hankins, JA and Adler, PR (1996) Water quality and waste production in a recirculating trout culture system with feeding of a higher energy or a lower energy diet. *Aquaculture Research*, 27: 699–710.

IEA (2012a) *International Energy Agency, Key World Energy Statistics*. OECD, Paris.

Klapsis, A and Burley, R (1984) Flow distribution studies in fish rearing tanks. Part 1—Design constraints. *Aquacultural Engineering*, 3: 103–118.

Larmoyeux, JD, Piper, RG and Chenoweth, HH (1973) Evaluation of circular tanks for salmonid production. *Progressive Fish-Culturist*, 35: 122–131.

Lawson, TB (1995) *Fundamentals of Aquacultural Engineering.* Chapman and Hall, New York.

Lindley, JA and Whitaker, JH (1996) *Agricultural Buildings and Structures.* Revised Edition. American Society of Agricultural and Biological Engineers (ASABE), St. Joseph, MI.

Meade, JW (1985) Allowable ammonia for fish culture. *Progressive Fish-Culturist*, 47(3): 135–145.

MWPS (1987) *Structures and Environment Handbook.* 11th Edition. Midwest Plan Service, Iowa State University, Ames, IA, USA. ISBN: 0-89373-057-2.

New Brunswick Department of Fisheries and Aquaculture (1995) Selection of a Trout Farming Site. New Brunswick Department of Fisheries and Aquaculture. Fredericton, NB, Canada.

NGMA (1996) National Greenhouse Manufacturers Association Standards for Design Loads in Greenhouse Structures. Structural Loads. Harrisburg, PA, USA.

NGMA (2004) National Greenhouse Manufacturers Association Structural Design Manual. Harrisburg, PA, USA.

OMAFRA (2001) *Growing Greenhouse Vegetables.* Ontario Ministry of Agriculture, Food and Rural Affairs. Publication 371. ISSN: 1492-6601. Toronto, Canada, p. 77.

OMAFRA (2003) *Production Recommendations for Greenhouse Floriculture. Ontario Ministry of Agriculture, Food and Rural Affairs.* Publication 370. ISSN: 1492-6512. Toronto, Canada, p. 39.

Piper, RG, McElwain, IB, Orme, LE, McCraren, JP, Fowler, LG and Leonard, JR (1982) *Fish Hatchery Management.* U.S. Fish and Wildlife Service, Washington, DC.

Roberts, WJ, Bartok, JW, Fabian, EE and Simpkins J (1989) *Energy Conservation for Commercial Greenhouses.* NRAES-3. Northeast Regional Agricultural Engineering Service. Cornell University, Ithaca, NY, USA.

Sase, S and Christianson, LL (1990) *Screening Greenhouses—Some Engineering Considerations.* Proceedings of the 1990 Northeast Agricultural/Biological Engineering Conference, pp. 1–13.

Statistics Canada (2019) CANSIM Tables: 32-10-0019-01; 32-10-0022-01; 32-10-0023-01; 32-10-0026-01; 32-10-0245-01. Ottawa, Canada.

Summerfelt, ST, Davidson, JT and Timmons, MB (2000b) Hydrodynamics in the 'Cornell-type' dual-drain tank. The Third International Conference on Recirculating Aquaculture, July 22–23, 2000, Roanoke, VA, Virginia Polytechnic Institute and State University, pp. 160–166.

Summerfelt, ST, Timmons, MB and Watten, BJ (2000a) Tank and raceway culture. In: Stickney, RR (Ed.), *Encyclopedia of Aquaculture.* Wiley, New York, pp. 921–928.

Timmons, MB (2000) Aquacultural engineering-experiences with recirculating aquaculture system technology. Part 1. *Aquaculture Magazine*, 26(1): 52–56.

Timmons, MB and Ebeling, JM (2010) *Recirculating Aquaculture.* 2nd Edition. NRAC Publication No. 401-2010. ISBN 978-0-9712646-2-5.

Timmons, MB, Summerfelt, ST and Vinci, BJ (1998) Review of circular tank technology and management. *Aquacultural Engineering*, 18: 51–69.

Tvinnereim, K and Skybakmoen, S (1989) Water exchange and self-cleaning in fish rearing tanks. In: De Pauw, N, Jaspers, E, Ackefors, H and Wilkens, N (eds.), *Aquaculture: A Biotechnology in Progress.* European Aquaculture Society, Bredena, Belgium, pp. 1041–1047.

Watten, BJ and Johnson, RP (1990) Comparative hydraulics in rearing trial performance of a production scale cross-flow rearing unit. *Aquacultural Engineering*, 9: 245–266.

Watten, BJ, Honeyfield, DC and Schwartz, MF (2000) Hydraulic characteristics of a rectangular mixed-cell rearing unit. *Aquacultural Engineering*, 24: 59–73.

Wheaton, FW (1977) *Aquacultural Engineering.* John Wiley and Sons, New York.

Yildiz, I (2021) *Greenhouse Engineering: Integrated Energy Management.* 1st Edition. CRC Press, Taylor and Francis Group, Boca Raton, FL. ISBN: 978-1-4987-4313-6.

4 Plant, Animal, and Aquatic Environments

4.1 INTRODUCTION

Environment is defined as the total of all external conditions surrounding plants and animals, which comprise physical, chemical, and biological factors; that is, no plant or animal can exist independently from an environment. Factors such as light and radiation, temperature, relative humidity (RH), oxygen, carbon dioxide (CO_2), water, nutrition, pollution, insect, and disease stresses are the key environmental components affecting optimum plant and animal growth, production, and reproduction. Controlled-environment greenhouses and livestock, poultry, and aquaculture buildings provide suitable environments for the intensive production of a variety of plants, livestock, poultry, and fish. They are designed to provide optimal control of solar radiation, temperature, RH, oxygen, CO_2, dissolved gases, nutrients, air movements, and pollutants in the aerial or aquatic environments.

From the engineering point of view, the environment provides the initial and boundary conditions in plant and animal biosystems; therefore, an engineer is concerned with i) how much food, feed, or fiber plants or animals will produce in a specific environment; ii) if the environment is not ideal, what needs to be changed and how much; iii) if the cost of changing the environment will be offset by a profitable increase in output; and iv) the pertinent (if any) social and environmental impacts. To answer these questions, an engineer needs to have a basic understanding of plants' and animals' physiological responses to their environments.

Modification of the abovementioned environmental parameters provides ample opportunities for engineering applications in controlled-environment buildings. For instance, low light at higher latitudes in winter is a limiting factor and not usually adjusted except where artificial supplemental lighting is used at seedling and plant-raising stages. Therefore, greenhouses are built with proper glazing materials to maximize natural light penetration so that plant growth can be maintained during low light seasons as well. Energy costs are typically high in both winter and summer. During colder months, supplemental heat is typically needed at night, and it may even be needed during the day. Due to high solar radiation loads, shade curtains, whitewashing, and other types of cooling must be provided to maintain appropriate indoor thermal environments.

This chapter first briefly covers basic plant, animal, and fish physiology and then moves to the impacts of major environmental parameters and their interactions on controlled-environment plant, animal, and fish production. The parameters covered include light and radiation, temperature, RH, oxygen, CO_2, various dissolved gases, pH, salinity, water hardness and alkalinity, solids balance, pollutants, air movement, and altitude.

4.2 ESSENTIALS OF PLANT AND ANIMAL PHYSIOLOGY

Physiology is the science of normal organism function at the cellular level, including the functioning of discrete organs and whole organisms. Physiologists study the structure and functions of the various parts of animals and plants and how the different parts work together to enable animals and plants to perform their normal functions and respond to their environments.

Acknowledging the presence of great diversity, we still observe a number of important commonalities among all physiological processes. The first of these common themes is that all physiological processes obey physical and chemical laws. The second theme is that physiological processes are regulated within acceptable ranges to maintain internal conditions. For instance, animals and plants must maintain water, oxygen, CO_2, temperature, nutrient, and other balance. This is termed homeostasis and is maintained though feedback loops that sense the existing conditions and in return activate appropriate responses.

The third theme is animal and plant phenotypes, which develop as the product of their genetic makeups (genotypes) and their interactions with the environment. An organism's genotype and its environment produce the phenotype of an adult organism interacting through development. Due to phenotypic plasticity, one genotype may produce many phenotypes, depending on the effects of the environment. Hence, genotype and phenotype are linked. The complex biological processes interact to produce complex behaviors and physiological responses. The environment, in turn, affects the adult phenotype. As a result of learning, organisms change their behavior, and modifying their phenotypes alters their physiological responses. Eventually, the phenotype of an animal or a plant determines its reproductive success.

The fourth theme is the genotype, which is a product of evolutionary change in a group of animals or plants over many generations in time. In other words, differential survival of animals or plants with particular phenotypes may result in evolutionary change in the genotype of an animal or a plant population over many generations. In short, phenotypes are the product of evolution.

Physiology can be divided into several subdisciplines that can be distinguished by the biological level of organization such as cell and molecular physiology, systems physiology, integrative physiology, and organismal physiology. It is not the intent of this book to get into the specifics of these subdisciplines; however, we still need to emphasize the importance of organismal physiology in particular due to its immediate relevance to controlled environment engineering as it is quite often concerned with the way an animal or a plant undertakes a specific

DOI: 10.1201/9781003349297-4

process or behavior in response to environmental variables. Organismal physiologists typically specialize in certain groups of animals or plants, such as avian physiology, crop physiology, swine physiology, dairy physiology, and fish physiology.

Furthermore, physiological subdisciplines can also be distinguished by the process that generates variation, that is, the mechanism by which changes in physiological processes arise. For instance, developmental physiology deals with how structures and functions change as animals or plants grow through the various life stages from embryo to reproductive maturity to senescence and death. Evolutionary physiology deals with explaining how specific physiological traits arise within lineages over the course of multiple generations, and environmental physiology is interested in how organisms develop physiological responses to different environmental challenges. For instance, changes in temperature influence many animal and plant physiological processes in complex ways; therefore, an environmental physiologist is concerned with the way an individual animal or a plant organizes its physiology in order to survive the environmental challenge.

Physiological research can further be distinguished on the basis of the ultimate goal (e.g., pure and applied science). An applied animal or plant physiologist intends to achieve a specific and practical goal. Some applied physiologists study certain groups of animals for their economic importance, for instance specializing in avian, dairy, swine, fish, and other physiology. Again, using the information generated by these highly specialized professionals and scholars as the initial and boundary conditions, engineers design and build unique systems for different agricultural animals, crops, or fish so that they can keep growing and producing in a healthy way.

4.2.1 The Basis of Physiology: Physics and Chemistry

To understand animal and plant physiology, we need to have a basic understanding of chemistry and physics. Animals and plants are obviously constructed from natural materials and therefore obey the same physical and chemical laws that apply to everything that we see around us. Temperature, for instance, exerts its effects on physiology by altering the nature of chemical bonds in biomolecules or the solubility of gases in solution. Hence, physiologists often borrow concepts and techniques from the physical and chemical sciences, including engineering, to help them understand how animals and plants work, and in turn, engineers design systems for animals and plants to grow, produce, and reproduce.

4.2.2 Physiological Regulation

Just like any other organisms, animals and plants face environmental variations. In natural environments, temperature, food availability, and the physiochemical environment around an animal or a plant may change with the time of day, the season, or the movement of an animal across the landscape. In controlled environments, these variables are fully controlled based on an animal's or a plant's needs.

Multicellular animals and plants can be classified according to the strategies they use to cope with changing conditions. Accordingly, conformers' internal environments change when they face changing external conditions, whereas regulators maintain relatively constant internal environments regardless of the conditions in the external environment. For instance, fish have low body temperatures in cold water and high temperatures in warm water. Hence, each of the cells in the body of a fish must cope with the effects of changes in ambient temperature. A swine's body temperature, on the other hand, is relatively constant whether it is in a warm controlled environment or standing outside on a very cold day. The body of a swine has mechanisms to maintain its internal temperature, and the vast majority of its cells do not have to cope with changes in external temperature.

4.2.3 Homeostasis and Feedback Loops

The maintenance of internal conditions within tolerable limits in the face of environmental disturbances is referred to as homeostasis. Homeostasis does not mean that there is no change in the organism, only that the animal or the plant initiates specific responses to control or regulate a particular variable. Homeostasis is maintained by what's called reflex control pathways—where a change in the internal or external environment provides a stimulus—including antagonistic controls, negative feedback, and positive feedback. A swine's body temperature stays relatively constant due to constantly changing physiological processes such as adjusting the rates of heat production and heat loss, for instance. To maintain homeostasis, animals and plants must i) detect external environmental conditions and ii) if needed, initiate adequate compensatory responses that iii) keep vital areas buffered against undesirable changes. Animals and plants most often maintain homeostasis using reflex control pathways.

4.2.3.1 Negative Feedback Loops

Maintaining homeostasis, negative feedback loops tend to minimize the changes in regulated variables. In a negative feedback loop, the response sends a signal back to the stimulus, reducing the intensity of the stimulus. For instance, when animals eat, the incoming food causes the stomach to swell. The change in stomach volume and early digestion triggers a negative feedback loop that acts through their brain to reduce their appetite.

Many physiological systems have a set-point, a preferred physiological state defended through feedback loops. For instance, different livestock body temperatures have different set points. The exact body temperature set point for a particular species varies between individual animals and changes throughout the day. When temperature rises in their external environments, they will seek shade or well-ventilated spaces, and their bodies may sweat to cool them down; in contrast, lower body temperatures will direct them to seek shelter or simply trigger shivering to warm them back to their set-point.

4.2.3.2 Positive Feedback Loops

Unlike negative feedback, which minimizes the change in the regulated variable, positive feedback tends to amplify and maximize the change in the regulated variable and cause explosive responses. As we can observe in many natural systems, some physiological systems are controlled by slowly starting but rapidly increasing intensity positive feedback loops. For instance, the stomach muscles are normally regulated to contract and relax in a pattern to gently mix food. When a toxin is detected, however, a positive feedback loop is activated to induce forceful contractions and hence vomiting. In such a loop, the animal also needs a signal to stop the process at a proper time to maintain control over the action taken.

4.2.4 Phenotype, Genotype, and the Environment

Overall, it must be noted that the physiological properties of an animal or a plant are aspects of the species' phenotype. However, physiological traits, like other characteristics of animals or plants, are determined mostly by the genes of the genome—the genotype—but are also influenced by the way the genes are regulated in response to external conditions.

Environmental heat and cold cause compensatory changes in animal and plant physiology and in morphology in many species. These changes help an individual organism to cope with temperature stress. Acclimation and acclimatization are types of phenotypic plasticity in which environmental variables result in reversible changes in phenotypes. Most animals and plants remodel their physiological processes in response to external environmental conditions.

Acclimation refers to the process of change in response to a specific variable in any controlled environment, whereas acclimatization refers to the process of change in response to variations in the natural environment. For instance, a fish that is moved from a controlled setting of water at 15°C to 5°C water will show a variety of physiological changes over time in the controlled environment via acclimation. That fish in a natural lake or pond environment with a mean temperature of 15°C will show many similar changes when observed in a 5°C lake or pond, but this process is acclimatization because the changes take place in the natural setting. In brief, animals make compensatory biochemical adaptations to changes in temperature.

Acclimatization refers to adaptation to not only temperature changes but also changes in day length, food availability, and many other environmental parameters; both acclimation and acclimatization are generally reversible physiological changes. This book deals with controlled environments, and therefore, discussions are confined to acclimation.

Furthermore, the genotype of an animal or a plant is the result of evolutionary processes including adaptation and genetic drift. Adaptation, however, has two different meanings within the context of physiology. The most common usage among physiologists refers to the product or process of evolution by natural selection, that is, changes in a population of organisms over evolutionary time. Having said that, however, physiologists quite often use the word adaptation in place of acclimation, and this author does so as well. For engineers, putting scientific knowledge into practical use requires understanding the basic physiology of animals and plants as influenced by the parameters of their environments.

4.3 PLANT ENVIRONMENTS

Green plants are photosynthetic autotrophs that manufacture (synthesize) their own food from inorganic raw materials (CO_2 from the air and water from the root zone) by photochemical and biochemical processes with the help of chlorophyll (a green pigment that absorbs light) and using the energy absorbed from light, which is called photosynthesis. The classical formulation of photosynthesis is provided as follows, which shows that the plant stores energy in the form of carbohydrates:

$$6CO_2 + 12H_2O \xrightarrow[(2{,}880\ \text{kJ/mol glucose})]{\text{Light}} C_6H_{12}O_6\ (\text{sugars}) + 6H_2O + 6O_2 \tag{4.1}$$

The stored sugars are used in the growth and production of biomass. Therefore, maintaining optimum light, temperature, CO_2, water, and nutrient regimes ensures desired yield or form of plant. A vegetative plant, for instance, produces more leaves than flowers and fruits, and its environmental requirements are different from those for a generative plant, which typically has smaller leaves and a high density of flowers and fruits for the size of the plant. However, for optimum plant development, the environmental parameters need to be adjusted so that the plant can have the vegetative phase at certain times of the year and then move to the generative phase.

Plants obtain their CO_2 and O_2 by diffusion of gas through their stomata (tiny openings on leaves), and the gas exchange (O_2 uptake and CO_2 release) with the environment with a release of energy, related to the breakdown of reserves (sugars), is called respiration. The classical formulation of respiration is provided as

$$C_6H_{12}O_6\ (\text{sugars}) + 6O_2 + 6H_2O \xrightarrow{\left(-2{,}880\frac{\text{kJ}}{\text{mol}}\text{glucose}\right)} 6CO_2 + 12H_2O \tag{4.2}$$

The released energy is used for supporting different plant processes such as root, leaf, and bud formation. Dark respiration is common to all plants and does not depend on light; it takes place during all 24 hours of the day. Some plants, however, have an additional mechanism known as photorespiration that occurs only when light is available. Green plants are classified into three major groups on the basis of their photosynthetic mechanisms: i) C_4 plants whose stomata are open during the day and dark respiration occurs all day and night (e.g., corn, sugar cane, sorghum); ii) C_3 plants whose stomata are open during the day where dark respiration occurs all day and night and photorespiration takes place during daytime; and iii) CAM plants whose stomata are open at night and dark respiration occurs all day and night (e.g., pineapple). When O_2 concentration increases, the rate of net photosynthesis in C_3 plants is inhibited due to the increased photorespiration (Warburg effect); therefore, enriching the CO_2 concentration in the aerial environment helps tremendously for developing a highly productive plant. This topic will be explored further in this chapter.

Under low light and temperature (early winter), for instance, assuming proper watering and nutrients, greenhouse plants typically have large leaves and flowers and relatively small fruits. In these conditions of high controlled temperature, plant growth is better with lower relative humidity. We can modulate the environment further by manipulating the day and night temperatures. During high daytime temperatures, for instance, plants manufacture more sugar and grow faster; at lower nighttime temperatures, the sugars manufactured during the day in the fruits, leaves, stems, and the growing points generate thicker, heavier plants. Engineers adjust not only temperature but other environmental parameters as well in designing control systems so that growers can obtain well-balanced, highly productive plants in their greenhouses.

4.3.1 Light and Radiation

Light energy is required directly to sustain the growth of all green plants. As described in the classical formulation of photosynthesis, light must always be present for green plants to perform photosynthesis, but temperature and CO_2 concentration may modify its influence.

4.3.1.1 Photosynthetically Active Radiation Range

Light intensity, quality, and duration all influence plant development, and plants respond primarily to the levels and duration of light within the photosynthetically active radiation (PAR) range (400–700 nm), the visible portion of the spectrum. Figure 4.1 shows the McCree curve, the photosynthetic response of typical plants. As we can see, blue (400–500 nm) and red (600–700 nm) are the most photosynthetically efficient wavelengths, but other colors such as green and yellow are quite useful as well. With the help of developments in LED technology, every passing day, we are learning more about how light quality affects plant growth, development, flowering, and antioxidant and vitamin content.

4.3.1.1.1 Red Light

Red light (600–700 nm) is quite effective at regulating plant growth and development. It is the most efficient light for photosynthesis as chlorophylls have absorbance peaks in the red band of the spectrum. Plants primarily use two photoreceptors termed cryptochromes and phytochromes that mediate flowering of plants with a photoperiodic flowering response (photoperiodic lighting is used at a low intensity to inhibit flowering of short-day plants and promote flowering of long-day plants), which are highly sensitive to the blue (400–500 nm) and UV-A (315 nm–400 nm) light regions and the red (600–700 nm) and far-red light (700–800 nm) regions, respectively, to determine the length of the day for photoperiodism (a response based on the portion of light to darkness in a 24-hour cycle).

Phytochromes are used as temperature sensors as well as for photomorphogenesis in seed germination. Plants are overly sensitive to low intensities of red light, which can inhibit flowering of short-day plants and promote flowering of some long-day plants when provided at night. In short, it can be stated that red light is the most dominant color for both photosynthetic and photoperiodic lighting. Longer wavelengths including red light can penetrate the growth media such as soil as well; and therefore, they are used by seedlings to determine in what direction to grow.

However, plants grown under red light only (e.g., LEDs) do not have adequate growth characteristics (e.g., thin and large leaves, taller plants) due to extension growth. When a small amount of blue light is added, however, the extension

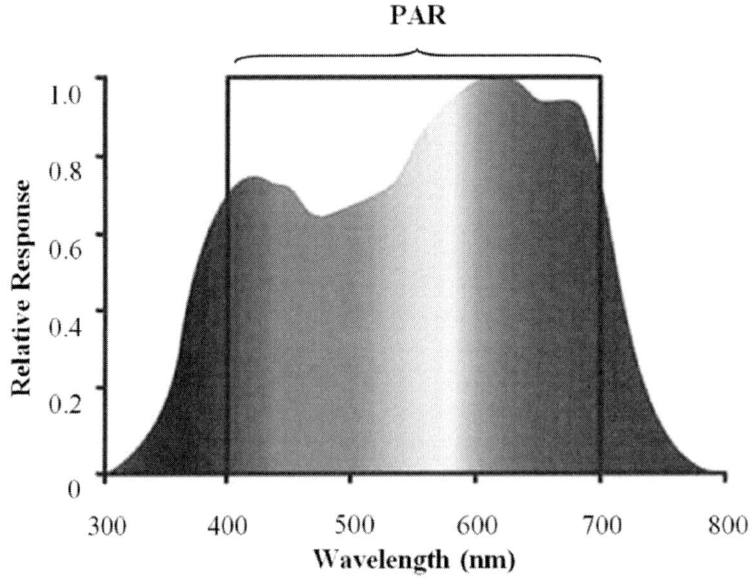

FIGURE 4.1 Relative photosynthetic response of typical plants.

Source: *Greenhouse Engineering*, Edn. 1 by İlhami Yıldız, Copyright © 2021 by Imprint. Reproduced by permission of Taylor & Francis Group (*Greenhouse Engineering*).

growth is inhibited. In greenhouses, on the other hand, plants receive the full solar spectrum and therefore the color provided by supplemental lighting does not have much impact on extension growth. The addition of far-red radiation also stimulates the flowering of many long-day plants.

4.3.1.1.2 Blue Light

Blue light (400–500 nm) is also one of the most essential light ranges for photosynthesis and high-quality vegetative growth and flowering. Blue light also promotes antioxidant and vitamin activity. Generally, low blue light is sufficient for fully functional photosynthesis. Extension growth is generally suppressed by blue light; plants grown with blue light have narrow internodal spacing and stronger stems along with smaller, thicker, and darker green leaves.

Moreover, blue light promotes flowering of long-day plants and inhibits flowering of short-day plants at higher intensities, although it does not regulate flowering of most daylength-sensitive crops at low light intensities. Blue light also regulates the stomatal openings (tiny pores on leaves), which control transpiration (water loss) and CO_2 uptake for growth. Blue light contributes to leaf coloration via anthocyanin production as well, and plants as a result produce appealing vegetation. Furthermore, blue light affects plant morphology through plant cryptochromes, which are sensitive to blue light and tell the plant to grow toward the light.

4.3.1.1.3 Green Light

Figure 4.1 shows that plants, even though not as efficiently as blue or red light, also use green light (500–600 nm) very well. Chlorophylls do not absorb green light that well, and plants reflect more green light than other wavelengths (this is why we see green plants as green). Green light mitigates the performance of lower leaves because it penetrates the canopy into the leaf mesophyll, energizing chloroplasts and enhancing photosynthesis. Green light has high energy transfer efficiency (quantum yield), which means that more energy reaches the reaction centers and be used for photosynthesis under high light intensity when the chlorophylls and pigments on the upper leaf surface are already saturated by blue and red lights, much of which is absorbed on the upper leaf surface by flavonoids and carotenoids before reaching the reaction centers and therefore leaving chlorophylls and pigments deeper in the leaf not saturated. Green light is particularly useful under high light intensity. As discussed earlier, phytochromes and cryptochromes mediate flowering of plants with a photoperiodic response. These pigments absorb green light as well, and recent research findings have shown that in many plants, green light is as effective at regulating flowering of long-day plants as the same intensity of red plus far-red radiation.

4.3.1.2 Non-PAR Range

Both solar radiation (see Figure 2.34 in Chapter 2) and supplemental lighting (see Chapter 5) provide PAR and non-PAR in greenhouses. Non-PAR from 750 to 3,000 nm, for instance, contributes to the air and soil temperatures in greenhouses while increasing the plant temperatures. Some non-PAR may contribute to crop responses as well.

4.3.1.2.1 Ultraviolet Radiation

At non-PAR range, UV radiation is defined as the part of the electromagnetic spectrum from 1 to about 400 nm (Figure 2.34), biologically active radiation that regulates plant growth. As discussed in Chapter 2, the Earth's atmosphere (molecular oxygen and ozone) filters out UV radiation at wavelengths smaller than 280 nm, which is called the UV-C range.

UV-A is wavelengths between 315 and 400 nm. Regular optical (glazing) materials such as glass and acrylic do not effectively transmit UV radiation with wavelengths smaller than 360 nm, and chlorophyll *a* has an absorption peak around 415 nm. Plants also have a photoreceptor called UVR8 (responsible for triggering the stress response in plants) that is highly sensitive to UV-B light from 280 to 315 nm and centered at about 285 nm. The stress response in plants can be manipulated for growing stronger plants if done carefully. For instance, low doses of UV-B can increase THC (tetrahydrocannabinol), the main active ingredient in cannabis followed by secondary metabolites such as terpenes.

Overall, UV light increases leaf coloration, thickness, and waxiness and promotes shorter stems and smaller leaves, thereby inhibiting extension growth. Recent research showed that some UV light suppresses insect feeding on leaves (increases pest resistance) and vulnerability to fungal pathogens, in contrast with little or no UV. UV-B and UV-C were particularly effective at killing pathogens on leaves. UV-C and sometimes UV-B are also used for disinfecting water, another important use of UV light in not only greenhouse but also aquacultural operations. Furthermore, as discussed earlier in Chapter 3, UV is also used by insects including bumblebees to help them navigate to flowers. If an environment is totally free of UV light, pollinators like bumblebees and pests may have problems locating flowers.

4.3.1.2.2 Far-Red Light

Another non-PAR range is far-red light, which covers the range from 700 to 800 nm. As discussed earlier, plant photoreceptors are quite sensitive to red and far-red light, using them to evaluate their growth environment. For instance, in full sunlight, the proportion of red light to far-red inside a greenhouse is high, and as is well known, plants respond to this very well.

When the solar daily light integral (DLI) inside a greenhouse is considerably low, then supplemental lighting is effective at increasing plant growth, and far-red light accelerates the flowering of long-day plants in particular; far-red light does not influence flowering in day-neutral and short-day plants. When the DLI is low, however, providing some red and far-red light during long nights can accelerate flowering of some species such as tomatoes. Under overcast or shade conditions, however, the proportion of far-red to red light is high, which initiates a phenomenon known as the shade avoidance response in plants.

Under such conditions, plants use their energy to grow tall toward the Sun and in the process neglect vegetative growth. This leads to undesirably tall, weak plants. The proportion of red to far-red light is manipulated by the photoperiodic plants for determining the time of year, so that they would know when to start flowering. As the Sun sets, the proportion of far-red increases, and using this information, photoperiodic plants determine the length of the day. Once the plant, either short-day or long-day, determines that the length of day is adequate, then it performs flowering. The effectiveness for photoperiod response in plants typically peak at wavelengths of 660 nm (red) and 730 nm (far-red).

4.3.1.3 Photoperiodism

At this point, it would be wise to talk about photoperiodism little further. Daylength (photoperiod) influences flowering in some plants along with fruit growth and stem elongation. There exist four basic daylength groups of plants: i) short-day plants require a period of darkness longer than a critical length, usually in fall (e.g., chrysanthemum, poinsettia, aster); ii) long-day plants require a period of darkness shorter than a critical length, usually in summer (e.g., lettuce, most grains); iii) daylength-intermediate plants flower only within a narrow range of daylength, usually 10 and 14 hours (e.g., sugarcane, coleus); and iv) day-neutral plants are insensitive to the length of the day (e.g., tomato, carnation, sunflower, corn). Depending on this classification, an artificial single light source (e.g., sodium lamps) as the sole source of lighting may not be adequate in greenhouse environments; the environment may require the added blue, red or infrared emissions, which will be discussed further in Chapter 5.

4.3.1.4 Light Intensity and Uniformity

Light intensity influences plant development, and as Figure 4.2 shows, the rate of photosynthesis increases as light intensity increases; however, light does not increase the photosynthetic rate beyond a maximum intensity. If the light available to plant is too low, its energy production from photosynthesis may be equaled by its energy output in respiration. This is called the light compensation point. At this point, the light becomes a limiting factor.

In mid-winter months, there is less light energy available for plant growth than in the summer, for instance; therefore, it takes twice as long for plants to perform the same amount of growth in mid-winter as in mid-spring or early summer. It is therefore particularly important to effectively schedule greenhouse crop production. For cost-effective production, the scheduling must be well aligned with the prevailing light conditions.

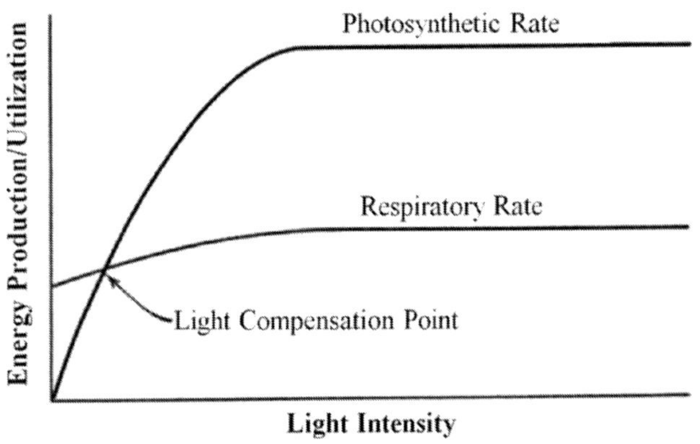

FIGURE 4.2 Schematic representation of photosynthetic and respiratory rates versus light.

Source: *Greenhouse Engineering* (2021).

Plants can well adapt to different light levels depending on the age of plant or leaves and the acclimatization length. Under low light intensity, plants develop larger leaf surface areas for intercepting more light to compensate for the adverse effects of low light intensity without impacting the photosynthetic rate per unit surface area of these leaves. Plants also arrange the light capturing chloroplasts accordingly depending on the light levels. Under low light conditions, they align chloroplasts perpendicular to the incoming radiation so that they can capture more light. Under high light conditions, however, the chloroplasts are aligned parallel to the incoming radiation.

As discussed in Chapter 3, the amount of light that is received by plants in a greenhouse is directly affected by the greenhouse covering material. Glass and double PE are the most used covering materials, and their transmissivities were provided earlier. Modern roof designs and elevated gutter heights enable improved light transmission to the crop canopy. Only 60% to 75% of incoming light reaching the cover's outside surface is transmitted into the greenhouse. There exists a theoretical maximum of transmissivity for each type of glazing because of the refractive index at each glazing–air interface (~4%) and absorptivity of each glazing material. Therefore, choosing the right glazing material in the first place is extremely important for providing adequate light intensity for the plant grown in a greenhouse.

Especially, in the production of flowering ornamental plants and some other specialty crops, the common goal of all growers is to consistently grow marketable high-quality and uniform plants. There might be different reasons for plant nonuniformity, and one of these reasons might be nonuniform light distribution in a greenhouse. Therefore, we must make sure that greenhouse lighting is evenly distributed to ensure uniform crop response.

4.3.1.5 Light and Carbon Dioxide Interaction

While several environmental stressors such as watering and nutrient availability also influence photosynthesis, particularly the CO_2 concentration and temperature affect photosynthesis dramatically (Figures 4.3 and 4.4).

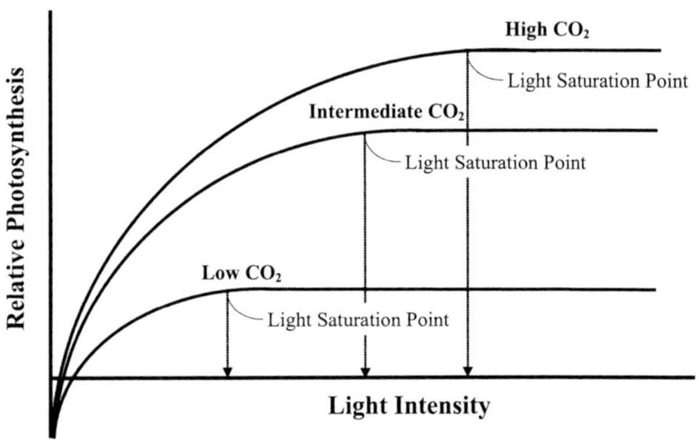

FIGURE 4.3 Schematic illustration of light and CO_2 interaction on photosynthesis. Plants use more light and perform more photosynthesis when the CO_2 concentration increases. Beyond the light saturation point at each CO_2 concentration, increased light does not influence photosynthesis.

Source: *Greenhouse Engineering* (2021).

Figure 4.3 indicates that, at low light intensities, the rate of photosynthesis depends on the quantity of incident light, whereas at high light intensities, i.e., beyond the light saturation point (the intensity where further increases in light do not increase photosynthesis), the amount of CO_2 is rate limiting. Figure 4.3 also illustrates that low CO_2 concentration decreases the light saturation point, reducing the rate of photosynthesis, while high CO_2 levels increase the light saturation point, resulting in more effective utilization of light by the plants.

4.3.1.6 Light Pollution

Today, more greenhouse operations are employing supplemental lighting to increase plant growth and development; however, this electric lighting can have undesirable and unintended consequences to neighboring areas at night, called light pollution. These neighboring areas might be other greenhouse sections within the same operation or nearby greenhouse complexes or residences.

Some plants are sensitive to extremely low light intensities; therefore, the stray light can have sufficient intensity to reach other sections or nearby operations to modify the photoperiod of plants grown there. Light pollution may come not only from lighting fixtures used in greenhouses but also from other nearby buildings and street lighting. Under such conditions, to prevent light pollution, greenhouse operators can use retractable blackout screens and curtains along the greenhouse sidewalls and roof. These systems might be expensive to install; however, a blackout curtain not only prevents light pollution but also serves as an excellent curtain that reduces heat losses from the greenhouse at night especially during the winter. Therefore, it is a justifiable investment.

4.3.2 Temperature

Temperature (and RH) affects a number of plant processes that influence plant growth including photosynthesis and respiration; the development of roots, leaves, and flowers; and leaf size and shape. It is a critical environmental parameter in greenhouse design.

For whole-plant development, 24 hour average temperature is important; leaf and flower formation increase by 10% for each degree increase between 10 and 30°C. The higher the light intensity, the higher the optimum temperature; this is why the 24 hour average temperature is higher under high light conditions. Under low light, a lower 24 hour average temperature might increase foliage quality, but plants develop much more slowly. Figure 4.4 shows that photosynthesis increases as temperature increases until a maximum and then decreases; however, as clearly seen in the figure, the temperature influence is also a function of CO_2 concentration. The temperature for maximum development increases when the CO_2 concentration increases.

Besides, too high or too low 24-hour temperature is no good for plant development. High temperatures promote thinner leaves and stems and small flowers, while low temperatures promote thicker leaves and stems as well as large flowers and damage flowers or delay harvest. Additionally, the day–night temperature differentials influence the development of thermoperiodic plant species, which have different optimum day and night temperatures for each stage of their growth. When there is no day–night temperature differential, plants produce more vegetation with shorter stems and thicker and larger leaves, and when there is a large temperature differential, then plants turn to generative production with longer stems and thinner and smaller leaves. Many plants grow better when the nighttime temperature is lower than the daytime temperature. In practice, generally a more even day-night temperature control is provided in early development of a plant, and then a temperature differential is used after the plants flower and move into the fruit-setting stages.

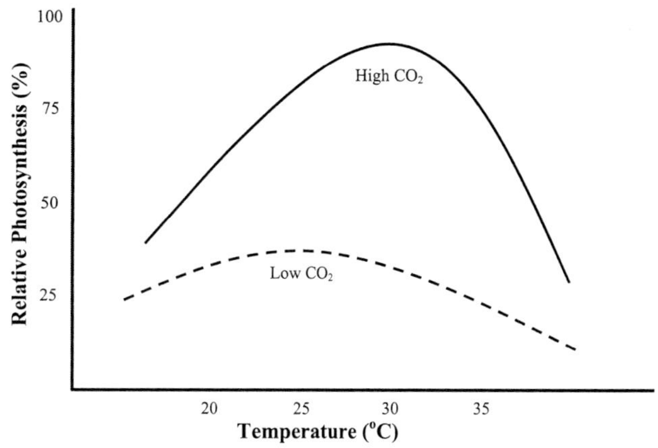

FIGURE 4.4 Schematic illustration of temperature and CO_2 interaction on photosynthesis. Photosynthesis increases as temperature and CO_2 increase until a certain point at which it then decreases.

Source: *Greenhouse Engineering* **(2021).**

Moreover, temperature and RH of the ambient air are intimately linked: RH decreases as temperature increases and vice versa. RH directly affects plant transpiration rate and overall water status. Plants also generally produce longer stems and larger leaves at high RH.

Overall, plants have significantly varying minimum, optimum, and maximum temperatures that depend heavily on light intensity, the stage of plant growth, CO_2 concentration, and season. The optimum temperatures for the major greenhouse plants lettuce, tomatoes, peppers, and cucumbers are generally reported to be 17–18°C, 19°C, 19°C, and 21°C, respectively; however, further specifics of temperature requirements for these and other crops are provided in Table 4.1 and Table A.51. As discussed earlier, many crops grow better when the night temperature is lower than the day temperature. They have different optimum day and night temperatures for each stage of their growth, and recommended night temperatures for a list of greenhouse crops can be found in Table A.51 as provided by ASHRAE (1997) and OMAFRA (2001).

4.3.3 Water and Relative Humidity

RH was defined in Chapter 2 as the ratio of the mole fraction of water vapor in moist air (the actual amount of water

TABLE 4.1
Recommended Air Temperatures for Three Major Greenhouse Crops

Plant		Temperature (°C)				
		Germination	**Transplanting**	**Raising**	**Harvesting Stage**	**Full Harvest**
Tomato	Day	25	24	19–21	19	20–22
	Night	25	24	19–21	19	17–19
Cucumber	Day	28	25	24	23	23
	Night	28	25	21	21	17–21
Pepper	Day	25	21	21	21	21
	Night	25	24	23	24	18

Source: OMAFRA (2001).

vapor present) to the mole fraction in the air saturated (the maximum amount of water vapor that can be held) at the same temperature and pressure. RH influences transpiration rate, latent heat transfer, and certain diseases, especially bacterial and fungal. Normal plant growth takes place at RH between 20% and 80% if plants have well-developed root systems; however, RH in most greenhouses ranges between 40% and 100%, which causes many problems. Figure 4.5 gives water vapor contents per unit volume of air at different RH.

The water holding capacity per unit volume of air (saturation humidity) also correlates with air temperature; as Figure 4.6 shows, it nearly doubles for every 10°C increase in air temperature.

It should be kept in mind that RH in a greenhouse changes with changes in air temperature if absolute humidity is not adjusted via ventilation; Figure 4.7 shows that RH increases or decreases by about 5% for each degree Celsius decrease or increase, respectively. This is particularly important when greenhouse set point temperatures are changed during day and night.

We typically use RH to express the moisture content of greenhouse air even though it gives the moisture status only within a narrow temperature range. It is not sufficient to know RH as it fluctuates quite a bit with temperature changes (see Figure 4.7). Figure 4.8 further clarifies why RH does not provide adequate information about the actual moisture content of greenhouse air. The figure presents water contents and total water-holding capacities of air with 70% RH at various temperatures; the meaning of 70% RH

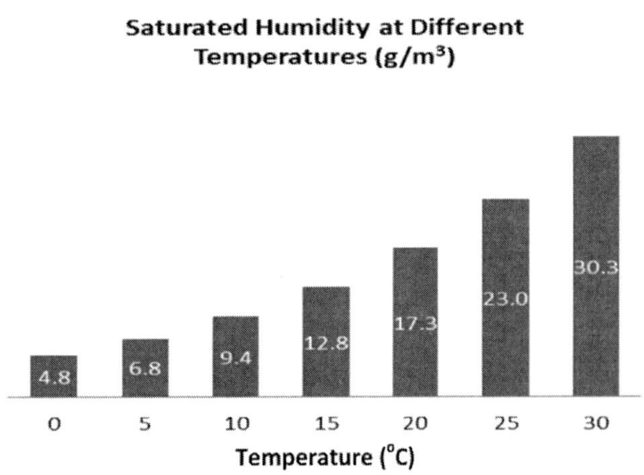

FIGURE 4.5 Actual water content per unit volume of air (g/m^3) at 18 °C with varying RH.

Source: *Greenhouse Engineering* (2021).

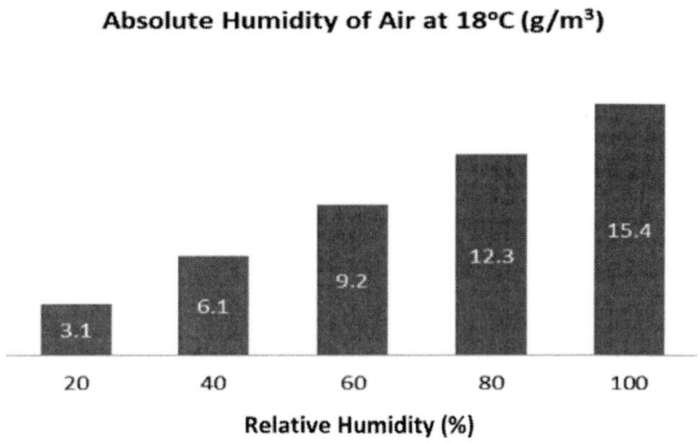

FIGURE 4.6 Effect of temperature on the water holding capacity (g/m^3) of air.

Source: *Greenhouse Engineering* (2021).

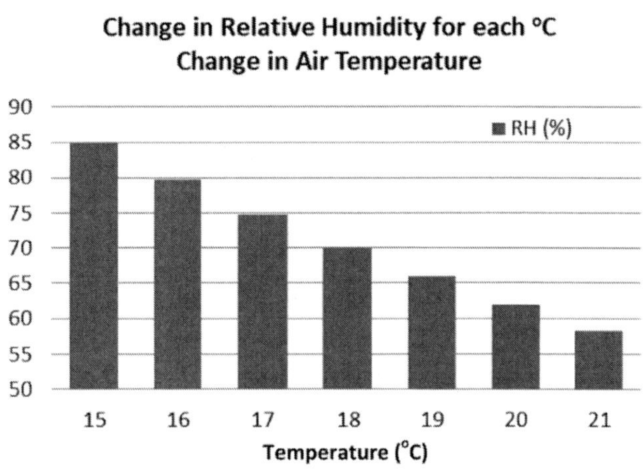

FIGURE 4.7 Change in relative humidity for each degree Celsius change in air temperature (T = 18°C and T_d = 12.5°C (i.e., 70% RH) was used as the starting point).

Source: *Greenhouse Engineering* (2021).

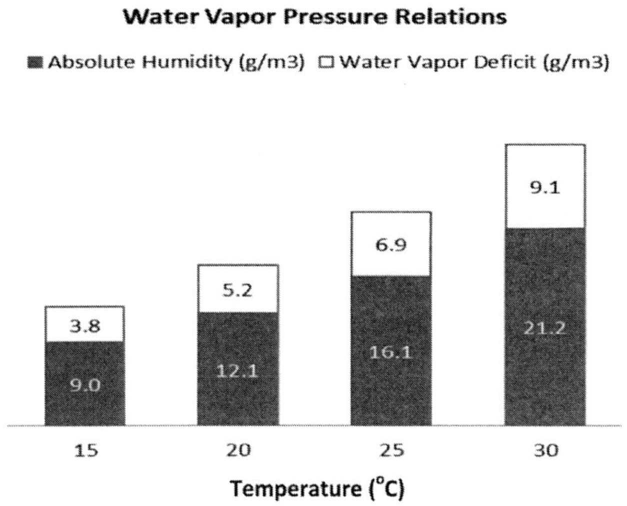

FIGURE 4.8 Actual water contents (g/m^3) and water-holding capacities (g/m^3) of air with 70% relative humidity at various temperatures. Note: The gray shaded area is the optimum range for most greenhouse crops. 1 mbar = 0.1; kPa = 0.0145 psi; 0.1 kPa = 0.7 g; m^5 = 3?i RH.

Source: Greenhouse Engineering (2021).

at each temperature with respect to the actual water content changes as the water-holding capacity increases with temperature.

For instance, while the air at 30°C and 70% RH contains much more moisture (21.2 g/m^3) than the lower temperatures presented in Figure 4.8, the air still has quite a bit of room (water vapor deficit of 9.1 g/m^3) before it gets saturated with water vapor. This means that the air at a lower temperature has lower water-holding capacity and therefore becomes saturated faster.

A high water vapor deficit, on the other hand, means that there exists a higher forcing function for transpiration and therefore that transpiration will increase. Tiny pores on plant leaves called stomatal openings biologically control transpiration by opening and closing. When there is abundant water or light or low internal CO_2 concentration, stomatal openings will open; in reverse, when there is a lack of water or light, or when internal CO_2 concentration is high, stomatal openings will close. Stomatal cavities in plant leaves are normally fully saturated with water vapor while the greenhouse air is usually not, and the water vapor on each side exerts pressure in the air whether it is greenhouse air or the air in the stomatal cavity. The difference between these two water vapor pressures, also known as vapor pressure deficit (VPD), regulates the plant water loss through transpiration. Table 4.2 presents the VPDs for various air temperature and RH combinations, and VPDs between 4 and 8 mbar are considered optimal for optimal nutrient uptake and transpiration supported by proper carbohydrate production and transfer, which generate healthy, productive plants. Too low or too high VPDs have detrimental effects on plant growth and productivity. Engineers manipulate VPD information in their designs for controlling the greenhouse environment for a safe, healthy and productive greenhouse operation.

Due to the incoming solar radiation, leaf temperatures are typically higher than the air temperatures during the day. However, at night, in the absence of solar radiation under especially clear sky conditions, leaf temperatures

TABLE 4.2
Vapor Pressure Deficit as Influenced by Temperature and Relative Humidity (mbar)

Temperature	Relative Humidity (%)									
<°C)	50	55	60	65	70	75	80	85	90	95
15	8.5	7.7	6.8	6.0	5.1	4.3	3.4	2.6	1.7	0.9
16	9.1	8.2	7.3	6.4	55	4.6	3.6	2.7	1.8	0.9
17	9.7	8.7	7.8	6.8	5.8	7.5	3.9	2.9	1.9	1.0
18	10.3	9.3	8.3	7.2	6.2	5.2	4.1	3.1	2.1	1.0
19	11.0	9.9	8.8	7.7	6.6	5.5	4.4	3.3	2.2	1.1
20	11.7	10.5	9.4	8.2	7.0	5.9	4.7	3.5	2.3	1.2
21	12.4	11.9	9.9	8.7	75	6.2	5.0	3.7	25	1.2
22	13.2	11.9	10.6	9.3	7.9	6.6	5.3	4.0	2.6	1.3
23	14.1	12.6	11.2	9.8	8.4	7.0	5.6	4.2	2.8	1 4
24	14.9	13.4	11.9	10.4	9 0	7.5	6.0	4.5	3.0	1.5
25	15.8	14.3	12.7	11.1	9.5	7.9	6.3	4.8	3.2	1.6
26	16.8	15.1	13.4	11.8	10.1	8.4	6.7	5.0	3.4	1.7
27				12.5	10.7	8.9	7.1	5.4	3.6	1.8
28				13.2	11.3	9.5	7.6	5.7	3.8	1.9
29					12.0	10.0	8.0	6.0	4.0	2.0
30					12.7	10.6	8.5	6.4	4.2	2.1

Source: OMAFRA (2001).

may drop several degrees due to radiational cooling depending on the greenhouse glazing and whether a night curtain is used or not. Under such conditions, if the inside RH is around 95%–96%, then condensation takes place on leaf and other surfaces. Lower leaves, which stay moist longer, are typically more susceptible to condensation, and therefore to bacterial and fungal diseases. Transpiration from the leaves raises the relative humidity in interfoliage air; that is, the still air surrounding the plant canopy is usually moister. However, manipulating rail pipe, underbench, and grow pipe heating systems in particulare coupled with ventilation for induced convection within the canopy can remove the risk of condensation on plant and other surfaces whenever it occurs.

Humidity extremes are not good for plants whether low or high. Rapid drops in RH due to excessive ventilation can burn foliage and reduce fruit quality and yield in some greenhouse crops. RH also impacts pollen transfer and germination for fertilization and proper fruit size. Tomatoes, for instance, require RH between 70% and 75% for proper pollination.

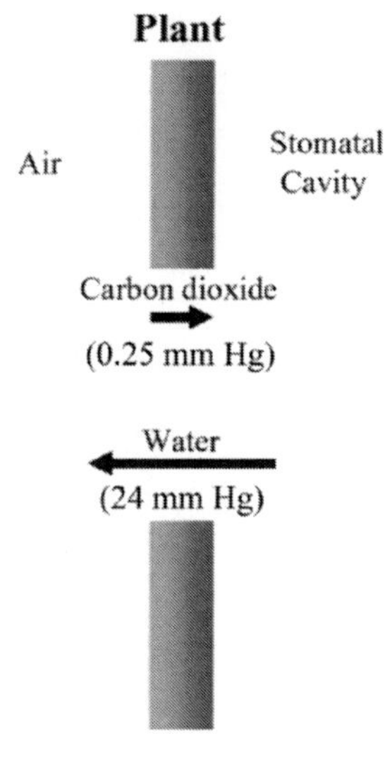

FIGURE 4.9 Partial pressure differences across stomatal openings. **Source:** ***Greenhouse Engineering*** **(2021).**

4.3.4 CO_2 and Air Pollutants

4.3.4.1 CO_2

Today, the atmospheric CO_2 concentration is about 0.04% (400 ppm), and burning fossil fuels continuously adds significant amounts of CO_2 to the atmosphere every day. As was discussed earlier, CO_2 taken up by plants through their stomatal openings is an essential ingredient for photosynthesis, which responds to the changes in CO_2 concentration as it is the main constituent of plant dry matter, approximately 45%, and is essential for the plant growth and yield.

The amount of CO_2 available to terrestrial plants is usually not limiting, although at high altitudes it can be a limiting factor. The amount of CO_2 available in greenhouses can quickly drop well below the ambient levels and become limiting, especially during sunny periods depending on the light, temperature, ventilation, and supplemental CO_2 if any. Plants use more light and perform more photosynthesis when the CO_2 concentration in the ambient air increases (Figure 4.3). At low light intensity, the rate of photosynthesis depends on the incident light, whereas at high light intensity, the amount of CO_2 becomes rate limiting assuming that other management practices are adequately performed. Elevated CO_2 concentrations coupled with elevated daytime temperatures and high light shorten the time between germination and harvest by as much as 50%.

As Figure 4.9 schematically shows, the driving force for CO_2 is no more than 0.25 mm Hg (0.13 in. H_2O) (0 inside of plant tissues). At 25°C, the outward driving force for loss of water to dry air from plant tissues is 24 mm Hg (12.9 in. H_2O) (about one hundred times as high as the inward driving pressure for CO_2). Due to the low diffusion gradient of CO_2 into the plant, normally it is difficult for plants to obtain the necessary CO_2 from the ambient air without incurring relatively high water loss by transpiration through the stomatal openings. Because of this unfavorable situation, plants have

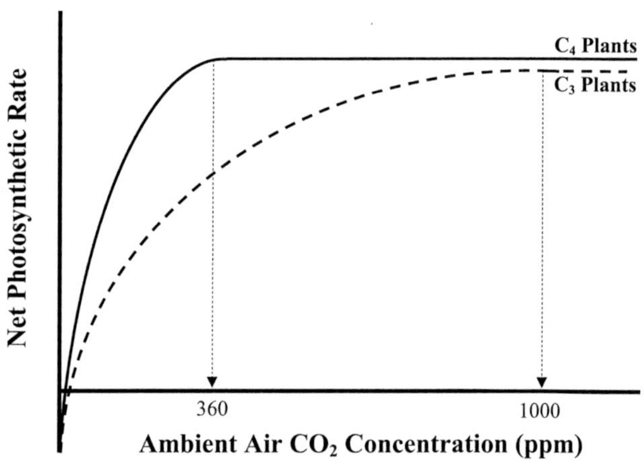

FIGURE 4.10 Schematic representation of photosynthetic response of different plant species to CO_2 concentration. **Source:** ***Greenhouse Engineering*** **(2021).**

very high water requirements. In most plants, the stomatal openings are located primarily in the lower epidermis—the side of the leaf usually turned away from the Sun's rays, where drying is less severe. When plants run short of moisture, they block most water loss by simply closing the stomatal openings at the expense of growth.

It is possible, however, to increase the rate of plant growth by increasing CO_2 concentrations several times above normal values. Plant species differ in their photosynthetic responses to CO_2, with C_3 plants exhibiting increases in net photosynthetic rate of about 1000 ppm (CO_2 saturation point) and C_4 plants exhibiting the maximum rate at about 360 ppm (Figure 4.10). As discussed earlier, C_3 plants have an additional light-dependent respiratory mechanism known as photorespiration; the normal O_2 concentration of 21% inhibits CO_2 uptake by the plant and increases the loss of carbon due to photorespiration.

Since the partial pressure of O_2 in the air at 1 standard atm (760 mm Hg) is 159 mm Hg, which is much greater than that of CO_2 (0.25 mm Hg), the rate of net photosynthesis in C_3 plants is inhibited due to the increased photorespiration (Warburg effect) as a result of the competition between ambient CO_2 and O_2 to be fixed by the enzyme ribulose diphosphate carboxylase. The oxygen inhibition of photosynthesis decreases with CO_2 enrichment and is almost eliminated at a CO_2 concentration of about 900 ppm; consequently, the net photosynthesis in plants increases (Figure 4.10). Elevated CO_2 concentrations also increase the optimal temperature for growth (Figure 4.4). Pot plants, cut flowers, vegetables, and forest plants expose quite positive responses to CO_2 enrichment with increased plant quality and number of flowers.

Research findings have shown that plant growth is improved by 15%–20% in greenhouses when CO_2 concentrations increase to 800–1,300 ppm during the day. Continuous CO_2 enrichment during daylight has been found to be superior to intermittent CO_2 application. The literature suggests that the optimal CO_2 concentration for plant growth is between 600 and 1,000 ppm for most species; in general, the optimum CO_2 coupled with temperature and PAR intensity is about 400 ppm on cloudy days or vents are opened more than 10% and 1,000 ppm on sunny days.

Furthermore, elevating the CO_2 concentration leads to higher internal CO_2 concentrations in plant tissues closing the stomatal openings and hence reduces the transpiration of plants by 20%–40%. Water consumption is therefore significantly reduced by CO_2 enrichment while photosynthesis is increased simultaneously. We know that increasing RH from, say, 60% to 95% decreases plant transpiration by about 50%; additionally, CO_2 enrichment at 95% RH decreases transpiration and water uptake by about 30% in some plants, also increasing the water-use efficiency (amount of water used per unit dry weight produced) by about 30%. The decreased transpiration could, however, cause higher leaf temperature due to the reduced evaporative cooling performed by plants themselves, especially under high light conditions.

4.3.4.2 Air Pollutants

The most common method used for CO_2 enrichment in greenhouses is burning hydrocarbons such as natural gas, propane, or low sulfur kerosene. Air pollution in greenhouses due to CO_2 enrichment by burning these hydrocarbons sometimes causes visible leaf injuries, which include chlorosis, necrosis and curling of the leaves, and invisible injuries such as reduced growth. The research findings explain the injurious effects of high CO_2 concentrations by i) too-high leaf temperatures at high light levels as affected by reduced transpiration at high CO_2 concentrations; ii) accumulation of starch, which breaks down the chlorophyll, promoted by high light levels and low temperatures; and iii) reduced nutrient uptake because of reduced transpiration, particularly at high air humidity levels.

When CO_2 is supplied by burning kerosene, then ethylene (C_2H_4) pollution can be observed due to incomplete combustion. Ethylene is a plant hormone that stimulates senescence of the plants, causing leaf chlorosis and abscission and flower drop; therefore, it has potential to cause substantial crop losses. The concentration in the greenhouse environment should not exceed 0.05 ppm, above which plant injuries might be observed. These long-term plant exposures to even quite low ethylene concentrations usually inhibit overall plant growth and can harm small flower buds. While new kerosene CO_2 dosing equipment has pretty good burners, adequate burner maintenance and care is still vital.

Carbon monoxide (CO) is a pollutant that is deadly for people at certain concentrations and exposure times (see Table A.16); it is produced by incomplete combustion and causes injuries similar to those caused by ethylene, but thousands of times higher concentrations of CO must be present to cause comparable damage. In practice, the CO concentration in greenhouses is somewhere between 0 and 5 ppm, which creates no problems. Sometimes, kerosene contains sulfur that might generate enough SO_2 to cause damage as well; however, using low-sulfur kerosene eliminates this problem.

Other hydrocarbons such as propane can also cause pollution. Commercial propane leakage has caused severe plant injuries in some greenhouse operations in the past as propane contains propylene, which is an analogue of ethylene. With the modern equipment and current safety measures, however, the probability of leakage is quite low. Natural gas, on the other hand, consists mainly of methane, and with complete combustion, it does not pose any harm to plants.

Along with ethylene, nitrogen oxides (NO_x) are other important pollutants sourcing from the burning of hydrocarbons. NO_x is produced during the combustion when O_2 reacts with atmospheric N_2, and in greenhouses, concentrations up to 1 ppm have been measured with kerosene as well as with propane burners. At a CO_2 dosing of 1000 ppm, the NO_x concentration is usually about 0.5 ppm. NO normally makes up about 75% of the oxides, and concentrations above 250 ppb sustained over 10 days are harmful to plants; some propane and kerosene burners produce mainly NO_2, and concentrations above 132 ppb sustained over 10 days are harmful. The injuries of these oxides are similar and species specific, ranging from visible to invisible injuries.

Since burning hydrocarbons can injure plants, probably the safest source is pure liquid CO_2 from containers, recommended for general use for greenhouse crops even though it can also contain some undesirable impurities such as ethylene. However, as mentioned earlier, along with liquid CO_2, the flue gas CO_2 coming out of natural gas boilers is commonly and successfully used for CO_2 supplementation in commercial greenhouses.

4.3.5 Air Movement and Altitude

Air movement affects evaporation, plant transpiration, and the CO_2 available to plants. Air speed of 0.5 to 0.7 m/s is reported to be optimum in greenhouses. Air speeds exceeding 1 m/s inhibit plant growth, and 4.5 m/s can cause physical damage. The direction of air movement does not seem to affect results.

Altitude, on the other hand, affects plant growth indirectly due to the decreased average temperatures, wavelength distribution of solar radiation, humidity, average wind speed, etc. At high altitudes, greenhouse heating costs for providing an adequate plant thermal environment rise due to decreased outside temperatures and increased wind speeds; partial pressures of atmospheric CO_2 and O_2 concentrations drop as well, and unless proper measures are taken, they may be limiting.

Furthermore, combustion equipment is designed and operated to ensure complete combustion. If insufficient air is provided to burners, incomplete combustion occurs, with non-oxidized components in the flue gas such as C, H_2, and CO. Incomplete combustion does not use fuel efficiently, which results in air pollution and can pose even further danger due to CO production and a potential for explosion. CO is odorless, tasteless, colorless, and toxic. As discussed earlier, every combustion process has the potential to produce this toxic gas as a result of incomplete combustion. Air at higher altitudes has lower density and therefore less oxygen per unit volume. Excess air must be adjusted accordingly; otherwise, we can have incomplete combustion, and combustion can eventually cease.

Plants, animals, insects, bacteria, fungi, and viruses form the biotic environment in a greenhouse, and these must be managed below an economic injury level in an integrated fashion using cultural, physical, biological, and chemical control strategies.

4.4 ANIMAL ENVIRONMENTS

An animal's environment is the total of all external conditions that affect it. Thermal conditions, air quality, lighting, ion concentration, noise, and crowding are important in designing facilities for animals. Thermal environment affects the rate of chemical processes and heat dissipation; it specifically covers the effects of air temperature, air moisture content, air velocity, and solar radiation on the regulation and balance of animal heat, and their influence on growth, feed conversion, production, and health. As in plants, lighting influences photoperiodism in animals; air quality, ion concentrations, noise, and crowding can influence growth, productivity, and health of animals. This section summarizes the physiological factors involved in controlling animal environments.

4.4.1 Thermal Environment and Animal Response to Temperature

As mentioned earlier, temperature has great influences on all physiological processes. Animal performance (growth, egg or milk production, wool growth, and reproduction) and conversion of feed to useful products are very much linked to the thermal environment. As a result of these influences, each animal develops and displays strategies such as combining behavioral, biochemical, and physiological responses that maintain their body temperatures (T_b) within acceptable limits. Ambient temperature (T_∞) has the most important environmental influence on the thermal strategy: Animals must be able to survive the thermal extremes in their environments.

4.4.1.1 Physiological Control Systems

Animal body temperature is regulated via homeothermy, which is a form of homeostasis. Livestock and poultry are homeothermic, meaning that they can maintain constant body temperatures when ambient temperature changes. The hypothalamus regulates animals' body temperature by controlling physiological responses to regulate heat production and loss.

Internal body temperature is dynamically controlled by an equilibrium between internal heat generation and environmental heat loss. Internal heat production rate varies quite a bit with animal's size, body weight, breed, age, growth stage, feed type and intake rate, health, production level, gestation, degree of activity, and environmental conditions. Approximately 25%–40% of the gross feed energy consumed is typically converted into heat and eventually lost to the environment. In a cold environment, heat loss is greater than normal heat production, and the net energy normally used for animal growth and production is used to maintain body temperature, which results in rising feed requirements per unit production. Animal performance is, therefore, greatly affected by the heat exchange between animals and their environments.

4.4.1.2 Thermal Comfort, Cold Stress and Heat Stress

Each homeothermic species has an optimum thermal environment that facilitates all necessary physiological functions with minimum energetic input. Figure 4.11 demonstrates this range as the thermal comfort zone, between A and A' in the figure (i.e., welfare is enhanced by maintaining environmental conditions within this zone). The thermal comfort zone is not easily defined for animals, but it can be approximated.

For a controlled environment system engineer, the thermal environment conditions (i.e., dry-bulb temperature, humidity, air movement, and radiation) are extremely important for achieving maximum animal performance, efficiency, well-being, and economic return for a given species. Even beyond the temperature range of optimal productivity for a particular species, performance and efficiency do not greatly decline until near the lower critical temperature (LCT) or upper critical temperature (UCT), and potential performance losses may be acceptable. Such response relationships permit adequate environmental selection and design criteria based on economic costs (i.e., penalties to performance) and animal well-being particularly when used with climatological information to evaluate risks (Hahn et al., 1983). It must also be noted that factors like individual versus group penning and floor type can affect the LCT by approximately 5°C; therefore, designing an adequate controlled-environment building for a particular livestock requires caution.

Overall, acceptable LCTs and UCTs depend on a number of variables, such as the species, breed, age, weight, sex, type of feed and level of feeding, disease and parasites, space allocation, lactation or gestation, and other physical features of the animal's environment. However, adequate conditions are typically established using temperature because an animal's sensible heat loss is mainly determined

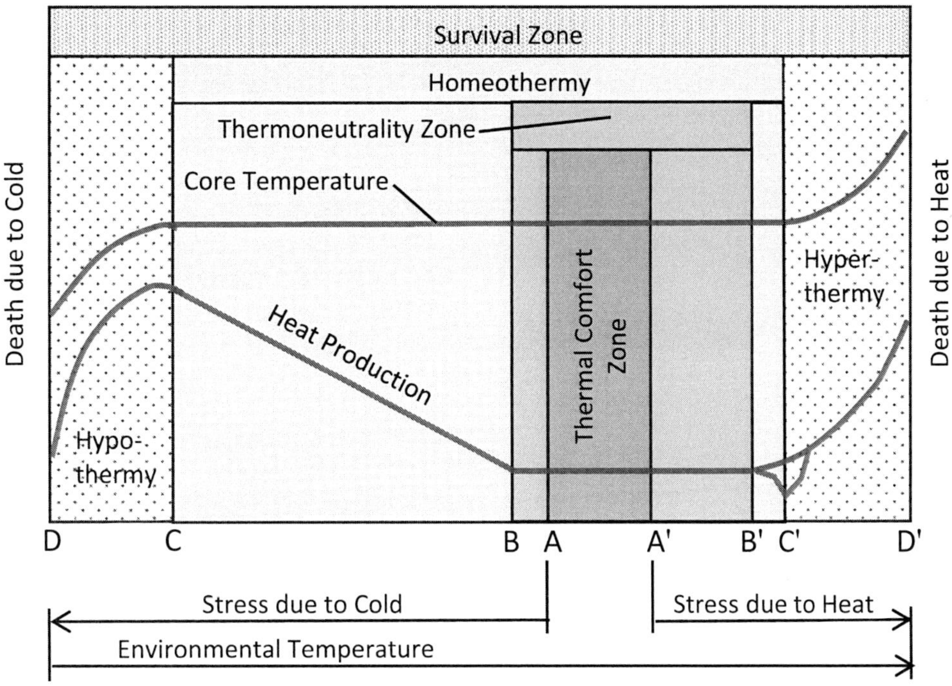

FIGURE 4.11 Schematic representation of homeothermic heat production and body temperature as affected by environmental temperature.

Source: ***Structures & Environment*** **(1987).**

by the temperature difference between the animal's body surface and ambient air. Air movement is another important factor in sensible heat dissipation. If the air temperature is lower than the animal's skin temperature, air movement helps heat loss by evaporation and by convection. When the air and skin temperatures approach each other, then animals feel more comfortable with air movements.

At low temperatures, however, air movement may lead to excess cooling of unprotected skin areas. In addition, it should be remembered that air movement is also needed to remove noxious and toxic gases. Furthermore, with increased environmental temperature, elevated humidity as well can restrict performance severely but is only moderately important at moderate ambient temperatures. During cold temperatures, RH has little effect on animals' heat dissipation. Extremely low humidity in livestock and poultry buildings cause irritation of the mucous membranes, while too high humidity can promote fungus infections. As side information, it must also be remembered that high humidity contributes to decay in building materials. Therefore, the RH in livestock and poultry buildings should be kept within the range of 40% to 80%.

When the ambient temperature drops below point A in Figure 4.11, hair or feather erection and vasoconstriction start conserving animals' metabolic heat. Blood flow to the skin is restricted, reducing heat transfer from the body core to the body surface. If the ambient temperature drops below the LCT (point B), animal heat production increases, and the additional heat generated protects the animal's core temperature from falling, reflecting homeothermy. At this point, increased feed intake can sustain the increased metabolic rate.

However, at point C in Figure 4.11, the extra heat produced by the animal can no longer balance the heat loss, and homeothermy simply fails; beyond this point, both the heat production and core temperature decrease, eventually leading to death from cold. At this point, we must emphasize that time is extremely important for a body to stand cold stress, and Figure 4.11 does not have such a scale to demonstrate the impact of time on survival.

Heat stress, on the other hand, begins when ambient temperature reaches point A' in Figure 4.11. Unlike the response of metabolism to combat cold stress, the response to heat stress (i.e., beyond point A') is physical. The surface vascular vessels enlarge (vasodilation), increasing blood flow to the surface and raising the skin surface temperature, which increases the temperature difference between the skin surface and the environment, therefore the heat transfer. In some species, increased respiration, sweating, or both is observed. In other species, water vapor diffusion through the skin increases. As a result of the increased heat production, appetite is depressed to reduce animal heat production.

The LCT, point B' in Figure 4.11, is the threshold for drastic changes in heat production, where the evaporative cooling mechanisms increase in intensity; as a result, heat production may decrease. Evaporative cooling is the most powerful form of heat dissipation as ambient temperature gets close to animal's skin surface temperature. However, it must be remembered that in a hot and dry climate, evaporation is rapid, but in a hot and humid climate, the ability of the air to take on further moisture is limited; therefore inadequate cooling may still result in heat stress.

At point C', the organism's homeothermy stops functioning because evaporative cooling cannot increase the heat dissipation through vapor loss from the skin any further. Poultry does not have sweat glands, so all evaporative heat loss from poultry has to originate from the respiratory tract, while other livestock species have varying abilities to sweat. Moreover, at point C', the respiration rate may keep rising; however, its cooling effect beyond this point becomes limited and hence does not make much difference. When evaporation rate from the skin reaches its maximum level, the body core temperature starts rising as a consequence, and thence increase biochemical reactions that actually further increase heat production by the body, eventually leading to death after a few hours of sustained heat stress in most animals.

4.4.1.3 Air Temperature and Livestock Production Efficiency

Animal's appetite is important for livestock production. Below thermoneutrality (see Figure 4.11), core temperature is maintained by increased appetite and feed intake at the expense of maximum production of milk, meat, eggs, and fiber. Above thermoneutrality, however, feed intake is reduced due to depressed appetite at the expense of productivity. We can determine the range of temperatures for the zone of nominal losses for each livestock species (see Figure 4.12a), which includes optimal productivity (growth and milk and egg production by mature animals) and efficiency levels, representing acceptable design conditions to achieve maximum performance.

Figure 4.12a shows the energetic and performance relationships as influenced by air temperature. The lower part of the figure shows the temperature range for maximum animal production efficiency. Between T_1 and T_2 is the thermal comfort zone—the nominal losses zone—where production efficiency changes only slightly.

The effect of air temperature on the productivity of some species of livestock and poultry is shown in Figure 4.12b. Most of the response data here was observed at moderate humidity under still air conditions. Obviously, the optimum temperature zone and the performance rates for each species vary with age, genetics, and air velocity and vary slightly with RH.

Animals not only respond to environmental conditions such as ambient temperature but also to changes in those conditions. Cycles of temperature frequently occur in animal environments as well, but we can calculate heat loss from cattle and swine, for instance, from average daily ambient temperatures accurately for design load calculations. Productivity slightly varies for averaged temperatures when a cyclic range less than approximately 10°C exists (Nienaber et al., 1987). Beyond this range, productivity is lower than that expected at the average temperature. However, normal egg production is maintained under daily varying ambient temperatures of 20 to 25°C under widely varying humidities.

4.4.2 Heat Production and the Environment

Heat production has been measured for many livestock animals and poultry. The rate of heat production is primarily a function of temperature, animal species, and animal size (Figure 4.13a). For relevant data, see ASABE Standards (2018) and MWPS (1987). Heat production also varies diurnally, based on animal activity and eating times, etc. Building type and existing waste-handling system as well can influence the percent distribution of sensible and latent heats; hence the ventilation system design of a livestock building requires careful attention. Since heat production is influenced by animal body weight and surface area, heat

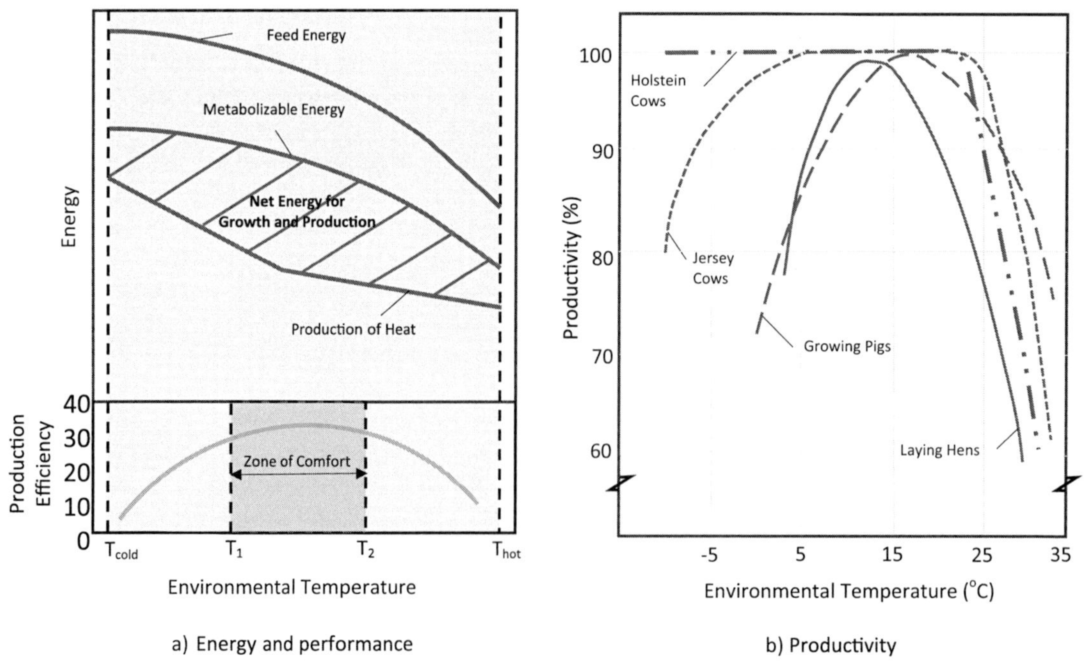

FIGURE 4.12 Energy, performance, and productivity of different animal species as affected by environmental temperature. Source: *Structures & Environment* (1987).

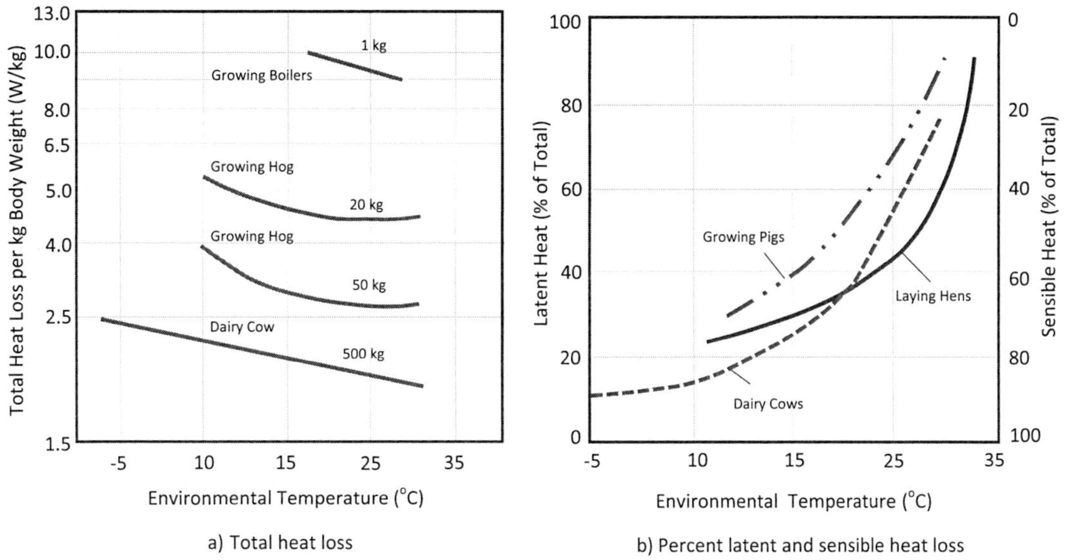

FIGURE 4.13 Schematic representation of mature animals' heat loss as affected by ambient temperature.
Source: ***Structures & Environment*** **(1987).**

transfer per unit weight is higher for smaller animals. Heat production under normal metabolic conditions is useful for a controlled environment system engineer. Figure 4.13b presents total, sensible, and latent heat losses of animals fed in typical housing conditions.

4.4.2.1 Heat Transfer to the Environment

As discussed, the total heat exchanged between an animal and its environment consists of both sensible and latent heat (ASABE Standards, 2018; MWPS, 1987). Most sensible heat of livestock animals and poultry is lost through the skin via conduction, convection, and radiation. All animals lose significant amounts of latent heat by vaporization of water from their respiratory tracts and body surfaces as well.

Sheep, swine, and poultry have evaporative losses largely from their upper respiratory tracts, and lose only a little moisture—and therefore latent heat—through their skin surface. Cattle and horses, on the other hand, transfer most of their latent heat through their skin. Blood transfers the heat produced by metabolism via conduction and convection to the skin or to the lung surfaces, where it is converted to latent heat by internal evaporation from the mucous layer coating of the lungs. Inhaled air reaching the lungs is heated close to the animal's body temperature and saturated with water vapor in the upper respiratory tract. Spraying an animal's skin with water also maximizes cooling by external evaporation. Overall, engineering design variables influencing animal heat loss are i) air temperature; ii) vapor pressure of the air; iii) air movement; iv) configuration, absorptivity, emissivity, and surface temperature of the building surfaces; and particularly v) temperature and conductivity of surfaces with which the animal might be in contact, such as floors.

For instance, water on building floors evaporates and gets converted into latent heat, which requires thermal energy either from animals or indoor air typically through the building heating system. After water vaporizes, the excess moisture in the air in the form of latent heat has be removed by adequately designed ventilation systems to prevent condensation on building surfaces and equipment. We also have to minimize the adverse health effects of moisture on animals. The negative consequences of excess moisture can typically occur when indoor humidity levels exceed 80% or so.

Moreover, water spillage by animals or floor flushing can increase latent heat production by approximately one-third from animal moisture production data, with a corresponding reduction in sensible heat production per animal (ASABE Standards, 2018). Excessive water spillage and floor flushing increase the moisture removal ventilation rate. The room moisture production, however, is lower in livestock buildings with partially or totally slatted floors with under-floor waste storage by approximately 35% (and up to 50%) than in buildings with solid floor systems. As a consequence, the ventilation rate required for moisture removal is reduced as well while excessive water spillage increases the moisture removal ventilation rate.

4.4.2.2 Cattle and the Thermal Environment

Figure 4.12b shows the effect of ambient temperature on productivity of some dairy cattle. The ideal environment for Holstein cattle should normally not exceed 24°C, while Jerseys can tolerate an upper limit of around 27°C (Yeck and Stewart, 1959). At the lower end of the temperature scale, production decreases may be expected below 5°C for Jerseys and below −12°C for Holsteins and Brown Swiss. Temperature has a greater influence on cows with genetic potential for high milk production than on those with a lower potential. Most controlled-environment dairy buildings are kept at 4–15°C from November through April, while relative humidities of 60%–80% are common in ventilated dairy buildings. For beef cattle, cold housing (inside temperatures about the same as outside) is adequate.

Extended low temperatures do not seem to affect the reproductive performance of livestock animals, even below freezing. Under high-temperature conditions, however, breeding efficiency of both sexes decreases. Temperatures above 24°C decrease spermatogenesis, and sustained temperatures above 29°C decrease fertility and sperm quality in males as well as increasing anestrus and embryonic death in females. The temperature–humidity index combines air temperatures with relative humidity values, which is a better and more accurate predictor of the risk for heat stress than air temperature alone because changes in air temperature and RH affect heat stress risk additively. Figure 4.13 shows the total, sensible and latent heat loss of dairy cows as affected by ambient temperature.

4.4.2.3 Sheep and the Thermal Environment

Sheep vary among breeds and strains in their ability to adapt to environmental changes. At elevated temperatures, sheep grow slowly. The amount and quality of wool produced varies considerably among breeds; however, environmental factors such as photoperiod, nutritional level, and temperature also affect wool growth. High skin temperature is a predominant factor for high wool growth.

Sheep normally mate only during certain periods of the year, but ewes may breed if exposed to modified air temperature of 7°C before mating is attempted. However, high air temperatures beyond 32°C affect the spermatogenesis rate, while subjecting ewes to 38°C just before mating causes failure in fertilization. In a study, a temperature and humidity combination of 32°C and 65% at mating killed most embryos at an early stage. Generally speaking, all reproductive processes appear to be more adversely affected by high air temperature when combined with high RH. Here, it should be noted that early embryonic death is the biggest loss in potential sheep offspring conceived in a high-temperature environment. Near mating time, the risk of high temperature seems to be greatest; however, susceptibility generally decreases after mating as time goes on. High temperatures particularly during the last third of gestation produce weak lambs. Generally, ewes and feeder lambs are kept in cold housing, where inside temperatures are about the same as outside temperatures. In cold climates, using warm houses for lambing is common, where inside temperatures maintained above freezing, and often at 7–12°C.

4.4.2.4 Swine and the Thermal Environment

Ambient air temperature affects the feed conversion and daily growth of growing swine. A temperature range of 15 to 22°C produces maximum growth and feed conversion rates for 70 to 100 kg hogs, while a wider range of 10 to 24°C influences performance slightly (Kibler and Brody, 1956). Younger animals, on the other hand, necessitate a temperature range of 23 to 28°C for best performance, and piglets from three days to two weeks of age require 30 to 32°C temperature conditions (Hahn, 1983). Moreover, Nienaber et al. (1987) reports that daily air-temperature cycles of more than about 5°C on either side of 21°C results in reduced daily gain and an increased feed requirement per unit of gain. No matter how small the cycles are, pigs generally desire relatively constant air temperatures.

Swine gains and feed conversion rates are highly sensitive to air velocities. Swine less than eight weeks of age should not be exposed to velocities greater than 0.25 m/s; and even much lower velocities are preferred when temperatures are in the recommended range. Generally speaking, pigs benefit from air movement if the air temperature is somewhere between 32 and 39°C. Above 39°C, high wind speeds increases the animals' heat loads; they are negatively affected by air velocities greater than 1 m/s (Bond et al., 1965; Gunnarson et al., 1967; Riskowski and Bundy, 1986), preferring 0.15 m/s (30 ft/m) at optimal air temperatures.

Furthermore, Merkle and Hazen (1967) and Heard et al. (1986) demonstrated that some type of cooling in hot weather is beneficial for sows. In these studies, sprinkling the sow during high-temperature conditions at breeding and shortly after produced more live births than the sows without sprinkling. Sprinkling boars before mating also increased the number of live births. Conception rate varied from about 100% of normal at 21°C to only 70% at 32°C. Figure 4.13 shows the total, sensible, and latent heat losses of growing pigs as affected by ambient temperature.

4.4.2.5 Chickens and the Thermal Environment

Supplemental heating is usually required for newly hatched chicks. Broiler strains respond quite well to brooder temperatures of 33°C under the hovers. After that, air temperatures may be decreased at a rate of approximately 3°C per week until 21°C is reached. Good feathering and market-quality broilers are produced at RH of 65% to 70%; however, RH 70% to 75% is generally recommended to maintain conditions that are neither too dusty nor too restrictive of latent heat dissipation.

Drury (1966), on the other hand, observed some benefit from increased air velocities up to 2.5 m/s as well at air temperatures from 24 to 35°C; however, high-velocity air at a temperature above the feather temperatures causes more heat stress in chickens. Typically, air speeds greater than 0.3 m/s (60 ft/min) at broiler height are not recommended for broilers under two weeks of age. Increasing air speeds from 0.2 to 2.5 m/s (40 to 500 ft/min) around 1.4 kg (3 lb) broilers improve weight gain and water-use efficiency when the temperature fluctuated diurnally from 21 to 36°C. Similar increases in air speed reduce heat stress in eight-week-old broilers that were exposed to environment temperatures less than the chicken's body temperature of 40.6°C. At environmental temperatures above body temperatures, however, an increase in wind speed will exacerbate the heat stress. For turkeys, fans that moved the air approximately 1.8 m^3/h.kg (0.5 ft^3/min.lb) distributed throughout the house provided the most effective air circulation (ASABE Standards, 2018).

In poultry, temperatures over 29°C produce fewer eggs with lower weights and thinner shells. Hens older than a year are more negatively affected by high-temperature environments than younger hens (e.g., Rhode Island Reds and White Leghorns). Hatchability as well declines in high-temperature environments. While the recommended ideal environment is somewhere between 7 and 24°C, fertility is still good at 24°C, but it declines at 30°C. Feed requirements increase markedly in colder air, especially below 7°C, and activity and productivity decline below freezing.

Figure 4.13 shows the total, sensible, and latent heat losses of growing broilers and laying hens as affected by ambient temperature.

4.4.2.6 Turkeys and the Thermal Environment

Turkeys are reported to have the best growth rate and feed conversion efficiency at a brooder room temperature of 21 to 24°C for the first two weeks and 18°C thereafter at 70% RH. Initial brooder temperature is typically 38°C and is reduced by approximately 3°C per week thereafter until room temperature is reached. DeShazer et al. (1974) reported that mature turkeys tolerate wide temperature and humidity ranges of at least −7 to 32°C and 35% to 85%, respectively; the authors provide heat loss data on large-breed growing turkeys, but estimates from heavy chickens (on a live-weight basis) would also be sufficient for design purposes.

4.4.3 Light Intensity, Quality, and Photoperiodism

Light is one of the earliest discovered and most important environmental variables affecting the growth, production, and reproductive processes of animals. As discussed earlier regarding plants, light intensity (amount), light quality (spectrum), and light duration (photoperiod) all are important influences on animal response.

Light can harm (phototoxicity) animal cells if unprotected, stimulate vision receptors, and control photoperiodism. Just like in plants, photoperiod has a direct influence on animal performance especially on the breeding season for sheep and poultry egg production. Photoperiodic control of reproduction in animals is a common practice; however, it is more complex than those in plants. As will be discussed further shortly, artificial light is used in the temperate and colder zones to equalize egg production throughout the year, for instance.

In hot climates, on the other hand, additional hours of light before sunrise and after sunset are recommended to encourage the hens to eat during relatively cooler hours. Having said that, however, the effects of light vary widely among animal species and age. While some animals grow and remain healthy with or without light (e.g., growing swine), for some others, lighting management is quite important. Effects in poultry, for instance, seem to be more important than in livestock animals. Research has shown that intermittent light and darkness (photoperiod) improves growth and conserve feed and energy (McDaniel and Brewer, 1975). Essentially, there exist three daylength groups of domestic animals: i) short-day breeders mate when daily photoperiod decreases in the fall; short photoperiods induce or accelerate estrus development in these animals (e.g., sheep and goats); ii) long-day breeders mate when daily photoperiod increases in the spring (e.g., birds and horses); and iii) day-neutral breeders mate readily during all seasons (e.g., cattle and swine).

Both the photoperiod and the quality of light source are important for the early growth of layer chickens (Manser, 1996; Prescott et al., 2003). Light quality, however, varies widely from one light source to another. Incandescent light, for instance, contains more red and fewer blue wavelengths than natural light, while natural light has higher intensity and a more complete spectrum than incandescent light (Prescott et al., 2003). Birds, in addition to the three types of cones that all together sense electromagnetic radiation between 400 and 730 nm, have another type of cone that senses UV-A light (315–400 nm; see Figure 2.34) (Hart et al., 1999; Prescott and Wathes, 1999). Little is known about the effect of different light sources on behavioral development of chicks, but lighting influenced cannibalism among chickens. Taylor et al. (2003) found that lower light intensities restricted the movement of birds; Brown Leghorn hens, for instance, showed a clear preference for increased light intensity (Taylor et al., 2001). Type of light also influences feeding and perching behavior (Gunnarsson, 2013).

Laying hens are active, gregarious birds under light. The red jungle fowl, its ancestor, evolved in the equatorial jungle and is reported to have a diurnal rhythm of 12 h of photoperiod and 12 h of darkness (Collias and Collias, 1996). Hence, lighting in egg production is crucial for laying hens. Within certain limits, increasing photoperiod stimulates egg production in poultry. UV-light, on the other hand, activates vitamin D in the animal's blood, promoting absorption of calcium from the gut; when intensive, however, UV-light may create pathological disturbances in animals. In another study, Shutze et al. (1962) observed that while the lighting treatments did not affect the percent hatchability, continuous white incandescent light during incubation of White Leghorn eggs caused eggs to hatch from 16 to 24 hours earlier than eggs incubated in darkness. Red, yellow, or blue lights gave comparable results.

Light is also manipulated to delay sexual maturity in hens, which enhances subsequent production. This is typically done by gradually decreasing photoperiods from hatching to 22 weeks of age, or by abruptly decreasing photoperiods to 9 hours at 14 to 16 weeks of age. If pullets are previously exposed to increasing photoperiods, the change of photoperiod typically takes place at the age of 14 weeks; if they are exposed to a constant or decreasing photoperiod schedule, then at this time, a change at the age of 16 weeks is found adequate. Light intensities of 11 to 22 lux (measured at bird height) were sufficient in all cases. Photoperiod is then suddenly increased to 14 or 16 hours at the age of 21 weeks. The economic value of increasing photoperiods beyond 14 hours in a windowless poultry house and 16 hours in an open poultry house has not been justified. Photoperiods such as 8 hours of light, 10 hours of darkness, 2 hours of light and 4 hours of darkness, and other cycles have been reported to improve feed utilization, egg production, and poultry growth (Buckland, 1975). Furthermore, under a light intensity of 11 to 32 lux (measured at bird height), continuous lighting from hatching through 20 to 21 weeks noticeably depressed subsequent egg production (but not egg biomass) and caused severe eye abnormalities in birds.

Proper control of photoperiod stimulates growth rate in turkeys as well. Mature birds are kept for fertile egg production, and therefore, lighting for off-season egg production is important. The young turkeys are raised to an age of 20 weeks on natural light; thereafter, females are placed in a totally darkened pen and given 8 hours of light per day at an intensity of 22 to 54 lux at bird height. At 30 weeks, the sexes are mixed, and the photoperiod is increased to 13 to 15 hours.

With respect to pigs, there is wide and sometimes contradictory literature about the effects of light on pig behavior and performance. Light intensity, in general, does not seem to be an important variable affecting the welfare of pigs. For instance, van Putten (1980) studied the behavior of pigs and indirectly their welfare in the presence and absence of light and observed no effects experimentally. However, they did find that tail biting decreased greatly when pigs were maintained in a warm and low light environment (van Putten, 1968, 1984). Furthermore, aggression among recently grouped unfamiliar pigs also decreased when pigs were reared in darkness (Barnett et al., 1994). Frederiksen et al. (2006), on the other hand, reported that male pigs grown in low light had increased daily weight gain and carcass weight while their androstenone levels lowered.

McGlone and Curtis (1985) observed that photoperiod had no particular benefit for nursery pigs. However, the situation was different for breeding sows. Stevenson et al. (1983) reported that exposing lactating sows to 16 h of supplemental light per day in farrowing rooms increased the weight of litters weaned at four weeks of age and resulted in more prompt return to estrus after weaning. The length of photoperiod is reported to have a clear physiological influence on reproduction, growth, lactation and health of dairy cows as well (Dahl et al., 2012).

Dahl and Petitclerc (2003) and Dahl et al. (2012) reported that extended photoperiods can result in increased milk yields compared to shorter photoperiods. A number of other studies have shown that cows exposed to a long photoperiod produced 5% to 15% more milk than cows held in a short photoperiod (Peters et al., 1978; Bodurov, 1979; Dahl et al., 2012). However, there was no effect of 24 h continuous lighting on milk yield of dairy cows when compared with a natural photoperiod (Marcek and Swanson, 1984). The intensity of light has also been reported to affect both behavior and physiology of cows (Stanisiewski et al., 1988; Phillips and Schofield, 1989; Muthuramalingam et al., 2006).

LEDs are being increasingly used every year, although incandescent and fluorescent bulbs are still widely used in modern livestock and poultry houses. LEDs are relatively new and more energy-efficient than incandescent and fluorescent light. Son et al. (2020), one of the groups working on LEDs, investigated the effects of different colored LEDs on milk production, milk composition, and physiology of Korean Holstein cows. They observed that blue and yellow LED groups demonstrated greater rates of decline in milk production than white LED groups, and that at the finish point, milk fat, protein, and lactose contents were the lowest in the blue LED group. Furthermore, extended exposure to blue LED light lowered antioxidant enzyme activity and insulin-like growth factor-1 levels, and the blue LED group had the highest cortisol level among all the groups studied. Overall, their findings suggest that blue LED light decreases milk production and generates more stress than white and yellow LED lights.

In another study, Asher et al. (2022) studied the effect of short-wavelength white LED illumination throughout the night on the milk fatty acid profile of high-yielding Holstein cows and showed no commercial benefit in using white LED light at night over the natural light–dark cycle during the autumn months, at 32°N latitude, consisting of 10 to 11 h light exposure period. Their findings supported using the natural light cycle to save energy, maintain the cows' wellbeing, and preserve the natural circadian variation in milk fat composition.

4.4.4 Air Composition, Pollutants, and Altitude

4.4.4.1 Oxygen

Energy-rich molecules are broken down (oxidation) into smaller components (CO_2 and H_2O) with the release of energy, which is required in all metabolic activities. Just like plants, animals' gas exchange (O_2 uptake and CO_2 release) with the environment, related to the breakdown of reserves, is called respiration. While plants obtain their O_2 by diffusion of gas through their stomata, most terrestrial animals have lungs for supplying O_2 to their tissues. As a result, animals housed in controlled environments modify air composition by reducing O_2 content and increasing CO_2 and vapor contents.

The inward driving force for O_2 in animals is 159 mm Hg (85.6 in. H_2O) in the air. At 25°C, the outward driving force for loss of water to dry air is 24 mm Hg (12.9 in. H_2O), only a small fraction of the inward driving pressure of O_2 (Figure 4.14). As a consequence, animals do not need to lose much water to obtain O_2. Gram for gram, plants require 15–20 times as much water as animals. When plants run short of moisture, they can block most water loss by simply closing their stomata. As a result, net photosynthesis ceases, but the plants are protected. In this regards, the problem for animals is way more severe. Animals cannot produce their own O_2 for use in metabolism; hence they must breathe, and as a consequence, water is lost.

Animals also have blood circulatory systems and respiratory pigments (hemoglobin) to facilitate the rapid distribution of O_2, and easy access to O_2 in the atmosphere permits a high rate of metabolism in animals. The first response to low partial O_2 pressure is increasing ventilation of the lungs and the rate of heartbeat. Long-term acclimatization

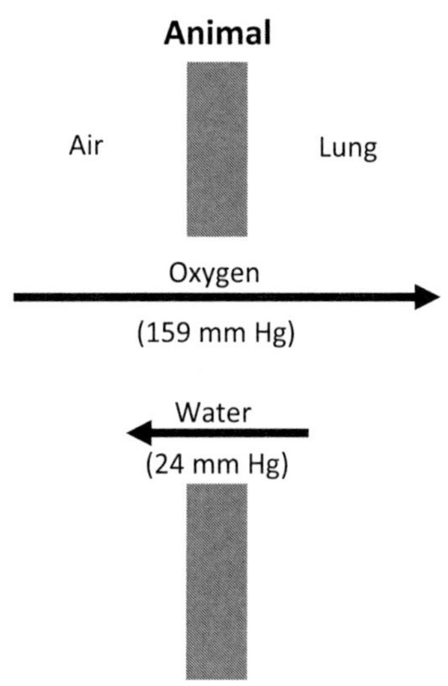

FIGURE 4.14 Partial pressure differences of oxygen and water across lungs in animals.

to reduced O_2 is the stimulation of the production of the hemoglobin carrying cells of the blood. All of these adaptive responses ensure better trapping and use of O_2 at low concentrations.

4.4.4.2 Carbon Dioxide

CO_2 is a product of animal respiration. Burning of fossil fuels adds significant amounts of CO_2 to the atmosphere as well. The atmosphere contains about 0.03% CO_2. Unlike O_2, CO_2 dissolving in water combines chemically with it to form carbonic acid (H_2CO_3), which influences the acid–base balance of the environment. Mammals and birds regulate the rate of ventilation of the respiratory surfaces according to the CO_2 concentration in the lung air. CO_2 is more important than O_2 in regulating respiration in animals.

4.4.4.3 Pollutants

Animals are sensitive to pollutants, and each pollution source is different. Direct or indirect effects on different animal species are also different; and young animals are more sensitive to pollution than adults. Typical contaminants in livestock housing are i) microbes, both pathogenic and nonpathogenic, usually hosted in the respiratory tracts or animal wastes; ii) gases from decomposing waste products such as ammonia, hydrogen sulfide (H_2S), and CO; and iii) dusts from air movement, animal activities, feed, bedding material, manure, and animal skins, hair, and feathers. Livestock animals and poultry can typically tolerate higher levels of most air contaminants without having any negative health consequences. However, young animals and others experiencing stress (e.g., hot or cold stress, nutritionally limited animals, and sick animals) are more sensitive, and even low levels of contaminants can have serious adverse effects. An adequately engineered ventilation rate sufficient to remove water vapor usually controls dusts and gases as well.

4.4.4.3.1 Ammonia

Ammonia levels in livestock buildings are affected by the sanitation practices, ventilation rate, and the waste-handling system employed. Curtis (1983) showed that ammonia levels of 50 ppm reduce weight-gaining rates in swine, and Donham (1987) showed that negative effects begin between 5 and 10 ppm. However, as mentioned earlier, properly designed ventilation for moisture control is generally sufficient to control ammonia levels where buildings are sanitized regularly and wastes are not allowed to accumulate for more than a month or so.

4.4.4.3.2 Hydrogen Sulfide

H_2S is mainly produced from decomposing wastes and is toxic only when liquid wastes in a pit below the animals are agitated; gases from the waste handling area can leak into the livestock housing area. It must be ensured that ventilation is properly designed and sufficient to remove the contamination.

4.4.4.3.3 Carbon Monoxide

CO is typically caused by the failure of gas-fired unit heaters or ventilation equipment. These heaters should be vented to the outdoors and properly equipped with sensors to ensure complete combustion.

4.4.4.3.4 Dust and Contaminants

Dust from air movement, animal activities, feed, bedding material, manure, and animal skins, hair, feathers, and litter particles are quite common in enclosed buildings. While some contaminants are dry and relatively nonadhesive, others are oily and adhere tightly on various surfaces. In poultry buildings, typical contaminants creating problems are feathers, fibers removed from the feather rib, skin flakes, and dust from feed and litter particles. In other livestock buildings, feathers are replaced by hair as a contaminant that is not only an irritant but also a carrier of pathogens that can affect animal performance. Pneumonia and other pathogenic bacteria can be transferred among animals through the air. Both pathogenic and nonpathogenic bacteria can create health complications overloading the animal's immune system.

Janni et al. (1984) reported that the physiological effect of dust on livestock animals is not fully known; however, animal growth rate was reduced in animals inoculated with an airborne pathogen without any significant effect on mortality, and this growth rate reduction intensified at low temperatures. Additionally, charging the indoor air with negative ions shows beneficial results of reduced air contamination and bacteria levels. However, some researchers suggest that negative ionization is not beneficial because although negative ions kill bacteria, they can increase particle retention in the respiratory tract (Janni et al., 1984).

Normally, ventilation that is adequate for moisture removal is also sufficient to prevent problems due to airborne microbes (Christianson and Fehr, 1983). It should however be remembered that when the ventilation equipment is stopped by a power outage or equipment failure, air contaminants can reach extremely dangerous or even lethal levels.

4.4.4.4 Altitude

As was discussed earlier, altitude affects plant growth indirectly due to decreased average temperatures, wavelength distribution of solar radiation, humidity, average wind speed, etc. At high altitudes, CO_2 concentration is a limiting factor for plants. Altitude also affects animals, primarily due to reduced partial O_2 and CO_2 pressures, which has already been discussed earlier.

At high altitudes, heating costs for providing adequate thermal environments for animals rise as well due to decreased outside temperatures and increased wind speeds; as discussed earlier, we have to make sure that combustion equipment is properly designed and operated to ensure complete combustion. If insufficient air is provided to a burner, combustion is incomplete, leaving nonoxidized components in the flue gas such as C, H_2, and CO. Not only does incomplete combustion use fuel inefficiently, it results in air pollution that can pose even more danger and increase the potential for explosion. Every combustion process has the potential to produce this toxic gas as a result of incomplete combustion; however, air at higher altitudes is less density and has less oxygen per unit volume. Therefore, excess air at high altitudes must be adjusted accordingly; otherwise, we can have incomplete combustion, or combustion can eventually cease. Overall, just as for plants, controlled-environment facilities for animals, insects, bacteria, fungi, and viruses must be managed below economic injury levels in an integrated fashion using cultural, physical, biological, and chemical control strategies.

4.5 AQUATIC ENVIRONMENTS

4.5.1 Introduction

Aquaculture has already been defined in Chapter 3 as the farming of aquatic organisms, and it is supplying more than half of all seafood consumed today. Based on the water source and intensity of production, aquaculture systems are classified as extensive (pond culture), semi-intensive (cage culture), or intensive (recirculating aquaculture). Extensive and semi-intensive systems are difficult to control in open air, where we have almost no control over their environments; these are highly susceptible to airborne and waterborne contamination and have long-term sustainability issues. Although controlled-environment recirculating aquaculture production currently represents just a small portion of the world aquacultural production, these systems have a very promising future primarily due to their higher production intensity, better environmental control, lower water usage and impact on the natural environment, control of market time and size, and site independence. This book focuses mainly on fully sustainable controlled environment recirculating aquaculture systems (RAS) while providing some essential information on the other systems.

In RAS, we have total control of the secure tanks where fish are grown in enclosed environments. All environmental variables, such as temperature, oxygen, pH, salinity, alkalinity, ammonia, nitrite, CO_2 and other dissolved gases, and suspended solids are continuously monitored and fully controlled. Water recirculates through the system components, and due to their low water makeup requirements (e.g., 10% per day or less), they can also capture 96% to 100% of the waste produced (Heinen et al., 1996). This means that the solid wastes are filtered and removed, adequate dissolved oxygen (DO) levels are maintained, and effluent water is sent through a biofilter for biological conversion of ammonia-nitrogen to nitrate-nitrogen.

This section covers basic physiological functions such as respiration and osmoregulation first and then the impacts of major environmental parameters and their interactions on fish in general and specifically on controlled-environment fish production. The environmental parameters discussed include temperature, oxygen and various other dissolved gases, nitrification and denitrification processes, solids balance and generation, light, and pollutants.

4.5.2 Basic Fish Physiology and Aquatic Environments

The physical, chemical, and biological parameters of aquatic environments influence the growth, reproduction, and survival of the aquatic organisms (Parker and Davis, 1981). Even though the anatomical and physiological systems of fish are, in many ways, similar to those of other animals, some of their life processes are unique and quite different from those of land animals.

Knowledge of these life processes, such as respiration, osmoregulation, feeding, digestion, and excretion, are particularly important for understanding and rationally designing rearing systems that meet the physiological requirements of fish. Physical environmental parameters such as temperature in an aquatic system change more slowly than chemical parameters such as dissolved gases, pH, nitrogenous metabolites, alkalinity, hardness, salinity, and toxic compounds. This quickly changing chemical environment can cause stress and even death to the cultured organisms. Therefore, the information provided here in this section on basic physiological functions and on the interactions among the chemical, physical, and biological components of controlled rearing environments for intensive recirculating aquaculture systems.

4.5.2.1 Respiration

As mentioned earlier, energy-rich molecules are broken down (oxidation) into smaller components (CO_2 and H_2O) with the release of energy, which is required in all metabolic activities. Just like plants and terrestrial animals, fish gas exchange (O_2 uptake and CO_2 release) with the environment, related to the breakdown of reserves, is called respiration.

In unicellular animals and plants, the required amount of gas exchange is accomplished by passive diffusion alone. In complex organisms such as fish and most terrestrial animals, however, both a dedicated anatomical structure (gills and lungs, respectively) for gas exchange and a gas transport system (blood and circulation) are required to supply sufficient O_2 to the tissues and remove CO_2. Fish gills also perform other functions that will be discussed later such as the ion exchange between water and the tissues needed for osmoregulation and acid/base regulation. The processes by which the respiratory system obtains, transports, and exchanges O_2 and CO_2 between water, blood, and tissues are particularly relevant to the engineering design and operation of intensive aquaculture systems that will meet physiological requirements and sustain fish health.

It should be noted that substantial O_2 must be transferred from the water across the gills and supplied to the fish tissues. Just as an example, for active cold-water fish such as salmonids, even the resting consumption rate of fish can be 100 mg O_2 per kilogram body weight per hour or more. The respiratory system of an actively swimming fish must provide O_2 at quite high rates, such as as 800 mg O_2 per kilogram per hour, while large amounts of CO_2 must also be removed correspondingly (Wedemeyer, 1996).

Having said that, however, water is an oxygen-poor environment where the maximum dissolved O_2 is usually less than 10–12 mg/L. The high dissolved salt concentration in saltwater, on the other hand, reduces the available DO to a maximum of only 8–9 mg/L. As a result, relatively large volumes of water must move over the fish gills to extract enough O_2 to support their life. The mouth and opercules (the bony flap that protects the gills) function as force and suction pumps so that a sufficient flow of water is produced when resting. Because of the high density and viscosity of water, the energy cost of gill ventilation is at least 10% of the oxygen consumed. Active fish species such as salmonids can maintain the required water flow over the gills by what is called ram ventilation—simply opening the mouth when swimming. In either ventilation, it is theoretically possible to reach 80% DO utilization due to the countercurrent flow anatomy of gills. The actual oxygen utilization is, however, species dependent, typically averaging 30%–40% in trout and 70%–80% in carp (Randall and Daxboeck, 1984).

As water passes over the gills, DO diffuses passively through the very thin gill epithelium into the blood. Once O_2 is in the blood, most of it diffuses and is bound by hemoglobin and circulated to the tissues by the cardiovascular system. Here, it must be emphasized that the hemoglobin molecule has a unique physicochemical characteristic: increased acidity [H+] decreases its binding affinity for O_2 molecules (the Bohr effect) and in some species, its maximal capacity to bind O_2 (the Root effect).

Therefore, as blood circulates through the tissues, the acidity caused by the CO_2 being produced weakens the hemoglobin-O_2 bond, facilitating the release of O_2 molecules that passively diffuse into the cells where the low O_2 tension is present. At the same time, CO_2 diffuses out of the tissues into the blood driven by its own concentration gradient. However, unlike O_2, most of the CO_2 dissolves in the plasma and is circulated back to the gills as bicarbonate (HCO_3^-). Then, as the blood passes through the gills, an enzymatic reaction dehydrates the HCO_3^- ions to molecular CO_2, which then diffuses out into the water.

Because the residence time of a unit volume of blood within the gills is only a few seconds, the abovementioned enzymatic reaction has to be extremely fast. The diffusion of CO_2 out through the gills is also rapid because of the high CO_2 tension gradient between the blood and the water. Thus, while the blood O_2 tension may vary by 100 mm Hg or more, the blood CO_2 concentration remains low and changes very little. The Bohr effect, particularly in active cold-water fish species, tends to be large (begins at low blood CO_2 levels). In aquaculture systems, for example, the Bohr effect will begin to impair the oxygen transport of salmonids if the dissolved CO_2 concentration in the water is allowed to rise to only 20 mg/L. The commonly cultured warmwater fish (tilapia, carp, channel catfish) are generally less sensitive to the CO_2 concentration, but it is still good practice to manage the fish-rearing conditions to prevent CO_2 from accumulating in the culture water.

In addition to the effect of CO_2, lactic acid production can also increase blood acidity and impair O_2 transport. The most common cause for this is the radical swimming activity as a result of excitement and stress, which results in lactic acid accumulation in fish blood and tissues due to an oxygen shortage in the white muscle.

In intensive aquaculture systems, the acidosis resulting from high blood lactic acid concentrations or high blood CO_2 concentrations leads to reduced O_2 transport to the tissues due to the Bohr or Root effects. Usually, excessive swimming due to excitement and stress or low DO concentrations are the typical reason for this. In addition, the use of pure aerated O_2 to achieve higher loading densities (during rearing or transportation) can result in supersaturated DO levels with high blood CO_2 concentrations as well. This is a side effect of this practice because the high DO in the end suppresses the gill ventilation rate, resulting in high blood CO_2 concentrations and increased arterial CO_2 tensions. However, it should also be remembered that high blood CO_2 concentrations will impair O_2 transport to the tissues only if the resulting acidity overwhelms the normal blood buffering capacity and respiratory acidosis occurs. In fish-culturing operations with adequate water quality, reduced O_2 transport due to the Bohr effect is typically a consequence of metabolic acidosis due to the excessive swimming lactic acid production.

Basu (1959) first documented the magnitude and relationship of Bohr effect respiratory stress to actual CO_2 and DO concentrations in the water. Accordingly, the minimum amount of DO to support moderate swimming increases from about 6 mg/L if little or no CO_2 is present to more than 11 mg/L if the dissolved CO_2 concentration rises to 30 mg/L. If dissolved CO_2 levels are not kept well below 30–40 mg/L, the blood O_2-carrying capacity of salmonids will be depressed to the point that even high DO concentrations are insufficient to prevent tissue hypoxia.

Careful fish handling reduces fish excitement and swimming; as a result, respiratory stress from the Bohr and Root effects is minimized. Adequate water exchange or aeration that quickly removes dissolved CO_2 and provides DO will do the same. These can be listed as two of the most important practical considerations in meeting the physiological needs of fish in controlled environment intensive RAS.

The O_2 consumption of fish in intensive aquaculture systems increases due to fish culture procedures and natural processes. Stress resulting from handling, increased swimming resulting from excitement, and the natural processes of feeding and digestion are the most important reasons for increased O_2 consumption. For instance, when juvenile steelhead trout are stressed due to handling, their O_2 consumption can immediately double and remain high for at least an hour (Barton and Schreck, 1987). Also, increased O_2 consumption caused by excitement and stress commonly occurs immediately after fish are loaded into transport tanks. When O_2 aeration is available, the hauling tank water is typically supersaturated, in some cases using liquid O_2, to a DO level of 14–16 mg/L (hyperoxic conditions) in order to compensate for this initial O_2 demand.

In addition, the natural processes of feeding and digestion increase fish O_2 consumption because the energy costs of digestion, absorption, and assimilation can reach almost 40% of the resting metabolic rate (Brown and Cameron, 1991). It is for this reason that for channel catfish, salmonids, and tilapia, an O_2 consumption increase of 40%–50% for several hours should be planned each time the fish are fed. That is, fish should not be handled or transported immediately after feeding because the excitement and stress can increase their O_2 consumption such that even aeration cannot overcome the effects.

The common practice of withholding food for 24–48 hours prior to handling or transportation will prevent this undesirable effect and reduce the fish O_2 consumption rates. Swimming increases O_2 consumption as well, the swimming activity levels of fish in intensive aquaculture systems are typically much lower. Another important factor in intensive RAS that can be managed to influence O_2 consumption is water temperature. As will be discussed later in depth, warmer water temperatures increase O_2 consumption rates by increasing the overall metabolic rate.

In controlled-environment salmonid culture, for instance, the water exchange is controlled so that the fish O_2 consumption does not reduce the DO concentration below approximately 6.0 mg/L at the outflow end. If the DO is not replaced as quickly as it is being consumed by the fish, O_2 will be depleted. The breathing rate of fish is stimulated by falling DO concentrations, not by rising CO_2, which is the case for terrestrial animals. Therefore, species such as carp, catfish, and trout respond physiologically to declining DO levels by first increasing the gill ventilation rate using the mouth and opercules and the blood flow through the gills by increasing the cardiac output and blood pressure.

In the case of salmonids, even moderate DO depletion results in dramatic increases in the gill ventilation rate. These physiological responses increase O_2 uptake; however, the faster water flow also reduces the proportion of the DO that can be extracted during each pass over the gills. As the extracted DO drops, the amount of O_2 transferred into the blood during each pass will also drop from the theoretical maximum of 80% to as low as 15% (Smith, 1982).

It must also be considered that the energy cost of moving more water over the gills increases dramatically—from the resting rate of about 10% of the extracted O_2 to almost 70%. As a result, the cost to obtain O_2 increases as the amount of O_2 dissolved in the water and the arterial blood O_2 tension decrease. However, when the O_2 tension in arterial blood decreases to the point that hemoglobin in the red blood cells is less than 60% or so saturated, then certain hormones are released so that the branchial vasculature is dilated and Na^+/H^+ exchange by the red cells' membrane is stimulated, increasing the intracellular pH. As a result, through a series of events, the changes in hemoglobin-O_2 affinity and capacity mediated by the Bohr and Root effects improve both O_2 transfer and O_2 unloading at the gills and the tissues, respectively (Thomas and Perry, 1992).

If the DO falls below about 5 mg/L, salmonids become anorexic presumably—a behavioral response to prevent the normal increase in O_2 consumption that would occur due to feeding and digestion. In salmonids, the bioenergetic cost of obtaining and utilizing O_2 begins to exceed the energy that can be derived from its consumption at a DO of about 2 mg/L, and unconsciousness, and eventually death will occur. In contrast, many of the warmwater fish species important to aquaculture can survive for a couple of hours even if the DO concentration decreases below 1 mg/L, although tissue hypoxia will be inevitable and will eventually result in unconsciousness followed by death (Wedemeyer, 1996). However, in aquacultural practice, we can temporarily decrease the rate of O_2 consumption by simply manipulating the same biological and environmental parameters we use to increase O_2 consumption. For instance, reducing the water temperature (hypothermia), fasting, and using anesthetics to reduce swimming activity, excitement, and stress during hauling are the most common practices employed.

4.5.2.2 Osmoregulation

Similar to that of other vertebrates, the chemical composition of fish blood is about 80% water, 18% proteins, and 2% other solutes, about half of which are inorganic salts such as Na^+ and Cl−; the other half is small molecular weight organic molecules such as glucose. To achieve the homeostasis required to maintain life, fish have to regulate the water and most of the solutes in the blood within quite narrow concentration limits (Wedemeyer, 1996). The blood-buffering system in salmonids, for instance, maintains a narrow pH range of approximately 7.4–7.6. Significant shifts in blood chemistry from normal can cause problems in fish health and physiological condition.

Moreover, with a few exceptions, the fish blood and tissue composition are usually different from the dissolved materials in both freshwater and seawater. The dissolved substances present in the blood are separated from the aquatic environment by the very thin gill epithelium. As discussed earlier, the gill itself has a permeable structure that facilitates O_2 diffusion, but water molecules too are small to easily pass through the gills, even though their diffusion rates are different from dissolved substances. A concentration gradient (either inward or outward) for certain substances in the blood and tissues is always present. This gradient, across a semipermeable membrane between the blood and the aquatic environment, results in a continuous influx of water by osmosis in freshwater fish and a continuous efflux in saltwater fish (Figure 4.15). If osmosis is not regulated, it fatally reduces the concentration of normal blood constituents in freshwater fish species and dehydration in marine fish species. This means that the success of osmoregulation within an existing environmental concentration gradient is dependent on fish species; usually fish cannot survive in environmental salinity conditions much different from what they are used to. A number of fish species (e.g., the striped bass), in contrast, are able to maintain relatively constant numbers of particles of solute in their blood over a wide range of environmental salt concentrations, but to maintain

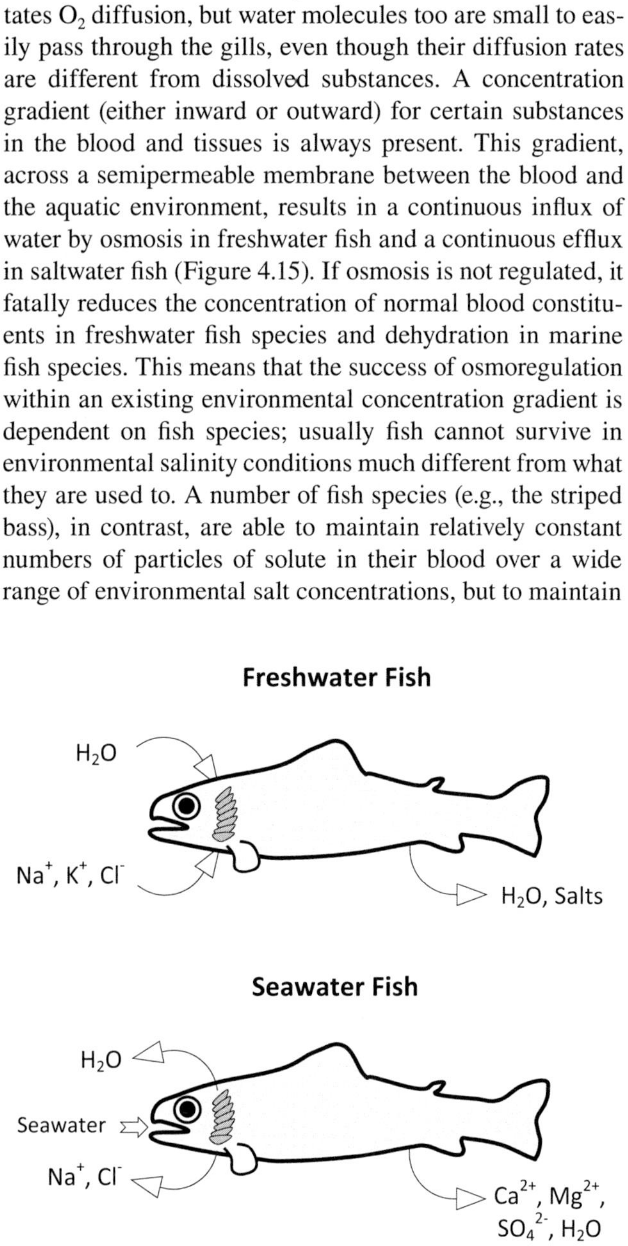

FIGURE 4.15 Comparative features of osmoregulation depicted for freshwater and saltwater fish.

osmoregulation, they use considerable amounts of dietary energy when they are grown under high salinity conditions.

The primary area of osmotic water influx in fish is the gills because they are the largest permeable surface on a fish body (Smith, 1982). Consequently, changing the respiratory surface area of the gills is one form of osmoregulation. If a fish is stressed, then certain hormones are produced to dilate the branchial vasculature, which increases the respiratory surface area and therefore the O_2 uptake. The increased water influx increases the osmoregulatory load as well. The balance required between the O_2 requirement and the need to minimize osmosis is termed the osmorespiratory compromise.

The structure of fish skin also has important implications for the effects of environmental conditions on fish health. Unlike terrestrial vertebrates, in which the outer layer of the epidermis consists of dead cells, the skin of most fish species is composed of living cells (Yasutake and Wales, 1983). The living skin layers are easily abraded by handling and infections, and problems can occur very quickly that increase the normal osmoregulatory load.

Additionally, ionoregulation is an important aspect of osmoregulation in both marine and freshwater fish species. Chloride cells, clustered around the bases of the gill filaments, transport Na^+ and Cl^- from the water into the blood of freshwater fish. The chloride cells help maintain homeostasis by transporting monovalent ions (sodium, potassium, chloride) from the water into the fish blood to replace those lost by diffusion through the gills and in the large amounts of urine that is produced. The lower the mineral content of the water, the larger the diffusion gradient between the fish blood and the water.

This is important for aquaculture because large concentration gradients of up to 3,000-fold can occur in very soft water (20 mg/L total hardness, as $CaCO_3$; see Table 2.1), and as a result, the transport of these ions can consume a considerable percentage of the energy provided by the fish diet. In contrast, the energy costs are lower in hard water because the concentration gradient is much narrower. In saltwater, on the other hand, the concentration gradient is reversed because the environmental salt concentration is normally 30–35 ppt salinity, and ions tend to diffuse through the gills into the fish blood instead. In this case, the direction of active transport turns into the opposite direction of the freshwater situation.

It should also be noted that the gill epithelium is the site of coupled Na^+/H^+ and Cl^-/HCO_3^- exchange between the blood and the water as well. This function of the gill epithelium not only regulates the acid–base balance of the fish but also facilitates the excretion of ammonia and CO_2. We should also remember that heavy metal exposure in fish inhibits these important processes. The resulting ionoregulatory dysfunction and the increase in blood ammonia can seriously impact fish health and condition.

Additionally, the involvement of the gill ion exchange mechanism in the excretion of blood ammonia by the exchange of Na^+ from the water has implications for the health of fish reared in naturally low mineral content water. For instance, mineral salt supplements have been successfully used to raise dissolved Na^+ concentrations to 20 mg/L. This elevation helps reduce blood ammonia levels in smelting steelhead trout and prevents swollen gills and related mortalities that are thought to impair ammonia excretion by the branchial Na^+/NH_4^+ ion exchange mechanism, which relies on low environmental Na^+ concentrations (Bradley and Rourke, 1984).

We should also remember that the water that continuously diffuses by osmosis through the gills of freshwater fish must also be continuously excreted by the kidneys at the same rate. As a consequence, urine production is relatively high; however, the implication for fish health is not the large volumes of water that are excreted but the fact that the urine starts having the same electrolyte concentration as the blood. Even though extensive salt re-absorption takes place in the kidneys, blood electrolytes are continuously lost because the volume of the urine is quite large. These electrolytes must somehow be replaced, either from the water through the gill chloride cells or from the fish diet; otherwise, life-threatening ionoregulatory disturbances will develop quickly.

In contrast, as discussed earlier and Figure 4.15 shows, osmotic forces cause water to move out through the gills in marine fish species, dehydrating the fish, whereas salts diffuse inward driven by the steep concentration gradient between saltwater and the fish blood. To prevent dehydration, marine fish must drink relatively large amounts of saltwater and then physiologically distill it to remove the salts it contains. After marine fish drink saltwater, most of the ingested water and monovalent ions pass into the blood through the gut epithelium, and these ions are later excreted by the gills.

Most of the divalent and trivalent ions (magnesium, calcium, sulfate, phosphate, etc.) coming from the saltwater remain in the gut and are excreted rectally with the feces. Small quantities of these ions, however, pass through the intestinal epithelium into the blood and are excreted by the kidneys. Urination represents a loss of water that must be made up from other sources. Marine fish produce much less urine than freshwater species, and the gut epithelium here serves not only osmoregulation but also nutrient absorption.

4.5.2.3 Dissolved Oxygen

All aquatic life with the exception of anaerobic microorganisms necessitate a supply of DO for their metabolism. High-intensity aquatic life requires even more DO. DO and temperature are in fact the most critical aquatic environmental parameters for cultured organisms, and both require continuous monitoring in intensive controlled-environment production systems.

A given aquatic environment has an ongoing biological requirement for oxygen. To maintain biomass growth and reproduction in an aquatic system, there is a continuous requirement to replace the DO that is continuously extracted from the water by aerobic organisms. This continuing demand for DO in an aquatic environment should not, however, be confused with biological oxygen demand (BOD), which is commonly used in the field of water pollution and wastewater management.

BOD is a widely used measurement of the pollution potential of wastewater and the concentration of biologically degradable wastes in wastewater. In essence, BOD measures the concentration of waste in terms of the total microbial oxygen demand that would be required for aerobic microorganisms to completely use the decomposable organic portion of waste as metabolic substrates. BOD, just like DO, is measured in units of mg DO/L; in wastewater, BOD is a one-way process; it is a measure of the total DO required to biologically degrade a given concentration of organic wastes. The DO requirement for living aquatic systems is, however, an ongoing demand.

The metabolic processes of living aquatic organisms require a continuous resupply of sufficient DO in their environments to meet their respiration requirements. For instance, the BOD for a given wastewater can be as high as several thousand mg of DO. This is the total oxygen needed per liter of wastewater to completely degrade the organic waste, which has no meaning in an aquaculture system. Water can, on the other hand, only hold a few mg/L of DO. What is important in aquaculture production is that the DO concentration in water is maintained above a critical level to sustain the life of aquatic animals and, preferably, that the DO is kept at an optimal level to promote rapid growth and reproduction and maximum resistance to disease.

In extensive aquaculture operations, natural means of oxygenation can be sufficient while controlled-environment intensive RAS usually require artificial input to maintain optimal levels of DO. Typically, a minimum DO concentration of 5 mg/L is recommended for warmwater species, while it is 6 mg/L for cold-water species (Wheaton, 1977; Lawson, 1995). This means that warmwater species feed best, grow fastest, and are healthiest when DO concentrations are above 5 mg/L, and the same is true for cold-water species when the concentrations are above 6 mg/L.

The response of fish to low DO is not an immediate life-or-death issue but a continuum of physiological effects; lower concentrations than these will cause the fish be stressed, resulting first in reduced fish growth and eventually in mortality depending on exposure time, size, fish health, water temperature, CO_2 concentration, and other environmental conditions. Fish respiration and therefore O_2 consumption vary in relation to a number of factors such as species, size, and age (young require more O_2 per body weight than the same species adults); physical activity level (e.g., fish swimming in strong currents requires more O_2 than resting fish); available O_2 in the water; whether the fish are well-fed or fasting, and the water temperature (e.g., metabolic rates of cold-blooded animals increase with increasing temperatures). Furthermore, fish respiration can increase with other environmental stresses.

Fish O_2 consumption increases with available O_2 up to a critical concentration above which the fish engage in normal physiological activity. McLarney (1984) defined three important thresholds of DO concentrations for fish in aquatic environments: i) the limiting level, the level below which fish cannot survive; ii) the critical level, the level at which survival is good but growth and activity are restricted; and iii) the full-activity level, the level at which growth and activity are not restricted by O_2. These levels vary by species, average size, and a number of interrelated environmental conditions.

Table 4.3 shows O_2 consumption rates of some common fish species and DO concentrations corresponding to the fish activity levels. In principle, the partial pressure (tension) of O_2 in water governs the rate at which fish can load and unload their blood and hemoglobin. Although the atmospheric air contains 21% O_2, O_2 is only slightly soluble in

TABLE 4.3
Oxygen Consumption Rates of Some Common Fish Species and DO Concentrations Based on Fish Activity Levels

	Oxygen Consumption Rate		Dissolved Oxygen Level and Activity		
Species	**kg O_2/1000 kg fish-day**	**Size (g)**	**ppm**	**Temperature (°C)**	**Limiting**
Rainbow trout	7.2 at 15°C and unknown feeding level	100	2.4–3.7	16	Limiting
	7.2 at 15°C and production feeding	100	1.05–2.06	18.5	Limiting
			2.5	13–20	Limiting
			5–5.5	20	Limiting
			9	20	Full
Sockeye salmon	5.6 at 15°C and unfed	28.6			
	6.6 at 15°C and 3% feeding level	28.6			
Channel catfish	13.4 at 30°C and unfed	100	1.0		Limiting
	19.5 at 30°C and satiation feeding	100	3.0–5.0		Full
Largemouth bass	2.7 at 20°C and unfed	35	2.3–4.8	0–4	Limiting
	2.7 at 20°C and unfed	150			
	15.3–19.9 at 20°C and standard feeding	1.6			
Blue tilapia			1.0	25	Critical
Malaysian prawn	36.0 at 24°C and unfed	0.5			
	43.0 at 24°C and satiation feeding	0.5			

Sources: Beamish (1970), McLarney (1984), Colt and Tchobanoglous (1981).

water. As a consequence, aquatic species must expend more energy to receive DO from water than terrestrial animals expend to obtain O_2 from the atmosphere.

In addition, the solubility of O_2 decreases as salinity increases. Table A.47 provides the saturation O_2 concentration in water as a function of salinity and temperature; both altitude and barometric pressure directly influence oxygen concentration in water as well (see Equation (2.252) in Chapter 2). As mentioned in Chapter 2, the oxygen-holding capacity of water decreases as temperature increases above 0°C. Increasing salt concentration in water decreases the dissolved saturation O_2 capacity of water as well. Table A.47 provides the solubility of O_2 in water exposed to air at 760 mm Hg (1 atm) pressure for temperatures from 0 to 30°C and for increasing salt concentration.

At the surface of fish gills, the partial pressure of O_2 in the water is higher than the partial pressure of O_2 in the blood so that O_2 is loaded from the water to the blood hemoglobin. The blood carries the O_2 to the body tissues, and the tissue cells rapidly use O_2 for metabolism so that partial pressure of O_2 in the tissue is low and O_2 is unloaded from hemoglobin into the tissue. The tissue cells produce CO_2 as a byproduct of metabolism, so that partial pressure of CO_2 is high in the tissue. Additionally, while O_2 is being unloaded from the blood, CO_2 is loaded back from the tissues to the blood. The blood returning to the gills has a higher partial pressure of CO_2 than that of the water surrounding the fish, and CO_2 is unloaded from the gill surfaces back into the water.

High levels of CO_2 in blood reduce its carrying capacity for O_2. If water is high in CO_2, the O_2-loading capacity of hemoglobin is reduced; therefore, fish require higher DO concentrations when CO_2 levels are high in water. Warmwater species have better capacity to load and unload O_2 in their hemoglobin than do cold-water species; as a consequence, cold-water fish species require higher DO concentrations than warmwater species.

The rearing unit effluent for salmonids should, for instance, contain from 6.0 to 8.0 mg/L DO, whereas the allowable minimum levels for warmwater species such as catfish and tilapia are much lower, usually 2.0 to 3.0 mg/L, and it is recommended to maintain levels closer to 5.0 to 6.0 mg/L (Timmons and Ebeling, 2010). Here, we see great variation in the allowable minimums when we use the unit of mg/L to express DO; therefore, it is more appropriate to use the O_2 tension, the partial O_2 pressure for determining the lower limits. For instance, a partial pressure of 90 mm Hg is a reasonable target for salmonids (Downey and Klontz, 1981).

As Table A.45 shows, the atmospheric air contains 20.95% O_2, and at standard pressure of 760 mm Hg at 20°C (68°F), the partial pressure of O_2 is 155.5 mm Hg = 0.2095 × [(760 mm Hg − 17.54 mm Hg)], corresponding to a DO saturation of 9.1 mg/L in freshwater (see Table A.47), where 17.54 mm Hg is the water vapor pressure at 20°C (68°F). Therefore, an O_2 partial pressure of 90 mm Hg at 20°C (68°F) corresponds to 5.2 mg/L DO. If the temperature is 5°C (41°F), then the partial O_2 pressure is 158.2 mm Hg = 0.2095 × [(760 mm Hg − 6.54 mm Hg)] and corresponds to DO saturation of 12.8 mg/L in freshwater (see Table A.47). In this case, O_2 partial pressure of 90 mm Hg at 5°C (41°F) corresponds to 7.3 mg/L DO. This shows that the warmer the water, the lower the effluent DO in mg/L can go at a partial pressure of 90 mm Hg. Applying 90 mm Hg O_2 partial pressure at higher temperatures than 20°C (68°F) would, on the other hand, set target values of less than 5.2 mg/L DO for culture tank oxygen levels.

4.5.2.3.1 DO Supersaturation

In some circumstances, natural waters and controlled-environment aquaculture systems can become supersaturated with DO; that is, the O_2 levels actually become greater than the normal saturation concentration. Supersaturation can be caused by high turbulence in streams or by water falling over falls and dams, or it can happen in lakes and coastal waters during periods of high photosynthetic activity by plankton. O_2 supersaturation in water can lead to supersaturation of O_2 in the blood of fish as well and can cause problems even including mortality.

The formation of gas bubbles in the blood, which is commonly known as gas bubble disease, results in bubbles on the fish gills and behind their eyes that cause their eyes to bulge (Timmons and Ebeling, 2010). Because fish continuously utilize O_2 from their blood for respiration, supersaturation is not as great a concern as a cause of gas bubble disease; some O_2 supersaturation is acceptable. In fact, some controlled-environment intensive aquaculture operations even provide initial O_2 supersaturation in their incoming water to ensure a good supply of O_2 to very dense fish populations. In these operations, O_2 supersaturation may be reduced by stopping bubble aeration or vigorously stirring water so that the excess O_2 is lost to the air.

4.5.2.4 Temperature

Water temperature, like dissolved oxygen, is an equally critical environmental parameter for aquatic organisms that requires continuous monitoring in controlled-environment intensive production systems. Temperature directly affects the physiological processes of aquatic organisms, such as respiration rate, efficiency of feeding and assimilation, growth, behavior, and reproduction. Most aquatic organisms have a narrow temperature range for optimum growth.

Fish have traditionally been classified into three main groups by temperature: cold-water, cool water, and warmwater species. Generally speaking, warmwater species such as tilapia, can tolerate a wider range of thermal conditions than cold-water species such as trout (Parker and Davis, 1981). Stickney (1994) defines cold-water species as those with temperature optima below 20°C (68°F) and warmwater as those with temperature optima at or above 25°C (77°F). Species that have their temperature optima within these two temperatures are considered cool-water or mid-range species. This is a general classification, and the specific temperature preference is influenced by organism's species, age, size, and thermal background. Table 4.4 provides optimum temperature ranges for several representative species of interest for aquacultural production.

Fish are not homeothermic but rather are poikilothermic, or cold-blooded, aquatic animals. This means that their body temperature is approximately the same as their

TABLE 4.4
Optimum Temperature Ranges for Common Aquatic Species

Species	Temperature (°C)	Reference
Brook trout	7–13	Piper et al. (1982)
Brown trout	9–16	Petit (1990) and Piper et al. (1982)
Rainbow trout	10–16	Petit (1990) and Piper et al. (1982)
Atlantic salmon	15	Petit (1990)
Chinook salmon	10–14	Piper et al. (1982)
Coho salmon	9–14	Piper et al. (1982)
Sockeye salmon	15	Petit (1990)
Sole	15	Petit (1990)
Turbot	19	Petit (1990)
Plaice	15	Petit (1990)
European eel	22–26	Petit (1990)
Japanese eel	24–28	Petit (1990)
Common carp	25–30	Petit (1990)
Mullet	28	Petit (1990)
Tilapia	28–30	Petit (1990)
Channel catfish	21–29	Tucker and Robinson (1990); Piper et al. (1982)
Striped bass	13–23	Piper et al. (1982)
Red swamp crawfish	18–22	Romaire (1985)
Freshwater prawn	30	Romaire (1985)
Brine shrimp	20–30	Romaire (1985)
Brown shrimp	22–30	Romaire (1985)
Pink shrimp	>18	Romaire (1985)
American lobster	20–24	Hedgecock et al. (1976)
American oyster	>8	Romaire (1985)

Sources: Hedgecock et al. (1976), Piper et al. (1982), Petit (1990), Tucker and Robinson (1990), Romaire (1985).

ambient environment temperature, and each species has an optimum temperature range that maximizes growth and upper and lower limits beyond which they cannot survive. The growth rates within the tolerable temperature range, however, increase as the water temperature increases until the optimum growth temperature is reached.

It must, however, be remembered that water has a high specific heat (4,186 J/kg°C; see Table A.4), and therefore it is an excellent heat storage medium. This property of water protects and sustains aquatic life against large temperature fluctuations in natural environments. However, it is exactly because of this property that controlled-environment (warm-water) aquacultural systems require large amounts of heat for initial raising of water temperature. Therefore, above the optimum temperature, the energy requirements for fish food conversion and other metabolic processes increase and follow the law of diminishing returns. Furthermore, beyond optimum temperatures, the food conversion ratios decrease and can eventually reach upper lethal limits. As a result, we must maintain system temperatures as close as possible to the optimum value so that fish growth can be maximized while the stress is minimized.

4.5.2.5 Carbon Dioxide

As discussed in depth earlier, CO_2 is about 41 times more soluble in water than O_2 on a weight per volume basis; however, since CO_2 is a small component (0.0033% by vol.) of atmospheric air (Table A.45), it is not usually present in high concentrations in water. For instance, the partial pressure of O_2 in air at 1 standard atm (760 mm Hg) is 159 mm Hg = (760 mm Hg × 0.2095), and the partial pressure of CO_2 in air at 1 standard atm (760 mm Hg) is 0.25 mm Hg = (760 mm Hg × 0.00033). That is, the atmospheric partial pressure of O_2 is 634 times that of CO_2. At 20°C, when DO concentration is 9.2 mg/L (Table A.47) = 2.88×10^{-8} mol/L, the dissolved CO_2 concentration is 0.56 mg/L = 1.22×10^{-5} mol/L. This shows that the moles of O_2 is approximately 24 times that of the dissolved CO_2.

Additionally, as we can see in Table A.47, the solubility of CO_2 in water decreases as temperature increases and with increasing salinity (Table A.47). In addition, Table A.46 lists the solubility of the major atmospheric gases in water at 15°C. The concentration of CO_2 in water is determined by a gas–liquid equilibrium relationship and a series of acid–base reactions (bicarbonate buffering); therefore, CO_2 is different from oxygen, nitrogen, and other gases. Gas–liquid equilibrium, first affects the transfer of CO_2 between air and water; then, the abovementioned acid–base reactions determine the chemical form in which dissolved inorganic carbon is present in water. Dissolved CO_2 concentration is therefore a function of pH or the alkalinity of water, and the total dissolved inorganic carbon present in water: CO_2, H_2CO_3, HCO_3^-, and carbonate (CO_3^{2-}). Changes in temperature and salinity can also have minor influences since they affect the equilibrium constants (Piedrahita and Seland, 1995). For detailed discussions of the sources and sinks, the solubility of CO_2, and bicarbonate buffering in aquatic systems, the readers are referred to Sections 2.4.3.2.4 and 2.4.9.2.2.

CO_2, in contrast, has little direct toxicity to fish. In general, high CO_2 levels are associated with low DO levels. It is usually not necessary to monitor CO_2 if O_2 is being monitored. Dissolved CO_2 in excess of 20 mg/L may be harmful to fish (Piper et al., 1982), but most species of fish survive for several days at dissolved CO_2 levels up to 60 mg/L (Boyd, 1982). CO_2 displaces O_2 in fish hemoglobin, so that if DO levels are low when CO_2 levels are high, O_2 uptake in fish may be hindered.

Unfortunately, CO_2 levels tend to be high when DO levels are low; if respiration continues in the absence of photosynthesis in aquatic environments, CO_2 is produced as a byproduct of respiration; when photosynthesis activity is high, the CO_2 uptake by plants and algae is greater than their O_2 production—a photosynthesis byproduct—and exceeds O_2 utilization by respiration. At night, CO_2 levels tend to increase and DO levels decrease in natural aquatic systems with high plankton and algae populations.

When fish are exposed to high CO_2 concentrations, their respiration efficiency and tolerance to low DO concentrations decrease. High dissolved CO_2 concentration in aquatic systems influences O_2–CO_2 exchange at fish gills. As a result of the reduced CO_2 excretion, the CO_2 concentration in the fish blood increases and thereby lowers the blood pH, which causes a condition termed respiratory acidosis. As a consequence, the oxygen-carrying capacity of hemoglobin is reduced and respiration distress can occur even at high DO concentrations. This is called the Bohr–Root effect (Timmons and Ebeling, 2010).

Methods of CO_2 degassing are presented and discussed later in Chapter 5. In aquaculture systems, we can add bases to increase pH and hence shift the carbonate chemistry equilibrium (Grace and Piedrahita, 1994; Summerfelt, 1996; Loyless and Malone, 1997). However, adding bases to the culture water does not result in the removal of dissolved inorganic carbon from water; as discussed in Chapter 2, it causes a decrease in the CO_2 concentration by simply shifting the carbonate carbon balance to bicarbonate and carbonate ions as the pH is increased. Usually, two types of chemicals are used in aquaculture to control pH and decrease CO_2 concentrations: i) strong bases not containing any carbon (e.g., sodium hydroxide) and ii) bases containing carbon (e.g., sodium bicarbonate). In each case, the alkalinity of water is increased by the addition of bases. In the case of the carbon-containing bases, however, the increase in alkalinity is accompanied by an increase in total carbonate carbon.

4.5.2.6 Nitrogen

Nitrogen is an essential nutrient for all living organisms; however, it is required in relatively small quantities, and physiological needs are easily satisfied. Nitrogen in aquatic systems can be in either organic or inorganic form, and it is recycled between the inorganic and organic forms. Nitrogen is present in large amounts in organic matter, and there is a rapid exchange in the form of inorganic and organic nitrogen compounds. Inorganic forms of nitrogen in aquatic systems include free nitrogen gas (N_2), NH_4^+-N (ionized ammonia nitrogen), NH_3-N (unionized ammonia nitrogen), NO_2^-–N (nitrite nitrogen), and NO_3-N (nitrate nitrogen), while organic forms are amino acids, proteins, nucleic acids, and enzymes.

Ammonia, nitrite, and nitrate are all highly soluble in water, and as listed above, ammonia exists in two forms: unionized NH_3 and ionized NH_4^+. The sum of the two ammonia forms (i.e., $NH_4^+ + NH_3$) is termed as total ammonia nitrogen (TAN) or just ammonia. The relative concentration of the two forms is primarily a function of water hydrogen ion concentration (pH), salinity, and temperature.

As mentioned earlier in Section 2.4.9.2.3, molecular nitrogen (N_2) forms 78% (by vol.) of the gas mixture in atmospheric air (Table A.45). It is relatively inert and does not take part as a nutrient directly in aquatic plant or animal growth. The main source of dissolved N_2 gas in water is the atmosphere, and the main mechanism for removal of dissolved N_2 from water is by transfer back to the atmosphere or by exposure to air/water interfaces under conditions of supersaturation. Natural waters contain more dissolved N_2 than O_2 because the partial pressure of N_2 in the atmosphere is 3.73 times greater than that of the partial pressure of O_2. Molecular N_2 enters water by N_2 fixation as well, mainly by blue-green algae.

Additionally, the major inorganic N_2 source in an aquatic environment is ammonification, which removes ammonia from organic molecules. The main sources of ammonia in natural waters are the bacterial decomposition of proteins and the deamination of amino acids, which are organic constituents of all natural detrital organic matter. Fish also directly excrete ammonia.

Ammonia is subject to nitrification, a bacterially mediated that can be inhibited by light. Nitrate can be converted back to molecular N_2 or NO_x by specialized types of bacteria that use nitrate instead of oxygen as an electron acceptor. This process is called denitrification, another bacterially mediated process. Nitrification and denitrification will be discussed further in this section. Since N_2 compounds are very dynamic, and many N_2 sources are difficult to measure, the N_2 budget of an aquatic system is difficult to establish.

N_2 is not toxic to fish, but it poses a threat at levels above saturation. Like O_2, N_2 is absorbed by blood in fish. O_2 is continuously removed by the breathing tissues, but N_2 in blood is not taken up by cells. N_2 concentrations in blood tend to build until the partial pressure of N_2 in the blood is equal to the partial pressure of N_2 in the water. As long as N_2 in the blood of fish remains below the saturation concentration, there will be no problem. However, if the blood becomes supersaturated with N_2, it can come out of solution to form gas bubbles in the blood, which is a dangerous situation for fish.

Small N_2 bubbles in the blood can block arteries or veins and stop the blood circulation. Sometimes fish develop obvious gas bubbles under the skin or in the gills or fins, or N_2 bubbles can form behind the eyes causing the eyes to bulge out of their sockets. As discussed earlier, this condition is known as gas bubble disease. N_2 supersaturation occurs when air is introduced into water under high pressure that is subsequently lowered. Heating water can also cause supersaturation because the solubility of N_2 in water decreases with temperature.

N_2 supersaturation can also occur at the bottom of waterfalls or humanmade spillways because turbulence can cause air bubbles to reach depths of more than a meter. The bubbles at this depth are under higher pressure than the surface atmospheric pressure. As this water returns to the surface, the pressure drops, and the nitrogen may become supersaturated. Moreover, in cases where well-aerated cold water is heated or mixed with warm water, N_2 supersaturation can occur because N_2 solubility is lower at higher temperatures. Both freshwater and saltwater fish are susceptible to gas bubble disease. Owsley (1981) reported that any level of N_2 supersaturation above 100% can be a problem for fish, and levels over 110% in a fish culture operation should receive remedial intervention (Piper et al., 1982).

Moreover, as briefly mentioned earlier, fish create and release a number of nitrogenous waste products through gill diffusion, gill cation exchange, and urine and feces excretion. Organic debris of dead and dying organisms, uneaten feed, and N_2 gas from the atmosphere accumulate nitrogenous waste in aquatic environments. Needless to say, due to the toxicity of ammonia, nitrite, and to some extent, nitrate, decomposition of these compounds is particularly important in controlled environment intensive aquaculture. Unionized ammonia (NH_3-N) is the most toxic form of ammonia due to its ability to move across cell membranes, and the toxicity of TAN mainly depends on the percentage present in the unionized form of ammonia. As the ammonia concentration in water increases, fish ammonia excretion decreases with a consequent increase in the levels of ammonia in blood and tissues (Colt and Armstrong, 1979). This causes a rise in blood pH, which affects membrane stability and interferes with enzyme catalyzed reactions. Ammonia increases O_2 consumption by tissues, damages gills, and reduces the ability of blood to transport O_2. Exposure to sublethal concentrations of ammonia decreases disease resistance in fish (Boyd, 1982).

Equilibrium Equation (2.261) shows that as pH increases, the concentration of OH^- increases, so that the concentration of the unionized form of ammonia (NH_3) increases as well. Table A.50 presents the percentage of NH_3 in aqueous solutions at different temperatures as a function of pH variation (Boyd, 1982). For instance, water containing 3.0 mg/L of total NH_3 at pH 7 and 20°C would contain 0.012 mg/L of NH_3 = (3.0 mg/L × 0.004), which has a negligible impact on most fish; if pH is raised to 9.4, the NH_3 concentration becomes 1.5 mg/L of NH_3 = (3.0 mg/L × 0.50), killing most species of fish in hours. This shows that NH_4^+ transforms into NH_3 at alkaline pH. The concentration of NH_3 can also be calculated knowing the temperature and pH of the water using the following equation (Lawson, 1995):

$$\frac{NH_3 - N}{TAN} = \frac{1}{1 + 10^{(10.068 - 0.033T - pH)}} \tag{4.1}$$

where NH_3-N is the computed concentration of unionized ammonia (mg/L), TAN is the measured concentration of total ammonia nitrogen (mg/L), T is the water temperature (°C), and pH is the measured pH of the water.

Now, let's calculate the concentration of NH_3 using Equation (4.1), given a TAN 5.0 mg/L in freshwater at 20°C and pH 7.0. In this case, the concentration of NH_3 is only 0.020 mg/L, which has a negligible impact on most fish. However, at pH 9.0, the NH_3 concentration again increases to 1.41 mg/L, killing most species of fish in hours.

As mentioned, NH_3 is the most toxic form of ammonia to fish and has a direct effect on the growth of aquatic animals. Toxic concentrations of NH_3 were reported by European Inland Fisheries of Advisory Commission (1973) even at low concentrations, with 96-hour LC_{50} varying widely by species, for instance from as low as 0.08 mg/L NH_3-N for pink salmon to 2.2 mg/L NH_3-N for common carp. Warmwater species are more tolerant of ammonia toxicity than coldwater species, and freshwater species are more tolerant than saltwater species. Table 4.4 provides LC_{50} values for various fish species for unionized ammonia nitrogen, nitrite nitrogen and nitrate nitrogen. In general, for commercial production, it is recommended that unionized ammonia be kept below 0.05 mg/L and TAN concentrations below 1.0 mg/L for long-term exposure.

As shown next, nitrite is the ionized form of the relatively strong nitrous acid and is the intermediate product in the nitrification of ammonia to nitrate. Nitrite is generally found in low concentrations in unpolluted lakes, and in very low concentrations in seawater where sunlight does not penetrate. Large nitrite concentrations are often indicative of sewage pollution. Nitrifying bacteria convert ammonia to nitrite, while some algae can reduce nitrate to nitrite as well. Nitrite concentrations follow the cycle of nitrate, and although nitrite is quickly converted to nitrate by ozone or by the nitrifying bacteria in a properly balanced biofilter, it is still a problem in recirculating systems because it can build up easily if the second bacteria-driven stage of the nitrification process does not work properly. The two-stage microbial conversion of ammonia to nitrate takes place is shown as follows:

$$NH_4^+ + OH^- + 1.5O_2 \rightarrow H^+ + NO_2^- + 2H_2O + 59.4 \text{ kcal}$$

and

$$NO_2^- + 0.5O_2 \rightarrow NO_3^- + 180 \text{ kcal}$$

Consequently, nitrite is an important environmental parameter in aquatic environments to monitor and mediate if the concentrations go beyond acceptable limits (see Table 4.4). Beyond these limits, nitrite enters the fish's bloodstream, oxidizing the iron in the hemoglobin molecule from the ferrous state to the ferric state, causing "brown-blood disease" and affecting the hemoglobin's ability to carry oxygen. The nitrite-to-chloride ratio in the aquatic environment influences the amount of nitrite entering the blood of fish, where increased levels of chloride reduce the amount of nitrite absorption. NaCl (ordinary salt) or $CaCl_2$ can be used to increase chloride levels in the water; for this purpose, at least a 20:1 chloride to nitrite-nitrogen (Cl: NO_2^--N) ratio is recommended for channel catfish, tilapia, and rainbow trout, for instance (Timmons and Ebeling, 2010).

As shown earlier, nitrate is the end-product of the bacterially mediated nitrification process and is the least toxic form of nitrogen compounds, with a 96-hour LC_{50} usually beyond 1,000 mg NO_3-N/L (Table 4.4). In controlled-environment intensive aquaculture systems, nitrate levels are typically controlled by daily water exchanges. In systems with low water exchange or high hydraulic retention times, the denitrification process becomes important. The two-stage process where microbial conversion of nitrate to nitrous oxide takes place is

$$4NO_3^- + 3CH_4 \rightarrow 2N_2 + 3CO_2 + 6H_2O + 475 \text{ kcal}$$

or

$$2NO_3^- + 2H^+ + 21 \text{ kcal} \rightarrow N_2O + 2O_2 + H_2O$$

Nitrate concentrations in eutrophic lakes are often lowest in the upper layers because of plankton utilization. The deepest layers in lakes are often depleted in nitrate due to bacterial reduction under anaerobic conditions. In oligotrophic lakes, on the other hand, there is little nitrate stratification. In seawater, however, nitrate is present in higher concentrations than any of the other inorganic forms of nitrogen. Maximum nitrate concentrations are found in the ocean at depths below several hundred meters, and deep ocean waters can have up to 40 mg/L of nitrate. Surface waters in tropical oceans have almost no nitrate at all. Upwelling currents bring deep ocean waters to the surface and supply nitrates, which creates a very fertile environment for fish and other aquatic organisms.

4.5.2.7 Methane and Hydrogen Sulphide

As discussed earlier in Section 2.4.9.2.5, anaerobic decomposition of organic matter by microorganisms produces both CH_4 and CO_2 as carbon-based byproducts. Hydrogen sulfide (H_2S) is another byproduct gas formed during anaerobic decomposition of organic matter, and in water, H_2S dissociates into HS^- ions, which can further dissociate into S^{2-}. Further insights about CH_4 and H_2S can be found in the abovementioned section in Chapter 2.

Low concentrations of CH_4 do not appear to be toxic to fish (Wheaton, 1977), but the presence of CH_4 is indicative of low O_2 somewhere in the system. Low O_2 is an extremely serious problem for aquaculture operations. In culture systems with heavy biological loading, small pockets of anaerobic decomposition can spread rapidly following small upsets to the system such as heavy algae blooms or reduced aeration.

The ionized forms of H_2S (HS^- and S^{2-}) are not very toxic to fish, but the unionized H_2S is very toxic. Piper et al. (1982) reported that H_2S is toxic to fish even at just a few parts per billion (ppb). Smith et al. (1976) reported that the concentration of H_2S that caused 50% lethality (LC_{50}) in bluegill fish in water at 20°C was 0.048 mg/L (48 ppb). H_2S is very toxic to humans as well. It has an easily detectable rotten egg smell and can be detected by taking a handful of bottom mud from a pond, let's say, and mixing it with the fingers to see if the typical rotten egg odor is detectable (Blakely and Hrusa, 1989).

4.5.2.8 Hydrogen Ion Concentration

As discussed earlier in Section 2.4.3.2, pH is a convenient means of expressing the concentration of H^+ ions and is defined as the negative logarithm to the base 10 of the hydrogen ion concentration. In the case of pure water, where $[H^+] = 10^{-7}$ mol / litre, then pH = 7, and the water is said to be neutral; that is, the concentration of $[H^+]$ balances the concentration of $[OH^-]$. If $[H^+]$ are added to water, the $[H^+]$ concentration increases and pH becomes less than 7, in which case the water is said to be acidic. Conversely, if $[H^+]$ are added to water, the $[H^+]$ concentration will decrease so that pH becomes greater than 7; the water in this case becomes basic.

However, the pH of most surface waters and groundwater is buffered by the bicarbonate-carbonate system, and usually, therefore, pH ranges from 5 to 9. Groundwater with high concentrations of dissolved CO_2 and sometimes water draining from mines can be cited as exceptions. The pH of natural lakes varies between 6 and 9, while the pH of seawater is quite stable and typically varies between 7.8 and 8.3 because of the buffering capacity of saltwater (Wheaton, 1977).

It is important to note that the pH of water has a large influence on aquatic organisms, typically through its indirect effects resulting from interactions with other chemicals in the aquatic environment. The pH of water controls a wide variety of solubility and equilibrium reactions, and the most important of them is the relationship between the unionized and the ionized form of ammonia and nitrite. For instance, as already discussed earlier, when pH rises into the basic range, NH_4^- dissolved in water combines directly with $[H^+]$ to convert to NH_3, which exists in water as a dissolved gas and is toxic to fish even at low concentrations. The optimum pH for the growth and health of most freshwater aquatic animals is in the range of 6.5 to 9.0. The pH of water affects the toxicity of H_2S and metals such as copper, cadmium, zinc, and aluminum (Timmons and Ebeling, 2010).

4.5.2.9 Salinity

Natural waters always contain other compounds either dissolved or suspended as particles. As covered in Section 2.4.3.2.2, salinity is a term usually used in reference to seawater and is defined as the concentration of total dissolved solid material in water when all the carbonate (CO_3^{2-}) has been converted to oxide (CO_2), the bromine and iodine replaced by chlorine, and all the organic material completely oxidized. Salinity is expressed as grams of dissolved solid material per kilogram (ppt) of seawater.

Earlier, it was discussed that fish maintain dissolved salts in their body fluids through a process called osmoregulation that regulates the ion uptake from the environment through restricting the ion loss (see Section 4.5.2.1). Freshwater fish, for instance, tend to accumulate water because they have body fluids that have higher ion concentrations than those of the surrounding water. Each aquatic species has an optimum range of salinity for reproduction and growth, although the salinity tolerance of most aquaculture species is quite broad.

Commercially cultured milkfish can, for instance, perform osmoregulation over salinity ranges of 0–100 ppt or more, but considerable amounts of energy from their diets must be used for osmoregulation when these fish are reared outside of their optimum range under high salinity conditions. If salinity deviates too much from a species' optimum, osmoregulation eventually fails, the animal cannot maintain homeostasis, and it dies. Freshwater aquaculture systems are generally maintained at 2–3 ppt salinity to reduce stress in the fish and the amount of energy required for osmoregulation, thereby increasing growth rates. Most commercially important freshwater fish species reproduce and grow well at salinity of up to 4–5 ppt (Timmons and Ebeling, 2010). Readers are referred to Section 4.5.2.1 for further information on osmoregulation, its mechanisms, and other relevant topics.

4.5.2.10 Hardness

As discussed in Section 2.4.3.2.3, the concept of hardness is closely related to salinity. The term hardness is generally applied to freshwater while salinity is used when discussing saltwater. Hardness is defined as the sum concentration ($CaCO_3$ equivalent) of all polyvalent cations in a water sample, such as calcium (Ca^{2+}), magnesium (Mg^{2+}), iron (Fe^{2+}), manganese (Mn^{2+}), strontium (Sr^{2+}), and aluminum (Al^{3+}).

Among the polyvalent cations, calcium carbonate and magnesium carbonate are very soluble in water and therefore produce the majority of hardness in most water sources.

The total hardness of natural waters ranges widely from less than 5 to over 10,000 mg/L $CaCO_3$. Table 2.1 shows the qualitative classification of water according to hardness. Aquifers in regions of basalt and granite often have waters with very low polyvalent cation concentrations.

Additionally, as discussed in Section 4.5.2.1, the lower the mineral content of the water, the larger the diffusion gradient between the fish blood and the water. This is important for aquaculture because large concentration gradients of up to 3,000-fold can occur in very soft water (20 mg/L total hardness, such as $CaCO_3$); as a result, the transport of these ions can consume some of the energy provided by the fish diet. In contrast, the energy costs are lower in hard water because the concentration gradient is much smaller. If the water used for freshwater aquaculture has low hardness, then the water must be "hardened" by adding dissolved calcium to support newly fertilized fish eggs and for calcification of larval skeletal structures. Ca^{2+} and Mg^{2+} cations reduce the toxicity of dissolved metals as well, and the overall recommendations for total hardness range from 20 to 300 mg/L $CaCO_3$ equivalent (Timmons and Ebeling, 2010).

4.5.2.11 Alkalinity and Buffering Capacity

Alkalinity is a measure of the total anion concentration (opposite of hardness) in water that will react to neutralize added H^+ ions; i.e., a measure of the pH-buffering capacity or the acid-neutralizing capacity of water. Alkalinity is obviously different from hardness, but it is reported in the same units as milligrams per liter of basic anions converted to equivalent mg/L of $CaCO_3$. The majority of alkalinity in natural waters is produced by the presence of CO_3^{2-} and HCO_3^- ions; OH^- also contributes to alkalinity; and H^+ reduces alkalinity (see Equation 2.189).

Alkalinity of freshwater is determined by the geology of the aquifer or watershed and ranges from less than 5 mg/L in soft water to over 500 mg/L, whereas the alkalinity of seawater is about 120 mg/L $CaCO_3$ (Timmons and Ebeling, 2010). Aquifers in regions of basalt and granite often have low-alkalinity waters. Obviously, the required alkalinity concentrations are directly linked to system pH and CO_2 concentrations. Maintaining CO_2 at less than 15 mg/L and pH between 7.0 and 7.4 requires alkalinity less than 190 mg/L $CaCO_3$ (for low pH) to 70 (for high pH).

The relationship between pH and alkalinity is a significant issue for controlled-environment intensive aquaculture, which requires careful monitoring and intervention in both alkalinity and CO_2 levels to maintain optimum pH for both the cultured aquatic species and the biofilters. Alkalinity can easily be adjusted through the addition of sodium bicarbonate (NaHCO), common baking soda; other materials can be used, but NaHCO is readily available globally, inexpensive, and safe. It is also easy to apply and has very high water solubility at ambient temperature. Typically, for every kg of feed, approximately 250 g of NaHCO is applied to the water, and the concentrations of CO_2 are routinely controlled through degasification systems.

4.5.2.12 Solids Balance and Generation

Solids are usually classified into three categories: i) settleable solids, ii) suspended solids, and iii) dissolved solids. Settleable solids settle out in less than an hour in cone-shape cylinders, while suspended solids do not and require a different separation process from conventional gravity basins. Dissolved solids are difficult to remove because they are already dissolved in water. Waste solids are important in controlled-environment intensive RAS because they affect the efficiency of all processes in these systems. In an aquaculture system, waste typically comes from uneaten feed and feed fines, fish fecal matter, biofilm cell mass from biological filters, etc. Fish produce between 0.3 and 0.4 kg of total suspended solids (TSS) for every kilogram of feed fed; they are a major source of carbonaceous oxygen and nutrient input in water. TSS can directly affect fish health within recirculating systems by damaging fish gills and harboring pathogens.

On the other hand, while an upper limit of waste solids for freshwater fish is tentatively reported as 25 mg TSS/L, a level of 10 mg TSS/L is usually recommended for normal operations. For instance, tilapia performs well with TSS levels of 80 mg/L if all the other water quality parameters are ideal (Timmons and Ebeling, 2010). Waste solids need to be removed from the aqueous rearing environment as soon as possible without creating much turbulence or mechanical shearing.

4.5.2.13 Light

Just as for plants and terrestrial animals, light is important for aquatic animals as well. First of all, some aquatic animals are nocturnal feeders while others feed diurnally, which affects the feeding regimes in aquaculture operations. Nocturnal feeding animals can generally easily adapt to daylight feeding, whereas it may be difficult for diurnal feeders to adapt to nocturnal feeding (McLarney, 1984); therefore, light intensity is important. Lighting effects on fish production can generally be classified as i) growth rate effects, ii) maturation effects, and iii) vigor (ASABE Standards, 2018).

Photoperiod and light intensity in many fish species affect growth rate. Usually, a very low light difference (<1 lux) is needed to induce photoperiod responses in fish; in addition, fish can be extremely sensitive to sudden changes in light level. The resulting stress can negatively affect feeding and growth of the fish. Therefore, dimming circuits or banked lamps on a slow ramp-up and down might be useful to prevent stress on fish and ensure proper growth.

Usually, lighting requirements for an aquacultural facility's operational needs are not compatible with the needs of the fish. Hence, special care must be taken to make sure that lighting for the operational needs does not negatively impact the fish population. Reduced light intensity can be helpful in mitigating transport stress in anadromous salmonids. Reducing light intensity during all stages of smolt transport is a commonly used practice (Wedemeyer, 1996). In addition, manipulating the photoperiod can control the speed of maturation of some fish species. For this purpose, one of the methods used is varying the photoperiod above the fish tanks from day to day to simulate the natural fluctuations in day length from season to season. Fish vigor, or survival rate, also depends on lighting: Generally, longer photoperiods result in greater fish vigor (ASABE Standards, 2018).

Current research is, however, still limited for many species, although Sigholt et al. (1989) reported that long photoperiods during the freshwater period result in higher growth than shorter photoperiods in Atlantic salmon and that continuous light results in higher growth than simulated daylight (Krakenes et al., 1991). It was also reported that a long photoperiod (16 h light/day) results in increased growth and feed conversion in rainbow trout (Mason et al., 1991). Furthermore, Duston and Saunders (1995) showed that long photoperiods interrupted by two months of short photoperiods resulted in faster maturation and higher survival rates than with constant long photoperiods.

In the case of coho salmon, however, absence of a dark period (24 h light/day) resulted in reduced growth, silvering index, and seawater adaptability (Thorarensen et al., 1989), whereas pink salmon mature faster by accelerating the seasonal photoperiod (Beacham and Murray, 1993). Hart et al. (1996) studied greenback flounder and found that lack of lighting resulted in death of the species; an 18 to 24 hour photoperiod was recommended for optimum production. Moreover, striped bass reproductive cycles can be compressed to six or nine months by compressing the photoperiod cycle, but reproductive success diminishes as a result (Blythe et al., 1994).

4.5.2.14 Altitude

Altitude affects fish primarily due to reduced partial O_2 pressure in the atmosphere and therefore reduced DO concentration in water, which was already discussed earlier. At high altitudes, heating costs are higher for providing adequate thermal environments for controlled-environment RAS rise due to decreased outside temperatures and increased wind speeds.

We must ensure that combustion equipment is properly designed and operated so that we can ensure complete combustion. If air provided to a burner is not sufficient, then incomplete combustion will occur, leaving nonoxidized components in the flue gas such as C, H_2, and CO. Incomplete combustion not only uses fuel inefficiently but also results in air pollution that can pose even further danger due to CO production and increase the potential for explosion. Every combustion process has the potential to produce this toxic gas as a result of incomplete combustion; however, air at higher altitudes has lower density and hence less oxygen per unit volume. Therefore, excess air at high altitudes must be adjusted accordingly; otherwise, we can have incomplete combustion, or combustion can eventually cease.

4.5.3 Water Pollutants and Their Effects

This chapter has already defined the environment as the total of all external conditions surrounding plants and animals, comprising physical, chemical, and biological factors. Environmental degradation can negatively impact the sustainability of food and biomass production systems that affect our day-to-day standard of living. In the context of aquatic environments, an unpolluted environment is essential for productive aquaculture.

Pollution is therefore another important term to define. Most people accept that the term pollution has a negative connotation and that it is generally used to describe degraded alterations of our environments. Pollution is defined as undesirable changes in the physical, chemical, or biological characteristics of the air, water, or land that can harmfully affect the health, survival, or activities of humans or other living organisms; it can be caused not only by human activities but also by natural processes such as volcanic eruptions. Now we can talk about pollutants and their effects in aquatic environments.

4.5.3.1 Toxicity of Pollutants in Aquatic Environments

The severity of the effects of a pollutant released into the environment can be measured by the following characteristics: i) chemical properties—how harmful the pollutant is to living organisms; ii) concentration—the amount of pollutant dispersed in the environment per volume of air, water, or soil; and iii) persistence—how rapidly can natural systems degrade the pollutant, that is, is the pollutant readily degradable or will it last for a long time in the environment? Some pollutants can be rapidly reduced by physical, chemical, and biological processes, and these are termed biodegradable.

For example, sewage dumped into a river is biodegradable in that it will be readily assimilated if it is not added in excessive amounts that overwhelm the natural biosystem. Other pollutants are persistent, however, such as insecticides and chlorfluorocarbons (CFCs), which break down very slowly in the environment. Other pollutants are nondegradable, such as lead and mercury, and cannot be broken down by natural processes; these will persist in the environment virtually forever.

All chemicals are potentially harmful when present in high concentrations. For instance, NaCl in seawater is toxic to freshwater fish, causing failure of osmoregulation and cellular dehydration; even distilled water is lethal to fish since they cannot replace Na^+ and Cl^- in their cellular fluids. At low concentrations, toxic chemicals produce a variety of effects in fish that negatively impact their growth rates and reproductivity; at high concentrations, toxic chemicals even result in death. Therefore, to evaluate the severity of different chemical pollutants in aquatic systems, it is necessary to have specific data on the limiting concentrations of different pollutants that can cause death or other significant toxic effects.

4.5.3.2 Standard Toxicity Tests

For standard toxicity tests, multiple concentrations of a chemical are tested in aquaria. These concentrations are usually chosen after preliminary trials to establish suitable ranges for the toxicity tests. From 7 to 10 fish are placed in each aquarium, and then the survival times for all fish are recorded.

A control tank with no chemical addition is also set up for the fish. If any fish die in the control tank, then the test results are discarded because some other environmental factor(s) different from the chemical may be affecting the fish survival. The times at which each individual fish dies in each chemical concentration are recorded. With this data, the average survival time for each aquaria of fish is calculated against the corresponding concentrations of chemical, and the results are plotted on a time versus concentration graph. The time taken for 50% of fish to die at a given specified toxin concentration is LT_{50}. An alternate means of plotting the result is to determine the toxic concentrations causing 50% mortality at specified times of exposure; this is called LC_{50}. LC_{50} can also be plotted on a time concentration graph. The typical time for conducting standard LC_{50} toxicity tests for fish species is 96 hours.

A great many other factors influence the susceptibility of a given species of fish to a toxic pollutant. It is also important to determine whether fish exposed to potentially lethal concentrations of chemicals for time periods less than what would cause mortality will suffer irreversible health effects following short-term exposure or not.

4.5.3.3 Factors Affecting Pollutant Toxicity

A number of physical, chemical, and biological factors can influence the severity of toxicity imposed by given chemicals on fish. Different fish species have different sensitivities to different toxic pollutants. An active fish species with a high rate of metabolism in cold water, such as trout, will have a corresponding rapid exchange of chemicals through the gills, and the uptake of toxic chemicals will be quick as well. As a result, the metabolic processes that enable detoxification and excretion will be faster. In contrast, warmwater fish species tend to have a slower metabolic rate and will therefore absorb toxins more slowly.

4.5.3.3.1 Temperature Effects

In natural waters, temperature fluctuates on a cyclic basis. Although water temperature varies much less than air temperature, depending on the geographic location and local climate, daily fluctuations can be as much as 5°C, and seasonal fluctuations can be up to 30°C. Temperature changes influence fish sensitivity to pollutants. Since fish are cold-blooded animals, the internal physiology of fish is influenced by changes in water temperature.

When the water temperature changes, fish adjust by producing different enzymes that function optimally at the new temperature. This change places stress on the fish until it can acclimate to the new temperature. As discussed earlier, at higher body temperatures, the overall metabolism is higher. This also means that at higher metabolic rates, chemicals are absorbed at a faster rate, and the toxicity curve tends to shift downwards. In some cases, the delayed uptake of chemicals by fish in cold water is a factor in assessing the severity of an accidental chemical spill into a fish habitat. If the spill happens in winter, aquatic systems can flush the chemicals before the fish can absorb lethal doses. Water temperature also has a direct effect on the toxicity of some chemicals. For instance, as discussed earlier, water temperature has an effect on the proportion of toxic NH_3 in equilibrium with NH_4^+. Under controlled-environment aquaculture production systems, however, the influence of temperature change on fish sensitivity to pollutants is not an issue.

4.5.3.3.2 Dissolved Oxygen Effects

Cold water has a greater capacity to carry DO than warm water. As water temperatures warm, the respiration rate of fish increases at the same time the DO concentration in the water decreases. Therefore, at higher water temperatures, fish have to pump more water over their gills and are thus exposed to more toxins. Low DO levels, for any reason, cause stress on fish, and therefore toxin effects are more severe. The toxicity is generally lowered by low DO levels.

4.5.3.3.3 Other Factors

As also discussed earlier, the alkalinity of natural waters acts as a buffer to control pH to near neutral levels. A number of chemicals may be toxic in one ionic state and not toxic in another. It should be very well known by now that pH has a great influence on the equilibrium concentration between ionic components in the aquatic environment. Again, a good example is NH_3 that is highly toxic to fish, while NH_4^+ is not nearly as toxic. With a pH increase from 7.0 to 8.0, NH_3 as a mole fraction of TAN increases ten-fold. Water hardness also affects the toxicity of metals. As we know, hardness is due in large part to calcium and magnesium, and calcium competes with toxic metals such as copper, zinc, and lead for uptake at fish gill surfaces; that is, high hardness reduces the uptake of toxic heavy metals by fish. Chlorine, on the other hand, reduces the toxicity of nitrite in fish.

Humic acids in water also reduce the toxicity of some heavy metals by forming soluble complexes that reduce the bio-availability of heavy metals in fish. Suspended solids in water also reduce harmful toxin exposures in fish because many low-solubility toxic chemicals are absorbed by suspended solids and become unavailable for uptake by fish.

4.5.3.4 Concentration Limits of Toxins

General recommendations for maximum allowable limits of toxins in aquaculture water may be given as percents of LC_{50}. To apply these recommendations, it is also necessary to have the appropriate information for the specific toxin in question. The values for 96 hour LC_{50} will of course vary with different toxicants and fish species.

4.5.3.4.1 Recommendations

Avault (1996) provided the following recommendations:

(1) Concentrations of non-persistent toxicants or toxicants that have non-cumulative effects should not exceed 0.1 of the 96 hour LC50 for any sample taken at any time or place. The 24 hour average of the concentration of the toxicant should not exceed 0.05 of the LC50 after mixing.

(2) For persistent or cumulative toxicants, concentrations should not exceed 0.05 of 96 hour LC_{50} at any time or place for a single sample, and nor should the 24 hour average concentration exceed 0.01 of 96 hour LC_{50}.

(3) When two or more toxic materials are present at the same time, it is assumed that the toxic effects are additive.

Examples of some experimentally observed LC_{50} concentrations for a number of commercially used agricultural pesticides on rainbow trout are provided by Avault (1996).

NOMENCLATURE

DO	dissolved oxygen
LC	lethal concentration
LCT	lower critical temperature
LED	light-emitting diode
LT	lethal time
nm	nanometer
PAR	photosynthetically active radiation
ppb	parts per billion
ppm	parts per million
ppt	parts per thousand
RH	relative humidity
T	temperature
THC	tetrahydrocannabinol
UCT	upper critical temperature
UV	ultraviolet
VPD	vapor pressure deficit

Subscripts

b	body
d	dew point
∞	ambient
1,2	numbers
50	50 percent

BIBLIOGRAPHY

Plant Environments

Abeles, FB (1973) *Ethylene in Plant Biology*. Academic Press, New York, London.

Aldrich, RA and Bartok, JW (1994) *Greenhouse Engineering*. NRAES-33. Northeast Regional Agricultural Engineering Service. Cornell University, Ithaca, NY, USA.

Anderson, L (1978) Gas-fired heaters cause crop damage. *Grower*, 89: 574–575.

Anderson, LS and Mansfield, TA (1979) The effects of nitric oxide pollution on the growth of tomato. *Environmental Pollution*, 20: 113–121.

Ashenden, TW, Mansfield, TA and Harrison, RM (1977) Generation of air pollutants from kerosene combustion in commercial and domestic glasshouses. *Environmental Pollution*, 14: 93–100.

ASHRAE (1997) *Handbook of Fundamentals*. American Society of Heating, Refrigeration and Air Conditioning Engineers, Inc., Atlanta, GA 30329, pp. 10.25–10.29.

Calvert, A and Slack, G (1975) Effect of carbon dioxide enrichment on growth, development and yield of glasshouse tomatoes. I. Responses to controlled concentrations. *Journal of Horticultural Sciences*, 50: 61–71.

Calvert, A and Slack, G (1976) Effect of carbon dioxide enrichment on growth, development and yield of glasshouse tomatoes. II. The duration of daily periods of enrichment. *Journal of Horticultural Sciences*, 51: 401–409.

Capron, TM and Mansfield, TA (1975) Generation of nitrogen oxide pollutants during CO_2 enrichment of a glasshouse atmosphere. *Journal of Horticultural Sciences*, 50: 233–238.

Capron, TM and Mansfield, TA (1976) Inhibition of net photosynthesis in tomato in air polluted with NO and NO_2. *Journal of Experimental Botany*, 27: 1181–1186.

Capron, TM and Mansfield, TA (1977) Inhibition of growth in tomato by air polluted with nitrogen oxides. *Journal of Experimental Botany*, 28: 112–116.

Dutta Gupta, S (2017) *Light Emitting Diodes for Agriculture*. Springer, Singapore.

Ehleringer, JR (1979) Photosynthesis and photorespiration. Biochemistry, physiology, and ecological implications. *HortScience*, 14: 217–222.

Enoch, HZ and Hurd, RG (1977) Effect of light intensity, carbon concentration, and leaf temperature on gas exchange of spray carnation plants. *Journal of Experimental Botany*, 28: 84–95.

Enoch, HZ and Hurd, RG (1979) The effect of elevated CO_2 concentrations in the atmosphere on plant transpiration and water use efficiency. A study with potted carnation plants. *International Journal of Biometeorology*, 23: 343–351.

Enoch, HZ, Rylski, I and Samish, Y (1970) CO_2 enrichment to cucumber, lettuce and sweet pepper plants grown in low plastic tunnels in a subtropical climate. *Israel Journal of Agricultural Research*, 20: 63–69.

Enoch, HZ, Rylski, I and Spigelman, M (1976) CO_2 enrichment of strawberry and cucumber plants grown in unheated greenhouses in Israel. *Scientia Horticulturae*, 5: 33–41.

Forrester, ML, Krotkow, G and Nelson, CD (1966) Effect of oxygen on photosynthesis, photorespiration and respiration in detached leaves. I. Soybean. *Plant Physiology*, 41: 422–427.

Freeman, R (1983) Better crops through carbon dioxide injection. *Grower Talks*, 47: 42–48.

Goldammer, T (2019) *Greenhouse Management: A Guide to Operations and Technology*. Apex Publishers, Centreville, VA, USA.

Hanan, JJ (1973) Ethylene pollution from combustion in greenhouses. *HortScience*, 8: 23–24.

Hand, DW (1971) CO_2 from hydrocarbon fuels. *A.D.A.S.Q. Review*, 1: 18–23.

Hand, DW (1979) Injury to crops in glasshouse polluted with nitrogen oxides. *A.D.A.S.Q. Review*, 33: 134–143.

Hand, DW (1982) CO_2 enrichment, the benefits and problems. *Scientia Horticulturae*, 33: 14–43.

Hand, DW and Cockshull, KE (1975) Roses. I: The effects of CO_2 enrichment on winter bloom production. *Journal of Horticultural Sciences*, 50: 183–192.

Hand, DW and Postlethwaite, JD (1971) The response to CO_2 enrichment of capillary-watered single-truss tomato at different plant densities and seasons. *Journal of Horticultiral Sciences*, 46: 461–470.

Hand, DW, Slack, G and Sweeney, DG (1981) Lettuce: Crop responses to controlled levels of CO_2 enrichment. *Representative of Glasshouse Crops Research Institute*, 1979: 58–59.

Haukeness, MO, Maginnes, EA, Green, GH and Brooks, EE (1978) Using the heat and carbon dioxide from turbine exhaust gases for the production of greenhouse tomatoes. *Horticultural Science*, 13: 292–293.

Heij, G and Uffelen, JAM van (1984) Effects of CO_2 concentration on growth of glasshouse cucumber. *Acta Horticulturae*, 162: 29–36.

Hesketh, J (1967) Enhancement of photosynthetic CO_2 assimilation in the absence of oxygen, as dependent upon species and temperature. *Planta* (Berl.), 76: 371–374.

Hill, AC and Bennett, JH (1970) Inhibition of apparent photosynthesis by nitrogen oxides. *Atmospheric Environment*, 4: 341–348.

Holley, WD and Goldsberry, KL (1963) CO_2 and temperature recommendations. *Color Flower Growers Association Bulletin*, 164, p. 3.

Holley, WD, Goldsberry, KL and Juengling, C (1964) Effects of CO_2 concentration and temperature on carnation. *Color Flower Growers Association Bulletin*, 174: 1–5.

Janes, BE (1970) Effect of carbon dioxide, osmotic potential of nutrient solution, and light intensity on transpiration and resistance to flow of water in pepper plants. *Plant Physiology*, 45: 95–103.

Joliffe, PA and Tregunna, EB (1968) Effect of temperature, CO_2 concentration, and light intensity on oxygen inhibition of photosynthesis in wheat leaves. *Plant Physiology*, 43: 902–906.

Jones, P, Allen, LH, Jones, JW and Valle, R (1985) Photosynthesis and transpiration responses of soybean canopies to short- and long-term CO_2 treatments. *Agronomy Journal*, 77: 119–126.

Kimball, BA and Mitchell, ST (1979) Tomato yields from CO_2 enrichment in unventilated and conventionally ventilated greenhouses. *Journal of the American Society for Horticultural Science*, 4: 515–520.

Kozai, T, Fujiwara, K and Runkle, ES (2016) *LED Lighting for Urban Agriculture*. Springer, Singapore.

Laing, WA, Ogren, WL and Hageman, RH (1974) Regulation of soybean net photosynthetic CO_2 fixation by the interaction of CO_2, O_2, and ribulose 1,5-diphosphate carboxylase. *Plant Physiology*, 54: 678–685.

Lopez, R and Runkle, ES (2017) *Light Management in Controlled Environments*. Meister Media Worldwide, Willoughby, OH, USA.

Matsumaru, T, Yoneyama, T, Totsuka, T and Shiratori, K (1979) Absorption of atmospheric NO_2 by plants and soils. (I) Quantitative estimation of absorbed NO_2 in plants by 15N method. *Soil Science and Plant Nutrition*, 25: 255–265.

McCree, KL (1972) The action spectrum, absorptance and quantum yield of photosynthesis in crop plants. *Agricutural Meteorology*, 9: 191–216.

Monson, RK, Stidham, MA, Williams, GJ III, Edwards, GE and Uribe, EG (1982) Temperature dependence of photosynthesis in Agropyronsmithii Rydb. *Plant Physiology*, 69: 921–928.

Morison, JIL (1985) Sensitivity of stomata and water use efficiency to high CO_2. *Plant, Cell & Environment*, 8: 467–474.

Morison, JIL and Gifford, RM (1984) Ethylene contamination of CO_2 cylinders. *Plant Physiology*, 75: 275–277.

Mortensen, LM (1983a) Air pollution and plant injuries in greenhouses. Dep. Floriculture and Greenhouse Crops, Agricultural University of Norway, Rep. 281: 1–22 (in Norwegian).

Mortensen, LM (1983b) Growth responses of some greenhouse plants to environment. VIII. Effect of CO_2 on photosynthesis and growth of Norway spruce. *Meldinger fra Norges Landbrukshoegskole*, 62: 1–13.

Mortensen, LM (1983c) Growth responses of some greenhouse plants to environment. IX. Effect of CO_2 on photosynthesis of Chrysanthemum morifolium Ramat. at different light, temperature and O_2 levels. *Meldinger fra Norges Landbrukshoegskole*, 62: 1–12.

Mortensen, LM (1985a) Effects of CO_2 enrichment and supplementary light on growth and flowering of poinsettia, Euphorbia pulcherrima Willd. *Meldinger fra Norges Landbrukshoegskole*, 64: 1–8.

Mortensen, LM (1985b) Nitrogen oxides produced during CO_2 enrichment. I. Effects on different greenhouse plants. *New Phytologist*, 101: 103–108.

Mortensen, LM (1985c) Nitrogen oxides produced during CO_2 enrichment. II. Effects on different tomato and lettuce cultivars. *New Phytologist*, 101: 411–415.

Mortensen, LM (1986a) CO_2 during ventilation to cucumber and roses. *Gartneryrket*, 76: 140 (in Norwegian).

Mortensen, LM (1986b) Effect of intermittent as compared to continuous CO_2 enrichment on growth and flowering of Chrysanthemum × morifolium Ramat. and Saintpauliaionantha H. Wendl. *Scientia Horticulturae*, 29: 283–289.

Mortensen, LM (1986c) Nitrogen oxides produced during CO_2 enrichment. III. Effects on tomato at different photon flux densities. *New Phytologist*, 104: 653–660.

Mortensen, LM (1987) Review: CO_2 enrichment in greenhouses: Crop responses. *Scientia Horticulturae*, 33: 1–25, Elsevier Science Publishers B.V., Amsterdam, The Netherlands.

Mortensen, LM and Moe, R (1983a) Growth responses of some greenhouse plants to environment. IV. Effects of carbon dioxide on photosynthesis and growth of Chrysanthemum morifolium Ramat. *Meldinger fra Norges Landbrukshoegskole*, 61: 1–11.

Mortensen, LM and Moe, R (1983b) Growth responses of some greenhouse plants to environment. V. Effect of CO_2, O_2 and light on net photosynthetic rate in Chrysanthemum morifolium Ramat. *Scientia Horticulturae*, 19: 133–140.

Mortensen, LM and Moe, R (1983c) Growth responses of some greenhouse plants to environment. VI. Effect of CO_2 and artificial light on growth of Chrysanthemum morifolium Ramat. *Scientia Horticulturae*, 19: 141–147.

Mortensen, LM and Moe, R (1983d) Growth responses of some greenhouse plants to environment. VII. The effect of CO_2 on photosynthesis and growth of roses. *Meldinger fra Norges Landbrukshoegskole*, 62: 1–11.

Mortensen, LM and Ulsaker, R (1985) Effect of CO_2 concentration and light levels on growth, flowering and photosynthesis of Begonia × hiemalis Fotsch. *Scientia Horticulturae*, 27: 133–141.

OMAFRA (2001) *Growing Greenhouse Vegetables*. Ontario Ministry of Agriculture, Food and Rural Affairs. Publication 371. ISSN: 1492-6601. Toronto, Canada, p. 77.

OMAFRA (2003) *Production Recommendations for Greenhouse Floriculture*. Ontario Ministry of Agriculture, Food and Rural Affairs. Publication 370. ISSN: 1492-6512. Toronto, Canada, p. 39.

Rogers, HH, Sionit, N, Cure, JD, Smith, JM and Bingham, GE (1984) Influence of elevated carbon dioxide on water relations of soybeans. *Plant Physiology*, 74: 233–238.

Runkle, ES (2011) The R to FR ratio. *Greenhouse Product News Magazine*, May. Sparta, MI, USA., p. 46.

Runkle, ES (2012) Maximizing photosynthesis to improve growth. *Greenhouse Product News Magazine*, December. Sparta, MI, USA, p. 66.

Runkle, ES (2015a) Interactions of light, CO_2 and temperature on photosynthesis. *Greenhouse Product News Magazine*, June. Sparta, MI, USA, p. 54.

Runkle, ES (2015b) Light wavebands and their effects on plants. *Greenhouse Product News Magazine*, March. Sparta, MI, USA., p. 42.

Runkle, ES (2017a) Effects of blue light on plants. *Greenhouse Product News Magazine*, February. Sparta, MI, USA., p. 38.

Runkle, ES (2017b) Growing plants with green light. *Greenhouse Product News Magazine*, June. Sparta, MI, USA., p. 58.

Runkle, ES (2019) Is green light useful to plants? *Greenhouse Product News Magazine*, June. Sparta, MI, USA., p. 50.

Salisbury, FB and Ross, C (1969) *Plant Physiology*. Wadsworth, Belmont, CA, p. 194.

Saxe, H (1981) Growth reduction by NO_x during CO_2 enrichment in the greenhouses. *Ugeskr Jordbruk*, 28: 406–409 (in Danish).

Saxe, H and Christensen, OV (1985) Effects of carbon dioxide with and without nitric oxide pollution on growth, morphogenesis and production time of pot plants. *Environmental Pollution*, 38: 159–169.

Schapendonk, AHCM and Gaastra, P (1984) A simulation study on CO_2 concentration in protected cultivation. *Scientia Horticulturae*, 23: 217–229.

Slack, G and Hand, DW (1985) The effect of winter and summer CO_2 enrichment on the growth and fruit yield of glasshouse cucumber. *Journal of Horticultural Sciences*, 4: 507–516.

Tolley, LC and Strain, BR (1984) Effects of CO_2 enrichment and water stress on growth of Liquidambar stipaciflua and Pinus taeda seedlings. *Canadian Journal of Botany*, 62: 2135–2139.

Yildiz, I (2021) *Greenhouse Engineering: Integrated Energy Management*. 1st Edition. CRC Press, Taylor and Francis Group, Boca Raton, FL. ISBN: 978-1-4987-4313-6.

Yoneyama, T, Sasakawa, H, Ishizuka, S and Totsuka, T (1979) (II) Nitrite accumulation, nitrite reductase activity and diurnal change of NO_2 absorption in leaves. *Soil Science and Plant Nutrition*, 25: 267–275.

Zeevaart, AJ (1976) Some effects of fumigating plants for short periods with NO_2. *Environmental Pollution*, 11: 97–108.

Animal Environments

ASABE Standards (2018) Standards, Engineering Practices, and Data. V21(1). St. Joseph, MI, USA.

ASHRAE (1997) *Handbook of Fundamentals*. American Society of Heating, Refrigeration and Air Conditioning Engineers, Inc. Atlanta, GA 30329, pp. 10.25–10.29.

Asher, A, Fialko, M, Fares, F, Moallem, U, Yaacoby and Gutman, R (2022) The effect of short-wavelength white LED illumination throughout the night on the milk fatty acid profile of high-yielding dairy cows. *Biology*, 11(12): 1799.

Barnett, JL, Cronin, GM, McCallum, TH and Newman, EA (1994) Effects of food and time of day on aggression when grouping unfamiliar adult pigs. *Applied Animal Behaviour Science*, 39(3–4): 339–347.

Blatchford, R, Klasing, K, Shivaprasad, H, Wakenell, P, Archer, G and Mench, J (2009) The effect of light intensity on the behavior, eye and leg health, and immune function of broiler chickens. *Poultry Science*, 88(1): 20–28.

Bodurov, N (1979) Effect of supplementary artificial illumination with visible rays on biochemical indices in the blood serum, milk yields and fertilization during lactation. *Veterinary Medicine Science*, 16: 58–65.

Bond, TH, Heitman, H and Kelly, CF (1965) Effects of increased air velocities on heat and moisture loss and growth of swine. *Transactions of the ASAE*, 8: 167. ASAE, St. Joseph, MI.

Buckland, RB (1975) The effect of intermittent lighting programmes on the production of market chicken and turkeys. *Worlds Poultry Science Journal*, 31(4): 262.

Christianson, LL and Fehr, RL (1983) Ventilation energy and economics. In: Hellickson, MA and Walker, JN (eds.), *Ventilation of Agricultural Structures*. ASAE Nomograph No. 6, ASAE St. Joseph, MI 49085, p. 336.

Christison, GI (1996) Dim light does not reduce fighting or wounding of newly mixed pigs at weaning. *Canadian Journal of Animal Science*, 76(1): 141–143.

CIGR (1999) *Handbook of Agricultural Engineering*, Volume II: *Animal Production & Aquacultural Engineering*. Edited by CIGR—the International Commission of Agricultural Engineering. ASAE, St. Joseph, MI. ISBN 0-929355-98-9.

Collias, NE and Collias, EC (1996) Social organization of a red jungle fowl, Gallus gallus, population related to evolution theory. *Animal Behaviour*, 51: 1337–1354.

Curtis, SE (1983) *Environmental Management in Animal Agriculture*. Iowa State University Press, Ames, IA.

Dahl, GE, Buchanan, BA and Tucker, HA (2000) Photoperiodic effects on dairy cattle: A review. *Journal of Dairy Science*, 83: 885–893.

Dahl, GE and Petitclerc, D (2003) Management of photoperiod in the dairy herd for improved production and health. *Journal of Animal Science*, 81: 11–17.

Dahl, GE, Tao, S and Thompson, IM (2012) Lactation biology symposium: Effects of photoperiod on mammary gland development and lactation. *Journal of Animal Science*, 90: 755–60.

DeShazer, JA, Olson, LL and Mather, FB (1974) Heat losses of large white turkeys—6 to 36 days of age. *Poultry Science*, 53(6): 2047.

Donham, KJ (1987) Human health and safety for workers in livestock housing. *CIGR Proceedings*. ASAE Special Publication, pp. 6–87.

Drury, LN (1966) The effect of air velocity of broile growth in a diurnally sisling hot humid environment. *Transactions of the ASAE*, 9: 329. ASAE, St. Joseph, MI.

Esmay, ME and Dixon, JE (1986) *Environmental Control for Agricultural Buildings*. AVI Publications, Westport, CN.

Frederiksen, B, Nafstad, O, Lium, BM, Marka, CH, Dahl, D and Choinski, JU (2006) Artificial light programmes in entire male pig production—effects on androstenone, skatole and animal welfare. *Acta Veterinaria Scandinavica*, 48(Suppl. 1): 2.

Gunnarson, HJ et al. (1967) Effect of air velocity, air temperarture and mean radiant temperature on performance of growig-finishing swine. *Transactions of the ASAE*, 10: 715. ASAE, St. Joseph, MI.

Gunnarsson, S (2013) Lighting for laying hens: The effect of environmental factors on bird behaviour. In: Aland, A and Banhazi, T (eds.), *Livestock Housing: Modern Management to Ensure Optimal Health and Welfare of Farm Animals*. Wageningen Academic Publishers, The Netherlands, pp. 431–438. ISBN: 978-90-8686-217-7.

Hahn, GL (1983) Management and housing of farm animals environments. In: *Stress Physiology in Livestock*. CRC Press, Boca Raton, FL.

Hahn, GL, Nygaard, A and Siemsen, E (1983) Toward establishing rational criteria for selection and design of livestock environments. ASAE Paper No. 83-4517. ASAE, St. Joseph, MI.

Hart, NS, Partridge, JC and Cuthill, IC (1999) Visual pigments, cone oil droplets, ocular media and predicted spectral sensitivity in the domestic turkey, Meleagris gallopavo. *Vision Research*, 39: 3321–3328.

Heard, L, Froehlich, D, Christianson, L, Woerman, R and Witmer, W (1986) Snout cooling effect on sows and litters. *Transactions of the ASAE*, 29(4): 1097. ASAE, St. Joseph, MI.

Jacobs, GH, Deegan, JF and Neitz, JAY (1998) Photopigment basis for dichromatic color vision in cows, goats, and sheep. *Visual Neuroscience*, 15(3): 581–584.

Janni, KA, Redig, PT, Newman, J and Mulhausen, J (1984) Respirable aerosol concentration in turkey grower building. ASAE Paper No. 84-4522, ASAE, St. Joseph, MI.

Jiménez, NV, Hernández, KR, Cerón, JH, Avila, HRV, Hernández, GN and Godoy, AV (2015) 16 hours photoperiod in holstein heifers in the subtropics: Effects in development and age to first estrus. *Ecosistemasy Recursos Agropecuarios*, 2(4): 53–67.

Kibler, HH and Brody, S (1956) Influence of diurnal temperature cycles on heat production and cardioresprativities in Holstein and Jersey cows. University of Missouri

Agricultural Experiment Station Research Bulletin No. 601, Columbia, MO.

Manser, CE (1996) Effects of lighting on the welfare of domestic poultry: A review. *Animal Welfare*, 5: 341–360.

Marcek, JM and Swanson, LV (1984) Effect of photoperiod on milk production and prolactin of Holstein dairy cows. *Journal of Dairy Science*, 67: 2380–2388.

McDaniel, GR and Brewer, RN (1975) Intermittent light speeds broiler growth and improves efficiency. *Highlights of Agricultural Research*, 4: 9. Auburn University, Auburn, AL.McGlone, JJ and Curtis SE (1985) Behavoir and performance of weanling pigs in pens equipped with hide areas. *Journal of Animal Science*, 60(1): 20–24.

Merkle, JA and Hazen, TE (1967) Zone cooling for lactating sows. *Transactions of the ASAE*, 10:444. ASAE, St. Joseph, MI.

Moreland, AF (1975) Characteristics of research animal bioenvironment. *ASHRAE Transactions*, 81(2): 542.

Moyes, CD and Schulte, PM (2014) *Principles of Animal Physiology*. 2nd Edition. Pearson Education Limited, Harlow, Essex, England.

Muthuramalingam, P, Kennedy, AD and Berry, RJ (2006) Plasma melatonin and insulin-like growth factor-1 responses to dim light at night in dairy heifers. *Journal of Pineal Research*, 40: 225–229.

MWPS (1987) *Structures and Environment Handbook*. 11th Edition. Midwest Plan Service, Iowa State university, Ames, IA, USA. ISBN: 0-89373-057-2.

Nienaber, JA, Hahn, GL, Klencke, HG, Becker, BA and Blecha, F (1987) Cyclic temperature effects on growing-finishing swine. CIGR *Proceedings. ASAE Special Publication*, 687: 312.

OSHA (1985) OSHA Safety and Health Standard. U.S. Dept. of Labor Code 1910.1000, pp. 653–659.

Penev, T, Radev, V, Slavov, T, Kirov, V, Dimov, D, Atanassov, A and Marinov, I (2014) Effect of lighting on the growth, development, behaviour, production and reproduction traits in dairy cows. *International Journal of Current Microbiology and Applied Sciences*, 3(11): 798–810.

Peters, RR, Chapin, LT, Leining, KB and Tucker, HA (1978) Supplemental lighting stimulates growth and lactation in cattle. *Science*, 199: 911–912.

Phillips, CJC and Schofield, SA (1989) The effect of supplementary light on the production and behaviour of dairy cows. *Animal Science*, 48: 293–303.

Phillips, CJC and Lomas, CA (2001) The perception of color by cattle and its influence on behavior. *Journal of Dairy Science*, 84(4): 807–813.

Prescott, NB and Wathes, CM (1999) Spectral sensitivity of the domestic fowl (Gallus g. domesticus). *British Poultry Science*, 40: 332–339.

Prescott, NB, Wathes, CM and Jarvis, JR (2003) Light, vision and the welfare of poultry. *Animal Welfare*, 12: 269–288.

Rao, TKS, Kumar, B, Singh, A, Sriranga, KR, Patel, VA and Chaurasia, S (2017) Photoperiod management in dairy herd: A review. *International Journal of Science, Environment and Technology*, 6(1): 669–683.

Riskowski, GL and Bundy, DS (1986) The effect of air velocity and temperature on growth performance and stress indicators of weanling pigs. ASAE Paper No. 86-4531. ASAE, St. Joseph, MI.

Rozenboim, I, Biran, I, Chaiseha, Y, Yahav, S, Rosenstrauch, A, Sklan, D and Halevy, O (2004) The effect of a green and blue monochromatic light combination on broiler growth and development. *Poultry Science*, 83(5): 842–845.

Schmidt-Nielsen, K (1997) *Animal Physiology: Adaptation and Environment*. 5th Edition. Cambridge University Press, Cambridge. ISBN: 9781107268500.

Scott, NR, DeShazer, JA and Roller, WL (1983) Effects of the thermal and gaseous environment on livestock. ASAE Monograph No. 6. ASAE, St. Joseph, MI 49085.

Shutze, JV, Lauber, JK, Kato, M and Wilson, WO (1962) Influence of incandescent and colored light on chicken embriyos during incubation. *Nature*, 196(4854): 594.

Son, J, Park, J, Kang, D, Belal, SA, Cha, J and Shim, K (2020) Effects of white, yellow, and blue colored LEDs on milk production, milk composition, and physiological responses in dairy cattle. *Animal Science Journal*, 91(1): e13337.

Stanisiewski, EP, Chapin, LT, Ames, NK, Zinn, SA and Tucker, HA (1988) Melatonin and prolactin concentrations in blood of cattle exposed to 8, 16 or 24 hours of daily light. *Journal of Animal Science*, 66: 727–734.

Stanisiewski, EP, Mellenberger, RW, Anderson, CR and Tucker, HA (1985) Effect of photoperiod on milk yield and milk fat in commercial dairy herds. *Journal of Dairy Science*, 68(5): 1134–1140.

Stevenson, JS, Pollmann, DS, Davis, DL and Murphy, JP (1983) Influence of supplemental light on sow performance during and after lactation. *Journal of Animal Science*, 56(6): 1282–1286.

Taylor, PE, Coerse, NCA and Haskell, M (2001) The effects of operant control over food and light on the behaviour of domestic hens. *Applied Animal Behaviour Science*, 71: 319–333.

Taylor, PE, Scott, GB and Rose, P (2003) The ability of domestic hens to jump between horizontal perches: Effects of light intensity and perch colour. *Applied Animal Behaviour Science*, 83: 99–108.

van Putten, G (1968) Een onderzoek naar staarbijten bij mestmarkens. Ph.D. Thesis. University of Amsterdam, Amsterdam, the Netherlands.

van Putten, G (1980) Objective observations on the behaviour of fattening pigs. *Animal Regulatory Studies*, 3: 105–108.

van Putten, G (1984) The influence of three levels of light on the behaviour of fattening pigs. *Applied Animal Behaviour Science*, 13(1–2): 18.

Xie, D, Wang, Z, Dong, Y, Cao, J, Wang, J, Chen, J and Chen, Y (2008) Effects of monochromatic light on immune response of broilers. *Poultry Science*, 87(8): 1535–1539.

Yeck, RG and Stewart, RE (1959) A 10 year summary of the pyschroenergetic laboratory dairy cattle research at the University of Missouri. *Transactions of the ASAE*, 2(1): 71. ASAE, St. Joseph, MI.

Yousef, M (ed.) (1985) *Stress Physiology in Livestock*. 3 vol. CRC Press, Boca Raton, FL.

Aquatic Environments

ASABE Standards (2018) Standards, Engineering Practices, and Data. V21(1). St. Joseph, MI, USA.

Avault, JW Jr (1996) *Fundamentals of Aquaculture—A Step by Step Guide to Commercial Aquaculture*. AVA Publishing Company, Inc., Baton Rouge, Louisiana, p. 255.

Barton, BA and Schreck, CB (1987) Metabolic cost of acute physical stress in juvenile steelhead. *Transactions of the American Fisheries Society*, 116: 257–263.

Basu, SP (1959) Active respiration of fish in relation to ambient concentrations of oxygen and carbon dioxide. *Journal of the Fisheries Research Board of Canada*, 16: 175–212.

Beacham, TD and Murray, CB (1993) Acceleration of maturity of pink salmon (Oncorhynchus gorbuscha) using photoperiod control. *Aquaculture*, 109: 315–325.

Beamish, FWH (1970) Oxygen consumption of largemouth bass, Micropterus salmoides, in relation to swimming speed and temperature. *Canadian Journal of Zoology*, 48: 1121–1128.

Bisogni, JJ and Timmons, MB (1994) Control of pH in closed cycle aquaculture systems. In: Timmons, MB and Losordo, TS (eds.), *Aquaculture Water Reuse Systems: Engineering Design and Management*, Elsevier Science, Amsterdam, NY, pp. 235–245.

Blakely, DR and Hrusa, CT (1989) *Inland Aquaculture Development Handbook*. Fishing News Books, Oxford, Great Britain, Chap. 4.

Blythe, WG, Helfrich, LA, Libey, G and Beal, WE (1994) Induced maturation of striped bass morone saxatilis exposed to 6, 9, and 12 month photothermal regimes. *Journal of the World Aquaculture Society*, 25(2): 183–192.

Boyd, CE (1982) *Water Quality Management for Pond Fish Culture*. Elsevier Scientific Publishing Company, New York, NY, Chap. 2.

Boyd, CE (1990) *Water Quality in Ponds for Aquaculture*. Auburn University, Agricultural Experiment Station, Auburn, AL.

Boyd, CE and Tucker, CS (1998) *Pond Aquaculture Water Quality Management*. Kluwer Academic Publishers, Boston.

Bradley, TB and Rourke, AW (1984) The influences of addition of minerals to rearing water and smoltification on selected blood parameters of juvenile steelhead trout, *Salmo gairdneri* Richardson. *Physiological Zoology*, 58: 312–319.

Brown, CR and Cameron, N (1991) The induction of specific dynamic action in channel catfish by infusion of essential amino acids. *Physiological Zoology*, 64: 276–297.

CIGR (1999) *Handbook of Agricultural Engineering*, Volume II: *Animal Production & Aquacultural Engineering*. CIGR—The International Commission of Agricultural Engineering (ed.). ASAE, St. Joseph, MI. ISBN 0-929355-98-9.

Colt, J (2006) Water quality requirements for reuse systems. *Aquacultural Engineering*, 34: 143–156.

Colt, J and Armstrong, D (1979) *Nitrogen Toxicity to Fish, Crustaceans, and Molluscs*. Department of Civil Engineering, University of California, Davis, CA, p. 30.

Colt, JE and Tchobanoglous, G (1981) Design of aeration systems for aquaculture. In: Allen, LJ and Kinney, EC (eds.), *Proceedings of the Bio-Engineering Symposium*. American Fisheries Society, Bethesda, MD, pp. 138–148.

Downey, PC and Klontz, GW (1981) Aquaculture techniques: Oxygen (pO2) requirements for trout quality. Publication status unknown.

Duston, J and Saunders, RL (1995) Advancing smolting to autumn in age 0+ atlantic salmon by photoperiod, and long term performance in sea water. *Aquaculture*, 135: 295–309.

European Inland Fisheries of Advisory Commission (1973) Water quality criteria for European freshwater fish. *Report on Ammonia and Inland Fisheries*, 7: 1011–1022.

Grace, GR and Piedrahita, RH (1994) Carbon dioxide control. In: Timmons, MB and Losordo, TS (eds.), *Aquaculture Water Reuse Systems: Engineering Design and Management*. Elsevier Science, Amsterdam, NY, pp. 209–234.

Hart, PR, Hutchinson, WG and Purser, GJ (1996) Effects of photoperiod, temperature and salinity on hatchery-reared larvae of the greenback flounder (Rhombosolea tapirina Gunther, 1862). *Aquaculture*, 144: 303–311.

Hedgecock, D, Nelson, K and Shleser, RA (1976) Growth differences among families of the lobster Homerus americanus. *Proceedings of World Mariculture Society*, 7: 347–361.

Heinen, JM, Hunkins, JA and Adler, PR (1996) Water quality and waste production in a recirculating trout culture system with feeding of a higher-energy or a lower-energy diet. *Aquaculture Research*, 27: 699–710.

Henry, JG and Heinke, GW (1989) *Environmental Science and Engineering*. Prentice Hall, Englewood Cliffs, NJ, Chap. 1.

Krakenes, R, Hansen, T, Stefansson, SO and Taranger, GL (1991) Continuous light increases growth rate of atlantic salmon (Salmo salar L.) postmolts in sea cages. *Aquaculture*, 95: 281–287.

Lawson, TB (1995) *Fundamentals of Aquacultural Engineering*. Chapman & Hall, New York.

Lloyd, R (1992) *Pollution and Freshwater Fish*. Fishing News Books, Oxford University Press, Don Mills, ON, p. 37.

Loyless, CL and Malone, RF (1997) A sodium bicarbonate dosing methodology for pH management in freshwater—recirculating aquaculture systems. *Progressive Fish-Culturist*, 59: 198–205.

Mason, EG, Gallant, RK and Wood, L (1991) Productivity enhancement of rainbow trout using photoperiod manipulation. *Bulletin of the Aquaculture Association of Canada*, 91(3): 44–46.

McGraw-Hill Book Company (1984) *McGraw-Hill Dictionary of Scientific and Technical Terms*. 3rd Edition. Parker, SP (ed.). McGraw-Hill Book Company, New York, NY.

McLarney, W (1984) *The Freshwater Aquaculture Book*. Hartley and Marks Publishers, pp. 450–451.

Meade, JW (1985) Allowable ammonia for fish culture. *Progressive Fish-Culturist*, 47(3): 135–145.

Owsley, DE (1981) Nitrogen gas removal using packed columns. In: Allen, LJ and Kinney, EC (eds.), *Proceedings of the Bio-Engineering Symposium for Fish Culture*. Fish Culture Section of the American Fisheries Society, Bethesda, MD, pp. 71–82, 307.

Parker, NC and Davis, KB (1981) Requirements of warmwater fish. In: Allen, LJ and Kinney, EC (eds.), *Proceedings of the Bio-Engineering Symposium for Fish Culture*. American Fisheries Society, Bethesda, MD.

Petit, J (1990) Water supply, treatment and recycling in aquaculture. In: Barnabe, G (ed.), *Aquaculture*. Vol. 1, Ellis Horwood, New York, NY.

Piedrahita, RH and Seland, A (1995) Calculation of pH in fresh and seawater aquaculture systems. *Aquaculture Engineering*, 14(4): 331–346.

Piper, RG, McElwain, IB, Orme, LE, McCraren, JP, Fowler, LG and Leonard, JR (1982) *Fish Hatchery Management*. U.S. Fish and Wildlife Service, Washington, DC.

Poxton, MG and Allouse, SB (1982) Water quality criteria for marine fisheries. *Aquacultural Engineering*, 1: 153–191.

Radenko, VN and Alimov, IA (1992) Significance of temperature and light for growth and survival of Larvae of Silver Carp, Hypophthalmichthus molitrix. *Journal of Ichthyolog*, 32(1): 16–27.

Randall, DI and Daxboeck, C (1984) Oxygen and carbon dioxide transfer across fish gills. In Hoar, W and Randall, D (eds.), *Fish Physiology*. Vol. X, Academic Press, San Diego.

Romaire, RP (1985) Water quality. In: Huner, JV and Brown, EE (eds.), *Crustacean and Mollusk Aquaculture in the United States*. AVI Publishing, Westport, CT.

Rousseau, RC and Thurston, RV (1991) Aquaculture and water quality. In: *Advances in World Aquaculture*. Vol. 3, The World Aquaculture Society, Baton Rouge, Louisiana.

Smith, LL Jr, Oseid, DM, Kimball, GL and El-Kandelgy, SM (1976) Toxicity of hydrogen sulphide to various life history stages of bluegill (Leponis macrochirus). *Transactions of American Fisheries Society*, 105: 442–449.

Smith, LS (1982) *Introduction to Fish Physiology*. TFH Publications, Neptune, NJ.

Stickney, RR (1979) *Principles of Warm Water Aquaculture*. John Wiley & Sons, New York, NY, Chap. 4.

Stickney, RR (1993) Channel catfish. In: Stickney, RR (ed.), *Culture of Nonsalmonid Freshwater Fishes*. 2nd Edition, CRC Press, Boca Raton, FL, pp. 33–80.

Stickney, RR (1994) *Principles of Aquaculture*. John Wiley and Sons, New York, NY.

Stumm, W and Morgan, JJ (1981) *Aquatic Chemistry*. 2nd Edition. Wiley and Sons, New York, NY, p. 780.

Summerfelt, ST (1996) Engineering design of a water reuse system. In: Summerfelt, RC (Ed.), *Walleye Culture Manual*. North Central Regional Aquaculture Center Publication Office, Iowa State University, Ames, IA, pp. 277–309.

Summerfelt, ST, Davidson, J, Waldrop, T, Tsukuda, S (2000) A partial-reuse system for coldwater aquaculture. In: Libey, GS and Timmons, MB (eds.), *Proceedings of the Third International Conference on Recirculating Aquaculture*. Virginia Polytechnic Institute and State University, Blacksburg, VA, pp. 167–175.

Tchobanoglus, G and Schroeder, ED (1985) *Water Quality: Characteristics, Modelling, Modification*. Addison-Wesley Publishing Company, Don Mills, ON, Chap. 3.

Thomas, S and Perry, SF (1992) Control and consequences of adrenergic activation of red blood cell Na+/H+ exchange on blood oxygen and carbon dioxide transport in fish. *Journal of Experimental Zoology*, 263: 160–175.

Thorarensen, H, Clarke, WC and Farrell, AP (1989) Effects of photoperiod and various intensities of night illumination on growth and seawater adaptability of Juvenile Coho Salmon (Oncorhynchus kisutch). *Aquaculture*, 82: 39–49.

Timmons, MB (2000) Aquacultural engineering experiences with recirculating aquaculture system technology, Part 1. *Aquaculture Magazine*, 26(1): 52–56.

Timmons, MB and Ebeling, JM (2010) *Recirculating Aquaculture*. 2nd Edition. NRAC Publication No. 401-2010. ISBN 978-0-9712646-2-5.

Tucker, CS and Robinson, EH (1990) *Channel Catfish Farming Handbook*. Van Nostrand Reinhold, New York, NY.

Wedemeyer, GA (1996) *Physiology of Fish in Intensive Culture Systems*. Springer-Science+Business Media, B.V. ISBN 978-1-4613-7754-2.

Wedemeyer, GA, Meyer, FP and Smith, L (1976) *Environmental Stress and Fish Diseases*. TFH Publications, Neptune, NJ.

Wheaton, FW (1977) *Aquaculture Engineering*. John Wiley & Sons, New York, NY, Chap. 4 and 6.

Wickins, JF (1981) Water quality requirements for intensive aquaculture: A review. In: Tiews, K (ed.), *Aquaculture in Heated Effluents and Recirculation Systems*. Vol. 1, Heenemann Verlagsgesellschaft, Berlin, pp. 17–37.

Williamson, JH, Carmichael, GJ, Graves, KG, Simcoe, BA and Tomasso, JR (1993) Centrarchids. In: Stickney, RR (ed.), *Culture of Nonsalmonid Freshwater Fishes*. 2nd Edition. CRC Press, Boca Raton, FL, pp. 145–198.

Yasutake, WT and Wales, JH (1983) *Microscopic Anatomy of Salmonids: An Atlas*. U.S. Fish and Wildlife Service Resource Publication 150.

5 Controlled-Environment Systems Engineering

5.1 INTRODUCTION

As should be clear by now, the main purpose of building controlled environments is to convert the solar energy and the energy in feed into useful products utilized by humans, such as vegetables, fruits, flowers, eggs, fish, meat, milk, wool, hides and skins, and manure. The life spans of controlled environments for plant and animal production vary considerably between a few years and a couple of decades, but in economic terms, usually 15 to 20 years maximum is considered depending on the type of building. This actually indicates that an increase in production or a decrease in the frequency of plant, animal, or fish injury and disease, waste of production inputs or labor requirements for plant, animal, and fish handling will indeed return all the investment and care that have been put into the design and construction layout of the controlled environment building.

Extensive plant, livestock, poultry, and fish production involving indigenous species and low-cost feeding typically have low performance and hence only justify minimal investment, if any, in controlled-environment housing. However, it usually becomes more economical to increase the production intensity under some kind of environmental control, where improved varieties/breeds, management, and feeding are available. For instance, it is difficult to justify economic investment in sheep production using controlled-environment engineering even using the least expensive buildings. A relatively expensive farrowing, poultry, or aquatic building that provides a high level of environmental control, however, usually improves the survival rate in piglets, poultry, fish sufficiently to justify the cost and add to the profitability of the production operation.

The planning and design of any structure for a controlled-environment plant, livestock, poultry, or fish production system involve many alternative systems for controlling each of the environmental variables and can therefore easily turn into a complex issue. However, every facet of the design, including the production system itself, equipment, building materials, layout, and location, will therefore play an important role in determining the profitability of production; any variation in one of them can significantly affect the profitability of the whole operation.

Labor and energy are the major costs of controlled-environment operations, particularly in temperate and cold climates. High energy costs are frequently incurred to maintain the required thermal and radiant environments in greenhouses and similar structures during both winter and summer. While heating is mostly required at night during colder months, it is sometimes needed during the day as well. Just to have a sense of the influence of various production inputs in controlled environment systems, in greenhouse operations, for instance, we can easily say that the annual labor costs typically range from 30% to 35% of the total cost of production, while heating only costs from 20% to 35% of the total production cost depending on the location and the local fuel prices. There exist many commercial heating systems that are currently being used successfully in controlled-environment operations, some of which include old steam heating, modern hot water heating, radiant heating, spaced unit heaters, central heating, floor heating, and perforated tube air distribution.

In warmer weather, heat builds up inside the buildings due to high solar radiation loads and sensible heat gains from animals; if quality crops, livestock, poultry, or fish are to be grown year-round, it is necessary to provide buildings with appropriate cooling. Cooling can be provided by different means starting from the most cost-effective options such as whitewashing the building exterior, shading screens, and natural ventilation toward more costly options such as mechanical ventilation, evaporative cooling, and even refrigerated air-conditioning if its cost is justified.

As discussed in Chapter 4, along with temperature, environmental factors such as light and radiation, relative humidity, O_2, CO_2, other dissolved gases, air movements, water, nutrient levels, and pollutants in the aerial or aquatic environments are the key environmental components affecting optimum plant, animal, or fish growth, production, and reproduction. Modifying these environmental parameters provides ample opportunities for engineering applications in controlled-environment buildings, and indeed, this is exactly what we are going to be doing in this chapter.

Greenhouse, livestock, poultry, and fish growers successfully modulate these parameters for better plant, animal, and fish growth, production, and reproduction, including artificial lighting. As discussed in depth in Chapter 4, light (i.e., intensity, quality, and duration) is required directly to sustain the growth, production, and reproduction of all green plants, livestock, poultry, and fish. For instance, greenhouses are engineered with proper glazing materials to maximize natural light penetration so that healthy plant growth, production, and reproduction can be maintained during low light as well.

When the intensity and/or duration of natural light is not sufficient under normal operating conditions (i.e., temperature, CO_2 concentration, water, nutrients), then artificial lighting might be considered for increasing the intensity of light and extending the duration of light (photoperiod). Some of the lighting sources used in engineering designs include incandescent (INC) bulbs, fluorescent and compact fluorescent lamps (CFL), high-intensity discharge (HID) lamps such as high-pressure sodium (HPS) and metal halide (MH), and light-emitting diodes (LEDs).

This chapter first goes over estimating heating loads in greenhouses, livestock, poultry, and aquatic buildings and then moves to passive and unheated controlled environments. The chapter then briefly introduces the reverse

DOI: 10.1201/9781003349297-5

return principle and covers in depth the spaced unit heaters and central hot water (hydronic) heating systems using step-by-step approach. Actual boiler room (including the boiler loops, manifolds, transport groups, expansion systems, vent stack and fresh air inlets) and heat distribution system designs (i.e., the main supply and return lines, pipe rail heating, under-bench heating, in-floor heating, sidewall and gable-end heating, and snowmelt and overhead heating, etc.) are included. Moreover, flue gas condensers and low-grade heating systems, water heating, combined heat and power system integration, heat buffer system design, and carbon dioxide dosing system integration are covered as well.

The chapter then goes over and briefly introduces estimating cooling loads in greenhouses, livestock, poultry, and aquatic buildings, and then moves to the engineering design of passive and active cooling systems for controlled-environment buildings. The systems covered include white-washing the building exterior, shading screens, and natural ventilation, as well as more costly options such as mechanical ventilation, evaporative cooling, and heat pump cooling. The chapter then covers essential light concepts in depth including lighting systems, strategies, and sources used in greenhouses and other controlled environments. Finally, the chapter covers the water supply, loading rates and production terms, solids removal, biofiltration systems, denitrification reactor types and designs, ozone and UV disinfection systems, aeration, oxygenation, and degasification systems in controlled environment aquatic systems, and concludes with the aquaponics: integrated fish and crop production systems.

Real-world design examples are provided in each and every step in order to fully demonstrate and grasp the above-mentioned controlled environment engineering concepts.

5.2 PASSIVE AND UNHEATED SYSTEMS

Where possible, controlled environment operators or potential operators should definitely consider unheated or passively heated systems. Many times, especially at lower latitudes, the amount of incoming solar radiation to a controlled environment facility exceeds the total energy required to keep the indoor temperature at or above design temperatures. Since it is readily available and free, passive solar heating is the most cost effective and environmentally friendly method of heating not only greenhouses but also livestock and aquatic buildings.

Using proper building envelope materials (including insulation) and installing adequate curtain systems form the basis of successful passive controlled environment system designs. Insulating modified environment buildings is important for preventing condensation and maintaining the desired inside temperature while it is important for supplementally heated buildings for both minimizing heating costs and preventing condensation. It is also important to insulate cold barns for controlling condensation, frost, and high summer temperatures. In addition, passive solar greenhouse, livestock, and aquatic systems can include solar heat storage systems (rocks, concrete, water, etc.) and keep these buildings warm enough during cold nights and cloudy days.

5.3 HEATING SYSTEMS

Forced-air spaced-unit heaters are simple, convenient, and the most common heating systems for all controlled-environment buildings including small greenhouse complexes, livestock buildings and milking centers as well as aquatic buildings. They can be vented gas fired, unvented gas fired, oil fired, electric, steam or hot water unit heaters. However, for large operations (particularly greenhouse operations), typically and logically, central heating systems are used.

Central heating systems usually include a standard bare steel, black iron, copper or aluminum pipe or finned pipe (with minimal footage and water volume) containing hot water or steam, hot air distributed in sheet metal or plastic tubing, and electrical resistance strips. Steam heating is no longer preferred due to the required excessive maintenance, initial cost, lack of installers, etc. Today, the most popular controlled-environment heating systems are gas-fired unit heaters with one-way gas piping and hot water (hydronic) heating systems; therefore, this chapter focuses on gas-fired unit heaters and central hot water heating systems. Central hot water heating systems, however, are not commonly used in controlled livestock environments as they do not usually require much supplemental heating except for large operations in cold climates.

Moreover, uniform temperature distributions are needed in controlled environment buildings; therefore, horizontal air circulation fans (HAFs) or turbulators are integral, particularly spaced-unit heaters. Even though HAFs are desirable with hot water heat distribution systems, they readily provide uniform temperature distribution especially in greenhouses, which require custom engineering designs and complex installations for successful applications. They are designed to operate at different flow rates and temperatures. Bare piping, finned piping or even spaced-unit heaters can be used in central hot water heating systems.

In greenhouses, the piping or tubing in central heating systems can be placed as pipe rails between the crop rows, underbench, in-ground, along the outside walls, overhead or directly above plant bed surfaces for root zone heating. Pipe rail heating alone can usually provide approximately 75%–80% or more of the total heat loss of a greenhouse depending on the location and number of heating lines provided per bay. The remaining heat requirement is typically provided by sidewall, gable end, and overhead heating.

Induced air circulation usually provides uniform temperatures within the plant canopy or an animal micro-environment. Usually though, horizontal air circulation fans, the specifics of which will be discussed later, are desirable to improve the naturally induced circulation patterns. Overhead perforated plastic tubes with unit heaters are also used in greenhouse, livestock and aquatic buildings. In these systems, the discharge holes are arranged in such a way that air is not blown directly onto plants or animals. Several other discharge units are available with or without additional circulation fans to obtain better temperature distribution within the building. The unit employed may also differ depending on the architectural characteristics of the plants grown or the characteristics of the reared animal in a greenhouse, livestock, or poultry building.

Another common practice for heat distribution in controlled environment plant, animal, and aquatic buildings is floor heating, if plant containers can be set directly on the floor in a greenhouse, for instance. Floor heating can be provided by electric resistance cables or hot water tubes. Even though it varies from one location to another and from one building type to another, this system can for instance provide up to 25% of the heat requirement for a double-glazed greenhouse, and sometimes all the heat required in a livestock or poultry building. A number of studies reported that soil, floor, and bench-top heating systems for greenhouses conserve energy compared to more conventional natural or forced convection air heating systems (Boersma and Rykbost, 1975; Gosselin and Trudel, 1982; Seager, 1982; Zeroni and Gale, 1982).

The amount of heat input provided per unit floor surface area or the slab surface temperature for floor heating varies from one controlled environment building to another, or from one application to another within a plant or an animal controlled environment building. For instance, the floor under the saw is not heated in furrowing buildings. In floor heating, uniform concrete thickness and perimeter insulation are particularly important. And in livestock and poultry buildings, it should also be remembered that too much heated floor area results in wasted heat while too little heat encourages livestock and poultry to pile. Also, while floor heating systems are coupled with overhead, sidewall and gable end heating systems in greenhouses and greenhouse type aquatic building systems, they are used together with heat lamps and hovers in livestock and poultry buildings.

Another heating method available today is radiant infrared heating; when properly designed and installed, they transfer the heating energy quite well even without horizontal air circulation fans, and provide adequate temperatures while permitting the use of lower air temperature set points. Several studies compared infrared radiant heating systems with hot water systems in terms of the energy consumption. Knies et al. (1984) reported a net energy savings of 6% and 12% for two different configurations of infrared systems compared to hot water systems. Casault et al. (1989), on the other hand, compared three gas fired greenhouse heating systems (infrared emitters, condensing boilers, and unit heaters) in cold climates and reported that each system produced relatively uniform air temperature distributions both horizontally and vertically. However, the infrared emitters permitted equivalent leaf temperatures with lower air temperature set points (at least 3°C), while condensing boilers providing the best energetic performance because of their high steady-state efficiency and better sizing with respect to the heating load.

It should also be noted that, every passing year, alternative energy sources such as solar and wind energy are receiving greater attention for heating, cooling and power generation applications. In addition, the environmental concerns and the future of fossil fuels have advanced the use of heat pumps for heating and cooling applications. In order to improve the heat pump coefficient of performance (COP) and replace the fossil fuel energy sources, the idea of combining the heat pump and solar energy in beneficial ways has also been developed.

The principal advantage of using solar radiation as a heat pump energy source is that, it provides heat at a higher temperature level than other sources, resulting in an increase in COP. Solar heat pumps can typically be operated in two basic configurations—direct and indirect. In a direct or conventional system, the refrigerant passes through the evaporator tubes in a solar collector (usually a flat plate type), where it is evaporated by incident solar energy. The heat pump extracts energy and delivers it to the load at the required temperature. In this system, the collector and evaporator functions are combined into one unit. Substantial energy savings can be realized using solar heat pump systems; however, the economics may currently not always be favorable when initial and operating costs are considered together.

5.3.1 Estimating Heating Loads

Applying energy balance Equation (2.1) from Chapter 2 to a controlled-environment building such as a greenhouse, livestock, or aquatic building and ignoring the internal storage term gives the following energy balance equation:

$$\sum E_{gains} = \sum E_{loss} \tag{5.1}$$

And Equation (5.1) takes the following form:

$$\dot{Q}_t + \dot{Q}_{internal} = \dot{Q}_{cond} + \dot{Q}_{inf} + \dot{Q}_{evap} \tag{5.2}$$

where $\dot{Q}_t$ is the rate of supplemental heat provided by the furnace, $\dot{Q}_{cond}$ is the rate of heat loss from the building envelope, $\dot{Q}_{inf}$ is the rate of heat loss due to air exchange between the inside and outside air, $\dot{Q}_{evap}$ is the rate of heat loss by evaporating water, and $\dot{Q}_{internal}$ is the rate of total internal heat gains.

The rate of total internal heat gains, $\dot{Q}_{internal}$, for a greenhouse can be written as follows:

$$\dot{Q}_{internal} = \dot{Q}_{equipment} + \dot{Q}_{lighting} + \dot{Q}_{solar} \tag{5.3}$$

In a steady-state analysis, heat gains must be the average for the period in question, not the peak values. In particular, solar radiation must be based on averages, not peak values. Only when the outdoor temperature drops below the balance point temperature (see Chapter 7) is supplemental heat required to maintain the indoor design temperature. Above the balance point temperature, internal heat gains (solar, equipment and lights, when available) are able to maintain the indoor temperature at or above the indoor design temperature; therefore, no supplemental heating is required in such cases.

Supplemental heat is, however, called for at night when there is no sun. Furthermore, lighting or other equipment gains should never be included in the furnace capacity sizing because full heating may be required during the times when lighting or other equipment may not be operational. The rate of heat loss by evaporating water is ignored in Equation (5.2) as well. Therefore, the supplementary heating equation for a greenhouse takes the following form:

$$\dot{Q}_t = \dot{Q}_{cond} + \dot{Q}_{inf} \tag{5.4}$$

where $\dot{Q}_{cond}$ is the rate of heat loss from the greenhouse building envelope, and $\dot{Q}_{inf}$ is the rate of heat loss due to air exchange between the inside and outside air.

$$\dot{Q}_{cond} = \sum UA \times (T_i - T_o) \quad (5.5)$$

$$\sum UA = U_c A_c + U_{wi} A_{wi} + U_d A_d + U_w A_w + U_f A_f + U_p L_p \quad (5.6)$$

where A is the area of a greenhouse surface (m² in SI or ft² in I-P), T_i is the nighttime temperature of the greenhouse (°C in SI or °F in I-P; see Table A.51 for recommended night temperatures for a list of greenhouse crops), and T_o is the winter design dry-bulb temperature (99%) (°C in SI or °F in I-P; see Tables A.52–A.55 for climatic conditions for selected locations), and ΣUA is the total conduction heat loss term (W/°C in SI or Btu/h.°F in I-P); U_c, U_{wi}, U_d, U_w, and U_f are the heat transfer coefficients (W/°C.m² in SI or Btu/h.°F.ft² in I-P) for the greenhouse ceiling/roof, window (nonexistent for greenhouses), door, frame walls, and concrete walls, respectively; A's with the subscripts are the corresponding surface areas (m² in SI or ft² in I-P); U_p and L_p are the heat transfer coefficient (W/°C.m² in SI or Btu/h.°F. ft² in I-P) for the greenhouse perimeter and the corresponding perimeter length (m in SI or ft in I-P), respectively. See Table A.60 for U values of different greenhouse glazing and building materials, and Table 1 of Chapter 26 in the *ASHRAE Handbook* (ASHRAE, 2021) for typical thermal properties (density, thermal conductivity, thermal resistance, and specific heat) of common building and insulating materials.

As discussed in Section 2.3.5.1, it should be remembered that heat transfer from a surface is affected by the surface temperature and emissivity, air velocity, and temperature difference between the surface and air. The surface resistances therefore already combine the effects of conduction, convection, and radiation. The total resistance $R_t = 1/U$ and U includes the inside and outside boundary layer resistances, R_i and R_o, respectively, the values of which are provided in Table A.9 for different flow directions and surface emissivities.

Table A.9 shows that surface resistance increases as surface emissivity decreases. Usually, the resistance for inside surfaces is for still air, and the outside surface resistances for winter and summer are based on 6.6 m/s (15 mph) and 3.3 m/s (7.5 mph) wind speeds, respectively. For moving air, on the other hand, R is independent of surface slope and direction of heat flow because heat transfer is predominantly by forced convection under such conditions. While thermal conductivities of some common solid materials are provided in Table A.7, typical thermal properties (density, thermal conductivity, thermal resistance, and specific heat) of building and insulating materials can, on the other hand, be found in Table 1 of Chapter 26, *ASHRAE Handbook* (ASHRAE, 2021).

$$\dot{Q}_{inf} = \dot{m}_a c_p (T_i - T_o) \quad (5.7)$$

where $\dot{m}_a$ is the mass flow rate of air exchange between indoor and outdoor air (kg$_{da}$/s in SI or lb$_{da}$/h in I-P), c_p is the specific heat of air (J/kg$_{da}$.°C in SI or Btu/lb$_{da}$.°F in I-P; see Table A.13), T_i is the nighttime temperature of the greenhouse (°C in SI or °F in I-P; see Table A.51 for recommended night temperatures for a list of greenhouse crops), and T_o is the winter design dry-bulb temperature (99%) (°C in SI or °F in I-P; see Tables A.52–A.55 for climatic conditions for selected locations).

Since the mass flow rate of air exchange, $\dot{m} = \text{ACH}\,\rho V$, when the number of air changes per hour (ACH) is given, then it may be more convenient to use Equation (5.6) in I-P in the following form:

$$\dot{Q}_{inf} = \text{ACH}\,\rho V c_p (T_i - T_o) \quad \text{(I-P)} \quad (5.8a)$$

where ACH is the number of air changes per hour (see Table A.61 for greenhouses), ρ is the density of air (lb$_{da}$/ft³), c_p is the specific heat of air (Btu/lb$_{da}$ °F), and V is the volume of the building (ft³).

When ACH is given, it may be more convenient to use Equation (5.6) in SI in the following form:

$$\dot{Q}_{inf} = \text{ACH}\,\rho V \left(\frac{1}{3{,}600}\right) c_p (T_i - T_o) \quad \text{(SI)} \quad (5.8b)$$

where ACH is the number of air changes per hour, V is the volume of the greenhouse building (m³), 1/3,600 is for unit conversion (1 h = 3,600 s), ρ is the density of air (kg$_{da}$/m³), and c_p is the specific heat of air (J/kg$_{da}$ °C).

Similarly, when Q is given in I-P, then it may be more convenient to use Equation (5.6) in the following form:

$$\dot{Q}_{inf} = \frac{Q}{v}\left(60\frac{min}{h}\right) c_p (T_i - T_o) \quad \text{(I-P)} \quad (5.9a)$$

where Q is the rate of minimum winter ventilation (ft³/min or cfm), v is the specific volume of air (ft³/lb$_{da}$), 60 is for unit conversion (1 h = 60 min), and c_p is the specific heat of air (Btu/lb$_{da}$.°F).

When Q is given in SI, then it may be more convenient to use Equation (5.6) in the following form:

$$\dot{Q}_{inf} = \frac{Q}{v} c_p (T_i - T_o) \quad \text{(SI)} \quad (5.9b)$$

where Q is the rate of minimum winter ventilation (m³/s), v is the specific volume of air (m³/kg$_{da}$), and c_p is the specific heat of air (J/kg$_{da}$ °C).

To determine the overall heat loss from a greenhouse or the required supplementary heating for a greenhouse at a particular location, Equation (5.4) should be multiplied by an applicable wind factor provided in Table 5.1.

Heating loads from a greenhouse can be reduced using opaque and transparent insulation, retractable night curtains, reduced ventilation, etc. as covered in depth in Chapter 8. However, when determining the furnace capacity, these energy-saving measures are not taken into account and instead are totally ignored.

Moreover, as covered in Section 3.6.2, supplementally heated livestock and poultry buildings have thermal insulation and mechanical ventilation for better environmental control and more comfortable working conditions. Hence, the energy balance equation provided in Equation (5.1) is valid for livestock and poultry buildings as well; however,

TABLE 5.1
Recommended Wind Factors as a Function of Wind Velocity

Wind Velocity (m/s)	(mph)	Factor
7	15	1.00
8	18	1.02
9	20	1.04
10	22	1.06
11	25	1.08
12	27	1.10
13	30	1.12
14	31	1.14
15	35	1.16

Source: Greenhouse Engineering, Edn. 1 by İlhami Yıldız, Copyright © 2021 by Imprint. Reproduced by permission of Taylor & Francis Group (*Greenhouse Engineering*).

the rate of total internal heat gains, $\dot{Q}_{internal}$, takes the following form:

$$\dot{Q}_{internal} = \dot{Q}_{animal\_sensible} + \dot{Q}_{equipment} + \dot{Q}_{lighting} + \dot{Q}_{solar} \quad (5.10)$$

where $\dot{Q}_{equipment}$ is the rate of mechanical heat gain provided by the equipment, which can be estimated but is typically neglected; $\dot{Q}_{lighting}$ is the rate of heat gain provided by the lighting equipment such as incandescent and fluorescent bulbs, which can be estimated but is typically neglected as well; $\dot{Q}_{solar}$ is the rate of heat gain from solar radiation, which is also neglected due to the reasons discussed earlier; and, $\dot{Q}_{animal\_sensible}$ is the rate of total sensible heat gains provided by animals, which are tabulated for each livestock species.

For livestock and poultry buildings then, again ignoring the rate of heat loss by evaporating water from wet surfaces, Equation (5.2) takes the following form:

$$\dot{Q}_t + \dot{Q}_{animal\_sensible} = \dot{Q}_{cond} + \dot{Q}_{inf} \quad (5.11)$$

where $\dot{Q}_t$ is the supplemental heating provided (or to be provided) by the heating system, $\dot{Q}_{animal\_sensible}$ is the rate of sensible heat gain from the specific livestock or poultry (see Table A.63, or ASABE Standards (2018)), $\dot{Q}_{cond}$ is the rate of heat loss from the livestock building envelope, and $\dot{Q}_{inf}$ is the rate of sensible heat loss due to air exchange with the outside air.

$$\dot{Q}_{cond} = \sum UA \times (T_i - T_o) \quad (5.12)$$

where A is the area of a livestock building surface, T_i is the indoor design temperature of the livestock building (see Chapter 4 of this book, ASABE Standards (2018) or MWPS (1987)), and T_o is the winter design dry-bulb temperature (99%) (see Tables A.52–A.55 for climatic conditions for selected locations), and ΣUA is the total conduction heat loss term (W/°C in SI or Btu/h.°F in I-P; see Equation (5.6)). Also see Table 1 of Chapter 26, *ASHRAE Handbook* (ASHRAE, 2021) for typical thermal properties (density, thermal conductivity, thermal resistance, and specific heat) of common building and insulating materials.

It should further be noted that in Equation (5.7) for livestock buildings, $\dot{m}_a$ is the mass flow rate of air exchange between indoor and outdoor air (kg_{da}/s in SI or lb_{da}/h in I-P), c_p is the specific heat of air (J/kg_{da} °C in SI or Btu/lb_{da} °F in I-P; see Table A.13), T_i is the design temperature of the livestock building (°C in SI or °F in I-P; see Chapter 4 of this book, ASABE Standards (2018), or MWPS (1987)), and T_o is the winter design dry-bulb temperature (99%) (°C in SI or °F in I-P; see Tables A.52–A.55 for climatic conditions for selected locations).

Again, when ACH is given, then it may be more convenient to use Equation (5.8a) for I-P and Equation (5.8b) for SI. Similarly, when the minimum winter Q is given in I-P, then it may be more convenient to use Equation (5.9a), and if Q is given in SI, then it may be more convenient to use Equation (5.9b).

Furthermore, the energy balance equation, Equation (5.1), is also valid for aquatic buildings; however, the rate of total internal heat gains, $\dot{Q}_{internal}$, takes the following form:

$$\dot{Q}_{internal} = \dot{Q}_{fish\_sensible} + \dot{Q}_{equipment} + \dot{Q}_{lighting} + \dot{Q}_{solar} \quad (5.13)$$

where $\dot{Q}_{equipment}$ is the rate of mechanical heat gain provided by the mechanical equipment, which can be estimated but is typically neglected; $\dot{Q}_{lighting}$ is the rate of heat gain provided by the lighting equipment such as incandescent and fluorescent bulbs, which can be estimated but is typically neglected as well; $\dot{Q}_{solar}$ is the rate of heat gain from solar radiation, which is also neglected for the reasons discussed earlier; and, $\dot{Q}_{fish\_sensible}$ is the rate of total sensible heat gains provided by fish.

For aquatic buildings then, again ignoring the rate of heat loss by evaporating water from wet surfaces, Equation (5.2) takes the following form:

$$\dot{Q}_t + \dot{Q}_{fish\_sensible} = \dot{Q}_{cond} + \dot{Q}_{inf} \quad (5.14)$$

where $\dot{Q}_t$ is the supplemental heating provided (or to be provided) by the heating system, $\dot{Q}_{fish\_sensible}$ is the rate of sensible heat gain from the reared fish (0.61 W per kg body mass or 1 Btu/h per lb body mass), $\dot{Q}_{cond}$ is the rate of heat loss from the aquatic building envelope, and $\dot{Q}_{inf}$ is the rate of sensible heat loss due to air exchange with the outside air.

$$\dot{Q}_{cond} = \sum UA \times (T_i - T_o) \quad (5.15)$$

where A is the area of an aquatic building surface, T_i is the indoor design temperature of the aquatic building (see recommended design temperatures for each fish species and Table 4.3), and T_o is the winter design dry-bulb temperature (99%) (see Tables A.52–A.55 for climatic conditions for selected locations), and ΣUA is the total conduction heat loss term (W/°C in SI or Btu/h°F in I-P; see Equation (5.6)). Also see Table A.60 for U values of different greenhouse glazing and building materials, and Table 1 of Chapter 26, *ASHRAE Handbook* (ASHRAE, 2021) for typical thermal properties (density, thermal conductivity, thermal resistance, and specific heat) of common building and insulating materials.

Many times, greenhouse type buildings are used for aquatic production; in such cases, we can easily use the rate of heat loss equation, $\dot{Q}_{cond}$, developed for a greenhouse building envelope. If the aquatic building is a tight concrete block, we follow the same procedures in calculating $\dot{Q}_{cond}$ using different building and insulation materials listed in Table 1 of Chapter 26, *ASHRAE Handbook* (ASHRAE, 2021).

Furthermore, in Equation (5.7) for aquatic buildings, $\dot{m}_a$ is the mass flow rate of air exchange between indoor and outdoor air (kg_{da}/s in SI or lb_{da}/h in I-P), c_p is the specific heat of air ($J/kg_{da}\cdot°C$ in SI or $Btu/lb_{da}\cdot°F$ in I-P; see Table A.13), T_i is the design temperature of the aquatic building (°C in SI or °F in I-P; see recommended design temperatures for each fish species and Table 4.3), and T_o is the winter design dry-bulb temperature (99%) (°C in SI or °F in I-P; see Tables A.52–A.55 for climatic conditions for selected locations).

Moreover, when ACH is given, then it may be more convenient to use Equation (5.8a) for I-P and Equation (5.8b) for SI. If a greenhouse building is used as an aquatic building, then the tabulated ACH provided in Table A.61 might be too small because typically there exists plenty of space for free water surface evaporation. Therefore, if the tabulated air change is less than 3.0, then it is safer to use minimum 3.0 air changes per hour.

If the building is very tight, on the other hand, such as a concrete block building with no windows, etc., then the air change would probably be somewhere around 0.5 or so; again, in this case as well, a minimum 3.0 air change per hour should be supplied to remove excess moisture. Similarly, when the minimum winter ventilation rate (Q) is given in I-P, then it may be more convenient to use Equation (5.9a), and if Q is given in SI, then it may be more convenient to use Equation (5.9b). Sometimes, natural infiltration of air may not be enough to remove excess moisture in the building; in such cases, minimum winter ventilation rates to remove excess moisture are calculated, and these ventilation rates are used in Equations (5.9a) or (5.9b) as will be shown later.

5.3.2 Reverse Return Principle

In piping systems with multiple branches and loops, the flow takes the path with the least resistance. In direct return systems with no controls, the return header is connected to the load closest to the pump, and there exists an inherent difference in flows. This can be due to many factors, including pipe length and size, pipe roughness, material, fittings, and bends. A different yet a simple approach is reverse return. A reverse-return system is a closed-loop system where the return header is connected to the most hydraulically remote load. The basic principle is last in, first out.

Compared with the direct return system, reverse return distributes the flows and pressures more evenly across the system, making it inherently balanced. The reverse return principle is employed extensively in the heating, ventilation, and air conditioning (HVAC) industry to help balance flows, and we will be using this principle in this chapter extensively as well. HVAC has many identical devices that require equal amounts of flow, such as boilers, unit heaters, and chillers along with their respective heat emitters such as pipe rail, underbench, in-floor, overhead, and sidewall and gable end heat distribution systems.

5.3.3 Unit Heater Systems

As mentioned earlier, unit heaters are the most commonly used heating systems in controlled environments (see Figure 5.1). They are located within the space and heat the space directly. They can be vented gas fired, unvented gas fired, oil fired, electric, steam or hot water unit heaters. Controlled-environment operations usually start small and expand, and as they expand, the operators keep adding more unit heaters to meet the heating requirement of their operations. Unit heaters are simple and cost effective and normally rated on their input rather than output capacity (i.e., the capacity they deliver). When sizing an adequate unit heater, the output capacity must be considered in order to meet the heating demand, and the unit heaters typically have 80% efficiency.

The two main types are vented and unvented. Unvented gas-fired unit heaters use the indoor air for combustion and exhaust the flue gas back into the indoor environment; therefore, all the energy after the natural gas or propane is combusted stays inside and is used to heat the indoor air. Some greenhouse operators prefer this type of heaters assuming that they could also provide CO_2 enrichment in their greenhouses; however, this argument has no justification because most of the time heating is required when CO_2 is not.

Another byproduct of combustion is water vapor that leads to condensation, which is not desirable in

FIGURE 5.1 Installed unit heaters with fresh-air louvers on the back.

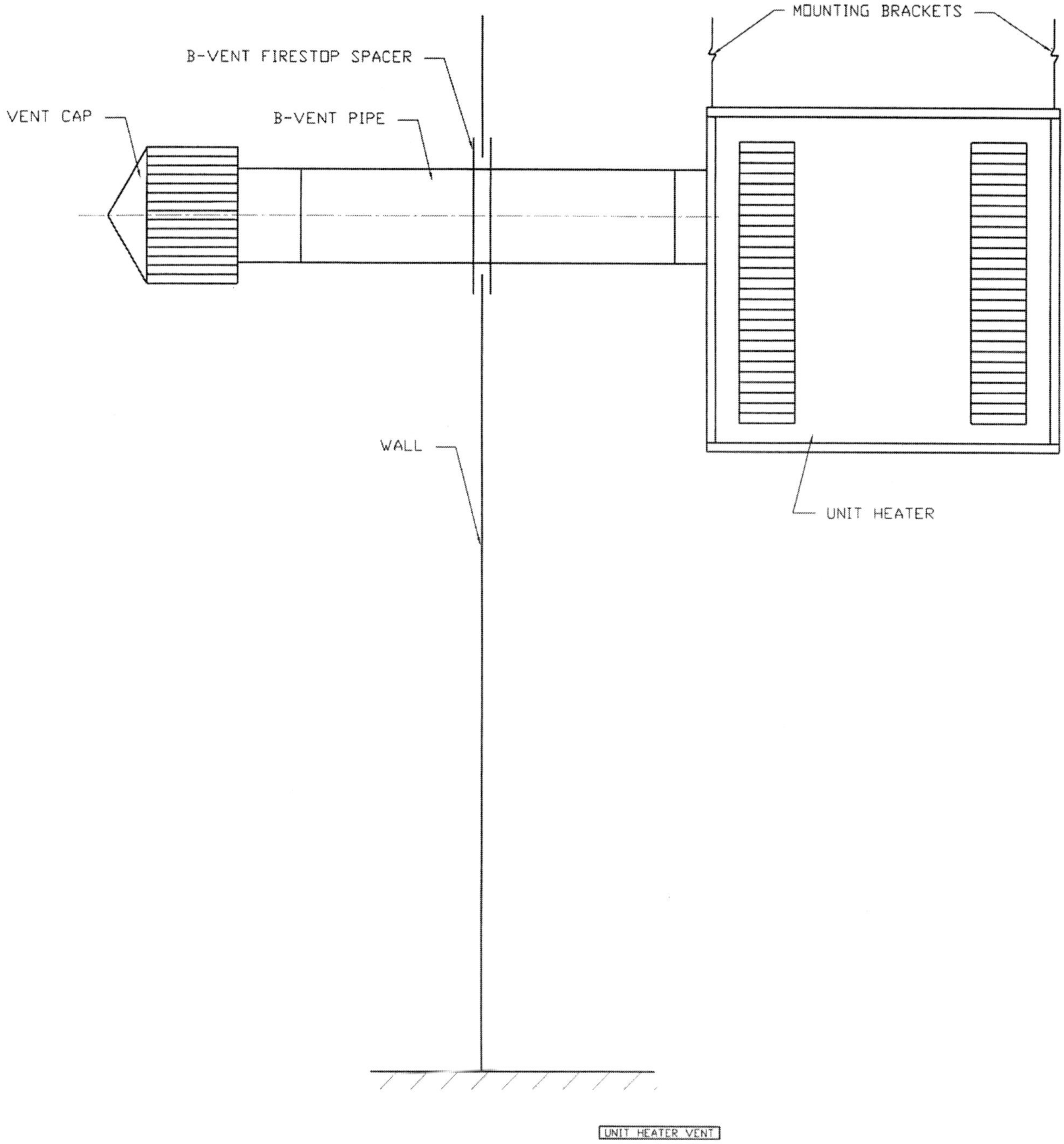

FIGURE 5.2 Unit heater with straight B-Vent connection.

controlled-environment buildings. Additionally, when the combustion is incomplete, toxic CO is released that must be monitored closely. Vented gas-fired unit heaters, on the other hand, exhaust the flue gas along with some heat to the outside air. When possible, the unit heaters are placed along the coldest wall to reduce cold drafts and radiant heat losses to the wall. With more than one unit heater, particularly in livestock and poultry buildings, the heaters should be arranged in such a way that a circular air pattern is created within the room.

For proper combustion, each unit heater must be provided with sufficient fresh air supply so that its maximum efficiency can be achieved (see Figure 5.1 for the fresh air inlet louver on the back of the unit heater; Figure 5.2 shows a vented gas-fired unit heater with straight B-Vent connection). As discussed in Chapter 2, the ASME states that 1 square inch free area per 2,000 Btu/h maximum fuel input of combined burners must be provided and kept clear at all times. Adequate inlet louver sizing exercises will be performed later in this chapter.

Figure 5.3 shows the connection details of another type of unit heater, a hot water heater. These heaters are hooked up to a hot water or steam boiler system and require further engineering expertise for piping, balancing, and controls.

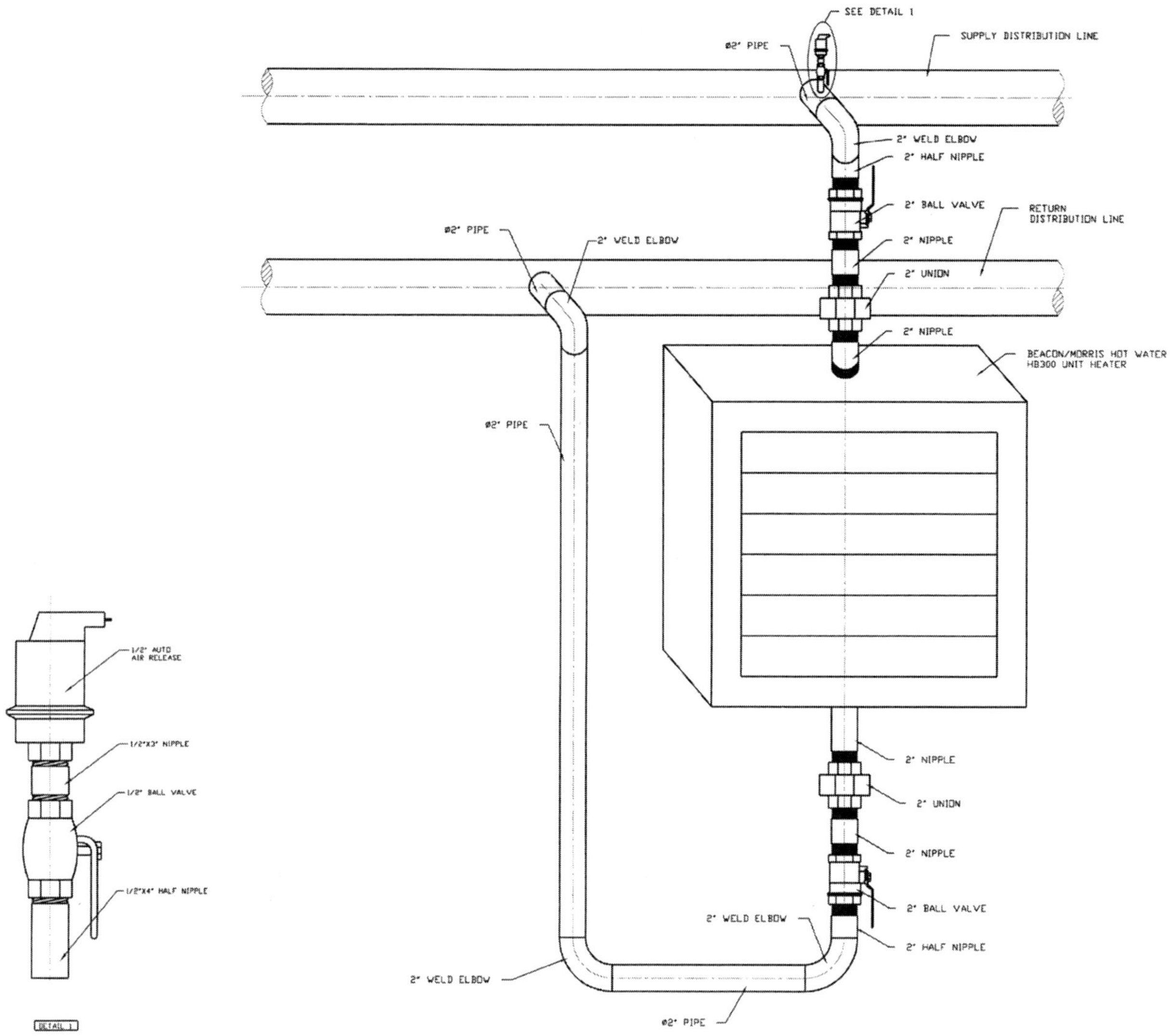

FIGURE 5.3 Hot water unit heater connection details.
Source: ***Greenhouse Engineering*** **(2021).**

Example 5.1: Sizing Unit Heaters for a Gutter-Connected Single-Glass Greenhouse Complex

Now let's consider a 10-bay gutter-connected single-glass gable roof greenhouse complex in Columbus, Ohio, USA, with an uninsulated floor, 21 ft (6.4 m) span, 130 ft (39.6 m) long, 16 ft (4.9 m) wall height, and 5.25 ft (1.6 m) gable height. Assuming that the head-house is preheated, determine the heating load and design the heating system for the greenhouse using natural gas-fired unit heaters. See the layout of the greenhouse complex in the next figure.

Solution

GREENHOUSE DIMENSIONS

COMPONENT	(ft)	(m)
Wall Height	16	4.9
House Width	21	6.4
House Length	130	39.6
Rafter Length	11.5	3.5
Gable Height	5.25	1.6
Number of Bays	10	10

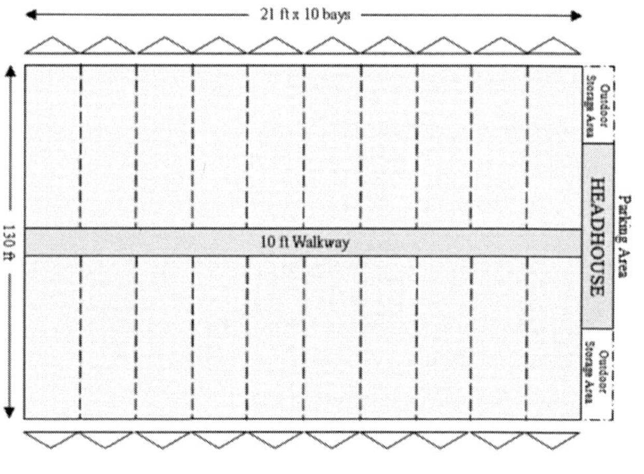

Layout of the gutter connected glass greenhouse for Example 5.1.

OTHER PROVIDED or CALCULATED INPUTS		
PARAMETER	**(I-P)**	**(SI)**
T_i	65 °F	18 °C
T_o	6 °F (99%); Table A.52	−14 °C (99%); Table A.53
ΔT	59 °F	32 °C
Greenhouse Area	27,300 ft²	2,536 m²
	0.63 acre	0.25 ha
Greenhouse Volume	508,463 ft³	14,389 m³
Air Change per Hour	1.00 (Table A.61)	
Wind Factor	1.00 (Table 5.1)	
Location	Columbus, Ohio, USA	

	AREA		*U* VALUES*		*UA* VALUES & TRANSMISSION LOSSES			
COMPONENT	*A* (ft²)	*A* (m²)	*U* (Btu/h.°F.ft²)	*U* (W/°C.m²)	*UA* (Btu/h.°F)	*UA ΔT* (Btu/h)	*UA* (W/°C)	*UA ΔT* (W)
Single Material Wall	10,880	1,011	1.1	6.2	11,968	706,112	6,267	200,533
Gable End	1103	102	1.1	6.2	1,213	71,552	635	20,321
Gable Roof	29,900	2,778	1.1	6.2	32,890	1,940,510	17,222	551,098
Perimeter Length	680	207	0.8	1.4	544	32,096	290	9,285
TOTAL					**46,615**	**2,750,270**	**24,414**	**781,237**

*See Table A.60 for determining *U*.

HEAT LOSSES	Btu/h	W
$Q_{cond} = UA \times \Delta T$	2,750,270	781,237
$Q_{inf} = \rho \times C_p \times \Delta T \times V \times \text{ACH}$	599,986	155,023
$Q_t = Qcond + Qinf$	**3,350,256**	**936,260**
	100 BHP	

The required heating load of 3,350,256 Btu/h (936 kW) can be supplied by using 20 80% efficient gas-fired heaters, 10 GHE200 (156,000 Btu/h) and 10 GHE230 (184,000 Btu/h), providing a total of 3,400,000 Btu/h). The unit heaters come with 5 in. (12.7 cm)) round flue gas connection on the back and can be installed with straight B-Vent connection as shown in Figure 5.4 and diagonally in each bay as shown. GHE200 and GHE230 have ½ in. and ¾ in. gas piping sizes, respectively, and natural gas plumbers can easily connect the individual unit heaters to a single natural gas supply pipe along each gable end.

In this project, we have used 10 (ten) 80% efficient GHE200 (156,000 Btu/h output each) and 10 (ten) 80% efficient GHE230 (184,000 Btu/h output each) providing a total of 3,400,000 Btu/h). This makes (10 × 156,000 Btu/h + 10 × 184,000 Btu/h)/0.8 = 4,250,000 Btu/h (input total). As mentioned earlier, the ASME states that *1 square inch free area per 2,000 Btu/h* maximum fuel input of combined burners must be provided and kept clear at all times. Then,

Total Free Area Required = (4,250,000 Btu/h input total) × (1 in²/2,000 Btu/h)

Total Free Area Required = 2,125 in²

Free Area per Unit Heater = 2,125/20 heaters ≈ 106 in²

Inlet louvers typically have 60% free area; therefore, 106 in²/0.6 =177 in² louver area per unit heater needs to be provided. Then, for the purpose of providing fresh air here, 20 GHE13 motorized inlet shutters with opening dimensions of 13 ½ in. × 13 ½ in. each (which provides a louver area of 182 in²) would meet the requirement. Instead, we could have sized the inlet louvers separately for the GHE200 and GHE230 heaters. Please note that while the specifications are real, the manufacturer model is generic.

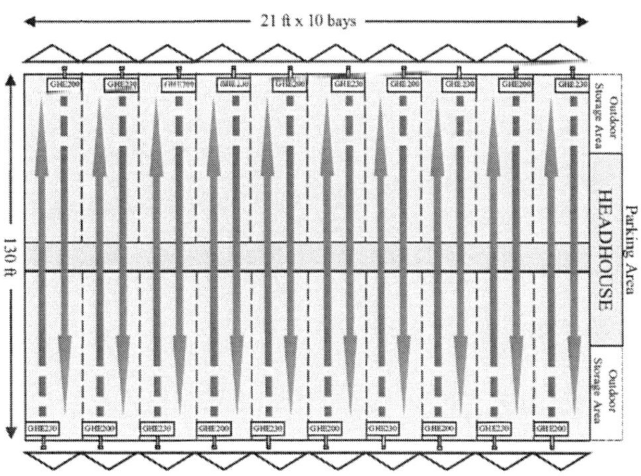

Layout of the unit heater heating system for Example 5.1.

For this heating system, we also need to install an adequate number of properly sized horizontal air flow (HAF) fans, which will be dealt with in Example 5.6 in Section 5.7 in this chapter.

Example 5.2: Sizing Unit Heaters for a Livestock Building

See Examples 5.13 and 5.14 in this chapter for heating a swine finishing building and a poultry confinement building.

Example 5.3: Sizing Unit Heaters for a Gutter-Connected Double-Poly Aquaculture Complex

This time, let's consider a 10-bay gutter-connected, IR-inhibited, double-poly, curved-roof greenhouse-type aquatic building complex in Columbus with an insulated floor, 21 ft (6.4 m) span, 130 ft (39.6 m) long, 16 ft (4.9 m) wall height, and 5.25 ft (1.6 m) gable height. Assuming that the headhouse is preheated, determine the heating load and design the heating system for this building complex using natural gas-fired unit heaters. See the layout in the next figure. NOTE: Even though the heat produced by fish is approximately 1 Btu/h per lb (2.2 kJ/h per kg) body mass, it is assumed that it contributes to water heating rather than space heating, and the rearing water is controlled at the indoor design temperature of 25°C (77°F).

Solution

AQUATIC BUILDING DIMENSIONS

COMPONENT	(ft)	(m)
Wall Height	16	4.9
House Width	21	6.4
House Length	130	39.6
Length of Arch	24.4	7.4
Gable Height	5.25	1.6
Number of Bays	10	10

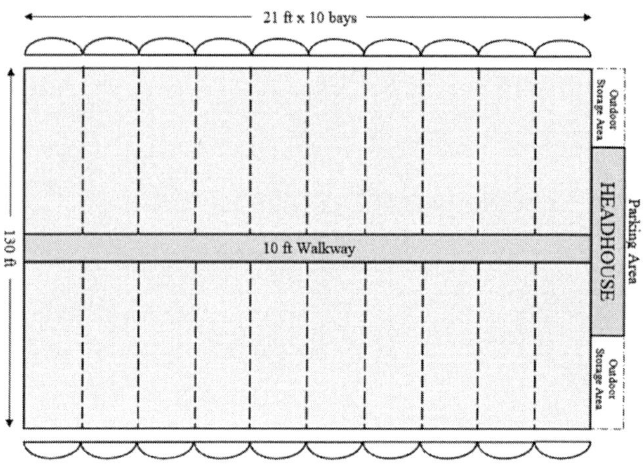

Layout of the gutter connected double poly aquatic building for Example 5.3.

OTHER PROVIDED or CALCULATED INPUTS

PARAMETER	(I-P)	(SI)
T_i	77 °F	25 °C
T_o	6 °F (99%); Table A.52	−14 °C (99%); Table A.53
ΔT	71 °F	39 °C
Aquatic Building Area	27,300 ft²	2,536 m²
	0.63 acre	0.25 ha
Building Volume	532,350 ft³	15,066 m³
Air Change per Hour	Use 3.0 minimum (Table A.61)	
Wind Factor	1.00 (Table 5.1)	
Location	Columbus, Ohio, USA	

	AREA		U*		UA & TRANSMISSION LOSSES			
COMPONENT	A (ft²)	A (m²)	U (Btu/h.°F.ft²)	U (W/°C.m²)	UA (Btu/h.°F)	$UA\,\Delta T$ (Btu/h)	UA (W/°C)	$UA\,\Delta T$ (W)
Single Material Wall	10,880	1,011	0.5	2.8	5,440	386,240	2,830	110,370
Gable End	1,433	133	0.5	2.8	717	50,907	373	14,547
Gable Roof	31,720	2,947	0.5	2.8	15,860	1,126,060	8,251	321,789
Perimeter Length	680	207	0.4	0.7	272	19,312	145	5,655
TOTAL					**22,289**	**1,582,519**	**11,599**	**452,361**

*See Table A.60 for determining U.

HEAT LOSSES

	Btu/h	W
$Q_{cond} = UA \times \Delta T$	1,582,519	452,361
$Q_{inf} = \rho \times C_p \times \Delta T \times V \times \text{ACH}$	2,267,810	593,471
$Q_t = Qcond + Qinf—Qfish\_sensible$	**3,850,329**	**1,045,832**
	115 BHP	

The required heating load of 3,350,256 Btu/h (1046 kW) can be supplied by using 20 80% efficient gas fired GHE250 (200,000 Btu/h) providing a total of 4,000,000 Btu/h. The unit heaters come with 5 in. (12.7 cm) round flue gas connections on the back and can be installed with straight B-Vent connections as shown in Figure 5.2 and diagonally in each bay as shown in the next figure. The GHE250 units have ¾ in.

pipes, and natural gas plumbers can easily connect the individual heaters to a single natural gas supply pipe along each gable end.

The project used 20 80% efficient GHE250 (200,000 Btu/h) unit heaters. This makes (20 × 200,000 Btu/h)/0.8 = 5,000,000 Btu/h (input total). Again, 1 square inch free area per 2,000 Btu/h maximum fuel input of combined burners must be provided and kept clear at all times. Then,

Total Free Area Required = (5,000,000 Btu/h input total) × (1 in^2/2,000 Btu/h)

Total Free Area Required = 2,500 in^2

Free Area per Unit Heater = 2,500/20 heaters ≈ 125 in^2

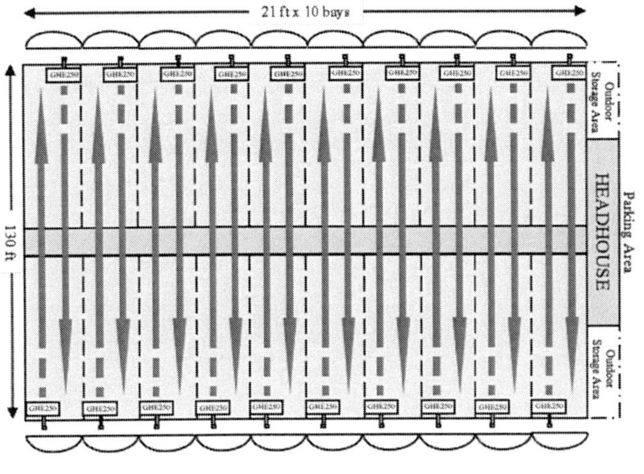

Layout of the unit heater heating system design for Example 5.3.

Inlet louvers typically have 60% free area; therefore, 125 in^2/0.6 = 208.3 in^2 louver area per unit heater needs to be provided. Then, for the purpose of providing fresh air here, 20 GHE14 motorized inlet shutters with opening dimensions of 14 ½ in. × 14 ½ in. each (which provides a louver area of 210 in^2) would meet the requirement in this project. Please note that while the specifications are real, the manufacturer model is generic. Additionally, for this heating system, for uniform temperature distribution and adequate moisture removal and management, it is also desirable to install an adequate number of properly sized HAF fans, which will be dealt with in Section 5.7.

5.3.4 Central Hot Water Heating Systems

Central heating systems have a central hot water or steam boiler and a heat distribution system made up of standard black pipe (natural convection), finned-pipe (natural convection), or hot water or steam (forced convection). Boilers come in many different shapes and sizes and are classified by operating temperature and pressure, fuel type (fuel oil, natural gas, coal, wood, etc.), and whether they condense flue gas or not. They are also described by their heat exchanger designs: cast-iron (oldest designs), fire-tube, steel water tube, and copper water tube. In a fire-tube boiler, the hot flue gas moves through the tubes surrounded by hot water, whereas in a water tube boiler, whether it is steel or copper, water moves through tubes, which are surrounded by the hot flue gas. Condensing boilers can either be fire tube or water tube boilers, which condense the water vapor in the flue gas and extract the waste heat before the flue gas is exhausted into the atmosphere.

Hot water systems are also classified by temperature and pressure, piping arrangement, and mode of circulation. Low-temperature systems are generally open to the atmosphere through vent pipes and expansion tanks and are called low-pressure systems. They typically operate with flow water temperatures of less than 90°C (195°F). Modern systems today use lightly pressurized diaphragm-type or nitrogen expansion vessels and have no open vent pipes as in the previous low-temperature systems. These systems can operate at temperatures up to 120°C (250°F) and are known as medium-temperature or pressure systems. The systems that operate at temperatures above 120°C (250°F) are called high-temperature or pressure systems. As mentioned, natural gas, fuel oil, coal, or wood is burned in boilers to heat water and generate transportable heat. Boilers are rated in boiler horsepower based on the output capacity (1 bhp = 33,479 Btu/h). The hot water generated by a boiler is pumped and distributed in controlled-environment buildings for heating purposes.

5.3.4.1 Boiler Room

5.3.4.1.1 The Heat Source

The heat source in a hydronic heating system is typically a hot water boiler housed in the boiler room. The heating load of a controlled environment building, as calculated using the estimating methodology introduced earlier in this chapter, determines the size of the heat source, a boiler for instance. As mentioned earlier, boilers are rated based on the output bhp; therefore, we first need to convert the building heating load to bhp.

In greenhouse and aquaculture applications, we divide the total bhp requirement (e.g., 200) by 2 to determine the capacity of each boiler (100 bhp) to be installed if we decide to install at least two boilers. In these applications, it is important to provide a backup capacity of approximately 50%. This means, if one of the boilers fails, then we would still have a 50% capacity to keep the building inside air temperature above freezing and protect the crop or fish while the system is being fixed. If future expansion is considered, then the boiler room should be designed accordingly to accommodate the future boilers, pumps, mixing groups, etc. Figure 5.4 shows the connection details of a three-boiler central heating system where potential future expansion is considered.

The boiler lines, the manifold and the transport groups and mains are insulated with at least 30 mm thick insulation and jacketed with 0.8 mm aluminum sheet. The temperature and pressure sensors must be located where the volumetric flow is fully developed, which means that the sensors are installed at a minimum distance of five times the diameter of a pump or a valve. The air within the system collects at the highest point, typically where the flow direction changes from horizontal to vertical downward or from vertical upward to horizontal. The air should be bled by auto air release systems or other forms of air release systems at these locations (see Figures 5.4, 5.7, and 5.8). Should it be needed, a water filtering assembly needs to be installed as well. Thermal expansion of transport pipes, especially the long lines, should also be accommodated adequately so that transport line and/or structural damages can be prevented. Overall, the boiler room must comply with the applicable local codes and regulations.

5.3.4.1.2 *The Manifold*

In a hot water heating system, the mixing groups, which will be discussed shortly, sit on two large diameter pipes (supply and return) called the manifold (see Figure 5.4, right side, and Figure 5.5 for the details). As shown in Figure 5.6, the supply and return manifolds are connected via a check valve preventing the back flow. The manifold simply provides extra water volume for pumps so that they have access to enough hot water to transport heat. The manifold size is determined using the following relationship:

$$D = \sqrt{2\left(M_1^2 + M_2^2 + M_3^2 + ... + M_n^2\right)} \quad (5.16)$$

where D is the manifold diameter in mm and M_1, $M2$, . . ., and M_n are the size of mixing groups in mm.

5.3.4.1.3 *Boiler Main Lines*

The boiler main lines are the pipelines that connect the boilers to the manifold and are designed according to the

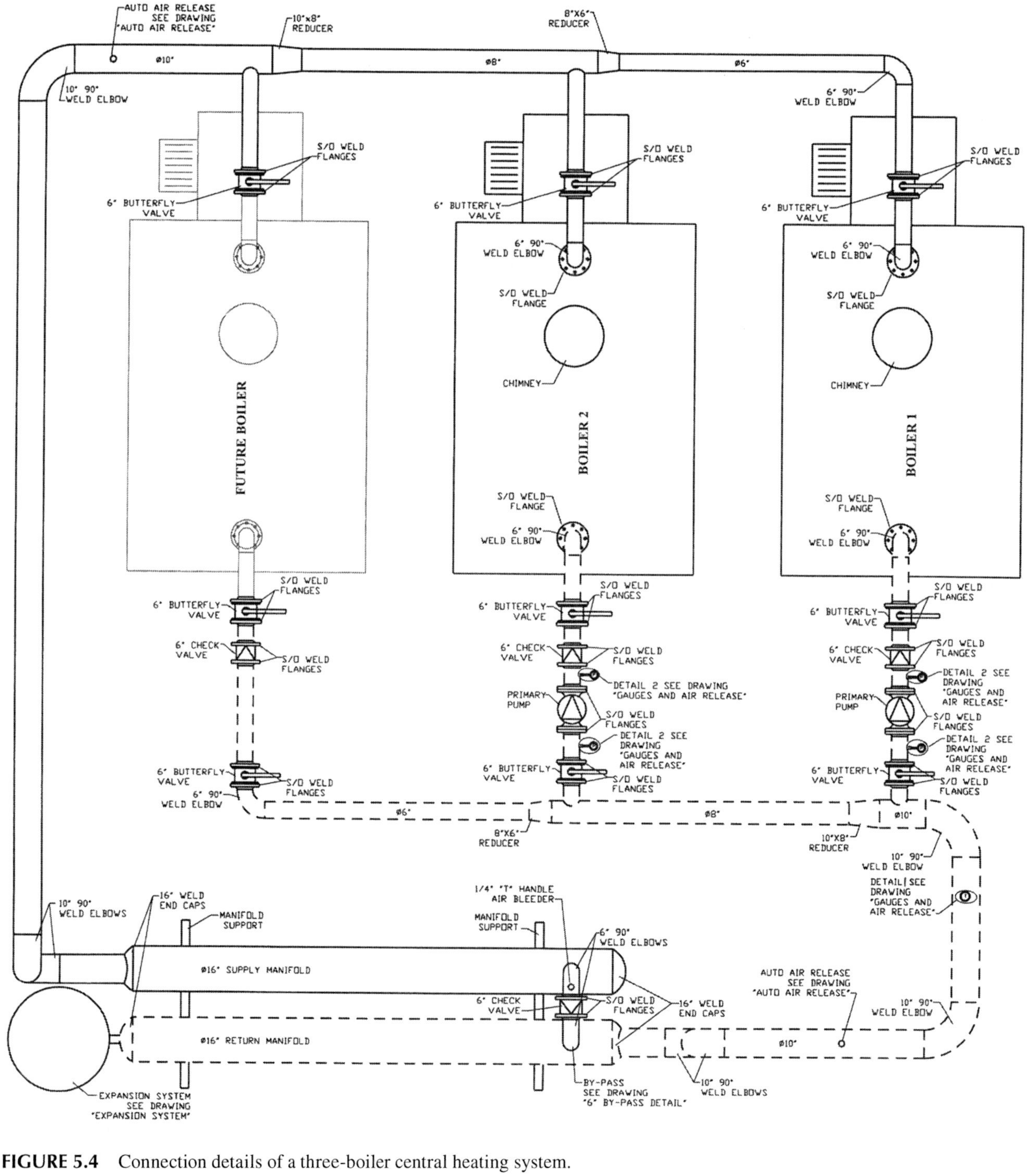

FIGURE 5.4 Connection details of a three-boiler central heating system.

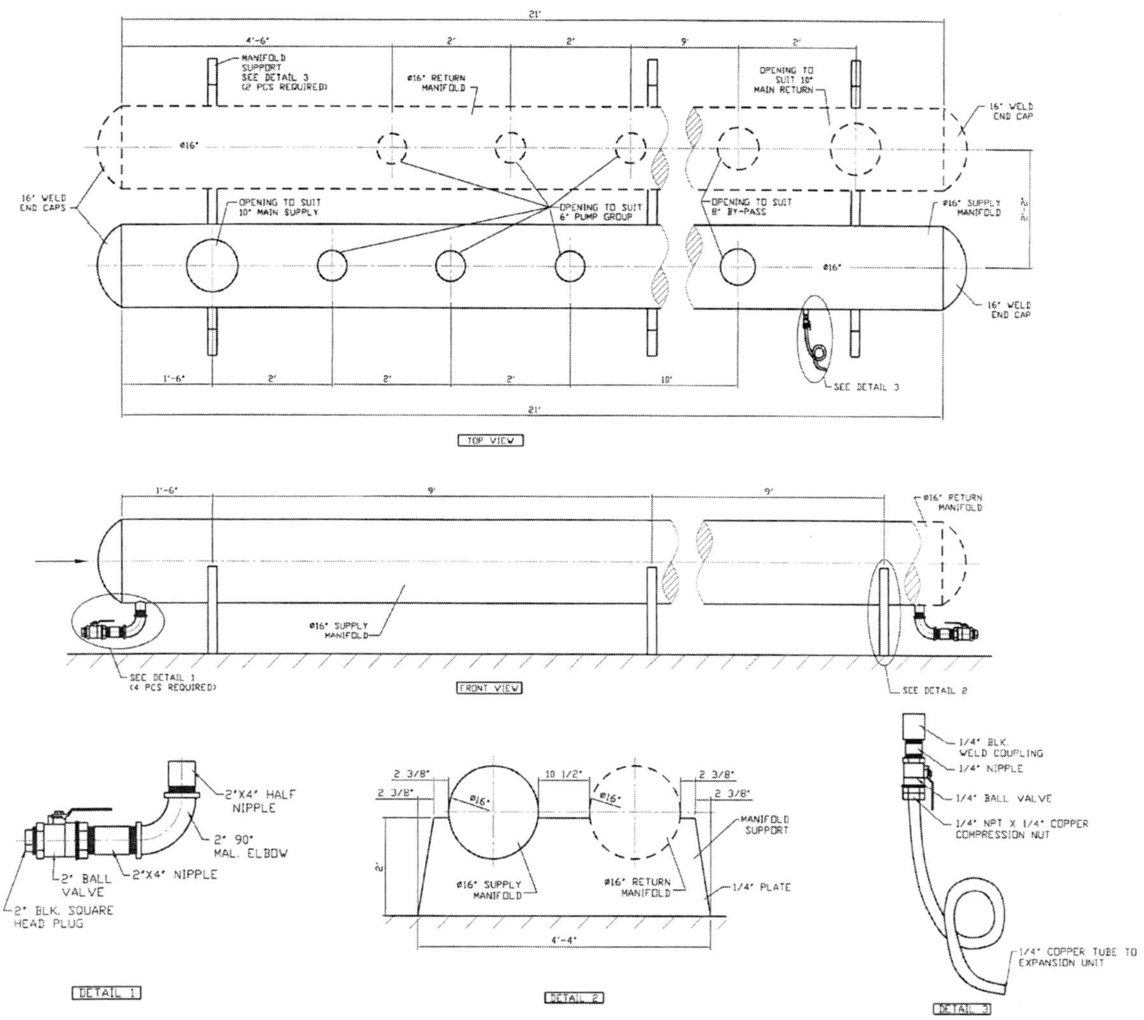

FIGURE 5.5 Details of a 16 in. manifold with three 6 in. pump connection openings.

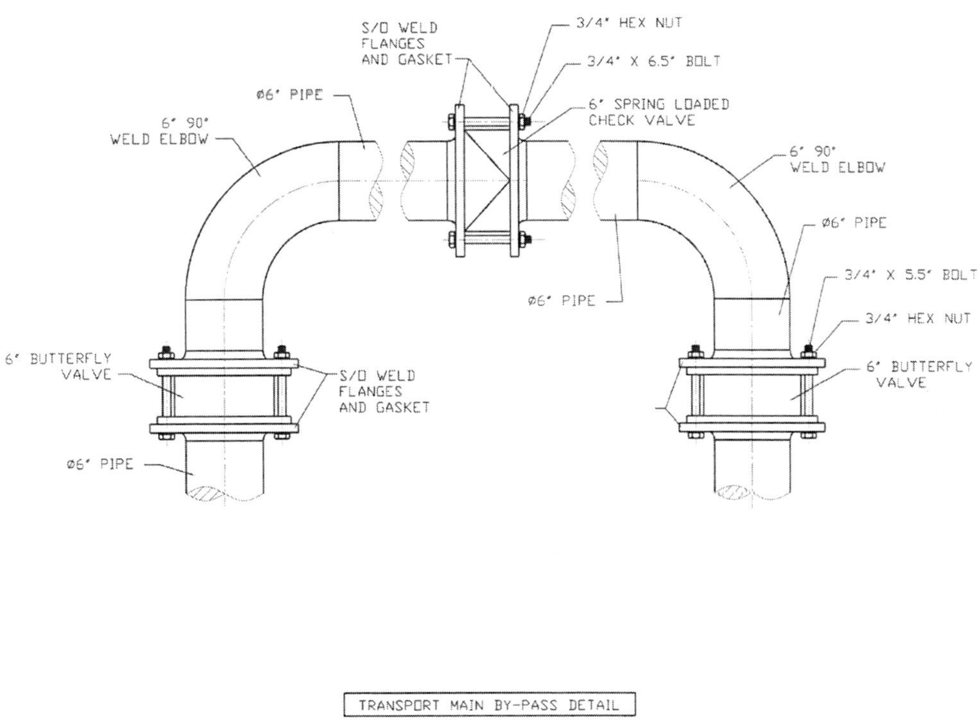

FIGURE 5.6 Connection details of a 6 in. transport main bypass system.

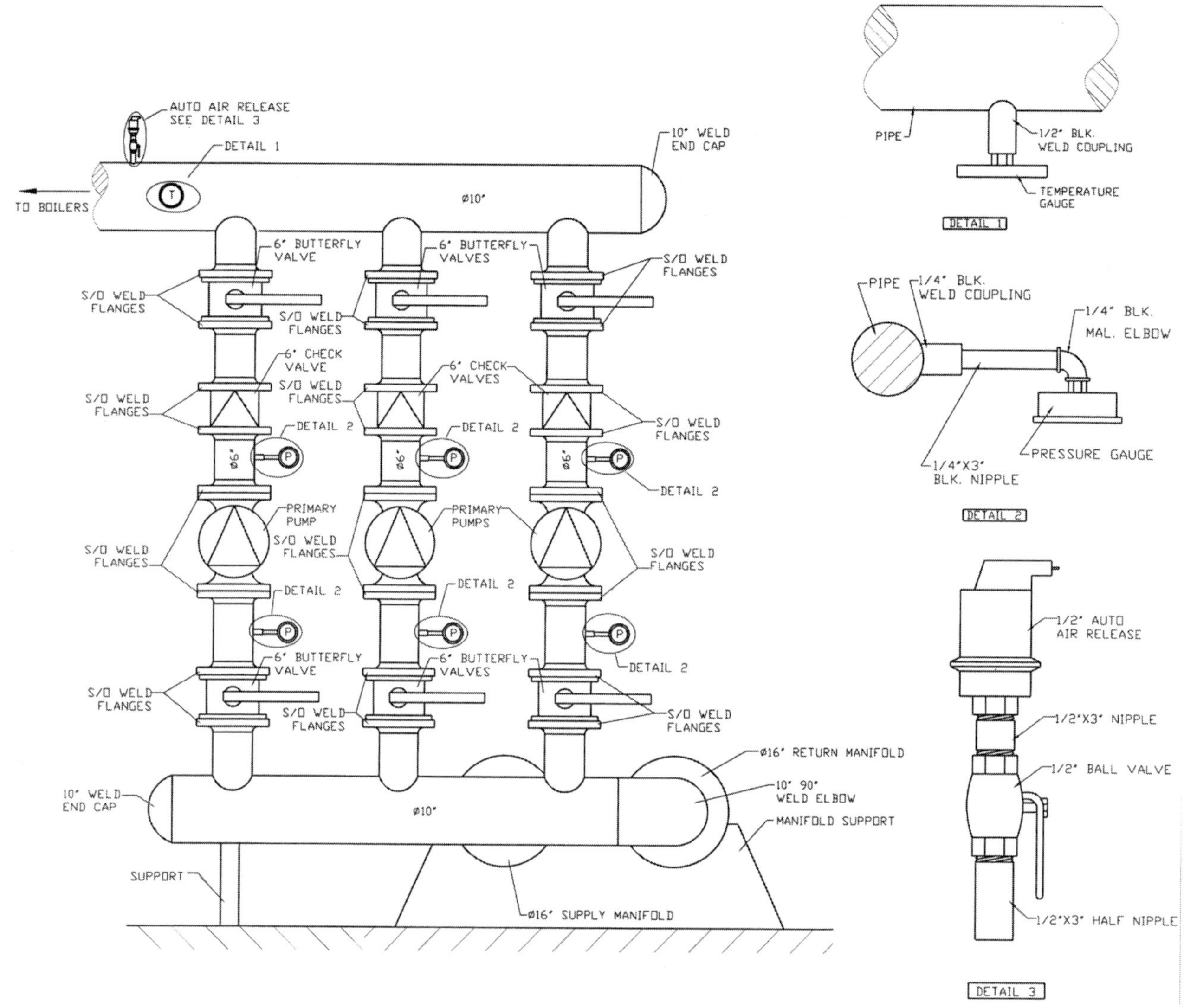

FIGURE 5.7 Connection details of a three-pump primary group.

reverse return principle discussed earlier in this chapter. Remember that the basic principle is last in, first out (see Figures 5.4 and 5.7). The boiler main lines can be designed using the methods introduced in Section 2.4 in Chapter 2.

5.3.4.1.4 The Primary Loop

Most boilers require continuous flow over, and this is provided by the primary loop system. Circulation pumps are installed in the main return line, and they keep circulating the water through the boilers and the supply and return manifolds, which are typically connected with a check valve to prevent backflow in the case of a pump failure. The loop is equipped with temperature and pressure gauges along with an automated air release system installed at the loop's highest point to release the air within the loop. If one of the circulation pumps fails, then the other pump will circulate enough water so that the greenhouse can still be provided with some heat in order to prevent the crops from freezing. Figure 5.7 shows the connection details of a three-pump primary group.

The capacity of the pump is calculated using Equation (2.24) and keeping in mind that the hot water boiler typically delivers 95°C (203°F) hot water with a return temperature of 80°C (176°F), that is, ΔT of 15°C (27°F). After determining the pump capacity in L/h (or in gpm), the boiler main line pipe sizes and fittings and therefore, as stated, the resistances are determined, and the main line is designed. It should be kept in mind that the main line resistance includes not only the pipe, valve, and fitting resistances but also the boiler's internal resistance.

Boiler manufacturers typically provide the required resistance (head loss) information of the boiler; however, if the information is not provided, the experience has shown that a boiler head loss of 750 mm H_2O (2.5 ft H_2O) can be assumed and used safely, and it should be added to the line resistance in the primary loop. The selected circulation pump should obviously be able to deliver the required volumetric capacity (L/h or gpm) at the designed head (mm H_2O or ft H_2O). This author gives an example of a real-world step-by-step primary-loop design in his *Greenhouse Engineering* (Yildiz, 2021).

5.3.4.1.5 *The Transport or Mixing Group*

As shown in Figure 5.8, insulated and normally jacketed pipes and manifolds deliver the hot water to the greenhouse zone being heated. The mixing group connected to the manifold adjusts the water temperature based on the need by mixing the supply water (typically 90°C or 194°F) with the return water (typically 75°C or 167°F) with ΔT of 15°C (27°F). If the heating zone calls for more heat, then the supply water is mixed with less return water, and if the zone requires less heat, then the supply water is mixed with more return water coming at a lower temperature.

The transport pumps circulate the hot water through the main transport lines in the controlled environment and brings back to the mixing group located on the manifold. Figure 5.8 shows all the connection details of a three-pump transport group, each connected to the return line with a dedicated three-way mixing valve. As discussed in Chapter 2, the mixing valve is sized using Equation (2.216) or Equation (2.218) for I-P or SI, respectively, and the relationship between the flow coefficients C_v and K_v is given by Equation (2.220). To determine the flow coefficient for the mixing valve, the flow rate (in proper units) and a pressure drop of 1.45 psi (or ~1,020 mm H_2O) are used, and the mixing valve is sized based on the calculated flow coefficient (C_v or K_v). Then, using the flow coefficient of the actual mixing valve, the actual pressure loss across the mixing valve is calculated and used in determining the transport pump's head. This author gives an actual design in *Greenhouse Engineering* (Yildiz, 2021).

The capacity of the transport pump is calculated using Equation (2.24) and typically a supply water temperature of 90°C (194°F) with a return water temperature of 75°C (167°F). After determining the pump capacity either in L/h or in gpm, the main transport line pipe sizes and fittings, and therefore the resistances, are determined, and the transport line is designed. In designing the transport mains, the maximum allowable flow rates for the steel pipes are used, and the head losses are determined accordingly. To take the special resistance into account, typically 15% resistance added to the total line resistance excluding the mixing valve resistance. An actual transport design will be shown later in this chapter. The author gives a real-world example in *Greenhouse Engineering* (Yildiz, 2021).

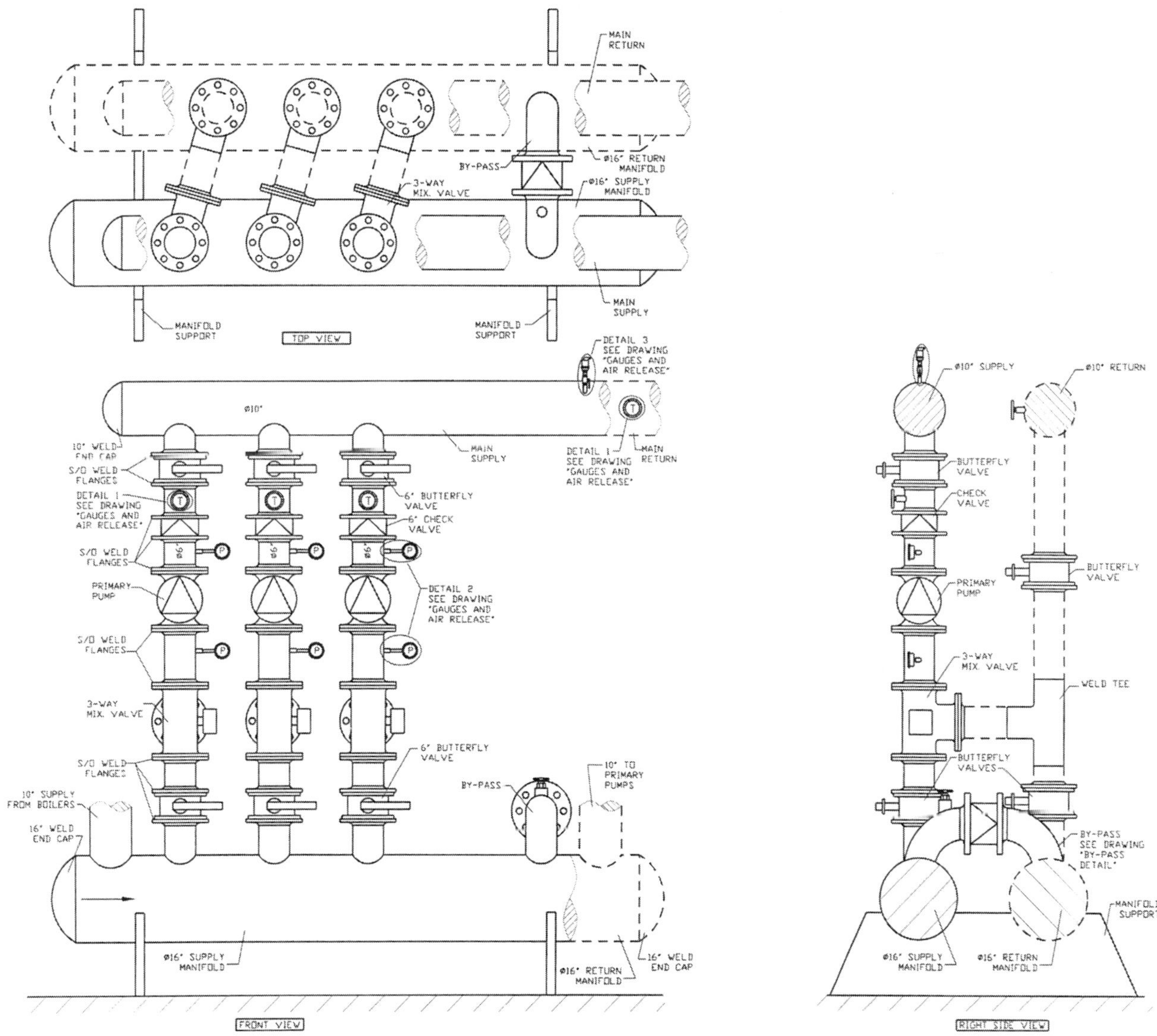

FIGURE 5.8 Connection details of a three-pump transport group.

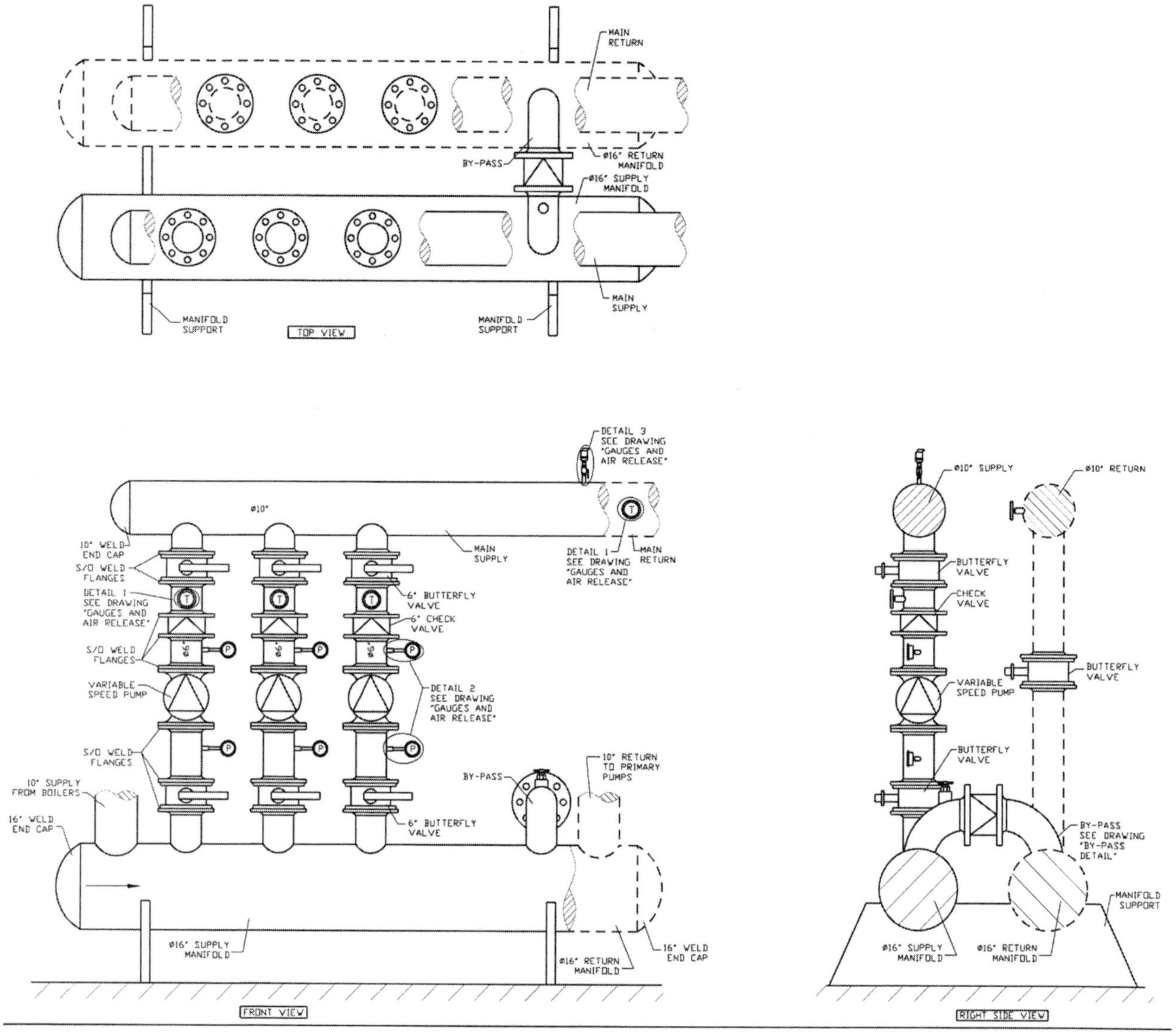

FIGURE 5.9 Connection details of a frequency-controlled variable speed three-pump transport system.

Figure 5.9 shows the connection details of a frequency-controlled variable speed three-pump transport system. The design procedure is the same as in the previous case except for the mixing valve. No mixing valve is needed or used in a frequency-controlled transport line circuit; therefore, the resistance or the pump head is now smaller. The benefits of using a frequency-controlled variable speed pumps are discussed with numerical examples later in Chapter 8.

5.3.4.1.6 Expansion System

The expansion and contraction of the water in a hot water heating system is controlled by the expansion vessel. When the system is heated, the expansion vessel stores the increased volume of the system's water content and prevents the piping system from bursting. When it cools and shrinks down, the stored water is sent back to the system. The size and location of the expansion vessel are important. The mass of the water in the installation does not change; however, the volume of the water increases as temperature increases, that is, the density of the water decreases. Table 5.2 lists the density of water at different temperatures. See Figure 5.10 for connection details of three-tank membrane type expansion vessels and auto fill system.

As discussed earlier in Chapter 2, water has its maximum density at 4°C or 39.2°F (ρ = 1,000 kg/m^3). When water is heated from 4 to 90°C, the volume of water increases by 1000/965 = 1.036 times with 3.6% expansion. In practice, 4% to 5% expansion is used.

Expansion systems are typically categorized under six different types: i) open expansion vessel; ii) expansion vessel with nitrogen cushion of varying pressure; iii) expansion vessel with constant pressure nitrogen cushion; iv) open pump-expansion vessel based on the overflow principle; v) closed pump-expansion vessel with pressure-controlled pump; and vi) membrane expansion vessel.

An open expansion vessel openly communicates with the environment. A secondary function of the open expansion vessel is to protect the heating system against overpressure. The vessel is connected to the hydronic heating system by a non-closable safety pipe. Excess water is discharged outside through an overflow. A glass gauge indicates the water level in the vessel, and the position of the vessel determines the pressure in the

TABLE 5.2
Density of Water as a Function of Temperature

Temperature °C (°F)	**4 (39.2)**	**10 (50)**	**20 (68)**	**30 (86)**	**40 (104)**	**50 (122)**	**60 (140)**	**70 (158)**	**80 (176)**	**90 (194)**
Density kg/m^3 (lb/ft^3)	1,000 (62.43)	999.8 (62.42)	998.3 (62.32)	995.8 (62.17)	992.3 (61.95)	988.1 (61.69)	983.2 (61.38)	977.7 (61.04)	971.4 (60.64)	965.1 (60.25)

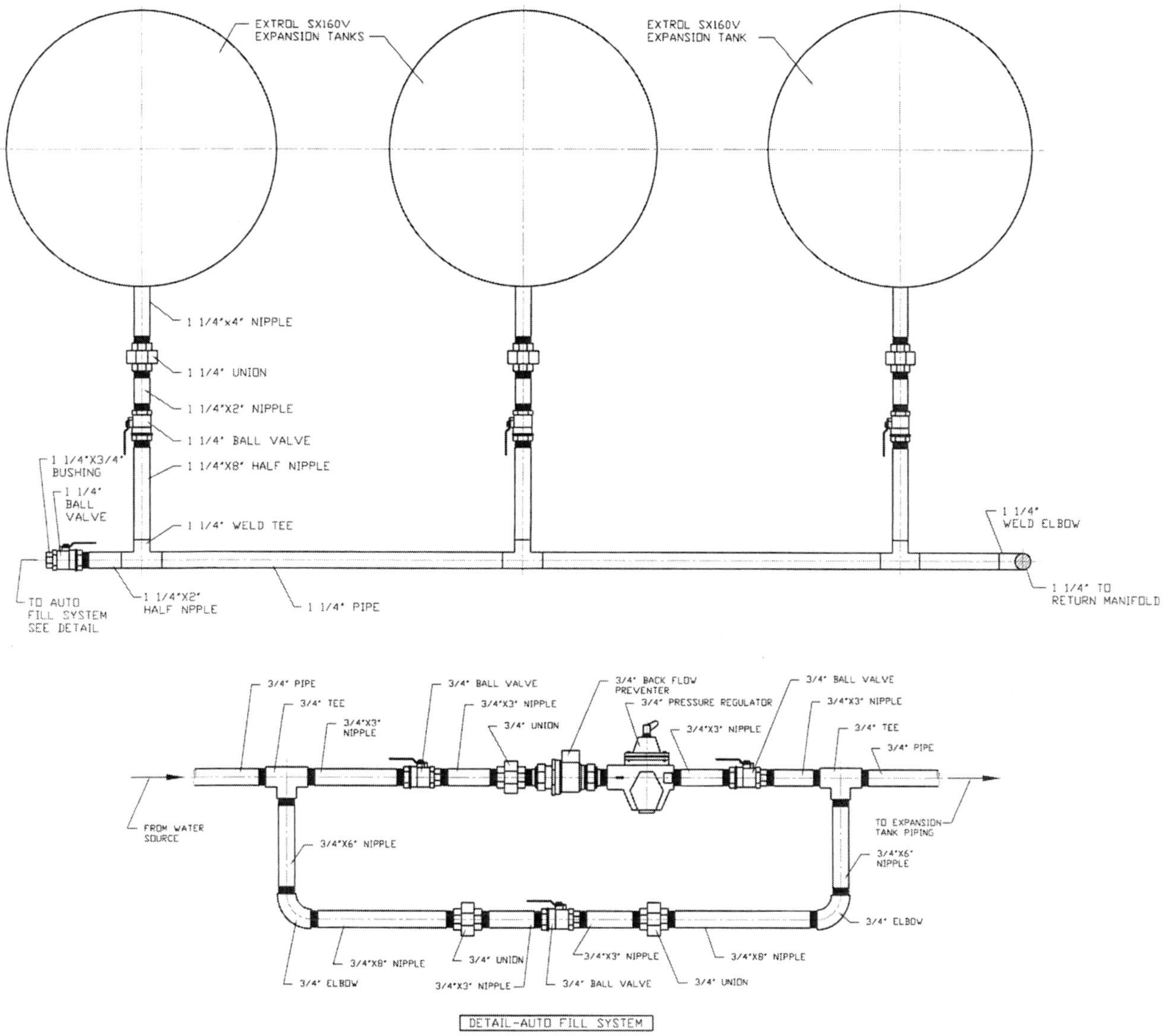

FIGURE 5.10 Connection details of three-tank membrane type expansion vessels and autofill system.

installation. Obviously, the pressure variations are caused by the varying water level in the vessel, and the higher the vessel is placed, the higher the static pressure in the system.

When the boiler room height is considered, installing this kind of expansion vessel can be problematic. Therefore, the open expansion vessels are installed horizontally, creating a very large water-air separation surface. It means that the water is in contact with air; hence, the water absorbs plenty of air, and the absorbed oxygen leads to corrosion in the hydraulic system. Unless it is released properly, the air accumulates at the highest point of a plumbing section and partially blocks the circulation.

In the expansion vessel with nitrogen cushion of varying pressure, the nitrogen volume above the water surface is compressed as the water expands, and when the water shrinks, the nitrogen cushion expands. The pressure of the nitrogen cushion and therefore the heating system pressure vary between 0.5 and 2 bar. The lost nitrogen is supplemented from a permanently connected nitrogen bottle.

In the expansion vessel with constant pressure nitrogen cushion, when the system's water is warmed up, the water expands, the water level in the vessel rises, and the nitrogen pressure increases. The compressor starts running and presses the excess nitrogen back into the expansion vessel.

The maximum nitrogen pressure in the storage tank is 5 bar. When the water in the system cools down, the water shrinks and the pressure in the expansion vessel drops. An on/off control valve installed between the expansion valve and the nitrogen supply vessel is opened, and the nitrogen pressure in the expansion vessel increases until the pressure goes back to the set point pressure. This expansion system includes a nitrogen bottle as well to fill the expansion system and compensate for any losses.

The open pump expansion vessel based on the overflow principle stores and replenishes the water in a hot water heating system. The vessel is connected to the heating system via a pressure booster pump with a pressure regulator. Above the vessel, there exists air at about atmospheric pressure. If the pressure is too high, the relief valve discharges the air, and a vacuum valve prevents the observance of underpressure in the vessel. A continuously running booster pump creates the overpressure in the system, and the pressure increase is set and controlled by a pressure regulator. This regulator is incorporated in a shunt or overflow pipe and maintains the necessary continuous flow through the overflow pump. In this system, provisions must be made to ensure that the booster pump remains in continuous operation.

Instead of a continuously operating pressure boosting pump and a safety valve, in a closed-pump expansion vessel with pressure-controlled pump, the pump and solenoid valve are controlled by a pressure switch. When the pressure in the heating system increases, the solenoid valve opens and the water flows to the expansion vessel. When the system pressure drops, this time the water is pumped from the expansion vessel to the hot water heating system.

In a membrane or diaphragm type expansion tank, a membrane separates the water and the compressed air that accommodates the expanding water. When the system's water expands, the excess water moves into the membrane and the air in the tank is compressed; when it shrinks during the cooling process, the air volume in the closed section expands.

The content of a membrane type expansion reservoir can be determined as follows: If the content of the expansion reservoir is expressed as I, ΔV is the maximum volume increase of water due to expansion, and the minimum and maximum pressures in the system are p_{min} and p_{max} (absolute pressures), respectively; then from Boyle's law, we have the following relationship for the air volume:

$$I \times p_{min} = (1 - \Delta V) \times p_{max} \tag{5.17}$$

Then,

$$I = \Delta V \left(\frac{p_{max}}{p_{max} - p_{min}} \right) \tag{5.18}$$

The expansion vessels are located on the return side (see Figures 5.4 and 5.10) of the hydronic heating system so that the whole system can operate at or above the system pressure. Hydronic systems typically operate at a pressure of 12–15 psi (0.83–1.03 bar), which is controlled by expansion vessels.

Example 5.4: Determining the Size of a Membrane-Type Expansion Tank

Determine the size of a membrane-type expansion vessel if the water volume increase or expansion is 2,000 liters.

Solution

Using Equation (5.18), we can easily determine the tank size as follows:

The initial and final pressures are 0.5 bar (1.5 bar absolute) and 3 bar (= 4 bar absolute), respectively. Then the content of the expansion tank becomes

$$I = (2{,}000 \text{ L}) \times \left(\frac{4}{4 - 1.5} \right) = 3{,}200 \text{ L}$$

5.3.4.1.7 Boiler Room Air Supply

As discussed in Chapter 2, for complete combustion and safety, boiler rooms must be provided with proper openings so that enough fresh air is supplied. Temporary air intake openings such as doors and windows should be avoided as they may be closed and hence cut off the required air supply. From experience, the ASME states that 1 square inch free area per 2,000 Btu/h maximum fuel input of combined burners located in the boiler room works, and a minimum of two air supply openings must be provided and kept clear at all times. It should also be kept in mind that inlet louvers typically have 60% free area; therefore, the actual (gross) louver area per unit heater will be larger. Properly designed motorized inlet shutters will do the job perfectly fine, and an actual design example will be shown later in this chapter.

5.3.4.2 Heat Distribution Systems

As mentioned earlier, the insulated transport mains and manifolds deliver the hot water to the controlled environment zone that is being heated. The heat distribution systems in greenhouses, for instance, are normally thin-walled 51 mm OD pipe rails and 33 mm OD grow pipes for the row vegetable crops or cut flowers; 51 mm OD underbench pipes for seed germination, propagation, or general growth on the bench tops; 25 mm OD (1 in.) in-floor or on-floor PE tubes for the plants in containers directly on the floor; 51 mm OD overhead and snowmelt pipes, if needed; and again 51 mm OD pipe sidewall and gable end heating systems.

In supplementally heated livestock and poultry buildings, on the other hand, unit heaters are usually used for space heating. For large operations and when multiple purpose heating is required, then central heating systems can also be used in such situations. Again, there exists a boiler room and main transport (supply and return) lines as discussed earlier. Depending on the heating zone characteristics, one or more heat distribution forms can be employed including hot water unit heaters (see Figure 5.3) directly hooked up to the boiler or thin-walled 51 mm OD overhead and snowmelt (in greenhouse type livestock and poultry buildings) pipes, if needed; 51 mm OD pipe sidewall and gable end heating systems; and/or 25 mm OD (1 in.) in-floor PE tubes.

Furthermore, if central hot water heating system is used in an aquatic building, the heat distribution system there consists of either hot water unit heaters (see Figure 5.3) again directly hooked up to the boiler or thin-walled 51 mm OD pipe overhead and snowmelt (in greenhouse type aquatic buildings) systems, if needed; 51 mm OD pipe sidewall and gable end heating systems; and/or 25 mm OD (1 in.) in-floor PE tubes. It should be kept in mind that the zone requirements might be different for each zone in any of the controlled environment buildings; however, design procedures are exactly the same whether the building is a greenhouse or a livestock, poultry, or aquatic building.

In each of these cases, the temperature of the circulating water is controlled via a zone mixing valve to match the heat requirement of a particular heating zone. What is critical here is that the heating zones are typically determined by the ventilation zones; that is, the number of ventilation zones determines the number of heating zones. Hot water heating systems, when engineered and installed properly, provide even heat distribution, which results in more uniform temperature.

Again, when designed and installed properly, observing both the energy and mass balances, maximum spatial (horizontal) temperature difference observed even in large controlled environments such as greenhouses is approximately 1°C. To keep the spatial temperature differences at minimal levels, the building heating zones are designed accordingly. For instance, in small applications, if the length or width of a greenhouse section to be heated is less than 100 m, a single heating zone is used, where the mixing groups and transport lines are directly connected to the manifold in the boiler room.

In contrast, in large applications, if the length or width is greater than 100 m, typically the section is split into more heating zones, where the main transport line is connected to the pre-mix group in the boiler room and the zone mixing groups are typically connected to the main transport lines within the greenhouse space. Other considerations are also taken into account for uniform heat distribution within a controlled environment. For instance, aluminum painted pipes typically emit 30% less heat than black pipes; vertical pipes emit 50% less heat than same-size horizontal pipes; and plastic hoses half horizontal plus half vertical emit 3/4th of a horizontal plastic hose. Considering all these fine differences and taking advantage of them, we can design a very well-functioning gable end or sidewall heating system.

5.3.4.2.1 Pipe Rail Heating Systems

As mentioned, rail pipe heating systems are used quite effectively for greenhouse row vegetable crops or row cut flowers. To distribute heat in these systems, thin-walled 51 mm OD pipe loops are used (Figure 5.11). Depending on the bay width, four, five, or six loops of 51 mm OD pipes are installed. The rail loops start 0.7 m (2.3 ft) from the gable ends, run to the walkway, and are connected to the distribution headers along the gable end via 1.5 to 2 m (5 to 6 ½ ft) 19 mm or 25 mm ID black hot water heating hoses (Figures 5.11).

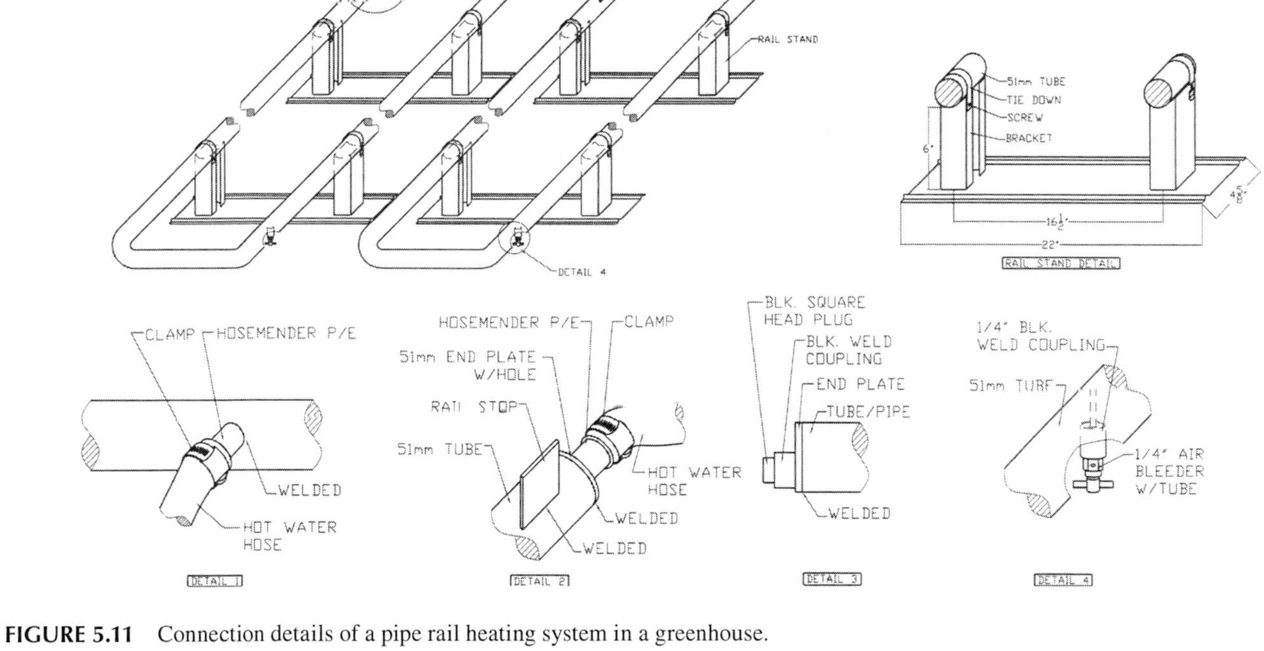

FIGURE 5.11 Connection details of a pipe rail heating system in a greenhouse.

Source: *Greenhouse Engineering* (2021).

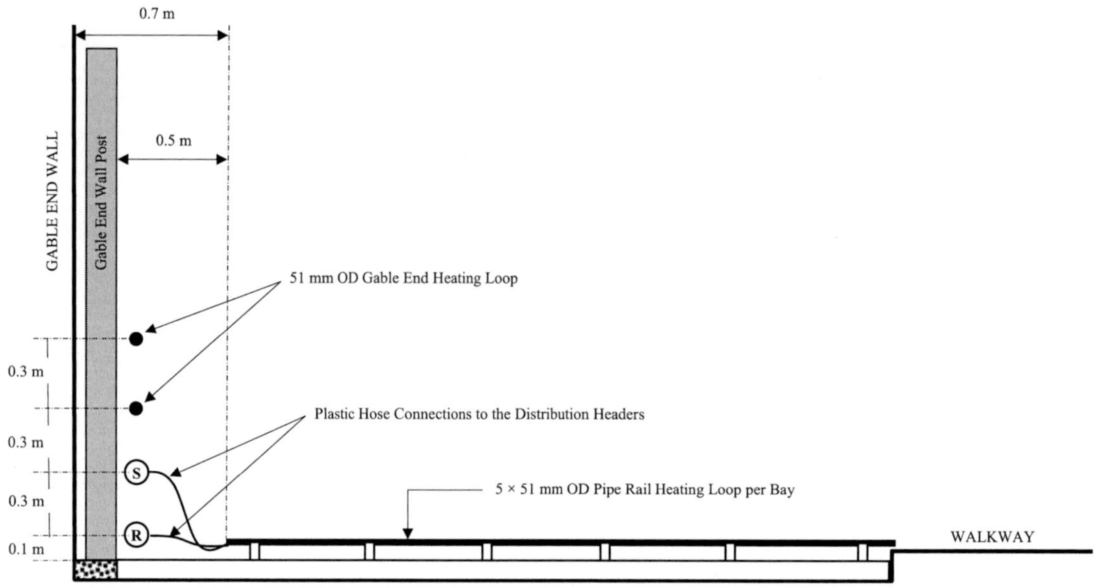

FIGURE 5.12 Distribution headers and the pipe rail heating loop (side view).
Source: *Greenhouse Engineering* (2021).

If the water flow in the loop is less than 800 L/h (3.5 gpm), a 19 mm ID black hot water hose is used, whereas a 25 mm ID black hot water hose is used if the flow is greater than 800 L/h (3.5 gpm). Each pipe rail loop has a return bend, and the older versions have 42 cm c.t.c. (50 cm total) widths; newer ones have 55 cm c.t.c. (60 cm total) widths. Each loop is supplied with an air bleeder with tube, and each distribution header is supplied with a regular air bleeder at the highest points (Figures 5.11). The distribution headers are designed according to the reverse return principle with a maximum flow resistance of 10 mm H_2O (0.033 ft H_2O) and connected to a rail pipe heating zone mixing group as discussed earlier. The average pipe size of the distribution header, whether it is supply or return, is 0.8 times the maximum pipe size in the distribution line. See Example 5.6 in *Greenhouse Engineering* (Yildiz, 2021) for a complete real-world rail heating system design example.

With respect to the heat loss from the greenhouse, the heat loss from the space 0.7 m (2.3 ft) from the gable end wall in the bay is treated as part of the heat loss from the gable end, and the heat demand is supplied by the distribution headers, the plastic hose connections, and the additional 51 mm OD pipes, if needed (see Figure 5.12). Maximum pipe rail heating system temperature is typically 90°C (194°F), and the return temperature is 75°C (167°F).

To obtain these system temperatures, the flow temperatures to/from the pipe rail heating zone mixing group are 91°C (~196°F) and 74°C (~165°F) for supply and return, respectively. The maximum flow resistance used in the heat distribution lines as well as in the loops is 10 mm H_2O (0.033 ft H_2O), and a minimum of 250 mm H_2O (0.83 ft H_2O) loop resistance from the supply header to return header is typically used over the 51 mm OD loop, including the connections. Special resistance is built in within this magnitude for the hose, the hose menders, the nipple, and the valve.

5.3.4.2.2 Grow Pipe Heating Systems

Grow or vegetation pipe heating systems are used with pipe rail heating systems to eliminate excess moisture within the canopy of row crops and minimize disease risks. The grow pipes not only eliminate excess moisture but contribute a modest amount of heat around the plants at lower temperatures than those of the rails. The heat released by the connecting arching pipes and the plastic hoses of grow pipe heating can also be used to cover the deficiency of the rail pipe heating along the walkways; therefore, the need for on-ground 51 mm OD heating along the walkway on both sides is reduced. Depending on the climate, sometimes one or two lengths of 51 mm OD pipe on each side of the walkway are required for balanced heating along the walkways as well.

As shown in Figures 5.13 and 5.14, grow pipe heating systems have separate distribution headers along the gable end walls and therefore separate mixing groups as well. A total of 2 ½ loops (5 × 33 mm OD, 1 in. or ¾ in. pipes) per bay, that is, one pipe length per pipe rail loop, are used. The pipes are installed about 2.1 m (7 ft) above the floor and 0.3 m (1 ft) below the plant level.

Just like pipe rail loops, the grow pipe loops start 0.7 m (2.3 ft) from the gable ends and run to the walkway and are connected to the distribution headers along the gable end via 1.5 to 2 m (5 to 6 ½ ft) 19 mm or 25 mm ID black hot water heating hoses (see Figures 5.13). As with pipe rail heating, for connecting to the distribution headers, a 19 mm ID black hot water heating hose is used if the water flow in the grow pipe loop is less than 800 L/h (3.5 gpm), whereas a 25 mm ID black hot water heating hose is used if the flow is greater than 800 L/h (3.5 gpm). A same-size pipe is used for the arch connecting the 33 mm OD, 1 in. or ¾ in. pipe lengths via 2.5 m long 19 mm or 25 mm OD black hot water heating hoses on each side in between (see Figure 5.14), and each loop and distribution header is supplied with a regular air bleeder at the highest points.

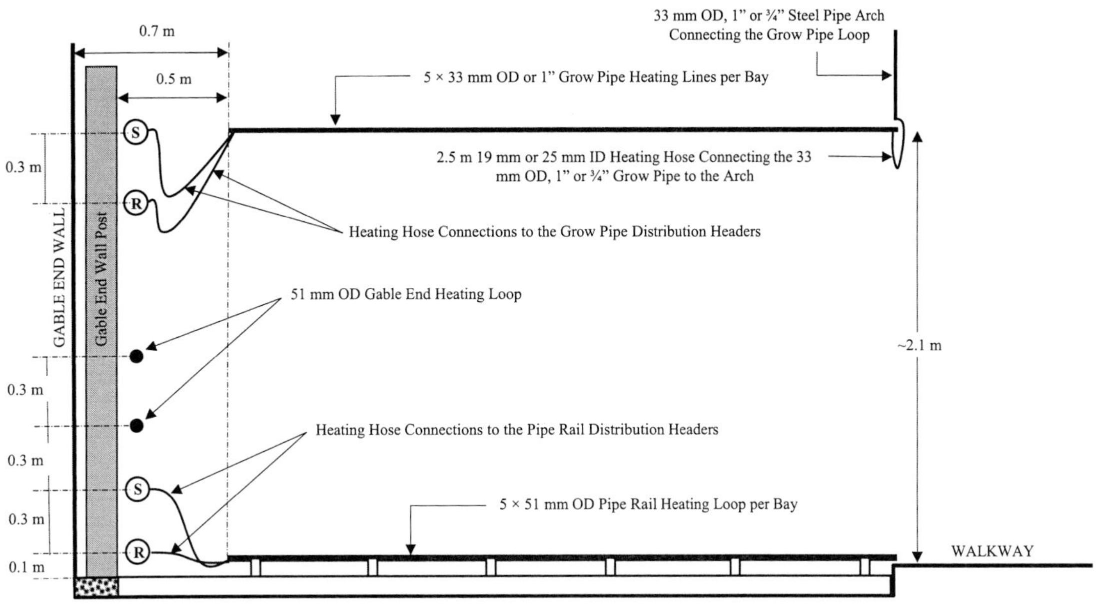

FIGURE 5.13 Distribution headers and the grow pipe heating loop along with the pipe rail (side view) (Yildiz, 2021).

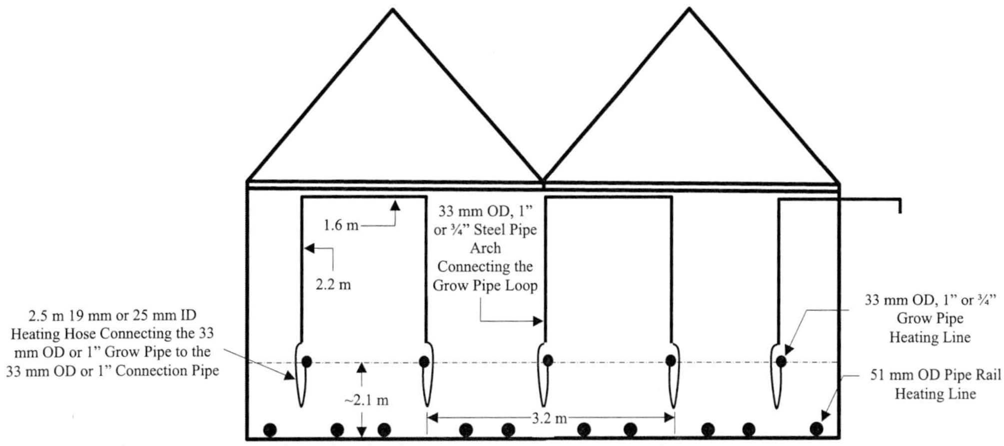

FIGURE 5.14 2 ½ loops (5 pipes) per bay grow pipe heating connections along with 5 x 51 mm OD pipe rail loops per bay (view from the walkway).

Source: ***Greenhouse Engineering*** **(2021).**

The distribution headers are designed according to the reverse return principle with a maximum flow resistance of 10 mm (0.033 ft) H_2O and connected to a grow pipe heating zone mixing group. The average pipe size of the distribution header, whether it is supply or return, is 0.8 times the maximum pipe size in the distribution line. The maximum grow pipe heating system temperature used is typically 70°C (158°F), whereas the return temperature is 55°C (131°F). To obtain these system temperatures, the flow temperatures to the grow pipe heating zone mixing group are 71°C (~160°F) and 54°C (~129 °F) for supply and return, respectively. See Example 5.7 in *Greenhouse Engineering* (Yildiz, 2021) for a complete grow pipe heating system design example.

Since the horizontal pipe has 100% of the horizontal pipe heat emission, this rate should be used for the 1.6 m horizontal section, and the vertical pipe has 50% of the horizontal pipe emission, so 50% of the horizontal pipe emission should be used for the two 2.2 m vertical pipes. Since the total pipe length over the arch is 6 m, the heat emission factor for the arch is calculated as follows:

$$\text{Factor for the Arch} = \left(\frac{1.6\,\text{m} \times 100/100 + 2 \times 2.2\,\text{m} \times 50/100}{6\,\text{m}}\right)$$

$$= 0.63$$

5.3.4.2.3 Underbench Heating Systems

As mentioned earlier in this section, thin-walled 51 mm OD pipe underbench heating systems are used for seed germination, propagation, or general growth on bench tops (see Figures 5.15). Four or five loops of 51 mm OD pipes per bay are installed under benches depending on the bay width and the need along with two or three lengths of 51 mm OD pipes. As in pipe rail heating, the underbench loops start 0.7 m (2.3 ft) from the gable ends and run to the walkway and are connected to the distribution headers along the gable end via 1.5 to 2 m (5 to 6 ½ ft) 19 mm or 25 mm ID black hot water heating hoses (see Figures 5.15).

If the water flow in the loop is less than 800 L/h (3.5 gpm), a 19 mm ID plastic hose is used, whereas a 25 mm ID plastic hose is used if the flow is greater than 800 L/h (3.5 gpm). Each underbench loop has a return bend and ball vales for supply and return lines, and each loop and distribution header is supplied with a regular air bleeder at the highest point. Underbench heating pipes (51 mm OD) emit 20% less heat than the same size uncovered horizontal pipes; therefore, just like pipe rail and grow pipe heating, it is vital for underbench heating systems to have an engineered design for adequate performance. The distribution headers are designed according to the reverse return principle with a maximum flow resistance of 10 mm H_2O (0.033 ft H_2O) throughout and connected to a dedicated underbench heating zone mixing group. The average pipe size of the distribution header, whether it is supply or return, is again 0.8 times the maximum pipe size in the distribution line.

5.3.4.2.4 (Low-Grade) Floor Heating Systems

Floor heating systems typically use in-floor or on-ground plastic hot water tubes or electric resistance cables buried directly in the floor; electric floor heating costs more than hot water heating. Hot water in-floor or on-ground heating systems are very effective for plants in bags and containers grown on the floor or even for vegetable crops grown directly in the soil. To distribute heat in such systems, 25 mm OD (1 in.) thermoplastic tubes are used (see Figure 5.16).

For on-ground heating, 25 mm OD smooth polyethylene tubes with a wall thickness of 1.9 mm are used, while 25 mm OD smooth PE tubes with a wall thickness of 2.7 mm are used for in-floor heating whether it is in sand, soil, or concrete. Depending on the purpose, either 20 cm c.t.c. (8 in. c.t.c.) (for propagation and growth) or 30 cm c.t.c. (12 in. c.t.c.) (for packing sheds, pathways, etc.) tube spacing is used. Maximum loop lengths are maintained between 125 m (415 ft) and 150 m (500 ft). The floor heating loops start almost immediately, 0.2 m (~0.65 ft) from the gable ends and run to the walkway, and both supply and return lines are directly connected via ball valves to the distribution headers along the gable end (see Figure 5.16). Each PE tube is secured to a steel mesh with tie wraps every 30 cm (12 in.) for integrity and longevity. When concrete floors are used, to get uniform floor temperatures, a minimum concrete slab thickness of 10 cm (4 in.) is used, and the floor should be insulated around the perimeter and as deep as the front line with typically 5 cm (2 in.) extruded polystyrene board insulation.

A zone supply temperature of 35°C (95°F) and a return temperature of 30°C (86°C) are used providing a ΔT of 5°C

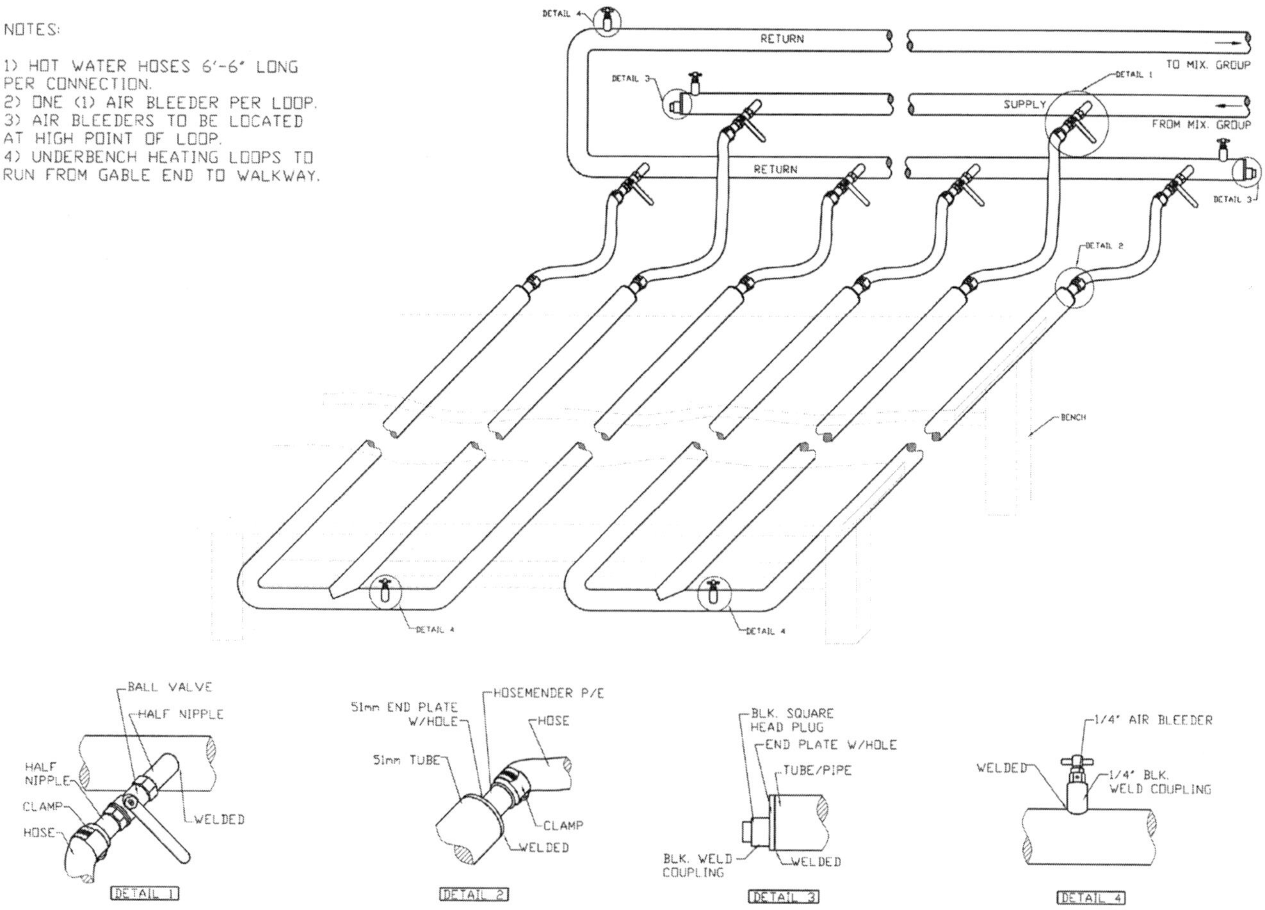

FIGURE 5.15 Connection details of a three-tube underbench heating system.

Source: *Greenhouse Engineering* (2021).

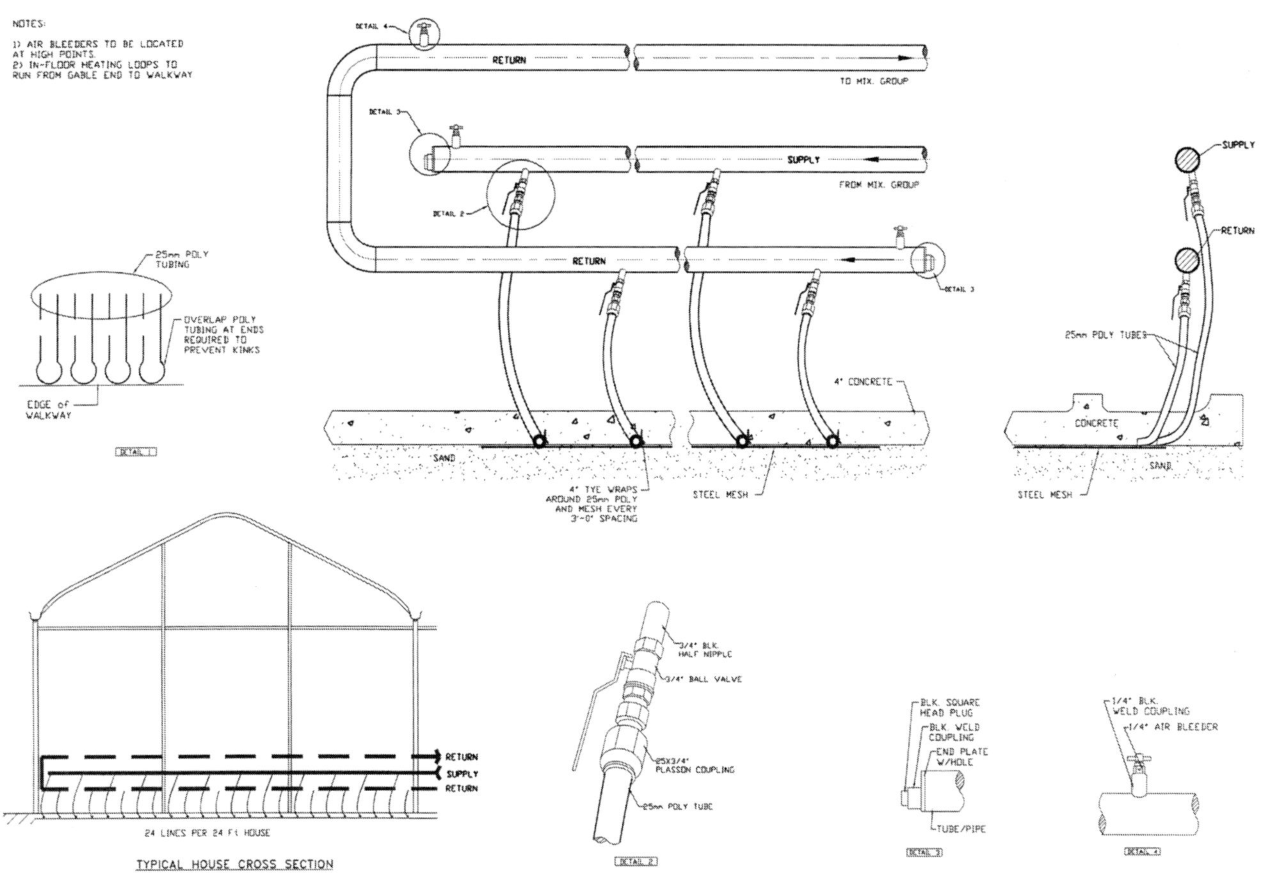

FIGURE 5.16 Connection details of in-floor heating system.
Source: ***Greenhouse Engineering*** **(2021).**

(9°F). The distribution headers are designed according to the reverse return principle with a maximum flow resistance of 10 mm (0.033 ft) H_2O and connected to a dedicated floor heating zone mixing group with a bypass. The average pipe size of the distribution header, whether it is supply or return, is 0.8 times the maximum pipe size in the distribution line. The bypass in the mixing group is essential because in the absence of the bypass, if the mixing valve is opened by hand or if the mixing valve actuator fails, then the water temperature in the PE tubes can exceed the critical temperature of 60°C (140°F) and melt the tubes (see Example 5.5 in *Greenhouse Engineering* (Yildiz, 2021)).

As an industry standard, typically 25 kcal/h.m^2 = 29 W/m^2 (100 Btu/h.m^2) floor and 50 kcal/h.m^2 = 58 W/m^2 (200 Btu/h.m^2) floor heating are provided for plant-covered floors and uncovered floors (sheds, walkways, etc.), respectively. During the PE tube installation, before they are covered with concrete, the tubing should be filled with water to check for and fix leakages, to make it more impact resistant, and to prevent it from floating. When designed properly, hot-water floor heating systems provide the most uniform heat distribution horizontally and within the crop canopy as well. Except for cool-season crops—lettuce, spinach, etc.—in addition to the floor heating, some type of supplemental heating will be required, and it is typically provided by overhead, gable end wall, and sidewall heating systems in greenhouses or greenhouse-type livestock, poultry, or aquatic buildings and by heat lamps or other radiant heaters in livestock or poultry buildings. See Example 5.8 in *Greenhouse Engineering* (Yildiz, 2021) for an in-floor heating system design example.

Floor heating is a convenient and commonly used heating system in livestock, poultry, and aquatic buildings not only in their production areas but also in the work areas and office spaces of such buildings. Again, the requirements for different zones in any of these buildings might be different, however, the design procedures are exactly the same for every custom design as shown in Examples 5.5 and 5.6 in *Greenhouse Engineering* (Yildiz, 2021).

5.3.4.2.5 Overhead and Snowmelt Heating Systems

In overhead heating systems, just like in pipe rail and underbench heating systems, thin-walled 51 mm OD pipes are used to compensate for deficiencies in pipe rail, underbench, and floor heating systems whether it is in a greenhouse, or a greenhouse type livestock, poultry or aquatic building. The number of 51 mm OD pipe lengths per bay is a function of the climate and the type of greenhouse installed.

The OD pipes are installed in trusses and run from gable end to gable end or from supply into return from alternating sides, or sometimes they are installed in loops and run from gable end to gable end from the supply header on gable end to the return header on the same gable end using a return bend. Overhead heating lengths start 0.15 m (6 in.) from the gable ends, run to the other gable end, and are connected to the distribution headers along the gable end via an average of 3.5 m (~12 ft) 19 mm or 25 mm ID clamped black hot

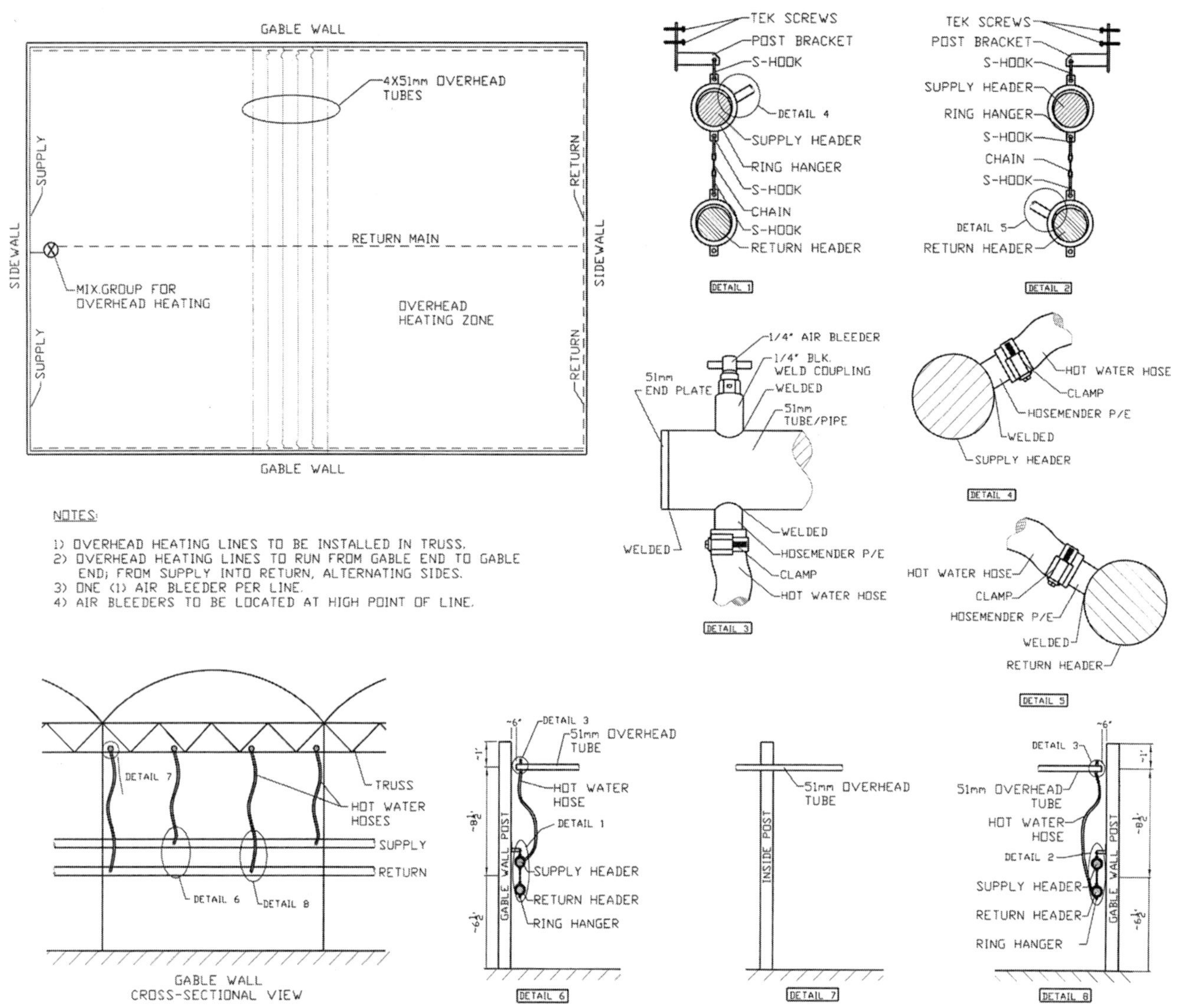

FIGURE 5.17 Connection details of a plastic hose connected overhead heating system.
Source: ***Greenhouse Engineering*** **(2021).**

water heating hoses or 27 mm (¾ in.) welded steel pipes (see Figure 5.18).

If the water flow in the loop is less than 800 L/h (3.5 gpm), a 19 mm ID black hot water heating hose is used, whereas a 25 mm ID black hose is used if the flow is greater than 800 L/h (3.5 gpm). Each 51 mm OD pipe length is supplied with a regular air bleeder at the highest point (see Figures 5.18). The distribution headers are designed according to the reverse return principle with a maximum flow resistance of 10 mm H_2O (0.033 ft) H_2O throughout and connected to a dedicated overhead heating zone mixing group (see Figure 5.20). The average pipe size of the distribution header, whether it is supply or return, is again 0.8 times the maximum pipe size in the distribution line. See Example 5.8 in *Greenhouse Engineering* (Yildiz, 2021) for a combined overhead and in-floor heating system design example.

In addition to the overhead heating, snowmelt systems are installed in greenhouse-like controlled-environment buildings as well at locations where considerable snow loads are expected. Typically, two lengths of thin-walled 51 mm OD pipes per bay are used for snow melting; they are installed immediately under gutters and rest on trusses and run from gable end to gable end or from supply into return; sometimes they are installed in loops and run from gable end to gable end from the supply header on gable end to the return header on the same gable end using a return bend.

The lines are connected to the headers via plastic hoses or 27 mm (¾ in.) welded steel pipes. Each OD pipe length is supplied with a regular air bleeder at the highest point. The distribution headers are designed according to the reverse return principle with a maximum flow resistance of 10 mm (0.033 ft) H_2O throughout and connected to a dedicated snowmelt zone mixing group, which is a separate mixing group from the overhead heating and provides further flexibility. The average pipe size of the distribution header, whether it is supply or return, is again 0.8 times the maximum pipe size in the distribution line.

5.3.4.2.6 Sidewall and Gable End Heating Systems

In sidewall and gable end heating, just like in pipe rail and underbench heating, thin-walled 51 mm OD pipes are used when there is still a deficiency after heating provided by the distribution headers.

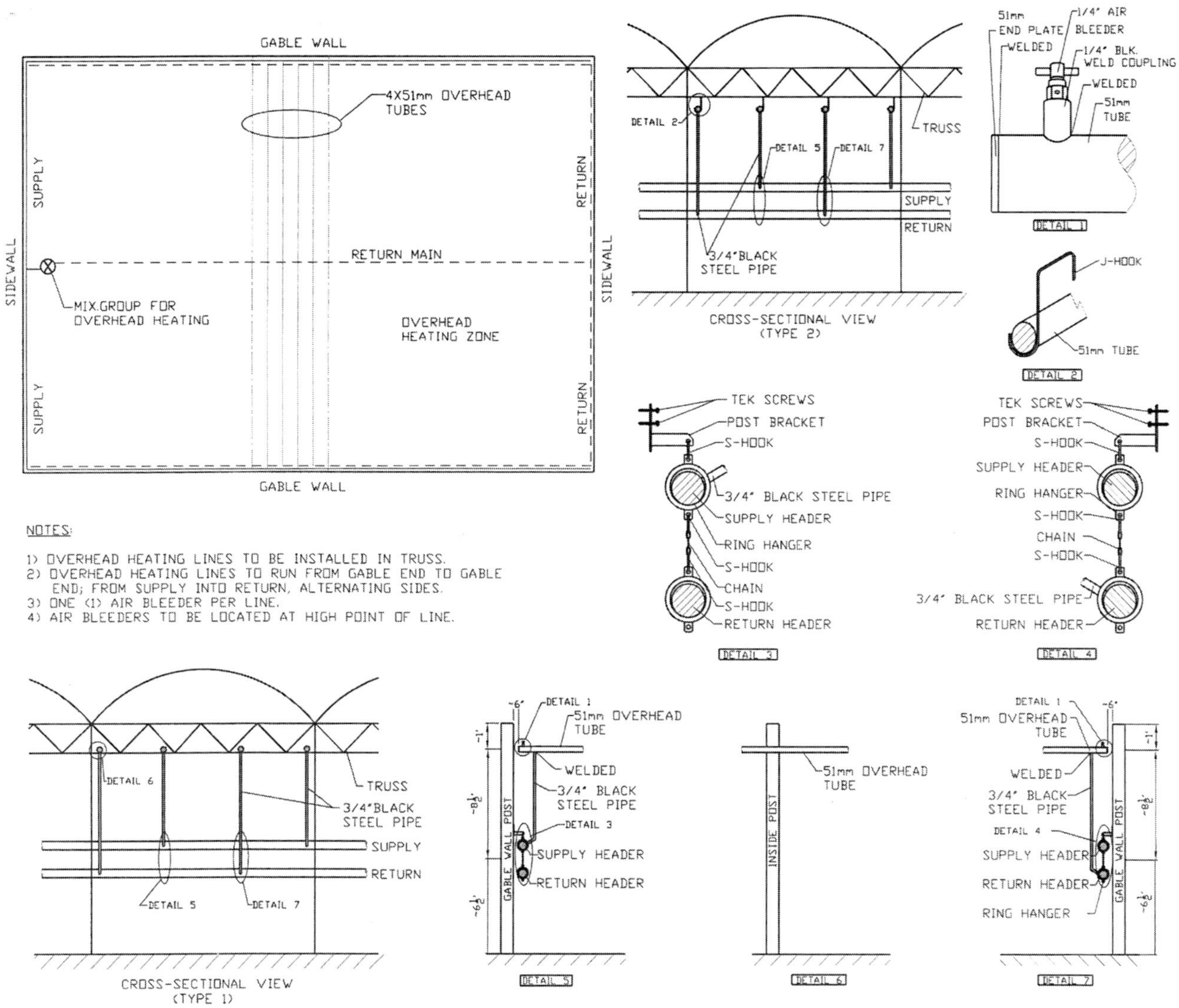

FIGURE 5.18 Connection details of a steel pipe connected overhead heating system.
Source: ***Greenhouse Engineering*** **(2021).**

The number of 51 mm OD pipe lengths per sidewall or gable end is a function of the climate and the type of building installed. The pipes run from the supply header on the gable end to the return header on the same gable end. Sidewall or gable end heating lengths start almost immediately 0.15 m (6 in.) from the gable end or the sidewall, respectively, run along the gable end or sidewall, and are connected to the distribution headers, typically via 19 mm or 25 mm ID clamped black hot water heating hoses or welded steel pipes. They are typically connected to the main hot water heating zone mixing group, that is, the pipe rail, underbench, or overhead heating in greenhouse-like controlled-environment buildings.

5.3.5 Radiant Heating Systems

As has been shown repeatedly, conventional controlled-environment building heating systems first heat the space (air), and then the heated air heats the plants or animals. Many types of radiant heaters are available today, from heat lamps to low infrared heaters and sophisticated fan-driven pipe units.

A low infrared heater, for instance, operates with heated pipes—up to 595°C (1100°F) for piglets—and first heats the objects in the building (including crops, animals, floor, or surfaces) directly through infrared radiation heat exchange; then the objects exchange this heat with the environment through conduction, convection, and again radiation heat transfer. As a result, the energy is absorbed, and the leaf, animal, or other surface is warmed. Infrared heaters typically emit 70% of their heat by radiation and the remaining 30% by convection. As mentioned, the radiation component directly heats the plants or animals as well as the floor surface while the remaining convection heats the air. That's why the temperature of the interior air is typically lower, and this low inside air temperature results in lower heat losses from the controlled environment. However, radiant heating creates less uniform thermal environments than properly designed space heating systems.

Infrared heaters eliminate the pumping energy requirements and the heat losses through distribution lines typically seen in conventional heating systems; when installed properly, infrared heating can provide good enough uniform surface temperatures, particularly plant and vertical air temperatures in greenhouses, and are a cost-effective option for controlled-environment heating. Radiant heating and other space heating systems can be combined used together or coupled particularly with the floor heating systems covered earlier in this section to improve the indoor thermal environments; however, this can lead to increased evaporation from wet floor surfaces, particularly in livestock buildings, and a rise in the indoor air moisture content and therefore ventilation requirements.

5.3.6 Other Heating Systems

In addition to the heating systems already discussed, other systems are available for controlled-environment heating such as air make-up heaters, heat supplied by cooling milk and equipment in milking centers, solar heaters, exhaust air reuse (not recommended in livestock or poultry buildings), heat exchangers, and earth tempering.

Air make-up heaters are typically unvented outside units that only heat the incoming ventilating air. They essentially supply ventilating air much as winter exhaust fans in controlled environments, and they provide constant temperature near the air inlet via modulating the fuel flow. Additionally, the heat supplied by cooling milk and equipment in milking centers can be easily directed to milking parlors, for instance, to dry the floor and provide operator comfort in winter.

Solar heating is already part of passive controlled-environment building heating. Solar heating can also provide some of the heating needed in livestock or poultry buildings via solar preheated ventilating air, which works very well with other space and floor heating systems. Various design approaches are available depending on the type of building.

Heat exchangers and earth tempering systems extract heat from exhaust air, other heat sources, or the soil through pipes or other heat-exchanging surfaces. When designed properly, these systems can easily handle cold and mild weather ventilation rates in livestock and poultry buildings. Hot weather ventilation, which will be discussed shortly in this chapter, must be provided additionally when these systems are employed.

5.4 CARBON DIOXIDE DOSING SYSTEMS

5.4.1 Plant Responses to CO_2 Enrichment

As discussed in depth in Chapter 4, research findings have shown that plant growth is improved by 15%–20% in greenhouses when CO_2 concentrations are increased to 700–1,200 ppm during the day. Continuous CO_2 enrichment during the light period has been found to be superior to intermittent CO_2 application. The literature indicates that the optimal CO_2 concentration for plant growth is between 600 and 1,000 ppm for most species; in general, the optimum CO_2 coupled with temperature and PAR intensity is about 400 ppm on cloudy days or vents are opened more than 10% and 1,000 ppm on sunny days.

Furthermore, elevating the CO_2 concentration leads to higher internal CO_2 concentrations in plant tissues, closing the stomatal openings and thereby reducing the transpiration of plants by 20%–40%. Water consumption is therefore significantly reduced by CO_2 enrichment while photosynthesis is increased simultaneously. It is well known that increasing the air relative humidity, let's say from 60% to 95% decreases plant transpiration by about 50%, and CO_2 enrichment at 95% relative humidity further decreases transpiration and water uptake by some plants by about 30% increasing the water-use efficiency (amount of water used per unit dry weight produced) by about 30%.

Regardless, because of the overall benefits of CO_2 as one of the basic nutrients for plants, CO_2 dosing has been used on a large scale in greenhouse horticulture for many years to increase production. CO_2, a natural component of the air, enters the plant through tiny openings in the leaves known as stomata. In that respect, CO_2 is different from all other nutrients, which are all absorbed through the roots.

Again as mentioned earlier, the effect of CO_2 on plant growth and production is different for different crops, but it is possible to generalize the relative effects of CO_2 on production as a rule of thumb based on experiments on vegetable crops using a base concentration of 350 ppm; in principle, the rule is valid for almost all crops. A rule of thumb is obviously not precise, but it is simple and generally applicable. This rule of thumb is expressed as follows:

$$X = \left(1{,}000 / CO_2\right)^2 \times 1.5 \quad (5.19)$$

where X is the percent production increase per 100 ppm CO_2 increase (% per 100 ppm). For instance, at 350 ppm increase (relative to the 350 ppm base concentration of CO_2) the effect of 100 ppm increase (up to 700 ppm) is determined as follows:

$$\left(1{,}000 / 350\right)^2 \times 1.5 = 12\% \text{ per } 100 \text{ ppm}$$

At 550 ppm increase (900 ppm in the greenhouse environment), the effect would be 5% and at 1,000 ppm the increase would only be 1.5% per 100 ppm.

However, increasing the CO_2 concentration has its limits, and maintaining too high concentrations are not necessary. In summertime, when the light level is high, CO_2 concentration should average 1,000 ppm and not exceed 1,200 ppm; in winter, when there is little light and no ventilation, average of 700 ppm and a maximum of 1,000 ppm are the recommended CO_2 concentrations. These levels are optimal from the crop perspective; however, economic side must also be considered when making decisions. In the summer months, a high CO_2 concentration in a greenhouse is desirable; however, it is not feasible due to high ventilation requirements. With a lot of light available and no CO_2 dosing, the concentration can quickly drop below the outside levels, and CO_2 then becomes the limiting factor, and production is greatly reduced by the CO_2 deficiency. In such cases, it is always profitable to maintain the CO_2 concentration at least at the outside levels.

5.4.2 Installation of CO_2 Dosing Systems

CO_2 dosing systems include i) supply of liquid (pure) CO_2; ii) small kerosene or natural gas burners, atmospheric of fan assisted, placed in the growth areas in small greenhouse operations; and iii) natural gas fired boilers typically placed in boiler rooms as part of a central heating system, which may have a cogeneration unit along with a heat storage tank. This book focuses on CO_2 dosing systems with natural gas-fired boilers placed in boiler rooms as part of a central heating system.

In properly engineered central CO_2 dosing systems, greenhouse growers take a portion of the flue gases from their natural gas-fired boilers into the greenhouse to increase the CO_2 concentration in the greenhouse aerial environment. Such an installation consists of a properly designed PVC piping system that transports the flue gases from the boiler through the CO_2 pipes. CO_2 distribution lines, on the other hand, are plastic branches (layflats) with a diameter of 40 or 60 mm. Today, 90 mm diameter layflats are also available. These branches are provided with tiny holes through which the flue gases flow into the greenhouse. Because the flue gases do not flow the CO_2 into the greenhouse by itself, a blower is placed at the beginning of the piping system. The primary duty of the blower is to suck the desired amount of flue gases from the boiler vent stack and bring it into the CO_2 transport line with adequate pressure.

5.4.3 Determining the Capacity of a CO_2 Blower

For optimal CO_2 dosing, the blower must have sufficient capacity to bring the desired amount of flue gases through the piping system into the greenhouse. When choosing a blower, it is important to determine the correct quantity of dosing gas and what resistance this gas will undergo through the piping system. An adequate blower can be selected based on i) the amount of dosing gases and ii) the dosing value.

The amount of dosing gases is determined by the desired dosing value and the temperature of the flue gases, and the dosing value is the amount of natural gas that the grower wants to burn per hectare for the CO_2 dosing. The dosing value is strongly dependent on the type of crop and whether a heat buffer is present or not. For greenhouse vegetables, the dosing value without a heat buffer is 40 to 80 m^3 gas/(h.ha), and with a heat buffer, the dosing value is 80 to 120 m^3 gas/(h.ha). For floriculture, a dosing value of 40 to 60 m^3 gas/(h.ha) is used when no heat buffer is present, and if a heat buffer is present, then the dosing value is 40 to 80 m^3 gas/(h.ha). For potting plants, there is no specific guideline or recommendation. It should be kept in mind that to provide CO_2 dosing, we can also operate burners at low speeds or with variable-speed fans.

When there is a heat buffer, it is particularly important that the capacity of the installed CO_2 dosing system be accurate and aligned with the content of the heat buffer system. If it is not designed properly, the heat buffer will heat up either too quickly or too slowly. As a general rule, the shortest filling time used for the heat buffer is six hours; based on the filling time and the buffer tank capacity, the desired dosing value can easily be calculated when a heat buffer is used.

5.4.4 Temperature of Flue Gas

A CO_2 dosing system is composed of PVC pipes; therefore, the temperature of the flue gases may not exceed 60°C. When there exists a flue gas condenser installed on the boiler, the heat recovered can be used for low-grade heating (such as workspaces, walkways, and in-floor heating) and the flue gas temperature stays under 60°C; however, without a flue gas condenser, the flue gas temperature goes up to 180°C. The flue gases at these temperatures are definitely not suitable for CO_2 dosing. Cooling the flue gases down to 60°C is still possible by using the outside air to make it suitable for CO_2 dosing. When flue gas temperature rises beyond 60°C, the amount of outside air to be mixed in and moved by the CO_2 blower obviously increases. The quantity of the air to be moved can be calculated using the dosing value and the temperature of the flue gases.

5.4.4.1 Resistance

If the amount of dosing gases is known, then the resistance in the pipe system can be calculated. The head of the blower for the calculated quantity of dosing gases must be at least equal to the resistance in the piping system. The head of a blower depends on the temperature of the air to be moved. The higher the temperature, the lower the head is for a given volume of air. Since most blower manufacturers specify the delivery head at an air temperature of 20°C, the delivery head of the blower must be adjusted to 60°C.

5.4.4.2 Blower Connection to the Boiler Vent Stack

The connection of the CO_2 blower to the boiler chimney is important when designing a CO_2 dosing system. The flue gases face resistance both before and after the blower. For an optimally functioning CO_2 dosing system, it is very important to keep the suction side resistance as small as possible, and this can be done in several ways: i) by placing the fan as close to the chimney as possible because the length of the suction side line (maximum 10 m) and the number of fittings is then kept at a minimum; ii) by preventing flow velocities in the suction pipe above 10 m/s; and iii) by placing so-called guide plates in the boiler chimney.

These partitions guide the flue gases from the chimney into the suction pipe of the CO_2 dosing system installation. It should be kept in mind that typically, 100 mm H_2O head loss is observed on the suction side for sucking the flue gas, and 70 mm H_2O head loss is observed just through the layflat connection to the distribution piping; this leaves the remaining blower head for the transport line and along the layflat. To have reasonable pressure drops, a maximum flue gas velocity of 10 m/s is used in the transport main and the distribution lines. The maximum velocity used in the layflats, on the other hand, is limited to 7.0 m/s. CO concentration in the flue gas is continuously monitored as well, and if the concentration is 30 ppm or higher, it is not considered for CO_2 dosing. Typically, however, when the CO concentration at the CO_2 blower reaches 25 ppm, then the CO_2 dosing system automatically shuts off.

5.4.4.3 CO_2 Piping System

The CO_2 piping system consists of three components: i) the transport pipes; ii) the distribution pipes; and iii) the branches. The transport pipe carries the CO_2 dosing gases from the blower to the distribution pipe. To keep the resistance in a transport line within acceptable limits, the number of bends in the transport line should be kept at a minimum level. The diameter of transport line PVC piping can be 125, 160, 200, 250, 315 or 400 mm (see Table A.28), and pipes are available with different wall thicknesses. The wall thickness is indicated by a class designation (Class 65, 51, 41, etc.). The lower the class, the thicker the wall. At least Class 51 must be used for CO_2 dosing systems; Class 65 is considered too thin.

The distribution pipes distribute the dosing gases evenly around the greenhouse, and in order to keep the resistance in the manifold within acceptable levels, a maximum dosing gas flow velocity of 10 m/s is used. Because the amount of dosing gases in the manifold decreases further along the line, smaller diameter pipes can be used as the amount of gas gets smaller. PVC pipes with diameters of 90, 110, 125, 160, 200, 250, and 315 mm are used for the CO_2 manifold, which will be shown in a design example in this section. In terms of the PVC class, the same applies to distribution pipes as to transport pipes.

The branch forms the connection between the distribution pipe and the CO_2 dosing layflats. A branch consists of at least a layflat holder with a throttling plate with which the layflat holder is screwed into the distribution pipes. Possible options for branching are shown in Figure 5.19; however, exceptions to these branch connections are possible. Depending on the amount of dosing gases flowing through a branch, a 32 mm or a 40 mm branch connection can be used. The distance between the branches on the distribution pipe is usually 1.60 or 3.20 m; however, this distance can be modified depending on the type of cultivation system being used in a greenhouse.

The layflats connected to the distribution pipe are provided with holes through which the dosing gases are released into the greenhouse environment. The number of holes in the CO_2 layflats determines the resistance that the dosing gases face in the layflats. Due to this resistance, pressure builds up in the layflats, which is called the layflat pressure. The more holes there are on a layflat, the more dosing gases a layflat can release into the greenhouse at the same layflat pressure. It means, the greater the dosing value and/or the flue gas temperature, the greater the quantity of dosing gases, and hence the more holes the layflats must have.

The number of holes on the CO_2 layflat is determined by the prick (puncture) distance, which is the distance between the holes along the length of layflat. By default, four holes with a diameter of 0.8 mm are pricked per prick distance. Instead of 4 × 0.8 mm holes, we can also use 1 × 2 mm hole per prick distance. For an optimal CO_2 distribution over a layflat, the design of a CO_2 dosing installation is based on a layflat pressure of 70 mm H_2O column. At this pressure, the CO_2 layflats are not easily squeezed by stems and/or fruits, and the condensate can easily get out of the layflats. Using Equations (5.23) and (5.24) provided at the end of this chapter, the prick distance on the CO_2 layflats can be determined for a layflat pressure of 70 mm H_2O column.

5.4.4.4 Pressure Drop across the Layflat

Pressure loss also takes place in a CO_2 layflat. As the amount of dosing gas through a layflat increases, the resistance increases as well. As a result, the pressure drop at the beginning of a layflat is larger than at the end of a layflat. When the pressure drop across a layflat exceeds 20% of the pressure at the beginning of the connection, the difference in CO_2 release between the beginning and the end of the layflat is not acceptable. The pressure drop can then be reduced by reducing the layflat distance or by using a larger-diameter layflat. In practice, excess pressure differences across the layflat can be prevented by adhering to the following guideline

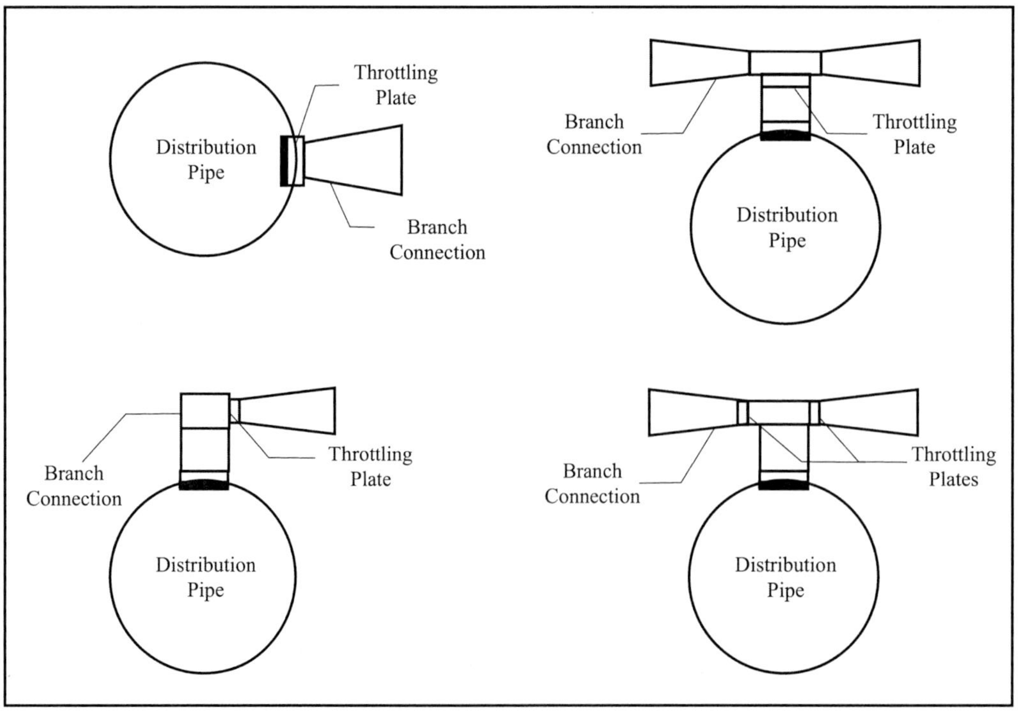

FIGURE 5.19 Possible options for branch connections.
Source: *Greenhouse Engineering* (2021).

values. With a branch distance of 3.20 m (c.t.c.) and layflat diameters of 41 and 60 mm, the maximum layflat lengths are 40 and 75 m, respectively. With a branch distance of 1.60 m (c.t.c.) and layflat diameters of 41 and 60 mm, the maximum layflat lengths are 60 and 120 m, respectively.

5.4.4.5 Throttling Plates (Discs)

The distribution of the dosing gases over the greenhouse is optimal if the pressure in all CO_2 lines is the same. However, the same line pressure is only possible if the dosing gases in the CO_2 piping system experience the same resistance, which is in practice not possible. The dosing gases that flow into the CO_2 layflats at the end of the distribution pipe obviously travel a much longer distance than the dosing gases that enter the CO_2 layflat at the beginning of the distribution pipe.

By using throttling plates, we can build in resistance and throttle the pressure to the desired layflat pressure, and as a consequence, an equal layflat pressure can be achieved everywhere using plates. Throttling plates are round plates with a hole of a certain diameter. The ratio between the diameter of the hole in the throttle plate and the diameter of the branch determines the degree to which the pressure is throttled. The smaller this ratio, the more pressure is throttled away. The available diameters of the holes in the throttle plates are for a 32 mm branch 11 to 18, 20, 22, and 24 mm. For a 40 mm branch, these are 20, 22, 24, 26, 28, 30, and 32 mm. Designing the correct throttle plates is not easy. Incorrectly designed throttle plates can have negative consequences on the dosing, and they should never be glued.

5.4.4.6 Condensate Drainage

One of the biggest problems in CO_2 dosing installations is flooding. The dosing gases in the CO_2 pipes have a higher temperature than the greenhouse air, which cools the dosing gases. Since the relative humidity of the dosing gases is around 100%, water condenses from the flue gases due to cooling. If this condensate water is not properly discharged, it will remain in the CO_2 pipes and result in poor CO_2 distribution. Flooding can be prevented by installing condensate discharge points in the pipes, laying the pipes with a slope and preventing the pipes from subsiding.

Condensation water can be discharged from the CO_2 pipes in two ways: through holes drilled in the bottom of the pipes or through a 'gooseneck' connection (see Figure 5.20). The first solution is quite simple and cheap. However, the disadvantage of this method is that significant leakages can occur across the holes. That is why this condensation discharge method is not recommended. The second solution is discharging the condensate water via a gooseneck connection, which is a more expensive but more secure way of condensate water removing. Since water remains in a gooseneck, no dosing gases can escape through the gooseneck. It is of course important that the gooseneck has sufficient length. Otherwise, the blower will blow the water out of the gooseneck, causing dosing gases to escape.

The gooseneck must also be properly connected to the CO_2 line. This is usually done with a pointed piece, as a result of which a layer of water will always remain in the pipe. This can be prevented by connecting the gooseneck

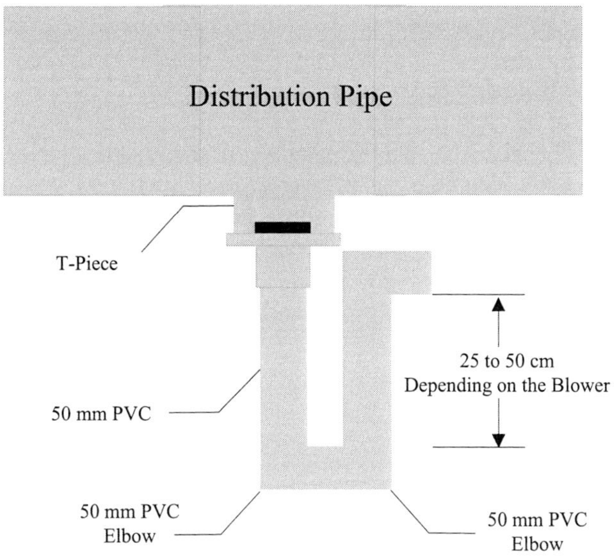

FIGURE 5.20 Condensate discharge through a gooseneck connection.

Source: *Greenhouse Engineering* (2021).

to the CO_2 pipe with the help of an insert inlet or a T-piece. With pipe diameters of 200 mm and larger, an insert inlet is used, but for smaller diameters, a T-piece is used.

For optimal condensate discharge, it is necessary to install sufficient discharge points in the CO_2 piping system. It is therefore recommended to place a condensate discharge point at least at the end of a transport pipe and in a distribution pipe every 25 meters. The condensate water from the discharge point can be discharged into the drainage via a gravel bed. It is also possible to drain the collected condensate water with a pipe to the recirculation system or the ditch. A minimum slope of 1 mm per meter of pipe length (0.1%) can be used safely. It is important that the slope is always in the flow direction of the dosing gases. Subsidence can occur in underground CO_2 pipes if they are not properly buried. As a result of the subsidence, water remains in the pipe, which will adversely affect the CO_2 distribution. Therefore, it is important to install the underground CO_2 pipes on a solid and level surface.

Designing a proper CO_2 piping system is a difficult task that requires thorough knowledge of fluid dynamics. The CO_2 dosing system installation must therefore be designed and installed by an experienced engineer who has proper experience in fluid dynamics.

5.4.4.7 Calculating Dosing Value and Dosing Gases for a Heat Buffer

The dosing value, D_s, is the quantity of natural gas to be burned per hour in the floor surface area (ha) of the greenhouse.

$$D_s = \frac{\left[V_{wb} \times c_w \times \rho_w \times \left(T_{max} - T_{min}\right)\right]}{A_{gh} \times \Delta t_{fill} \times H_b \times \eta_{boiler}} \tag{5.20}$$

where D_s is the dosing value in m^3 gas/h.ha; V_{wb} is the water content of the heat buffer in m^3, c_w is the specific heat of water, 4.2 kJ/kg.K; ρ_w is the density of water in kg/m^3; T_{max}

is the maximum water temperature in buffer in °C; T_{min} is the minimum water temperature in °C; A_{gh} is the greenhouse floor surface area in ha; Δt_{fill} is the minimum filling time of the heat buffer full load hours boiler (hour); H_b is the upper calorific value of natural gas, 35,170 kJ/m³ (varies regionally; therefore, local value must be used); and η_{boiler} is the efficiency of the boiler in fraction.

The quantity of dosing gases V_{dgas} can be calculated using D_s and the volume of combustion gases per m³ of natural gas to be burned V_{cg} as follows:

$$V_{dgas}\left(\frac{\text{m}^3}{\text{h.ha}}\right) = D_s \times V_{cg} \tag{5.21}$$

$$V_{cg}\left(\frac{\text{m}^3}{\text{m}^3 \text{ natural gas}}\right) = \frac{\left[(8.3871 \times n - 0.7171) \times T_{cg}\right]}{273} = 10\frac{\text{m}^3}{\text{m}^3 \text{ natural gas}} \tag{5.22}$$

where n is the excess air in the combustion process. For $n = 1.17$ and the combustion gas temperature of $T_{cg} = 27$°C, $V_{cg} = 10$ m³/m³ natural gas.

5.4.4.8 Prick Distance of CO_2 Layflat

The prick distance depends on the amount of dosing gases and the layflat (branch) distance.

At a branch distance of 3.20 meters (c.t.c.):

$$\text{Space}_{3.2} = \frac{450}{V_{dgas}} \text{ with a minimum of 1.2 m and a minimum of 0.2 m} \tag{5.23}$$

At a branch distance of 1.6 meters (c.t.c.):

$$\text{Space}_{1.6} = \frac{900}{V_{dgas}} \text{ with a minimum of 1.2 m and a minimum of 0.2 m} \tag{5.24}$$

See Examples 5.9 and 5.10 in *Greenhouse Engineering* (Yildiz, 2021) for determining the prick distance and a complete CO_2 dosing system design for a vegetable greenhouse in Canada, respectively.

5.5 HEAT STORAGE TANKS

When heat is produced but not consumed at the time of its production, it is stored in a tank for later use when there is heat demand. When the hot water is pumped to the top of the storage tank, a hot and a cold section are created at the top and the bottom of the tank, respectively. These sections are normally separated by a thin boundary layer in between, and this layer gets thicker when the hot water is stored longer. The heat storage tank is unloaded based on this line.

Heat loss occurs during the storage depending on the method of filling, the storage time, and the insulation of the storage system. After a period of heat surplus, typically a period of heat demand follows, and heat is stored exactly in between these two periods that follow each other within 24 hours. Heat storage tanks can be either horizontal (maximum temperature 100°C) or vertical (maximum temperature less than 93°C) and are used in greenhouses in the following situations; when they are used, they become the heart of a whole system.

1. **CO_2 Dosing**

   As discussed earlier, CO_2 is an essential ingredient for crop growth and development. In summer months, crops can take up a lot of CO_2, and therefore, for proper crop production, at least the outside CO_2 level has to be maintained in a greenhouse environment. In spring months, however, higher CO_2 levels can be maintained, and extra crop production can be achieved as a result. In a mass crop production facility, CO_2 can be generated in a cost-effective way by simply burning natural gas.

   The heat generated during CO_2 dosing is, however, quite often not used entirely in the greenhouse in the spring and summer months. For instance, there is a heat surplus during the time greenhouses are being ventilated to get rid of the excess heat accumulated inside. The heat in the storage tank is easily unloaded at night in the summer months; however, even though it is hoped for, it is a challenge to empty the tank during the days in summer. A heat storage (buffer) system for CO_2 dosing only is an old and proven system (see Figure 5.21 for heat storage [loading/filling] cycle and Figure 5.22 for heat recovery [unloading] cycle), for which a design example will be given at the end of this section. In these systems, the tank capacity is a function of CO_2 dosing norm. If the dosing norm is 100 m³ natural gas/h.ha, then the tank capacity is 100 m³/ha, and if the norm is 125 m³ natural gas/h.ha, then the tank capacity is 125 m³/ha. The dosing time has to be 10 hours.

   During the loading/filling cycle, the primary goal of the system is to perform CO_2 dosing in the greenhouse and store the hot water generated in the storage tank. This system must, during the CO_2 dosing, be able to respond to the heating needs of the greenhouse whether for primarily heating or humidity control. As shown in Figure 5.21, in the loading mode, hot water coming from the boiler is directed into the top of the storage tank while colder water is drawn from the bottom of the storage tank by the load/fill pump through a control valve. When the CO_2 dosing is finished, the loading cycle is complete, the hot water in the storage tank is unloaded as shown in Figure 5.22.

2. **Open Buffer Systems and Combined Heat and Power Generation**

   Peak shaving is the prime reason for the development of open buffer systems, and we must have a high number of operation hours for such systems. In these systems, the boiler is operated during the off-peak hours when there is not much or no heat demand at all in the greenhouse, and the generated heat is stored in heat storage tanks for later use; in this way, the overall natural gas consumption at the greenhouse operation is reduced. Hourly natural gas use in an open buffer system differs

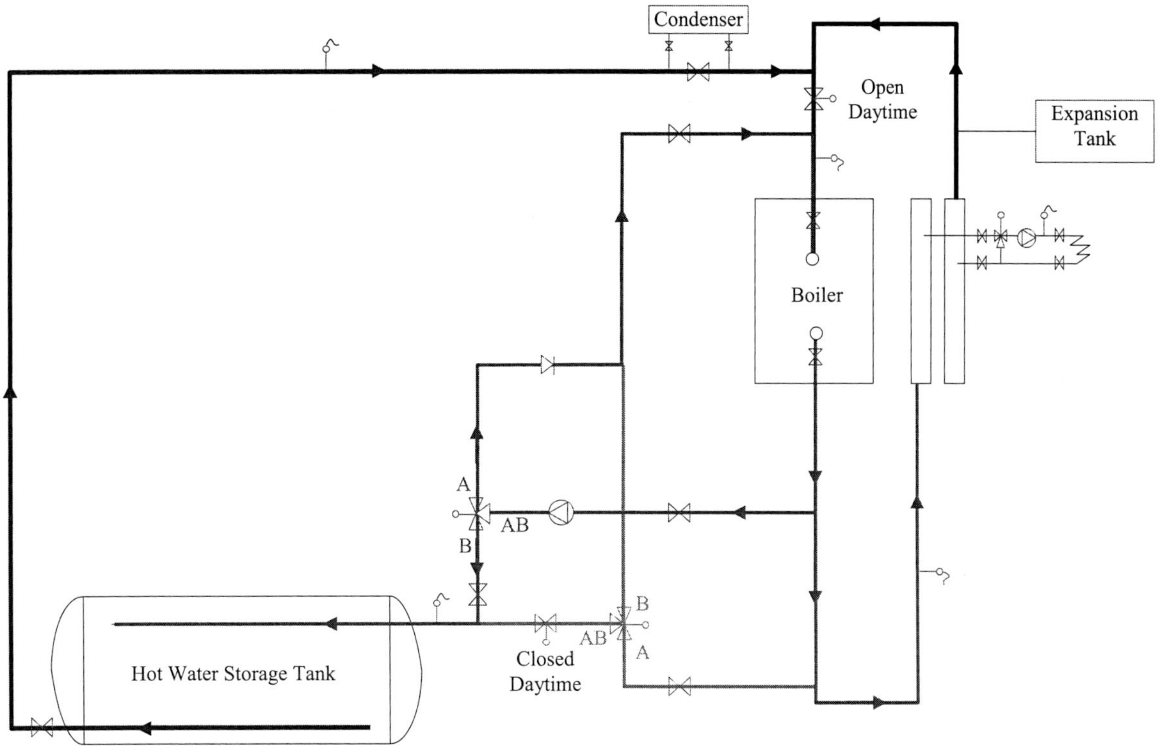

FIGURE 5.21 Hot water storage system with a horizontal tank (loading cycle).
Source: *Greenhouse Engineering* (2021).

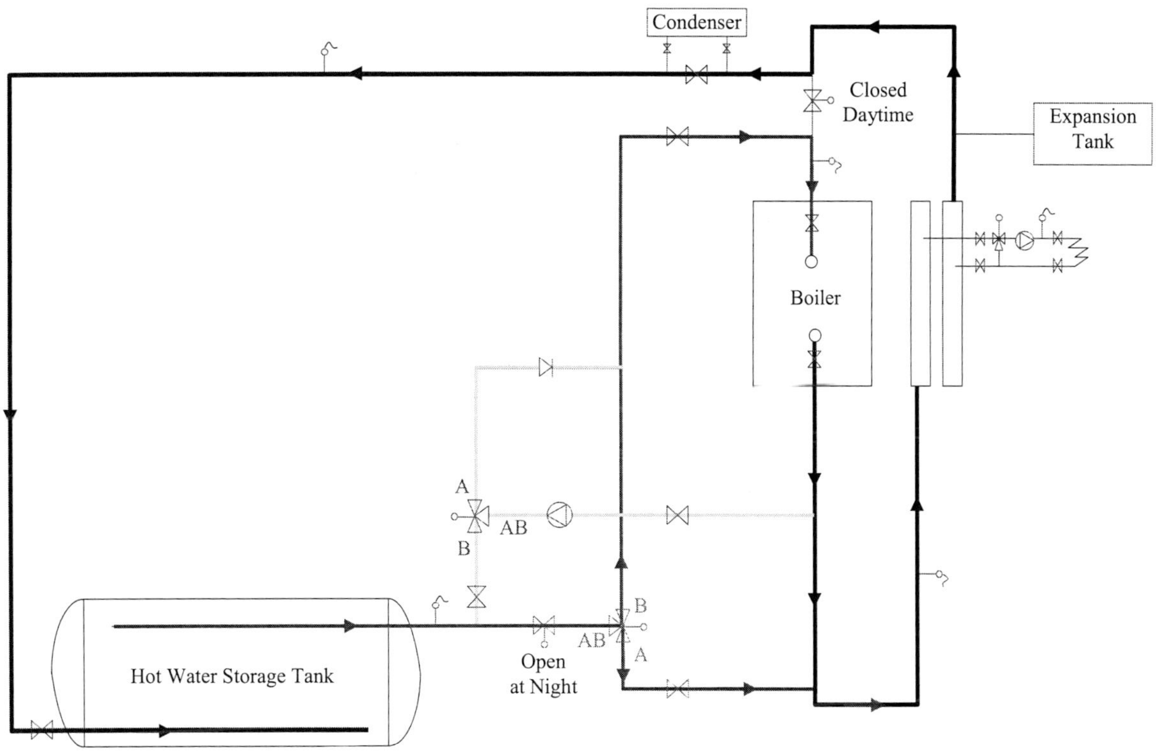

FIGURE 5.22 Hot water storage system with a horizontal tank (unloading cycle).
Source: *Greenhouse Engineering* (2021).

entirely from gas use in a system without an open buffer; the buffer saves a considerable amount of natural gas and help conserve our resources and make our greenhouse operations more sustainable. An example of an open buffer system will be given at the end of this section.

In an effort to conserve energy and make facilities more sustainable, greenhouse operations are increasingly using combined heat and power (CHP) generation systems every year. These localized systems are more environmentally friendly, more flexible, more efficient, and more cost-effective than traditional centralized systems, especially when power transmission savings are taken into account. Conventional generation of electricity in large central power stations is normally only 30%–40% energy efficient. Recent CHP cycle generation can improve this to 55%, excluding the transmission losses of electricity, which can be as great as 20% of the delivered efficiency.

Through the utilization of the heat, the efficiency of a cogeneration unit can reach 90% or more (see Figure 5.23 for loading and Figure 5.24 for unloading). Cogeneration of heat and power in

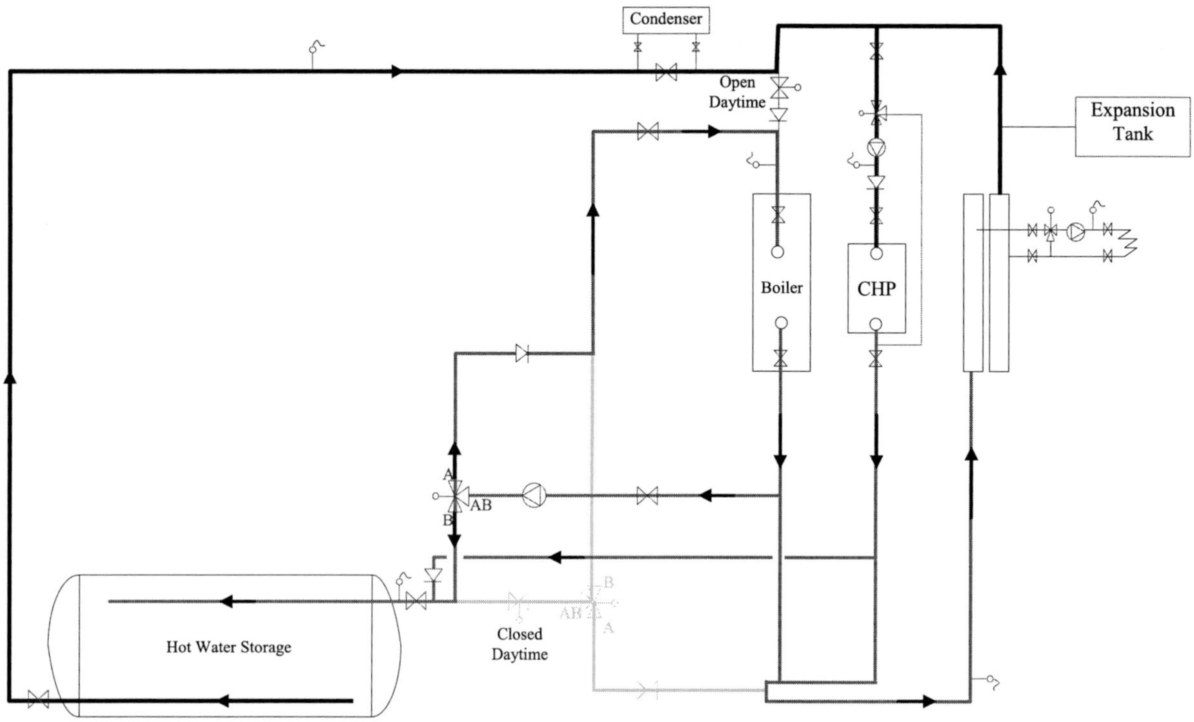

FIGURE 5.23 Hot water storage system with CHP in parallel connection (loading cycle).
Source: *Greenhouse Engineering* (2021).

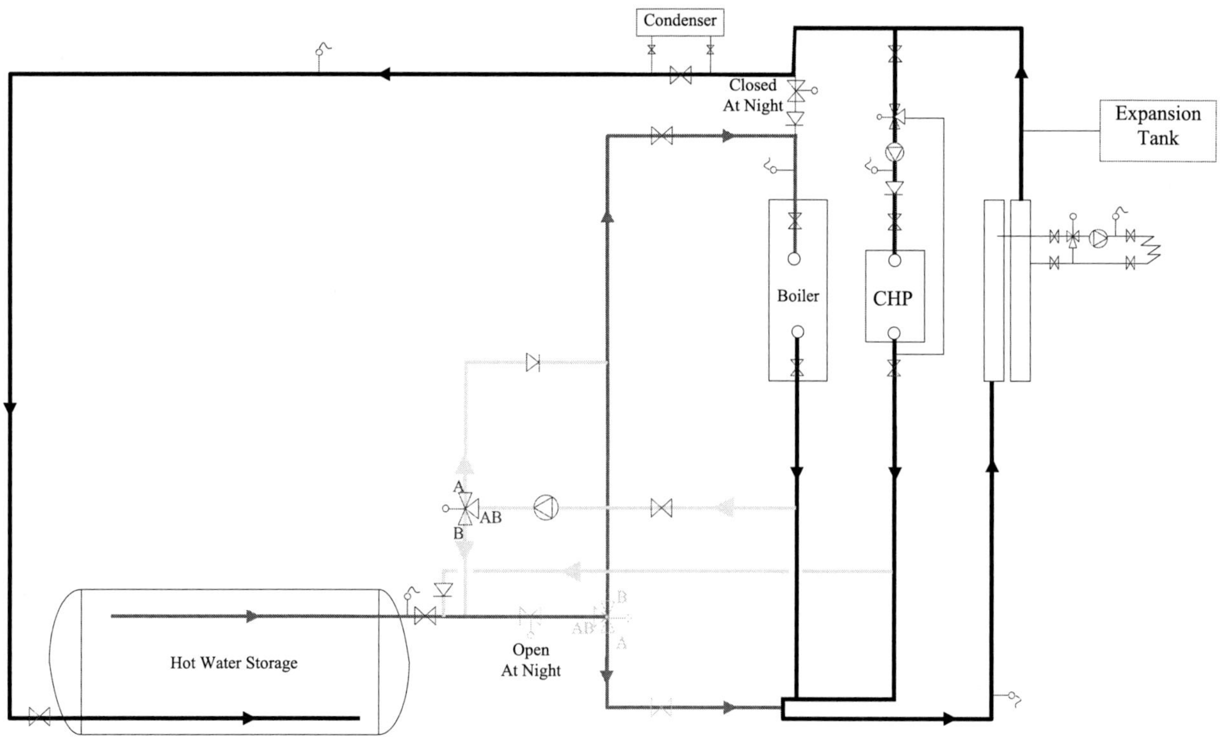

FIGURE 5.24 Hot water storage system with CHP in parallel connection (unloading cycle).
Source: *Greenhouse Engineering* (2021).

greenhouse operations not only utilizes the heat and power generated on site but also essentially eliminates the CO_2 emissions to the environment. In essence, these systems serve as multi-generation systems in greenhouse operations, generating heat, power, CO_2 and water, all precious resources. CHP systems can be connected to the system either in series or in parallel connection.

5.5.1 Tank Size, Heat Surplus, and Demand

A hot water storage tank must be used as efficiently as possible. To avoid unnecessary investments, the size of the tank will have to be sized and optimized in advance. Special attention must also be given to both filling and emptying the tank.

How long the burner only runs for CO_2 dosing can often be taken from the computer recordings in a greenhouse operation. The number of hours, in combination with the burner setting for CO_2 dosing (often the minimum setting), gives us the excess heat produced. The determination of heat surplus will be numerically shown in an example later in this section. Similarly, the exhausted heat (too large) from a CHP installation can also be calculated.

The stored heat has to be depleted in the following period with heat demand, usually at night. It is highly desirable to start the next day with an empty heat tank, especially when CO_2 dosing is practiced. In short, the heat demand during the night sets a limit for heat storage; that is, no further heat can be stored than what can be used in the short term. In spring months, the nights are still cold enough to be able to use much of the stored heat; hence, it is not a good practice to size the heat storage tank based solely on the heat levels in the summer nights.

5.5.2 Setting Up and Other Considerations

The heat storage efficiency decreases in proportion to the increase in the surface of the boundary layer. That is why the vertical arrangement of storage tanks is particularly important for large tanks (see Figure 5.25). Furthermore, with a vertical arrangement, the load per m^2 of ground surface increases, which obviously increases the cost of building a tank foundation. This can be prevented by placing several tanks in series; however, the consequence of having more tanks is to have a larger exposed surface and hence more heat losses. A horizontal tank that can be quickly unloaded is therefore simply the best option where interfacial mixing can hardly occur (see Figure 5.26). With the use of more than just two supports, the load of the tank on the ground can no longer be an issue; hence, no special foundation design is required. To reduce the load on the ground, it is recommended to use more supports, for instance every 2.5 m or so. It should also be kept in mind that considerable heat loss occurs through the supports.

To make optimum use of the heat storage tank, it is necessary to select sufficiently large pipe diameters. The stored heat must cover the entire heat demand of the greenhouse as much as possible. This means that to empty the heat tank, a pipe diameter that is at least the same size as the supply from the boiler is required. Both pipes to and from the heat storage tank can be the same size. As long as the pipe size is suitable for emptying, filling is never a problem. All the methods used in practice to connect the boiler and heat storage tank to the manifold are based on two methods: i) in series connection with the boiler or ii) in parallel connection with the boiler.

Occasionally, hot water storage tanks are damaged due to high pressure caused by improper use of valves or pipes that are too small. A safety valve is required to prevent major damages to the tank; and to prevent corrosion, an air vent must also be installed on the tank.

Moreover, we should always choose a tank that has more capacity than the calculated capacity for summer. The larger tank will not be more expensive than the one for summer storage, and it will provide more room for CO_2 dosing, especially in the spring. And remember that the tank does not have to be completely full in the summer. Other useful considerations include the following.

1. **Distance to Boiler Room**
   In practice, emptying the tank can be a problem. Sometimes, even though there is a lot of heat demand in the greenhouse, the heat storage tank cannot be used effectively because the hot water

FIGURE 5.25 A vertical hot water storage tank used in a greenhouse facility.

Source: *Greenhouse Engineering* (2021).

FIGURE 5.26 A horizontal hot water storage tank used in a greenhouse facility.

Source: *Greenhouse Engineering* (2021).

from the storage tank is not released quickly enough. This is because the pipe diameters are too small with respect to their lengths. In the event of immediate heat demand, sufficient heat can be supplied from the tank because the tank is full of heat, and the boiler is prevented from starting. The rule of thumb stating that the pipes to the heat storage tank should be the same size as the pipes to the boiler is only applicable when the heat storage tank is close to the manifold. The heat storage tank is, however, typically farther away from the manifold; therefore, larger diameters than those used for the boiler are often recommended. Additionally, due to the potential energy losses, the heat storage tank is placed as close as possible to the manifold.

2. **Insulating Supports**

   The supports of the hot water storage tank can become quite hot. However, insulating these supports is very easy and reduces heat losses during the storage. Additionally, rather than insulating these supports, rubber pieces can be placed between the tank and supports.

See Examples 5.11 and 5.12 in *Greenhouse Engineering* (Yildiz, 2021) for real-world design examples of a horizontal hot water storage tank system for CO_2 dosing and a vertical tank open buffer system, respectively.

5.6 WATER PREHEATING SYSTEMS

In greenhouse crop production as well as aquatic systems, irrigation or culture water is typically preheated to reduce the shock of the cold water to plants or fish and maintain water temperatures in culture tanks. Figure 5.27 shows the details of a water preheating system.

Example 5.5: Designing a Water Preheating System

See Examples 2.57 and 2.58 in Chapter 2 for water preheating for a greenhouse and an aquaculture operation.

5.7 HORIZONTAL AIR FLOW SYSTEMS

HAF fans, also known as circulation fans, provide a more uniform controlled environment and reduce the difference

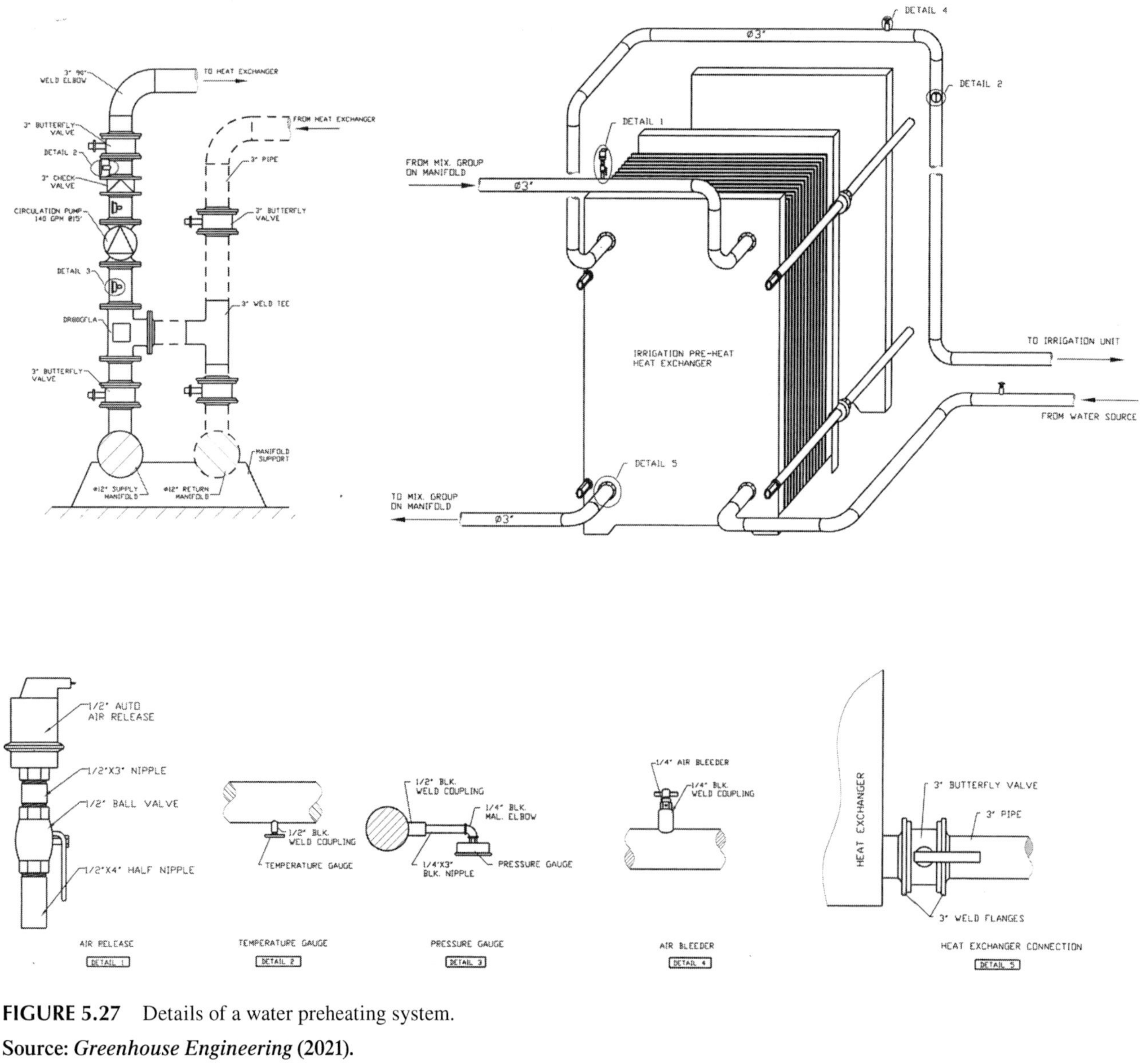

FIGURE 5.27 Details of a water preheating system.
Source: *Greenhouse Engineering* (2021).

FIGURE 5.28 A horizontal air flow fan installed in a greenhouse. Source: *Greenhouse Engineering* (2021).

between internal air temperature and plant temperature (see Figure 5.28) in greenhouses, for instance. They improve not only uniform temperature distribution but also humidity and CO_2 distributions in greenhouse operations costing marginal energy expenditure. The flowing air removes moisture and prevents condensation on the leaves, reducing foliar fungal diseases as well.

Circulation fans are also suitable for year-round ventilation within livestock and poultry buildings providing temperature uniformity throughout the building and removing condensation by circulating the air, thereby improving the animal environment at low cost and energy consumption. There are a number of other specific benefits of using circulation fans for livestock (e.g., pigs) and poultry applications, such as controlling insects because many insects are discouraged by a constant airflow throughout the livestock or poultry building and overcoming dunging problems; specifically, where the ventilation rates become too low or the fresh air is not circulated evenly, animals tend to congregate wherever the oxygen level is high in their controlled environment building.

Fans circulate fresh air evenly throughout the entire building, which creates a healthier environment for the animals; they reduce growth loss from heat stress. The constant airflow from fans keeps livestock and poultry buildings dry by evaporating moisture in the air, and the reduced heat stress by circulation fans can enhance animal reproductive efficiency. Exhaust fans move dust, pathogens, and manure gases away from the animals and out of buildings more efficiently.

5.7.1 HAF Fans in Greenhouse Environments

In greenhouses, typically, a total air flow of 0.01 $m^3/s.m^2$ (2 cfm/ft^2) of floor area is provided keeping the greenhouse air mass moving at a velocity of 0.5 m/s (100 ft/min) to 0.7 m/s (140 ft/min) and not exceeding 1.0 m/s (200 ft/min) in the vicinity of crop canopy. HAF systems come with 30 cm (12 in), 40 cm (16 in), 45 cm (18 in), and 50 cm (20 in) diameter blades, and generally move the air approximately 30 times the blade diameter. A 30 cm (12 in) diameter HAF fan moves the air 9 meters (30 ft).

In a well-designed HAF fan system, the fans are installed in a way that air flow is horizontally directed down the greenhouse bay and aligned with similar fans pushing the greenhouse air in the opposite direction down the next bay. To make sure that the air within the canopy is moved rather than the air above the canopy, fans are mounted at between 0.6 m (2 ft) to 0.9 m (3 ft) above the plants. The first and last fans are placed about 4.5 m (15 ft) to 6.0 m (20 ft) from the gable ends to pull the air that is coming from the adjacent bay. Crop canopy and hanging baskets resist horizontal air flow; therefore, the fans should be spaced properly. After the first fan, they are typically placed 9 m (30 ft) to 15 m (50 ft) apart to keep the greenhouse air mass moving smoothly.

Example 5.6: Sizing Horizontal Air Flow Fans for a Greenhouse Complex

Let's consider the 10-bay gutter-connected single glass gable roof greenhouse complex in Columbus, Ohio, for which we designed a heating system using natural gas-fired unit heaters in Example 5.1.

Solution

Remember that the unit heaters chosen for the greenhouse complex were models GHE200 and GHE230, which come with one 16 in. and two 16 in. diameter fans, respectively; as discussed, a 16 in. (40 cm) diameter fan can move the air approximately 30 times the blade diameter, 40 ft (12 m). The remaining distance in each bay one way can be handled by two 18 in. (45 cm) diameter fans, each moving the air additional 45 ft (13.5 m). Then, for each bay we will use four GHE18 fans, two fans in each direction as shown.

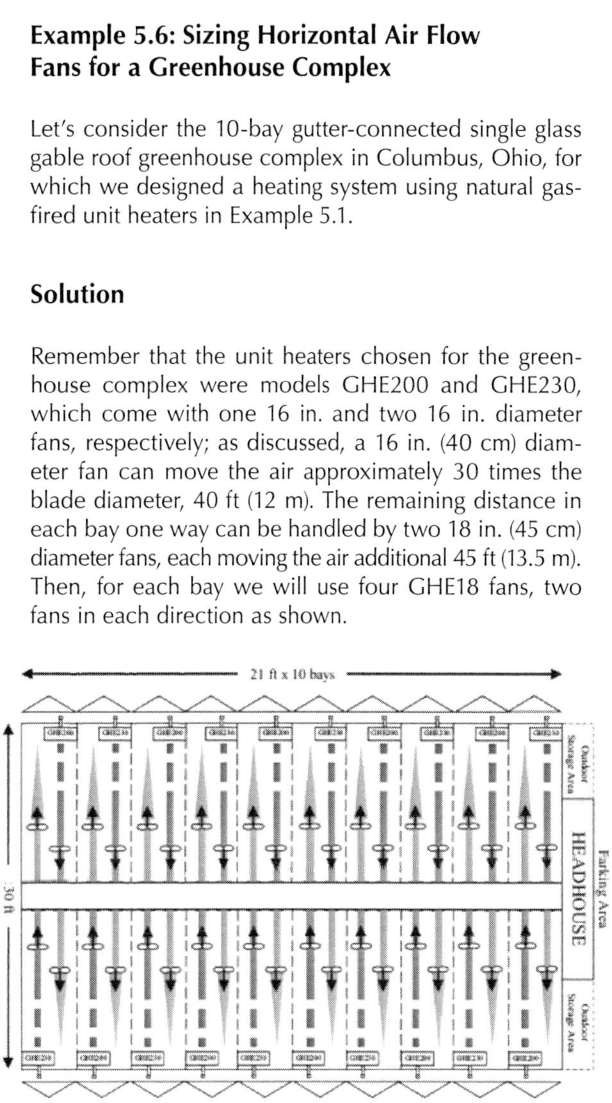

Layout of the horizontal air flow fan system for Example 5.6.

5.7.2 HAF Fans in Livestock and Poultry Environments

Air movement within livestock and poultry buildings providing temperature uniformity, reducing heat stress, increasing reproduction efficiency, and moving pathogens and manure gases away from the animals is usually accomplished by properly locating the air inlets and by varying intake cross-sectional area to fit the animal distribution. Automated motorized inlet controllers have been developed to adjust the slot intake opening outside depending on the pressure difference between inside and outside the building.

The slot is usually made a little wider than calculated, and the opening width is adjusted for desired pressure differential and distribution.

Another method of intake distribution in animal buildings uses a ventilation fan connected to a perforated PE air distribution tube or a rigid duct, which typically allows a combination of heating, circulation, and ventilation designed into one system. Furthermore, circulation fans improve air distribution, particularly in winter months to help dry or heat isolated corners. In summer, they also improve animal comfort by increasing air velocity over the skin. Circulation fans come in different sizes but can be 1750 rpm unrated units with sealed motors.

As mentioned in Chapter 4, swine gains and feed conversion rates are highly sensitive to air velocities. Swine less than eight weeks of age should not be exposed to velocities greater than 0.25 m/s (50 ft/min), and much lower velocities are preferred when temperatures are in the recommended range. Generally speaking, pigs benefit from air movement if the air temperature is between 32 and 39°C (90–103°F). Above 39°C (103°F), high wind speeds increase heat load, and pigs are negatively affected by air velocities greater than 1 m/s. Pigs prefer still air or at most 0.15 m/s (30 ft/min) at optimal air temperature. These circulation rates are typically provided with fans.

Again, as discussed in Chapter 4, some benefits are observed from increased air velocities up to 2.5 m/s (500 ft/min) at air temperatures from 24 to 35°C; however, high-velocity air at a temperature above the feather temperatures causes more heat stress in birds. Typically, air speeds greater than 0.3 m/s (60 ft/min) at broiler height are not recommended for broilers under two weeks of age. Increasing air speeds from 0.2 to 2.5 m/s (40 to 500 ft/min) around broilers improved the weight gains and water-use efficiencies of 1.4 kg (3 lb) broilers when the temperature fluctuated diurnally from 21 to 36°C (70 to 96°F).

Similar increases in air speed reduced heat stress in eight-week-old broilers that were exposed to environment temperatures less than the bird's body temperature of 40.6°C (105°F). At environmental temperatures above body temperatures, however, an increase in wind speed will exacerbate the heat stress. Similarly, for turkeys, air circulation provides some benefits, and the use of circulating fans distributed throughout a poultry house is considered the most efficient means to provide air stirring.

5.8 VENTILATION AND COOLING SYSTEMS

5.8.1 Introduction

Plants are biological air conditioners that cool their own immediate surroundings. However, during warm weather, heat builds up inside greenhouses due to high solar radiation loads; if quality crops are to be grown, it is required to provide greenhouses with appropriate cooling. Cooling may be provided by different means starting from the most cost-effective options such as whitewashing the building exterior, shading cloths or screens, and natural ventilation, and moving toward more costly options such as mechanical ventilation, evaporative cooling, heat pumps, refrigerated-air conditioning, or Watergy.

Similarly, during warm weather, heat builds up inside livestock and poultry buildings due to the heat generated by animals and high solar radiation loads on the walls and roofs. If proper animal growth, production and reproduction are expected, it is required to provide livestock and poultry buildings with appropriate cooling. Therefore, ventilation in livestock and poultry buildings is extremely important for continuously removing not only excess heat but also excess moisture, airborne dust and disease organisms, combustion products of unvented heaters, and odors and harmful gases from the building as well as providing fresh air for animals. It should be remembered that adequate air movement reduces the risk of moisture condensation on interior wall and roof surfaces in cold weather as well while maintaining acceptable inside temperatures in warm and hot weather.

The amount of moisture and excess heat to be removed, on the other hand, depends on animal species (see Figure 4.13), animal size, and manure handling system. Larger animals usually give off more moisture and require more ventilation for moisture control. Also, as discussed in Section 4.4.2.1, partly and totally slotted floors drain most of the liquids rather than evaporating, reducing the moisture load on the ventilation system by about 35% to 50%, respectively, when compared with solid floors. In order to prevent condensation and airborne bacteria risks and reduce dust levels within livestock and poultry buildings, 50% to 80% relative humidity levels are used as the target boundary design conditions for ventilation system designs.

Odor levels are heavily dependent on the animal species, the manure handling system, and air temperature. In temperate and cold climates, except for the coldest part of the year, the amount of ventilation required for removing excess animal heat and moisture is usually enough for controlling the odor levels. However, cold-weather-only ventilation controls only the moisture buildup within the building and might not be sufficient for proper odor control. Therefore, it is recommended to ventilate indoor manure storage pits as well for the purpose of reducing gases at the animal level. Exhausting the cold weather ventilation air from the pit would definitely help reduce the gas levels. It should also be kept in mind that when the indoor storage pits are agitated or being emptied, high enough volume ventilation rates are required, and if this is not possible, at least people should stay outside and the animals should be removed from the building.

Furthermore, livestock and poultry buildings can be cooled by a number of different means starting from the most cost-effective options, such as whitewashing the building exterior, when applicable, shading cloths or screens, natural ventilation, cool drinking water, and wet-skin cooling (i.e., fogging which cools the air, and sprinkler cooling which cools the animal directly), and moving towards relatively costly options such as mechanical ventilation, evaporative cooling, zone cooling, and refrigerated air-conditioning systems.

In aquatic buildings, on the other hand, open tank water surfaces cool the indoor environment via evaporative cooling. However, during hot and even sometimes warm weather, not only heat but also, in all weather conditions, moisture and other gas concentrations, such as CO_2, ammonia, and methane build up inside aquatic buildings as well; therefore, it is required to provide aquatic buildings with proper ventilation, which can also provide required fresh air to the building. In addition, when ventilation only cooling systems are not able to handle the excess heat build up within the building, other cooling mechanisms may also be required.

In conclusion, ventilation is a necessary component of conventional greenhouse, livestock, poultry, and aquatic buildings for all kinds of reasons, and it can essentially be classified into two systems: i) natural (passive) and ii) mechanical (fan or forced-air). Regardless of the type, ventilation is a fundamental instrument to control not only the indoor air temperature—especially in mild and hot summer months, but also the humidity and other gas concentrations such as CO_2, ammonia, and methane throughout the year. In greenhouse operations, in addition to removing the excess heat and moisture, ventilation is a vital element for bringing the CO_2 rich outside air to indoors, if additional CO_2 enrichment is not being utilized.

This chapter briefly goes over estimating cooling loads in controlled environment buildings first and then moves to the passive and active cooling techniques. The techniques covered include whitewashing the building exterior, permanent and retractable shading systems, natural and mechanical ventilation, evaporative cooling, fogging, sprinkler cooling, zone cooling, heat pump and other refrigerated-air conditioning, and Watergy.

5.8.2 Estimating Cooling Loads

Using the energy balance equation, Equation (5.2), and the other pertinent equations provided earlier in this chapter, we can determine the cooling load in a controlled environment building such as a greenhouse, livestock, poultry, or aquatic environment. Controlled-environment building cooling load can be reduced using permanent or retractable shading materials and devices before taking further cooling measures as discussed below.

5.8.3 Passive Cooling by Shading and Insulation

Shading is mostly used during warm sunny days to reduce cooling loads and therefore indoor temperatures; as will be discussed later in Chapter 8, it is also an effective energy conserving technique commonly used in commercial greenhouses, livestock, poultry, and aquatic buildings.

These simple and justified energy conservation techniques sometimes include an external spray (also known as whitewash) applied on the exterior surface of the building (Figure 5.29), insulation, and when applicable, variable retractable shading (internal or less commonly external) (Figure 5.30) using different materials. Whitewashing is an inexpensive method to provide shading and solar radiation reflection and hence reduce cooling loads; however, the proportion of shading it provides is relatively fixed and permanent even though it is washed away slowly in time. It can be applied on the whole or only a portion of the building in alternating strips (e.g., greenhouses). The degree of shading of whitewashing application depends on the coating thickness or the frequency of alternating strips. Also, the whitewash must be washed off especially from greenhouses when the solar radiation becomes limiting again in the fall.

FIGURE 5.29 Whitewash applied on the outside surface of the glazing.

Source: *Greenhouse Engineering* **(2021).**

Shades are quite effective for cooling livestock buildings as well. High shades at least 3 m high increase exposure to the cooler sky temperatures, increasing radiant heat loss from the animals.

Moreover, thermal insulation may be used in a controlled environment building for a number of reasons including i) conserving heat, which reduces the amount of supplemental heat required, ii) reducing heat gain during warm seasons, which reduces the amount of ventilation and cooling required—we must remember that the temperatures of walls and roofs exposed to direct sunlight can be as much as 25 to 30°C (77°F to 86°F) above air temperature, iii) providing warmer surface temperatures on the inside of exterior walls during cold seasons—warmer surface temperatures mean less condensation risk on the inside surfaces and lower the amount of heat loss from objects in the building to cold walls and roofs. See Section 2.3.5 for further information on thermal insulation and condensation and relevant examples. The effect of using and changing thermal insulation on energy conservation in controlled environment buildings will further be examined in numerical examples later in Chapter 8.

Retractable shading, as shown in Figure 5.30, on the other hand, is a relatively expensive shading technology used in greenhouses or greenhouse-type structures; however, when controlled properly, it dynamically and precisely reduces cooling load. When whitewashing or external shading curtains are used, solar radiation can be reflected before it enters and increases the cooling load in the greenhouse. These systems should not be controlled based on only inside temperatures but also inside light levels for not to hinder crop or animal growth and development. Both external and internal shading cloths come in different colors and materials having different costs and shading percentages. Another variable shading system that reduces the cooling loads in greenhouses or greenhouse-type structures combines polystyrene pellets that provides precise shading in greenhouses; however, it is not commercially available yet due to its relatively expensive channelized double acrylic glazing and control system.

FIGURE 5.30 Internal retractable shading system: open (left) and closed (right).
Source: ***Greenhouse Engineering*** **(2021).**

5.8.4 Passive Cooling by Natural Ventilation

In warm seasons, cooling is required to reduce the indoor air temperature and to alleviate plant and animal heat stress. Ventilation is a necessary component of conventional greenhouse, livestock, poultry, and aquatic systems and can be classified into two systems: i) natural (passive) or ii) mechanical (forced air).

Whether it is natural or mechanical, ventilation is a fundamental instrument to control not only the indoor air temperature, especially in mild and hot summer months, but also the humidity and other gas concentrations such as CO_2, ammonia, and methane throughout the year. Ventilation removes the excess heat and moisture from the greenhouse, livestock, poultry, and aquatic environments reducing both the cooling and moisture loads while improving the other air quality parameters. In greenhouses, it is also a vital environmental control component in terms of bringing the outside air, which has a higher CO_2 concentration than the inside air, if CO_2 enrichment is not being utilized. As will be discussed later in Chapter 8, in energy-conserving greenhouses, livestock, poultry, and aquatic buildings, it is desirable to reduce ventilation rates to reduce the heat loss due to ventilation in colder months. That being said, however, some other greenhouses may have entirely closed-systems that require no ventilation at all, which will also be covered later in this chapter as well as in Chapter 8.

5.8.4.1 Natural Ventilation in Plant Environments

In natural (passive) ventilation, obviously, a pressure difference has to be created by wind or temperature gradient for natural ventilation to occur, and this type of cooling heavily depends on the crops' cooling themselves via transpirational cooling. In order to take advantage of wind-created pressure differences, it is recommended that vent openings be present on the sides and on the ridge of the controlled-environment building. See Figures 5.31 to 5.34 for different configurations of natural ventilation in conventional greenhouse systems. Most greenhouse operations depend only on natural ventilation even for summer cooling. In relatively dry climates where wet-bulb depression permits efficient evaporative cooling, then evaporative cooling systems become handy to support both commercial and institutional greenhouse operations as a cost-effective option.

Greenhouses with only sidewall vents depend on wind pressure to force air exchange and are usually ineffective, especially in large commercial greenhouse operations. Therefore, greenhouses typically have vent openings on both sides of the ridge and on both sidewalls running continuously the full length of the greenhouse. The total ridge vent area should be equal to the total sidewall vent area, and each should be designed to have at least 15%–20% of the greenhouse floor area. Depending on the wind direction, the leeward (downwind side) vents are opened automatically to create a negative pressure at the top of the ridge so that effective natural ventilation can be achieved. Vent openings are adjusted depending on the rain and wind conditions as well.

Efficient greenhouse ventilation systems are desperately needed for providing comfortable and uniform greenhouse environments in different greenhouse structures. Much research effort has been spent on providing uniform greenhouse ventilation; however, the outcomes of ventilation studies performed in research-scale greenhouses are many times questionable with respect to their effectiveness in much larger commercial greenhouses. Many thanks to the powerful computational fluid dynamics modeling tools that we are now able to determine the weakest distribution areas of the traditional ventilation systems and increase their performances.

FIGURE 5.31 Sidewall (left) and ridge (right) ventilation at a Venlo greenhouse complex.
Source: *Greenhouse Engineering* (2021).

FIGURE 5.32 Roof ventilation with retractable shading system at a greenhouse complex.
Source: *Greenhouse Engineering* (2021).

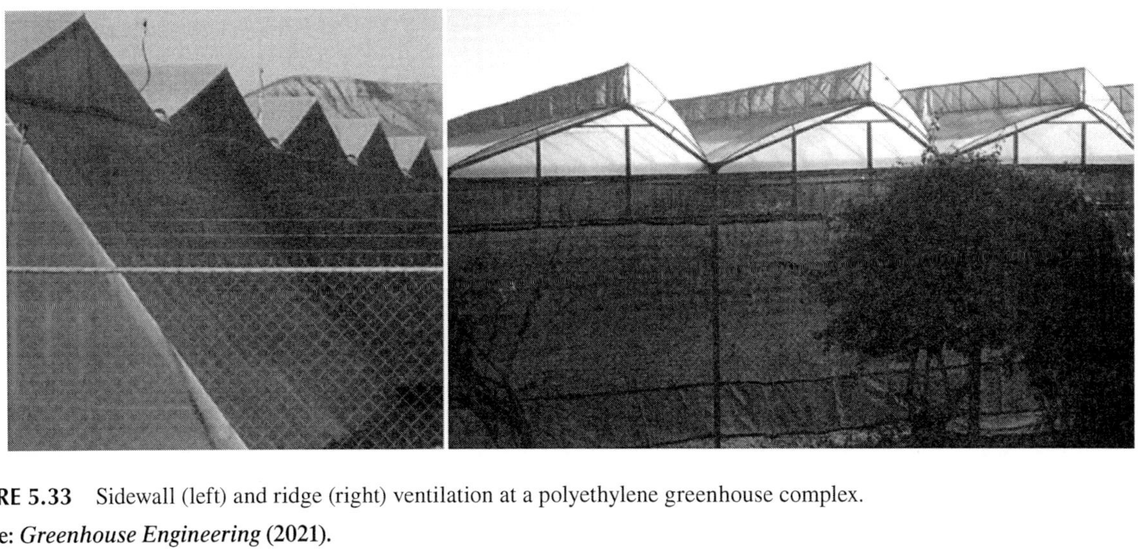

FIGURE 5.33 Sidewall (left) and ridge (right) ventilation at a polyethylene greenhouse complex.
Source: *Greenhouse Engineering* (2021).

FIGURE 5.34 Sidewall and roof ventilation with insect screen installed at a research facility.

Source: ***Greenhouse Engineering*** **(2021).**

Example 5.7: Sizing Natural Ventilation for a Greenhouse

Now let's consider a 3 bay single-glass gable roof greenhouse with 24 ft (7.2 m) span, 150 ft (45 m) long, and 14 ft (4.2 m) wall height. Determine the required sidewall (both sides) and ridge (both sides) opening sizes for natural ventilation running continuously the full length of the greenhouse.

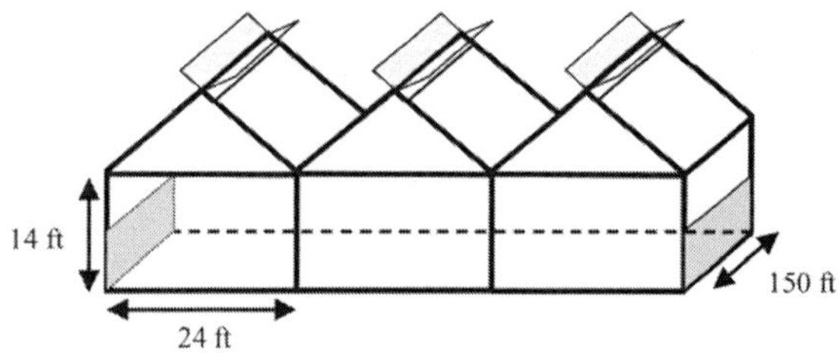

Solution

Remember that the total ridge vent area will be equal to the total sidewall vent area, and each will be designed to have at least 15%–20% of the greenhouse floor area. Here, let's assume that we are going to apply 20%.

$$\text{Greenhouse Floor Area} = (3\,\text{bays})(24\,\text{ft})(150\,\text{ft})$$
$$= 10{,}800\,\text{ft}^2$$

$$\text{Total Ridge Vent Area} = (0.20)\left(10{,}800\,\text{ft}^2\right) = 2{,}160\,\text{ft}^2$$

(This area is the total sidewall vent area as well.)

$$\text{Ridge Vent Area per Bay} = \left(2{,}160\,\text{ft}^2\right)/(3\,\text{bays}) = 720\,\text{ft}^2$$

$$\text{Ridge Vent Area (each side) per Bay} = \frac{720\,\text{ft}^2}{2} = 360\,\text{ft}^2$$

$$\text{Sidewall Vent Area (each side)} = \frac{2{,}160\,\text{ft}^2}{2} = 1{,}080\,\text{ft}^2$$

Both the ridge and sidewall openings will be running continuously the full length of the greenhouse; then,

$$\text{Ridge Vent Area (each side) Dimension} = 2.4\,\text{ft} \times 150\,\text{ft}$$
$$\text{Ridge Vent Area (each side) Dimension} = 360\,\text{ft}^2\left(\approx 34\,\text{m}^2\right)$$

$$\text{Sidewall Vent Area Dimension} = 7.2\,\text{ft} \times 150\,\text{ft}$$
$$\text{Sidewall Vent Area Dimension} = 1{,}080\,\text{ft}^2\left(\approx 100\,\text{m}^2\right)$$

And this will be installed at the ground level on each side.

5.8.4.2 Natural Ventilation in Animal Environments

As mentioned earlier, a pressure difference has to be created by wind or temperature gradient for natural ventilation to occur. The difference between inside and outside wind pressure and temperature moves the air through the building. In order to make use of the wind or temperature created pressure differences, just like in greenhouses, we must design adjustable and fixed-vent openings on the sides and on the ridge of livestock and poultry buildings so that natural ventilation can be accomplished. For this purpose, adjustable openings may include windows, wall panels, vents, and ridge ventilators, while fixed openings include building open fronts and continuous eave and ridge slots.

As discussed in Section 3.6.2, there are two types of naturally ventilated buildings: i) cold buildings—these buildings are usually uninsulated or lightly insulated, enclosed buildings with unregulated or very limited controlled natural ventilation, and remain cold in winter; hence, they are normally used for machinery storage, and only for large animals, such as cattle or finishing and gestating swine; and ii) modified environment buildings—these buildings are well insulated buildings and designed for controlled natural and fan ventilation, and typically used for warm free stall livestock buildings, poultry production buildings, and swine finishing and gestating buildings. Generally speaking, in economical terms, natural ventilation is much more economical for mature animals than forced or mechanical ventilation. It can also be used for young animals on bedded floors or in well-insulated modified environment buildings with zone heat.

It should also be noted that building orientation is an integral part of both natural and mechanical ventilation. Hence, open-front buildings should be oriented with the open side away from prevailing wind direction; and when possible, in order to provide best summer shade and winter solar light penetration, they should be built with an east-west roof orientation in mid and high latitudes. In well-insulated modified environment buildings, on the other hand, the solar load on the north-south oriented roofs is not a great concern as for the cold buildings; therefore, these buildings can be built perpendicular to winter and summer prevailing winds for better ventilation. As is clearly seen here, winter and summer time ventilation requirements and dynamics of buildings are quite different (Figure 5.35), and therefore, they should be treated differently.

In winter ventilation of the buildings, an air suction is created across the open ridge that draws warm and moist air out through the ridge openings, while fresh air sucked in through the eave openings. In summer ventilation of the buildings, on the other hand, large sidewall openings (typically one-third to one-half of the sidewalls) in each sidewall allow an air cross-flow within the building. The ridge openings have, though they still contribute, not much effect in summer.

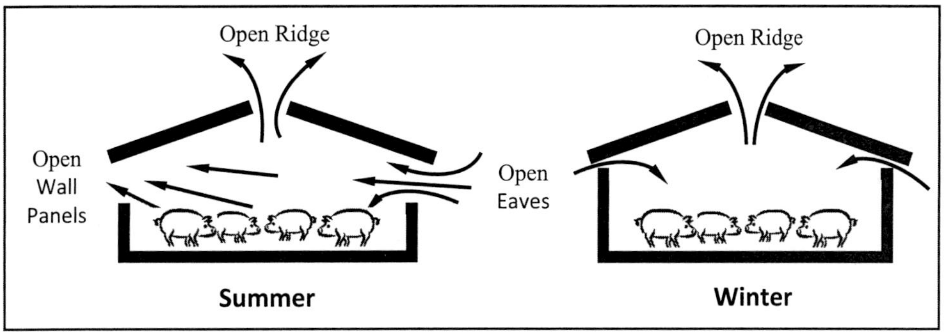

FIGURE 5.35 Summer (*left*) and winter (*right*) air flow with natural ventilation.

TABLE 5.3
Natural Ventilation Openings for Cattle Housing

Building Width (m)	**9**	**10**	**11**	**12**	**13**	**14**	**15**	**16**	**17**	**18**	**19**	**20**	**21**	**24**
Eave and Ridge Opening (cm)[a]	15	16.7	18.3	20	21.7	23.3	25	26.7	28.3	30	31.7	33.3	35	40
Sidewall Opening (cm)[b]	70	76.7	83.4	90	96.7	103.3	110	116.7	123.3	130	136.7	143.3	150	170

[a] Ridge openings are sized for 4.5 m/s (10 mph) wind.
[b] Sidewall openings are sized for a 0.9 m/s (2 mph) wind and are a minimum. Opening the back sidewall more increases summer cooling.

Source: Structures & Environment (1987).

5.8.4.2.1 Natural Ventilation Design in Animal Environments

Considering there exists two types of ventilation requirements for livestock buildings, we have to provide two different sizes of adjustable openings for summer and winter ventilation: large and smaller slot-type air inlets, respectively. For summer ventilation in temperate climates, the air inlets are usually 7.5 cm × 15 cm or 10 cm × 20 cm hinged, pivoting, or sliding doors placed in the sidewalls for effective cross ventilation. For winter ventilation, on the other hand, the air inlets typically consist of openings at the eaves of the building and the air outlets are placed at the ridge or high point of the roof.

Air inlets in winter ventilation are usually placed between rafters or under overhangs in order to prevent air drafts directly at animal level. These openings are baffled in climates with severe winters so that the inlets can be closed in stormy weather in order to prevent snow from getting into the building. Air outlets in winter ventilation are typically baffled slot type openings, which control the airflow. In order to avoid large temperature fluctuations inside the building due to changing weather, vent openings need to be adjusted, and this can easily be performed using a thermostat controller located at the center of the building for automatic door adjustment, for instance. It can also be convenient to disconnect all unwanted doors during cold weather and control only the doors needed at that particular time. Moreover, sometimes rather than using baffled slots, roof ventilators are used that have adjustable openings depending on the weather conditions while others rotate from wind forces. It should be noted that roof ventilators are not as effective as an open ridge ventilation due to the birds' nests, dust, or frost.

5.8.4.2.1.1 Ridge Openings As mentioned earlier, snow buildup is important at ridge openings and can be prevented by using a ridge opening of at least 7.5 cm wide. For cattle, a continuous slot opening sized at about 5 cm of width for each 3 m of building width is used (see Table 5.3). For swine in temperate climates, ridge openings are typically sized to provide 30 to 40 cm² per finishing pig and 45 cm² per gestating sow. In the case swine are relatively evenly distributed in the livestock building with adequate space, then one can comfortably choose opening sizes using Table 5.4.

In cold buildings, the ridge opening is simply constructed by removing the ridge cap and covering each truss with adequate flashing. In modified-environment buildings, ridge openings are constructed with closures or commercial ridge ventilators with adjustable openings so that airflow can be reduced conveniently during cold weather. In addition, employing adjustable closures, one can increase the ridge opening width 50% or more to improve airflow on relatively calm days.

It should also be kept in mid that in general, better airflow regimes are obtained when the roofs are steeper and smoother underside. Also, exposed purlins deeper than 10 cm should be avoided. In temperate and colder climates, roof slopes less than 3/12 for gable roofs and 2/12 for monoslope roofs should, for all kinds of reasons, be avoided as well. Moreover, a properly sized ridge opening (i.e., the air exiting through) normally prevents most precipitation from entering the building, hence the raised ridge caps are not recommended as they disrupt air flow and may trap snow as well. In the case birds create a problem, then installing a 2 cm × 2 cm (¾" × ¾") hardware cloth should be considered over the ridge openings; however, screened ridge openings need to be maintained properly in cold climates due to freezing shut.

5.8.4.2.1.2 Eave Openings Typical eave openings include the openings between rafters and the openings at the top of pivot doors. They can easily be provided with slightly opened sidewall doors or at the open spaces between trusses. Also, continuous eave openings are built along both sides of the building, and are deliberately located high on the sidewall so that incoming air has a chance to warm up before reaching the animals. Eave openings should also be provided with facia boards to reduce drafts on animals, and with eave baffles so that airflow can be reduced without completely closing them. It must be noted that, for an effective natural ventilation, each eave inlet should be sized to have at least as much open area as the ridge opening. Again, if birds are a problem, then $1.25\,\text{cm} \times 1.25\,\text{cm} \left(\frac{1}{2}" \times \frac{1}{2}"\right)$ hardware cloth should be considered as a screen to keep birds out.

5.8.4.2.1.3 Sidewall Openings Summer vent doors such as plastic curtains, pivot doors, and top- or bottom-opening doors perform very well. When these systems are used, usually 1/3 to 1/2 of the sidewall is kept open in order to provide a non-disturbing airflow at animal level. Sidewall openings in swine buildings are sized per Table 5.4, and the vent doors are constructed to open fully to a horizontal position so that none of the opening is constricted. Increasing door size to meet the recommended open area is always an option.

With respect to the management of openings, in gable roof modified-environment buildings, only the eave doors are adjusted to maintain building air conditions in cold weather, while the ridge is kept fully open except during very cold weather or blowing snow conditions. The large, sidewall vent doors, on the other hand, are kept closed in cold weather unless they are used for the eave openings. When the weather starts warming up, the first thing to do is gradually opening up the downwind vent doors, and when downwind doors are about half way open, then we start opening the upwind doors, and eventually open all the vent doors completely when the weather gets hot. We should be doing exactly the reverse of this sequence as temperature decreases again.

Moreover, open fronts in gable roof open-front buildings have limited ventilation control, and animals must adjust to quickly changing temperatures in the building. In order to have better control and provide the animals with some kind of adjustment opportunity to the rapidly changing temperatures, the back sidewall doors are opened in spring and closed in fall, and the eave vents are closed when outside air temperatures drop below freezing. In addition, the ridge vents are kept fully open and rarely closed, regardless of the weather.

TABLE 5.4
Natural Ventilation Openings for Swine Housing: Slot Widths for Cold and Mild Weather

	Winter Air Inlet Height (H_i)		Winter Air Outlet Constant (C_t)	
Building Width (m)	**Finishing (cm)**	**Gestation (cm)**	**Finishing**	**Gestation**
3.0–4.5	1.9	1.3	9	6
4.6–6.0	2.5	1.9	12	9
6.1–7.5	3.2	2.5	15	12
7.6–9.0	3.8	3.2	18	15
9.1–10.5	4.5	3.8	21	18
10.6–12.0	5.0	3.8	24	18

Install a continuous winter air inlet along each side of the building.
Make each inlet the height (H_i) in this table.
Make winter air outlets at least 7.5 cm wide to prevent them from freezing shut. It is usually necessary to install several shorter vents that are 7.5 cm or wider, instead of a continuous narrow opening.
Provide an outlet at least 4.8 m o.c. to determine the length of the outlet opening required per 16 ft of building length (L_o), and divide the table constant (C_t) by the outlet opening width (W_o is 7.5 cm or wider).

Source: Structures & Environment (1987).

Example 5.8: Designing Natural Ventilation Openings for a Gable Roof Swine Finishing Building

Assuming a ridge opening width (W_o) of 7.5 cm (3 in.), design the natural ventilation openings for a given 12 m (40 ft) wide gable roof swine finishing building.

Solution

Step 1: From Table 5.4, each continuous eave inlet is 5.0 cm high (H_i).

Step 2: Assume a 7.5 cm (3 in.) wide ridge opening (W_o). Divide the table constant (C_t) by the ridge opening width (W_o) to get the length of 7.5 cm (3 in.) opening required for each 4.8 m (16 ft) of building length (L_o). From Table 5.4, the table constant (C_t) is 24, then the opening for each 4.8 m (16 ft) of building length (L_o) can be calculated as:

$$L_o = \frac{C_t}{W_o} = \frac{24}{7.5\ \text{cm}} = 3.2\ \text{m}$$

TABLE 5.5
Natural Ventilation Openings for Swine Housing: Sidewall Openings for Warm Weather

	Gable Roof	Monoslope Roof	
Building Width (m)	**Each Sidewall (cm)**	**Backwall (cm)**	**Frontwall (cm)**
3.0–4.5	60	30	90
4.6–6.0	75	40	120
6.1–7.5	90	45	150
7.6–9.0	105	60	165
9.1–10.5	120	-	-
10.6–12.0	150	-	-

Heights given are for one opening the entire building length. Provide an opening in each sidewall.

Source: Structures & Environment (1987).

This means the opening for each 4.8 m (16 ft) of building length (L_o) is 3.2 m (10 ft, 8 in.).

Step 3: From Table 5.5, install a continuous 150 cm (60 in.) high opening along each sidewall.

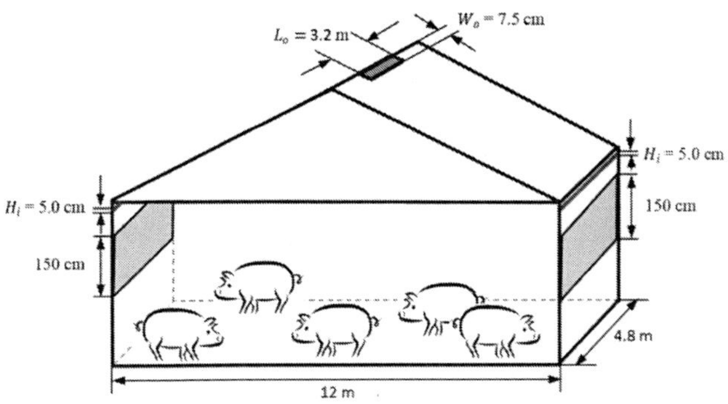

5.8.4.3 Natural Ventilation in Aquatic Environments

Just like in greenhouses, most greenhouse-type aquatic buildings depend only on natural ventilation for excess moisture and gas removal, and even for summer cooling. While some evaporative cooling already takes due to the existing free water surfaces in aquatic buildings, in relatively dry climates where wet-bulb depression permits further evaporative cooling, then commercial evaporative cooling systems become handy to support both the cooling of commercial and institutional aquatic buildings as a cost effective option.

Greenhouse-type aquatic buildings with only sidewall vents depend on wind pressure to force air exchange, and while they are effective for small operations, they are usually ineffective in large operations. Therefore, these buildings typically have vent openings on both sides of the ridge and on both sidewalls running continuously the full length of the building. The rule of thumb is that the total ridge vent area should be equal to the total sidewall vent area, and each should be designed to have at least 15%–20% of the aquatic building floor area. Depending on the wind direction, the leeward (downwind side) vents are opened automatically to create a negative pressure at the top of the ridge so that effective natural ventilation can be achieved. Vent openings are adjusted depending on the rain and wind conditions as well. In aquatic buildings other than greenhouse-type buildings, adequate natural ventilation is designed based on the type of building, when possible, providing proper sidewall and ridge openings so that sufficient natural ventilation can be provided for the building. Example 5.9 demonstrates a natural ventilation system design for a greenhouse-type aquatic building.

Example 5.9: Sizing Natural Ventilation for an Aquatic Building

Now let's consider a single glass gable roof greenhouse type aquatic building with 24 ft (7.2 m) span, 150 ft (45 m) long and 14 ft (4.2 m) wall height. Determine the required sidewall (both sides) and ridge (both sides) opening sizes for natural ventilation running continuously the full length of the building.

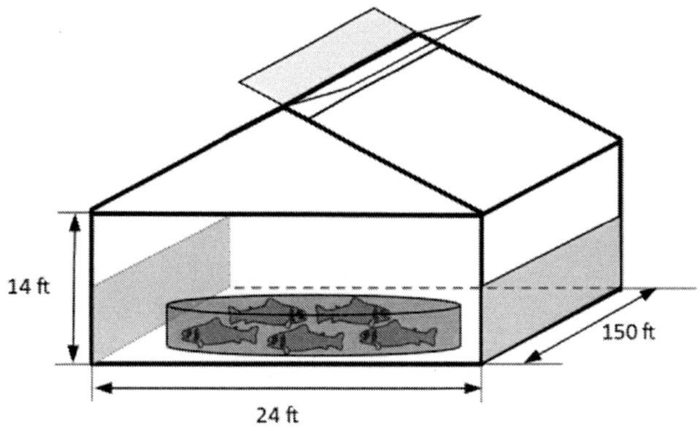

Solution

Remember that the total ridge vent area will be equal to the total sidewall vent area, and each will be designed to have at least 15%–20% of the building floor area. Here, let's assume that we are going to apply 20%.

$$\text{Greenhouse Floor Area} = (1\text{bay})(24\,\text{ft})(150\,\text{ft}) = 3{,}600\,\text{ft}^2$$

$$\text{Total Ridge Vent Area} = (0.20)\left(3{,}600\,\text{ft}^2\right) = 720\,\text{ft}^2$$

(This area is the total sidewall vent area as well.)

$$\text{Ridge Vent Area per Bay} = \left(720\,\text{ft}^2\right)/(1\text{bay}) = 720\,\text{ft}^2$$

$$\text{Ridge Vent Area (each side) per Bay} = \frac{720\,\text{ft}^2}{2} = 360\,\text{ft}^2$$

$$\text{Sidewall Vent Area (each side)} = \frac{720\,\text{ft}^2}{2} = 360\,\text{ft}^2$$

Both the ridge and sidewall openings will be running continuously the full length of the building; then,

$$\text{Ridge Vent Area (each side) Dimension} = 2.4\,\text{ft} \times 150\,\text{ft}$$

$$\text{Ridge Vent Area (each side) Dimension} = 360\,\text{ft}^2 \left(\approx 34\,\text{m}^2\right)$$

$$\text{Sidewall Vent Area Dimension} = 7.2\,\text{ft} \times 150\,\text{ft}$$

$$\text{Sidewall Vent Area Dimension} = 1{,}080\,\text{ft}^2 \left(\approx 100\,\text{m}^2\right)$$

This will be installed at the ground level on each side.

5.8.5 Active Cooling by Mechanical Ventilation

As mentioned earlier, the other ventilation system is mechanical, or fan or forced-air ventilation. It is the most commonly used tool to control the indoor air temperature (especially in mild and hot summer months), the humidity and other gas concentrations in controlled environments, such as CO_2, ammonia, and methane throughout the year.

5.8.5.1 Mechanical Ventilation in Plant Environments

Mechanical ventilating systems have fans, air inlets or outlets, and pertinent control systems. They provide better control over indoor air temperature, air movement, and moisture build up than natural ventilation. And mechanical ventilation systems, as will be discussed further in Section 5.8.5.2.2, can be positive pressure, negative pressure (exhaust), or neutral pressure systems.

FIGURE 5.36 Mechanical ventilation systems installed on a gable end (left) and a sidewall (right).
Source: *Greenhouse Engineering* (2021).

In greenhouses, the fans are usually located on the leeward side or on the gable end of the greenhouses (Figure 5.36), and incoming air is introduced into the greenhouse so that it mixes and is warmed by interior air before contacting the plants. If the fans need to be placed on the windward sides, then air exchange capacity should be increased at least 10%. Ventilation fans are sized to deliver required air exchange for a minimum static pressure of 0.015 kPa (0.060 in. H_2O); however, a static pressure of 0.025 kPa (0.10 in. H_2O) is typically used in designs.

The fans are placed no more than 7.6 m (25 ft) along the sidewalls or gable ends. It is recommended that mechanical ventilation rates be 0.005–0.01 $m^3/s.m^2$ (1–2 cfm/ft^2) for minimum winter air exchange and 0.06 $m^3/s.m^2$ (12 cfm/ft^2) for summer cooling. The fans should be activated in stages by the control system based on the air exchange needs of the greenhouse, and in the heating season, the first stage of ventilation should be activated at least 4–6°C (8–10°F) above the heating set-point so that the exhaust fans will not operate at the same time. In greenhouses where the air inlet to fan distance is less than 30 m (100 ft), ventilation flow rates should be increased to improve the heat transfer based on the following equation:

$$Q_{adj} = \frac{C \cdot Q_v}{\sqrt{D}} \tag{5.25}$$

where Q_{adj} is the amount of adjusted ventilation rate (m^3/s in SI or cfm in I-P), Q_v is the amount of minimum winter ventilation requirement (m^3/s in SI or cfm in I-P), C is a coefficient (5.5 in SI or 10.0 in I-P), and D is the inlet to fan distance (m in SI or ft in I-P).

For an efficient fan operation, a distance of four to five fan diameters should also be maintained between the fan discharge and any nearby obstruction such as another greenhouse or building. Air inlet louvers or shutters, on the other hand, have an inlet area of minimum 1.25 times the fan area, and should open outward. In cold regions where outside temperatures go below 5°C (40°F), incoming air is mixed and warmed by the warmer interior air before reaching the plants. In such cases, continuous louvers can be installed high along the sidewall, and the inlet air velocity is maintained at 3.6 m/s (700 ft/min) at the required ventilation flow rate. They are motorized and controlled in proportion to the number of active fans or in response to the pressure drop generated across the inlet louver.

Example 5.10: Sizing Mechanical Ventilation for a Greenhouse

Using the same greenhouse that was used in Example 5.7, now design a mechanical ventilation system and determine the fan sizes required. Assume that the static pressure required for the fans is 0.10 in H_2O.

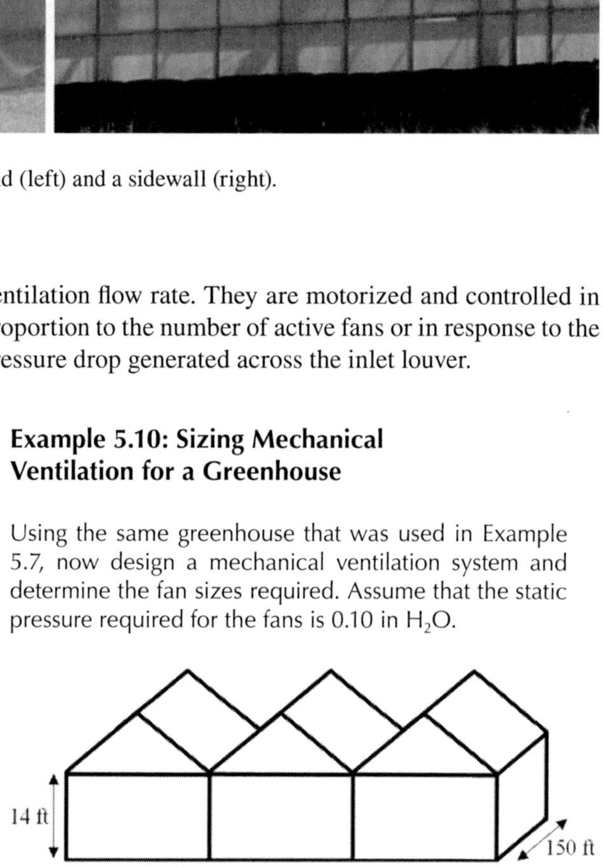

Solution

Remember that the air exchange rate for summer cooling is 0.06 $m^3/s.m^2$ (12 cfm/ft^2) while the minimum winter air exchange requirement is between 0.005–0.01 $m^3/s.m^2$ (1–2 cfm/ft^2).

$$\text{Greenhouse Floor Area} = (3\,\text{bays})(24\,\text{ft})(150\,\text{ft}) = 10{,}800\,\text{ft}^2$$

$$\text{Air Exchange Required for Summer Cooling} = \left(12\frac{\text{cfm}}{\text{ft}^2}\right)\left(10{,}800\,\text{ft}^2\right) = 129{,}600\,\text{cfm}$$

Since the inlet-to-fan distance is 72 ft (21.6 m), which is less than 100 ft (30 m), we need to adjust the ventilation rate using Equation (5.25) as follows:

$$Q_{adj} = \frac{(10.0)\,(129{,}600\,\text{cfm})}{\sqrt{72\,\text{ft}}} = 152{,}735\,\text{cfm}$$

(This is the adjusted maximum summer ventilation rate.)

$$\text{Minimum Winter Air Exchange Requirement} = \left(2\frac{\text{cfm}}{\text{ft}^2}\right)\left(10{,}800\,\text{ft}^2\right)$$

$$\text{Minimum Winter Air Exchange Requirement} = 21{,}600\,\text{cfm}$$

Again, we need to adjust the minimum ventilation rate as follows:

$$Q_{adj} = \frac{(10.0)\,(21{,}600\text{ cfm})}{\sqrt{72\text{ ft}}} = 25{,}456\text{ cfm}$$

(This is the adjusted minimum winter ventilation rate)

Minimum winter ventilation can uniformly be supplied by four fans installed each delivering 6,438 cfm at the assumed static pressure of 0.10 in. (model: GHE24; fan size: 24 in.; motor: 1/2 hp; fan RPM: 891). Please note that while fan specifications are real, the manufacturer model used in this example is generic.

These four fans can deliver a total of 25,752 cfm when they operate all together; therefore, in order to determine the spring/fall and summer ventilation fans, this total needs to be subtracted from the grand total (152,735 cfm).

$$152{,}735\,\text{cfm} - 25{,}752\,\text{cfm} = 126{,}983\,\text{cfm}$$

(This much more to be provided for spring/fall and summer ventilation)

We may have other locally available options; however, for this example, we will pick six fans, each delivering 21,283 cfm at the assumed static pressure of 0.10 in. (model: GHE54; fan size: 54 in.; motor: 1 hp; fan RPM: 374). Again while fan specifications are real, the manufacturer model is generic.

		Manufacturer Model
1) Winter Ventilation	4 fans at 6,438 cfm each	GHE24 (single speed)
2) Spring/Fall Ventilation	3 fans at 21,283 cfm each	GHE54 (two speed)
3) Summer Ventilation	3 fans at 21,283 cfm each	GHE54 (two speed)
Total Fan Capacity:	**153,450 cfm (installed)**	

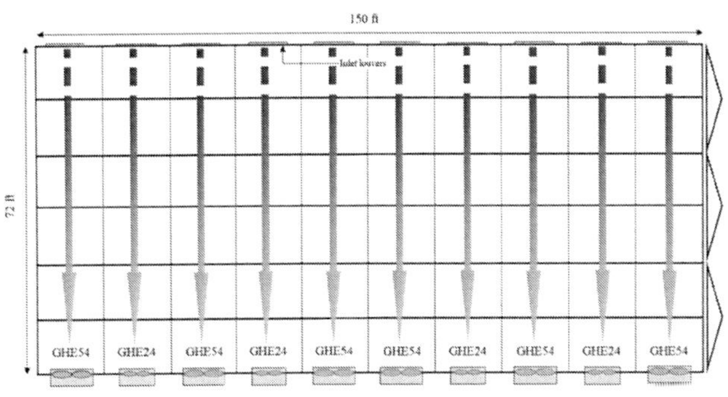

In this design, the fans and inlet louvers are placed within the distance of 7.6 m (25 ft) along the sidewalls. Also note that the inlet-to-fan distance is less than 100 ft, therefore, we have made an adjustment in the ventilation rate. In the following example, we are going to use five bays, which will give us an inlet-to-fan distance of 120 ft that is greater than 100 ft, where no adjustment will be required. For an efficient fan operation, keep in mind that a distance of four to five fan diameters should also be maintained between the fan discharge and any nearby obstruction such as another greenhouse or building.

Our design for this greenhouse has a total installed capacity of 153,450 cfm, which will provide little more than what is required (152,735 cfm); therefore, the overall ventilation system will be in good shape providing uniform air exchange with the provided configuration in the greenhouse as shown above. The fans are nicely sized and located to provide uniform air exchange and prevent oscillations when controlled properly in stages. Remember that we have used four fans (6,438 cfm each) for minimum (winter) ventilation distributed along the sidewall, which will operate continuously. These four fans can as well operate in steps or stages providing further (proportional) control over minimum ventilation when needed.

Furthermore, we have split the total summer (the maximum) air exchange requirement into two stages (low and high) using two-speed fans, providing half the capacity in stages for spring/fall ventilation, and the other half to top up in stages for summer ventilation. This means that all the GHE24 fans (minimum winter ventilation fans) and all the GHE54 fans (low stage only) would be operating for spring/fall ventilation in stages as needed; and all the GHE24 and GHE54 fans (low and high stages) would be operating in stages and eventually providing the total air exchange of 153,450 cfm when needed in summer.

This configuration will provide excellent spatial and temporal ventilation for this specific greenhouse. It should however be kept in mind that larger fan diameter and lower HP fans consume less operational energy for delivering the same air exchange; therefore, alternative fans with larger diameter and lower HP motors should be chosen, when possible. Reducing fan rotational speed is the most efficient way to reduce fan energy use in variable flow, and this can be efficiently achieved with frequency controlled electric motors. For instance, if we reduce the fan rotational speed by 10%, based on the fan/pump affinity laws covered in Chapter 2 (and this is further discussed in Chapter 8), the fan power will reduce 27%. It means, not only the initial costs but also the operational costs of fans should be considered for making a justified fan decision.

Here, we will pick motorized air inlet louvers based on 1.5 times the areas of the fans. They open outward during the fan operation and are controlled based on the number of active fans or in response to the pressure drop generated across the inlet louver. Total installed ventilation capacity in this example is 153,450 cfm. Using the inlet air velocity maintained at about 3.6 m/s (700 ft/min) at the inlet louver, we can calculate the cross-sectional area of inlet louver as follows:

$$\text{Total Cross-Sectional Area of Inlet Louver} = 1.5\left(\frac{153{,}450\,\text{ft}^3/\text{min}}{700\,\text{ft}/\text{min}}\right)$$

$$\text{Total Cross-Sectional Area of Inlet Louver} \approx 329\,\text{ft}^2$$

If a continuous inlet louver is to be installed on the opposite sidewall, since the total length of the greenhouse sidewall is 150 ft, then the width of the inlet louver becomes 329 ft^2 ÷ 150 ft = 2.2 ft ≈ 2¼ ft. Hence, the inlet louver dimension becomes 2¼ ft × 150 ft. We can also install multiple louvers distributed along the opposite sidewall for a uniform air flow distribution in the greenhouse. As shown on the figure above, for instance, 10 inlet louvers can be used with a cross-sectional area of approximately 33 ft^2 each.

Example 5.11: Sizing Mechanical Ventilation for a Greenhouse

Maintaining the floor area of the greenhouse that was used in Example 5.7, however, this time with five bays, design a mechanical ventilation system and determine the fan sizes. Again, assume that the static pressure required for the fans is 0.10 in. H_2O.

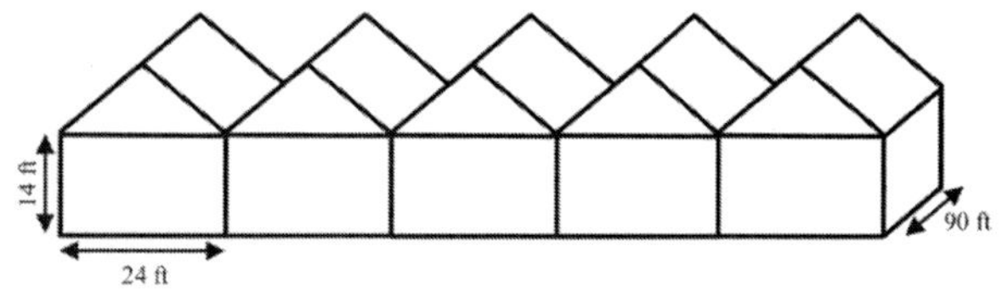

Solution

Again, remember that the air exchange rate for summer cooling is 0.06 m^3/s.m^2 (12 cfm/ft^2) while the minimum winter air exchange requirement is between 0.005–0.01 m^3/s.m^2 (1–2 cfm/ft^2).

$$\text{Greenhouse Floor Area} = (5\,\text{bays})(24\,\text{ft})(90\,\text{ft}) = 10{,}800\,\text{ft}^2$$

$$\text{Air Exchange Required for Summer Cooling}:$$

$$= \left(12\frac{\text{cfm}}{\text{ft}^2}\right)\left(10{,}800\,\text{ft}^2\right) = 129{,}600\,\text{cfm}$$

$$\text{Minimum Winter Air Exchange Requirement}:$$

$$= \left(2\frac{\text{cfm}}{\text{ft}^2}\right)\left(10{,}800\,\text{ft}^2\right) = 21{,}600\,\text{cfm}$$

In this case, minimum winter ventilation can uniformly be supplied by four fans each delivering 5,421 cfm at the assumed static pressure of 0.10 in. (model: GHE24; fan size: 24 in.; motor: 1/3 HP; fan RPM: 769). Again, the manufacturer model used in this example is generic.

These four fans can deliver a total of 21,684 cfm maximum when they operate all together; therefore, in order to determine the spring/fall and summer ventilation fans, this total need to be subtracted from the grand total, 129,600 cfm:

$$129{,}600\,\text{cfm} - 21{,}684\,\text{cfm} = 107{,}916\,\text{cfm}$$

(This much more to be provided for spring/fall and summer ventilation)

We may have other locally available options; however, for this example, we will pick six fans each delivering 18,238 cfm at the assumed static pressure of 0.10 in. (model: GHE48; fan size: 48 in.; motor: 1 HP; fan RPM: 380). Again, while the fan characteristics are real, the manufacturer models are generic.

		Manufacturer Model
1) Winter Ventilation	4 fans at 5,421 cfm each	GHE24 (single speed)
2) Spring/Fall Ventilation	3 fans at 18,238 cfm each	GHE48 (two speed)
3) Summer Ventilation	3 fans at 18,283 cfm each	GHE48 (two speed)
Total Fan Capacity:	**131,112 cfm (installed)**	

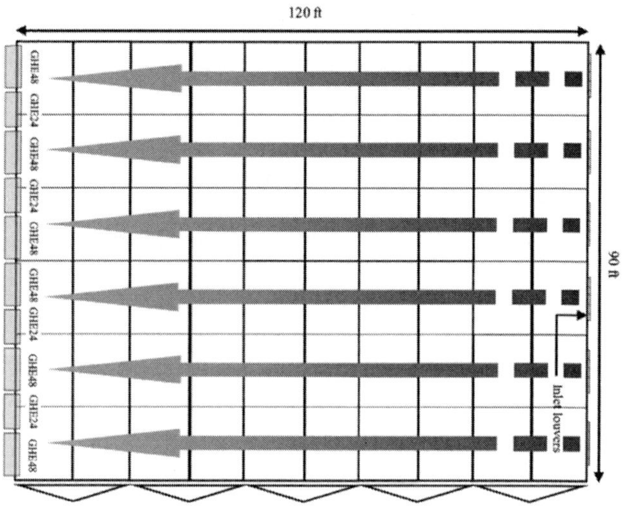

In this design, the fans and inlet louvers are placed within 7.6 m (25 ft) along the sidewalls. Also note that the inlet-to-fan distance is less than 100 ft; therefore, no adjustment was required in the ventilation rate. For an efficient fan operation, keep in mind that a distance of four to five diameters should also be maintained between the fan discharge and any nearby obstruction such as another greenhouse or building.

Our design for this greenhouse has the total installed capacity of 131,112 cfm, which will provide little more than what is required (129,600 cfm); therefore, the overall ventilation system will be in good shape providing uniform air exchange with the provided configuration in the greenhouse as shown. The fans are nicely sized and located to provide uniform air exchange and prevent oscillations when controlled properly. Remember that we have used four fans (5,421 cfm each) distributed along the sidewall for minimum (winter) ventilation that will operate continuously. These four fans can operate in steps or stages providing further (proportional) control over minimum ventilation when needed.

Furthermore, we have split the total summer (the maximum) air exchange requirement into two stages (low and high) using two-speed fans, providing half the capacity in stages for spring/fall ventilation, and the other half to top up in stages for summer ventilation. This means that all the GHE24 fans (minimum winter ventilation fans) and all the GHE48 fans (low stage only) would be operating for spring/fall ventilation in stages as needed; and all the GHE24 and GHE48 fans (low and high stages) would be operating in stages and eventually providing the total air exchange of 131,112 cfm when needed in summer. This configuration will provide excellent spatial and temporal ventilation for this specific greenhouse. It should however be kept in mind that larger fan diameter and lower HP fans consume less operational energy for delivering the same air exchange; therefore, alternative fans with larger diameter and lower HP motors should be chosen, when possible. Again, reducing the fan rotational speed is the most efficient way to reduce fan energy use in variable flow, and this can be efficiently achieved with frequency controlled electric motors.

Here, we will again pick motorized air inlet louvers based on 1.5 times the areas of the fans. They open outward during the fan operation and are controlled based on the number of active fans or in response to the pressure drop generated across the inlet louver.

Total installed ventilation capacity in this example is 131,112 cfm. Using the inlet air velocity maintained at about 3.6 m/s (700 ft/min) at the inlet louver, we can calculate the cross-sectional area of inlet louver as follows:

$$\text{Total Cross-Sectional Area of Inlet Louver} = 1.5\left(\frac{131{,}112\,\text{ft}^3/\text{min}}{700\,\text{ft}/\text{min}}\right)$$

$$\text{Total Cross-Sectional Area of Inlet Louver} \approx 281\,\text{ft}^2$$

If a continuous inlet louver is to be installed on the opposite sidewall, since the total length of the greenhouse sidewall is 150 ft, then the width of the inlet louver becomes 281 ft² ÷ 90 ft = 3.12 ft ≈ 3¼ ft. Hence, the inlet louver dimension is 3¼ ft × 90 ft. We can also install multiple louvers distributed along the opposite sidewall for a uniform air flow distribution in the greenhouse. As shown in the figure, for instance, six inlet louvers can be used with a cross-sectional area of approximately 48 ft² each.

5.8.5.2 Mechanical Ventilation in Animal Environments

As briefly introduced in Section 5.8.5.1, mechanical ventilation systems have fans, air inlets or outlets, and pertinent control systems. They provide better control over indoor air temperature, air movement, moisture build up and odors in livestock and poultry buildings compared to natural ventilation. And as already covered in depth in Chapter 4, livestock and poultry produce best when the environment provided is within their optimum growth and production zones. Therefore, we need to provide the best environmental conditions and control systems in these buildings.

5.8.5.2.1 Heat and Moisture Balances

Design of the environmental control in these buildings begins with analyzing the heat and moisture flows in the building. Energy and mass balances were already discussed in Chapter 2. An energy balance in livestock and poultry buildings employs sensible heat gains and heat losses; a moisture balance, on the other hand, employs latent heat sources and losses. Hence, two basic equations will be presented below for determining ventilation requirements: one based on available sensible heat and the other on latent heat.

First, using energy balance Equation (2.1) and ignoring the internal storage term, we earlier developed Equations (5.1) and (5.2) for a livestock and poultry building. $\dot{Q}_{evap}$, the rate of sensible heat loss by evaporating water, in Equation (5.2) is typically ignored because some of the heat generated by animals is used to evaporate within the building. Furthermore, $\dot{Q}_{equipment}$, the rate of sensible heat gain provided by the mechanical equipment, $\dot{Q}_{lighting}$, the rate of sensible heat gain provided by the lighting equipment, and $\dot{Q}_{solar}$, the rate of sensible heat gain from solar radiation in Equation (5.10) are all neglected due to the reasons discussed earlier. Then, Equation (5.11) was obtained, leaving the heat sources of $\dot{Q}_t$, the rate of supplemental heat provided by the furnace, and $\dot{Q}_{animal\_sensible}$, the rate of total sensible heat gains provided by animals (tabulated for each livestock species), and the heat loss terms of $\dot{Q}_{cond}$, the rate of heat loss from the building envelope, and $\dot{Q}_{inf}$, the rate of sensible heat loss due to air exchange between the inside and outside air, i.e., the ventilation.

5.8.5.2.1.1 Sensible Heat Balance Sensible heat gains, the terms on the left side of Equation (5.11), in a livestock and poultry building, raise the building temperatures, while the losses (on the right side of the equation), drop the temperatures. Hence, if we want to hold the building temperature constant, then sensible heat gains in that building have to be removed at the same rate it is produced, and this is done by properly controlled ventilation. If too much sensible heat is removed by ventilation, then temperatures start dropping, and if not enough sensible heat is removed, then this time, temperatures rise.

Rearranging Equation (5.11), the sensible heat required to be removed by ventilation for temperature control in a livestock and poultry building can be written as follows:

$$\dot{Q}_{inf} = \dot{Q}_t + \dot{Q}_{animal\_sensible} - \dot{Q}_{cond} \tag{5.26}$$

where $\dot{Q}_{inf}$ is the rate of sensible heat loss due to ventilation, $\dot{Q}_t$ is the rate of supplemental heat provided by the furnace, $\dot{Q}_{animal\_sensible}$ is the rate of total sensible heat gains provided by animals (tabulated for each livestock species), and $\dot{Q}_{cond}$ is the rate of heat loss from the building envelope.

As already expressed in Equation (5.7), $\dot{Q}_{inf}$, the rate of sensible heat loss due to ventilation, can also be written as follows:

$$\dot{Q}_{inf} = \dot{m}_a c_p \Delta T \tag{5.27}$$

or

$$\dot{Q}_{inf} = \dot{m}_a (h_2 - h_1) \tag{5.28}$$

where $\dot{Q}_{inf}$ is the rate of sensible heat loss due to ventilation (J/s in SI or Btu/h in I-P), $\dot{m}_a$ is the mass flow rate of dry air, i.e., the amount of ventilating air (kg_{da}/s in SI or lb_{da}/h in I-P), c_p is the specific heat of air at constant pressure (J/kg.°C in SI or Btu/lb.°F in I-P), ΔT is the temperature difference between the outside and exhaust air (°C in SI or °F in I-P), h_1 and h_2 are the enthalpies of incoming and exhaust air (J/kg_{da} in SI or Btu/lb_{da} in I-P), respectively.

Then, the ventilation rate for temperature control can be calculated as follows:

$$Q = m_a\, v/60 \quad \text{(I-P)} \tag{5.29a}$$

where Q is the ventilation rate (ft³/min or cfm), $\dot{m}_a$ is the mass flow rate of dry air, i.e., the amount of ventilating air (lb_{da}/h), and v is the specific volume of air (ft³/lb_{da}), which is the inside conditions for exhaust systems, and outside conditions for positive pressure systems, and 60 is for unit conversion (1 h = 60 min).

$$Q = m_a\, v \quad \text{(SI)} \tag{5.29b}$$

where Q is the ventilation rate (m^3/s), $\dot{m}_a$ is the mass flow rate of dry air, i.e., the amount of ventilating air (kg_{da}/s), and v is the specific volume of air (m^3/kg_{da}), which is the inside conditions for exhaust systems, and outside conditions for positive pressure systems.

Then, ventilation rate required to maintain a desired livestock and poultry building temperature can easily be calculated using Equation (5.26). Assuming that no supplemental heating is provided (i.e., $\dot{Q}_t$ is zero), the animal sensible heat production rate ($\dot{Q}_{animal\_sensible}$) must be in balance with the rate of heat loss through building surfaces ($\dot{Q}_{cond}$).

$$\dot{Q}_{animal\_sensible} = \dot{Q}_{cond} + \dot{Q}_{inf} \quad (5.30)$$

After substituting Equation (5.12) and Equation (5.27) into Equation (5.30), the ventilation rate (Q) equation takes the following forms:

$$Q = \frac{v \times \left[\dot{Q}_{animal\_sensible} - \sum UA \times (T_i - T_o)\right]}{\left[60 \times c_p \times (T_i - T_o)\right]} \quad \text{(I-P)} \quad (5.31a)$$

where Q is the ventilation rate for controlling the indoor temperature (ft^3/min or cfm), v is the specific volume of air (ft^3/lb_{da}), $\dot{Q}_{animal\_sensible}$ is the rate of sensible heat gain from animals (Btu/h; see Table A.63 or ASABE Standards (2018)), ΣUA is the total building heat loss factor (Btu/h °F; see Equation (5.6)), T_i is the indoor design temperature of the building (°F; see ASABE Standards (2018) or MWPS (1987)), T_o is the outside design dry-bulb temperature (99%; see Tables A.52–A.55 for climatic conditions for selected locations), 60 is for unit conversion (1 h = 60 min), and c_p is the specific heat of air (Btu/lb_{da}°F; see Table A.13). See Table 1 of Chapter 26, *ASHRAE* (2021) for typical thermal properties (density, thermal conductivity, thermal resistance, and specific heat) of common building and insulating materials.

$$Q = \frac{v \times \left[\dot{Q}_{animal\_sensible} - \sum UA \times (T_i - T_o)\right]}{\left[c_p \times (T_i - T_o)\right]} \quad \text{(SI)} \quad (5.31b)$$

where Q is the ventilation rate for controlling the indoor temperature (m^3/s), v is the specific volume of air (m^3/kg_{da}), $\dot{Q}_{animal\_sensible}$ is the rate of sensible heat gain from animals (J/s; Table A.63 or ASABE Standards (2018)), ΣUA is the total building heat loss factor (W/°C; see Equation (5.6)), T_i is the indoor design temperature of the building (°C; see ASABE Standards (2018) or MWPS (1987)), T_o is the outside design dry-bulb temperature (99%; see Tables A.52–A.55 for climatic conditions for selected locations), and c_p is the specific heat of air (J/kg°C; see Table A.13). See Table 1 of Chapter 26, *ASHRAE Handbook* (ASHRAE, 2021) for typical thermal properties (density, thermal conductivity, thermal resistance, and specific heat) of common building and insulating materials.

5.8.5.2.1.2 Calculating Heat Loss from Building Envelope
As discussed earlier and expressed in Equation (5.5), the rate of heat loss through each building component is proportional to its area and the difference between the inside and outside temperatures, $\dot{Q}_{cond} = \sum UA \times (T_i - T_o)$. The rate of heat flow is also determined by the total resistance (R_t) of the building component; the higher the total resistance, the lower the rate of heat flow. The total resistance $R_t = 1/U$ and the overall heat transfer coefficient U include the inside and outside boundary layer resistances, R_i and R_o, respectively, the values of which are provided in Table A.9 for different flow directions and surface emissivities.

It should be noted that all heat loss components are given per unit area except for the floor perimeter, which is the only special case the resistance value is given per length (Table 5.6). In the case of floor perimeter, the area, A, in Equation (5.5) is replaced by the length of the exterior wall. In order to obtain the total heat loss from a building, the losses through each building component are simply added together. Example 5.12 illustrates the procedure in a step-by-step fashion.

TABLE 5.6
R Values for Selected Building Components

Component	R_T (h.ft^2.°F/Btu)	R_T (m^2.K / W)
Roofs		
Metal roofing, 25/32" insulating sheathing	2.91	0.51
Metal roofing, 0.4" expanded polystyrene	3.95	0.70
Metal roofing, 1" molded polystyrene, 1 pcf	5.85	1.03
Ceiling		
2" expanded polyurethane	13.35	2.35
1/2" plywood, 4" glass or mineral wool fill insulation	13.47	2.37
Metal roofing, R-19 blanket insulation	19.85	3.49
1/2" plywood, 8" glass or mineral wool fill insulation	25.47	4.48
Doors		
1/2" plywood, R-2 blanket insulation, 3/4" air space, 1/2" plywood	4.99	0.88
1/2" plywood, 1" polysterene molded, 1 pcf 3/4" air space, 1/2" plywood	7.99	1.41
Floor Perimeter (per foot (I-P) or meter (SI) of exterior wall)		
Concrete	1.23	0.70
Concrete, with 2" × 24" of rigid insulation around perimeter	2.22	1.26

Source: Structures & Environment (1987).

Example 5.12: Calculating Heat Loss from Building Envelope

Calculate the amount of heat loss from the building envelope of a 24 ft × 48 ft building constructed without any windows as shown below. The height of the wall is 6 ft, and the height of the foundation wall is 2 ft. Also, the building has two 3 ft × 7 ft doors, insulated with 1 in., 1 pcf of molded polystyrene. The building is located near Columbus, Ohio, and the required inside design temperature is provided as 50°F.

Solution

Step 1: We first list the length, width, wall height, and foundation height of the building. Then, we calculate the perimeter, frame wall area (excluding windows and doors), concrete wall, ceiling, window, and door areas.

BUILDING DIMENSIONS & SURFACE AREAS

BUILDING DIMENSIONS	**(ft)**	**(m)**
House Width	24	7.2
House Length	48	14.4
Frame Wall Height	6	1.8
Concrete Wall Height	2	0.6
Perimeter	144	43.2
SURFACE AREAS	**(ft^2)**	**(m^2)**
Ceiling Area	1,152	103.7
Window Area	0	0
Door Area	42	3.8
Frame Wall Area (excluding windows and doors)	822	74
Concrete Wall Area	288	25.9

Step 2: We then list the inside and outside design temperatures of the building, and calculate and list R (see Table 5.6) for the building sections.

Frame Wall Resistance		**Concrete Wall Resistance**	
Outside surface (15 mph wind)	0.17	Outside surface (15 mph)	0.17
Sheet metal	0.00	Concrete (6 in.)	0.48
Batt insulation	11.00	Extruded polystyrene (2 in.)	10.00
Supplemental vapor barrier	0.00	Hardboard, tempered	0.25
Plywood (1/2 in.)	0.62	Inside surface (still air)	0.68
Inside surface (still air)	0.68	Total resistance, $(R_T)_f$	11.58
Total resistance, $(R_T)_w$	12.47		

OTHER PROVIDED or CALCULATED INPUTS

PARAMETER	**(I-P)**	**(SI)**
T_i	60 °F	15.5 °C
T_o	6 °F (99%); Table A.52	−14 °C (99%); Table A.53
ΔT	54 °F	29.5 °C
Location	Columbus, Ohio, USA	
R-Values		
Ceiling	13.47 $h.ft^2.°F / Btu$; Table 5.6	2.37 $m^2.K / W$; Table 5.6
Window	- $h.ft^2.°F / Btu$;	- $m^2.K / W$;
Door	7.99 $h.ft^2.°F / Btu$; Table 5.6	1.41 $m^2.K / W$; Table 5.6
Frame Wall	12.47 $h.ft^2.°F / Btu$; Calculated	2.20 $m^2.K / W$; Calculated
Concrete Wall	11.58 $h.ft^2.°F / Btu$; Calculated	2.04 $m^2.K / W$; Calculated
Perimeter	2.22 $h.ft^2.°F / Btu$; Table 5.6	1.26 $m^2.K / W$; Table 5.6

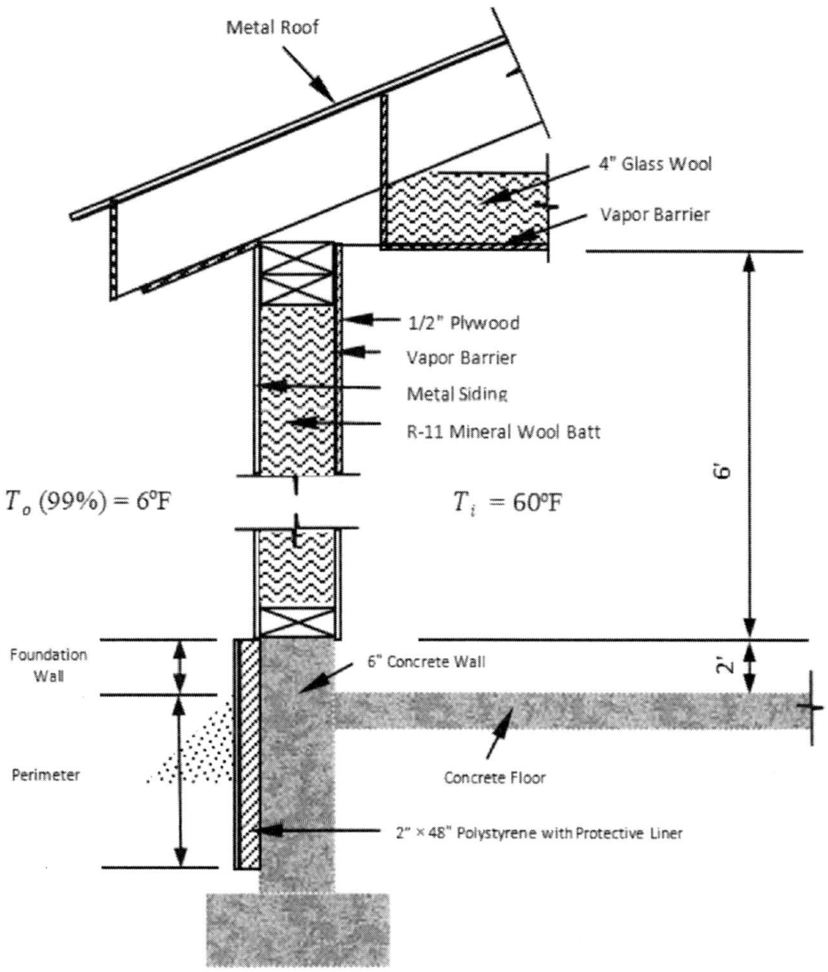

Building components for Example 5.12.

Step 3: Then, using Equations (5.5) and (5.6), we calculate the heat loss from the building sections and building.

HEAT LOSS FROM BUILDING SECTIONS

Building Section	$(1/R_T)_i$ (Btu/h.ft^2.°F)	A_i (ft^{2}of ft)	$(T_i - T_o)$ (°F)	$(\dot{Q}_{cond})_i$ (Btu/h)	$(\dot{Q}_{cond})_i$ (W)
Ceiling Area	1/13.47	1,152	54°F	4,618.3	1,354.3
Window Area	0	0		0	0
Door Area	1/7.99	42		283.9	83.3
Frame Wall	1/12.47	822		3,559.6	1043.9
Concrete Wall	1/11.58	288		1,343.0	393.8
Perimeter	1/2.22	144		3,502.7	1,027.2
Total Heat Loss, $\dot{Q}_{cond}$	246.5 Btu/h.°F			**13,311**	**3,904**

5.8.5.2.1.3 Moisture Balance As mentioned earlier, a moisture balance employs latent heat sources and losses in a livestock and poultry building. Moisture sources in these buildings include water vapor produced by animals that is calculated from latent heat production of animals. Moisture evaporation from wet building surfaces is, on the other hand, typically included in the latent heat production data of the animals. Moisture losses include water vapor losses taking place in the ventilating air (as colder and drier outside air replaces warmer and moister indoor air). Also, since surface condensation is not desirable in livestock and poultry housing, moisture loss from the indoor air by condensation on cold building surfaces is small, and therefore, neglected.

Using Equation (2.14) in steady-flow processes, the total moisture inflow rates are equal to the total moisture outflow rates; or in other words, if the indoor humidity ratio is to remain constant (typically below 70%–80%), the total rate of moisture loss must be equal to the total rate of moisture production as follows:

$$\dot{m}_a (W_2 - W_1) = W_a - W_e \qquad (5.32)$$

where $\dot{m}_a$ is the mass flow rate of dry air, i.e., the amount of ventilating air (kg_{da}/s in SI or lb_{da}/h), W_1 is the humidity ratio (water vapor) in incoming air (kg_{water}/kg_{da} in SI or lb_{water}/lb_{da} in I-P), W_2 is the humidity ratio (water vapor) in exhaust air (kg_{water}/kg_{da} in SI or lb_{water}/lb_{da} in I-P), W_a is the water vapor production by animals (kg_{water}/s in SI or lb_{water}/h in I-P; tabulated for each livestock species), and W_e is the water vapor production from surface water evaporation (kg_{water}/s in SI or lb_{water}/h in I-P; typically included in W_a).

Then, ventilation rate required to maintain a desired building moisture can easily be calculated using Equation (5.32). Considering that the water vapor production from surface water evaporation is usually included in the water vapor produced by animals, and after substituting Equation (5.29a) for I-P and Equation (5.29b) for SI in Equation (5.32), Q is calculated as

$$Q = \left(\frac{v}{60}\right) \times \left(\frac{W}{W_2 - W_1}\right) \quad \text{(I-P)} \qquad (5.33a)$$

where Q is the ventilation rate for controlling the indoor moisture (ft^3/min or cfm), v is the specific volume of air (ft^3/lb_{da}), which is the inside conditions for exhaust systems, and outside conditions for positive pressure systems, $W = W_a - W_e$, is the total moisture to be removed (lb_{water}/h) (determined from animal latent heat production data), 60 is for unit conversion (1 h = 60 min), W_1 is the humidity ratio (water vapor) in incoming air (lb_{water}/lb_{da}), and W_2 is the humidity ratio (water vapor) in exhaust air (lb_{water}/lb_{da}).

$$Q = v \times \left(\frac{W}{W_2 - W_1}\right) \quad \text{(SI)} \qquad (5.33b)$$

where Q is the ventilation rate for controlling the indoor moisture (m^3/s), v is the specific volume of air (m^3/kg_{da}), which is the inside conditions for exhaust systems, and outside conditions for positive pressure systems, $W = W_a - W_e$, is the total moisture to be removed (kg_{water}/s) (determined from animal latent heat production data), W_1 is the humidity ratio (water vapor) in incoming air (kg_{water}/kg_{da}), and W_2 is the humidity ratio (water vapor) in exhaust air (kg_{water}/kg_{da}).

5.8.5.2.2 Ventilation System Design in Livestock and Poultry Buildings

Two basic equations have been presented for determining ventilation requirements: one based on available sensible heat and the other on latent heat. Once the appropriate calculations are performed for the various operating conditions (winter design, fall-spring, and summer) and the feasible alternatives are analyzed, an appropriate ventilation system can be specified. The appropriate calculations that must be considered are as follows.

5.8.5.2.2.1 Minimum Continuous Ventilation Using both ventilation rate equations for sensible heat and moisture removals—Equations (5.31a) or (5.31b) and Equation (5.33a) or Equation (5.33b), respectively—the minimum continuous ventilation rate should be calculated. The winter outside design dry-bulb temperature (99%) with the relative humidity at 80% and the minimum inside design temperature with the desired inside relative humidity should be used in these equations. This will provide two ventilation rates: one for moisture control and one for temperature control. The higher of the two values should be specified as the minimum continuous ventilation rate.

If the larger value is obtained from the sensible heat balance equation and this fan size is specified, the inside temperature will be at the selected value. This ventilation rate will be larger than required to maintain the selected inside relative humidity resulting in a lower than selected value. This reduced relative humidity is normally desirable and no additional provisions need to be made. If the larger value is obtained from the latent heat (moisture) balance equation and this value is specified, the inside relative humidity will be at the selected value. Because this rate will provide more ventilation than required to maintain the selected inside design temperature, the inside temperature would fall below the selected value if no supplemental heat were provided. Consequently, an additional calculation to determine the required amount of supplemental heat is necessary.

In order to calculate the amount of supplemental heat ($\dot{Q}_t$) required, the energy balance equation is solved for $\dot{Q}_t$ using the specified winter design conditions. The quantity of supplemental heat should be set and the controls should be set to turn this heat on whenever the inside temperature falls below the specified inside design condition. Special care should be exercised in this portion of the calculations to anticipate management practices that might significantly alter the solutions. These practices include changes in livestock and poultry stocking density, variation in animal size during the production cycle, and management practices, which could significantly alter the assumed sensible and latent heat production values.

For instance, if the minimum continuous ventilation rate is determined by the moisture balance equation under the worst conditions (i.e., high stocking density and large size animals), this will result in too much ventilation if a lower stocking density is used or if smaller animals occupy the building. This excess ventilation will normally have a profound influence on the supplemental heat energy required to maintain the selected temperature. It should be kept in mind that a ventilation rate only 1.5 times greater than what is necessary for moisture removal may easily quadruple the energy consumption for supplemental heat. Whenever large changes in stocking density or animal size are expected, several minimum continuous ventilation rates should be determined and provisions made to achieve them.

5.8.5.2.2.2 Fall-Spring Ventilation Rate As the outside temperature increases above the minimum design dry-bulb temperature on warm winter days and during the spring and fall months, it will be necessary to increase the ventilation rate above the minimum continuous level to maintain desirable inside conditions. In practice, this increase in the ventilation rate is obtained by sensing the inside dry-bulb temperature with a thermostat, which increases the ventilation rate as the inside temperature increases.

In order to determine the fall–spring ventilation rate, the sensible heat balance equations, Equations (5.31a) or Equation (5.31b), are used. Since it is warmer outside and the inside temperature has increased above the minimum inside design temperature, a second inside design temperature is selected. Often, this higher inside design temperature is chosen to be approximately 5.5 to 11°C (10 to 20°F) above the minimum inside design temperature. The outside temperature is usually taken as 3 to 8°C (5 to 15°F) below the interior temperature, ΔT = 3 to 8°C or ΔT = 5 to 15°F. This additional ventilation capacity is then obtained incrementally with additional or variable speed fans controlled by thermostats. Since these fans are controlled by thermostats that sense only temperature, there is little point in performing a moisture (latent heat) balance.

5.8.5.2.2.3 Maximum Ventilation Rate During the summer and other periods when the outside and inside temperatures are above those used for the spring–fall ventilation rate, additional ventilation must be provided in livestock and poultry buildings. In many cases, this additional capacity is most economically provided by manually opening large ventilation panels and simply relying on natural ventilation. If mechanical ventilation is desired during these periods, additional fan capacity must be provided. The additional capacity is calculated using the sensible heat balance equations, Equation (5.31a) or (5.31b). The summer design dry-bulb temperature (1%) is used for the outside temperature, and the desired inside temperature is assumed to be 1.5 to 3°C (3 to 5°F) higher than the outside design temperature. Fans to provide this additional capacity are then installed and controlled by thermostats.

5.8.5.2.2.4 Procedure for Ventilation System Design In this section, a step-by-step procedure is used for developing a ventilation system design for a swine finishing building near Columbus, Ohio, covering minimum (winter) continuous ventilation, mild-weather (fall–spring) ventilation, and hot weather (summer) ventilation requirements.

Step 1: Determine indoor and outdoor design temperature and humidity.

Indoor Design Temperature, T_i

- For cold and cool weather, use an indoor design temperature based on the species optimum temperature requirement of 50–70°F (10–21°C). Here, let's use 50°F (10°C).
- For summer, use 3–5°F (1.5–3°C) above the outside design temperature, that is, the summer design dry-bulb temperature (see Tables A.52–A.53). Then, the summer design temperature is 88°F + 5°F = 93°F (31°C + 3°C = 34°C).

Outdoor Design Temperature, T_o

- For minimum winter ventilation, use 99% winter design dry-bulb temperature (see Tables A.52–A.53). For Columbus, the outside design temperature at 99%, T_o (99%) = 6°F (–14°C).
- For maximum fall–spring ventilation, use the average temperature for January, 5–15°F (3–8°C) below maximum indoor design temperature, T_i. Then, 70°F – 10°F = 60°F (21°C – 5.5°C=15.5°C).
- For summer ventilation, use 1% summer design dry-bulb temperature (see Tables A.52–A.53). For Columbus, the summer design temperature at 1%, T_o (1%) = 88°F (31°C).

Indoor Relative Humidity, ϕ_i

- For cold weather, use an indoor relative humidity of 70%–80%.

Outdoor Relative Humidity, ϕ_o

- For cold weather, assume 80% outdoor relative humidity.

Step 2: Determine animal sensible and latent heat generations. Use Table A.63 or ASABE Standards (2018).

Step 3: Calculate the minimum winter ventilation rate using the ventilation rate equations for sensible heat and moisture removals, Equations (5.31a) or (5.31b) and Equations (5.33a) or (5.33b), respectively. Choose the larger ventilating rate as the minimum continuous winter ventilation rate. If more air is needed for moisture removal than that of temperature control, then supplemental heating is required to maintain the desired temperature levels. It should be remembered that the moisture removal ventilation rate is usually enough for controlling odors in warm buildings; however, if this calculated rate is lower than the recommended ventilation rates provided in Table A.64, then the removal of odorous air may become a concern, which can easily be handled by increasing the minimum winter ventilation rate above what is required for moisture control.

Step 4: Calculate the maximum ventilation rate for fall–spring using Equations (5.31a) or (5.31b).

Step 5: Specify fans and controls to achieve the additional ventilation as T_o increase causes T_i to increase. For constant speed fans, we normally use at least two fans to prevent oscillation.

- Turn on the first fan when T_i is 5–10°F (3–5°C) above the minimum T_i, (e.g., T_i > 60°F or T_i > 15°C).
- Turn on the second fan when T_i increases another 5°F (3°C) (e.g., T_i > 65°F or T_i > 18°C).
- Turn on the third fan when T_i increases another 5°F (3°C) (e.g., T_i > 70°F or T_i > 21°C).

Step 6: Calculate the summer ventilation rate, if fans are used, using the sensible heat removal Equation (5.31a) or (5.31b) and specify the fans.

Step 7: If required, size the supplemental heating using Equation (5.11) based on the worst case—99% winter design dry-bulb temperature (6°F or −14°C) for the outdoor design temperature—the chosen minimum winter ventilation rate, and the lowest possible total sensible heat generation ($\dot{Q}_{animal\_sensible}$) by the livestock or poultry.

Step 8: Size and locate inlets and outlets.

Step 9: Check against standard recommendations, and if possible, reevaluate.

Example 5.13: Mechanical Ventilation for a Swine Building

Calculate the ventilation rates in a swine finishing building with a solid floor and 200 head of 175 lb (80 kg) pigs near Columbus. The finishing building is a 24 ft × 48 ft building constructed without any windows as shown in Example 5.12. Assume inside design conditions of 60°F (15°C), and 80% inside and outside relative humidity. Use 218 Btu/h per 175 lb pig (64 W per 80 kg pig) latent heat emission and 409 Btu/h per 175 lb pig (120 W per 80 kg pig) sensible heat emission. These values can be obtained from Table A.63 or ASABE Standards (2018).

Solution

Step 1: Determine indoor and outdoor design temperature and humidity.

Indoor Design Temperature, T_i

- For cold and cool weather, use an indoor design temperature based on the species optimum temperature requirement of 50–70°F (10–21°C). Here, we will use 60°F (15°C).
- For summer, use 3–5°F (1.5–3°C) above the outside design temperature, the summer design dry-bulb temperature (see Tables A.52–A.53). Then, the summer temperature is 88°F + 5°F = 93°F (31°C + 3°C = 34°C).

Outdoor Design Temperature, T_o

- For minimum winter ventilation, use 99% winter design dry-bulb temperature (see Tables A.52–A.53). For Columbus, Ohio, the outside design dry-bulb temperature at 99%, T_o (99%) = 6°F (−14°C).
- For maximum fall-spring ventilation, use the average January temperature, 5–15°F (3–8°C) below maximum indoor design temperature, T_i. Then, 70°F − 10°F = 60°F (21°C–5.5°C=15.5°C).
- For summer ventilation, use 1% summer design dry-bulb temperature (see Tables A.52–A.53). For Columbus, Ohio, the summer design dry-bulb temperature at 1%, T_o (1%) = 88°F (31°C).

Indoor Relative Humidity, ϕ_i

- For cold weather, assume indoor relative humidity of 80%.

Outdoor Relative Humidity, ϕ_o

- For cold weather, assume outdoor relative humidity of 80%.

Step 2: Determine animal latent and sensible heat generations.

- Use 221 Btu/h per 175-lb pig (65 W per 80 kg pig) latent heat emission. This value is the difference of total and sensible heat generations for 175 lb (80 kg) pig provided in Table A.63 or ASABE Standards (2018).
- Use 409 Btu/h per 175 lb pig (120 W per 80 kg pig) sensible heat emission. Relevant data is available in Table A.63 or ASABE Standards (2018).

Step 3: Calculate the minimum winter ventilation rate using the equations for sensible heat and moisture removals, Equations (5.31a) or (5.31b) and Equations (5.33a) or (5.33b), respectively. Choose the larger rate as the minimum continuous winter ventilation rate. If more air is needed for moisture removal than that of temperature control, then supplemental heating is required to maintain the desired temperature levels. It should be remembered that the moisture removal ventilation rate is usually enough for controlling odors in warm buildings; however, if this calculated rate is lower than the recommended ventilation rates provided in Table A.64, then the removal of odorous air may become a concern, which can easily be handled by increasing the minimum winter ventilation rate above what is required for moisture control.

The moisture production via latent heat generation for growing-finishing pigs were obtained from pig housing systems, therefore, no additional allowance is needed for water being evaporated. Hence,

$$W = \left(221 \frac{\text{Btu}}{\text{h.pig}}\right) / \left(1{,}044 \frac{\text{Btu}}{\text{lb}_{\text{water}}}\right) \times 200 \text{ pigs} = 42 \frac{\text{lb}_{\text{water}}}{\text{h}}$$

Please note that moisture production provided in Table A.63 or ASABE Standards (2018) is 1.2 g_{water}/h.kg pig for 80 kg pig is equal to 42 lb_{water}/h for 200 pigs weighing 80 kg.

At $T_i = 60°F$ and $\phi_i = 80\%$, from the psychrometric chart, the humidity ratio, $W_i = 0.00884 \, lb_{water} / lb_{da}$ and specific volume, $v = 13.34 \, ft^3 / lb_{da}$. And at $T_o = 6°F$ and $\phi_o = 80\%$, from the psychrometric chart, the humidity ratio, $W_o = 0.00086 \, lb_{water} / lb_{da}$ and specific volume, $v = 11.8 \, ft^3 / lb_{da}$.

Now using Equation 5.33a (or 5.33b), we can calculate the ventilation rate (Q) required for moisture control (exhaust system) as follows:

$$Q = \left(\frac{v}{60}\right) \times \left(\frac{W}{W_2 - W_1}\right)$$

$$Q = \left(\frac{13.34 \, \text{ft}^3 / \text{lb}_{\text{da}}}{60}\right) \times \left(\frac{42 \frac{\text{lb}_{\text{water}}}{\text{h}}}{0.00884 \frac{\text{lb}_{\text{water}}}{\text{lb}_{\text{da}}} - 0.00086 \frac{\text{lb}_{\text{water}}}{\text{lb}_{\text{da}}}}\right)$$

$$Q = 1{,}170 \, \text{cfm} (\approx 0.55 \, \text{m}^3 / \text{s})$$

And now assuming the pigs are confined in the building described in Example 5.12, and using Equation 5.31a (or (5.31b)), we can calculate the ventilation rate (Q) required for indoor temperature control (exhaust system) as follows:

$$Q = \frac{v \times \left[\dot{Q}_{animal\_sensible} - \sum UA \times (T_i - T_o)\right]}{\left[60 \times c_p \times (I_i - T_o)\right]}$$

where $v = 13.34 \, ft^3 / lb_{da}$, $c_p = 0.24 \, Btu / lb_{da}.°F$ (see Table A.13), and

$$\dot{Q}_{animal\_sensible} = (409 \, \text{Btu} / \text{h.pig}) \times 200 \, \text{pigs} = 81{,}800 \, \text{Btu} / \text{h}$$

$$\dot{Q}_{cond} = \sum UA \times (T_i - T_o) = 246.5 \, \text{Btu} / \text{h.°F} \times (60 - 6) \, \text{°F}$$

$$Q = \frac{13.34 \, \text{ft}^3 / \text{lb}_{\text{da}} \times \left[81{,}800 \, \text{Btu} / \text{h} - 246.5 \, \text{Btu} / \text{h.°F} \times (60 - 6) \, \text{°F}\right]}{\left[60 \times 0.24 \, \text{Btu} / \text{lb}_{\text{da}}.\text{°F} \times (60\text{°F} - 6\text{°F})\right]}$$

$$Q = 1{,}175 \, \text{cfm} (\approx 0.55 \, \text{m}^3 / \text{s})$$

Theoretically, the minimum ventilation rate is the larger of the two rates just calculated. However, we need to compare the calculated minimum ventilation rate with the recommended rates provided in Table A.64.

The recommended minimum ventilation (cold weather) rate for 200 lb finishing pig is provided as 10 cfm per pig; thence, the total minimum ventilation rate for 200 head is 2,000 cfm (or 0.94 m³/s). So we will use one two-speed commercially available fan (2,080/1,190 cfm) (model: CCE14-2; fan size: 14 in.; motor: 1/4 HP; fan RPM: 1725/1140) for minimum winter ventilation along the sidewall, which will operate continuously with an option to reduce the rate based on the need. This much ventilating air is enough for moisture and odor removal, but it should be remembered that supplemental heating is required to maintain the desired temperature levels. See Step 7 for determining the required supplemental heating.

Step 4: Calculate the maximum ventilation rate for fall-spring using Equations (5.31a) or (5.31b).

Using sensible heat balance Equation 5.31a (or (5.31b)), we can now calculate fall–spring Q required for indoor temperature control (exhaust system) as follows:

$$Q = \frac{v \times \left[\dot{Q}_{animal\_sensible} - \sum UA \times (T_i - T_o)\right]}{\left[60 \times c_p \times (T_i - T_o)\right]}$$

where T_i is the maximum allowed cool weather indoor temperature, i.e., 70°F, and $T_o = \text{Maximum } T_i - 10°F = 70°F - 10°F = 60°F$.

At $T_i = 70°F$ and $\phi_i = 80\%$, from the psychrometric chart, the specific volume, $v = 13.68 \, ft^3 / lb_{da}$; and from Table A.13, $c_p = 0.24 \, Btu / lb_{da}.°F$. From Table A.63, we can estimate by interpolation the sensible heat emission for 175 lb (80 kg) pig at $T_i = 70°F$ as 308 Btu/h per 175 lb pig (90 W per 80 kg pig). Then,

$$\dot{Q}_{animal\_sensible} = (311 \, \text{Btu} / \text{h.pig}) \times 200 \text{ pigs} = 62{,}200 \text{ Btu} / \text{h}$$

$$\dot{Q}_{cond} = \sum UA \times (T_i - T_o) = 246.5 \, \text{Btu} / \text{h.°F} \times (70 - 60) \, \text{°F}$$

$$\dot{Q}_{cond} = 2{,}465 \, \text{Btu} / \text{h}$$

Then, the fall–spring ventilation rate is

$$Q = \frac{13.68 \, \text{ft}^3 / \text{lb}_{\text{da}} \times \left[62{,}200 \, \text{Btu} / \text{h} - 2{,}465 \, \text{Btu} / \text{h}\right]}{\left[60 \times 0.24 \, \text{Btu} / \text{lb}_{\text{da}}.\text{°F} \times (70\text{°F} - 60\text{°F})\right]}$$

$$Q = 5{,}674 \, \text{cfm} (\approx 2.68 \, \text{m}^3 / \text{s})$$

We now need to compare this calculated fall–spring ventilation rate with the recommended rates provided in Table A.64. The recommended fall–spring ventilation rate for 200-lb finishing pig is provided as 35 cfm per pig; therefore, the recommended fall–spring ventilation rate for 200 head would be 7,000 cfm (or 3.30 m³/s). Since the recommended fall–spring ventilation rate is more than the calculated rate, no adjustment is needed for odor removal or so.

Step 5: Specify fans and controls to achieve the additional ventilation as T_o increase causes T_i to increase. For constant speed fans, we normally use at least two fans to prevent oscillation.

Based on these calculations, we can choose the following fans, although other fans may be chosen as appropriate.

- One two-speed CEE14-2, 2080/1190 cfm fan operating continuously
- One two-speed CEE30-2, 5200/4100 cfm fan to turn on at about 65°F and the second speed at about 70°F

That is,

- first fan is operating continuously.
- Turn on the second fan at first speed when T_i increases 5°F (3°C) (e.g., T_i > 65°F or T_i > 18°C).
- Turn on the second fan at second speed when T_i increases another 5°F (3°C) (e.g., T_i > 70°F or T_i > 21°C).

Remember that we have used one two-speed fan (2,080/1,190 cfm) (model: CCE14-2; fan size: 14 in.; motor: 1/4 HP; fan RPM: 1725/1140) for minimum winter ventilation along the sidewall, which will operate continuously with an option to reduce the rate based on the need. Furthermore, we have used one two-speed fan (5,200/4,100 cfm) (model: CCE14-2; fan size: 14 in.; motor: 1/2 HP; fan RPM: 1080/980) for the additional fall–spring air exchange requirement providing 5,200/4,100 cfm ventilation at two different speeds as needed. This configuration will provide excellent spatial and temporal ventilation for this specific swine-finishing building. Please note that while the fan specifications are real, the manufacturer models are generic.

It should be kept in mind that larger fan diameter and lower HP fans consume less operational energy for delivering the same air exchange; therefore, when possible, alternative fans with larger diameter and lower HP motors should be chosen. Also, reducing fan rotational speed is the most efficient way to reduce fan energy use in variable flow, and this can be efficiently achieved with frequency controlled electric motors. For instance, if we reduce the fan rotational speed by 10%, based on the fan/pump affinity laws (this is further discussed in Chapter 8), the fan power will reduce 27%. This means that not only the initial costs but also the operational costs of fans should be considered for making a justified fan decision.

Step 6: Calculate the summer ventilation rate, if fans are used, using the sensible heat removal Equations (5.31a) or (5.31b).

The number of fans installed in Step 5 would only be enough for mild weather, fall to spring; we need to provide additional ventilation for hot weather. Normally, the additional capacity for hot weather is most economically provided by opening up the large ventilating panels used for natural ventilation. However, if mechanical ventilation is required or desired for some reason, we can simply use additional fans to provide the additional ventilating capacity, which is calculated using the sensible heat balance equation. For this purpose, the summer design temperature is used for the outside temperature and the desired inside temperature is assumed to be 3–5°F higher than the outside design temperature. For summer ventilation, we need to use 1% summer design dry-bulb temperature (see Tables A.52–A.53); and for Columbus, Ohio, the summer design temperature at 1%, T_o (1%) = 88°F or 31°C. Then, the summer ventilating rate (exhaust system) for our swine finishing building can be calculated using Equation (5.31a) or (5.31b) as follows:

$$Q=\frac{v\times\left[\dot{Q}_{animal\_sensible}-\sum UA\times(T_i-T_o)\right]}{\left[60\times c_p\times(T_i-T_o)\right]}$$

where T_i is the hot weather inside temperature, $T_i = 88°F + 5°F = 93°F$.

At $T_i = 93°F$ and $\phi_i = 80\%$, from the psychrometric chart, the specific volume, $v = 14.5\,ft^3/lb_{da}$; and from Table A.13, $c_p = 0.24\,Btu/lb_{da}.°F$. From Table A.63, we can estimate the sensible heat emission for 175 lb (80 kg) pig at $T_i = 93°F$ as 134 Btu/h per 175 lb pig (39 W per 80 kg pig). Then,

$$\dot{Q}_{animal\_sensible} = (134\,Btu/h.pig)\times 200\,pigs = 26{,}800\,Btu/h$$

$$\dot{Q}_{cond} = \sum UA\times(T_i-T_o) = 246.5\,Btu/h.°F\times(93-90)°F$$

$$\dot{Q}_{cond} = 739.5\,Btu/h$$

Then, the summer ventilation rate is

$$Q=\frac{14.5\,ft^3/lb_{da}\times[26{,}800\,Btu/h-739.5\,Btu/h]}{\left[60\times 0.24\,Btu/lb_{da}.°F\times(93°F-90°F)\right]}$$

$$Q = 8{,}747\,cfm(\approx 11.3\,m^3/s)$$

We now compare the calculated value with the recommended rate in Table A.64. The table value is much higher than needed for sensible heat removal, in order to provide high air velocities around the pig. The recommended hot weather ventilation rate for 200 lb finishing pig is provided as 120 cfm per pig; therefore, the recommended summer ventilation rate for 200 head would be 24,000 cfm (or 3.30 m^3/s). If mechanical ventilation is desired, we should also consider circulation fans, which can replace exhaust fans for one half of the hot weather ventilation.

Based on these calculations, we can choose the following fans, although other fans may be chosen as appropriate

- One two-speed CEE14-2, 2080/1190 cfm fan operating continuously.
- One two-speed CEE30-2, 5200/4100 cfm fan to turn on at about 65°F and the second speed at about 70°F.
- Two single-speed CEE36-1, 8500 cfm each fan, one at 75°F and the other at 80°F.

That is,

- the first fan is operating continuously.
- Turn on the second fan at the first speed when T_i increases 5°F (3°C) (e.g., T_i > 65°F or T_i > 18°C).
- Turn on the second fan at the second speed when T_i increases another 5°F (3°C) (e.g., T_i >70°F or T_i >21°C).
- Turn on the third fan when T_i increases another 5°F (3°C) (e.g., T_i > 75°F or T_i > 24°C).
- Turn on the fourth fan when T_i increases another 5°F (3°C) (e.g., T_i >80°F or T_i >27°C).

Remember that we have used one two-speed fan (2,080/1,190 cfm) (model: CCE14-2; fan size: 14 in.; motor: 1/4 HP; fan RPM: 1725/1140) for minimum winter ventilation along the sidewall, which will operate continuously with an option to reduce the rate based on the need. Furthermore, we have used one two-speed fan (5,200/4,100 cfm) (model: CCE14-2; fan size: 14 in.; motor: 1/2 HP; fan RPM: 1080/980) for the additional fall–spring air exchange requirement providing 5,200/4,100 cfm ventilation at two different speeds as needed. And we have used two single speed CEE36-1, 8500 cfm each (model: CEE36-1; fan size: 36 in.; motor:

1/2 HP; fan RPM: 850) for the additional hot weather air exchange requirement. This configuration will provide excellent spatial and temporal ventilation for this specific swine finishing building. Please note that while the fan specifications are real, the manufacturer models are generic models.

Again, it should be kept in mind that larger fan diameter and lower HP fans consume less operational energy for delivering the same air exchange; therefore, when possible, alternative fans with larger diameter and lower HP motors should be chosen. Also, reducing fan rotational speed is the most efficient way to reduce fan energy use in variable flow, and this can be efficiently achieved with frequency-controlled electric motors. For instance, if we reduce the fan rotational speed by 10%, based on the fan/pump affinity laws (this is further discussed in Chapter 8), the fan power will reduce 27%. This means that not only the initial costs but also the operational costs of fans should be considered for making a justified fan decision.

Step 7: Now we can size the supplemental heating using Equation (5.11) based on the worst case of 99% winter design dry-bulb temperature (6°F or −14°C) for the outdoor design temperature, the chosen minimum winter ventilation rate, and the lowest possible total sensible heat generation ($\dot{Q}_{animal\_sensible}$) by the pigs. Here, we will assume that the lowest possible number of pigs in the building is only half-stocked with 100 pigs weighing 130 lb (60 kg). Then, the supplemental heating is

$$\dot{Q}_t = \dot{Q}_{cond} + \dot{Q}_{inf} - \dot{Q}_{animal\_sensible}$$

where $\dot{Q}_t$ is the supplemental heating provided to be provided by the heating system, $\dot{Q}_{cond}$ is the rate of heat loss from the livestock building envelope (13,311 Btu/h), and $\dot{Q}_{inf}$ is the rate of sensible heat loss due to 2,000 cfm air exchange with the outside air, and $\dot{Q}_{animal\_sensible}$ is the rate of sensible heat gain from 100 pigs weighing 130-lb (see Table A.63, or ASABE Standards (2018)).

Using Equation 5.9a (or (5.9b)), $\dot{Q}_{inf}$, the rate of sensible heat loss due to 2,000 cfm air exchange with the outside air can be calculated as follows:

$$\dot{Q}_{inf} = \frac{Q}{v}\left(60\frac{\text{min}}{\text{h}}\right)c_p\left(T_i - T_o\right)$$

$$\dot{Q}_{inf} = \frac{2{,}080\,\text{cfm}}{13.34\,\text{ft}^3/\text{lb}_{da}}\left(60\frac{\text{min}}{\text{h}}\right)(0.24\,\text{Btu}/\text{lb}_{da}.{}^\circ\text{F})\left(60^\circ\text{F} - 6^\circ\text{F}\right)$$

$$\dot{Q}_{inf} = 121{,}225\,\text{Btu/h}$$

From Table A.63 or ASABE Standards (2018), the sensible heat emission is found as 580 Btu/h per 130 lb pig (170 W per 60 kg pig). Then, the total sensible heat emission by 100 pigs weighing 130 lb would be

$$\dot{Q}_{animal\_sensible} = (580\,\text{Btu}/\text{h.pig})\times 100\,\text{pigs}$$
$$= 58{,}000\,\text{Btu}/\text{h}$$

Then, the required supplemental heating, $\dot{Q}_t$ is

$$\dot{Q}_t = 13{,}311\,\text{Btu}/\text{h} + 121{,}225\,\text{Btu}/\text{h} - 58{,}000\,\text{Btu}/\text{h}$$
$$\dot{Q}_t = 76{,}536\,\text{Btu}/\text{h}\left(\approx 22{,}445\,\text{W}\right)$$

The required heating load of 76,536 Btu/h (22,445 W) can be supplied by using one 82% efficient propane-fired CEE100 (100,000 Btu/h input) providing a total of 82,000 Btu/h. The unit heater will turn on at about 60°F. The unit heater comes with 4 in. (10.2 cm) round flue gas connection on the back and can be installed with straight B-Vent connection and diagonally. CEE100 has ½ in. gas piping size, and natural gas plumbers can easily connect the individual unit heater to a propane supply tank.

In this project, we have used one 82% efficient CEE100 (100,000 Btu/h input). As mentioned earlier, the ASME states that 1 square inch free area per 2,000 Btu/h maximum fuel input of combined burners must be provided and kept clear at all times. Then,

$$\text{Free Area Required} = (100{,}000\ \text{Btu/h input}) \times (1\ \text{in}^2/2{,}000\ \text{Btu/h})$$
$$\text{Free Area Required} = 50\ \text{in}^2$$

Inlet louvers typically have 60% free area; therefore, 50 in^2/0.6 = 83 in^2 louver area per unit heater needs to be provided. Then, for the purpose of providing fresh air here, one CEE10 motorized inlet shutter having opening dimensions of 9.5 in. × 9.5 in. (which provides a louver area of 90 in^2 and a free area of 54 in^2) would meet the requirement.

Please note that while the specifications are real, the manufacturer models are generic models. For this heating system, we also need to install an adequate number of properly sized HAF fans.

Step 8: Size and locate inlets and outlets. See Examples 5.15 and 5.16. Same protocols applied.

Step 9: Check against standard recommendations, and if possible, reevaluate. It has already been checked, and necessary adjustments were made accordingly.

Example 5.14: Mechanical Ventilation for a Poultry Building

Calculate the ventilation rates in a poultry confinement building with hanging cages over pits and 690, 4 lb laying hens near Columbus, Ohio. The confinement building is a 24 ft × 48 ft building constructed without any windows as shown in Example 5.12. Assume inside design conditions of 65°F (18°C), and 70% inside and 80% outside relative humidity. Use 0.012 lb_{water}/h per bird (5.28 g/h per bird) latent heat emission and 24.2 Btu/h per bird (7.1 W per bird) sensible heat emission. These values can be obtained from Table A.63 or ASABE Standards (2018).

Solution

Step 1: Determine indoor and outdoor design temperature and humidity.

Indoor Design Temperature, T_i

- For cold weather, use an indoor design temperature based on the species optimum temperature requirement. Here, we will use 65°F (18°C). Also, the maximum allowed cool weather indoor temperature is 85°F (29°C).
- For summer, use 3–5°F (1.5–3°C) above the outside design temperature, i.e., the summer design dry-bulb temperature (see Tables

A.52–A.53). Then, the summer design temperature is 88°F + 5°F = 93°F (31°C + 3°C = 34°C).

Outdoor Design Temperature, T_o

- For minimum winter ventilation, use 99% winter design dry-bulb temperature (see Tables A.52–A.53). For Columbus, Ohio, the outside design dry-bulb temperature at 99%, T_o (99%) = 6°F (−14°C).
- For maximum fall-spring ventilation, use the average temperature of January, or 5–15°F (3–8°C) below maximum indoor design temperature, T_i. Maximum indoor design temperature is assumed to be 85°F; then, T_o = 85°F − 10°F = 75°F (30°C − 5°C = 25°C).
- For summer ventilation, use 1% summer design dry-bulb temperature (see Tables A.52–A.53). For Columbus, Ohio, the summer design temperature at 1%, T_o (1%) = 88°F (31°C).

Indoor Relative Humidity, ϕ_i

- For cold weather, assume indoor relative humidity of 70%.

Outdoor Relative Humidity, ϕ_o

- For cold weather, assume 80% outdoor relative humidity.

Step 2: Determine animal latent and sensible heat generations:

- Use 0.012 lb_{water}/h.bird (5.3 g/h.bird) latent heat emission.
- And use 24.2 Btu/h.bird (7.1 W/bird) sensible heat emission.
- These values are obtained from Table A.63 or ASABE Standards (2018).

Step 3: Calculate the minimum winter ventilation using the equations for sensible heat and moisture removals, Equations (5.31a) or (5.31b) and Equations (5.33a) or Equation (5.33b), respectively. Choose the larger ventilating rate as the minimum continuous winter rate. If more air is needed for moisture removal than for temperature control, then supplemental heating is required to maintain the desired temperature levels. It should be remembered that the moisture removal ventilation rate is usually enough for controlling odors in warm buildings; however, if this calculated rate is lower than the recommended ventilation rates provided in Table A.64, then the removal of odorous air may become a concern, which can easily be handled by increasing the minimum winter ventilation rate above what is required for moisture control.

The moisture production via latent heat generation for laying hens were obtained from calorimetric studies. Therefore, it should be kept in mind that the moisture production in poultry housing systems will be increased with a corresponding decrease in sensible heat loss. The moisture production by the total number of birds in the confinement building is as follows:

$$W = \left(0.012\frac{lb_{water}}{h.bird}\right)\times 920\text{ birds} = 11.0\frac{lb_{water}}{h}$$

At $T_i = 65°F$ and $\phi_i = 70\%$, from the psychrometric chart, the humidity ratio, $W_i = 0.00923\,lb_{water}/lb_{da}$ and specific volume, $v = 13.47\,ft^3/lb_{da}$. And at $T_o = 6°F$ and $\phi_o = 80\%$, from the psychrometric chart, the humidity ratio, $W_o = 0.00086\,lb_{water}/lb_{da}$ and specific volume, $v = 11.8\,ft^3/lb_{da}$.

Now using Equation 5.33a (or 5.33b), we can calculate the ventilation rate (Q) required for moisture control (exhaust system) as follows:

$$Q = \left(\frac{v}{60}\right)\times\left(\frac{W}{W_2 - W_1}\right)$$

$$Q = \left(\frac{13.47\,ft^3/lb_{da}}{60}\right)\times\left(\frac{11\frac{lb_{water}}{h}}{0.00923\frac{lb_{water}}{lb_{da}} - 0.00086\frac{lb_{water}}{lb_{da}}}\right)$$

$$Q = 295\,cfm(\approx 0.14\,m^3/s)$$

Now, assuming the layers are in the building described in Example 5.12 and using Equation 5.31a (or (5.31b)), we can calculate Q required for indoor temperature control (exhaust system) as follows:

$$Q = \frac{v\times\left[\dot{Q}_{animal\_sensible} - \sum UA\times(T_i - T_o)\right]}{\left[60\times c_p\times(T_i - T_o)\right]}$$

where $v = 13.47\,ft^3/lb_{da}$, $c_p = 0.24\,Btu/lb_{da}.°F$ (see Table A.13), and

$$\dot{Q}_{animal\_sensible} = (24.2\,Btu/h.bird)\times 920\,birds = 22,264\,Btu/h$$

$$\dot{Q}_{cond} = \sum UA\times(T_i - T_o) = 246.5\,Btu/h.°F\times(65-6)°F$$

$$\dot{Q}_{cond} = 14,544\,Btu/h$$

$$Q = \frac{13.47\,ft^3/lb_{da}\times[22,264\,Btu/h - 14,544\,Btu/h]}{[60\times 0.24\,Btu/lb_{da}.°F\times(65°F - 6°F)]}$$

$$Q = 122\,cfm(\approx 0.06\,m^3/s)$$

Theoretically, the minimum ventilation rate is the larger of the two rates just calculated. However, we need to compare the calculated minimum ventilation rate with the recommended rates provided in Table A.64.

The recommended minimum ventilation (cold weather) rate for layers is provided as 0.5 cfm per bird; therefore, the total minimum ventilation rate for 920 birds is 460 cfm (or 0.22 m³/s). Therefore, we use one single-speed commercially available fan (530 cfm) (model: CCE10-1; fan size: 10 in.; motor: 1/3 HP; fan RPM: 1625) for the minimum winter ventilation along the sidewall, which will operate continuously. This much ventilating air is enough for moisture and odor removal, but it should be remembered that with this winter ventilation rate, supplemental heating is required to maintain the desired temperatures. See Step 7 for determining the required supplemental heating.

Step 4: Calculate the maximum ventilation rate for fall–spring using sensible heat removal Equations (5.31a) or (5.31b).

Using Equations (5.31a) or (5.31b), we can now calculate fall–spring Q required for indoor temperature control (exhaust system) as follows:

$$Q = \frac{v\times\left[\dot{Q}_{animal\_sensible} - \sum UA\times(T_i - T_o)\right]}{\left[60\times c_p\times(T_i - T_o)\right]}$$

where T_i is the maximum allowed cool weather indoor temperature—85°F—and $T_o = \text{Maximum}\,T_i - 10°F = 85°F - 10°F = 75°F$.

At $T_i = 75°F$ and $\phi_i = 70\%$, from the psychrometric chart, the specific volume, $v = 13.81\,ft^3/lb_{da}$; and from Table A.13, $c_p = 0.24\,Btu/lb_{da}.°F$. From Table A.63, we can estimate by interpolation the sensible heat emission for 4 lb (1.8 kg) bird at $T_i = 75°F$ as 21.6 Btu/h per 4 lb bird (6.3 W per bird). Then,

$$\dot{Q}_{animal\_sensible} = (21.6\,Btu/h.bird) \times 920\,birds = 19{,}872\,Btu/h$$

$$\dot{Q}_{cond} = \sum UA \times (T_i - T_o) = 246.5\,Btu/h.°F \times (85-75)°F$$

$$\dot{Q}_{cond} = 2{,}465\,Btu/h$$

Then, the fall–spring ventilation rate is,

$$Q = \frac{13.81\,ft^3/lb_{da} \times [19{,}872\,Btu/h - 2{,}465\,Btu/h]}{[60 \times 0.24\,Btu/lb_{da}.°F \times (85°F - 75°F)]}$$

$$Q = 1{,}669\,cfm (\approx 0.79\,m^3/s)$$

We now need to compare this calculated fall–spring ventilation rate with the recommended rates provided in Table A.64. The recommended mild weather ventilation rate for layers is provided as 2 cfm per bird; therefore, the recommended fall–spring ventilation rate for 920 birds would be 1,840 cfm (or 0.87 m^3/s). Since the recommended fall–spring rate is more than the calculated rate, no adjustment is needed for odor removal.

Step 5: Specify fans and controls to achieve the additional ventilation as T_o increase causes T_i to increase. For constant-speed fans, we normally use at least two fans to prevent oscillation.

Based on these calculations, we can choose the following fans, although other fans may be chosen as appropriate.

- One single-speed CEE10-1, 530 cfm fan operating continuously.
- One variable-speed CEE12-VD, 1,540 cfm fan to turn on at about 65°F.

That is,

- the first fan is operating continuously.
- Turn on the second fan when T_i increases 5°F (3°C) (e.g., T_i > 65°F or T_i > 18°C).

Remember that we have used one single fan (530 cfm) (model: CCE10-1; fan size: 10 in.; motor: 1/3 HP; fan RPM: 1625) for the minimum winter ventilation along the sidewall, which will operate continuously. Furthermore, we have used one variable-speed fan (1,540 cfm) (model: CCE12-VD; fan size: 12 in.; motor: 1/3 HP; fan RPM: 1625) for the additional fall-spring air exchange requirement. This configuration will provide excellent spatial and temporal ventilation for this specific swine finishing building. Please note that while the fan specifications are real, the manufacturer models are generic.

It should also be kept in mind that larger fan diameter and lower HP fans consume less operational energy for delivering the same air exchange; therefore, when possible, alternative fans with larger diameter and lower HP motors should be chosen. Also, reducing fan rotational speed is the most efficient way to reduce fan energy use in variable flow, and this can be efficiently achieved with frequency controlled variable speed electric motors. For instance, if we reduce the fan rotational speed by 10%, based on the fan/pump affinity laws (this is further discussed in Chapter 8), the fan power will reduce 27%. This means that not only the initial costs but also the operational costs of fans should be considered in making a justified fan decision.

Step 6: Calculate the summer ventilation rate, if fans are used, using the Equations (5.31a) or (5.31b).

The number of fans installed in Step 5 would only be enough for mild weather, fall–spring. We need to provide additional ventilation for hot weather ventilation. As mentioned earlier, normally, the additional capacity for hot weather is most economically provided by opening up the large ventilating panels used for natural ventilation. That is what we will be doing for this poultry confinement building. However, if mechanical ventilation is required or desired for some reason, we can simply use additional fans to provide the additional ventilating capacity, which is calculated using the sensible heat balance equation, as in Example 5.13.

Step 7: Now we can size the supplemental heating using Equation (5.11) based on the worst case, 99% winter design dry-bulb temperature (6°F or –14°C) for the outdoor design temperature, the chosen minimum winter ventilation rate, and the lowest possible total sensible heat generation ($\dot{Q}_{animal\_sensible}$) by the birds. Here, we will assume that the lowest possible stocking density of birds in the building is only half-stocked with 460 birds weighing 4 lb (1.8 kg). Then, the supplemental heating is

$$\dot{Q}_t = \dot{Q}_{cond} + \dot{Q}_{inf} - \dot{Q}_{animal\_sensible}$$

where $\dot{Q}_t$ is the supplemental heating provided to be provided by the heating system, $\dot{Q}_{cond}$ is the rate of heat loss from the poultry confinement building envelope ($14{,}544\,Btu/h$), and $\dot{Q}_{inf}$ is the rate of sensible heat loss due to 460 cfm air exchange with the outside air, and $\dot{Q}_{animal\_sensible}$ is the rate of sensible heat gain from 460 birds weighing 4 lb (see Table A.63 or ASABE Standards (2018)).

Using Equation 5.9a (or (5.9b)), $\dot{Q}_{inf}$, the rate of sensible heat loss due to 460 cfm air exchange with the outside air can be calculated as follows:

$$\dot{Q}_{inf} = \frac{Q}{v}\left(60\frac{min}{h}\right)c_p(T_i - T_o)$$

$$\dot{Q}_{inf} = \frac{530\,cfm}{13.47\,ft^3/lb_{da}}\left(60\frac{min}{h}\right)(0.24\,Btu/lb_{da}.°F)(65°F - 6°F)$$

$$\dot{Q}_{inf} = 33{,}429\,Btu/h$$

From Table A.63 or ASABE Standards (2018), the sensible heat emission is 24.2 Btu/h per 4 lb bird (7.1 W per bird). Then, the total sensible heat emission by one-half the stocking density for 460 birds weighing 4 lb would be

$$\dot{Q}_{animal\_sensible} = (24.2\,Btu/h.bird) \times 460\,birds = 11{,}132\,Btu/h$$

Then, the required supplemental heating, $\dot{Q}_t$ is

$$\dot{Q}_t = 14{,}544\,Btu/h + 33{,}429\,Btu/h - 11{,}132\,Btu/h$$

$$\dot{Q}_t = 36{,}841\,Btu/h (\approx 10{,}804\,W)$$

The required heating load of 36,841 Btu/h (10,804 W) can be supplied by using one 80% efficient propane-fired CEE60 (60,000 Btu/h input) providing a total of 48,000 Btu/h. The unit heater will turn on at about 65°F. The unit heater comes with 5 in. (12.7 cm) round flue gas connection on the back and can be installed with straight B-Vent connection and horizontally. CEE60 has ½ in. gas piping size, and natural gas plumbers can easily connect the individual unit heater to a propane tank.

In this project, we have used one 80% efficient CEE60 (60,000 Btu/h input). As mentioned earlier, the ASME states that 1 square inch free area per 2,000 Btu/h maximum fuel input of combined burners must be provided and kept clear at all times. Then,

$$\text{Free Area Required} = (60{,}000\ \text{Btu/h input}) \times (1\ \text{in}^2/2{,}000\ \text{Btu/h})$$

$$\text{Free Area Required} = 30\ \text{in}^2$$

Inlet louvers typically have 60% free area; therefore, 30 in^2/0.6 = 50 in^2 louver area per unit heater needs to be provided. Then, for the purpose of providing fresh air here, one CEE9-1/2 motorized inlet shutter with opening dimensions of 7.5 in. × 7.5 in. (which provides a louver area of 56.3 in^2, and a free area of 33.8 in^2) would meet the requirement.

Please note that while the specifications are real, the manufacturer models are generic models. For this heating system, we also need to install adequate numbers of properly sized HAF fans.

Step 8: Size and locate inlets and outlets. See Examples 5.15 and 5.16. Same protocols applied.

Step 9: Check against standard recommendations, and if possible, reevaluate. It has already been checked, and necessary adjustments are made accordingly.

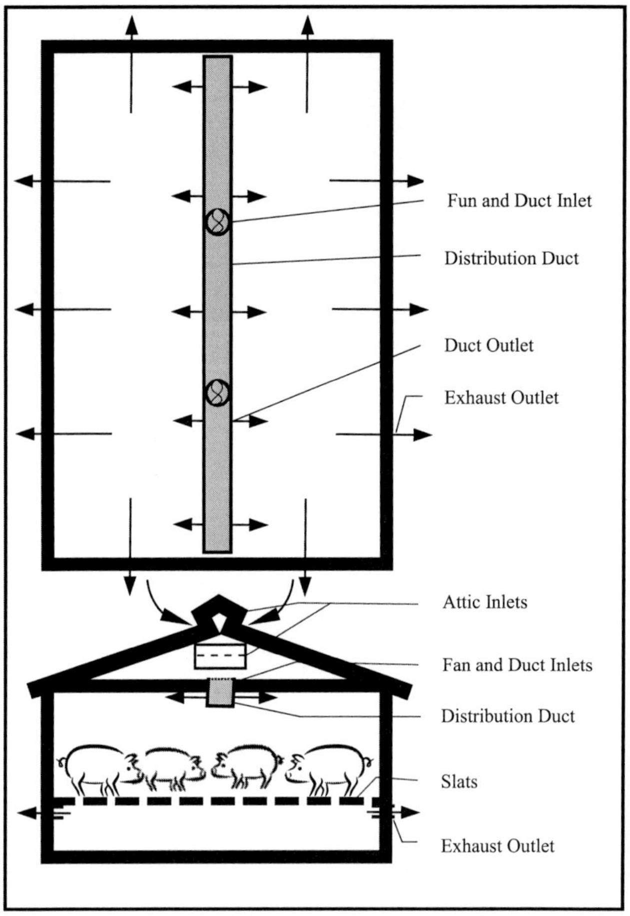

FIGURE 5.37 Positive pressure ventilating system.
Source: ***Structures & Environment* (1987).**

5.8.5.2.3 Ventilation Systems

In mechanical ventilation systems, if the ventilation fan forces fresh air into the controlled-environment building through the air inlets, the static pressure in the building becomes greater than the outside static pressure, and this pressure flows the inside air out through the outlet openings. This is commonly termed a positive pressure system. If the fan removes air from the building, the static pressure in the building will be lower than the outside pressure, and outside fresh air will flow in through the inlets. This system is referred to as a negative pressure system. In neutral pressure systems, however, while dedicated fans are forcing fresh air into the building, exhaust fans concurrently remove indoor air out of the building creating neither positive nor negative indoor pressure. This section covers all these systems further in depth, providing a number of numerical examples.

5.8.5.2.3.1 Positive Pressure Systems As mentioned, in positive pressure systems, ventilation fans force fresh outside air into the building, creating a positive indoor pressure that forces the humid and odorous inside air out of the building through adequately designed outlet openings and potential leaks in the walls and ceiling, if any (see Figure 5.37 for a possible configuration).

Because the air is forced from the building in positive pressure systems, these systems are ideal for applications where ventilation air must be filtered in order to prevent contaminants and pathogens from entering the building. However, this is also why the forced moist air infiltrates into the walls and attic; therefore, installing a good vapor barrier is extremely important in buildings having these systems. On the practical side, frost typically forms at air leaks around doors, etc., freezing and shutting them as a result. In addition, 20% to 30% of the energy used by fan motors that is rejected as waste heat is added to the building interior, an advantage in winter but a disadvantage in summer. The air in positive pressure systems is distributed with a baffle in front of the fan or a duct with carefully designed outlets that are insulated to prevent condensation.

Positive pressure exhaust outlets are sized at about 4,000 to 4,100 cm^2 per 1.0 m^3/s (or about 300 in^2 per 1000 cfm) fan capacity. For instance, one pair of 5 cm (2 in.) diameter holes for each 0.01 m^3/s (20 cfm) or one pair of 7.5 cm (3 in.) diameter holes for each 0.025 m^3/s (45 cfm) of airflow are provided and uniformly spaced along the duct. For small buildings up to 12 m wide or so, positive pressure fans with baffles work very well; for wider buildings, ducts should be spaced about 6 m apart from each other. In addition, the ceiling positive pressure fans with baffles should be 4.8 to 5.0 m apart from each other, while maintaining an approximately 2.5 m distance between the gable end wall and the closest fan. Also, while each of the positive, negative, and neutral pressure systems has its own characteristics that are considered when deciding which system to build, positive pressure systems are usually preferred in remodeled buildings because pressure outlets are relatively easier to construct than inlets for exhaust ventilation systems.

5.8.5.2.3.2 Exhaust Systems Exhaust or negative pressure systems are the most common ventilating system in livestock and poultry buildings. As mentioned earlier, exhaust fans in these systems create negative pressure inside the building, pulling the moist inside air out of the building. As a result of the pressure difference between outside and inside, the fresh outside air is forced into the building smoothly through baffled inlets and distributed uniformly throughout the building (Figure 5.38) as long as the inlets are adequately sized and properly located. Therefore, providing good air distribution in these systems is not challenging and is less costly because simple openings and slots constructed in walls function quite well to control and distribute air in the building.

Exhaust systems do not always provide uniform air distribution at low airflow rates due to potential air leaks and wind pressure effects. It should also be remembered to provide proper allowances for reduced fan efficiencies as dust accumulation takes place on fan blades due to the fact that the fans are located in the air exhaust, and hence are continuously exposed to dust and high humidity conditions.

5.8.5.2.3.3 Neutral Pressure Systems Neutral pressure systems have air inlet shutters, a duct fan, and an air distribution tube or duct as well as wall exhaust fans (Figure 5.39). In cold and mild weather, neutral pressure systems distribute ventilating air through the tube or duct and exhaust air through wall fans. The thermostatically controlled shutter is a motorized shutter located in the wall right across the duct fan that then allows fresh outside air to enter the tube or duct from the outside or from the attic, as shown in Figure 5.39.

In neutral pressure systems, the duct fan typically moves more fresh air than what is normally required in cold weather, so the shutter usually opens intermittently to allow in the required amount of fresh air. An exhaust fan, on the other hand, only operates when the shutter opens to pull fresh outside air in through the shutter. When the intake shutter closes, the exhaust fan stops and the duct fan keeps circulating only indoor air through the distribution tube or duct. Should a unit space heater is required, then it is placed at the tube or duct intake so that the tube or duct can distribute the heat as well.

Neutral pressure duct and duct fan systems are not designed to handle hot weather ventilation but at most the mild weather. To provide hot weather ventilation, additional inlets and exhaust fans are required. The hot weather inlets, typically motorized shutters, are located in the wall opposite the wall exhaust fans.

While neutral systems provide satisfactory ventilation, they have some inherent problems such as dust build-up in the duct, shutter freeze-up, and moreover, using the same air duct from one group of animals to another can transfer dust and bacteria., which must be seriously considered during such systems are designed. The only solution to dust buildup in the duct is periodic cleaning of the duct, which may require partial disassembly. Shutters can also freeze if they are too far from the duct fan to move warm air past the shutters, or if no supplemental heating is provided at

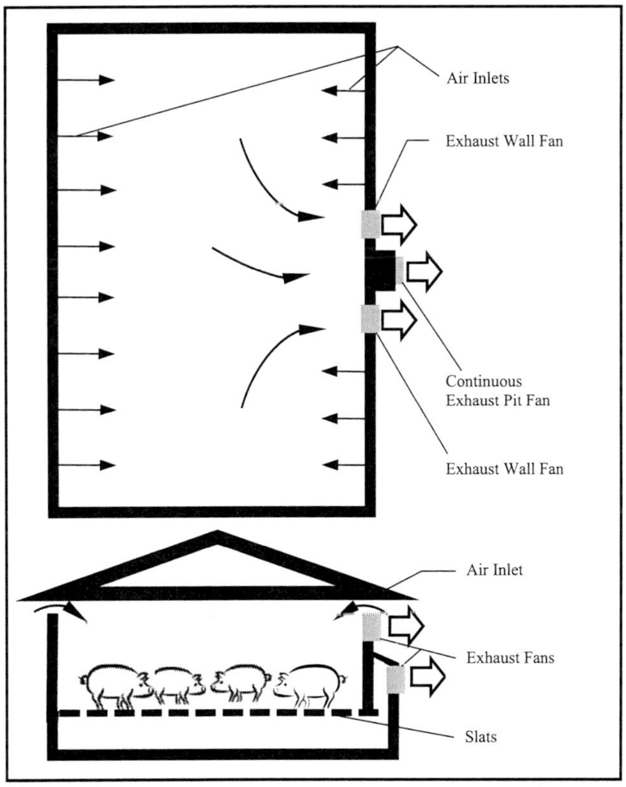

FIGURE 5.38 Exhaust (negative pressure) ventilating system.
Source: *Structures & Environment* (1987).

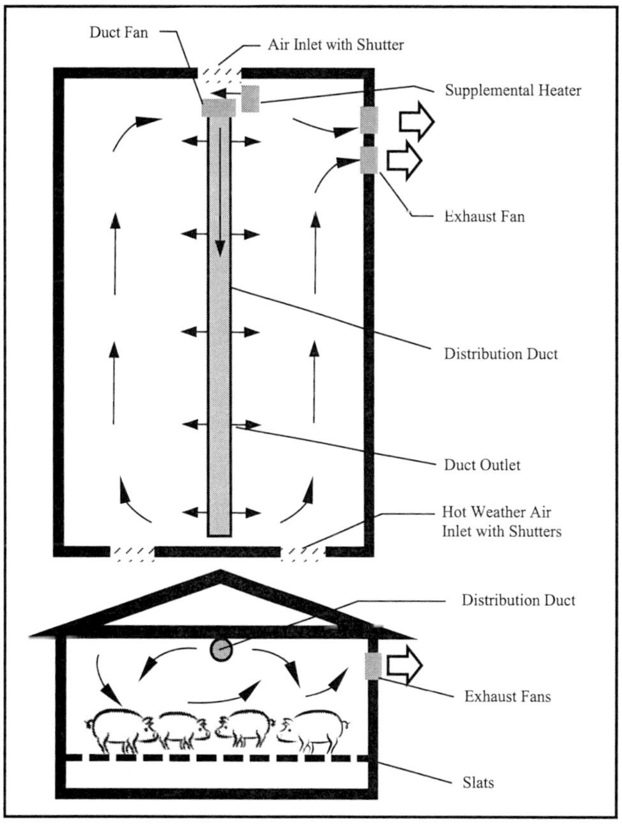

FIGURE 5.39 Neutral pressure ventilating system.
Source: *Structures & Environment* (1987).

all. Shutter freezing is a serious problem because it might eventually restrict ventilating air. Persisting shutter freeze-up problems can easily be solved by keeping the shutters warm with a heat lamp.

5.8.5.2.3.4 Ventilation Design and Air Distribution
In exhaust systems, ventilation is achieved through reducing the pressure within the ventilated building below outside pressure, which causes outside fresh air to enter the building wherever openings present. Pressure differences across walls in ventilated livestock or poultry buildings should typically range from 0.02 to 0.12 in. H_2O (W.G) (5 and 30 Pa). The distribution of the fresh air, on the other hand, is determined by the location, number, and cross-sectional area of the openings, and distribution of horizontal circulation fans.

As shown in Examples 5.13 and 5.14, the amount of fresh air entering a livestock or poultry confinement building is a function of the amount of heat and moisture produced by the animals in that particular building. Adequate air distribution can be achieved by properly locating the inlets and varying intake cross-sectional area to fit the animal distribution within the building. In order to obtain satisfactory air distribution within livestock and poultry buildings, motorized inlet controllers have been developed to adjust the air intake opening depending on the static pressure difference between the inside and outside air (Figures 5.38). In these units, the inlet baffle openings are adjusted in relation to the pressure difference between inside and outside, so that the fresh air distribution within the building is better controlled and managed.

Figures 5.37 and 5.39 show another method for proper fresh air distribution, which basically uses a fan connected to a perforated polyethylene tube or a rigid air distribution duct, which allows a combination design of heating, circulation, and ventilation as one system. This kind of air distribution can be used in either exhaust or positive pressure systems. However, for proper application and custom design of these systems, relevant equipment manufacturers should be consulted.

5.8.5.2.3.4.1 Air Flow through Openings

The ventilation rate in a livestock or poultry confinement building is determined by the ventilation capacity; however, the uniformity of air distribution primarily depends on the location, design, and the size of air inlets. This statement is true for both exhaust and pressure systems, provided that the longest path between inlet to outlet is less than about 75 ft (or 25 m).

The slot inlet dimensions and volume of air flow permitted at various static pressures are empirically related and expressed as shown in Table 5.7; the airflow rate through a slot at a constant static pressure difference is proportional to the width of the slot. For instance, the average rate through an entire 4 in. (10 cm) wide slot at 0.04 in. H_2O (10 Pa) is 200 cfm (0.09 m^3/s), four times the flow through a 1 in. slot at the same static pressure difference. The same relationship holds at the higher static pressure of 0.125 in. H_2O (30 Pa). If the static pressure across the slot is increased about three times, say 0.04 in. (10 Pa) to 0.125 in. H_2O (30 Pa), it only doubles the airflow rate through the slot. For instance, the average rate through an entire 4 in. (10 cm) wide slot at 0.125 in. H_2O (30 Pa) is 400 cfm (0.19 m^3/s) double the flow through a 4 in. (10 cm) slot at 0.04 in. H_2O (10 Pa) static pressure difference (Table 5.7).

In Table 5.7, a common rule of thumb is used for air inlets at 1 ft^2 (144 in.) of area (A) for each 600 to 1200 cfm of fan capacity (Q), and air velocity through the inlet is assumed to be the average velocity (V) through the full slot width ($Q = V.A$). Using this rule of thumb, the average inlet velocity through the full slot width is $V = Q / A = 600$ ft / min (3.0 m/s) at 0.04 in. (10 Pa) and 1,200 ft / min (6.0 m/s) at 0.125 in. (30 Pa).

In practice, it should be remembered that inlets gradually fill with debris and dirt and constrict the slot opening over time; therefore, inlets are not designed for the full static pressure rating of the fan. Otherwise, the fans are overloaded. Instead, the slots are designed slightly oversized, slightly wider than calculated, and then the opening width is simply adjusted for the desired pressure difference. Moreover, HAF fans are important for air distribution in livestock and poultry buildings. As discussed earlier in Chapter 4, in winter, they improve air distribution to help dry or heat corners; and in summer, they improve animal comfort by increasing air velocity over the skin.

TABLE 5.7
Airflow through 1 ft (0.3 m) Slots

Slot Width inches (cm)	CFM (m^3/s) per foot (0.3 m) length of slot opening — Static Pressure: 0.04 in. H_2O (10 Pa)	0.125 in. H_2O (30 Pa)
1.0 (2.5)	50 (0.024)	100 (0.047)
1.5 (3.8)	75 (0.035)	150 (0.071)
2.0 (5.0)	100 (0.047)	200 (0.094)
2.5 (6.4)	125 (0.059)	250 (0.118)
3.0 (7.5)	150 (0.071)	300 (0.142)
3.5 (8.9)	175 (0.083)	350 (0.165)
4.0 (10.0)	200 (0.094)	400 (0.189)

Source: Structures & Environment (1987).

5.8.5.2.3.4.2 Air Inlets for Exhaust Ventilation

When air inlets are designed properly, there are no damp corners, drafts, or dead air spots in a livestock or poultry building. One of the best air inlets commonly used today is a continuous slot with an adjustable hinged baffle (Figure 5.40a). Hinged rigid baffles are typically insulated to prevent condensation and freezing, and are used to adjust the inlet opening to increase air velocity and improve air mixing. They also prevent warping and uneven air distribution.

Intermittent, that is, center-ceiling flat baffle slotted, inlets (Figure 5.40b) are common as well, especially in remodeled livestock and poultry buildings, for their ease in building compared with continuous slots. For more uniform air distribution, intermittent openings are uniformly spaced in buildings at least 8 ft (2.5 m) from walls and at least 16 ft (5 m) apart. Obviously, the intermittent inlet sizes depend on the required airflow; however, for better control of the incoming air, the baffles are typically constructed to extend 4 in. (10 cm) beyond all sides of the inlet opening. Please note that the air enters through two sides of the center-ceiling slot inlet system. Hence, the airflow per unit of slot inlet length is obtained by multiplying the airflow rate per foot (or per meter) length of the slot opening by two.

While most exhaust ventilating systems with continuous slot inlets force air across the ceiling (Figure 5.40a, b), a "down-the-wall" pattern shown in Figure 5.40c also works. Regardless of the type of inlets, in addition to the airflow through the slot inlets, infiltration airflow through doors, windows, and unplanned openings should also be included in order to have a true description of the air inlet system.

5.8.5.2.3.5 Inlet Location Slot inlets in many exhaust systems are typically located around the perimeter of the ceiling (excluding the gable walls and the space near the fan or group of fans). The slots are continuous or intermittent, and a wide slot placed around the perimeter of the ceiling for high flow requirements during the summer can partially be restricted with an adjustable hinged baffle during the winter.

In addition, during cold weather, air inlets within about 8 ft (2.5 m) of the winter exhaust fans are closed and sealed to prevent short circuiting. As mentioned earlier, the longest path between the inlet and outlet (the fan) should be less than about 75 ft (or 25 m). For livestock or poultry confinement buildings up to about 35 ft (10 m) wide, continuous slot inlets are typically placed at the ceiling along the two sidewalls; for wider buildings, in addition to the continuous slot inlets, adding one or more intermittent ceiling inlets would be sufficient.

The fresh air coming from the attic space is desirable for winter ventilation due to the warming effect and wind protection; however, it would not be that desirable for hot weather ventilation, especially when the flow rate is low and the air experiences considerable heat gain through the attic. In hot weather, fresh air is directly pulled from the outside through the eaves instead of the attic (see Figure 5.40a, c), and part of this entering air also ventilates the attic and helps lowering the ceiling temperature. Additionally, the minimum winter ventilation fans, the moisture control fans, run continuously to control the moisture within the building, and if these fans do not operate, the warm and moist indoor air rises through the slot openings and condenses on the underside of the roof and drips on ceiling insulation, which is definitely not a desirable situation.

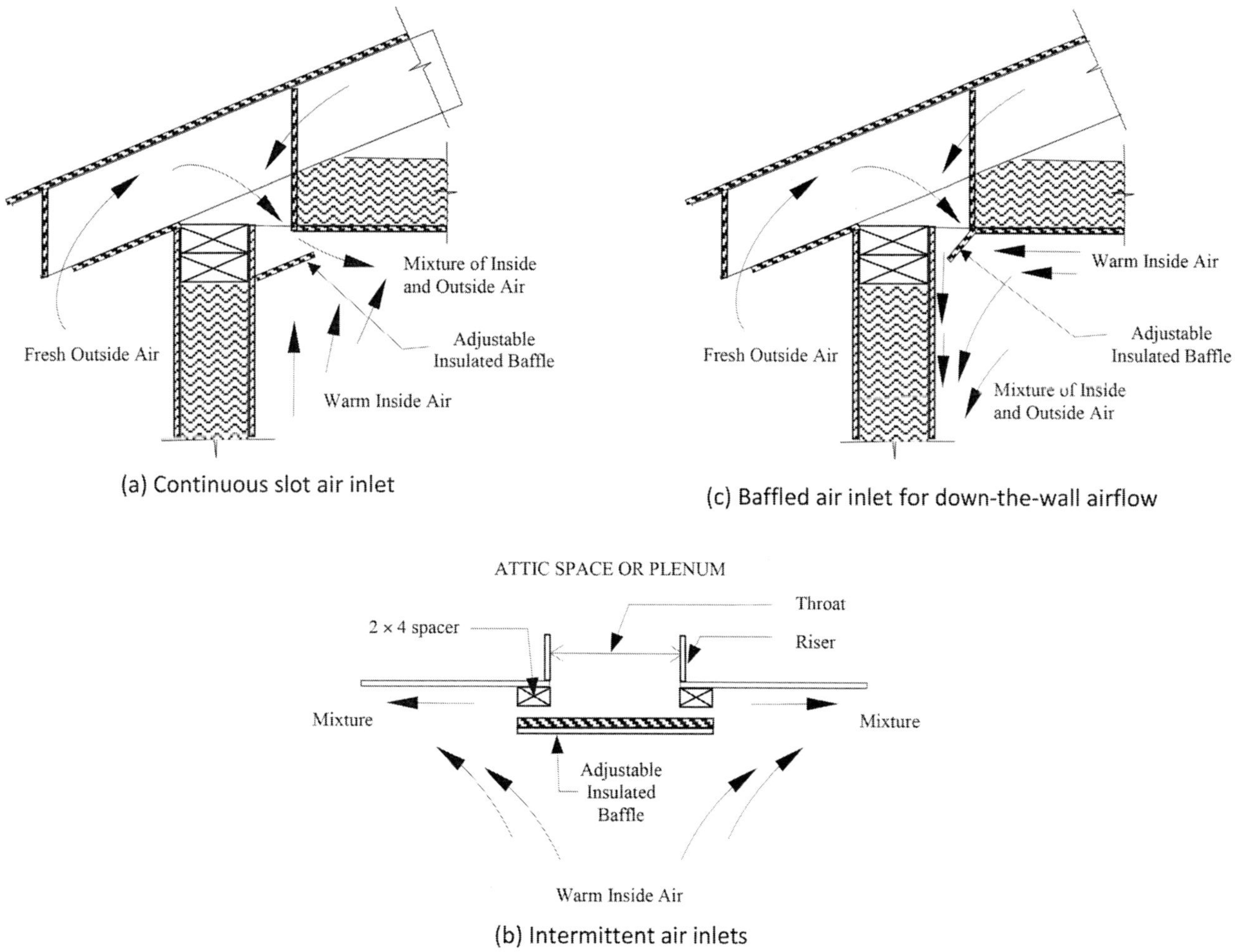

FIGURE 5.40 Slotted inlet configurations.

Sources: *Structures & Environment* (1987); and ASAE EP270.5 FEB03 Design of Ventilation Systems for Poultry and Livestock Shelters, Copyright © 2003 by American Society of Agricultural and Biological Engineers, St. Joseph, MI, USA).

5.8.5.2.3.6 Inlet Size As it is shown below in Examples 5.15 and 5.16, Table 5.7 can be used to design slotted inlet systems. In order to get a well mixing of cold and warm air, the inlets are sized to maintain a high average velocity (600 to 1200 fpm, or 3.0 to 6.0 m/s) (see Table 5.7). If the incoming air velocity is lower than these figures, the fresh and cold air settles as soon as it gets into the room, and hence may chill the livestock and poultry, affecting their health and well-being. If, on the other hand, the air velocity is higher than these figures, then the resulting high static pressure decreases the fan capacity and efficiency. Inlets are typically designed for the hot weather ventilation rate, and then reduced with an adjustable baffle for lower ventilating rates in cold and mild weather.

Example 5.15: Sizing a Slot Air Inlet for a Swine Building

Size a slot air inlet for a maximum fan delivery of 24,000 cfm at 0.125 in. H_2O (30 Pa) for the swine finishing building with 200 pigs near Columbus, Ohio, in Example 5.13. The finishing building is 24 ft × 48 ft. Assume a maximum slot width of 2.5 in. from Table 5.7.

Solution

By interpolation, Table 5.7 provides that each foot of a 2.5 in. wide slot allows 250 cfm at 0.125 in. H_2O. The needed slot length is then

$$24{,}000 \text{ cfm} / 250 \text{ cfm} = 96 \text{ ft of 2.5 in. slot is needed.}$$

Since the building width is less than 35 ft, we locate continuous slot inlets at the ceiling along both sidewalls. The incoming air should travel less than 75 ft from the inlet to the fan. During cold weather, we close and seal the air inlets within 8 ft of the winter exhaust fans to prevent short circuiting.

Example 5.16: Sizing a Slot Air Inlet for a Swine Building-2

Size a slot air inlet for the 24 ft × 48 ft swine finishing building with 200 head of 175-lb (80 kg) pigs near Columbus, Ohio, in Example 5.13. Assume that the group exhaust fans are placed in the center of the south wall and locate inlets at the ceiling on the long sides of the building. Design the slot air inlet for 0.125 in. H_2O static pressure.

Solution

Step 1: Total slot inlet length is

$$\text{Building Length} \times 2 = 48 \text{ ft} \times 2 = 96 \text{ ft}$$

As mentioned earlier, in cold weather, we close the inlets over the continuous exhaust fans to prevent short circuiting of air. Winter slot opening is about 80 ft.

Step 2: Maximum ventilating rate per foot of inlet is

$$(\text{Hot Weather Rate} \times \text{Number of Pigs}) / \text{Total Slot Length}$$

And the recommended hot weather rate for finishing pig is 120 cfm from Table A.64. Then,

$$(120 \text{ cfm} \times 200 \text{ pigs}) / 96 \text{ ft} = 250 \text{ cfm/ft}$$

Step 3: A 1 in. wide slot provides 100 cfm per foot of length with a 0.125 in. H_2O static pressure drop (Table 5.7), so the hot weather slot width is

$$(250 \text{ cfm/ft}) / (100 \text{ cfm/in.}) = 2.5\text{" wide}$$

Step 4: In cold weather, the ventilating rate is much lower:

$$(\text{Cold Weather Rate} \times \text{Number of Pigs}) / \text{Total Slot Length}$$

And the recommended cold weather rate for finishing pig is 10 cfm from Table A.64. Then,

$$(10 \text{ cfm} \times 200 \text{ pigs}) / 80 \text{ ft} = 25 \text{ cfm/ft}$$

Step 5: To provide this cold weather rate, we partly close the slot with a baffle, and obtain a slot width of

$$(25 \text{ cfm/ft}) / (100 \text{ cfm/in.}) = 0.25 \text{ in.}$$

However, in practice, a 0.25 in. slot opening might be quite small to control precisely. Therefore, we close every other 4 ft section of inlet and double the slot opening instead; that is, we make it 0.5 in. instead of 0.25 in. in the alternate 4 ft sections.

5.8.5.2.3.6.1 Inlets Controls

Air inlet control is extremely important for air distribution and a well-functioning ventilation system. Typically and preferably, the size of the slot inlet is adjusted automatically as a function of the changes in ventilating rates. The inlet sizes can also be controlled easily by manually adjusted baffles using a winch and cable system form one location. For precise adjustment, one can also place a manometer right next to the winch. Here, the manometer indicates the flow resistance that the fans must overcome to move air through the building. As shown in Table 5.7, ventilating systems for livestock or poultry buildings are typically designed for 0.04 in. H_2O (10 Pa) to 0.125 in. H_2O (30 Pa) static pressures. As mentioned earlier and also performed in Examples 5.15 and 5.16, a slot width sized to deliver air at 600 to 1,200 fpm (3.0 to 6.0 m/s) creates static pressure of 0.04–0.125 in. H_2O (10–30 Pa) for continuous slot inlets in exhaust systems. Sometimes, plastic gravity curtain slot inlets are used as well, but they do not provide as precise controls as the rigid baffles. It should also be kept in mind that condensation and frost on the curtain inlets and ceiling are often a problem.

5.8.5.2.3.6.2 Fan Location and Controls

Fan location does not have much effect on air distribution as long as the rigid baffle slot inlets are properly designed and operated. As mentioned earlier, fans are placed no more than about 75 ft (or 25 m) from the farthest part of the inlet, and downwind from prevailing winds, when possible. If the fans must exhaust against prevailing winter winds, they must be protected with fan windbreaks 5–10 ft (or 1.5–3 m) from the building. Fan hoods can be used to reduce wind effects as well. Moreover, as discussed in Chapter 3, mechanically ventilated livestock and poultry buildings should also be spaced at least 35 ft (or 10 m) apart, so that fans do not blow exhausted air into the intakes of neighboring buildings. Also, shutters must be used to prevent back-drafts on all non-continuous fans and on the animal side of fans to decrease freezing problems. Typically, thermostats with an on-off range of 3–5°F (5–10°C) developed for livestock and poultry applications are used for controlling ventilating systems in these buildings, and they are placed about 4–5 ft (1.2–1.5 m) off the floor, near the center of building width and away from furnace exhausts, inlet air, and direct sunlight.

5.8.5.3 Mechanical Ventilation in Aquatic Environments

As mentioned earlier, mechanical ventilation systems provide better control over indoor air temperature, air movement, moisture build up and odors in aquatic buildings compared to natural ventilation. And as already covered in depth in Chapter 4, fish produce best when the environment provided is within their optimum growth and production zones. Therefore, just like in greenhouses, livestock and poultry buildings, we need to provide the best environmental conditions and control systems in aquatic buildings as well.

5.8.5.3.1 Heat and Moisture Balances

Just like in livestock and poultry buildings, design of the environmental control in aquatic buildings begins with analyzing the heat and moisture flows in these building. Heat and moisture balances in livestock and poultry buildings have already been discussed in depth earlier in this chapter. A heat balance in aquatic buildings employs sensible heat gains and heat losses, while a moisture balance employs latent heat sources and losses. Since moisture generation and removal is the primary concern in aquatic buildings, a ventilation system designed using a moisture balance in the building will be necessary.

A moisture balance in an aquatic building employs all the latent heat sources and losses in the building. Moisture source in these buildings is mainly the water evaporated from free water surfaces in culture tanks. Moisture losses, on the other hand, include the water vapor losses taking place in the ventilating air (as colder and drier outside air replaces warmer and moister indoor air). Since surface condensation is not desirable in aquatic buildings, moisture loss from the indoor air by condensation on cold building surfaces is small, and therefore, neglected.

Then, using Equation (2.14) in steady-flow processes, the total moisture inflow rates are equal to the total moisture outflow rates; or in other words, if the indoor humidity ratio is to remain constant (typically below 70%–80%), the total rate of moisture loss must be equal to the total rate of moisture production as follows:

$$\dot{m}_a\left(W_2 - W_1\right) = W_e \tag{5.34}$$

where $\dot{m}_a$ is the mass flow rate of dry air, i.e., the amount of ventilating air (kg_{da}/s in SI or lb_{da}/h in I-P), W_1 is the humidity ratio (water vapor) in incoming air (kg_{water}/kg_{da} in SI or lb_{water}/lb_{da} in I-P), W_2 is the humidity ratio (water vapor) in exhaust air (kg_{water}/kg_{da} in SI or lb_{water}/lb_{da} in I-P), and W_e is the water vapor production from free water surfaces in culture tanks (kg_{water}/s in SI or lb_{water}/h in I-P).

Then, ventilation rate required to maintain a desired aquatic building moisture can easily be calculated using Equation (5.34). After substituting Equations (5.29a) for I-P and (5.29b) for SI into Equation (5.34), the Q equation for moisture control in aquatic buildings takes the following form:

$$Q = \left(\frac{v}{60}\right) \times \left(\frac{W_e}{W_2 - W_1}\right) \quad \text{(I-P)} \tag{5.35a}$$

where Q is the ventilation rate for controlling the indoor moisture (ft³/min or cfm), v is the specific volume of air (ft^3/lb_{da}), which is the inside conditions for exhaust systems, and outside conditions for positive pressure systems, W_e is the total moisture to be removed (lb_{water}/h) (evaporation from free water surfaces in culture tanks), 60 is for unit conversion (1 h = 60 min), W_1 is the humidity ratio (water vapor) in incoming air (lb_{water}/lb_{da}), and W_2 is the humidity ratio (water vapor) in exhaust air (lb_{water}/lb_{da}).

$$Q = v \times \left(\frac{W}{W_2 - W_1}\right) \quad \text{(SI)} \tag{5.35b}$$

where Q is the ventilation rate for controlling the indoor moisture (m³/s), v is the specific volume of air (m^3/kg_{da}), which is the inside conditions for exhaust systems, and outside conditions for positive pressure systems, W_e is the total moisture to be removed (kg_{water}/s) (evaporation from free water surfaces in culture tanks), W_1 is the humidity ratio (water vapor) in incoming air (kg_{water}/kg_{da}), and W_2 is the humidity ratio (water vapor) in exhaust air (kg_{water}/kg_{da}).

5.8.5.3.2 Ventilation System Design in Aquatic Buildings

Once the calculations are performed for the various operating conditions (cold and warm weather) and the feasible alternatives are analyzed, an appropriate ventilation system can be specified.

5.8.5.3.2.1 Minimum Continuous Ventilation Using the ventilation rate equation for moisture removal, Equations (5.35a) or (5.35b), the minimum continuous ventilation rate should be calculated. The winter outside design dry-bulb temperature (99%) with at 80% relative humidity and the minimum desired inside design temperature with the desired inside relative humidity (70%–80%) should be used in these equations.

The ventilation rate for moisture removal will provide good enough ventilation for moisture removal; however, as a result of this ventilation rate, the inside temperature might fall below the selected indoor design temperature if no supplemental heat were provided. Consequently, an additional calculation to determine the required amount of supplemental heat is necessary. To calculate the amount of supplemental heat ($\dot{Q}_t$) required, the energy balance equation is solved for $\dot{Q}_t$ using the specified winter design conditions. The quantity of supplemental heat should be installed and the controls should be set to turn this heat on whenever the inside temperature falls below the specified inside design condition.

5.8.5.3.2.2 Fall-Spring Ventilation Rate As the outside temperature increases on warm winter days and during the spring and fall months, it will be necessary to increase the ventilation rate above the minimum continuous rate to maintain desirable inside conditions. In practice, this increase in the ventilation rate is obtained by sensing the inside dry-bulb temperature with a thermostat, which increases the ventilation rate as the inside temperature increases. Since the moisture removal is still the primary concern, in order to determine the fall-spring ventilation rate, we still use the ventilation rate equation for moisture removal, Equations (5.35a) or (5.35b). Since it is warmer outside and the inside temperature has increased above the minimum inside design temperature, a second inside design temperature is selected. Often, this higher inside design temperature is chosen to be approximately 10 to 20°F (5.5 to 11°C) above the minimum inside design temperature. The outside temperature is usually taken as 5 to 15°F (3 to 8°C) below this higher interior temperature, ΔT = 5 to 15°F or 3 to 8°C. This additional ventilation capacity is then obtained incrementally with additional or variable speed fans controlled by thermostats.

5.8.5.3.2.3 Maximum Ventilation Rate During the summer and other periods when the outside and inside temperatures are above those used for the spring-fall ventilation rate, additional ventilation must be provided in aquatic buildings. In many cases, this additional capacity is most economically provided by manually opening large ventilation panels and simply relying on natural ventilation, which is what we will be doing here in this book.

5.8.5.3.2.4 Procedure for Ventilation System Design In this section, a step-by-step procedure is used for developing a ventilation system design for an aquatic building near Columbus, Ohio, covering minimum (winter) continuous ventilation, mild weather (fall–spring) ventilation, and hot weather (summer) ventilation requirements.

Step 1: Determine indoor and outdoor design temperature and humidity.

Indoor Design Temperature, T_i

- For cold and cool weather, use an indoor design temperature based on the species optimum temperature requirement of, for instance, 55–75°F (13–24°C).
- For summer, use 3–5°F (1.5–3°C) above the outside design—summer design—dry-bulb temperature (see Tables A.52–A.53). Then, the summer design temperature for Columbus is 88°F + 5°F = 93°F (31°C + 3°C = 34°C).

Outdoor Design Temperature, T_o

- For minimum winter ventilation, use 99% winter design dry-bulb temperature (see Tables A.52–A.53). For Columbus, Ohio, the outside design temperature at 99%, T_o (99%) = 6°F (14°C).
- For maximum fall–spring ventilation, use the average temperature of January, 5–15°F (3–8°C) below the maximum indoor design temperature, T_i. Then, 75°F − 10°F = 65°F (24°C − 6°C = 18°C).
- For summer ventilation, use 1% summer design dry-bulb temperature (see Tables A.52–A.53). For Columbus, Ohio, the summer design temperature at 1%, T_o (1%) = 88°F (31°C).

Indoor Relative Humidity, ϕ_i

- For cold weather, use indoor relative humidity 80%.

Outdoor Relative Humidity, ϕ_o

- For cold weather, assume 80% outdoor relative humidity.

Step 2: Determine latent heat generations in the building. Assume 0.25 in. (6 mm) water is evaporated per day from a free water surface or can be predicted as a function of inside temperature, relative humidity, and air speed.

Step 3: Calculate the minimum winter ventilation rate using the ventilation rate equation for moisture removal, Equations (5.33a) or (5.33b).

Step 4: Calculate the maximum ventilation rate for fall–spring moisture removal again using Equations (5.35a) or (5.33b).

Step 5: Specify fans and controls to achieve the additional ventilation as increasing T_o causes T_i to increase.

- Turn on the first fan when T_i is 5–10°F (3–5.5°C) above the design T_i, (e.g., T_i > 60°F or T_i > 15.5°C).
- Turn on the second fan at the first speed when T_i increases another 5°F (3°C) (e.g., T_i > 65°F or T_i > 18°C).

  Turn on the second fan at the second speed when T_i increases another 5°F (3°C) (e.g., T_i > 70°F or T_i > 21°C).

  Turn on the second fan at the third speed when T_i increases another 5°F (3°C) (e.g., T_i > 75°F or T_i > 24°C).

Step 6: Calculate the summer ventilation rate. The number of fans determined above would be adequate for only mild weather. Additional ventilation is required for hot weather ventilation. In many cases, this additional capacity is most economically provided by manually opening large ventilating panels for natural ventilation.

Step 7: If required, size the supplemental heating using Equation (5.14) based on the worst case, 99% winter design dry-bulb temperature (6°F or −14°C) for the outdoor design temperature, the chosen minimum winter ventilation rate, and the lowest possible total sensible heat generation by fish. Heat produced by fish is approximately 1 Btu/h.lb$_{fish}$ (2,200 W/kg$_{fish}$) body mass.

Example 5.17: Mechanical Ventilation for an Aquatic Building

Calculate the ventilation rates in an aquatic building with a solid floor near Columbus, Ohio. The building is the same 24 ft × 48 ft building with no windows from Example 5.12. Assume inside design conditions of 55–75°F (13–24°C), and 80% inside and outside RH.

Total free water surface area is 828 ft² (m²). Also assume that the heat produced by fish is zero and that 0.25 in. (6 mm) of water is evaporated per day from a free water surface.

Solution

Step 1: Determine indoor and outdoor design temperature and humidity.

Indoor Design Temperature, T_i

- For cold and cool weather, use an indoor design temperature based on the species optimum temperature requirement of, for instance, 55–75°F (13–24°C). Here, we will use an indoor design temperature of 55°F (13°C), and a maximum allowable indoor temperature of 75°F.
- For summer, use 3–5°F (1.5–3°C) above the outside design temperature, the summer design dry-bulb temperature (see Tables A.52–A.53). Then, the summer design temperature is 88°F + 5°F = 93°F (31°C + 3°C = 34°C).

Outdoor Design Temperature, T_o

- For minimum winter ventilation, use 99% winter design dry-bulb temperature (see Tables A.52–A.53). For Columbus, Ohio, the outside design temperature at 99%, T_o (99%) = 6°F (−14°C).
- For maximum fall-spring ventilation, use the average temperature for January, or 5–15°F (3–8°C) below maximum indoor design temperature, T_i. Then, 75°F − 10°F = 65°F (24°C − 5°C = 18.5°C).
- For summer ventilation, use 1% summer design dry-bulb temperature (see Tables A.52–A.53). For Columbus, the summer design temperature at 1%, T_o (1%) = 88°F or 31°C.

Indoor Relative Humidity, ϕ_i

- For cold weather, assume indoor relative humidity of 80%.

Outdoor Relative Humidity, ϕ_o

- For cold weather, assume outdoor relative humidity of 80%.

Step 2: Determine the latent heat generations in the building.

- Use 0.25 in. (6 mm) of water evaporation per day from the free water surface.

Step 3: Calculate the minimum winter ventilation rate using Equations (5.35a) or (5.35b).

The moisture production from 0.25 in. (6 mm) of water evaporation per day from a free water surface,

$$W_e = \left(828\,\text{ft}^2_{\text{water}}\right)\left(0.25\frac{\text{in}_{\text{water}}}{\text{day}}\right)\left(\frac{1\text{ ft}}{12\text{ in}}\right)\left(\frac{62.4\,\text{lb}_{\text{water}}}{\text{ft}^3_{\text{water}}}\right)\left(\frac{1\text{ day}}{24\text{ h}}\right)$$

$$W_e = 44.9\frac{\text{lb}_{\text{water}}}{\text{h}}$$

At $T_i = 55°\text{F}$ and $\phi_i = 80\%$, from the psychrometric chart, the humidity ratio, $W_i = 0.00736\,\text{lb}_{\text{water}}/\text{lb}_{\text{da}}$ and specific volume, $v = 13.18\,\text{ft}^3/\text{lb}_{\text{da}}$. And at $T_o = 6°\text{F}$ and $\phi_o = 80\%$, from the psychrometric chart, the humidity ratio, $W_o = 0.00086\,\text{lb}_{\text{water}}/\text{lb}_{\text{da}}$ and specific volume, $v = 11.8\,\text{ft}^3/\text{lb}_{\text{da}}$.

Now using Equation 5.33a (or (5.33b)), we can calculate the ventilation rate (Q) required for moisture control (exhaust system) as follows:

$$Q = \left(\frac{v}{60}\right)\times\left(\frac{W}{W_2 - W_1}\right)$$

$$Q = \left(\frac{13.18\,\text{ft}^3/\text{lb}_{\text{da}}}{60}\right)\times\left(\frac{44.9\frac{\text{lb}_{\text{water}}}{\text{h}}}{0.00736\frac{\text{lb}_{\text{water}}}{\text{lb}_{\text{da}}} - 0.00086\frac{\text{lb}_{\text{water}}}{\text{lb}_{\text{da}}}}\right)$$

$$Q = 1517\,\text{cfm}(\approx 0.72\text{ m}^3/\text{s})$$

This is the theoretical minimum ventilation (cold weather) rate for the given building and operating conditions. In this case, we will provide little more room for the minimum winter ventilation and use a commercially available fan delivering 1,540 cfm (0.73 m^3/s) at 0.10 in. static pressure (model: CCE12; fan size: 12 in.; motor: 1/4 HP; fan RPM: 1625). Again, while fan specifications are real, the manufacturer model used in this example is a generic model.

This much ventilating air is enough for moisture and odor removal, but it should be remembered that supplemental heating is also required to maintain the desired temperature levels. See Step 7 for determining the required supplemental heating.

Step 4: Calculate the maximum ventilation rate for fall-spring using Equations (5.35a) or (5.35b).

At $T_i = 75°\text{F}$ and $\phi_i = 80\%$, from the psychrometric chart, the humidity ratio, $W_i = 0.01497\,\text{lb}_{\text{water}}/\text{lb}_{\text{da}}$ and specific volume, $v = 13.86\,\text{ft}^3/\text{lb}_{\text{da}}$. And at $T_o = 65°\text{F}$ and $\phi_o = 80\%$, from the psychrometric chart, the humidity ratio, $W_o = 0.01057\,\text{lb}_{\text{water}}/\text{lb}_{\text{da}}$ and specific volume, $v = 13.5\,\text{ft}^3/\text{lb}_{\text{da}}$.

Now using Equation 5.33a (or 5.33b), we can calculate the ventilation rate (Q) required for moisture control (exhaust system) as follows:

$$Q = \left(\frac{v}{60}\right)\times\left(\frac{W}{W_2 - W_1}\right)$$

$$Q = \left(\frac{13.86\,\text{ft}^3/\text{lb}_{\text{da}}}{60}\right)\times\left(\frac{44.9\frac{\text{lb}_{\text{water}}}{\text{h}}}{0.01497\frac{\text{lb}_{\text{water}}}{\text{lb}_{\text{da}}} - 0.01057\frac{\text{lb}_{\text{water}}}{\text{lb}_{\text{da}}}}\right)$$

$$Q = 2{,}347\,\text{cfm}(\approx 1.11\text{m}^3/\text{s})$$

Step 5: Specify fans and controls to achieve the additional ventilation as T_o increase causes T_i to increase.

- Turn on the first fan when T_i is 5–10°F (or 3–5.5°C) above the design T_i, (e.g., T_i > 60°F or T_i > 15.5°C).
- Turn on the second fan at the first speed when T_i increases another 5°F (3°C) (e.g., T_i > 65°F or T_i > 18°C).

Turn on the second fan at the second speed when T_i increases another 5°F (3°C) (e.g., T_i > 70°F or T_i > 21°C).

Turn on the second fan at the third speed when T_i increases another 5°F (3°C) (e.g., T_i > 75°F or T_i > 24°C).

Based on these calculations, we can chose the following fans, although other fans may be chosen as appropriate.

- One single speed CEE12, 1,540 cfm fan operating continuously.
- One 3-speed CEE12-3, 1100/900/800 cfm fan to turn on at 65°F and beyond.

Our design for this aquatic building has a total installed capacity of 2,640 cfm, which will provide little more than what is required (2,347 cfm); therefore, the overall ventilation system will be in good shape removing the excess moisture in mild weather and providing uniform air exchange provided a proper configuration. The fans are nicely sized, and once located properly, will provide uniform air exchange and prevent oscillations when controlled properly in stages as shown above. Remember that we have used one single speed fan (1,540 cfm) (model: CCE12; fan size: 12 in.; motor: 1/4 HP; fan RPM: 1625) for minimum winter ventilation along the sidewall, which will operate continuously. Furthermore, we have used one three-speed fan (1100/900/800 cfm) (model: CCE12-3; fan size: 12 in.; motor: 1/12 HP; fan RPM: 1600/1420/1150) for the additional fall-spring air exchange requirement providing 1100/900/800 cfm ventilation at three different speeds as needed. This configuration will provide excellent spatial and temporal ventilation for this specific aquatic building. Please note that while the fan specifications are real, the manufacturer models are generic models.

It should again be kept in mind that larger fan diameter and lower HP fans consume less operational energy for delivering the same air exchange; therefore, when possible, alternative fans with larger diameter and lower HP motors should be chosen. Also, reducing fan rotational speed is the most efficient way to reduce fan energy use in variable flow, and this can be efficiently achieved with frequency controlled electric motors. For instance, if we reduce the fan rotational speed by 10%, based on the fan/pump affinity laws (this is further discussed in Chapter 8), the fan power will reduce 27%. It means, not only the initial costs but also the operational costs of fans should be considered for making a justified fan decision.

Step 6: Fall-spring ventilation is just enough for mild weather ventilation; hence additional ventilation is required for summer ventilation. Hot weather ventilation is most economically provided by manually opening large ventilating panels for natural ventilation. So, open the natural ventilation panels for the required hot weather ventilation.

Step 7: Now we can size the supplemental heating using Equation (5.14) based on the worst case, i.e., 99% winter design dry-bulb temperature (6°F or −14°C) for the outdoor design temperature, the chosen minimum winter ventilation rate, and the lowest possible total sensible heat generation ($\dot{Q}_{fish\_sensible}$) by the fish. Here, we will assume that the sensible heat generated by the fish is zero. Then, the supplemental heating is

$$\dot{Q}_t = \dot{Q}_{cond} + \dot{Q}_{inf} - \dot{Q}_{fish\_sensible}$$

where $\dot{Q}_t$ is the supplemental heating provided to be provided by the heating system, $\dot{Q}_{cond}$ is the rate of heat loss from the livestock building envelope (246.5 Btu/h°F $\times(55°\text{F} - 6°\text{F}) = 12{,}079$ Btu/h), and $\dot{Q}_{inf}$ is the rate of sensible heat loss due to 1,000 cfm air exchange with the outside air, and $\dot{Q}_{fish\_sensible}$ is assumed to be zero.

Using Equation 5.9a (or 5.9b), $\dot{Q}_{inf}$, the rate of sensible heat loss due to 1,000 cfm air exchange with the outside air can be calculated as follows:

$$\dot{Q}_{inf} = \frac{Q}{v}\left(60\frac{\text{min}}{\text{h}}\right)c_p\left(T_i - T_o\right)$$

$$\dot{Q}_{inf} = \frac{1{,}600\,\text{cfm}}{13.18\,\text{ft}^3/\text{lb}_{\text{da}}}\left(60\frac{\text{min}}{\text{h}}\right)(0.24\,\text{Btu}/\text{lb}_{\text{da}}\cdot{}^\circ\text{F})\left(55^\circ\text{F} - 6^\circ\text{F}\right)$$

$$\dot{Q}_{inf} = 85{,}657\ \text{Btu/h}$$

And

$$\dot{Q}_{fish\_sensible} = 0\ \text{Btu}/\text{h}$$

Then, the required supplemental heating, $\dot{Q}_t$ is:

$$\dot{Q}_t = 12{,}079\ \text{Btu}/\text{h} + 85{,}657\ \text{Btu}/\text{h} - 0\ \text{Btu}/\text{h}$$

$$\dot{Q}_t = 97{,}736\ \text{Btu}/\text{h}\left(\approx 28{,}662\ \text{W}\right)$$

The required heating load of 97,736 Btu/h (28,662 W) can be supplied by using one 80% efficient propane-fired CEE145 (143,750 Btu/h input) providing a total of 115,000 Btu/h. The unit heater will turn on at about 55°F. The unit heater comes with 4 in. (10.2 cm) round flue gas connection on the back and can be installed with straight B-Vent connection and diagonally. CEE145 has ½ in. gas piping size, and natural gas plumbers can easily connect the individual unit heater to a propane supply tank.

In this project, we have used one 80% efficient CEE145 (143,750 Btu/h input). As mentioned earlier, the ASME states that 1 square inch free area per 2,000 Btu/h maximum fuel input of combined burners must be provided and kept clear at all times. Then,

Free Area Required = (143,750 Btu/h input) × (1 in²/2,000 Btu/h)
Free Area Required = 71.9 in²

Inlet louvers typically have 60% free area; therefore, 57.5 in²/0.6 = 120 in² louver area per unit heater needs to be provided. Then, for the purpose of providing fresh air here, one CEE11 motorized inlet shutter having opening dimensions of 11 in. × 11 in. (which provides a louver area of 121 in², and a free area of 72.6 in²) would meet the requirement.

Please note that while the unit heater and inlet louver specifications are real, the manufacturer models are generic models.

Step 8: Size and locate inlets and outlets. See Examples 5.15 and 5.16. Same protocols applied.

Step 9: Check against standard recommendations, if any; and if possible, re-evaluate. It has been checked and proper adjustments are made accordingly.

5.8.6 Active Cooling by Evaporative Cooling

Evaporative cooling, using current technology and available equipment, is an energy efficient and cost-effective method of cooling. It cools air by evaporation of water. When water evaporates into the air being cooled, it creates direct evaporative cooling, which is the oldest and the most common form of cooling. As discussed in Chapter 2 earlier, evaporative cooling is essentially an adiabatic saturation process, from an initial state at low humidity ratio and high dry-bulb temperature to a final state at higher humidity and lower dry-bulb temperature. The quantity of heat associated with a change in phase from liquid water to water vapor, termed as *latent heat of vaporization*, Q_L, that is, the energy required to evaporate 1 kg liquid water without a change in temperature is about 2,430 kJ (at ~30°C). In I-P units, 1 lb of water at 86°F requires ~1,044 Btu of energy to evaporate. This energy in the form of sensible heat is extracted from the air and this is how the evaporative cooling system cools the air as water evaporates and transforms the sensible heat into the latent heat. When the evaporation occurs separately and the air is cooled without humidity gain, the process is called indirect evaporative cooling. There are a number of techniques to cool the air using direct evaporative cooling. These include i) fan and pad systems, ii) humidifiers such as fog systems, iii) spray-filled and wetted-surface air washers, and iv) sprayed coil units.

Some evaporative cooling methods can provide cost-effective cooling in greenhouses, livestock, poultry and aquatic buildings. However, direct evaporative cooling adds moisture to the indoor air that may reduce plant transpiration in greenhouses and also increase the risk of diseases particularly fungal diseases if not managed adequately in greenhouses, livestock and poultry buildings. The amount of useful evaporative cooling obtained depends largely on the wet bulb temperature of the air entering the equipment. High ambient relative humidity levels severely limit the effectiveness of evaporative cooling and fog systems because of the reduced vapor pressure differential (also known as wet-bulb depression). The *wet-bulb depression*, which is the difference between the dry-bulb and corresponding web-bulb temperatures, provides the maximum limit the air can be cooled down via evaporative cooling. The drier the air, the bigger the wet-bulb depression is, and therefore, the bigger the potential for evaporative cooling.

Example 5.18: Determining Wet-bulb Depressions on Psychrometric Chart

Determine the wet-bulb depressions for Tucson, Arizona, USA; Columbus, Ohio, USA; and Montreal, Québec, Canada, and show the evaporative cooling potentials on the psychrometric chart.

Solution

Using Tables A.52–A.56, we first determine the summer climatic conditions at the given sites (1%) as follows:

LOCATION		°C	°F	Wet-bulb Depression (°C)	Wet-bulb Depression (°F)
Tucson, AZ	Design Dry-bulb Temperature (T)	39	102	21	37
	Coincident Wet-bulb Temperature (T_w)	18	65		
Columbus, OH	Dry-bulb Temperature (T)	31	88	8	15
	Coincident Wet-bulb Temperature (T_w)	23	73		
Montreal, QC	Dry-bulb Temperature (T)	28	83	7	13
	Coincident Wet-bulb Temperature (T_w)	21	70		

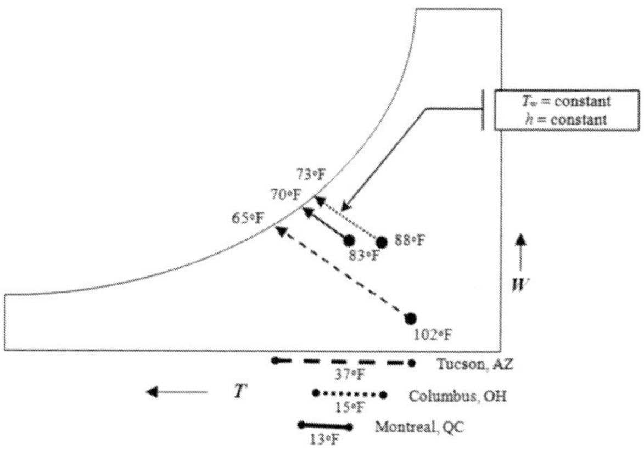

As shown above, Tucson, AZ, has very hot and dry summers, while Columbus, OH, has a milder and more humid summer climate. Montreal, QC, on the other hand, has relatively cooler and more humid summer climate compared to the other two locations. As it has been discussed earlier, the drier the air, the bigger the wet-bulb depression is, and therefore, the bigger the potential for evaporative cooling. Tucson, AZ, has the biggest wet-bulb depression (37°F) of all; and if we assume 80% efficiency for the evaporative cooler, then we can achieve 0.8 × 37°F = 29.6°F ≈ 30°F temperature drop with this cooler, while the same evaporative cooler can approximately provide 12°F and 10°F temperature drops in Columbus, OH, and Montreal, QC, respectively. This exercise clearly shows that evaporative cooling is quite an effective cooling method in hot and dry climates such as the one in Arizona.

5.8.6.1 Fan and Pad Evaporative Cooling Systems

The fan and pad type evaporative cooling system (see Figure 5.41) has been the standard cooling system particularly in greenhouses for many years. These systems consist of exhaust fans on the sidewall or gable end of the greenhouse and the pad assembly on the opposite side of the building. When the exhaust fans operate, they dump the hot and moist indoor air outside pulling the cooler and drier outside air through the wetted pads where the water evaporates and the outside air is cooled through the process. Each kilogram of water removes 2,430 kJ (and each pound of water removes 1,044 Btu) at ~30°C (86°F) of heat from the air by changing the phase from liquid to vapor, hence cools the outside air, which is being pulled to the building.

A fan and pad cooling system, as shown in Figure 5.42, has a pad, a water pump with valves and a water filter, a water distribution pipe, a water collection gutter at the bottom of the fiber pad (connected to a covered sump via a return pipe with a 50 mesh inclined screen), and a covered sump tank, and fan(s) on the opposite wall of the building. The water is pumped and distributed horizontally through 3 mm (0.125 in.) holes spaced 7.5 cm (3 in.) from each other along the distribution pipe at the top of the vertically mounted pads. For pad lengths less than 15 m (50 ft), a 32 mm (1¼ in.) PVC pipe handles the task, while for lengths up to 18 m (60 ft) a 38 mm (1½ in.) PVC pipe is typically used. Non-evaporated water is collected in the collection gutter at the bottom of the vertical pads and returned to a sump tank to be circulated again. The fans mounted on the opposite sidewall or gable end determine the total evaporative cooling pad area and pulls in fresh air through the pads, and cools the building interior.

FIGURE 5.41 A fan and pad evaporative cooling system. Only the pad system on gable wall is shown here.

Source: *Greenhouse Engineering* (2021).

There are a number of commercially available pad materials such as corrugated cellulose (most widely used), aspen pads, and other pads made from aluminum or plastic fibers. Corrugated pads are relatively inexpensive and, when maintained properly, have a lifetime of approximately ten years. Recommended air velocities through various evaporative cooling pads are presented in Table 5.8.

As mentioned earlier, the fans mounted on the opposite wall determine the total pad area, which can be calculated by dividing the volume of air exchange (m^3/s or ft^3/min) by the face velocity of air (m/s or ft/min, respectively) (Table 5.8) that can be moved through the pad. The pads are typically mounted on the interior sides of sidewalls or gable ends extending the entire length of the building wall. On the exterior side of the pad, air inlets are installed so that outside air can be sealed off during cold weather. When the pad height is more than that of the inlet, the pad should have a minimum of half the height difference space from the inlet. And maximum air velocities through the inlet should be limited to 1.8 m/s (350 ft/min).

Obviously, the cooled outside air moves from the pad side wall towards the opposite wall where the fans are mounted. And as a consequence of this distance, a temperature gradient forms across the building. Depending on the temperature tolerance of animals and crops—a temperature

TABLE 5.8
Recommended Air Velocities through Various Cooling Pad Materials

Material	Face Velocity of Air through Pad [a] (m/s)	(ft/min)
Corrugated Cellulose, 100 mm (4 in) thick	1.27	250
Corrugated Cellulose, 150 mm (6 in) thick	1.78	350
Aspen Fiber, 50–100 mm (2–4 in) thick, mounted vertically	0.76	150
Aspen Fiber, 50–100 mm (2–4 in) thick, mounted horizontally	1.00	200

[a] Where space is limiting, these velocities may be increased by no more than 25%. *Source:* Reprinted with permission from ANSI/ASAE EP406.2 (OCT95), Copyright © 2003 by American Society of Agricultural and Biological Engineers, St. Joseph, Michigan, USA.

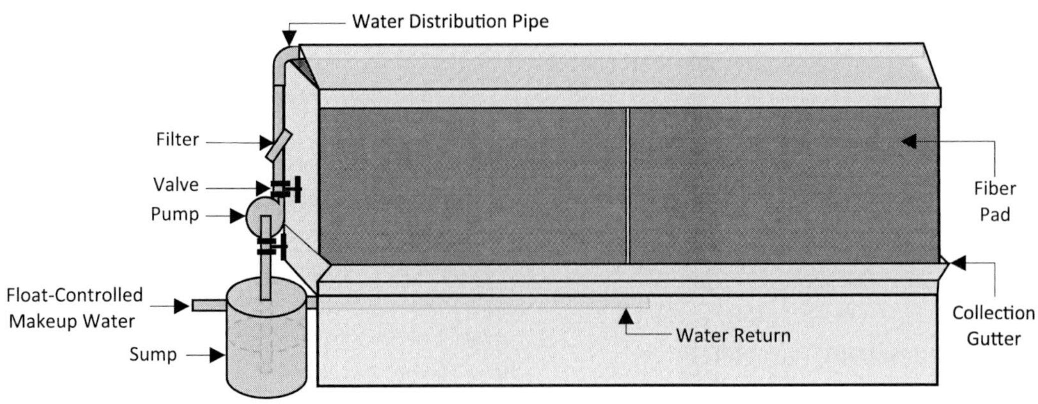

FIGURE 5.42 Schematic view of a fan and pad evaporative cooling system. Only the pad system is shown here.

Source: *Greenhouse Engineering* (2021).

TABLE 5.9
Recommended water flow rates and sump capacities for vertically mounted pads

Material	Minimum Water Flow Rate per Unit Length of Pad (L/min.m)	Minimum Water Flow Rate per Unit Length of Pad (US gal/min.ft)	Minimum Sump Capacity per Unit Pad Area (L/m^2)	Minimum Sump Capacity per Unit Pad Area (US gal/ft^2)
Corrugated Cellulose, 100 mm (4 in)	6.2	0.50	33	0.80
Corrugated Cellulose, 150 mm (6 in)	9.9	0.80	40	1.00
Aspen Fiber, 50–100 mm (2–4 in)	3.7	0.30	20	0.50
Aspen Fiber, 50–100 mm (2–4 in), desert conditions	5.0	0.40	20	0.50

Source: Reprinted with permission from ANSI/ASAE EP406.2 (OCT95), Copyright © 2003 by American Society of Agricultural and Biological Engineers, St. Joseph, Michigan, USA.

gradient of 4 to 5.5 °C (7 to 10 °F) is common, greater pad-to-fan distances can be used. However, a pad-to-fan distance of 30–45 m (100–150 ft) is typically utilized even though it is sometimes extended up to 60 m (200 ft).

Cooling pads are typically installed continuously or uniformly along the sidewall or gable end (when possible in the windward direction of prevailing summer winds) opposite the exhaust fans. For uniform wetting, therefore, an efficient cooling, vertical pad heights are kept at or below 2.4 m (8.0 ft), but not less than 0.60 m (2.0 ft). Algae growth and salt buildup should be monitored and controlled. We need to ensure that the pads remain wet from top to bottom during the cooling process, and when the pads are sized based on Table 5.8, the recommended water flow rates given in Table 5.9 are good enough. When algae growth becomes a problem, then it can be controlled via injecting chemicals through the circulation water. Where high mineral content water is used for evaporative cooling, a continuous water bleed-off rate of 0.02 L/min per m^3/s (0.05 gal/min per 1,000 cfm) of airflow is for minimizing mineral precipitation and buildup in the pads. The sump volumes given in Table 5.9 are sized to operate at the half depth of the tank, hence they will provide room for water returning from the pad when the system is turned off.

As mentioned earlier, exhaust fans are sized to provide the required airflow rate at an overall static pressure of 30 Pa (0.12 in. H_2O) and mounted on the opposing building wall. And they should not exhaust directly into the pads of the neighboring buildings unless the distance between fan and inlet is more than 15 m (50 ft). Also, the fans should not exhaust directly into each other unless they are more than four to five diameters apart.

Example 5.19: Fan and Pad Evaporative Cooling

Let's now consider a fan and pad evaporative cooling system used for air-conditioning an institutional greenhouse. Assume that the outside air enters the evaporative cooling pad at 35°C dry-bulb temperature and 25% relative humidity with a flow rate of 4 m^3/s. Assuming that the fan and pad evaporative cooler has a cooling efficiency of 80%, determine the supply T, ϕ, total cooling capacity provided, and the rate of water used for this evaporative cooling. Note: 1 ton of refrigeration equals a heat removal rate of 12,000 Btu/h.

Solution

See Example 2.74 in Chapter 2.

Example 5.20: Fan and Pad Type Evaporative Cooling

Now, using the same greenhouse that was used in Example 5.11, design a fan and pad type evaporative cooling system using 100 mm (4 in.) thick cellulose pads. Remember that the static pressure required for the fans is 30 Pa (0.12 in. H_2O). Assuming that the greenhouse is located in Tucson, Arizona, and the fan and pad evaporative cooler has a cooling efficiency of 80%, determine the supply T, ϕ, total cooling capacity provided, and the rate of water used for this evaporative cooling. Note: 1 ton of refrigeration equals a heat removal rate of 12,000 Btu/h.

Solution

Remember that in Example 5.11 the total fan capacity, 131,112 cfm, was installed for a static pressure of 0.10 in. H_2O. Since the static pressure required for the evaporative fans is 0.12 in. H_2O, we now need to pick new fans as follows:

Minimum Winter Air Exchange Requirement = 21,600 cfm

In this case, minimum winter ventilation can uniformly be supplied by four fans each delivering 5,648 cfm at a static pressure of 0.125 in. (model: GHE30; fan size: 30 in.; motor: 1/4 HP; fan RPM: 491). Again, the manufacturer model used in this example is a generic model.

These four fans can deliver a total of 22,592 cfm maximum when they operate all together; therefore, in order to determine the spring/fall and summer ventilation fans, this total need to be subtracted from the grand total, 129,600 cfm.

$$129{,}600 \text{ cfm} - 22{,}592 \text{ cfm} = 107{,}008 \text{ cfm}$$

(This much more to be provided for spring/fall and summer ventilation)

We may have other locally available options; however, for this example, we will pick five fans each delivering 17,456 cfm at a static pressure of 0.125 in. (model: GHE48; fan size: 48 in.; motor: 1 HP; fan RPM: 380); and another fan delivering 20,480 cfm again at the required static pressure (model: GHE54; fan size: 54 in.; motor: 1 ½ HP; Fan RPM: 414). Both manufacturer models are generic.

		Manufacturer Model No
1) Winter	4 fans at 5,648 cfm each	GHE24 (single speed)
2) Spring/Fall/ Summer	5 fans at 17,456 cfm each	GHE48 (two speed)
3) Spring/Fall/ Summer	1 fan at 20,480 cfm	GHE54 (two speed)
Total Fan Capacity:	130,352 cfm (installed)	

Now, using the total installed fan capacity of 130,352 cfm, we can calculate the face area of the 6 in. pad as follows:

$$\text{Face Area of the Pad} = \frac{130,352\,\text{cfm}}{250\,\dfrac{\text{ft}}{\text{min}}} = 521\,\text{ft}^2$$

(see Table 5.8 for 4 in. thick cellulose)

Now, calculate the pad height if pad extends full length of 90 ft sidewall:

$$\text{Pad Height} = (\text{Area of the Pad}) / (\text{Length of the Greenhouse})$$

$$\text{Pad Height} = \frac{521\,\text{ft}^2}{90\,\text{ft}} \approx 6\,\text{ft}$$

Then, the overall 4 in. pad dimension is 6 ft × 90 ft.

Now, we can calculate the pump flow rate as follows:

$$\text{Water Flow Rate} = (\text{Pad Length})(\text{Flow Rate per Unit Length of Pad})$$

(see Table 5.9)

$$\text{Water Flow Rate} = (90\,\text{ft})(0.5\,\text{gal/min.ft})$$

(see Table 5.9 for 4 in. thick cellulose pad)

$$\text{Water Flow Rate} = 45\,\text{gpm}$$

And, the minimum sump size as follows:

Minimum Sump Capacity

$$= (\text{Pad Face Area})(\text{Required Sump Capacity per unit Pad Area})$$

$$\text{Minimum Sump Capacity} = (6 \times 90)\,\text{ft}^2\,(0.80\,\text{gal/ft}^2)$$

(see Table 5.9 for 4 in. thick cellulose pad)

Minimum Sump Capacity = 432 gal

Note that in greenhouses where the air inlet-to-fan distance is less than 30 m (100 ft), ventilation flow rates should be increased to improve the heat transfer based on Equation (5.25). In our example, the distance is 36 m (120 ft); therefore, no adjustment was needed.

Now considering the above fan and pad type evaporative cooling system has a cooling efficiency of 80%, and assuming that the greenhouse is located in Tucson, AZ, and the outside air enters the evaporative cooling pad at the summer design dry-bulb temperature of 102°F (39°C) (Tables A.51 and A.56) and the coincident wet-bulb temperature of 65°F (18°C) (Tables A.51 & A.56) with a flow rate of 130,352 cfm (~61.3 m³/s), we can determine the supply T, ϕ, and total cooling capacity provided as follows:

State 1 is the outside air condition. If the cooling system were 100% efficient, then the temperature at State 2 (after the cooling pad) would be the T_w of State 1 (102°F), that is, State 1 is located at the intersection of T = 102°F and T_w = 65°F, which would at the same time be the temperature at State 2. With 80% cooling efficiency, however, T_2 would be T_1 minus the 80% of $(T_1 - T_w)$, i.e., T_2 = 102°F – 0.8 (102°F – 65°F) = 72.4°F, which is the supply temperature to the greenhouse.

Using the intersection of T_2 = 72.4°F and the constant wet-bulb temperature line of T_w = 65°F on the psychrometric chart or using psychrometric equations (Chapter 2), we can now determine the humidity ratio at State 2 (after the pad) as W_2 = 0.01150 $\text{lb}_{\text{water}}/\text{lb}_{\text{da}}$, the specific volume v_2 = 13.66 $\text{ft}^3/\text{lb}_{\text{da}}$, and the relative humidity at State 2 as $\phi \approx$ 69%.

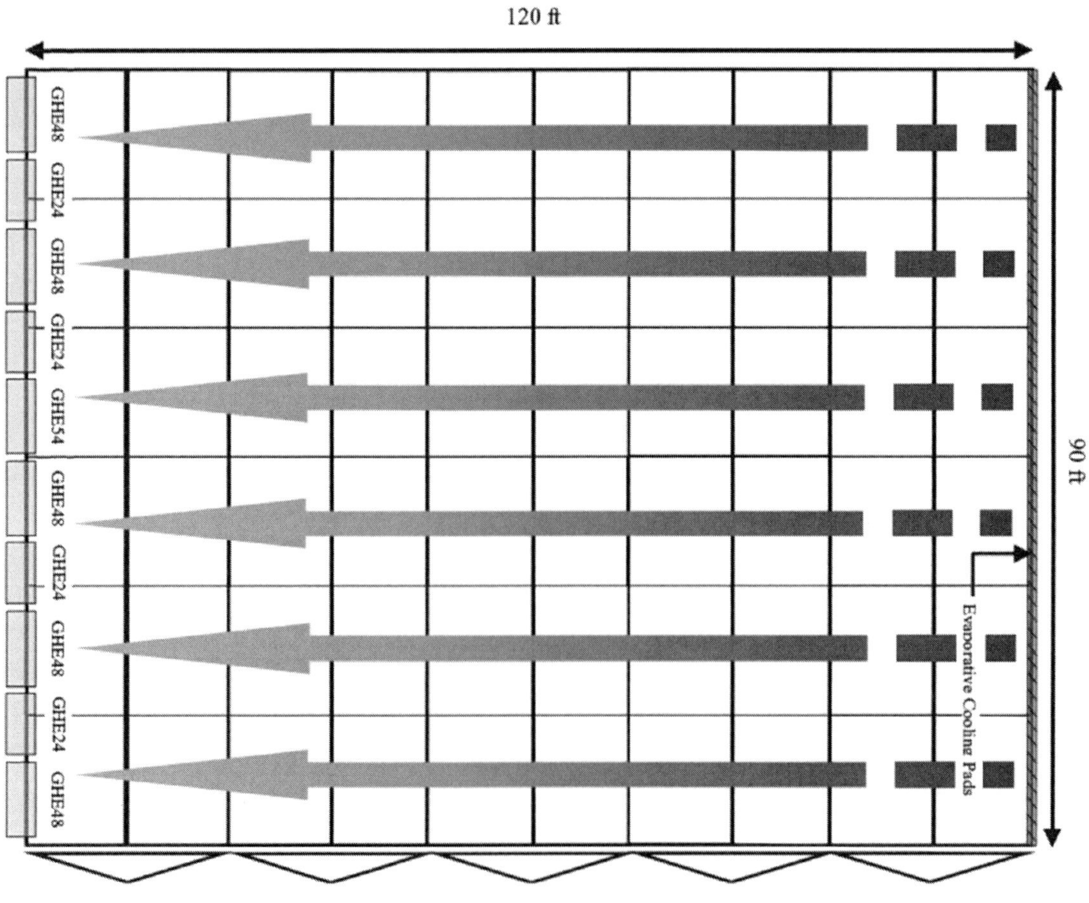

Now, we can first calculate the mass flow rate of dry air using the specific volume v_2 at State 2 as the fan is positioned on the opposite wall (negative pressure or suction type ventilation). The mass flow rate of dry air is then:

$$\dot{m}_a = \left(130{,}352\frac{ft^3}{min}\right) / \left(13.66\frac{ft^3}{lb_{da}}\right) = 9{,}543\frac{lb_{da}}{min}$$

And now using Equation (2.140) and the specific heat of air $0.24\ \frac{Btu}{lb_{da}\,°F}$ (Table A.13),

$$\dot{Q} = \dot{m}_a c_p (T_2 - T_1)$$

$$\dot{Q} = \left(9{,}543\frac{lb_{da}}{min}\right)\left(60\frac{min}{h}\right)\left(0.24\frac{Btu}{lb_{da}}\right)(72.4 - 102)$$

$$\dot{Q} = -4{,}067{,}608\frac{Btu}{h}\ (\approx -1{,}192\ kW)$$

And since 1 ton of refrigeration is equal to 12,000 Btu/h, the amount of cooling provided by this fan and pad evaporative cooling system is approximately equal to 340 ton.

And using Equation (2.136), the theoretical rate of water use for the evaporative cooling process can be calculated as follows:

$$\dot{m}_w = \dot{m}_a (W_2 - W_1)$$

$$\dot{m}_w = \left(9{,}543\frac{lb_{da}}{min}\right)(0.01150 - 0.00543)\frac{lb_{water}}{lb_{da}} = 57.93\frac{lb_{water}}{min}$$

$$\dot{m}_w = \left(57.93\frac{lb_{water}}{min}\right)\left[0.12\frac{gpm}{\frac{lb}{min}}\right] = 6.95\ gpm \approx 7 gpm$$

Example 5.21: Fan and Pad Evaporative Cooling

Now, using the same greenhouse from Examples 5.11 and 5.20, design a fan and pad evaporative cooling system, this time, using 150 mm (6 in.) thick cellulose pads.

Solution

Remember that the total fan capacity is 130,352 cfm. Now we can calculate the face area of the 6 in. pad as follows:

$$\text{Face Area of the Pad} = \frac{130{,}352\,cfm}{350\frac{ft}{min}} = 372\,ft^2$$

(see Table 5.8 for 6 in. thick cellulose)

Now, calculate the pad height if pad extends full length of 90 ft sidewall:

$$\text{Pad Height} = (\text{Area of the Pad}) / (\text{Length of the Greenhouse})$$

$$\text{Pad Height} = \frac{372\,ft^2}{90\,ft} \approx 4\frac{1}{2}\,ft$$

Then, the overall 6 in. cellulose pad dimension is 4 ½ ft × 90 ft.

Now, we can calculate the pump flow rate as follows:

$$\text{Water Flow Rate} = (\text{Pad Length})$$
$$(\text{Flow Rate per Unit Length of Pad})$$

(see Table 5.9)

$$\text{Water Flow Rate} = (90\,ft)(0.8\,gal/min.ft)$$

(see Table 5.9 for 6 in. thick cellulose pad)

$$\text{Water Flow Rate} = 72\,gpm$$

And, the minimum sump size as follows:

$$\text{Minimum Sump Capacity}$$
$$= (\text{Pad Face Area})\begin{pmatrix}\text{Required Sump}\\ \text{Capacity per unit Pad Area}\end{pmatrix}$$

$$\text{Minimum Sump Capacity} = \left(4\frac{1}{2}\times 90\right)ft^2\,(1.00\,gal/ft^2)$$

(see Table 5.9 for 6 in. thick cellulose pad)

$$\text{Minimum Sump Capacity} = 405\,gal$$

See the difference between Examples 5.20 and 5.21. The total amount of cooling is 329 ton here as well.

5.8.6.2 Fog Cooling Systems

Fog cooling uses exactly the same principle that is used in fan and pad evaporative cooling. Fog cooling systems however employ high-pressure pumps typically operating at 6.9 MPa to 10.3 MPa (1,000 psi to 1,500 psi), copper, stainless steel or reinforced flexible poly-tubes for water piping, and high-pressure ceramic, plastic or stainless-steel nozzles to form 0.5–50 μm (microns) water droplets for cooling the building (see Figure 5.43). The water consumption per nozzle is approximately 0.00105 to 0.00126 L/s (1.0 to 1.2 gallons per hour). The pipes and nozzles in greenhouses are located over aisles in order to prevent water dripping directly over plants. The water droplets almost vaporize immediately and cool the indoor air; however, fog cooling systems are not as effective as the fan and pad systems because these systems typically operate on the building interior air. Therefore, they must be

FIGURE 5.43 Fog cooling system installed together with a retractable shading system in a greenhouse.

Source: *Greenhouse Engineering* (2021).

monitored carefully with respect to the inside relative humidity levels. Even though they can be uniformly spaced and used with natural ventilation systems, for an effective fog cooling system, the conventional exhaust fans need to replace the humid air inside periodically with the drier outside air. In fan cooled buildings, most of the nozzles are located near the inlet with the remaining distributed evenly over rest of the building space. Fog nozzles can also be installed inside the air inlets when mechanical ventilation is used.

Fogging can be controlled with a computer or simply by a two-stage thermostat. Temperature settings of 3 to 5.5°C (5 to 10°F) apart would provide good enough control with this two-stage control. Since the nozzles have quite tiny openings, good quality water and filtration is extremely important for the performance of fog cooling systems.

5.8.7 Active Cooling by Heat Pumps

Several heat pump cycles are considered feasible: the reversed Rankine cycle, reversed Brayton cycle, reversed Stirling cycle, and Ericson cycle. Among these, the most commonly used cycle is the reversed Rankine cycle, which is also known as the vapor-compression cycle (see Figure 5.44). The driving energy of these cycles can be electricity generated in a power plant, an internal combustion engine, a Rankine or Brayton engine cycle, or an absorption system.

The vapor-compression cycle is the most common residential refrigeration/heat pump cycle used today. Current electric heat pump systems use a vapor-compression cycle. The refrigerant removes heat from the environment in the evaporator; it is, then, compressed in an electrically driven compressor or an internal combustion engine, and then sent to a condenser. The condenser rejects heat to the heat sink, and then the refrigerant flows back to the evaporator through an expansion valve. The energy consumer of this cycle is the electric motor driving the compressor. The COP (coefficient of performance) of heat pumps with respect to the compressor work varies in cooling and heating modes depending on the refrigerant used. However, if the energy source were considered, real efficiency is then depended on the electric plant or the engine efficiency. Colosimo (1987) reported that the typical efficiency of an electric power plant is about 32%, and approximately 3% (even more) of the electricity is lost during the transmission process, leaving only 29% of the energy supplied by fuel to be utilized as electrical energy. Consequently, the actual COP value of the electric heat pump system is less than a third of the value obtained with respect to the compressor work.

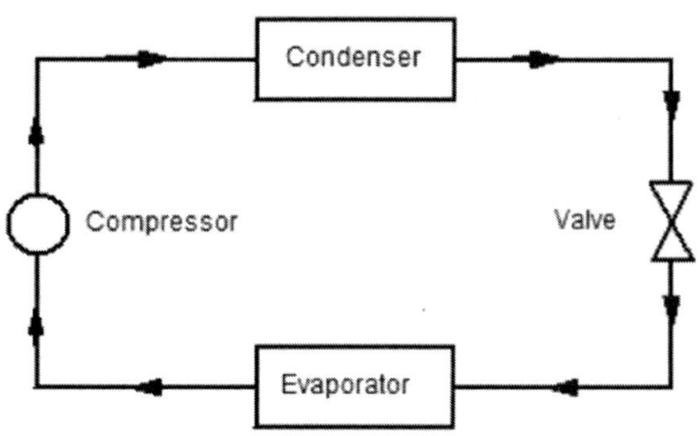

FIGURE 5.44 Vapor-compression cycle.

In order to improve COP, alternative cycles have been developed utilizing the energy source directly (oil or natural gas). These are the Brayton or Rankine cycle engine-driven heat pumps. The power cycles in these systems generate energy to drive the compressor for the vapor-compression cycle. Since both of these systems do not have an acceptable COP in both heating and cooling, Christensen and Santoso (1990) developed an alternative configuration of a Rankine cycle engine-driven heat pump system, which uses a hydraulically connected rolling diaphragm piston-cylinder device as the motor, compressor and pump. Initial evaluations of the system gave heating and cooling COPs of 2.01 and 1.06, respectively.

Another commonly used refrigeration cycle is the absorption refrigeration cycle (see Figure 5.45). Usually, natural gas is used as a source of energy, which supplies some heat to the generator. This cycle uses an absorption system (instead of a compressor), which consists of an absorber, a pump, and a generator to drive the whole system. Ammonia-water and water-lithium bromide solutions are the most commonly used refrigerants. Absorption refrigeration systems have several advantages compared to the vapor compression cycle. These are the quietness and the high efficiency at reduced load operations. However, they have disadvantages (such as lower COPs, more components, and ammonia toxicity) when compared to the vapor-compression systems. Santoso (1989) compared an absorption heat pump to a Rankine cycle engine-driven heat pump for the same operating conditions, and determined that the absorption system using ammonia-water had COPs of 0.73 and 0.63 in the heating and the cooling modes, respectively. For the same operating conditions, a Rankine power cycle, engine-driven heat pump system had better COP predictions using R22 and R113 refrigerants. The COPs were 2.01 and 1.06 in the heating and the cooling modes, respectively.

Yildiz (1993) developed a simulation model for a gas fired Rankine cycle engine-driven heat pump system, which was used for winter heating and warm weather cooling requirements of a greenhouse. Three energy sources are typically used for heat pump heating applications. These are the ambient air, ground water and the earth. The earth is considered to be the best energy source; however, since most heat pump applications today use ambient air as an energy source or sink, ambient air was used as the source for this system.

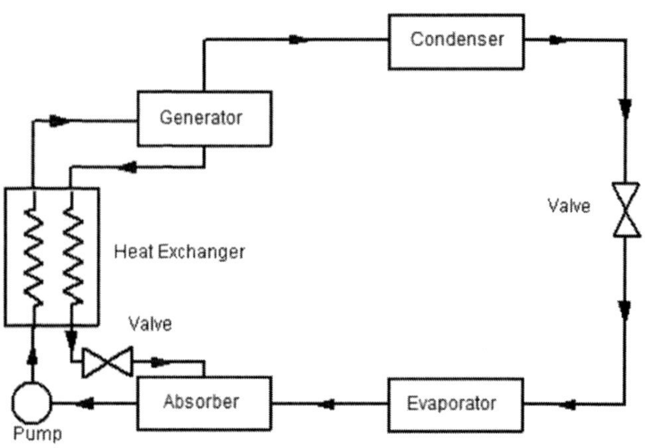

FIGURE 5.45 Absorption cycle.

The heat pump system was a 10,548 W (3 ton, 36,000 Btu/hr) unit consisting of a Rankine cycle and a vapor compression cycle. In the system, the power generated by the Rankine cycle replaces the electric power used to drive the compressor in Figure 6.12. Natural gas was used as the energy source, which supplied heat to the evaporator of the Rankine cycle. A fixed burner rate of 12,306 W (42,000 Btu/hr) and a burner efficiency of 0.9 were used. The vapor compression cycle removed the energy from the heat source through an evaporator attached to a hydronic cycle. The Rankine cycle used R123 (dichlorotrifluoroethane) as the working fluid; and the vapor compression cycle used R22 (chlorodifluoromethane) as its refrigerant.

Using this heat pump system, Yildiz and Stombaugh (2006a, 2006b) studied heating and cooling of a greenhouse at Delaware, Ohio, USA (latitude 40° 17'N, longitude 83° 05'W). January, April, and July weather files for were used to represent winter, spring, and summer in the simulations, respectively. Simulations were performed starting at the beginning of the fifth day and ended at the end of the 29th day of the month providing a 25-day simulation.

Table 5.10 presents some important greenhouse characteristics. The heat pumps evaluated were 3-ton gas fired units, and in order to provide multi-position proportional control, it was assumed that three heat pump units were used in each greenhouse. Along with variable shading and ventilation, an evaporative cooler was employed for cooling the conventional (CON) greenhouse. Outdoor air was used in the evaporative cooling system, and it was assumed that the air at the evaporative cooler outlet was fully saturated. In the open (OHP) and closed heat pump (CHP) systems, however, in addition to the variable shading system, heat pump units provided the cooling requirements. Indoor air was circulated and introduced back to the inside at a lower temperature using an overhead plastic tube for the air distribution. Two shading clothes with transmissivities of 0.75 and 0.50 were used to reduce the cooling loads. The use of these shading clothes provided shading levels of 25%, 50% and 62.5% by using them individually or together. In the open systems (CON and OHP), ventilation was provided by two fans, one with a fixed flow rate to provide a minimum level (0.01 $m^3/s.m^2$) of air exchange at all times, and the other one with a variable flow rate (with a maximum rate of 0.08 $m^3/s.m^2$).

TABLE 5.10
Greenhouse and Crop Characteristics Used in the Simulation Model

Greenhouse length	7.50 m (CON & OHP) and 25.0 m (CHP)
Greenhouse width	7.50 m
Greenhouse height at eaves	2.50 m
Greenhouse height at ridges	4.50 m
Greenhouse glazing	Double polyethylene
Floor surface	Reflective mulch with 60% reflectivity
Crop	Cucumber
Crop row orientation	North–South
Distance between plant rows	0.75 m
Average leaf length and width	0.30 m and 0.75
Plant height	2.0 m

Based on the indoor air temperature, the control system operated in either the heating or cooling mode. If the system was in heating mode and if heating was required, the ventilation rate was set to the minimum rate. The cooling mode operated in two steps. The first step was to reduce the cooling load using a variable shading system and to cool the inside air by increasing ventilation rates. If the first step in cooling could not handle the cooling load, then the second step was activated, in which the heat pump units (OHP) or evaporative cooling system (CON) provided the cooling. Inside relative humidity in the conventional system was controlled indirectly by the temperature control. In the open heat pump system, however, additional relative humidity control was provided. When the inside relative humidity levels exceeded 80%, additional ventilation was introduced to decrease inside relative humidity. In the cooling mode of the closed heat pump system, there was only one step unlike the conventional and open heat pump systems, which had two-step cooling systems. Here, no cooling was provided by ventilation; instead, it was done by the three heat pump units providing a multi-position proportional control, when possible, after reducing the cooling load using the variable shading system. The operation of the shading system was the same as in the other two systems. Here, either the cooling units or the dehumidifier (the first heating unit) controlled inside relative humidity.

Figure 5.46 shows the diurnal changes in solar radiation, relative humidity and temperature regimes in conventional, open and closed heat pump greenhouse systems on a spring day. Figure 5.47 presents the long-term operation of the simulation model and behavior of climatic variables, hourly water use, collection, and energy consumption rates per unit ground area in the conventional, open and closed heat pump greenhouse systems for 10 successive days in summer. In the conventional system, water was used for both transpiration and evaporative cooling (Figure 5.47a), whereas in the heat pump greenhouse systems water was used for transpiration only, some of it was recovered in the open system and almost all in the confined system (Figure 5.47b).

Table 5.11 summarizes the daily averages of energy consumption for cooling and water consumption in conventional, open and closed heat pump greenhouse systems in winter, spring, and summer. When both heating and cooling seasons are considered, it was determined that 99.5% of the energy consumption in the conventional system was used for heating, while the remaining portion (0.5%) was used for ventilation at this particular location. In the open heat pump system, 94.2% of the total energy consumption was used for heating, while 3.1% and 2.7%, respectively, were used for ventilation and heat pump cooling. In the closed heat pump system, 70% of the total energy consumption was used for heating, while 30% was used for cooling. The energy used for cooling was much less than the energy used for heating, because the shading system functioned very well and reduced the cooling loads. Also, the plants themselves are biological air-conditioners and it should be remembered that they can provide much of their own cooling.

Table 5.11 shows that, in addition to the transpiration, water was also used for evaporative cooling in the conventional system. In the closed heat pump system, however, almost all the transpired water was recovered on the

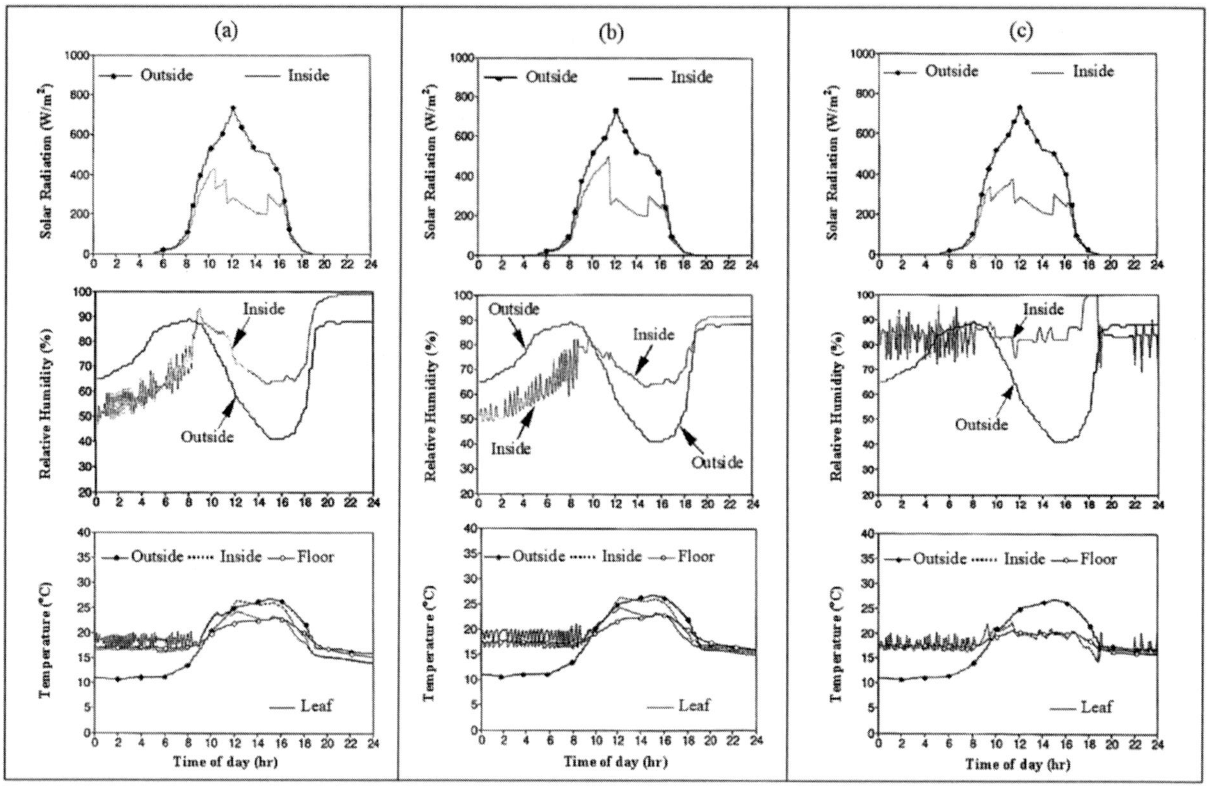

FIGURE 5.46 Diurnal changes in predicted greenhouse climatological quantities in the conventional (a), open heat pump (b), and closed heat pump (c) greenhouse systems on a spring day.

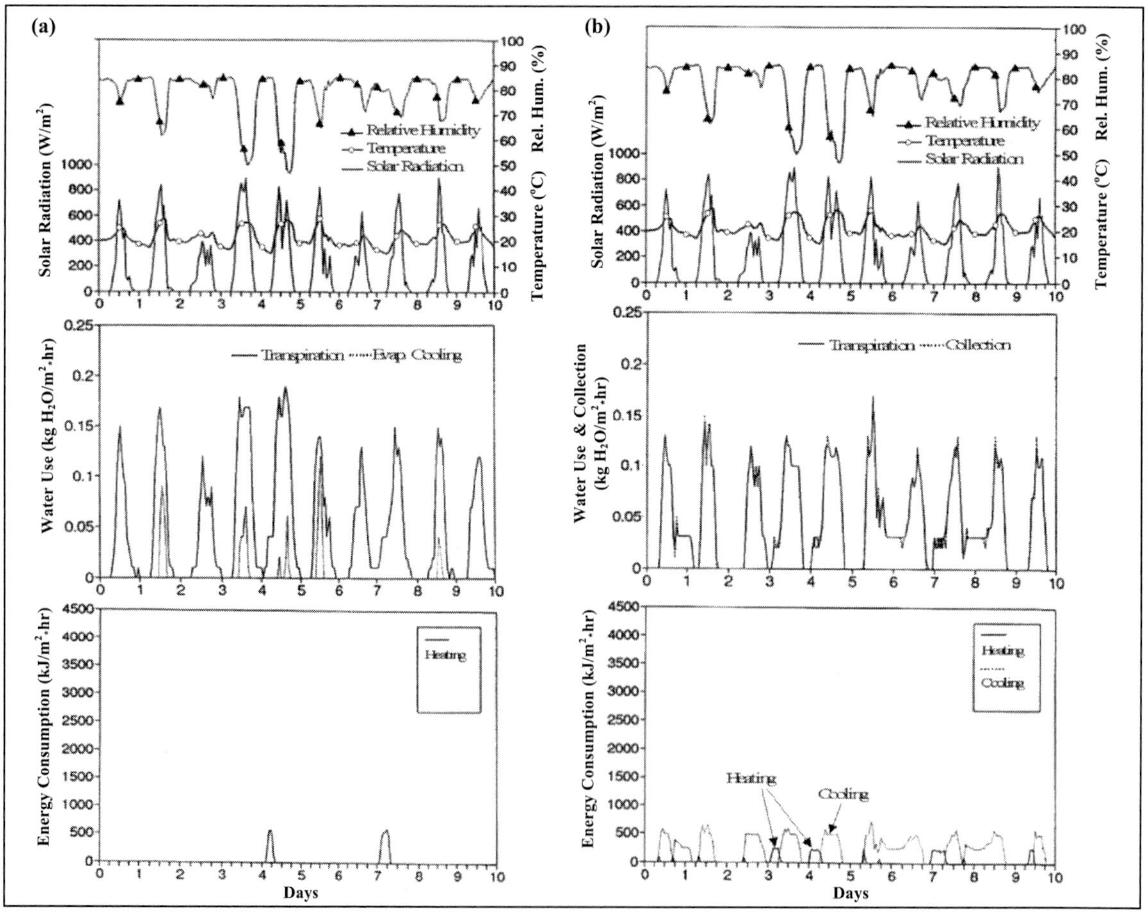

FIGURE 5.47 Predicted water use, collection and energy consumption per unit ground area in the conventional (a) and closed heat pump (b) greenhouse systems for 10 successive days in summer.

TABLE 5.11
Energy Consumption with Respect to Minimum Ventilation Rate in Winter, Spring, and Summer

	WINTER			SPRING			SUMMER		
	CHP	OHP	CON	CHP	OHP	CON	CHP	OHP	CON
ENERGY CONSUMPTION FOR COOLING (MJ/d.m^2)									
Natural Gas	0.02	–	–	1.94	–	–	5.19	0.91	–
Fan (Electric)	*	–	–	0.05	–	–	0.14	0.02	–
Pump (Electric)	*	–	–	0.03	–	–	0.08	0.01	–
Subtotal	0.02	–	–	2.02	–	–	5.40	0.94	–
Ventilation	–	0.11	0.11	–	0.23	0.11	–	0.71	0.13
TOTAL	**0.02**	**0.11**	**0.11**	**2.02**	**0.23**	**0.11**	**5.40**	**1.65**	**0.13**
WATER CONSUMPTION (kg H_2O/d.m^2)									
Transpiration	0.85	2.52	2.54	1.05	2.22	2.17	1.17	1.66	1.28
Water Recovered	0.83	–	–	1.04	–	–	1.17	0.35	–
Evaporative Cooling	–	–	–	–	–	–	–	–	0.29
TOTAL	**0.02**	**2.52**	**2.54**	**0.01**	**2.22**	**2.17**	**0.00**	**1.31**	**1.57**

* *Negligible amount.* CON, OHP and CHP*:* Conventional, open and closed heat pump greenhouse systems, respectively.

condensing coils, making the overall water consumption in this system essentially zero. Thus, the closed heat pump system was the most water conserving system, while the conventional system used the most water of all the systems in this study. When all three systems are compared with respect to the total energy consumption (both heating and cooling) the closed heat pump system was the most energy conserving system, while the conventional system used the most energy of all the systems. If the three systems are to be compared with respect to the energy consumption for cooling only, based on the average daily energy consumption (winter, spring, and summer values averaged together), the closed heat pump system had the highest energy consumption for cooling (2.48 MJ/day.m^2), while the open heat pump and conventional greenhouse systems had the consumptions of 0.66 MJ/day.m^2 and 012 MJ/day.m^2, respectively.

Yildiz and Stombaugh (2006a, 2006b) concluded that greenhouse operation with the heat pump system has a great potential for enormous energy and water savings, as well as for dehumidification. Closed greenhouse operation with a heat pump system would also make it easier to control the humidity and keep high CO_2 levels in greenhouses, while reducing the risk of insects and diseases within the confined greenhouse environment. The heat pump system performed very well with a variable shading system significantly conserving both energy and water when both heating and cooling are considered. A combination of open and closed heat pump systems would be the most energy and water conserving system, operating the system as a closed system in fall, winter and spring and as an open system in summer. However, when only the cooling season is considered, the commercial feasibility of using a heat pump system would depend on the market conditions of the produce and the other savings such as savings in water and CO_2 use.

It should be kept in mind that greenhouses, livestock, poultry, and aquatic buildings can be cooled not only by air-coupled heat pumps but also by earth- or water-coupled heat pumps; and cold water reservoirs can be underground water, oceans, lakes and rivers. Furthermore, we can directly cycle cold water from these reservoirs and provide ground cooling using PE tubes in such buildings.

5.8.8 Active Cooling by the Watergy Concept

For hot and arid regions, Buchholz et al. (2004) developed the Watergy concept—a novel, closed greenhouse system. The Watergy concept follows the development of a technology platform for decentralized basic supply of energy, water and food, which can be used within a number of different applications. The system has no external aeration and has almost no need for further water supply. The technological principle is actually proven at two different prototypes that were built within a European Union (Framework Program Promoting Energy, Environment, and Sustainable Development) funded research project. In 2004, the first prototype was built for the purpose of closed greenhouse operation in Almeria, Spain (see Figure 5.48), which holds promise for heating and cooling greenhouses. The second prototype was built in Berlin, Germany, as a living or office building with an attached façade-greenhouse. It has a great potential to reduce the demand for or to totally displace natural gas, to reduce the load on power grid by electricity savings, and the demand for irrigation water in greenhouse operations. It also allows plant protection without chemical insecticides and the accumulation of carbon dioxide without any aeration losses. The process generates heat, clear water and cooled/dehumidified air at minimal need for primary energy.

The main focus of Watergy is decreasing the temperature peaks in greenhouses by using water based thermal storage for alternate heating and cooling. A day-night balance of the temperatures is aimed at greenhouse daytime cooling for times of strong solar irradiation and for heating during winter nights. Day-night storage is a cheap system and can economically provide heating and cooling with temperature differences from about 10 K. Daytime cooling during the summer is provided from the storage that is unloaded through the night. In the cooling duct, the circulated air has to reach the dew-point temperature, as the greenhouse air has to be continuously dehumidified to enable a sufficient transpiration from the vegetation for providing enough evaporative cooling capacity in the closed system. The accumulated condensate in this process can be used for further irrigation.

Besides the production of food, closed greenhouses can also establish a basic module for decentralized systems of combined water supply and wastewater treatment. In order to exploit the full economical potential of the system, it is necessary to go beyond the concept of just recovering the evaporated water for re-use within the same greenhouse. Graywater can be used for irrigation in the greenhouse, and following the above-described heating/cooling process, can also supply drinking water. The idea is to substitute drinking water with cheaper and easier available sources of rainwater, graywater and saltwater. Rainwater is already commonly used in horticulture when water resources are limited. Through a further humidification device in the roof area of the greenhouse, even additional sea or brackish

FIGURE 5.48 First prototype of Watergy greenhouse built in Almeria, Spain.

Source: *Greenhouse Engineering* (2021).

water can be evaporated to provide desalination within the same process. In this way, greenhouse horticulture can turn from a water-consuming activity into a water-producing technology.

The ability of air to absorb water vapor is however dependent on its temperature. Due to the temperature gradients within the air cycle of the Watergy system, humidification is achieved within two steps: i) the first step is done through the transpiration of the vegetation. While using graywater, the waste components in the water are also used as plant nutrition; ii) in the second step at higher temperatures, the water gets into direct contact with the air by using sprinklers or sprayers. Here, rainwater, pretreated graywater or saltwater can be used. In the latter case, the brine can be concentrated within the component to the limit of solubility and after that, sent back to the sea or to an exterior system for complete desalination. This has not only a cleaning effect but also a passive cooling effect in the greenhouse. Evaporation acts as an interface between the water and energy management systems.

5.9 LIGHTING SYSTEMS

5.9.1 Introduction

Environmental factors such as light and radiation, temperature, relative humidity, carbon dioxide (CO_2) and a number of other gases in the environment, water, feed/nutrition, insect and disease stresses are the key environmental components affecting optimum animal and plant growth, production, and yield. And for instance, greenhouse growers have been successfully using a number of mechanisms for modulating each and every one of these components for better growth, production, and yield including artificial lighting. As discussed in depth earlier in Chapter 4, light energy is required directly to sustain the growth of all green plants in the first place while light intensity (quantity), light quality (wavelength), and light duration (photoperiod) all influence plant development, and plants respond primarily to the intensity and duration of light within photosynthetically active radiation (PAR) range, the visible portion of the spectrum. Light has very strong interactions on photosynthesis as well particularly with ambient temperature and CO_2 concentrations.

The greenhouse location, orientation and glazing determine the amount of free natural light that is available for plant growth and development in a greenhouse. The best greenhouse location therefore is the location where the free solar energy enters the greenhouse all day long with no shadows cast on the greenhouse. When the intensity and/or duration of natural light is not sufficient under normal operating conditions (temperature, CO_2 concentration, water, nutrients), then artificial lighting might be considered for increasing the intensity of light and extending the duration of light.

On the other hand, light is one of the earliest discovered and most important environmental variables affecting the growth, production, and reproductive processes of animals. As we have discussed earlier in plants, light intensity (amount), light quality (spectrum), and light duration (photoperiod) all are important when the influence of light on animal response is studied. Light can harm (phototoxicity) animal cells if unprotected, stimulate vision receptors, and control photoperiodism. Just like in plants, photoperiod has a direct influence on animal performance especially on the breeding season for sheep and poultry egg production; and photoperiodic control of reproduction in animals is quite a common practice. Essentially, there exist three daylength groups of domestic animals: i) short-day breeders—mate when daily photoperiod decreasing in the fall. Short photoperiods induce or accelerate estrus development in these animals (e.g., sheep and goats); ii) long-day breeders—mate when daily photoperiod increasing in the spring (e.g., birds and horses); and iii) day-neutral breeders—mate readily during all seasons (e.g., cattle and swine). Effects in poultry, however, seem to be more important than those in livestock animals. Artificial light is used in the temperate and colder zones to equalize egg production throughout the year, for instance. In hot climates, on the other hand, additional hours of light before sunrise and after sunset are recommended in order to encourage the hens to eat during relatively cooler hours.

As for plants and terrestrial animals, light is important for aquatic animals as well. First of all, some aquatic animals are nocturnal feeders while others are diurnal feeding, which may impact the feeding regime in aquaculture operations. Again, as already discussed in Chapter 4, lighting effects on fish production were classified as: i) growth rate effects, ii) maturation effects, and iii) vigour and survival rates. Photoperiod and light intensity in many fish species affect growth rate. Having said that, however, current research is still limited for many fish species.

Some of the lighting sources include INC, CFL, HID lamps such as HPS and MH, and LEDs. The use of LEDs is getting pretty close attention every passing year while incandescent and fluorescent bulbs are still widely used, for instance, in modern livestock and poultry houses.

Overall, this section covers essential light definitions and units used in lighting systems first, and then move to the lighting strategies and different light sources used particularly in greenhouses, and other controlled environments. And finally, the section concludes with the future perspectives of lighting in controlled environments.

5.9.2 Some Definitions, Light Units and Conversions

As discussed earlier in Chapter 2, light is considered as a stream of particles of radiant energy, also known as *photons*, and light waves require no medium to propagate. Photons have no mass, occupy no space, and travel at the speed of light. The amount of energy carried by an individual photon is, on the other hand, extremely small and inversely related to wavelength. The shorter the light's wavelength is, the higher the energy content of its photons. In the visible spectrum, for instance, blue light photons have more energy content than those of red light photons.

While sunlight is the most common and free source of radiation for animal and plant growth, production, and reproduction, the intensity and/or duration of this natural light may not be sufficient. In such cases, artificial lighting is provided. In order to understand what's happening under either natural or artificial lightings, we need to know and understand the following concepts pertinent to controlled environment lighting.

First of these concepts is *photosynthetically active radiation* (PAR), which stimulates photosynthesis and was defined earlier as the visible portion (400–700 nm) of the spectrum. And basically, three measuring systems are used for measuring radiation intensity:

1) Radiometric Units (Irradiance): W/m² or W/ft² with specific wavelength intervals (e.g., 400–700 nm).
2) Quantum Units (Photon Flux Density): μmol/s.m² (400–700 nm unless otherwise specified).
3) Photometric Units (Illuminance): lumen/m² (lux) or lumen/ft² (footcandle); 1 footcandle = 10.8 lux.

$$DLI\left(\frac{\text{mol}}{\text{m}^2.\text{d}}\right) = \left(\frac{\text{Photosynthetic Photon Flux Density}\left(\frac{\mu\text{mol}}{\text{s.m}^2}\right)\left(\text{Photoperiod}\left(\frac{\text{h}}{\text{d}}\right)\right)\left(3{,}600\,\frac{\text{s}}{\text{h}}\right)}{\left(1{,}000{,}000\,\frac{\mu\text{mol}}{\text{mol}}\right)}\right) \tag{5.36}$$

Only one of the above three, the quantum units, is appropriate for photosynthesis and therefore for plant production; and it will be discussed further below and becomes more meaningful. Also, since the spectral power distribution varies, it is important to use proper conversions (see Thimijan and Heins, 1983) between individual light sources for the wavelength interval considered so that meaningful comparisons can be made.

Photosynthetic photon flux (PPF) is the amount (μmol) of PAR (amount of photons available between 400 and 700 nm) emitted from a lamp per second. The unit of PPF is μmol/s. PPF is typically measured in laboratory settings with an integrating sphere that captures all the photons emitted from a particular lamp, for instance.

Photosynthetic photon flux density (PPFD) is the PPF (μmol/s) per unit surface area (m²); therefore, it has an overall unit of μmol/s.m².

Engineers and plant scientists have different preferences with respect to the PPF and PPFD; however, since the PPFD refers to the intensity of PAR at a surface and is typically measured at the top of a plant canopy by a quantum sensor, it is more meaningful than the PPF, which refers to the total amount of light emitted by a lamp spherically, and not necessarily the intensity arriving on a surface such as plant canopy. And there are three groups of PPFD provided as follows:

1) Low-light Intensity	: 100 μmol/s.m²
2) Medium-light Intensity	: 300 μmol/s.m²
3) High-light Intensity	: 600 μmol/s.m²

Microgreen crops are considered to be low-light intensity (100–200 μmol/s.m²) crops while tomatoes, for instance, are considered as high-light intensity (600 μmol/s.m²) crops. During peak hours, daylight PPFD can easily reach about 1,500 μmol/s.m² and 2,000 μmol/s.m² levels in mild and hot desert climates, respectively.

Photosynthetic photon efficacy (PPE) is the photosynthetic photon (400 to 700 nm) output (μmol/s) per unit electrical power input (W=J/s). In essence, PPE expresses how effectively a lighting fixture converts electrical power into photosynthetic light useful for photosynthesis, and the overall unit of PPE is therefore μmol/s.W or μmol/J.

Luminous efficiency is the number of lumens emitted per unit electrical power input (W). This efficiency is relevant when comparing the efficiency of lamps for human applications of visible light and not for plant applications such as greenhouses and other controlled environments.

Daily light integral (DLI) is the integrated photosynthetic photon flux density (400 and 700 nm) received on one square meter area per day. The unit is therefore mol/m²·d and the overall equation can be expressed as follows:

Table 5.12 was generated using Equation (5.36), and the readers may find it useful for quick visualization of the DLI as a result of the combination effect of PPFD and photoperiod. Using this table, one can quickly decide how long of a photoperiod must be provided for a given DLI requirement and PPFD value, or what sort of a photoperiod and PPFD combination can be used for providing a required optimum DLI value for a particular plant considering the plant's minimum and maximum PPFD limits.

There exists a correlation between the average DLI and plant growth, which increases almost linearly with increasing DLI. As a rule of thumb, a 1% increase in production is estimated for each 1% increase in DLI. Number of fruits and fruit sizes increase with the increasing DLI as well for fruiting vegetables such as tomatoes and cucumbers. And the DLI in a greenhouse is obviously influenced by the greenhouse glazing, overhead obstructions, shading curtains, and

TABLE 5.12
Daily Light Integral Values Obtained Using Equation (5.36) for Corresponding PPFDs and Photoperiods

DLI	Photosynthetic Photon Flux Density (μmol/s.m²)																			
Photoperiod (h/d)	50	100	150	200	250	300	350	400	450	500	550	600	650	700	750	800	850	900	950	1,000
1	0.2	0.4	0.5	0.7	0.9	1.1	1.3	1.4	1.6	1.8	2.0	2.2	2.3	2.5	2.7	2.9	3.1	3.2	3.4	3.6
2	0.4	0.7	1.1	1.4	1.8	2.2	2.5	2.9	3.2	3.6	4.0	4.3	4.7	5.0	5.4	5.8	6.1	6.5	6.8	7.2
3	0.5	1.1	1.6	2.2	2.7	3.2	3.8	4.3	4.9	5.4	5.9	6.5	7.0	7.6	8.1	8.6	9.2	9.7	10.3	10.8
4	0.7	1.4	2.2	2.9	3.6	4.3	5.0	5.8	6.5	7.2	7.9	8.6	9.4	10.1	10.8	11.5	12.2	13.0	13.7	14.4
5	0.9	1.8	2.7	3.6	4.5	5.4	6.3	7.2	8.1	9.0	9.9	10.8	11.7	12.6	13.5	14.4	15.3	16.2	17.1	18.0
6	1.1	2.2	3.2	4.3	5.4	6.5	7.6	8.6	9.7	10.8	11.9	13.0	14.0	15.1	16.2	17.3	18.4	19.4	20.5	21.6
7	1.3	2.5	3.8	5.0	6.3	7.6	8.8	10.1	11.3	12.6	13.9	15.1	16.4	17.6	18.9	20.2	21.4	22.7	23.9	25.2
8	1.4	2.9	4.3	5.8	7.2	8.6	10.1	11.5	13.0	14.4	15.8	17.3	18.7	20.2	21.6	23.0	24.5	25.9	27.4	28.8
9	1.6	3.2	4.9	6.5	8.1	9.7	11.3	13.0	14.6	16.2	17.8	19.4	21.1	22.7	24.3	25.9	27.5	29.2	30.8	32.4
10	1.8	3.6	5.4	7.2	9.0	10.8	12.6	14.4	16.2	18.0	19.8	21.6	23.4	25.2	27.0	28.8	30.6	32.4	34.2	36.0
11	2.0	4.0	5.9	7.9	9.9	11.9	13.9	15.8	17.8	19.8	21.8	23.8	25.7	27.7	29.7	31.7	33.7	35.6	37.6	39.6
12	2.2	4.3	6.5	8.6	10.8	13.0	15.1	17.3	19.4	21.6	23.8	25.9	28.1	30.2	32.4	34.6	36.7	38.9	41.0	43.2
13	2.3	4.7	7.0	9.4	11.7	14.0	16.4	18.7	21.1	23.4	25.7	28.1	30.4	32.8	35.1	37.4	39.8	42.1	44.5	46.8
14	2.5	5.0	7.6	10.1	12.6	15.1	17.6	20.2	22.7	25.2	27.7	30.2	32.8	35.3	37.8	40.3	42.8	45.4	47.9	50.4
15	2.7	5.4	8.1	10.8	13.5	16.2	18.9	21.6	24.3	27.0	29.7	32.4	35.1	37.8	40.5	43.2	45.9	48.6	51.3	54.0
16	2.9	5.8	8.6	11.5	14.4	17.3	20.2	23.0	25.9	28.8	31.7	34.6	37.4	40.3	43.2	46.1	49.0	51.8	54.7	57.6
17	3.1	6.1	9.2	12.2	15.3	18.4	21.4	24.5	27.5	30.6	33.7	36.7	39.8	42.8	45.9	49.0	52.0	55.1	58.1	61.2
18	3.2	6.5	9.7	13.0	16.2	19.4	22.7	25.9	29.2	32.4	35.6	38.9	42.1	45.4	48.6	51.8	55.1	58.3	61.6	64.8
19	3.4	6.8	10.3	13.7	17.1	20.5	23.9	27.4	30.8	34.2	37.6	41.0	44.5	47.9	51.3	54.7	58.1	61.6	65.0	68.4
20	3.6	7.2	10.8	14.4	18.0	21.6	25.2	28.8	32.4	36.0	39.6	43.2	46.8	50.4	54.0	57.6	61.2	64.8	68.4	72.0
21	3.8	7.6	11.3	15.1	18.9	22.7	26.5	30.2	34.0	37.8	41.6	45.4	49.1	52.9	56.7	60.5	64.3	68.0	71.8	75.6
22	4.0	7.9	11.9	15.8	19.8	23.8	27.7	31.7	35.6	39.6	43.6	47.5	51.5	55.4	59.4	63.4	67.3	71.3	75.2	79.2
23	4.1	8.3	12.4	16.6	20.7	24.8	29.0	33.1	37.3	41.4	45.5	49.7	53.8	58.0	62.1	66.2	70.4	74.5	78.7	82.8
24	4.3	8.6	13.0	17.3	21.6	25.9	30.2	34.6	38.9	43.2	47.5	51.8	56.2	60.5	64.8	69.1	73.4	77.8	82.1	86.4

high-intensity supplemental lighting (and distance from the canopy) when employed. Natural DLI in a commercial greenhouse rarely reaches 30 mol/m².d and based on their DLI requirements, crops can be classified as follows:

1) Low-light crops	: 3–6 mol/m².d
2) Medium-light crops	: 6–12 mol/m².d
3) High-light crops	: 12–18 mol/m².d
4) Very high-light crops	: 18+ mol/m².d

Runkle (2019) and Morgan (2013) provided a list of simplified DLI targets for different plants; however, the provided figures might change as a function of ambient air temperatures, CO_2 concentrations, and other environmental and economic conditions.

As it will be worked on numerically later in this section, for instance, having the long-term natural DLI figures available, one can easily determine how much extra light is required for maximizing crop yields in a greenhouse at a particular location; and then using this information, it can easily be decided whether installing and operating artificial lighting would be feasible or not. Knowing DLI also makes it possible to determine the intensity of artificial lighting and when and how long the artificial light sources need to be operated for providing optimal light conditions for maximum production in greenhouses where artificial lighting systems already exist. DLI knowledge in and outside of a greenhouse also let a greenhouse grower decide whether a shading system needs to be used or removed, or whether the time for cleaning or replacing a greenhouse glazing has come or not.

Greenhouse lighting systems designers and growers may find the DLI maps very useful, especially the monthly maps as more of those become publicly available around the world. Outdoor annual DLI maps around the world, and the outdoor monthly average DLI maps for the USA can be found in Korczynski et al. (2002), and Yildiz (2021).

Now, it is a good time to perform some numerical examples for the concepts and conversions introduced above.

Example 5.22: Radiometric to quantum and quantum to photometric units conversions and light comparisons

Determine and compare the photosynthetic photon flux densities (PPFD), footcandle per μmol/s.m², and lux per μmol/s.m² for 400 W/m² daylight (400–700 nm) and 400 W/m² high-pressure sodium lamp light (400–700 nm).

Solution

Using the conversion factors provided by Thimijan and Heins (1983), we can easily determine all the conversions as follows:

Photosynthetic photon flux density conversions:

For daylight,

$$\left(400\,\frac{\text{W}}{\text{m}^2}\right)\left(\frac{4.57\ \mu\text{mol/s.m}^2}{\text{W/m}^2}\right)=1{,}828\ \mu\text{mol/s.m}^2$$

For high-pressure sodium lamp,

$$\left(400\,\frac{\text{W}}{\text{m}^2}\right)\left(\frac{4.98\ \mu\text{mol/s.m}^2}{\text{W/m}^2}\right)=1{,}992\ \mu\text{mol/s.m}^2$$

As we see here, the light coming from the high-pressure sodium lamp provides higher photosynthetic photon flux density compared to the daylight at the same radiation intensity per unit surface area.

Footcandle conversions:

For daylight,

$$\left(1{,}828\,\frac{\mu\text{mol}}{\text{s.m}^2}\right)\left(\frac{5.0\ \text{footcandle}}{\mu\text{mol/s.m}^2}\right)=9{,}140\ \text{footcandle}$$

$$=9{,}140\ \text{lumen/ft}^2$$

For high-pressure sodium lamp,

$$\left(1{,}992\,\frac{\mu\text{mol}}{\text{s.m}^2}\right)\left(\frac{7.6\ \text{footcandle}}{\mu\text{mol/s.m}^2}\right)=15{,}139\ \text{footcandle}$$

$$=15{,}139\ \text{lumen/ft}^2$$

Lux conversions:

For daylight,

$$\left(1{,}828\,\frac{\mu\text{mol}}{\text{s.m}^2}\right)\left(\frac{54\ \text{lux}}{\mu\text{mol/s.m}^2}\right)=98{,}712\ \text{lux}$$

$$=98{,}712\ \text{lumen/m}^2$$

For high-pressure sodium lamp,

$$\left(1{,}992\,\frac{\mu\text{mol}}{\text{s.m}^2}\right)\left(\frac{82\ \text{lux}}{\mu\text{mol/s.m}^2}\right)=163{,}344\ \text{lux}$$

$$=163{,}344\ \text{lumen/m}^2$$

And further comparisons between different light sources can be made easily using the conversion factors provided by Thimijan and Heins (1983).

Example 5.23: Photometric light (footcandle of light) from high-pressure sodium lamp into quantum conversion

Determine the amount of PPFD provided by 1,500 footcandle of light from high-pressure sodium lamp (400–700 nm).

Solution

ain, using the proper conversion factor provided by Thimijan and Heins (1983), we can easily perform this conversion as follows:

$$\left(\frac{1{,}500\ \text{footcandle}}{7.6\ \dfrac{\text{footcandle}}{\mu\text{mol/s.m}^2}}\right)=197\ \mu\text{mol/s.m}^2$$

Example 5.24: Photometric light (lux of light) from metal halide lamp into quantum conversion

Determine the amount of PPFD provided by 14,000 lux of light from metal halide lamp (400–700 nm).

Solution

Once again, using the proper conversion factor provided by Thimijan and Heins (1983), we can perform this task as follows:

$$\left(\frac{14{,}000\ \text{lux}}{71\ \dfrac{\text{lux}}{\mu\text{mol/s.m}^2}}\right)=197\ \mu\text{mol/s.m}^2$$

Example 5.25: Quantum to radiometric conversion

Convert 1,992 μmol/s.m² of high-pressure sodium light (400–700 nm) to radiometric units (400–700 nm).

Solution

The quantum unit (μmol/s.m²) is based on the number of photons, and photons at different wavelengths have different energy contents. Therefore, in order to be able to calculate total energy, we have to know both the number of photons available at each wavelength and the energy content of photon at each wavelength. In reality, it is a complex calculation; however, Thimijan and Heins (1983) makes this process much simpler for known light sources. For this example, the light source is high-pressure sodium lamp; and for this lamp, Thimijan and Heins (1983) provides a conversion factor of 4.98. Therefore, we can perform this task easily as follows:

$$\left(\frac{1{,}992\ \dfrac{\mu\text{mol}}{\text{s.m}^2}}{\dfrac{4.98\ \mu\text{mol/s.m}^2}{\text{W/m}^2}}\right)=400\ \text{W/m}^2$$

One should also keep in mind that it is impossible to convert 24-hour average solar radiation values into, for instance, 8-, 16-, or 20-hour average solar power as the time distribution of solar power throughout the day is typically not known. Hence, we must convert lamp light data into 24-hour average values first, and then compare them with sunlight values. Let's now assume that the above 400 W/m² high-pressure sodium light is instantaneous and the irradiance is constant, and the lamp is used in a greenhouse lighting for 8 hours, and would like to convert this radiometric light output to 24-hour basis:

$$(\text{Irradiance})\left(\frac{\text{Hours of Irradiation}}{24\ \text{hours}}\right)$$

$$\left(400\,\frac{\text{W}}{\text{m}^2}\right)\left(\frac{8\ \text{h}}{24\ \text{h}}\right)=133\,\frac{\text{W}}{\text{m}^2}\qquad\text{(24-h basis)}$$

Furthermore, Runkle (2018a) reported the power consumption, quantity of photosynthetic photons (400 to 700 nm) (PPF) emitted per lighting fixture, and photosynthetic photon efficiency (PPE) for a number of selected lighting fixtures, i.e., HPS lamps and LEDs.

Runkle (2018a) also showed that the PPE is an important parameter for comparing lighting fixtures; however, it also shows that PPF and measured powers are also important. In addition to these considerations, other factors should also be considered before investing in a particular technology, such as light spectrum, spatial uniformity of light intensity, lighting fixture shading, durability, service, etc.

Runkle and Bugbee (2013), on the other hand, compared the photosynthetic photon efficacy (PPE) (for plants) and the luminous efficiency (for people) of four light sources. The reported that high PPE fixtures (blue and red LEDs) for plants do not necessarily have high luminous efficiency (which is appropriate for people) or high luminous efficiency fixtures do not necessarily have high photosynthetic photon efficacy (which is appropriate for plants). So, it is obvious that the measurement system based on photometric units (lux and footcandle) is quite misleading and inappropriate for plant applications while the measurement system based on quantum units is the most appropriate system for photosynthesis and therefore for plant production.

A double-ended 1,000-W HPS lamp with electronic ballast, for instance, has a PPE of approximately 1.7 μmol/J while the PPEs for LEDs vary quite a bit. The PPEs of many new LEDs today commonly exceed 2.0 μmol/J and even 3.0 μmol/J (Runkle,2018a). Obviously, the higher the PPE value of a light bulb is, the better it is at converting electric energy into photosynthetic photons.

Example 5.26: Calculating the DLI deficiency for supplemental lighting

Let's now consider that we have leafy greens in a greenhouse, which have a DLI requirement of 12 mol/m^2·d (Runkle, 2019; Morgan, 2013) and the natural light provides an average DLI of 7 mol/m^2·d inside the greenhouse. Determine the supplemental lighting intensity required to reach the DLI target of 12 mol/m^2·d if the grower wants to provide 20 hours of lighting per day.

Solution

We first need to calculate the DLI deficiency (the required supplemental DLI):

$$\text{Required Supplemental DLI} = \begin{bmatrix} (\text{Plant DLI Requirement}) - \\ (\text{Natural DLI Available}) \end{bmatrix}$$

$$\text{Required Supplemental DLI} = \begin{bmatrix} \left(12\,\text{mol}/\text{m}^2\cdot\text{d}\right) \\ -\left(7\,\text{mol}/\text{m}^2\cdot\text{d}\right) \end{bmatrix}$$

$$\text{Required Supplemental DLI} = 5\,\text{mol}/\text{m}^2\cdot\text{d}$$

The grower wants to provide 20 hours of lighting per day; so, the light intensity required to reach the DLI target can be calculated after rearranging Equation (5.36) as follows:

$$\text{Supplemental PPFD} = \left(\frac{\left(\text{Required Supplemental DLI}, \dfrac{\text{mol}}{\text{m}^2\cdot\text{d}}\right)\left(1{,}000{,}000\dfrac{\mu\text{mol}}{\text{mol}}\right)}{\text{Photoperiod}\left(\dfrac{\text{h}}{\text{d}}\right)\left(3{,}600\dfrac{\text{s}}{\text{h}}\right)}\right)$$

$$\text{Supplemental PPFD} = \left(\frac{\left(5\dfrac{\text{mol}}{\text{m}^2\cdot\text{d}}\right)\left(1{,}000{,}000\dfrac{\mu\text{mol}}{\text{mol}}\right)}{\left(20\dfrac{\text{h}}{\text{d}}\right)\left(3{,}600\dfrac{\text{s}}{\text{h}}\right)}\right) = 69\frac{\mu\text{mol}}{\text{s.m}^2}$$

And exactly the same magnitude (69 μmol/s.m^2) of supplemental PPFD can be found using Table 5.12 and the method of iteration for a DLI of 5.0 mol/m^2·d and a photoperiod of 20 hours per day.

Moreover, using this required light density (supplemental PPFD), we can now determine the number of particular lighting fixtures required for the high-density supplemental lighting taking into consideration the greenhouse dimension and the distance between the light source and the plants.

Example 5.27: Calculating how long to operate supplemental lighting lamps for DLI deficiency

Now, let's determine how long we need to operate the supplemental lighting to achieve a target supplemental DLI of 5 mol/m^2·d assuming that the supplemental lighting is provided by high-pressure sodium (HPS) lamps that provide 70 μmol/s.m^2 (532 footcandle or 5,740 lux (see Thimijan and Heins, 1983) for conversion factors) at plant height.

Solution

Rearranging Equation (5.36) for photoperiod this time, we can easily calculate the photoperiod, i.e., how long we need to operate the supplemental lighting:

$$\text{Photoperiod} = \left(\frac{\left(\text{Required Supplemental DLI}, \dfrac{\text{mol}}{\text{m}^2\cdot\text{d}}\right)\left(1{,}000{,}000\dfrac{\mu\text{mol}}{\text{mol}}\right)}{\text{Provided Light Intensity}\left(\dfrac{\mu\text{mol}}{\text{s.m}^2}\right)\left(3{,}600\dfrac{\text{s}}{\text{h}}\right)}\right)$$

$$\text{Photoperiod} = \left(\frac{\left(5\dfrac{\text{mol}}{\text{m}^2\text{·d}}\right)\left(1{,}000{,}000\dfrac{\mu\text{mol}}{\text{mol}}\right)}{\left(70\dfrac{\mu\text{mol}}{\text{s.m}^2}\right)\left(3{,}600\dfrac{\text{s}}{\text{h}}\right)}\right) = 20\frac{\text{h}}{\text{d}}$$

As we did in the previous example, we can achieve 20 h/d photoperiod using Table 5.12 and the method of iteration for a DLI of 5 mol/m^2·d and a PPFD of 70 μmol/s.m^2.

Example 5.28: Calculating how much of total DLI is coming from supplemental lighting lamps

Now, let's determine how much of DLI is coming from the supplemental lighting lamps in a greenhouse if metal halide (MH) lamps are operated from 6 a.m. to 11:00 a.m. and from 6 p.m. to 11:00 p.m. (10 hours of operation in total) providing 70 μmol/m^2·s.

Solution

Using Equation (5.36), now we can easily calculate how much of total DLI is coming from supplemental lighting lamps:

$$\text{DLI} = \left(\frac{\text{Provided PPFD}\left(\dfrac{\mu\text{mol}}{\text{s.m}^2}\right)\left(\text{Photoperiod}\left(\dfrac{\text{h}}{\text{d}}\right)\right)\left(3{,}600\dfrac{\text{s}}{\text{h}}\right)}{\left(1{,}000{,}000\dfrac{\mu\text{mol}}{\text{mol}}\right)}\right)$$

$$\text{DLI} = \left(\frac{70\left(\dfrac{\mu\text{mol}}{\text{s.m}^2}\right)\left(10\dfrac{\text{h}}{\text{d}}\right)\left(3{,}600\dfrac{\text{s}}{\text{h}}\right)}{\left(1{,}000{,}000\dfrac{\mu\text{mol}}{\text{mol}}\right)}\right) = 2.5\frac{\text{mol}}{\text{m}^2\cdot\text{d}}$$

And again, we can determine exactly the same DLI of 2.5 mol/m².d using Table 5.12 and the method of iteration for a PPFD of 70 μmol/m²·s and a photoperiod of 10 hours per day.

Example 5.29: Calculating the DLI from footcandle or lux measurements

Now, let's consider that the footcandle and lux measurements provided in the following table were made over a 24 hour period inside a greenhouse. Determine the DLI for each measurement units.

Time	Footcandle	Lux	Time	Footcandle	Lux
01:00	0	0	13:00	4,100	44,280
02:00	0	0	14:00	2,350	25,380
03:00	0	0	15:00	1,200	12,960
04:00	0	0	16:00	900	9,720
05:00	70	756	17:00	750	8,100
06:00	200	2,160	18:00	400	4,320
07:00	450	4,860	19:00	210	2,268
08:00	980	10,584	20:00	80	918
09:00	1,500	16,200	21:00	0	0
10:00	2,350	25,380	22:00	0	0
11:00	2,580	27,864	23:00	0	0
12:00	3,500	37,800	24:00	0	0

Solution

First, we need to calculate the daily total and 24-hour average footcandle and lux measurements, and then we can perform the following calculations as discussed below.

	Footcandle	Lux
Daily Total	: 21,625	233,550
24-hour Average (per hour)	: 901.0	9,731.3
PPFD (μmol/s.m²)	: 180.2	180.2
DLI (mol/m².d)	: 15.6	15.6

In the above calculations, 24-hour footcandle and lux averages were converted to PPDF using the conversion factors (1/5.0 = 0.2 μmol/s.m² per footcandle and 1/54 = 0.0185185 μmol/s.m² per lux, respectively) for daylight provided by Thimijan and Heins (1983). And then, DLIs were calculated using Equation (5.36).

5.9.3 Artificial Lighting Strategies

Artificial lighting is increasingly being used and now a very common practice in the commercial production of plants grown in controlled environments such as greenhouses. Different practices of greenhouse lighting have totally different goals, such as increasing the daily light integrals and/or extending the photoperiod. Photoperiodic lighting needs a low-intensity lighting (typically 10 footcandle or 110 lux or somewhere between 1–2 μmol/s.m2), which is only used in greenhouses after sunset or before sunrise, whereas supplemental (photosynthetic) lighting requires a high-intensity lighting (usually 400–500 footcandle or 4,320–5,400 lux or 60–120 μmol/s.m2) and is generally used in greenhouses from October through March in mid and high latitudes on cloudy days and during the night (Runkle, 2018b). Sole-source lighting, on the other hand, is used for indoor growing of plants, and typically requires 150–200 μmol/s.m² (Runkle, 2018b). For a majority of crops grown indoors, plants are usually provided light for 18–24 hours per day. Lighting is not a cheap investment, especially when the purchase, installation and operational cost of such systems are considered. Therefore, it is important to understand the differences between different lighting strategies in the first place and then install an adequate lighting system that can deliver the target DLI and/or the photoperiod to the plants in a greenhouse or other controlled environment plant production facilities. Hence, this section covers different lighting strategies and characteristics used in such plant production facilities.

Controlled environment growers sometimes use artificial lighting to create longer days (photoperiods) or to increase the DLI (supplemental lighting) or both. For instance, if we provide, let's say 16 hours of artificial lighting using high-pressure sodium lamps or any other artificial light sources on an overcast winter day, we will not only increase the daily light integral (supplemental lighting) but also provide a long-day photoperiod (photoperiodic lighting).

5.9.4 Photoperiodic Lighting

As discussed earlier in Chapter 2, daily path of the Sun is not constant throughout the year; rather it shifts slightly and steadily every day. And this has a huge impact on insolation (intensity) and daylight hours (natural photoperiods). The latitude and the season are a result of the solar geometry. The Northern Hemisphere, for instance, has more daylight hours in the summer, reaching a maximum value on summer solstice, June 21 (see Figures 2.36 and 2.37).

On summer solstice, the daily path is highest in the sky, and the length of daylight is the longest of the year. In the Southern Hemisphere, on the other hand, the situation is the reverse, and the hemisphere, on or about June 21, receives the least amount of sunlight and onset of winter. In the Northern Hemisphere's winter, the Sun is directly overhead at latitude 23.5°S, which is called the winter solstice (December 21); and at this time, the Southern Hemisphere starts enjoying the beginning of summer with long hours of the Sun high in the sky. At the extreme, the noontime Sun is directly overhead at either the Tropic of Cancer (summer solstice), or Tropic of Capricorn (winter solstice). At winter solstice, the areas north of the Arctic Circle have 24-hour darkness, while the areas south of the Antarctic Circle have 24-hour daylight. When the Sun is directly overhead at the equator (spring and autumnal equinoxes—on March 21 and September 21, respectively), everywhere on the Earth, there are 12 hours of daylight Figure 5.49); that is, all latitudes have equal amounts of day and night.

Figure 2.37 shows that as summer moves into fall, and then into winter, the sunrise and sunset times of the Sun's motion across the sky gradually move southward. As a result, the day lengths (natural photoperiods) get shorter and the solar path gets lower in the sky. So, the natural photoperiod changes by latitude and time of the year as shown in Figure 5.49. For instance, as we move from the equator towards higher latitudes, the annual range of natural photoperiods get much larger, and this phenomenon requires

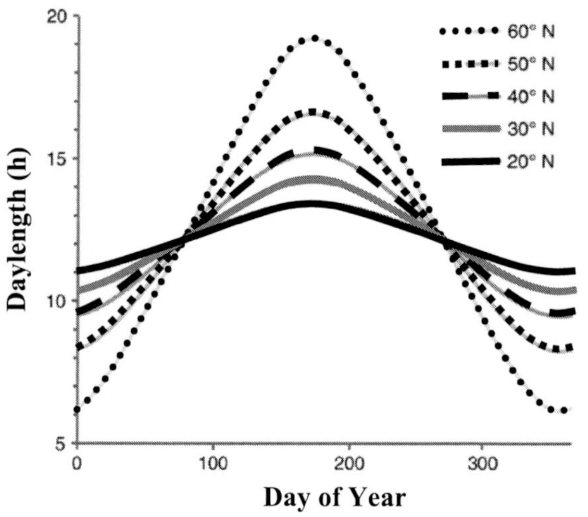

FIGURE 5.49 Calculated natural photoperiods by latitude and Julian day of year.

Source: *Greenhouse Engineering* **(2021).**

intervention in greenhouses so that the crops, livestock, poultry, and fish grown can have adequate photoperiods for their optimal growth, production and reproduction.

Photoperiodic lighting is called as the delivery of low-intensity lighting to artificially generate long days when the natural day lengths are short as discussed above. Once the natural photoperiod is long enough, there is no reason to provide artificial photoperiodic lighting. In plant controlled environments, typically, a light intensity of 1–2 μmol/s.m^2 is provided (Runkle, 2018b), which is not enough or negligible for plants to perform photosynthesis; but, it is bright enough for plants to sense and take proper action. Hence, this low-intensity (photoperiodic) lighting does not increase plant growth and development; rather, it is only used to regulate flowering, i.e., to accelerate and induce flowering of long-day plants (many annuals and perennials) so that the produce can make the target market dates, and to inhibit flowering of short-day plants (chrysanthemums, kalanchoes, poinsettias, etc.) again for marketing purposes.

We can extend the day length by simply turning on a HID light source such as HPS, MH, high-intensity LID lamps before the Sun sets, and keeping it on until the proper day length is achieved for a particular crop. Growers typically keep their stationary HPS lamps on for day-extension lighting, which indeed serves as both photoperiodic and supplemental lighting (increasing the DLI) depending on the light intensity and duration provided. Night interruption is another lighting strategy that has no effect on the DLI, and employs the photoperiodic lighting in the middle of night by interrupting the dark period for a four-hour period, typically from 10 p.m. to 2 a.m. because plants are extremely sensitive to light during this time. It is well known that continuous light during night interruption period is not absolutely required; hence an alternative lighting strategy called as cyclic or intermittent photoperiodic lighting for generating longer days can be employed and energy consumption can be reduced considerably. During the cyclic lighting period, lights are turned on for 10 minutes and off for the following 20 minutes (or lights are kept on 5 minutes minimum every 30 minutes). Since both extending the day and night interruption strategies are similarly effective for plant response, using the day-extension strategy with HPS lamps may cost additional energy if only operated for providing low-intensity photoperiodic lighting. The simplest way to deliver photoperiodic lighting is by installing low-output, screw-in lighting fixtures (CFL, INC or LED bulbs) and turning them on for 5 minutes minimum every half an hour. Using high-intensity fixtures (HPS, MH or high-intensity LED bulbs) with rotating reflectors or on a moving boom above crop canopy and moving back and forth for about four hours during night does the same job. Fluorescent lamps, red LEDs and white LEDs have been reported quite effective for inhibiting flowering of short-day plants. Lamps that emit red and far-red lights are, on the other hand, reported as the most effective lighting for promoting flowering of long-day plants.

For livestock and poultry controlled environments, the recommended light intensity and photoperiod levels are provided in Tables A.65, A.66, and A.67 for dairy, swine, and poultry, respectively. The influence of spectral distribution on different attributes of livestock and poultry are currently being studied extensively with the help of ever-growing LED technologies. Each color or combination of different colors has a distinct effect on livestock and poultry physiology.

And as discussed earlier in Chapter 4, the information on lighting needs of fish in controlled environment production are very limited because current research on many species is still limited. Specific requirements of some fish species, when available, were also provided in Chapter 4; however, it must be noted again that quite a low light level difference is typically required (less than 1 lux) to induce photoperiod responses in fish.

5.9.5 Supplemental Lighting

High-intensity photosynthetic (supplemental) lighting provides much higher light intensity compared to that of low-intensity photoperiodic lighting, typically around 60–120 μmol/s.m^2 (Runkle, 2018b) from HID lamps such as HPS and MH lamps or high-intensity LEDs, and increases photosynthesis and therefore plant growth and production. As a result of the increased DLI, plants having supplemental lighting typically flower one to two weeks earlier. As discussed earlier in the previous section, photosynthetic lighting can also be used to extend daylength (photoperiod); however, compared to low-intensity photoperiodic lighting, the installation and operational costs of high-intensity supplemental lighting are much higher. Therefore, supplemental lighting is generally used for high-value crops and when the harvestable yield of crops increases with light quantity (e.g., fruiting vegetables and cut flowers), and there is no need to have separate photosynthetic and photoperiodic lighting systems in the same greenhouse growing space. The choice of lighting technology is, on the other hand, primarily based on economics; therefore, the decision of providing lighting and the choice of lighting source will vary from one greenhouse operation to another.

For livestock and poultry controlled environments, the recommended light intensity and photoperiod levels are provided in Tables A.65, A.66, and A.67 for dairy, swine, and poultry, respectively. And the limited information on lighting needs of some fish species in controlled environment production facilities were provided and discussed in Chapter 4.

5.9.6 Sole-Source Lighting

Sole-source lighting is the delivery of light in controlled environment plant, livestock, poultry, or fish production facilities that do not receive natural light from the Sun, i.e., artificial lighting is provided for indoor production and provides the whole light required as the sole source of light. Inside plant factories and growth room facilities, environmental conditions are optimized for basically any desired crop, livestock, poultry or fish characteristics. And the most important environmental factor controlling particularly the plant growth and operational production costs in such kind of production facilities is lighting, that is, the sole-source lighting.

Relatively speaking, much higher intensity lighting (150–200 μmol/s.m^2) is provided for most plants grown in these production facilities, and the lighting fixtures operate for 18 to 24 hours per day mostly year round (Runkle, 2018b). As lighting fixtures generate a lot of heat, such facilities require properly designed cooling systems for providing and maintaining an adequate thermal environment for plants. Therefore, the lighting efficacy of light sources is critically important. New LEDs have relatively high lighting efficacies; hence, they generate less heat. As a consequence, we see plant factories today fully equipped with sole-source LED lighting systems, especially red LEDs with blue or white LEDs, or only a type of white LED.

5.9.7 Artificial Lighting Sources

In the previous sections, we have discussed the fundamental concepts of natural and artificial lighting, their relationships primarily with plant growth and development as well as manipulating different lighting strategies for better target attributes and maximum production. However, the ultimate question is how the light (natural, artificial, or both) required by plants, livestock, poultry and fish (preferably slightly over) will be deployed at the plant height, etc. The answer to this question requires an adequate engineering design so that the desired light intensity (PPFD) at the crop level (as a function of the crop type, the natural DLI, and the target DLI), uniformity and the duration of lighting can be designed effectively and sustainably. In their designs, engineers should also take the light depreciation, spectral changes, and material degradation into account so that the light output from the designed lighting systems can be maintained over their life cycles. Since each and every controlled environment lighting system is case-specific, it is extremely important for growers to work with a reputable engineer who has proper experience in controlled environment production lighting systems and can perform custom designs rather than using "a cookie cutter" approach.

This section covers different artificial lighting sources used successfully in controlled environment production facilities, and their characteristics.

5.9.7.1 High Intensity Discharge Lamps

HID lamps, such as HPS and MH lamps have been used in controlled environments effectively for a quite a long time. They are employed for providing daylight extension and night-interruption lighting. HPS and MH lamps are commonly suspended over the canopy, livestock, and poultry to provide lighting. However, as mentioned earlier in the previous section, they are sometimes mounted on booms that move back and forth above the canopy, and still provide the required lighting. Sometimes, these lamps are used with oscillating reflectors, a relatively more recent approach. Runkle (2018a) reported the PPE of different HPS bulbs. The higher the PPE is, the more effective the fixture is at converting electrical energy into photosynthetic photons. For instance, a 400-W single-ended HPS lamp with magnetic ballast has a PPE value of 0.94 μmol/J while a double-ended 1,000-W HPS lamp with electronic ballast has a PPE of 1.72 μmol/J. The PPE value for high-intensity LED products varies considerably, and many new LEDs now even approach and exceed 3.0 μmol/J.

5.9.7.2 Light-Emitting Diodes

Light-emitting diodes (LEDs) are solid-state, semi-conducting diodes, which can emit light from a wavelength of approximately 250 nm to over 1000 nm. They are a wonderful light source with extremely promising controlled environment production applications. There are several features of LEDs that make them attractive alternatives to HPS lamps. First, LEDs can produce quite a narrow-spectrum light at target wavebands useful particularly for plant growth and development, including blue (450 nm) and red (660 nm). Additionally, LEDs currently have a luminous efficiency of 38% (red) to 50% (blue) in converting energy to light with an estimated life of approximately 50,000 luminous hours or more. The theoretical maximum efficiency obviously occurs when all input energy is converted to energy in photosynthetic photons. Kusuma et al. (2020) reported that blue LEDs can be 93% efficient, phosphor-converted "whites" 76% efficient, and red LEDs 81% efficient. LED efficiency describes the ratio of the optical power output (W) to the electrical power input (W), using the same units. Therefore, the term efficiency is expressed as a percentage. As discussed earlier, the term efficacy applies to the same ratio using different units, i.e., μmol photon output per second, per watt of electrical power input, and therefore having an overall unit of μmol/J. Also, as discussed earlier in Chapter 2, the relationship between photon energy content and wavelength is expressed in Equation (2.270), the Planck's relationship. This equation is used to calculate the maximum possible photosynthetic photon efficacy for a given spectrum, and it is employed to convert between efficiency and efficacy. Based on Planck's relationship, energy content of photon is inversely proportional to wavelength.

In addition to efficacy, initial cost of lighting fixture per photon/s of output capacity is an important criterion. HPS fixtures (1 kW) range from US$200 to US$350 per kW (5–9 μmol/s per $) while LED fixtures ranging from US$1,000 to US$3,000 per kW (0.5–1.5 μmol/s per $). On a photon flux basis, the initial cost of LEDs is therefore reported to be somewhere between 3–18 times higher compared to those of HPS fixtures. The initial cost of both HPS and LED technologies has decreased, but the cost of LED fixtures is expected to decrease further and faster than HPS fixtures. While the initial cost is higher, LED fixtures have lower operational costs compared with those of HPS fixtures. Depending on the hours of usage and the cost of electricity, a payback period of 3–5 years for sole-source lighting and 5–8 years for supplemental lighting of LEDs can be expected (Nelson and Bugbee, 2014). The PPE of LEDs is increasing and initial costs are decreasing, and this makes the payback periods more acceptable every passing year. Their energy efficiency, long lifespan, and ability to target specific wavelengths make them an excellent option particularly for precision managing of photoperiod and some crop attributes. LEDs also provide the opportunity to manipulate the ratio of red (R) and far-red (FR) light for desired plant responses, for instance. As it was discussed earlier in Chapter 4, low red/far-red ratio promotes stem elongation (shade avoidance response) in many plant species. Overall speaking, green, blue, red, white light and their combinations have positive effects on different attributes of livestock and poultry as well at their different growth stages. The best combinations of different light mixtures and ratios are currently being studied and producing quite exciting findings.

5.9.7.3 Incandescent Lamps and Compact Fluorescent Lamps

Incandescent (INC) lamps have been used effectively in greenhouse, livestock, poultry and fish production for decades, and even though they are being dropped out of greenhouse production due to their low energy efficiency, they are still commonly employed for providing daylength extension and/or night-interruption lighting. Many growers have replaced most or all INC lamps with attractive alternative, compact fluorescent lamps (CFLs), which consume about one-fourth the amount of electricity as INC lamps. Since the frequent on-off operation required in cyclic lighting does not affect bulb life or fixture longevity, INC lamps are easily and adequately used for such lighting strategy. However, while CFLs have better energy efficiency and are quite effective for use in greenhouses, their spectrum is not as effective at controlling flowering of some long-day plants. CFLs are particularly low in far-red light, which is required for rapid flowering of some long-day crops (Runkle, 2007).

It is, however, possible to alternate CFL with INC bulbs to provide the required light quality for effective photoperiodic lighting so that some energy savings can still be achieved. A considerably more expensive alternative LEDs presented above, on the other hand, are much more energy efficient than both INC bulbs and CFLs, and last much longer.

5.9.8 Future Perspectives

The moderately efficient HPS lamp is currently the most widely used light source for high-intensity supplemental (photosynthetic) lighting in greenhouses. Efficiency, intensity and longevity of LEDs are being enhanced rapidly while INC lamps are being dropped out of greenhouse production due to their low energy efficiency. CFLs are cost-effective substitutes; however, as discussed earlier, they delay flowering in some long-day crops due to their very little far-red light emissions. Low-intensity LEDs are, on the other hand, slowly replacing INC lamps for low-intensity photoperiodic lighting.

As many greenhouses use supplemental lighting for only three or four months of the year, the payback periods are typically longer. However, as the production and sales of LEDs increase, their initial investment costs will drop further and faster. As they advance more and their costs decrease further, it will be a common practice to use them first in combination with HPS lamps, and then, they eventually become the only light source for both low-intensity and high-intensity lighting. As well, LED lighting fixture arrays will become much more compact for minimizing crop shading and making in-canopy lighting possible (for penetrating light deeper in the canopy) as they emit much less heat compared with HPS lamps. The flexibility of designing the spectral output of LEDs for specific plant and animal requirements will enable extremely exciting customized precision lighting applications for specific crops (and animals) and crop (and animal) attributes paving the way for biological engineering further. Responses to the spectrum of light have not been studied for livestock, poultry, and fish extensively. Future studies could, for instance, look at photoperiod applying different photoperiods to evaluate effects on young livestock, poultry, and fish.

Other lighting sources such as plasma lamps are also being developed for controlled environment plant production lighting; and companies are mostly focusing on high-intensity lighting to increase plant photosynthesis. In summary, we see a very bright light at the end of the tunnel, but more diverse and exciting developments are yet to come.

5.10 WATER SUPPLY IN CONTROLLED ENVIRONMENT AQUATIC SYSTEMS

5.10.1 Introduction

As discussed earlier in Chapter 3, one of the most important and critical factors to consider when choosing a proper site for a controlled environment operation is an adequate quality water source with continuous sufficient supply. Since surface waters are usually not used for aquaculture due to the risks mentioned earlier, the two most common water sources left are municipal water and ground water, assuming that they are economically available at required quantities, and their regulations and qualities permit their use for intensive aquaculture production under controlled environments.

5.10.2 Water Quality Requirements

A number of water quality concerns have been discussed earlier in different sections, however, the effect of certain organic compounds and high fine solid contents is still not known. And there is no single list of water quality standards for controlled environment aquatic systems, as the species, temperature regimes, and production techniques all affect the specific needs. This means that what might be considered good quality water for one type of organism may not be good for another organism. The effect of water quality on fish growth is typically the most important one, and a list of recommended water quality characteristics for aquaculture is provided in Table 3.4. Typically, the number one limiting water quality factor that controls the density and growth of fish reared in a system is the dissolved oxygen (DO) concentration. This is followed by the other two most important factors namely unionized ammonia and dissolved carbon dioxide concentrations. The fully recirculating systems typically reuse much greater portion of its total water volume; as a result, the depletion or accumulation rates of water quality parameters are affected and therefore must be treated accordingly. All these and other environmental factors, as well as their interrelationships and effects on fish growth are discussed in Chapter 4.

5.10.3 Water Quantity Requirements

How much water needed for an aquaculture operation is a function of several different factors such as species, density, production technology, management practices, etc. It is stated that, at a minimum, enough water should be available for filling production tanks within 24 to 48 hours, routine and emergency flushing of tanks, filter backwashing, facility clean up, and other domestic needs (Timmons and Ebeling, 2010). As a good rule of thumb, Timmons and Ebeling (2010) further suggests that it would be a wise thing to have sufficient water available to provide a 100% water exchange of total system volume per day. Beyond the minimum, the amount of additional new water for a given system is obviously a function of the degree of reuse and recycling of the water already in the system.

5.10.4 Freshwater Sources

5.10.4.1 Ground Water Sources and Characteristics

Ground water from springs, wells, and depressions is typically free from bacteria, viruses, and turbidity; but it has low or no dissolved oxygen, and may contain high levels of toxic dissolved gases such as nitrogen, carbon dioxide (CO_2), methane (CH_4), and hydrogen sulphide (H_2S), which are not desirable for aquaculture. Additionally, when pumped to the surface, ground water is often supersaturated with argon, carbon dioxide, and nitrogen gas. Therefore, before using, the ground water must be conditioned properly through aeration and degassing in order to add oxygen and remove the excess gasses. High concentrations of iron might be another problem with ground water; hence, it has to be removed via aeration. When the water is aerated, iron precipitates as iron oxide, which can be removed by granular filtration or settling basin. In limestone areas, ground water tends to have high levels of hardness and high alkalinity. In areas of granite bedrock, ground water tends to be soft and low in dissolved minerals, but it may have high dissolved carbon dioxide content. Also, ground water sources have more consistent year-round temperatures while surface waters may have wide seasonal temperature fluctuations, which may or may not be a desirable thing for a specific species under the controlled environment in mind.

5.10.4.2 Surface Water Sources and Characteristics

Surface waters coming from turbulent, well mixed sources such as small river and stream waters are fine; however, while they are well supplied with dissolved oxygen, they should not be the sole source of water due to the extensive damage they could possess in the case of a chemical or pathogenic contamination, or fish eggs, larvae, or predators or competitor organisms. Less turbulent, low circulation surface water sources include large rivers, lakes, ponds and reservoirs. Turbidity and suspended solids may also be problematic in surface waters. Surface waters typically have low dissolved minerals and low hardness, and their pH and mineral contents depend on the watershed area.

5.10.4.3 Municipal Water

Municipal water supplies are of very high quality for humans, plants and livestock; however, it is not the case for aquaculture while it is available in sufficient quantities. Municipal water is designed for human consumption, and therefore, treated to destroy pathogenic organisms; hence it often contains chemicals such as chlorine and fluorine, which are very toxic to fish. These chemicals must be neutralized before using municipal water in aquaculture. When municipal water is too expensive or not available, then surface water and ground water can definitely be used for aquacultural production, again depending on the availability and quality. Each of the above sources has different characteristics, and advantages and disadvantages for a specific controlled environment production.

5.10.5 Water Heating in Aquatic Systems

In aquatic systems, culture water is usually preheated to reduce the shock of the cold water to fish and maintain culture water temperatures in culture tanks depending on the species reared. Figure 5.27 shows the details of a water preheating system. See Examples 2.57 and 2.58 for water heating.

5.11 MASS BALANCES IN CONTROLLED ENVIRONMENT AQUATIC SYSTEMS

5.11.1 Introduction

Oxygen and waste transfer into or out of a fish culture tank are realized through the water flow. When we design a recirculating aquatic system, we must ensure that the essential parameters affecting water quality and fish productivity, such as ammonia, carbon dioxide, oxygen, and suspended solids are properly balanced. However, we should note that the minimum flow rate for maintaining any one of these parameters may be too low for maintaining other parameters. Mass balance concept for open systems has already been covered in Chapter 2. Using the mass balance concept presented in Equation (2.9) on any variable affecting water quality, we can rewrite the mass balance equation in word equation form as follows:

$$\text{Inflow Rate of "X"} + \text{Production Rate of "X"} = \text{Outflow Rate of "X"} \quad (5.37)$$

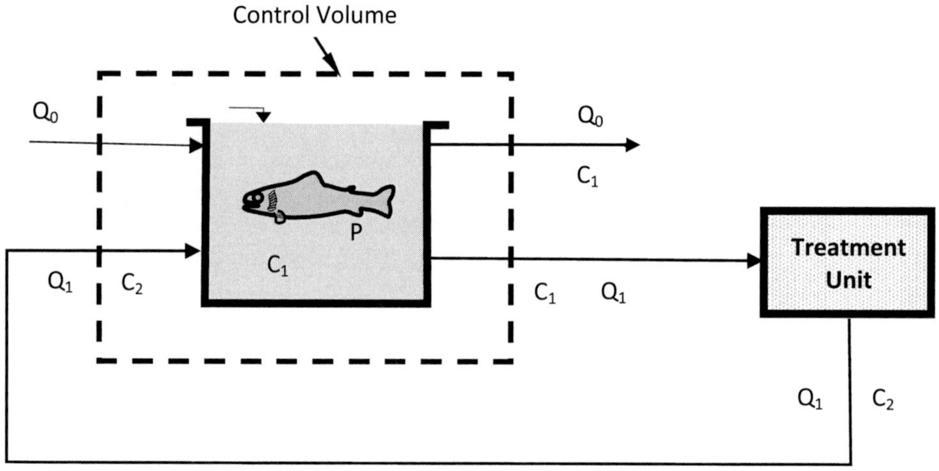

FIGURE 5.50 General mass balance on a culture tank.

The production rate of "x" in the above equation can, for instance, be the production of ammonia, CO_2, oxygen, or suspended solids, and in addition this term can be negative (e.g., oxygen). Figure 5.50 demonstrates a general mass balance for a culture tank where part of the flow is recirculated while the part of the flow is flow-through. The mass balance for the culture tank here assumes a well-mixed tank and that the tank has reached an equilibrium or steady state conditions.

As Figure 5.50 shows, water treatment process takes place outside the culture tank, which changes the concentration of the abovementioned parameter "x". Please note that there may be more than one treatment device lined up, each treating a different water quality parameter.

Applying Equation (5.37) on Figure 5.50 the mass balance equation can be rewritten as follows:

$$Q_1C_2 + Q_0C_0 + P = Q_0C_1 + C_1Q_1 \tag{5.38}$$

where Q_0 is the flow rate passing though the culture tank (discharge) (m^3_{water}/day as kg_{water}/day), Q_1 is the rate of recirculated water (m^3_{water}/day as kg_{water}/day), P is the production rate of parameter "x" (kg_x/day), and C_0, C_1, and C_2 are the concentration of parameter "x" crossing the control volume (kg_x/kg_{water}).

It is essential that each of the terms or products of terms in Equation (5.38) is represented by the same unit, kg_{oxygen}/day. For example, if we were doing a mass balance using (5.38) for oxygen, then all terms or products would need to have the same units of kg oxygen per unit time. It is convenient to use "day" as the time unit because growth rates and feeding rates are usually measured on per day basis. The unit balance example for a transport of oxygen flowing into the tank would then be QC = (kg_{water}/day)(kg_{oxygen}/kg_{water}) = kg_{oxygen}/day. In addition, flow is measured as volume per unit time or mass per unit time, and is generally provided in terms of gallons per minute (gpm), liters per second (L/s), kg per second (kg/s), or cubic meters per second (m^3/s). Also, most water quality parameters are typically expressed in terms of milligram per liter (mg/L, a.k.a. ppm).

Example 5.30: Calculating Available Oxygen for Fish Growth

Let's now calculate the available oxygen to support fish growth assuming a mass flow of water of 100 gpm of saturated inlet water at 60°F at an elevation 1,000 feet above sea level.

Solution

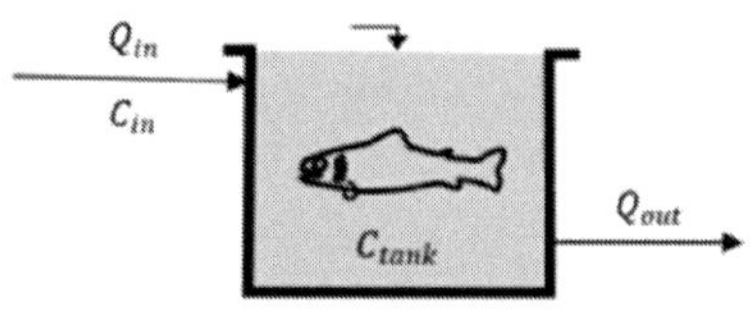

Based on the above diagram,

$$Q_{in}C_{in} = O_2\, transported\, in$$

$$Q_{out}C_{tank} = O_2\, transported\, out$$

Using Equation 5.39 we can estimate the oxygen concentration in non-saline waters as

$$C_{saturation}\left(\frac{mg}{L}\right) = \left(\frac{132}{T(°F)^{0.625}}\right)\left(\frac{760}{760 + \frac{Altitude(ft)}{32.8}}\right) \tag{5.39}$$

$$C_{saturation}\left(\frac{mg}{L}\right) = \left(\frac{132}{(60°F)^{0.625}}\right)\left(\frac{760}{760 + \frac{1,000\,ft)}{32.8}}\right)$$

$$C_{saturation} = 9.82\frac{mg}{L}$$

For the given flow rate of 100 gpm of water having a concentration of 9.82 mg/L, the mass of oxygen transported into the tank on a daily basis is:

$$Q_{in}(C_{in})_{oxygen} = \left(100\frac{gal}{min}\right)\left(9.82\frac{mg}{L}\right)$$

And now, after making the units consistent:

$$Q_{in}(C_{in})_{oxygen} = \left(100\frac{gal}{min}\right)\left(3.785\frac{L}{gal}\right)\left(9.82\frac{mg}{L}\right)\left(1,440\frac{min}{day}\right)$$

$$Q_{in}\left(C_{in}\right)_{oxygen} = \left(5{,}352{,}293\frac{mg}{day}\right)\left(\frac{kg}{10^6 mg}\right) = 5.35\frac{kgO_2}{day}$$

So, it means that this is the available oxygen on a daily basis to support fish growth and bacterial demand in the culture tank. However, since the water leaving the tank must still have minimum 5 mg/L oxygen concentration required to support fish growth, just a portion of the total available oxygen $(5.35 kgO_2 / day)$ can be used.

The available oxygen to support fish growth can then be calculated as follows:

$$Q_{in}\left(C_{in} - C_{out}\right)_{oxygen} = \left(100\frac{gal}{min}\right)\left(3.785\frac{L}{gal}\right)\left((9.82-5.0)\frac{mg}{L}\right)\left(\frac{kg}{10^6 mg}\right)\left(1{,}440\frac{min}{day}\right)$$

$$Q_{in}\left(C_{in} - C_{out}\right)_{oxygen} = 2.63\frac{kgO_2}{day}$$

If we go back to Equation (5.38), and apply the simplest recirculating system case where all the water flow is fully recirculated, hence there is no discharge. In such situation, the Q_0 terms in (5.38) would drop out and we would have the following form:

$$\begin{aligned} Q_1C_2 + P &= Q_1C_1 \\ Q_1(C_2 - C_1) &= -P \end{aligned} \tag{5.40}$$

We now need to determine the concentration of each particular water quality parameter leaving the treatment device (C_2), so that the mass flow of the water quality parameter into the culture tank can be determined and the mass balance solved. The treatment device shown in Figure 5.50 could be a biofilter, a CO_2 stripper, or a solids settling basin, each of which has its own treatment efficiency.

If we perform a mass balance on the treatment device control volume, then we can easily solve for the concentration of the water quality parameter of interest, C_{out}:

$$C_{out} = C_{in} + \frac{T}{100}\left(C_{best} - C_{in}\right) \tag{5.41}$$

where, C_{best} is the absolute best result that can be achieved by a treatment system (e.g., zero ammonia, saturated oxygen, or zero suspended solids), C_{in} and C_{out} are the water quality concentration entering and leaving the treatment device, respectively, and T is the treatment efficiency (%).

If the device is an oxygen addition unit, then the C_{best} term can be increased above atmospheric concentration values for oxygen by increasing the partial pressure above atmospheric oxygen partial pressure in the device. For instance, a pure oxygen supplying device would have a C_{best} value of approximately five times the C_{best} value that would be available if normal air is used at atmospheric pressure (e.g., trickling tower). C_{best} for ammonia and TSS are zero, but for CO_2 it is about 0.5 mg/L because there exists some CO_2 in the air.

5.11.2 Loading Rates and Production Terms

As fish density (carrying capacity) increases, minimum oxygen levels in culture tanks are better maintained by using pure oxygen suppliers, this is because pure oxygen systems allow inlet oxygen concentrations to be maintained at much higher levels (multiples) compared to dissolved atmospheric concentration levels. Having said that however, other water quality parameters start limiting the allowable fish density due to degradation in accumulated ammonia, carbon dioxide, or suspended solids. From basic stoichiometry and the production values for ammonia, solids, and carbon dioxide associated with feeding, 10 mg/L of oxygen consumed produces up to 1.4 mg/L of ammonia, 14 mg/L of carbon dioxide, and 10–20 mg/L of suspended solids.

The production term in Equation (5.38) refers to the production of some pollutant (e.g., ammonia, carbon dioxide, and suspended solids) or consumption term (e.g., oxygen). And these production terms in a recirculating system are proportionately related to the fish feeding rate; hence, we relate the production terms to feeding rates as follows:

P_{Oxygen}: −0.25 kg per kg feed consumed by fish (a negative production term)
−0.12 kg per kg feed consumed by nitrifying bacteria
−0.13 heterotrophic bacteria (estimate, can be as high as 0.5)

P_{Oxygen}: −0.50 kg per kg feed for system (including nitrifying and heterotrophic bacteria, and this is the minimum! Safe design value is 1.0 kg oxygen per kg feed fed)

P_{CO2}: 1.375 grams produced for each gram O_2 consumed (both fish and bacteria)

P_{TAN}: $F \times PC - 0.092$ (where F is the feeding rate (kg/day) and PC is the protein concentration of feed (decimal))

P_{TSS}: 0.25 kg feed fed (dry matter basis) (literature values are 20% to 40%)

5.11.3 Water Quality Targets

The controlled environment engineer must choose design or target operating conditions in a culture tank. These are the "C" terms in Equation 5.38 shown in Figure 5.50. These design numbers are obviously species dependent and typical recommended levels are provided in Chapter 4 where more of the water quality parameters are discussed in depth whether for warm- or cold-water fish species. The readers of this book are referred to Chapter 4 for further information.

5.11.3.1 Oxygen

Oxygen is consumed at quite a high rate due to high fish metabolism. Oxygen is transported by water flow, and loss of water flow in a recirculating system means lack of oxygen, resulting in death of fish. Even a partial loss of water flow generally results in insufficient oxygen for the fish, and the consequence is the death.

As discussed earlier in Chapter 4, the solubility of O_2 decreases as salinity and temperature increase. Table A.47 provides the saturation O_2 concentration in water as a function of salinity and temperature. And both altitude and barometric pressure directly influence oxygen concentration in water as well (see Equation (2.252) in Chapter 2). As also mentioned earlier in Chapter 2, the oxygen holding capacity of water decreases as temperature increases above 0°C. Considerably less oxygen is available at higher temperatures when metabolic rates, as well, are higher, affecting production capacity due to limited available oxygen. Increasing salt concentration in water decreases the dissolved saturation O_2 capacity of water, as well.

The rearing unit effluent for salmonids should, for instance, contain from 6.0 to 8.0 mg/L dissolved oxygen, whereas the allowable minimum levels for warmwater species such as catfish and tilapia are much lower than those for salmonids, usually 2.0 to 3.0 mg/L, while it is recommended to maintain levels closer to 5.0 to 6.0 mg/L (Timmons and Ebeling, 2010).

Finally, we should also remember that not only do the fish need oxygen, but also the biological filter, which is critically dependent on adequate oxygen supply for bacterial metabolism. The DO concentration within the filter must be maintained at minimum 2.0 mg/L in order to insure that the nitrification rate in the filter does not become limited due to insufficient levels of oxygen (Kumar, 1984; Manthe et al., 1988). We should always monitor the DO concentration coming off the biofilters; and if the concentration starts dropping down to 2.0 mg/L, then we must, for instance, increase the flow rate via increasing the hydraulic loading rate on the filter.

5.11.3.2 Ammonia and Nitrate

While ammonia was discussed in depth in Chapter 4, after reviewing the published literature on the effects of ammonia on fish, Meade (1985) concluded that a truly safe, maximum acceptable concentration of un-ionized, or of total ammonia is not known. The apparent toxicity of ammonia is extremely variable and depends on more than the mean or maximum concentration of ammonia.

As mentioned in Chapter 4, unionized ammonia (NH_3) is the most toxic form of ammonia to fish, and appears to have a direct effect on the growth of aquatic animals. Toxic concentrations of unionized ammonia were reported by European Inland Fisheries of Advisory Commission (1973), and it is toxic to fish even at low concentrations, with 96-hour LC_{50} values varying widely by species, for instance, from as low as 0.08 mg/L NH_3-N for pink salmon to 2.2 mg/L NH_3-N for common carp. In general, warmwater species are more tolerant to ammonia toxicity than coldwater species, and freshwater species are more tolerant than saltwater species. Table 4.4 provides values of the LC_{50} concentrations for various fish species for unionized ammonia nitrogen, nitrite nitrogen and nitrate nitrogen. In general, for commercial production, it is recommended that unionized ammonia should be kept below 0.05 mg/L, and TAN concentrations below 1.0 mg/L for long-term exposure. Toxicity concentrations of unionized ammonia-nitrogen, nitrite-nitrogen, and nitrate-nitrogen for several species of fin fish are provided in Table 4.5.

All the ammonia-nitrogen concentration numbers eventually depend on the estimate of ammonia generation, which is indeed based on the fish feeding rate and expressed as follows:

$$P_{TAN} = \frac{0.092 \times F \times PC}{t} \tag{5.42}$$

where F is the feeding rate (kg/day), PC is the protein concentration of feed (decimals), t is usually one day (i.e., feed fed uniformly over a 24 hour period), and the constant 0.092 in Equation (5.42) is coming from a series of approximations and estimates that when multiplied together result in: 0.092 = 0.16×0.80×0.80×0.90. Protein is 16% nitrogen, 80% nitrogen is assimilated, 80% assimilated nitrogen is excreted, and 90% of nitrogen excreted as TAN + 10% as urea (freshwater fish only).

Based on Equation (5.42), for simplicity, we can simply assume 10% of the protein in the feed becomes the ammonia-nitrogen generation rate. In (5.42), one day was used as the feeding time period. In recirculating systems, fish can be fed uniformly over 24 hours, hence distributing the ammonia-nitrogen load uniformly over the entire day. If this uniform 24-hour feeding is not used, then the equation should be adjusted and the time period should be the time between feedings. Many smaller feedings evenly spaced during the day would serve to minimize high values of P_{TAN}, which is in fact a strategy used for culture tank rearing systems through the use of automatic or demand type feeders.

As shown earlier in Chapter 4, nitrate (NO_3), on the other hand, is the end-product of the bacterially mediated nitrification process and is the least toxic form of nitrogen compounds, with a 96-hour LC_{50} value usually beyond 1,000 mg NO_3-N/L (Table 4.4); hence, it will not affect the controlling flow rates in the system. If need be, we can choose a value such as 500 mg/L to work with (Timmons and Ebeling, 2010). In intensive recirculating systems, nitrate levels are typically controlled by daily water exchanges. In systems where low water exchange or high hydraulic retention times is present, this time, denitrification process becomes important. In principle, if 1 kg/day of TAN is produced, then 1 kg/day of nitrate-N is produced as well. Hence, the equilibrium concentration of nitrate directly depends on the overall water exchange rate through the system.

5.11.3.3 Carbon Dioxide

Carbon dioxide (CO_2) has little direct toxicity to fish. In general, high CO_2 levels are associated with low DO levels. It is usually not necessary to monitor CO_2 if O_2 is being monitored. Dissolved CO_2 in excess of 20 mg/L may be harmful to fish (Piper et al., 1982); but; most species of fish are reported to survive for several days at dissolved CO_2 levels up to 60 mg/L. Colt and Watten (1988) as well recommend that the maximum level of CO_2 should not exceed 20 mg/L; however, Needham (1988) recommends the maximum not to exceed 10.0 mg/L. In the past, most recirculating systems relied upon aeration as the main means of supplying oxygen, and this type of management also kept CO_2 values at low levels, e.g., less than 20 mg/L. However, loading rates have in recent years increased; and instead of using aeration, it has become necessary to inject pure oxygen into these intensive recirculating systems. As a consequence, the stripping of CO_2 that takes place when employing aeration systems is no longer there; thence we are required to apply other means for CO_2 stripping.

The generation of CO_2 is based upon chemical stoichiometry by relating CO_2 production to oxygen consumption (1.375 is the ratio of molecular weights of the two gases, 44/32): CO_2 = 1.375 grams produced for each gram O_2 consumed (by both fish and bacteria).

As also discussed in Chapter 4, dissolved CO_2 concentration is a function of pH or the alkalinity of water, and the total dissolved inorganic carbon present in water, i.e., carbon dioxide (CO_2), carbonic acid (H_2CO_3), bicarbonate (HCO_3^-), and carbonate ions (CO_3^{2-}). Changes in temperature and salinity can also have minor influences, since they affect the equilibrium constants.

5.11.3.4 Suspended Solids

As already mentioned in Chapter 4, fish produce somewhere between 0.3 to 0.4 kg of total suspended solids (TSS) for every kilogram of feed fed, and hence, they are a major source of carbonaceous oxygen demand and nutrient input into the water. As a result, they can directly affect fish health and condition within recirculating systems by damaging fish gills and harboring pathogens. On the other hand, while an upper limit of waste solids for freshwater fish is tentatively reported as 25 mg TSS/L, a level of 10 mg TSS/L is usually recommended for normal operations. For instance, tilapia is reported to perform well with TSS levels of 80 mg/L if all the other water quality parameters are ideal levels (Timmons and Ebeling, 2010).

The quantity of suspended solids or TSS, generated per unit of feed being fed is estimated as follows: TSS = 0.25 kg feed fed (dry matter basis) (from 20% to 40% of feed fed dry basis). TSS in recirculating systems is treated as a dilute waste, and the TSS design concentrations in recirculating systems are reported to be in the range of 10 to 30 mg/L. For design purposes, controlled environment design engineers can assume that every kg of dry feed fed produces approximately 8 liters of liquid waste (i.e., one lb of dry feed fed produces one gallon of manure).

5.11.4 Estimating Circulation Requirements

In this section, it must be noted that we will work on a fish production example originally developed by Timmons and Ebeling (2010), assuming a target production goal of 100,000 lbs (45.5 tonnes) per year with a target market size fish of 750 g (1.65 lbs), whole on ice. Fingerlings are purchased from a biosecure facility at a mean size of 50 g. Example assumes no mortality or culling between stages.

5.11.4.1 Estimating Final Feed Rate

The daily feed rate as a percent of the fish weight is usually provided by the feed manufacturer or some other sources. If this information is unavailable, then we can estimate the final feed rate by calculating the weight of the fish the day before harvest, subtracting that from the weight at harvest and adjusting with an estimated feed conversion rate (FCR). In the example we will be working on in this book, however, we will assume that the feed conversion rate (FCR), i.e., the ratio of feed to weight gain, is 1.3 for the growout stage, reaching a maximum feeding level of 9.64 kg or 1.11 percent body weight per day. This is an extremely important information because all the other engineering designs are developed based on this feed rate, including circulation requirements, solids capture, biofiltration, aeration/oxygenation, gas stripping, and disinfection.

5.11.4.2 Determining Controlling Flow Rate

In this section, we will determine the flow rates to control dissolved oxygen, ammonia-nitrogen removal, carbon dioxide stripping, and suspended solids (TSS). It must be reiterated that these calculations are always based on a design feeding rate for a culture tank or collection of tanks that are placed on a shared controlled environment life support system, such as biofilter, gas transfer devices, solids removal units, etc.

In the following example, in order to keep the problem simple, it was assumed that as if the system had only one fish tank and that it reaches a maximum feeding level of 9.6 kg/day. Oxygen is transported into a culture tank and the generated waste products within the system are removed by water circulation. When a recirculating aquatic system is designed, a controlled environment engineer makes sure that the essential environmental parameters impacting water quality (e.g., dissolved oxygen, ammonia-nitrogen, carbon dioxide, and suspended solids) and hence fish productivity are adequately balanced. The design work requires determining the required flow rates for each of these four water quality parameters using mass balance equations in order to maintain the design water quality variables at or below (above for dissolved oxygen) their maximum tolerable (target) values. And then the system is operated at the highest flow rate calculated for these four essential water quality parameters. The determined maximum flow rate will normally be higher than the flow rates needed for the other three water quality parameters; however, it only means that these water quality parameters will indeed have better quality values than design targets.

5.11.4.3 Circulation Pumps and Piping Design

All recirculating systems use some kind of pump to move water and increase the system pressure for filtration, aeration, degassing, or other unit applications. In order for pumps to function efficiently and reliably, they must be carefully selected and incorporated into the overall system. A poor pump selection can increase operating and maintenance costs and even result in total system failure. There are many different types of commercially available pumps; and in recirculating systems, centrifugal pumps and axial pumps are the two most common types while centrifugal pumps forming the majority. Since fluid mechanics is the foundation of recirculating systems, the readers of this book are referred to Chapter 2 for the essentials of fluid mechanics required for water circulation and relevant numerical examples. Tables A.68, A.69, and A.70 present pressure losses and flow velocities for PVC pipes, pipe fittings and manual valves.

Example 5.31: Calculating the Controlling Flow Rate

Calculate the required design flow rate for a 100% recirculating flow for a growout tank assuming a feeding rate of 9.6 kg feed/day at 35% protein content.

Solution

In order to calculate the controlling flow rate, we must first calculate the required flow rate for each water quality parameter (oxygen, ammonia, carbon dioxide, and TSS) and then identify the controlling parameter.

Assuming a 90% efficiency for O_2 transfer, 35% for TAN removal, 70% for CO_2 stripping, and 90% for TSS removal, let's now calculate the required steady-state flow rate for maintaining the following water quality levels: 5 mg/L O_2, 2 mg/L TAN, 20 mg/L CO_2, and 10 mg/L TSS.

A good place to start is using the general mass balance equation, i.e., Equation (5.38), and the impact of a treatment device on the discharge water quality concentration can, on the other hand, be calculated using Equation (5.41).

c) Required Design Flow Rate for Dissolved Oxygen

Let's first calculate the effluent dissolved oxygen concentration C_{out} using a Speece cone with 90% oxygen transfer efficiency, and a 5 mg/L production tank DO level (target value for minimum) of C_{in} and an enriched oxygen stream at 16.0 mg/L from a VSA or PSA generator (Note: C_{best} could be as high as 100 mg/L, using pure oxygen and a 30 psi environment).

$$C_{out} = C_{in} + \frac{T}{100}(C_{best} - C_{in})$$

After solving for C_{out} we obtain:

$$C_{out} = 5\frac{mg}{L} + \frac{90}{100}\left(16\frac{mg}{L} - 5\frac{mg}{L}\right)$$

$$C_{out} = 15\frac{mg}{L}$$

Next, let's calculate the oxygen production (P) or in this case consumption (-P) equal to the sum of fish metabolism and bacterial oxygen consumption (heterotrophic and autotrophic) (trickling, RBC, and moving bed biofilters are self aerating systems and in these cases the oxygen load from the bacteria can be set to zero):

$$P = \frac{0.25\,kgO_2\,by\,Fish}{kg\,feed} + \frac{0.25\,kgO_2\,by\,Bacteria}{kg\,feed} = \frac{0.50\,kgO_2}{kg\,feed}$$

$$P = \left(\frac{9.6\,kg\,feed}{day}\right)\left(\frac{0.50\,kgO_2}{kg\,feed}\right)\left(\frac{10^6\,mg}{kg}\right) = \frac{4{,}800{,}000\,mgO_2}{day}$$

Now getting back to the general mass balance equation, i.e., Equation (5.38):

$$Q_1C_2 + P = Q_1C_1$$

$$Q_1\left(15\frac{mg}{L}\right) + \left(-\frac{4{,}800{,}000\,mg}{day}\right) = Q_1\left(5\frac{mg}{L}\right)$$

$$Q_1 = \frac{\left(\frac{4{,}800{,}000\,mg}{day}\right)}{(15.0-5.0)\frac{mg}{L}} = 480{,}000\frac{L}{day} = 5.56\frac{L}{s}$$

And this means that about 5.56 L/s (88.1 gpm) of influent tank water at 15.0 mg/L DO is required to satisfy the oxygen demand of the 9.6 kg of feed/day.

Required Design Flow Rate for Total Ammonia-Nitrogen

Let's first calculate the effluent dissolved oxygen concentration, C_{out} using a non-self aerating biofilter with a 35% removal efficiency and a production tank TAN level (target value for minimum) of $C_{in} = 2$ mg/L and a C_{best} value of 0.0 mg/L.

After solving for C_{out} we obtain:

$$C_{out} = 2\frac{mg}{L} + \frac{35}{100}\left(0\frac{mg}{L} - 2\frac{mg}{L}\right)$$

$$C_{out} = 1.3\frac{mg}{L}$$

The total ammonia-nitrogen (TAN) production (+P) is based on the feed rate and feed protein level (35%).

$$P = \left(\frac{9.6\,kg\,feed}{day}\right)\left(\frac{0.35\,kg\,Protein}{kg\,feed}\right)\left(\frac{0.092\,kg\,TAN}{kg\,Protein}\right)\left(\frac{10^6\,mg}{kg}\right) = \frac{309{,}000\,mg\,TAN}{day}$$

Now getting back to the general mass balance equation, i.e., Equation (5.38):

$$Q_1\left(1.3\frac{mg}{L}\right) + \left(-\frac{309{,}000\,mg}{day}\right) = Q_1\left(2.0\frac{mg}{L}\right)$$

$$Q_1 = \frac{\left(\frac{309{,}000\,mg\,TAN}{day}\right)}{(2.0-1.3)\frac{mg\,TAN}{L}} = 441{,}000\frac{L}{day} = 5.10\frac{L}{s}$$

And this means that about 5.10 (80.9 gpm) of influent water at 1.3 mg/L TAN is required to dilute the tank TAN concentration to 2.0 mg/L from the 9.6 kg of feed/day.

Required Design Flow Rate for Dissolved Carbon Dioxide

Now, let's calculate the required flow to strip the CO_2 production based on a maximum tank concentration of 20 mg/L CO_2, a stripping efficiency of 70%, and a C_{best} value of 0.5 mg/L assuming atmospheric air is used with 320 ppm concentration of CO_2 for stripping.

Again, after solving for C_{out} we obtain:

$$C_{out} = 20\frac{mg}{L} + \frac{70}{100}\left(0.5\frac{mg}{L} - 20\frac{mg}{L}\right)$$

$$C_{out} = 6.35\frac{mg}{L}$$

Carbon dioxide production (+P) is based on oxygen consumption, where 1.375 kg of CO_2 is produced for every kg of O_2 consumed. Then:

$$P = \left(\frac{9.6\,kg\,feed}{day}\right)\left(\frac{0.50\,kg\,O_2}{kg\,feed}\right)\left(\frac{1.375\,kg\,CO_2}{kg\,O_2\,consumed}\right)\left(\frac{10^6\,mg}{kg}\right) = \frac{6{,}620{,}000\,mg\,CO_2}{day}$$

Now getting back to the general mass balance equation, i.e., Equation (5.38):

$$Q_1\left(6.35\frac{mg}{L}\right) + \left(-\frac{6{,}620{,}000\,mg}{day}\right) = Q_1\left(20\frac{mg}{L}\right)$$

$$Q_1 = \frac{\left(\dfrac{6,620,000\,mg\,CO_2}{day}\right)}{(20-6.35)\dfrac{mg\,CO_2}{L}} = 482,000\frac{L}{day} = 5.58\frac{L}{s}$$

And this means that about 5.58 L/s (88.4 gpm) of influent water at 6.35 mg/L CO_2 is required to dilute the carbon dioxide concentration to 20 mg/L from the 9.6 kg of feed/day.

d) Required Design Flow Rate for Total Suspended Solids

And finally, let's now examine total suspended solids. Since this is directly based on feed fed, we can calculate the required flow to strip the suspended solids production based on a tank concentration of 10 mg/L, a stripping efficiency of 90%, and a C_{best} value of 0.0 mg/L. After solving for C_{out}, we obtain the following:

$$C_{out} = 10\frac{mg}{L} + \frac{90}{100}\left(0\frac{mg}{L} - 10\frac{mg}{L}\right)$$

$$C_{out} = 1.0\frac{mg}{L}$$

TSS production ($+P$) is based on feeding rate (assuming that the feed has 0% moisture content). Then:

$$P = \left(\frac{9.6\,kg\,feed}{day}\right)\left(\frac{0.25\,kg\,TSS}{kg\,feed}\right)\left(\frac{10^6\,mg}{kg}\right) = \frac{2,400,000\,mg\,TSS}{day}$$

And now getting back to the general mass balance equation, i.e., Equation (5.38) again:

$$Q_1\left(1.0\frac{mg}{L}\right) + \left(-\frac{2,400,000\,mg}{day}\right) = Q_1\left(10\frac{mg}{L}\right)$$

$$Q_1 = \frac{\left(\dfrac{2,400,000\,mg\,TSS}{day}\right)}{(10-1.0)\dfrac{mg\,TSS}{L}} = 267,000\frac{L}{day} = 3.09\frac{L}{s}$$

And this means that about 3.09 L/s (49.0 gpm) of influent water at 1.0 mg/L TSS is required to dilute the TSS concentration to 10 mg/L from the 9.6 kg of feed/day.

It is important to note that the design engineer must choose the design or target operating conditions in order to calculate the mass balances. This is why a decision was made to enrich the influent oxygen concentration to 16.0 mg/L (C_{best}) here in order to reduce the required flow rate for oxygen. This example particularly shows how sensitive the determined flow rates are to the values selected for the design values. A controlled environment engineer typically selects a realistic target value for the design value, then do the calculations, and then checks the economics of the calculated flow rate, and then make necessary adjustments in the target values, such as increasing oxygen availability with an enrichment device or using pure oxygen supplementation as we did in this example.

We have now calculated all four flow requirements, of which the highest flow rate is 5.58 L/s (88.4 gpm). For convenience, let's round this number to 6.30 L/s (100 gpm). It means the system is operated at the highest flow rate, i.e., 6.30 L/s (100 gpm). As it is seen, the determined maximum flow rate is much higher than the flow rates needed for the other three water quality parameters. As mentioned earlier, it only means these water quality parameters will have better quality values than design targets.

Before concluding this section, it must also be noted that we also need to determine the required hydraulic loading for the biofilter we would be using as in $m^3/h/m^2$ (gpm/ft) because some biofilters have minimum hydraulic loading requirement in order to function properly. Then this hydraulic loading may become the controlling flow rate over and above the flow rates required to maintain water quality conditions. We need to remember that, in such a case, the equilibrium concentrations for all those four water quality parameters would be slightly better.

5.12 SOLIDS IN CONTROLLED ENVIRONMENT AQUATIC SYSTEMS

5.12.1 Introduction

Developing an effective water treatment system is considered as the most important task in the design of controlled environment recirculating systems. These treatment processes, a.k.a. the unit processes, include the solids removal, conversion of toxic ammonia-nitrogen and non-toxic form of nitrite-nitrogen, dissolved oxygen addition to the water, and carbon dioxide removal from the water. All aspects of recirculating systems are negatively affected by suspended solids; therefore, it is extremely important to remove solid wastes from recirculating systems. Suspended solids, varying in size from microns to centimeters, are generated from uneaten feed, fish feces, and dead and living bacteria (biofloc). Hence, solids management through not only the removal of solids but also flow regulation, feed design and feeding management becomes vitally important as the production system intensifies (Cripps and Bergheim, 2000).

In controlled environment recirculating aquatic systems, main portion of the particles (by weight) are less than 100 μm in size; and in intensive recirculating systems they are 30 μm or less in size. And the problem is that these particles do not readily settle from the water column, and mechanical filtration is not really effective in removing these solids. Regardless of the size, all particles must be managed adequately by a properly selected solids removal technology within each size range, such as sedimentation and screening for removal of large particles and foam fractionation or ozone treatment for fine solids (<30 μm) removal. Granular media filters can, on the other hand, remove a wide range of particle sizes, and can be effective in removing solids down to about 20 μm. This and other solids removal technologies are briefly reviewed in this chapter.

Total suspended solids (TSS) concentration, having both inorganic and organic components, is defined as the mass of particles above 1 μm in diameter (APHA, 1995) observed in a known volume of water. The organic portion is termed as *volatile suspended solids* (VSS), which contributes to oxygen consumption and biofouling problems. Moreover, the inorganic components (ash) mainly form the sludge deposits. Suspended solids can further be partitioned into *settleable solids* (>100 μm) and *non-settleable suspended solids* (<100 μm) (EPA, 1975). The smaller and non-settleable suspended solids, on the other hand, are relatively more difficult to manage and the main reason of typical problems in controlled environment recirculating systems. A number of scholars reported that fish health, such as gill function can be impaired by the accumulated solids compromising oxygen transfer or by offering a suitable habitat for the pathogenic organisms (Braaten, 1992; Liltved and Cripps, 1999; Cripps and Bergheim, 2000). However, there is no agreement on an acceptable design value for TSS concentrations, which would then be used in recirculating system designs. For instance, the EIFAC (1980) recommended that TSS concentrations should be maintained below 15 mg/L in recirculating systems, while Muir (1982) suggested a limit of 20 to 40 mg/L for these systems.

5.12.2 Solids Generation and Physical Characteristics

In a recirculating aquatic system, all the waste generated essentially originate from the fish feed, which manifests itself in: i) uneaten feed, and ii) fish excrement in the form of solids, liquid, or gas. Of the feed that is eaten, 80% to 90% will eventually be excreted in some form primarily in the form of feces (Hopkins and Manci, 1989). The mass production rate of feces is proportional to the feeding rate; and as a rule of thumb, we typically use 25% of the feed fed to the fish as the mass that is produced as total suspended solids on a dry matter basis.

With respect to the solids management in recirculating aquatic systems, the two most important physical characteristics of suspended solids can be cited as the particle specific gravity and particle size distribution because the behavior of suspended particles in water is determined specifically by their particle size and specific gravity. APHA (1989) defines specific gravity as the ratio of the density of a wet particle to that of water, and is determined by the source of the particles. The size distribution is, on the other hand, determined by a number of factors, such as the source of particles, feed properties, fish size, solids removal process, water temperature, and turbulence in the system.

Chen et al. (1993) reported, from two recirculating systems, that the average specific gravity of the particulate matter was 1.19. Timmons and Ebeling (2010), on the other hand, reported a specific gravity of 1.05, which they think is more representative of suspended particles originating as tilapia feces. Regardless, the main issue is that the specific gravity of fish feces is not much different from that of water; and therefore, does not settle as rapidly.

And yes, feces are the main source of the suspended solids; however, uneaten fish feed is as well a significant source of total suspended solids in fish culture water. TSS from fish feed has different specific gravity and particle sizes than those originating from feces. As a consequence, they respond to waste management systems in different ways. Uneaten feed, even after several hours and reiterated passages through pumps, over 97% of the feed particles are reported to be greater than 60 μm in size, and 73% of which are typically larger than 500 μm. As mentioned earlier, in controlled environment recirculating aquatic systems, the focus of this book, particles less than 30 μm are the most prevalent in the water column. Chen et al. (1993) reported that 90% of the total particle mass in recirculating systems is from individual particles of less than 30 μm. Hence, it makes no sense to use, for instance, 60 μm drum filter screens to remove majority of the particles in recirculating systems. Sedimentation, i.e., gravitational removal methods, will not be able to remove the fine particles either from the water because the particles less than 30 μm have quite low settling velocities (Chesness et al., 1975; Chiang and Lee, 1986).

5.12.3 Solids Removal Mechanisms

It must again be emphasized that, if fish feces and waste fish feed are not removed from the aquatic system in a timely manner, they can easily harm the fish health and well-being as well as the biofilter performance, and hence increase the overall oxygen demand of the aquatic system in the first place. If the removal of solid waste from the system is poorly performed, all the other unit processes will eventually fail due to increased plugging within the aeration columns, screens, etc. Therefore, controlled environment recirculating systems must rapidly fractionate and remove settleable solids from the culture tanks to solids capture units positioned downstream.

Suspended solids are removed from culture waters in recirculating systems mainly by three methods namely gravity separation, filtration, and flotation. Gravity separation functions on the principle of sedimentation and settling velocities, and this technique of separation includes settling basins. However, the particle removal in gravity separation method is accomplished by screen, granular media, or porous media filters. Using gravity separation or even mechanical screen filtration, only particles larger than approximately 100 μm can be removed. Gravity separation or filtration methods cannot remove fine particles effectively. Granular filters can, for instance, only effectively remove particles larger than 20 μm. In a flotation process, on the other hand, particles attach onto air bubbles and that is how fine particles are separated from water. Foam fractionation, for instance, uses this process to remove fine particulate and dissolved organic compounds. Chen and Malone (1991) reported characterizations of solids removal techniques for controlled environment recirculating systems.

5.12.3.1 Sedimentation

Sedimentation, i.e., gravity separation, is one of the simplest solids removal technologies available, and it takes place because of the difference between the solid particles and water densities. This means a particle heavier than water will fall through the water until its speed reaches a terminal value for its settling velocity. It also means that the denser and larger particles will settle out of water faster than smaller and less dense particles. That's why large particle sizes should be maintained for more effective solids removal; and for this, the waste particles must be removed from the culture tanks quickly and particularly before any water pumping activity takes place. Warrer-Hansen (1982) reported settling velocities of 1.7–4.3 cm/s for fish feces while some other species produce even lower specific gravity fecal matter (Robertson, 1992). That's why these wastes have slower settling velocities. Juell (1991) reported that the feed pellets have settling velocities somewhere between 15–33 cm/s. In recirculating aquatic systems, however, even finer or less dense particles can be produced, which would obviously settle at much lower velocities such as 0.01 cm/s (IDEQ, 1988). And these finer and less dense particles cannot effectively settle in a settling basin, hence they have to be removed by other removal processes.

5.12.3.2 Settling Basins

When designed and operated properly, settling basins are relatively inexpensive and quite effective solids removal systems. In addition, they do not need much energy input and specialized labor skills. However, sedimentation basins have poor removal efficiency of small suspended solids (<100 μm), low loading rates, require more labor and floor space compared to microscreen filters or granular media filters. Henderson and Bromage (1988) states that settling basins are not effective solids removal systems in removing TSS when inlet concentrations are less than 10 mg/L. Since eliminating resuspension of TSS and dissolution is difficult in most settling basins, they usually need further TSS treatment system. All continuous rectangular flow sedimentation basins are divided into four zones according to function (Figure 5.51): i) the inlet zone, which uniformly distributes the suspension over the entire cross-section of the basin, ii) the settling zone where sedimentation takes place, iii) the sludge zone where the solids accumulate, and iv) the outlet zone where the clarified liquid is usually collected over the entire cross-section of the basin and discharged.

The volumetric flow of water per unit surface area of the basin or overflow rate (V_o) is the key design parameter for settling basins, which can be determined as follows:

$$V_o = Q / A_{sz} \tag{5.43}$$

where V_o is the overflow rate (m³/m².time), Q is the design flow rate (m³/time), and A_{sz} is the settling zone area (m²).

According to Hazen (1904), any particle having a settling velocity (V_s) greater than the overflow rate (V_o) will settle out. For other particles having $V_s < V_o$ will be removed in the ratio V_s/V_o depending on their vertical position in the tank at the inlet. The settling velocity (V_s) can be predicted using the following equation (Montgomery, 1985):

$$V_s = \sqrt{\frac{4g\left(\rho_p - \rho\right)D_p}{3C_d\,\rho}} \tag{5.44}$$

where V_s is the velocity of settling particles (m/s), g is the gravitational acceleration (m/s²), ρ_p is the density of particles (kg/m³), ρ is the density of water (kg/m³), D_p is the diameter of particles (m), and C_d is the dimensionless drag coefficient.

Since settling basin designs were originally developed for treating municipal waste waters, they give erratic performance for aquacultural waste when measured against the traditional approach of hydraulic retention time (τ) in the settling basin:

$$\tau = V_{basin} / Q \tag{5.45}$$

where τ is the hydraulic retention time, V_{basin} is the basin volume (m³), and Q is the design flow rate (m³/time).

A number of researchers recommended hydraulic retention times of 15 to 30 minutes for basin design (Liao and Mayo, 1974; Mudrak and Stark, 1981). It is also recommended that settling basins be designed with basin length to width ratios from 4:1 to 8:1 (Arceivala, 1983) and a minimum of 1 m (3.3 ft) of water depth (Liao and Mayo, 1974).

The overflow method developed by Stechey and Trudell (1990) provides more effective TSS removal compared to the hydraulic retention time approach. In intensive salmonid aquaculture, Stechey and Trudell (1990) recommend an overflow rate (V_o) for the design of settling basins to be between 40–80 m³/m².day (982–1,964 gpd/ft²). These overflow rates equal to particle settling velocities (V_s) of 0.046–0.092 cm/s. In short, for every 1.0 L/s of water flowing through the settling basin, 1.5 m² of surface area is required for settling (1.0 gpm flow per ft²). Mudrak (1981), on the other hand, employing several off-line settling basins used in intensive trout culture, reported that the removal of settleable solids was 90% or greater (although TSS removal was about 10% less) when the design overflow rate (V_o) was at approximately 60 m³/m².day.

5.12.3.2.1 Inlet and Outlets for Settling Basins

It is important to note that the rate of solids settlement is affected by the motion of water in the settling tank.

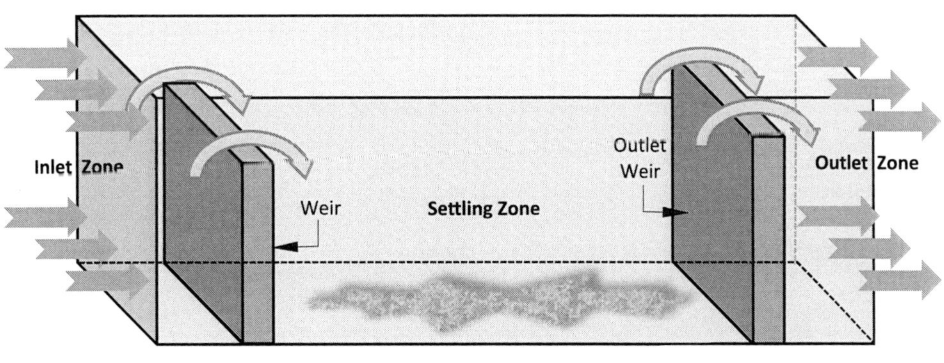

FIGURE 5.51 Schematic representation of the principal zones of a rectangular continuous flow sedimentation basin.

Therefore, the turbulence and mixing in the tank must be kept at an absolute minimum. The design of the inlet and outlets is critical for the effective performance of the water being handled during the sedimentation process; and when designing inlet structures, IDEQ (1988) emphasizes that i) the influent water should to be introduced evenly across the entire cross-section of the settling zone, ii) all flow through the settling zone should begin in even and horizontal path, and iii) the influent velocity to the settling zone should be slow so that excessive turbulence and mixing can be prevented.

5.12.3.2.2 Inlets

Inlets are designed with a submerged weir that separates the settling and inlet zones. The weir extends across the full width of the settling basin, and is submerged about 15% of the basin depth (Figure 5.51). The weir crest should have rounded edges to smooth the flow and the width should be about 20 to 30 cm (8 to 12 in.). As shown in Figure 5.52, for circular settling basins, the inlet is usually located at the center of the basin, and a baffle is placed surrounding the inlet pipe in order to reduce turbulence and distribute the flow in a radial pattern through the full depth of the basin.

5.12.3.2.3 Outlets

Rectangular settling systems are more efficient than are circular settling tanks, but they require considerably more floor space. The sub functions of rectangular systems are the inlet zone, the settling zone, and the outlet zone (Figure 5.51). The outlet weir divides the settling zone from the outlet zone, as it skims clear water from the surface of the settling zone. The outlet weir should be designed and constructed so that it distributes the water exiting the settling zone at a uniform depth and velocity across its width. This is necessary to avoid generating currents and the accompanying turbulence in the settling zone.

The outlet zone width area should be the same as the settling zone width, and the length should not be less than 1.5 times the depth of the settling zone as well. For instance, if the settling zone is 3.6 m (12 feet) wide, 9 m (30 feet) long, and 1.2 m (4 feet) deep, the outlet zone should be 3.6 m (12 feet) wide, 1.2 m (4 feet) deep, and at least 1.8 m (6 feet) long.

The outlet weir length, on the other hand, is governed by the weir discharge rate (volume of water discharged per unit length of weir per unit time) and should be extended as much as possible. And moreover, the weir discharge rates should be somewhere in between 400 and 600 m^3/d per meter length of the outlet weir (or 22 to 33 gpm/ft).

When we talk about the settling basin length, we typically mean the total length of the settling basin, which covers the length (area) required for the inlet zone, the actual settling zone, and the length (area) for the outlet zone. This total area requirement is often considered as the actual settling area; therefore, the settling basin will not perform effectively as intended. We should be careful and do not compromise on the size of the settling basin.

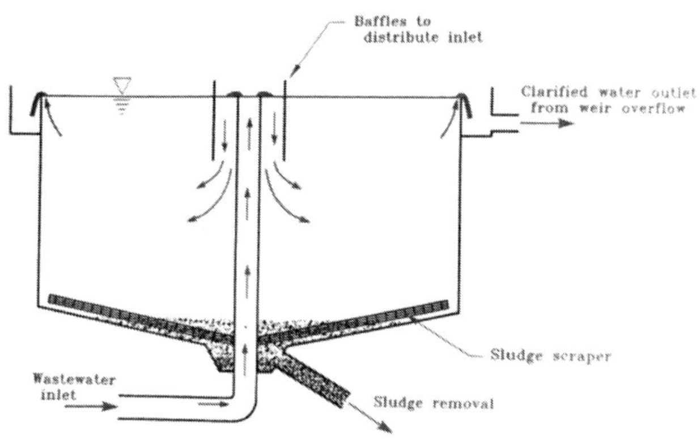

FIGURE 5.52 Circular settling basin.

Example 5.32: Designing a Settling Basin

Design a settling basin to remove 90% of the TSS particles greater than 0.5 mm in diameter with a design flow of 34 m^3/h (150 gpm).

Solution

Settling Basin Design Rules:

- Basin floor area of 0.68 $L/s.m^2$ (2.4 $m^3/m^2.h$ or 1 ft^2 per gpm of flow)
- 4.2 to 6.8 L/s per m width of weir for outflow (20 to 33 gpm per foot)
- Submerge inlet weir 15% of basin water depth
- Use 25 cm (10 in.) wide weirs and use rounded edges
- Maximize settling chamber length as much as possible

Using a design settling velocity (V_s) of 0.05 cm/s for an off-line settling basin (based on IDEQ, 1988) and converting to flow per unit area:

$$\left(0.05\frac{cm}{s}\right)\left(3{,}600\frac{s}{h}\right)\left(1\frac{m}{100\ cm}\right)\left(\frac{m^3}{m^2.h}\right)=1.8\frac{m^3}{m^2.h}$$

Now, we can determine the settling zone area using Equation (5.43),

$$A_{sz}=Q/V_o=\frac{34\frac{m^3}{h}}{1.8\frac{m^3}{m^2.h}}=18.9\,m^2\left(\text{or }210\,ft^2\right)$$

5.12.3.2.4 Settling Plates or Tubes in Settling Chambers

Normally the settling basins need a large floor area, and the required floor area can sometimes be expensive. However, the settling basin floor area can be decreased by adding obstructions, such as plastic settling plates or tubes, inside the settling basin in order to increase settling rates. The plate or tube settlers, a.k.a the settling chambers, is shown in Figure 5.53 where the incoming flow is brought into the

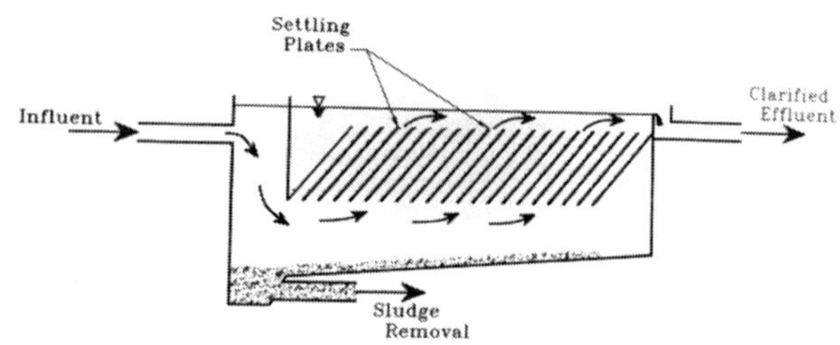

FIGURE 5.53 Settling chamber with settling plates to enhance solids removal efficiency.

settling basin under the settling chamber and forced to up-flow to exit the settlers while settling the solids settle within.

Gregory and Zabel (1990) reports typical separation distance of 5 cm (2 in.) between the inclined surfaces with a total inclined length of 0.9 to 1.8 m (3 to 6 ft). Usually the inclination angle is between 45° and 60° above horizontal, which facilitates the best gravity self-cleaning; however, settling plates or tubes do not properly self-clean, hence periodic cleaning is necessary by other means to prevent biofouling (Gregory and Zabel, 1990; Tchobanoglous and Burton, 1991). The water velocity through the plates or tubes should, on the other hand, be maintained in a laminar flow region (i.e., Re < 2,300) for preventing turbulence and re-suspending settled particles, which is quite important in settling basins. In addition, a number of different hydraulic loading rates have been suggested by different researchers; for instance, 1.5 m^2/m^2 per hour by McLaughlin (1981), 6.7 m^2/m^2 per hour by Parker (1981), and 7.4 m^2/m^2 per hour by Libey (1993). Quite an impressive efficiency of 80% for the removal of particles larger than 70 μm and that of 55% for the removal of all particles larger than 1.5 μm using a multi-tube settler are reported (Easter, 1992).

5.12.3.3 Microscreen Filters

We can comfortably say that all recirculating systems at the outlet water of culture tanks commonly use some type of commercially available microscreen filter such as drum filter, disc filter, or inclined belt filter equipped with microscreen filters of 40 to 100 microns mesh cloth for filtering solids. They are quite popular because they require minimal labor and floor space compared to settling basins, for instance. Microscreen filters clean the recirculating system water as suspended particles are removed, and stabilize the biofiltration processes as they reduce the organic solid load of the biofilter and improve conditions for nitrification due to the fact that the biofilter does not clog. Microscreen filter performance depends on screen size and the influent TSS concentration. Usually, the screen openings used in recirculating systems are some where between 40 to 100 μm, which provides a TSS removal between 30 and 80%. Vinci et al. (2001) provided the operational advantages and drawbacks of 100-micron mesh, 10 m^3/min capacity drum filters, disc filters and belt filters.

As schematically shown in Figure 5.54, microscreen drum filters, for instance, function on the principles including i) the outlet water of culture tanks enters the drum filter, ii) then the water is filtered through filtering microscreens. The driving force for the filtration is the water level difference between inside and outside the drum, iii) solids are captured on the microscreens and lifted to the backwash area by the drum's rotation, iv) rinse nozzles spray water from the outside of the microscreens. And the removed solids are simply washed out of the microscreens into the sludge tray underneath, and finally, v) the sludge flows together with water by gravity out of the drum filter and is transferred to external wastewater treatment.

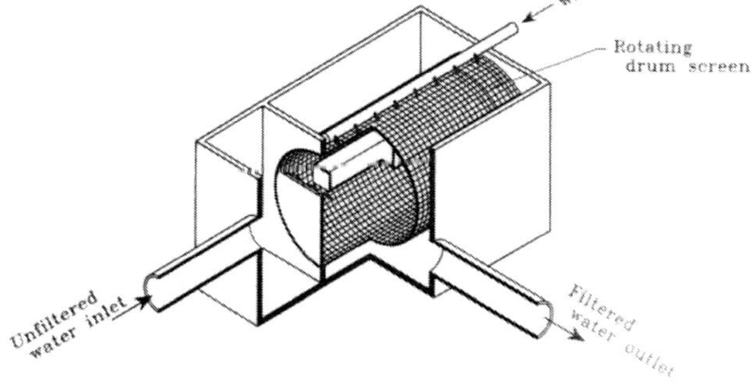

FIGURE 5.54 Microscreen drum filter.

5.12.3.4 Swirl Separators

Swirl separators, typically one- to two-meter diameter vessels, simply use water circulation to densify the solids in the center bottom of a tank and enhance separation (Figure 5.55). The rotational flow of water injected tangentially into a swirl separator, as shown on Figure 5.55, imposes a centrifugal force on the particles, causing the heavier particulate material to move towards the outer side of the tank. The particles are also affected by gravity, the main driver, causing them to fall through the water and move towards the bottom center drain. And it is right here at the bottom center drain, a small portion of the flow (known as underflow) is removed. Swirl separators are typically used to treat wastewater containing high specific gravity particles that would also have rapid settling velocities. However, it should be kept in mind that waste fish feed and fecal matter, as mentioned earlier, have specific gravities that are slightly greater than water, and hence relatively slow settling velocities.

Obviously, hydraulic loading rates of swirl separators depends on the size and specific gravity of the waste particles. Davidson and Summerfelt (2005) reported a loading rate of approximately 10 m^3/m^2 per hour (4 gpm per ft^3),

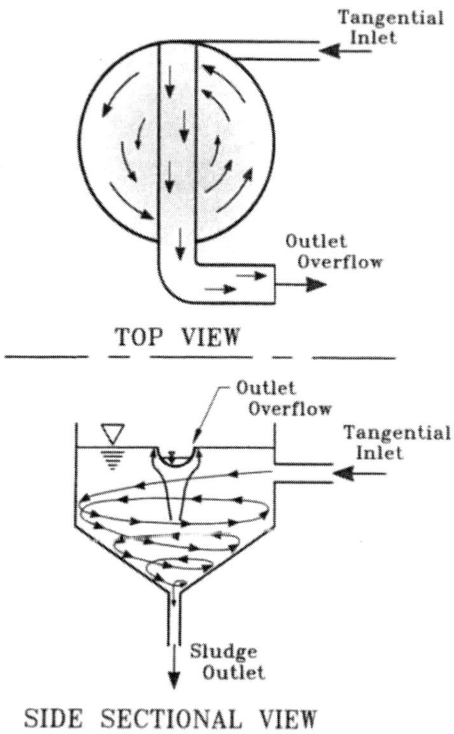

FIGURE 5.55 Swirl seperator.

which would be good for an effective solids removal. Note that m³/m² per hour is equivalent to m/h. It should also be noted that swirl separators are sometimes operated as a noncontinuous underflow. It means that the settled solids are kept in the base of the separator until they are manually flushed, several times daily. Just like settling basins and microscreen drum filters, while they are effective in TSS removal, swirl separators are not effective in fine solids (< 50 μm) removal either. Davidson and Summerfelt (2005) reported a solids removal efficiency of approximately 40%. Overall speaking, swirl separators are designed with tangential inlet flows where the inlet is typically submerged one third of the tank diameter, for a hydraulic loading of 2.0 to 3.4 L/s per m² (3 to 5 gpm per ft²) and an underflow of 5 to 15% of total flow.

5.12.3.5 Granular Media Filters

The process of granular media filtration involves passage of water through a bed of granular material (media), through which the deposition of solids onto this granular material takes place. These filters are quite often used together with some other treatment systems in order to increase the level of treatment. And there are a number of granular-medium filters, all of which use basically the same mechanisms of removal, and require a backwash of the filter to recondition the particulate removal capacity (Crites and Tchobanoglous, 1998).

Sand filtration is commonly used and employs the filtering properties of sand and produces clean water that is perfectly safe for recirculating systems. Furthermore, since sand filtration requires no replaceable parts but only periodic maintenance, it is considered as one of the best water treatment options for recirculating aquatic systems. The clean water produced by the filter is again totally safe for fish, and there are three types of such filters: rapid sand filters, slow sand filters, and up-flow sand filters. The most commonly used sand filter is the rapid sand filters (Figure 5.56), which can be cleaned (back flushed) by passing water upward through the filter after blowing compressed air up through the bottom for breaking up the compact media filter, and reused.

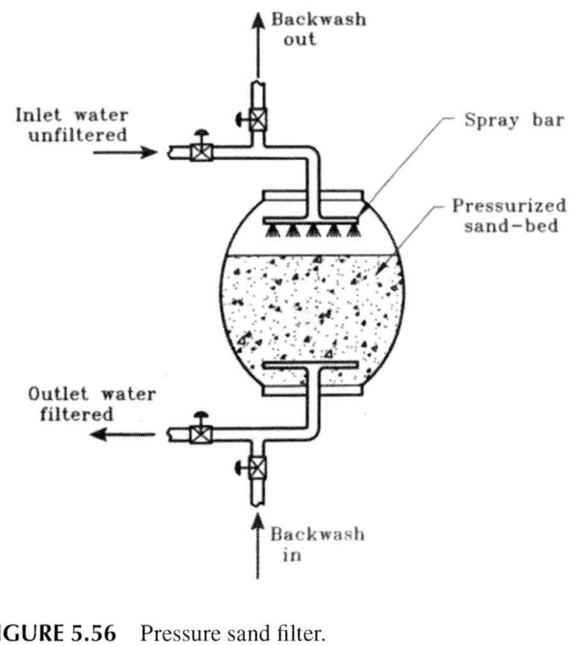

FIGURE 5.56 Pressure sand filter.

In addition, recirculating aquatic systems effectively use floating plastic media, a.k.a. the "bead" filters. These filtering systems as well require backwashing; however, the water loss in these systems is quite small compared to the traditional down-flow sand beds due to back wash for cleaning the clogged sand.

5.12.3.6 Porous Media Filters

Porous media filters are typically used to treat low flow rates due to their susceptibility to clogging even at low TSS concentrations. As a consequence, they are not used as the principal solids removal device in controlled environment recirculating systems due to the reconditioning or replacement requirement of the media filters as they get plugged and restrict the flow. Having that said, however, these filters are quite effective in removing fine solids. It means that porous media filters can be used to treat incoming water to a recirculating system as an additional particulate removal step in controlled environment recirculating system applications. The two common porous media filter systems are the cartridge filters and diatomaceous earth filters, both of which can remove very small solid particles (< 1 μm) due to their extremely fine pore sizes.

5.12.3.7 Foam Fractionators

Generally speaking, the traditional filtration systems cannot remove the fine suspended solids and dissolved organic solids within recirculation systems. Instead, foam fractionation systems are used to remove these solids; however, it must be emphasized that the removal of wastes by foam fractionation is not a substitute for effective primary TSS removal. Foaming results from the presence of surfactants, and surfactants are protein, which are polar with hydrophilic and hydrophobic ends. When the hydrophilic end is negatively charged and is sticking out into the water column, it will obviously attract positively charged materials in the water column; and these positively charged materials in turn will attract negatively charged particles. Besides, charged surfaces of an air bubble attract both negative and positive fine solids and bacteria. And consisting of high protein (approx. 70% on dry matter basis), a dead fish, when ground up through a pump and left in the tank, for instance, provides a huge amount of surfactant. And this will in turn result in noticeable foaming.

In the foam fractionation process, air is introduced into a column of water for generating foam at the surface of the column, and the resulting rising bubbles remove fine particulate and dissolved organic solids by adsorption. Eventually, the fine particulate and dissolved solids are trapped within the foam at the top of the water column, and then they are collected and removed as foam condensate, which consists mostly of dissolved organics and particles smaller than 30 μm. Foam fractionation method is a particularly suitable process in seawater systems compared to freshwater systems because the increased salinity levels enhance the efficiency of the system. That is, as the salinity concentration increases the surface tension increases, which results in permitting smaller air bubbles and therefore a higher filter area. Hence, we can easily say that foam fractionation is an integral part of recirculating seawater systems. Moreover, the pH has a considerable effect on foam creation. Higher the water pH is, higher the foam creation and the corresponding protein removal is. For instance, foam fractionators remove twice the amount of protein concentrations from water with a pH of 8.3 in comparison to water having a pH of 5.3 (Timmons and Ebeling, 2010).

Example 5.33: Foam Generation and Removal

Let's assume that in a tank of 20,000 fish, we will lose 5% of the animals over a 200-day growth period. It means that we would be losing an average of five fish each day. Use a carrying capacity of 120 kg/m², five fish at 1 kg each (~20% dry matter). Protein content of fish is approximately 70% on dry matter basis. Determine the number of days for a foam fractionator to remove the TSS.

Solution

Using a carrying capacity of 120 kg/m², 5 fish at 1 kg each (20% dry matter) would be the equivalent of:

$$20{,}000\,fish\left(1\frac{kg}{fish}\right)\left(\frac{m^3}{120\,kg}\right)=166.7\,m^3$$

$$=166{,}700\,L\approx 167{,}000\,L$$

$$5\,fish\left(\frac{1{,}000\,g_{flesh}}{fish}\right)\left(\frac{0.2\,g_{DM}}{1.0\,g_{flesh}}\right)\left(\frac{0.7\,g_{protein}}{1.0\,g_{DM}}\right)=700\,g_{protein}$$

Then, the concentration of 700 g of pure protein (a foaming agent) in the 167,000 L of water can be calculated as follows:

$$\frac{700\,g_{protein}\left(1{,}000\dfrac{mg}{g}\right)}{167{,}000\,L}=4.2\frac{mg}{L}$$

So, this much protein, if not removed, will create a lot of foaming. We now know that the amount of protein generated from dead fish per day is 700 grams. Chen et al. (1994) determined that only 1 mg of protein existed in 331 mg of VS collected as foam condensate. Then the 700 g protein (which effectively could become all surfactant material) would provide foaming potential for the following quantity of TSS or VS:

$$\theta_{TSS}=\left(700\,g_{protein}\right)\left(331\frac{g_{VS}}{g_{protein}}\right)=231{,}700\,g_{VS}$$

And as a rule of thumb, the generated TSS is:

$$\left(0.25\frac{g_{TSS}}{g_{feed}}\right)\left(\frac{0.01\,g_{feed}}{g_{fish}\cdot day}\right)\left(\frac{1{,}000\,g_{fish}}{fish}\right)$$

$$\left(20{,}000\,fish\right)=50{,}000\frac{g_{TSS}}{day}$$

And assuming that 100% of the VS is TSS type organic material, then the calculated $50{,}000\,g_{TSS}\,/\,day$ would require the following number of days for a foam fractionator to remove these TSS:

$$\frac{231{,}700\,g_{VS}}{50{,}000\dfrac{g_{TSS}}{day}}=4.6\,days$$

Thus, our calculations show that it would take for a foam fractionator approximately 5 days before the tank returned to normal foam levels. This example shows us the urgency of dead fish removal and other organic solids as quickly as possible.

5.13 BIOFILTRATION IN CONTROLLED ENVIRONMENT AQUATIC SYSTEMS

5.13.1 Introduction

In recirculating aquatic environments, the organic matter is not fully removed in the mechanical filter; as a result, dissolved compounds such as phosphate and nitrogen pass through the mechanical filter. Phosphate has no toxic effect to fish; however, decomposition of nitrogenous compounds is important in intensive recirculating systems because of the toxicity of ammonia, nitrite, and to some extent, nitrate, all of which are highly soluble in water. Different nitrogen and ammonia forms as well as their dynamics and effects on fish have already been discussed in depth in Chapter 4. We will recall that the nitrogen in the form of free ammonia (NH_3) is quite toxic to fish; therefore, is has to be transformed into non-toxic form in a biofilter. The removal process of ammonia through a biofilter is called nitrification, a two-stage process where microbial conversion of ammonia to nitrite and finally to nitrate takes place. The reverse of nitrification process is called denitrification, which is an anaerobic process where nitrate is converted to nitrogen gas with an intermediary step of nitrite. As stocking densities increase and water exchange rates are reduced in intensive recirculating systems, higher nitrate levels result in the culture systems; therefore, the denitrification process is becoming more important, particularly in marine systems.

Biological filtration is used as an effective means of controlling ammonia, instead of using water flushing to control ammonia levels. There are mainly two distinct groups of autotrophic bacteria that collectively perform nitrification in biofilters, and they derive their energy from inorganic compounds as opposed to heterotrophic bacteria that derive their energy from organic compounds. The nitrifying bacteria coexist with heterotrophic microorganisms, including heterotrophic bacteria, which grow significantly faster than nitrifying bacteria in biofilters, especially when dissolved and particulate organic matter concentrations are high. Biofilters are usually constructed with a very high surface area using plastic media, on which the bacteria grow as a thin film. It is also vitally important for biofilters to have a large free space for water to pass through, which is important for the backwash procedure as well.

Chen et al. (2006) classified numerous biological, chemical, and physical factors that influence the rate of nitrification into three main categories. The first of these categories are the factors that directly affect the microorganisms, such as pH, temperature, alkalinity, and salinity. The second of those are the ones that influence the supply of nutrients for the microorganisms, such as ammonia concentration, dissolved oxygen, and the mixing or turbulence that controls the mass transport of nutrients in the bioreactor. And finally, the third of the categories are the the ones that have effects on both the growth and nutrient supply, such as competition for nutrients and space with heterotrophic bacteria.

Jones and Morita (1985) along with Chen et al. (2006) reported that there is quite a wide range of optimum temperatures for nitrification, suggesting that nitrifying bacteria are able to adapt to a wide range of environmental temperature, as long as they are acclimated slowly. Moreover, in the case of no oxygen limitation, Zhu and Chen (2002) showed that temperatures from 14 to 27°C had no significant effect on the nitrification rate of a fixed film bioreactor. In reality, however, the water temperature at which the biofilter operates is normally determined by the requirements of the species reared, not by the needs of the biofilter bacteria; hence we do not have much control over the water temperatures for increasing nitrification efficiency of a biofilter.

Regulating pH with respect to biofilter efficiency is, however, important as low pH reduces the efficiency of the biofilter. Increasing pH, on the other hand, results in an increasing amount of an ionized ammonia (NH_3^-), which would enhance the ammonia toxicity. The main factors impacting the pH in recirculation systems are: i) production of CO_2 from the fish and from the biological activity in the biofilter, ii) acid produced through the nitrification process, and iii) as it is discussed below, consumption of alkalinity in the nitrification process. Haug and McCarty (1972) and Chen et al. (2006) suggested that the optimum range of pH for nitrification can range from 7.0 to 9.0.

As discussed in depth in both Chapters 2 and 4, alkalinity is a measure of the buffering capacity of an aquatic system. It was determined that for every gram of ammonium-nitrogen reduced to nitrate-nitrogen, 7.05 grams of alkalinity is consumed. So, nitrification is an acid-forming process, and if the biofilter system's water is poorly buffered, the system pH will decline, and in return, affect the biofilter performance. The loss of alkalinity is easily made up by the addition of sodium bicarbonate, referred to commonly as baking soda ($NaHCO_3$), or other bicarbonate supplement. Loyless and Malone (1997) provided a relevant rule-of-thumb that 0.25 kg of baking soda is required per 1 kg of feed (0.25 lbs of baking soda per pound of feed).

There is not much information on the impact of salinity on nitrification; however, providing enough time, nitrifying bacteria can acclimate to almost any salinity level. Chen et al. (2006) reported that pilot scale long term experiments performed with both freshwater and marine water recirculation systems suggest that the average removal rate is reduced by approximately 37% in saltwater compared to freshwater. Rusten et al. (2006), on the other hand, reported that commercial fish farms operating at a salinity of 21–24 ppt had a nitrification rate, which was approximately 60% of what would be expected in a freshwater system for moving bed bioreactors.

Moreover, it is important to know that the ammonia concentration directly impacts the nitrification rate. If ammonia concentration increases, then the biofilter performance as well increases linearly and proportionately over a range of concentrations. Ebeling and Wheaton (2006) graphically showed that this relationship exists from very low ammonia concentrations to between 2 and 3 mg/L TAN, and proportional increase decreases at some point, and then eventually levels off at a constant removal rate.

In addition, dissolved oxygen can become the nitrification rate-limiting factor in certain biofilters due to the low dissolved oxygen levels in the influent and the competing dissolved oxygen demands of the heterotrophic bacteria. It should be noted that 4.57 g of oxygen is required for every gram of ammonia-nitrogen oxidized to nitrate-nitrogen. Wheaton (1985) and Malone et al. (1998), on the other hand, stated that biofilter effluent levels of at least 2 mg/L of oxygen are probably enough to maintain maximum nitrification rates.

Recirculating aquaculture systems are extremely rich in both dissolved and particulate organic matter, which provides the required substrate for heterotrophic bacteria. It should also be noted that heterotrophic and nitrifying bacteria compete with each other for growing space, and heterotrophic bacteria have a five times bigger maximum growth rate than that of autotrophic nitrifying bacteria (Grady and Lim, 1980). Chen et al. (2006), on the other hand, observed an exponential decrease in nitrification rate with the increase in the COD/N ratio in a laboratory study using a series of biofilters, which means that organic solid must be removed immediately from the recirculating aquatic system.

5.13.2 Biofiltration System and Design

In a recirculation aquatic system, it is required to continuously treat the water in order to remove the waste products excreted by the fish and from the unused feed, and to add oxygen to keep the fish alive. The basic principle of recirculating aquatic system is that water first flows from the culture tank to a mechanical filter, and then to a biofilter before it is aerated, stripped of carbon dioxide, and returned to the culture tank. Depending on the requirements of the facility, a number of other treatment devices, such as oxygenation for adding higher oxygen concentrations, UV light or ozone disinfection, automatic pH regulation system, and heat exchanging system can also be added.

Not capturing solids, an ideal biofilter would obviously maximize media surface area and remove all of the inlet ammonia concentration, generating negligible nitrite, maximizing oxygen transfer, requiring relatively small footprint, having minimal head loss, using inexpensive media, and requiring very little maintenance. As one would expect, however, there is no single biofilter, which meets all these ideal expectations. Currently, there are numerous types of biofilters that are commonly used in commercial intensive recirculating systems, such as trickling biofilters, rotating biological contactors, floating bead biofilters, moving bed bioreactors, fluidized-bed biofilters, and many others. And we will be able to cover only a few of them here in this section.

Before we move into the specific biofilter types or designs, it would be helpful to have a basic set of biofilter definitions and terminology. Usually, the following terms are employed in biofilter design and characterization:

- ***Void space*** is the volume not occupied by biofilter media, and void ratio is that volume divided by the total volume of the biofilter. High void ratios reduce clogging by having large open spaces that allow solids to pass easily.
- ***Cross-sectional area*** refers to the area of the filter bed facing the direction of the water flow. Filter top

area is usually one of the last parameters selected in the filter design, to yield a desired hydraulic loading rate.

- ***Hydraulic loading rate*** is the volume of water pumped through the biofilter per unit of cross-sectional area of the filter per unit of time, typically expressed as gpm/ft^2 or m^3/m^2.day. Usually, there is a minimum and a maximum hydraulic loading rate for biofilters.
- ***Specific surface area*** is the surface area of the biofilter media per unit volume. The higher the specific surface area of a biofilter media, the more bacteria can grow on a unit volume, and the greater the total ammonia removal per unit volume of biofilter. The media size, void ratio, and specific surface area are all interrelated; the smaller the size, the smaller the void ratio, and the larger the specific surface area. Typical terms used to describe biofilter performance are based on either the volume of the biofilter media or its surface area. Although nitrification reaction rates are a surface area phenomenon, for fluidized sand beds and other granular media, because of the difficulty in measuring the actual media surface area, the rate is easier expressed per unit volume rather than per unit surface area.
- ***Volumetric TAN conversion rate*** is the grams of TAN per volume per day converted into nitrate (g TAN/ft^3 per day or kg TAN/m^3 per day).
- ***Areal TAN conversion rate*** is the grams of TAN per unit surface area per day converted into nitrate (g TAN/ft^2 per day or g TAN/m^2 per day).

Designing a biofilter for a recirculating system, on the other hand, is not a trivial job, in which the design engineer needs to make a number of choices and corresponding compromises in order to find an optimum solution for the system in hand. System volume, culture density, feed rates, temperature, and many other parameters constrain the biofilter design. All of the biofilters discussed below are designed to perform essentially the same function, i.e., oxidizing ammonia and nitrite to nitrate. Therefore, the biofilter must be designed to fully oxidize the produced ammonia-nitrogen, with some additional safety margin.

Generally speaking, we can say that one of the difficult design parameters to estimate is the oxygen demand of both the fish species being grown and the biofilter. Using stoichiometry, the oxygen demand could be as low as 0.37 kg of oxygen per kg of feed fed (0.25 kg for fish metabolism and 0.12 kg for nitrification). For design purposes, however, as a rule-of-thumb, using 1.0 kg oxygen per kg of feed would be a good starting point.

Each biofilter described in this section has advantages and disadvantages. For instance, one of the main advantages of both the trickling biofilter and the rotating biological contactor (RBC) is that they enrich the water flow with oxygen during their normal operation; and in addition, they do provide carbon dioxide stripping as well. The bead biofilters and fluidized-bed biofilters (FBF), on the other hand, are oxygen consumers and simply rely heavily on the the influent flow oxygen content to maintain aerobic conditions for the biofilm. Whenever the influent flow is poor in dissolved oxygen, then anaerobic conditions prevail within the biofilters. Furthermore, both the trickling biofilters and the RBC have low specific surface area media. At this point, it must be remembered that the capital cost of the biofilter is simply proportional to its total surface area, which means that the outcome is a physically large biofilter requiring larger floor space, and more costly biofilter. Besides, the bead biofilters and especially the FBF employ media with high specific surface areas, which means reduced cost and space requirements when compared to that in a trickling biofilter, for instance. Furthermore, if suspended solids are not properly controlled, then trickling biofilters and RBCs are already ready to biofoul. Organic matter consuming heterotrophic bacteria grow much faster than the autotrophic bacteria that perform nitrification. As a consequence of their extremely high growth rate and the relevant oxygen demand, heterotrophic bacteria simply suffocate the autotrophic bacteria starved for oxygen buried deeper in the biofilms, eventually resulting in death, and hence the loss of biofilter's nitrification capacity. Before moving on to the next section on design, this is telling us that we must actively plan for effective solids removal prior to the high ammonia water being moved into the biofilter. A simple recipe for disaster in a recirculating system is to have poor solids removal, which is the main cause for biofilter failure.

5.13.2.1 Trickling Tower

Trickling filters are one of the most commonly used biofiltration systems in recirculating aquatic systems. In trickling biofilters, water flows over suspended media and into a sump at the bottom of the biofilter. The advantage and disadvantages of trickling biofilters are already discussed above.

Before we move deeper into the biofilter design work, we must note that as a rule-of-thumb, using 1.0 kg of oxygen per kg of feed fed is recommended. In the following design example, however, since a trickling biofilter provides adequate oxygen levels for the autotrophic nitrifying bacteria on the trickling media, we will be using 0.50 kg of oxygen per kg of feed fed. By doing this, we increase the fish metabolism oxygen demand of 0.25 kg O_2/kg feed will be increased to 0.50 kg O_2/kg feed to allow for oxygen demand by heterotrophic bacteria in the water column).

In a trickling biofilter design, the following steps are used:

Step 1: Calculate the water quality loads on the system (oxygen, CO_2, TAN, and TSS).
Step 2: Calculate the controlling water flow requirement (Q_{tank}).
Step 3: Calculate TAN production by fish (P_{TAN}).
Step 4: Calculate the surface area of media (A_{media}) required to remove P_{TAN} from the areal TAN removal rate (ATR).
Step 5: Calculate the volume of media required (V_{media}) based upon the required area and the specific surface area (SSA) associated with the media being used.
Step 6: Calculate the biofilter cross-sectional area ($A_{biofilter}$).

Step 7: And finally, calculate the biofilter depth ($D_{biofilter}$) from the biofilter cross-sectional area ($A_{biofilter}$) and volume (V_{media}).

Example 5.34: Trickling Biofilter Design

Using a trickling tower biofilter whose media has a *SSA* of 200 m²/m³ (61 ft²/ft³), design a trickling biofilter for the recirculating system in Example 5.31.

Solution

The water flow requirements are based choosing the highest of the calculated flow rates for each of the control variables (oxygen, CO_2, TAN, TSS, and tank exchange). The biofilter sizing is based on the daily TAN production rate (0.309 kg TAN/day), which is directly proportional to the feeding rate of 9.6 kg per day.

Both Step 1 and 2 were demonstrated in Example 5.31 to calculate the controlling flow rate for the maintaining DO, CO_2, TAN, and TSS. In Example 5.31, using the mentioned design parameters for efficiency and tank water quality target values, we calculated a set of required flow rates. In this particular case, the calculated highest flow rate is 5.58 L/s (88.4 gpm). However, keeping in mind that we never compromise in the tank flow rates, and for convenience as well, we rounded this number to 6.30 L/s (100 gpm), which means the system is operated at the highest flow rate of 6.30 L/s (100 gpm).

Step 3: Calculate TAN production by fish (P_{TAN}) assuming the feed has a protein content of 35% at a design feeding load of 9.6 kg of feed per day.

$$P_{TAN} = a_{TAN} \times R_{feed}$$

$$P_{TAN} = (0.092 \times 0.35)\frac{kg_{TAN}}{kg_{feed}} \times 9.6\frac{kg_{feed}}{day}$$

$$P_{TAN} = 0.309\frac{kg_{TAN}}{day}$$

Step 4: Calculate the surface area (A_{media}) required to remove P_{TAN} from the Areal TAN removal rate (ATR). For trickling towers, we can use an estimated Areal TAN removal rate of 0.50 g TAN/m².day.

$$A_{media} = \frac{P_{TAN}}{ATR} = \frac{0.309\frac{kg_{TAN}}{day} \times \frac{1{,}000\,g}{kg}}{\frac{0.50 g_{TAN}}{m^2.day}}$$

$$= 618\,m^2 \left(\approx 6{,}650\,ft^2\right)$$

Please note, for coldwater applications (12–15 °C), an areal TAN removal rate of about 0.15–0.25 g TAN/m².day should be used when influent TAN concentrations flowing into a trickling column are less than 1–2 mg/L. Also, for saltwater applications at 24°C, the areal TAN removal rate drops to 0.1–0.2 g TAN/m².day when TAN concentrations entering the trickling filter are 1–2 mg/L.

Step 5: Calculate volume of media based on the specific surface area (SSA) associated with the media being used; here, just as an example, we will use a media having a specific surface are of 200 m²/m³ (61 ft²/ft³).

$$V_{media} = \frac{A_{media}}{SSA} = \frac{618 m^2}{200\frac{m^2}{m^3}} = 3.09 m^3 \left(\approx 109 ft^3\right)$$

Step 6: Calculate the biofilter cross-sectional area from required flow for the fish oxygen demand (Q_{tank} = 22.7 m³/h or 6.30 L/s) and the design hydraulic loading rate. Even though higher hydraulic loading rates can be used, a minimum *HLR* value of 255 m³/m².day (4.4 gpm/ft²), creating effective wetting of all surfaces, is required to prevent clogging of the media bed.

$$A_{biofilter} = \frac{Q_{tank}}{HLR} = \frac{22.7\frac{m^3}{h}\left(\frac{24\,h}{day}\right)}{255\frac{m^3}{m^2 day}} = 2.14\,m^2 \left(\approx 23 ft^2\right)$$

Then, the diameter of a single biofilter tank, $D_{biofilter}$ with this cross sectional area is:

$$D_{biofilter} = \sqrt{\left(\frac{4 \times A_{bed}}{\pi}\right)} = \sqrt{\left(\frac{4 \times 2.14\,m^2}{3.14}\right)} = 1.65\,m\left(\approx 5.5\,ft\right)$$

Step 7: Calculate the biofilter depth ($Depth_{media}$) from the biofilter cross-sectional area (A_{media}) and volume (V_{media}).

$$Depth_{media} = \frac{V_{media}}{A_{media}} = \frac{3.09\,m^3}{2.14\,m^2} = 1.44\,m\left(\approx 4.8\,ft\right)$$

As mentioned earlier, in addition to nitrification, trickling filter is used for the removal of carbon dioxide as well. A tricking filter can also be used for evaporative cooling in hot and dry climates. Moreover, Eding et al. (2006) reported that in order to remove suspended solids as much as possible, rotating drum filters with a mesh size of 30 to 60 μm are often used before the trickling tower.

5.13.2.2 Rotating Biological Contactor

Just like in trickling filter, in a rotating biological contactor (RBC), oxygen is obtained through the air and water, but is enhanced by a mechanical rotation of the media, of which about 40% is submerged at any point of time. Here as well, excessive heterotrophic bacterial growth can become an issue and result in clogging, which can lead to a decrease in nitrification and biofilter efficiency. Other advantage and disadvantages of RBCs are already discussed earlier. The design of a rotating biological contactor as well follows the same steps as that of the trickling tower with the nitrification rate based on the active surface area of the media. Using the initial design parameters of Example 5.34, the design process is the same until Step 4.

Employing several commercial scale RBC systems, van Gorder and Jug-Dujakovic (2005) determined an areal TAN removal rate of 1.2 g TAN/m².day, and each of these units was able to remove 1.13 kg TAN/day. This suggests that even one of these units would be more than enough to handle the ammonia load for this design example of 0.309 kg TAN/day; however, we would definitely check other commercially available RBCs having a closer TAN removal rates to 0.309 kg TAN/day and adopt that one instead. It should also be kept in mind that, in addition to nitrification, significant amount of carbon dioxide is degassed by the RBCs.

5.13.2.3 Floating Bead Biofilter

Floating bead filters (FBF) are similar to submerged filters; however, a floating polyethylene bead media is used instead. While the commercial applications usually use 3.0 mm-diameter beads, the bead diameters range from approximately 1.0 to 7.0 mm. FBFs exhibit advantages because of their ability to have high solids capture and low head loss during high organic loading rates. Other advantage and disadvantages of FBFs are already discussed earlier.

Assuming that the biofilter is used both as a bioclarifier to remove organic solids and as a biofilter for nitrification, Malone and Beecher (2000) reported that the primary method for sizing the bead volume of floating bead filters is based on an organic loading rate. The sizing criterion, v_f, for FBFs is based on the feed load as all organics in a recirculating system is from the feed. Then, the volume of bead media needed for an application, V_{media} (m^3), can be determined from the design load, R_{feed} (kg feed/day), and the floating bead filter sizing criterion, v_f (kg feed/m^3 of media.day) as follows:

$$V_{media} = R_{feed} / v_f \qquad (5.46)$$

Malone et al. (1993) reported another approach for sizing FBFs based on the areal nitrification capacity, which is based on the FBFs' areal conversion rates (ATR) of approximately 300 mg TAN/m^2.day, in systems with TAN and nitrite levels between 0.5 and 1.0 mg N/L (Malone and Beecher, 2000). Observed *VTR* values in these systems tend to increase with the increasing TAN tolerances, and hence, can be used to estimate the size of the FBFs:

$$V_{media} = (1.0 - I_s) \times \frac{P_{TAN}}{VTR} \qquad (5.47)$$

The *in situ* nitrification fraction, I_s, in the above equation is usually estimated at 30%, even though values higher than 50% have also been observed. Timmons and Ebeling (2010) in their designs ignored this contribution in favor of over designing the system to insure long term stability.

Example 5.35: Floating Bead Biofilter Design

Sizing the floating bead filter is based on the daily TAN production rate, which is directly dependent upon the feeding rate, design a floating bead biofilter for the recirculating system in Example 5.31.

Solution

The first three steps of the design process remain the same as in Example 5.34. Then,

Step 4: Calculate the volume (V_{media}) required to remove P_{TAN} from the Volumetric TAN removal rate (VTR). Assuming a growout application, and a VTR at the maximum TAN value of about 1 mg/L, 350 g N/m^3.day.

$$V_{media} = (1.0 - 0.0) \times \frac{0.309 \frac{kg_{TAN}}{day} \times \frac{1{,}000\,g}{kg}}{\frac{350\,g_{TAN}}{m^3\,day}}$$

$$= 0.88\,m^3\,(31.1\,ft^3)$$

In this example, we assumed an *in situ* nitrification fraction of 0.0, a conservative approach that produces the largest media requirement. For the operation in hand, we will then choose two commercially available 0.57 m^3 (20 ft^3) bead filters, which would be sufficient for the design, incorporating additional safety factors by having two separate biofilters.

5.13.2.4 Fluidized-Bed Sand Biofilter

A fluidized-bed reactor (FBR) is an up-flow biofilter that continuously flows, maintaining the media in suspension. In a FBR, nitrification rates are increased considerably, but may require high flow velocities and an active backwashing in order to maintain biofilter efficiency. Using low-density media can help mediate this problem, and hence minimize energy requirements. Other advantage and disadvantages of FBRs are already discussed earlier.

All biological filters are designed to perform the same function of oxidizing ammonia and nitrite to the fully oxidized form of nitrate. In a fluidized-bed sand biofilter design, the following steps are used:

Step 1: Determine the TAN load.
Step 2: Determine the sand volume required to match TAN load.
Step 3: Select the design depth of sand bed.
Step 4: Determine the cross-sectional area of the sand bed.
Step 5: Select the sand size in relation to flow rate available or desired or determine the required flow rate to fluidize the bed of the sand available.
Step 6: Compare to the other required flow rates for maintaining water quality parameters in the fish tank; make adjustments in the sand bed cross section if one of the Qs is larger than Q_{sand}.
Step 7: Design the water delivery system.

Example 5.36: Fluidized-Bed Sand Biofilter Design

Design a fluidized-bed sand biofilter for a warmwater system using 20/40 sized sand for a feeding rate of 9.6 kg feed per day.

Solution

Step 1: Determine the TAN load

$$TAN\,Load = 9.6 \frac{kg_{feed}}{day} \times (0.092 \times 0.35) \frac{kg_{TAN}}{kg_{feed}}$$

$$= 0.309 \frac{kg_{TAN}}{day}$$

Step 2: Determine the sand volume required to match TAN load

$$V_{media} = \frac{0.309 \frac{kg_{TAN}}{day}}{\frac{1 kg_{TAN}}{day.m^3_{sand}}} = 0.31 m^3_{sand}$$

Step 3: Select the design depth of sand bed

$$h_{sand} = 2m$$

Please note that this depth is the unexpanded sand bed height from floor level, considering the ceiling height and elevation required to permit gravity flow back to culture tank through aeration or oxygenation column or other equipment devices.

Step 4: Cross sectional area of bed

$$A_{bed} = \frac{V_{media}}{h_{sand}} = \frac{0.31\, m^3_{sand}}{2m} = 0.16 m^2$$

Step 5: Fluidization velocity

For a 20/40 sand, the effective size (the opening size, which will pass only the smallest 10%, by weight, of the granular sample. It provides an estimate of the smallest sand in the sample and is the size used to estimate the maximum expansion at a given velocity), D_{10} (40 mesh) = 0.42 mm. If the sand has a uniformity coefficient (UC) of 1.5 (a quantitative of the variation in particle size of a given media and is defined as the ratio of D_{60} to D_{10}), then we can estimate D_{90} (the sieve size for which 90% of the grains by weight are smaller), and the mean size, D_{50}, (the sieve size for which 50% of the grains by weight are smaller) as follows:

$$D_{90} = (0.42mm) \times 10^{1.67 \times log(1.5)} = 0.83mm$$

$$D_{50} = (0.42mm) \times 10^{0.83 \times log(1.5)} = 0.59mm$$

From Perry and Chilton (1973), we can obtain the required velocity to expand the above 20/40 sand by 50% with a D_{eq} = 0.59 mm as approximately 1.56 cm/s. In order to check if the largest sands will expand, the velocity required to expand a 0.83 mm sand 20% is 1.52 cm/s (again from Perry and Chilton (1973)). Because the velocity 1.52 cm/s is less than the bed design velocity used for the D_{50} sand at 50% expansion, the velocity selection is considered safe. If the D_{90} velocity had exceeded the D_{50} value, then we would have increased the overall velocity, and go with a larger expansion than 50%. In this design example, we will use 2.0 cm/s as a conservative safe estimate to ensure the large sands in the 20/40 sand used will fluidize. And now going back to Perry and Chilton (1973) and for the D_{50} value of 0.59 mm, we can say that the bed expansion is predicted as being around 75%, which is considered a very safe value to ensure full mixing and is therefore consistent with our recommended conservative approach.

Step 6: Required flow rate

Selecting an expansion velocity of 2.0 cm/s, and using the calculated cross-sectional area of the bed (0.16 m²), the required flow rate can be calculated as:

$$Q_{sand} = V_{sand} \times A_{bed} = 0.020 \frac{m}{s} \times 0.16 m^2 \times \frac{1{,}000 L}{m^3} = 3.2 \frac{L}{s}$$

Here, we see that the flow requirement for fluidization of the sand is much less than the controlling flow rate of 6.30 L/s for tank exchange rates. If we utilized that flow rate on our sand filters, we would easily blow the sand out of the biofilters. Instead, we only pumped part of the required flow (3.2 L/s) through the sand filter, while the rest bypassing the biofilter. Some other biofilters, such as trickling filter or moving bed biofilters can accept higher flow rates with no problem.

In a coldwater application, on the other hand, because of the high nitrification removal efficiency across the filter that can be achieved due to the high hydraulic retention time of the vessel, sands with D_{10} of around 0.25 mm are typically utilized. Since the sand selection locks in a velocity of near 1.0 cm/s and the recirculating flow rate is controlled by the oxygen requirements of the fish, coldwater biofilters are generally oversized. In the case of coldwater application, we would solve for A_{bed} (m²) knowing the flow rate required to maintain water quality design values in the fish rearing tank, Q_{wq} (m³/s), and the upflow velocity through the sand bed, V_{sand} (m/s):

$$A_{bed} = \frac{Q_{wq}}{V_{sand}}$$

5.13.2.5 Denitrification, Reactor Types and Design

5.13.2.5.1 Introduction

Being quite stable in the natural environment, nitrate (NO_3^-) can create a point source or non-point source pollution on the receiving waters, typically causing eutrophication and hence algal blooms and high dissolved oxygen consumption as the algae die and decompose. Denitrification process is not an inexpensive process, but it may be mandatory for many freshwater and most saltwater facilities in regions with either limited water supply or strict discharge requirements. In this subsection, the fundamentals of denitrification as well as the application of this process in recirculating aquatic systems are briefly discussed.

In both freshwater and especially marine recirculating aquaculture systems, nitrate reduction is becoming important because nitrate accumulations as high as 100–1,000 mg NO_3-N are quite common. Otte and Rosenthal (1979) and Honda et al. (1993) reported maximum nitrate concentrations in recirculating systems as high as 400–500 mg NO_3-N/liter. Tucker (1999) suggested that nitrate-nitrogen levels be maintained at concentrations less than 20 mg/L for marine larval and juvenile fish, and 50 mg/L for older fish. High nitrate concentrations can adversely affect the growth of reared aquatic organisms. Russo and Thurston (1991) emphasized the toxicity of nitrate to marine organisms due to its inhibitory effect of the animal's osmoregulatory ability, which can manifest itself in the form of inhibition of growth rates, reproductive cycles, poor egg development, delayed hatching times, and higher mortalities.

Biological denitrification is the microbial reduction of nitrate (NO_3^-) or nitrite (NO_2^-) to nitrogen gas (N_2) by heterotrophic and autotrophic facultative aerobic bacteria widely found in the natural environment. Quite a wide range of bacteria can use either nitrate or oxygen to oxidize organic material, shifting between oxygen respiration and nitrogen respiration, depending on the oxygen availability. Dissolved oxygen, however, is the preferred choice of denitrifying bacteria as the oxygen source over nitrate or nitrite. Accumulation of organic matter in an aerobic environment may, on the other hand, lead to anoxic conditions within the organic layer or biofilm, and may provide suitable conditions for proliferation and activity of denitrifying microorganisms. Under anoxic conditions, nitrate (NO_3^-) and nitrite (NO_2^-) replace oxygen as the electron acceptor for the oxidation of a wide variety of organic or inorganic electron donors. The optimal conditions for denitrification is defined by the term anoxic, i.e., low oxygen and high nitrate conditions. While denitrifying bacteria are quite common among the heterotrophic microorganisms, several autotrophic microorganisms are also able to perform denitrification. These organisms use reduced inorganic sulfur and iron compounds or hydrogen as their electron donor and derive carbon from an inorganic source are often prevalent in environments that are poor in organic matter. Denitrification process proceeds in a step-by-step fashion, in which nitrate (NO_3^-) is sequentially reduced to nitrite (NO_2^-), nitric oxide (NO), nitrous oxide (N_2O), and finally N_2 gas.

Reduced organic carbon, which is oxidized during heterotrophic denitrification can, for instance, be generally expressed as follows:

$$NO_3^- + 6H^+ + 5e^- \rightarrow \tfrac{1}{2}N_2 + 3H_2O \qquad (5.48)$$

And the overall transformation of nitrate (NO_3^-) yields:

$$NO_3^- + 6H^+ + 5e^- \rightarrow \tfrac{1}{2}N_2 + 3H_2O \qquad (5.49)$$

Park (2000) states that the above process can optimally be mediated by denitrifying bacteria under specific conditions, such as low redox potential, low oxygen levels, availability of an organic carbon source used as an electron and carbon donor, a nitrate source as an electron acceptor, an optimal pH range of 7.0–8.5, and an optimal temperature range of 25–32 °C. Moreover, Glass and Silverstein (1998) stated that no clear consensus exists on the effect of salinity on denitrification, while van Rijn et al. (2006) reporting no significant differences in denitrification rates between freshwater and marine recirculating systems.

5.13.2.5.2 Reactor Types and Design

Several denitrifying bioreactors are used in wastewater treatment: suspended growth systems and attached growth systems. Suspended growth systems are either completely-mixed or plug-flow reactors where a denitrifying biomass develops under anoxic and carbon-rich conditions. In attached growth systems, the bacteria is attached to a surface, which are called fixed-film or packed bed reactors, including fluidized bed reactors (FBR), and moving bed reactors (MBBR) (Ødegaard, 2006).

5.13.2.5.2.1 Moving Bed Bioreactor Processes for Denitrification The moving bed bioreactor (MBBR) was developed to remove nitrogen in a cost-effective manner through either pre- or post-denitrification processes. It has a smaller footprint and fewer maintenance requirements than trickling filters, for instance. Using an attached growth bioreactor, MBBR is widely applied in both small- and large-scale commercial operations for nitrification. Operating continuously, MBBR is a non-clogging biofilm reactor with low head loss, a high specific biofilm surface area, and has no requirement for backwashing. The reactor can be operated under either aerobic conditions for nitrification or anoxic conditions for denitrification. For nitrification, the media is maintained in constant circulation via air bubble aeration system creating aerobic conditions, and for denitrification via a submerged mixer to promote anoxic or anaerobic conditions. Media usually occupies up to 67% of the reactor volume (normally 50% fill), in that at higher percentage fill reduces mixing efficiency (Timmons and Ebeling, 2010).

5.13.2.5.2.2 Reactor Design For Denitrification An effective denitrification reactor can be designed by considering:

- Nitrate-nitrogen production in the system
- The maximum allowable nitrate concentrations in the culture water
- The volume of reactor required to match the nitrate load
- Retention time of the reactor
- The type of electron donor (organic or inorganic) and its stoichiometric requirements in the denitrification process
- Oxygen concentrations in the reactor

Example 5.37: Moving Bed Bioreactor (MBBR) Design

Design a denitrification system using a moving bed bioreactor (MBBR) for the recirculation system we have in hand, having a feeding rate of 9.6 kg feed per day, assuming the feed has a protein content of 35% (0.032 kg TAN/kg feed).

Solution

In this design, we will try to maintain the tank and the discharge water nitrate concentrations from the system as low as possible, e.g., <10 mg/L nitrate-N. The design process presented below is based on the success of the MBBR for nitrification in commercial aquaculture systems.

a) Nitrate-Nitrogen Production in the System

The daily maximum feed rate was estimated at 9.6 kg. Assuming a high-quality feed with a 35% protein content (0.032 kg TAN/kg feed), as determined before in the previous examples, 9.6 kg feed consumption per day suggests a daily production of 0.309 kg NH_3-N/day. It is difficult to determine the passive denitrification that can occur in any production system; therefore, we will assume that to be zero. Thence, if we assume that all of the TAN is converted into nitrate-nitrogen in the biofilter and that an insignificant quantity is lost in the discharge flow from the system, then 0.309 kg NO_3-N is generated per day.

b) The Maximum Allowable Nitrate Concentrations in the Culture Water

The nitrate-nitrogen will be maintained as low as possible in the tank and in the discharge by simply over sizing the removal system. If the denitrification rate is the same as the production rate of nitrate, which is the design goal, then the equilibrium concentration of nitrate in the fish system will approach zero if the recirculating system water discharge rate being added is free of nitrate.

c) The Volume of Reactor Required to Match the Nitrate Load

Volumetric denitrification removal rates (VDR) have been determined for several different types of MBBR media, and they are on the order of 2 g N/m^2.day. Then, the required surface area would be:

$$A_{media} = \frac{m_{NO_3-N}}{VDR} = \frac{\left(309\frac{g\,NO_3-N}{day}\right)}{\left(2\frac{g\,NO_3-N}{m^2.day}\right)} = 154\,m^2\left(\approx 1{,}658\,ft^2\right)$$

The effective biofilm surface area for a commercially available MBBR media is 500 m^2/m^3. Then, the volume of media required can be calculated as follows:

$$V_{media} = \frac{A_{media}}{SSA} = \frac{154\,m^2}{500\frac{m^2}{m^3}} = 0.31\,m^3\left(\approx 11\,ft^3\right)$$

Now, assuming a 67% fill factor for the MBBR, we can find a bioreactor volume of 0.46 m^3 (0.31 m^3/0.67) or 119 gallons.

d) Retention Time of the Reactor

Based on several municipal wastewater treatment systems, we can assume a hydraulic retention time (HRT) from 20 to 30 minutes.

e) The Type of Electron Donor (Organic or Inorganic) and Its Stoichiometric Requirements in the Denitrification Process

The available organic carbon source chosen for this design example is estimated at a maximum of 5:1 ratio, organic carbon to nitrate-nitrogen.

5.14 DISINFECTION IN CONTROLLED ENVIRONMENT AQUATIC SYSTEMS

5.14.1 Introduction

Disinfection means the removal, deactivation or killing pathogenic microorganisms. Bacterial and viral diseases are serious problems in both semi-intensive and intensive aquatic systems. As discussed earlier, using surface water in flow-through or recirculating systems poses a contamination risk by importing pathogenic microorganisms into the culturing system. In addition, commercial operations may also be required to disinfect their discharge waters before they are released into the natural environment. Disinfection by ozonation or UV-irradiation are the two main technologies often employed in aquatic systems. Other methods of disinfection are also briefly discussed in this section.

5.14.2 Ozonation

Ozone is known as an extremely reactive oxidant and a very effective bactericide and virucide. And, the ozone treatment is an efficient way of destroying unwanted organisms by the heavy oxidation of organic matter and biological organisms. In ozone treatment technology, microparticles are broken down into molecular structures, which will again bind together and form larger particles. By this form of flocculation, too small microscopic suspended solids can be caught and removed from the system instead of going through different types of filters in a recirculation system (Summerfelt and Hochheimer, 1997), which is particularly important for systems growing small fish that are sensitive to microparticles and bacteria in the water. Ozone has observed quite a wide employment in aquatic systems due to its rapid reaction rate, few harmful reaction byproducts in freshwater. Ozone treatment can also be used when the intake water to a recirculation system needs to be disinfected. Ozone application requires ozone generation, transfer, contact time to react and disinfect, and potentially ozone destruction in order to make sure no ozone residuals make into the culture tanks (Summerfelt and Hochheimer, 1997).

Using ozone in fish culturing systems has, however, been criticized due to the severe injury effects of over-dosage to the fish, and the harmful effects to the people working in the indoor environments. Therefore, utilizing the correct dosage and monitoring the loading together with proper ventilation in the aquatic building are critically important.

5.14.2.1 Ozone Generation and Transfer

The most efficient method of generating ozone is on-site generation by the electric corona discharge technique, which employs oxygen gas or air, and the generated ozone gas flow can be transferred into the aquatic system using any oxygen transfer device for deactivation of pathogenic microbiological organisms or other oxidative purpose (Summerfelt and Hochheimer, 1997). The transfer of ozone gas into the aquatic system is particularly because of the cost of producing ozone. This is especially true if the ozone is carried within purified oxygen feed gas that is either purchased or produced on site (Timmons and Ebeling, 2010).

5.14.2.2 Ozone Reaction and Pathogen Inactivation

In order to be effective, ozone oxidation necessitates maintaining certain amount of dissolved ozone concentration in the water for a specific contact time, which means disinfecting efficiency depends on the ozone residual concentration and its contact time. In this case, an ozone contact chamber facilitates the time required for the ozone residual to inactivate the target microorganisms. Wedemeyer (1996) reported that, depending on the target microorganism, the process of disinfecting waters may need a residual ozone concentration of 0.1–2.0 mg/L in a plug-flow type contact vessel for periods of 1–30 min. Right at this point, we must remember that it is extremely difficult to maintain residual concentrations above 1 mg/L in commercial aquatic applications.

It has been demonstrated that bacteria and viruses pathogenic to salmonids are quite sensitive to residual ozone in water. In distilled water, 99.9% or more inactivation is observed in the 0.01–0.10 mg/L residual ozone concentration range. As a rule-of-thumb, however, higher residual concentrations, i.e., 0.1–0.2 mg/L in natural sea-, brackish-, and freshwater, and 0.3–0.4 mg/L in aquacultural effluents, seem to be needed in order to obtain the legally required inactivation level (Timmons and Ebeling, 2010).

5.14.2.3 Ozone Destruction

While having an adequate concentration of residual ozone is required to inactivate bacteria, this same ozone has to be removed after inactivating bacteria but before the water reaches the fish in the aquatic system. Wedemeyer et al. (1979) reported that a de-ozonation unit must be integrated into the system because the residuals can, depending on the species and life stage, be lethal to fish at ozone concentrations as low as 0.01 mg/L. Ozone residuals can be eliminated by i) water retention within tanks immediately after ozonation, ii) applying small doses of a reducing agent, such as sodium thiosulphate, iii) passing the water through a forced-ventilation packed aeration column and stripping the dissolved ozone, iv) passing the water through a biofilter or bed activated carbon, v) reacting the dissolved ozone with low levels of hydrogen peroxide, or vi) contacting the dissolved ozone with high intensity UV-light (wavelengths ranging from 250 to 260 nm), catalyzing the conversion of O_3 to O_2.

5.14.3 UV Treatment

UV treatment works by applying light in certain wavelengths that destroy DNA of biological organisms. Pathogenic bacteria and single-celled organisms are the targeted organisms in aquatic systems. Since the UV treatment of water is performed outside the culture tanks (i.e., the fish production area), the treatment does not influence the fish. For the UV radiation to be effective, the organic matter content of the process water must be removed through mechanical filtration and biofiltration in order to eliminate supporting growth conditions for pathogenic microorganisms. The rule of thumb for the UV treatment in aquatic systems is that the water be prefiltered with a 50 μm mesh screen. One of the most widely used UV dosing unit is μWs/cm^2, and the efficiency of UV lighting system depends on the size and species of the target microorganisms and the turbidity of the water. We have to remember that the UV light must in the first place transmit through the water before even reaching the target organism. Since the UV transmission through the water column is quite small in recirculating aquatic systems, the UV applications in such kind of environments kill almost no organisms. Moreover, UV treatment does not produce any toxic residuals or form byproducts that pose a risk to fish (Oliver and Carey, 1976; de Veer et al., 1994).

5.14.3.1 UV Radiation

As already discussed in Chapter 4, UV radiation (190–400 nm) may damage microorganisms by directly and/or indirectly altering nucleic acids of organisms. Direct alteration is done due to absorption of UV radiation by DNA, resulting in photoproducts formation. DNA absorbance is high in the UV-C range (190–280 nm), but falls to one-third in magnitude in the UV-B range (280–315 nm), and is almost negligible in the UV-A band (315–400 nm) (Miller et al., 1999). The low-pressure mercury-vapor lamp emits approximately 85% of its energy at a wavelength of 253.7 nm, which falls in the optimum UV-C wavelength range of 250 to 270 nm for bactericidal effects. Solar UV-B irradiation is responsible for both direct and indirect DNA damage, while UV-A produces only indirect damage. And it should be remembered that photoreactivation can only repair direct DNA damage. As mentioned earlier, the UV irradiation doses do not produce any toxic residuals in the treated water because the low UV doses used do not generate significant amounts of photoproducts (Oliver and Carey, 1976; de Veer et al., 1994). UV treatment is commonly used in smolt farms for seawater and freshwater disinfection as well as for bacteriological control in recirculation systems (Rosenthal, 1981).

5.14.3.2 UV Disinfection and Pathogen Inactivation

A number of factors have been reported to influence the efficiency of bacterial inactivation by UV treatment (Kostenbauder, 1991; Bessems, 1998; Heinzel, 1998). Fish pathogenic bacteria are generally susceptible to UV doses of 1.5–3.4 μWs/cm^2 with a 99.9% or higher inactivation rate in high quality natural waters and buffers. Having that said, however, under continuous flow conditions with particles present in the flow, much higher doses are needed to reach a high degree of inactivation (Bullock and Stuckey, 1977; Liltved and Cripps, 1999). Hence, the inactivation doses obtained in laboratory conditions should be employed with extreme caution in predicting dose requirements under actual operating conditions.

5.14.3.3 UV Pathogen Inactivation Repair Mechanisms

The lethal effects of UV irradiation on bacteria may be repaired and reversed by two major mechanisms called photoreactivation and dark repair (a.k.a. liquid holding recovery). The mechanism of photoreactivation necessitates visible light. Light in the near UV and visible spectrum (330–480 nm) activates DNA photolyase, which reverts certain types of UV-induced damage (Jagger, 1967; Walker, 1984; Friedberg et al., 1995). And A. Kelner was the first who recognized the photoreactivation phenomenon in the late 1940s, discovering the reversibility of UV damage by visible light treatment (Kelner, 1961). In addition, bacteria use a complex cellular mechanism, known as the SOS regulatory system in dark repair processes.

Harm (1968) and Eker et al. (1991) reported that the photoreactivation rate is strongly correlated with light intensity. Several hours are normally required for complete reactivation at lamplight intensities, while sunlight intensities have been reported to reduce the time for photoreactivation completion down to fractions of an hour in irradiated *Escherichia coli* and halophilic bacteria. The photoreactivation mechanism is a widespread, but not universal trait among microorganisms. Viruses generally do not have the traits of photoreactivation (Asato, 1976). Moreover, the effect of photoreactivation is typically more prominent in aquacultural facilities compared to drinking water because UV treated water in aquacultural systems are exposed to artificial light or sunlight capable of initiating photoreactivation. As a rule-of-thumb, the applied UV dose should account for the worst case with respect to reactivation.

Liltved and Landfald (1996), on the other hand, found that dark repair mechanism in fish pathogenic bacteria was less efficient and slower than photoreactivation. Even at an elevated temperature of 22°C, at least 48 hours was required for the completion of repair. In aquacultural operations, such as salmon hatcheries that are typically characterized by relatively cold water and short hydraulic retention times, dark repair alone cannot have a strong influence on post-UV radiation survival. Furthermore, the combination of dark repair and photoreactivation generally leads to higher survival than dark repair alone (Moss and Davies, 1974).

5.14.4 Other Methods of Disinfection

There are numerous other methods of disinfecting water, equipment, and instruments used in aquaculture operations, each with its own advantage and disadvantages; however, getting deeper in this topic is not the particular focus of this book or this section.

5.15 AERATION/OXYGENATION AND DEGASSING IN CONTROLLED ENVIRONMENT AQUATIC SYSTEMS

5.15.1 Introduction

In a recirculating aquatic system, before the water runs back to the culture tanks, while the required dissolved oxygen concentration levels must be provided; in addition, the accumulated gases in the water that are detrimental to the fish must be removed. This degassing process, a.k.a. stripping, is performed by aeration of the water.

Carbon dioxide (CO_2) is introduced into the water through respiration of the fish and bacteria. For each mole of oxygen consumed, one mole of carbon dioxide is produced. In other words, on a mass basis, for one gram of oxygen consumed, 1.375 g of carbon dioxide is produced as the ratio of molecular weights is 44 for CO_2 to 32 for O_2). Measuring and calculating the concentration of dissolved CO_2 is complicated because, as discussed earlier in Chapters 2 and 4, CO_2 is part of a chemical equilibrium system that includes carbon dioxide (CO_2), carbonic acid (H_2CO_3), bicarbonate (HCO_3^-), and carbonate (CO_3^{2-}) ions (Figure 2.32). Hence, the concentration of any of them is pH dependent, so we must know the pH, alkalinity, salinity, and temperature of the water before attempting to determine their individual concentrations. It should also be noted that CO_2 concentration is proportional to alkalinity. The safe operating levels of CO_2 depend on the fish species, development stage, and overall water quality, and are provided in Chapter 4 for different fish species. It should be remembered that carbon dioxide reduces the capacity of the blood to transport oxygen. As the CO_2 concentration in the water increases, so does the blood CO_2 level. As also discussed in depth in Chapter 4, in intensive aquaculture systems, the acidosis resulting from high blood lactic acid concentrations or high blood CO_2 concentrations leads reduced O_2 transport to the the tissues due to what is called the Bohr or Root effects. In addition, carbon dioxide may also affect the biological filter's bacteria by reducing the overall system's pH. Degassing of CO_2 should be done when it is at the highest concentration level, which normally occurs after going through the biofilter.

Moreover, free nitrogen (N_2) as well gets accumulated in the water as a result of the ammonification and nitrification process. As a consequence, accumulation of carbon dioxide and nitrogen gases will have negative affect on the fish welfare and growth. Under anaerobic conditions, an extremely toxic gas hydrogen sulfide (H_2S) can also be produced, especially in saltwater systems. Aeration/oxygenation and degassing systems are presented and discussed in this section, providing some numerical examples as well.

5.15.2 Solubility of Gases in Water

For detailed discussions of the solubility of gases in water, the readers are referred to Section 2.4.9.2.

5.15.3 Dissolved Oxygen Requirements

For detailed discussions of the required dissolved oxygen (DO) concentrations for common fish species, the readers are referred to Section 4.5.2.3.

5.15.4 Oxygen Transfer to Water

For detailed discussions of the oxygen transfer to water, the readers are referred to Section 2.4.9.2.

5.15.5 Super Saturation and Degassing

As discussed above, gas supersaturation may occur in aquatic systems due to a variety of reasons. Dissolved carbon dioxide can easily be removed through a gas exchange process; however, predicting an accurate removal rate is a very difficult task because as CO_2 is removed, a shift in the carbonate buffering system occurs, affecting the pH and CO_2 concentration (see Chapter 2 and 4 for detailed discussions). Nitrogen gas supersaturation is of greatest concern for aquacultural operations because nitrogen is 78% of air and not removed from blood by metabolism demands of the tissues as oxygen does. Hence the danger of gas bubble disease in fish is consequently greater from nitrogen than it is from oxygen. Three types of degassing equipment are available for aquacultural systems: i) vacuum degassers, ii) oxygen injection degassers, and iii) packed column degassers. Vacuum degassers are comprised of a sealed column filled with packing material. Supersaturated water is applied to the top of the column and allowed to trickle down the packing material, which creates a large water surface area. A vacuum is also applied to the column to strip gas from the water flowing over the packing material. The vacuum must be less than the pressure head of water at the bottom of the column to allow water to flow out of the column. While stripping nitrogen, this method of degassing will also

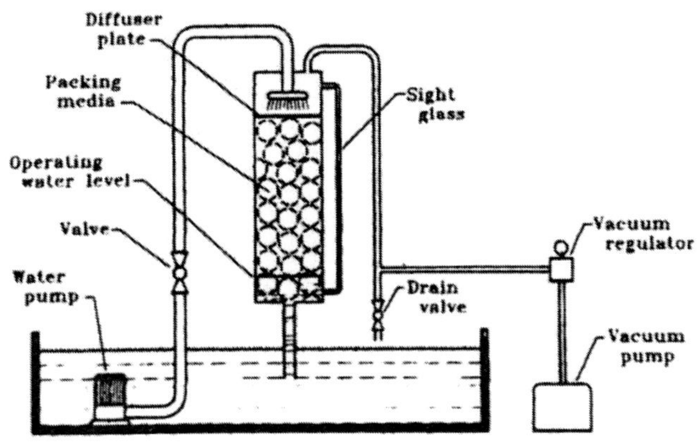

FIGURE 5.57 Vacuum degasser system.

cause a slight decrease in DO level. Figure 5.57 schematically shows a vacuum degasser.

An oxygen injection degasser is a packed column degasser with the addition of a system to inject pure oxygen into the column. The pure oxygen tends to displace the carbon dioxide and nitrogen dissolved in the water. This system has the advantage of removing dissolved carbon dioxide and nitrogen while increasing DO concentration at the same time. A simple packed column, the same as used for oxygenation of water, will also serve to remove dissolved carbon dioxide and nitrogen. Since the gas concentration is greater than the atmospheric partial pressures, the gas has a tendency to diffuse from the water into the air under a negative gas concentration gradient. These systems are only efficient when water has high levels of dissolved gases.

5.15.6 Types of Aerators

The water aeration process is the same physical process as degassing or stripping; but, it will add oxygen to the water instead of stripping. The DO content of water gets lowered once the water passes through culture tanks, and it is further reduced through the biofilter. Aeration of this water brings the DO saturation roughly up to 90% or so; and in some cases, 100% saturation can even be achieved. Oxygen saturation higher than 100% in the inlet water is preferred for a healthy fish growth; however, saturation levels beyond 100% requires a system using pure oxygen. Pure oxygen is often delivered in tanks in the form of liquid oxygen, but can also be produced on-site using an oxygen generator, such as a pressure swing adsorption (PSA) or a vacuum swing adsorption (VSA) unit. In general, aerators can be classified as: i) gravity aerators, ii) surface aerators, and iii) subsurface aerators.

5.15.6.1 Gravity Aerators

Gravity aerators make use of the kinetic energy in falling water to increase the air/water surface area in order to increase the rate of oxygen transfer to water. Falling water breaks into smaller units of volume under the turbulence caused by frictional forces as it falls through the air. Gravity aerators are primarily applicable for aquatic systems that inherently have an incoming water supply at an elevation raised above the fish growth area, which is typical with raceway systems. In recirculating systems, in which water has to be pumped through a biofilter and other treatment devices, the return flow may be raised to a sufficient elevation with little extra pumping energy to efficiently employ a gravity aerator. Since gravity aerators do not necessarily require motor power to supply energy, they are not usually rated for effectiveness and mechanical efficiency as other types of surface and submerged aerators that are rated in terms of field aerator efficiency (kg O_2/kWh), which gives the rate of oxygen supplied to water in terms of DO per kWh of motor energy. The efficiency of aerators can only be rated by Equation (5.50) with actual test runs using a given gravity aerator. Gravity aerators may be more conveniently rated in terms of aerator effectiveness as follows:

$$E = 100\frac{(C_b - C_a)}{(C_s - C_a)} \tag{5.50}$$

where E is the aerator efficiency (%), C_b is the outlet DO concentration (mg DO/L), C_a is the inlet DO concentration (mg DO/L), and C_s is the saturation DO concentration (mg DO/L).

A wide variety of types of gravity aerators have been found to effectively make use of free available gravitational energy; these aerator designs include: i) weir aerators, ii) corrugated inclined plane aerators, iii) lattice aerators, iv) cascade aerators, v) packed column aerators, vi) perforated plate or screen aerators, and vii) spray tower aerators. Aerators for raceways may also be constructed as an inclined plane instead of a single vertical drop. There are several variations of this design as well.

Figure 5.58 shows some examples of column type gravity aerators. The perforated-plate aerator is a column with a series of perforated horizontal plates to break the water flow into small trickles and droplets. The spray column is the simplest construction. Water is distributed as a spray by a nozzle at the top of the column and allowed to fall the height of the column exposed to air. Packed column aerators employ a vertical column packed with light-weight, high surface/area to volume media (Figure 5.58c). Water is uniformly distributed at the top of the packed column to trickle over the surfaces of the packing media. This creates a high surface area to volume ratio in the water. Air flows upwards through the column to supply air to the water surfaces. Packed column aerators usually require less volume for a given application than inclined surface aerators because packed columns create higher surface area to volume ratios. Packed columns provide nitrogen and CO_2 gas stripping as well.

As is the case with any aerator design, the performance of gravity aerators varies with water temperature and water quality. Overall speaking, gravity aerators are simple, have no moving parts, and can take advantage of elevated water supplies. But they only work when water is flowing, and the performance of some designs may be negatively effected by fouling.

5.15.6.2 Surface Aerators

Surface aerators agitate the water surface with a mechanical device or with a spray nozzle in order to increase the water surface exposed to air. There are several types of surface aerators including a floating propeller aerator, a paddle

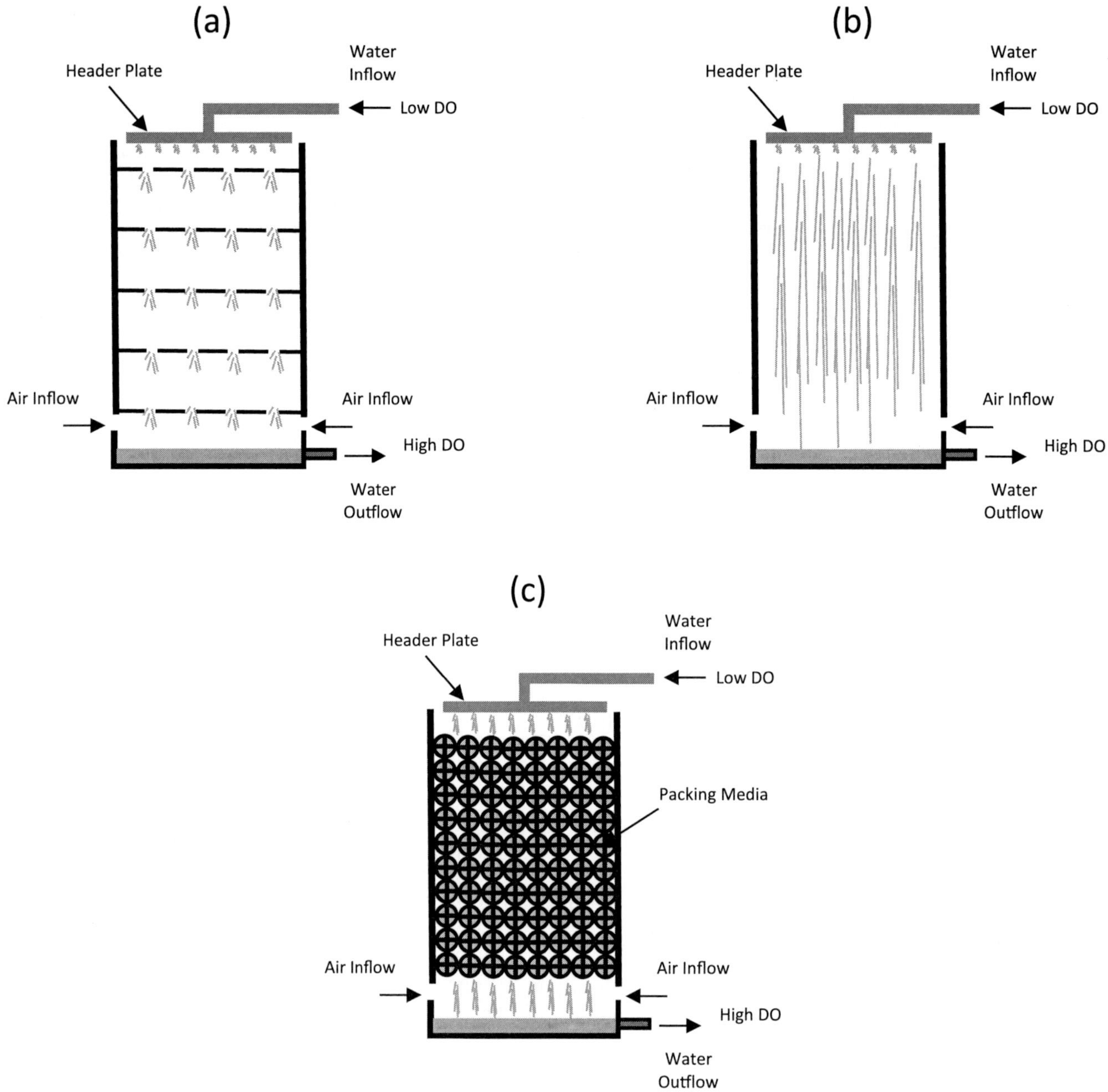

FIGURE 5.58 Examples of column type gravity aerators: (a) perforated plate aerator, (b) spray column aerator, and (c) packed column aerator.

wheel aerator, a propeller aerator with a draft tube, and a nozzle aerator.

The floating propeller aerators draw water from beneath the aerator by the mechanical work of the propeller and the deflectors positioned above the propellers spray water back onto the water surface. This creates a great deal of turbulence at the water/air interface area. Using a draft tube beneath the impeller draws water from further beneath the surface and also assists aeration by increasing water circulation.

In addition, the paddle wheel surface aerators are designed with a variety of paddle configurations. Paddle wheel aerators are turbine impellers partially submerged in water and rotated to agitate the water and spray it over the pond surface. In the past, paddle wheel aerators have been found to be one of the most energy efficient agitator designs.

Nozzle aerators, on the other hand, are often used to aerate round culture tanks. A single nozzle or series of nozzles are positioned radially (usually inclined 25° to 50° below the horizontal) over the tanks surface and directed tangentially at an inclined angle towards the water surface. The nozzle spray not only provides aeration but also sets up a useful water circulation in the tank.

5.15.6.3 Subsurface Aerators

Subsurface aerators may be classified as diffuser aerators, air lift pumps, and turbine aerators. Diffuser aerator designs are more common and there are many variations in the design of diffuser aerators. On the other hand, we must note that saturation levels above 100% requires a system using pure oxygen. Pure oxygen is often delivered in tanks in the form of liquid oxygen, but can also be generated on-site using an oxygen generator. Typically high pressure oxygen cone systems are used to prove supersaturated DO conditions in water. Water and pure oxygen are mixed under pressure whereby the oxygen is forced into the water. In the oxygen cone, the pressure is accomplished with a pump creating a high pressure of typically around 140 kPa (20 psi) in the cone.

5.15.6.3.1 Diffuser Aerators

Diffuser aerators inject air into water as bubbles. Oxygen is transported into the water through the air/water interface of the bubble surface. If the surface area of the bubbles is increased by forming small bubbles instead of large bubbles for a given air flow rate, oxygen transfer rates will increase. Also, lengthening the time that bubbles spend in water will increase the rate of oxygen transfer. The oxygen pressure gradient between the gas and liquid phase can be increased by injecting the bubbles at increased depths so that gas pressures are increased. Utilization of pure oxygen rather than air will also increase the efficiency of aeration since the saturation pressure of pure oxygen dissolved in water is much greater than the partial pressure of oxygen as a component of in air dissolved in water. Also, the use of pure oxygen eliminates the problem of nitrogen supersaturation that can result from using air under increased pressures. On the negative side, pure oxygen may be expensive; therefore, a cost/benefit analysis for the use of pure oxygen as compared to an air system should be done for any given application. There are a variety of different diffuser aerator designs, including simple diffusers, U-tube diffusers, venturi aerators, and aerator cones, as shown in Figure 5.59.

The simple diffuser aerator consists of an air supply pumped under pressure to a submerged diffuser that releases the air into the water as small bubbles. The diffuser could be as simple as a plastic coiled pipe with small holes to form air bubbles or it could be a porous solid material which creates very small bubbles, such as diffuser stones. The air supply for a diffuser must supply enough pressure to account for i) the friction losses due to flow in the pipes, fittings, and valves, ii) the pressure drop due to the diffuser, and iii) the pressure to overcome the liquid pressure at the depth of submergence. Although fine bubbles present a greater surface area for oxygen transfer, the formation of small bubbles with a fine pore diffuser utilizes a greater pressure drop than is required to form larger bubbles from a course pore diffuser.

Moreover, the U-tube aerator was designed to increase the residence time of air bubbles in water and to increase the gas pressure in the bubbles, thus increasing the overall gas transfer rate. Both parameters increase the aerator effectiveness. A diffuser introduces air bubbles into the flowing water at the top of a U-tube. The water downward velocity is greater than the rising velocity of the air bubbles so the bubbles travel downward with the water to the bottom of the U-tube and up the outlet side of the U-tube. This increases the bubble residence time in the water and the air pressure in the bubbles also increases as depth increases. As a rule-of-thumb, U-tubes are designed for a down flow velocity between 2 m/s to 3 m/s (Timmons and Ebeling, 2010). The oxygen transfer rate depends on the depth of water in the U-tube, the water velocity, the submergence depth of the diffuser, and the average bubble sizes. They can produce supersaturation due to the increased pressures at the bottom of the U-tube.

Venturi diffusers are, on the other hand, based on an aspirator system. Water at high pressure is pumped through a restricted orifice or venture nozzle; liquid velocity increase in the venturi reduces the liquid static pressure so that, if an opening to the atmosphere is made in the nozzle at a point

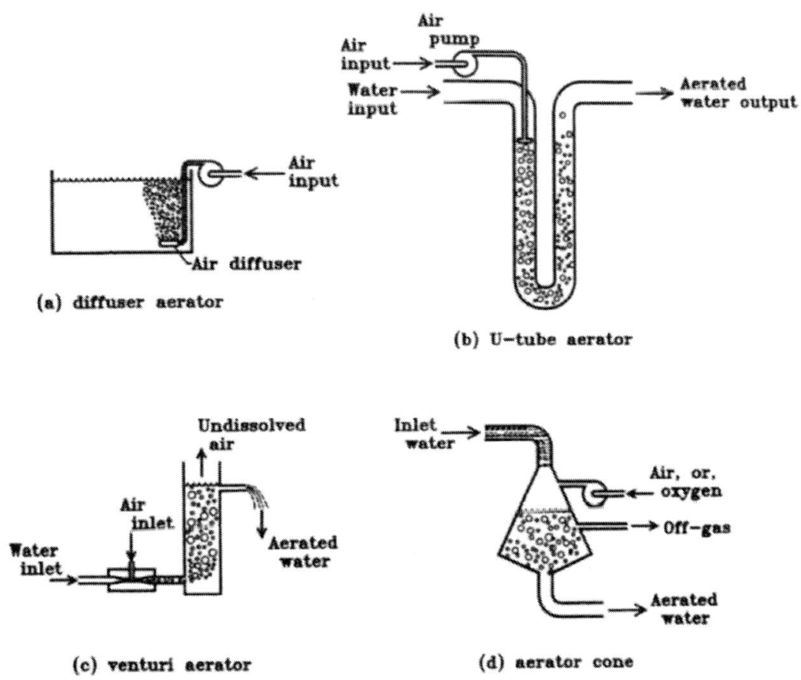

FIGURE 5.59 Sketches of different designs of diffuser aerators: (a) simple diffuser aerator, (b) U-tube aerator, (c) venturi aerator, and (d) aerator cone.

of pressure drop, air will be drawn into the water to form bubbles.

Finally, the aerator cone is like the U-tube, a variation on the diffuser design to increase the residence time of bubbles in water. Water is pumped into the top of an inverted cone at a velocity greater than the rising velocity of the air bubbles that are also injected into the top of the cone. The bubbles are forced to flow downward in the cone with the water. As the cross-section of the cone expands, the water velocity decreases; and, once the water velocity becomes less than the rising velocity of the air bubbles, the air bubbles can no longer travel downwards and they become trapped within the cone. This gives the bubbles an extended exposure time in water so that more of the oxygen in the bubbles is absorbed into the water.

Choice of a diffuser aerator system depends on a number of factors that balance cost of equipment, efficiency of oxygen transfer, and maintenance requirements. Simple diffusers require a source of compressed air or a high-pressure blower; venture aerators require a high-pressure water pump, but no air pump; while U-tubes and aerator cones require both an air pump and a water pump.

5.15.6.3.2 Air Lift Pumps

Air lift pumps consist of a vertical tube placed in the water column with an air diffuser placed near the bottom of the tube. The bottom of the tube is open to the water so that as the air bubbles rise, they draw water up the tube with them. In an air lift pump, the draft tube may be above the water surface if the air lift is used for a pumping application; or if the air lift is intended to provide only mixing and aeration, the outlet of the draft tube is kept below the water surface. An air lift pump may be useful if there is a need to circulate and mix bottom water with surface water as well as to provide aeration.

5.15.6.3.3 Turbine Aerators

A turbine aerator is a submerged impeller with a means of air or oxygen injection at the submerged impeller blade. The impeller increases circulation in the tank and thus increases the rate of surface aeration. The size and speed of the propeller and its position in the tank govern the rate of circulation while the oxygen gradient between the atmosphere and water governs the rate of oxygen transfer. In order to increase the rate of aeration with a turbine aerator, air may be injected near the rotating impeller. The air does not have to be supplied as fine bubbles since the propeller will itself break the air into smaller bubbles and disperse them in the water. Figure 5.60 demonstrates a turbine aerator with a hollow propeller shaft. The centrifugal action of the propeller blade rotation causes a low pressure near the centre of rotation; an air intake port in the hollow shaft above the water surface provides a source of air to the propeller where the air is broken into small bubbles and dispersed.

The oxygen transfer efficiency for turbine aerators with a source of pumped air is best for a system, in which the power input to the propeller is equal to the power input to the air compressor. Increasing the air compressor power over the propeller power creates larger bubbles that have a lower oxygen transfer rate. Moreover, increasing the propeller drive power above the air supply power results in more energy expended for water mixing than required for efficient oxygen absorption. Turbine aerators are, however, not widely used in fish culture tanks due to the obvious danger of physical injury to the fish; hence, turbine aerators are more applicable to non-mobile species.

5.15.7 Standard Test for Aerators

The simple gas transfer model presented earlier in Equation (2.257) is acceptable when the gas transfer coefficient is independent of the dissolved gas concentration, which is satisfied in systems where the gas bubbles do not change concentration significantly as they rise through the water column. We have this condition when the water column is shallow, the bubbles have large diameters, or the difference between the concentration of dissolved gas and the equilibrium concentration is small.

Equation (2.257) was run using linear regression so that the volumetric oxygen transfer rate coefficient, $K_L a$, can be determined from the slope of the straight line as shown in Equation (2.258). In actual aeration systems, however, it may be difficult to directly measure K_L and A (the

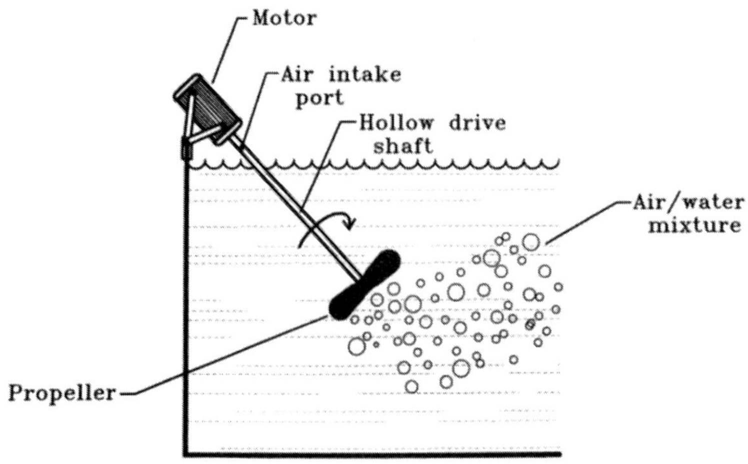

FIGURE 5.60 A turbine aerator with an air intake port in the hollow propeller shaft.

surface area of air/water interface area). As discussed earlier in Chapter 2, in practice, the value for the volumetric oxygen transfer rate, $K_L a$, can be measured in a test apparatus under standard conditions for a given aerator system. The calculated $K_L a$ can then be adjusted to apply to non-standard conditions for actual operating aeration systems. If the change in DO concentration is measured over time in a standard test system, Equation (2.258) plotted as a graph will produce a straight line with slope of ($-K_L a$). The graph will have a y-axis showing the natural logarithm of oxygen deficit [$\ln(C_s - C)$] as a function of time (x-axis). The slope of the straight line will be $K_L a$.

Aerators and aeration systems are normally tested under a set of standard conditions to calculate $K_L a$. The standard aerator test is made using tap water at 20°C with 0 mg DO/L at the start of the test. A series of measurements of DO is made in the tank with the aerator operating. About six measurements of DO are made as the DO increases from 10% to 70% of DO saturation capacity. These values are graphed to determine ($K_L a$), the standard volumetric oxygen transfer rate coefficient, for the given aeration system.

In practice, the value for $K_L a$ is adjusted to account for non-standard conditions typical of the actual aeration application. Corrections are made for temperature, if it is different than 20°C; corrections are also made if the starting DO concentration is greater than 0 and if impurities in water change the rate of oxygen absorption and/or lower the oxygen saturation concentration (Boyd, 1982).

The change in volumetric oxygen transfer coefficient due to impurities in water is adjusted by a factor called the oxygen transfer correction factor (α) (Colt and Orwicz, 1991):

$$\alpha = \frac{(K_L' a)_{20}}{(K_L a)_{20}} \qquad (5.51)$$

where α is the oxygen transfer correction factor, $(K_L' a)_{20}$ is the volumetric oxygen transfer coefficient for water being aerated (h^{-1}), $(K_L a)_{20}$ is the volumetric oxygen transfer rate coefficient for tap water at standard conditions (h^{-1}).

In addition, the oxygen saturation capacity for the water being aerated may be different than it is for the tap water used for the standard test because of potential impurities exist in the water. Here, a correction factor, called the oxygen saturation correction factor (β) is used (Colt and Orwicz, 1991):

$$\beta = \frac{(C_s')_{20}}{(C_s)_{20}} \qquad (5.52)$$

where β is the oxygen saturation capacity correction factor $(C_s')_{20}$ is the oxygen saturation capacity of water being aerated at 20 °C (mg DO/L), and $(C_s)_{20}$ is the oxygen saturation capacity of tap water at 20°C used for standard test (mg DO/L).

A temperature correction factor for the volumetric oxygen transfer coefficient can be made with the following equation (Colt and Orwicz, 1991):

$$(K_L a)_T = (K_L a)_{20}\, \theta^{(T-20^\circ\text{C})} \qquad (5.53)$$

where $(K_L a)_T$ is the volumetric oxygen transfer coefficient at temperature T (h^{-1}), $(K_L a)_{20}$ is the volumetric oxygen transfer coefficient at 20 °C (h^{-1}), and θ is the temperature adjustment factor, and T is the temperature at field conditions (°C)

The standardized oxygen transfer rate ($SOTR$) is the maximum rate of oxygen transferred in a test basin when the dissolved oxygen concentration in the basin is zero; and the $SOTR$ can be calculated as follows:

$$SOTR = (K_L a)_{20} \times C_s \times V \times 10^{-3} \qquad (5.54)$$

where $SOTR$ is the standard oxygen transfer rate in clean water at standard conditions (kg O_2/h), $(K_L a)_{20}$ is the volumetric oxygen transfer coefficient at 20°C (h^{-1}), C_s is the saturation dissolved oxygen concentration under standard conditions of 20 °C and 1 atm (mg DO/L), V is the water volume (m^3), and 10^{-3} is the conversion factor to convert to the proper units (L kg DO)/(m^3.mg DO).

And then, using the correction factors to convert to field conditions from standard test conditions, the field oxygen transfer rate (OTR_f) can be determined by

$$OTR_f = SOTR\left[\frac{\beta\, C_e - C_d}{C_s}\right] \theta^{(T-20^\circ\text{C})}\ \alpha \qquad (5.55)$$

where OTR_f is the field oxygen transfer rate (kg O_2/h), $SOTR$ is the standard oxygen transfer rate (kg O_2/h), β is the oxygen saturation capacity correction factor, C_e is the saturation dissolved oxygen concentration under field conditions of temperature and atmospheric pressure (mg DO/L), C_d is the existing dissolved oxygen concentration in the water (mg DO/L), C_s is the saturation dissolved oxygen concentration under standard conditions of 20 °C and 1 atm (mg DO/L), , is the temperature adjustment factor, T is the temperature (°C), and ± is the oxygen transfer correction factor (ratio of $K_L' a$ in dirty water to $K_L a$ in clean water).

It must further be noted that air is normally about 20.9% O_2; therefore, the normal partial pressure of O_2 is about one-fifth of standard atmospheric pressure. If pure O_2 gas is supplied instead of air, the saturation capacity of DO under given field conditions, C_e, is greatly increased. The use of pure oxygen can substantially increase the rates of oxygen transfer by an aerator system. However, buying bottled oxygen is expensive and if the pure oxygen is supplied under pressure, special care must be taken not to cause supersaturation of DO.

In order to evaluate the field efficiency of an aerator, and standard efficiency of oxygen transfer for a given aerator is defined as:

$$N_o = \frac{SOTR}{\text{Power Input to Aerator}} \qquad (5.56)$$

And using N_o, we can now calculate the field aeration efficiency for an aerator as follows:

$$N = N_o\left[\frac{\beta C_e - C_d}{C_s}\right]\theta^{(T-20^\circ\text{C})} \qquad (5.57)$$

where N is the field aerator efficiency (kg O_2/kWh), and N_o is the standard aerator efficiency (kg O_2/kWh).

Equation (5.55) shows clearly that the aerator efficiency decreases as the oxygen deficit in the water decreases. The value of N_o (standard aerator efficiency) is for tests run under specific standard conditions. The value for N_o is generally supplied by the aerator manufacturer, so that aerator efficiency for field conditions can be evaluated based on values for β, θ, α, C_e, and C_d.

Values of α, β, and θ have been determined for wastewater treatment systems. Although not much work has been done to define these values for pond water, it could be expected that the values of α, β, and θ would be similar to those for wastewaters. The values of α (the oxygen transfer correction factor) depends primarily on the surfactant concentration in water. In catfish ponds, α ranged from 0.66 to 1.07 with an average of 0.94 (Shelton and Boyle, 1983). Low values of a appeared to be related to the concentration of soluble compounds from feeds leaching into the water. Values of β (oxygen saturation capacity correction factor) typically range between 0.95 and 1.00 in wastewater. Colt and Orwicz (1991) recommended that in aquaculture systems a β value of 0.95 be used for ponds and recirculating systems and a β value of 1.00 be used for raceways. The value recommended for θ (temperature adjustment factor for determining volumetric oxygen transfer coefficient) in fresh water is 1.024, while values ranging from 1.008 to 1.047 have been reported (Colt and Orwicz, 1991).

Example 5.38: Calculation of Standard Aerator Efficiency

A manufacturer of a submerged aerator tested its equipment under standard conditions in a tank of 20 m^3 capacity. A series of tests were performed starting with a DO concentration of 0 mg/L. A number of DO measurements were made over time with the aerator running until the DO concentration in the test tank reached about 70% of saturation. The data gave a straight line on graph paper when plotting the values of [ln (C_s– C)] versus time (t) from the start of test. At time t_1 = 5 min from the start of the test, C_1 = 0.5 mg/L. At time t_2 = 40 min from the start of the test, C_2 = 7.20 mg/L. The saturation DO concentration of pure water, C_s, at 20 °C is equal to 9.17 mg DO/L. The power requirement for the aerator motor was 0.50 kW. Determine the following:

a) $(K_La)_{20}$, the standard volumetric oxygen transfer rate (h^{-1})
b) *SOTR*, the standard oxygen transfer rate for the aerator (kg O_2/h)
c) N_o, the standard aerator efficiency (kg O_2/kWh)

Solution

Using Equation (2.258), the standard volumetric oxygen transfer rate can be calculated as

$$(K_La)_{20} = \frac{\ln(C_s - C_1) - \ln(C_s - C_2)}{(t_2 - t_1)}$$

$$(K_La)_{20} = \frac{\ln(9.17 - 0.5) - \ln(9.17 - 7.20)}{\left(\frac{40}{60} - \frac{5}{60}\right)} = 2.54 h^{-1}$$

Then, using Equation (5.54), we can calculate SOTR as follows:

$$SOTR = (K_La)_{20} \times C_s \times V \times 10^{-3}$$

$$SOTR = 2.54\frac{1}{h} \times 9.17\frac{mg\,DO}{L} \times 20 m^3 \times (10^{-3}\frac{L.kg\,DO}{m^3.mgDO})$$

$$= 0.466\frac{kg\ DO}{h}$$

And then, using Equation (5.56), the standard aerator efficiency can be determined as follows:

$$N_o = \frac{SOTR}{Power\,Input\,to\,Aerator}$$

$$N_o = \frac{0.466\frac{kg\ DO}{h}}{0.5 kW} = 0.932\frac{kg\,O_2}{kWh}$$

5.15.8 Oxygen Transfer Efficiency for Submerged Aerators

The field aeration efficiency (N) for a submerged diffuser system can as well be calculated using Equation (5.57). However, the value of the saturation dissolved oxygen concentration under field conditions of temperature and atmospheric pressure (C_e) will be increased with increasing submergence depth. A one metre depth of water exerts a pressure of 0.0968 atm; or in other words, each metre depth of water increases pressure above standard atmospheric air pressure at the surface by 9.68%. The saturation DO concentration at a given depth would be approximately equal to C_s (1.00 + 0.968 Z), where Z is the depth of submergence. This value cannot be used as the value for C_e in (5.57) because the actual effective field dissolved oxygen concentration, C_e, is not equal to the saturation oxygen concentration at the diffuser depth. The actual increase in saturation dissolved oxygen concentration due to diffuser submergence depends on the type of diffuser, the bubble size, the bubble contact time and the amount of turbulence. The bubbles will only be exposed to the full hydrostatic effect of pressure while they are at the depth of the diffuser, as they rise towards the surface, the pressure decreases and so does the value of the saturation dissolved oxygen concentration. Huguenin and Colt (1989) suggested that values for C_e (effective saturation DO concentration at field conditions) be estimated by either Equation (5.58) for course and medium bubble diffusers or (5.59) for fine bubble diffusers:

$$C_e = C_s(1.00 + 0.0024Z) \quad (5.58)$$

$$C_e = C_s(1.00 + 0.0049Z) \quad (5.59)$$

where C_e is the effective saturation DO concentration for use in Equation (5.57) (mg DO/L) C_s is the saturation DO concentration under standard conditions of 20°C and 1 atm (mg DO/L), and Z is the diffuser submergence depth (m).

Example 5.39: Calculation of a Submerged Aerator Requirement

An recirculating aquaculture system operator wants to add oxygen to a round culture tank with the submerged aerator (let's make it a fine bubble diffuser) described in

Example 5.38 with a diffuser submergence depth, Z, of 1.4 m. The saturation DO concentration of pure water, C_s, at 20 °C is equal to 9.17 mg DO/L. Aeration will be used if the DO drops to 3.0 mg/L. Use the following values for correction factors:

α (oxygen transfer correction factor) : 0.85
β (oxygen saturation correction factor) : 0.95
θ (temperature correction factor) : 1.024

The water temperature in the culture tank is 17 °C, and the DO saturation capacity of pure water at 17 °C is 9.65 mg DO/L. Calculate the field aerator efficiency under culture tank conditions described in Example 5.38.

Solution

Let's first estimate the effective saturation DO concentration at field conditions, C_e, using Equation (5.59):

$$C_e = C_s(1.00 + 0.0049Z)$$

$$C_e = \left(9.17\frac{mg\,DO}{L}\right)(1.00 + 0.0049 \times 1.4m) = 9.23\frac{mg\,DO}{L}$$

Now, we can calculate the field oxygen transfer efficiency using the standard oxygen transfer efficiency from the previous problem using Equation (5.57):

$$N = N_o\left[\frac{\beta C_e - C_d}{C_s}\right]\cdot\,\theta^{(T-20^\circ C)}\,\alpha$$

$$N = 0.932\left[\frac{0.95(9.23) - 3.0}{9.17}\right]\cdot 1.024^{(17-20^\circ C)}(0.85)$$

$$= 0.464\frac{kg\,O_2}{kWh}$$

So, the field aerator efficiency of the given submerged aerator with a fine bubble diffuser is 0.464 kg O_2/kWh.

5.15.9 Aerator Selection and Controls

Aerator selection for an aquacultural operation depends upon whether it is intensive or extensive, the hydraulics of the system, the average water depth, whether the cultured organisms could be physically injured by the aerator system, how much noise the aerator system would produce, the capital and operating costs of the system, the reliability of the system, and whether aeration would be on an emergency, intermittent or continuous demand basis.

Fluid mechanics of a system has already been discussed in depth in Chapter 2. In aeration applications as well, the total pressure drop that must be overcome to supply air to a diffuser in a given application depends upon: i) the depth of submergence of the diffuser, ii) the pressure loss due to frictional flow losses in the air supply pipes and fittings, and iii) the pressure drop through the diffuser; and it can be calculated as follows:

$$h_{total} = h_{submergence} + h_{pipe+fittings} + h_{diffuser} \qquad (5.60)$$

where h_{total} is the total discharge pressure that is required at the outlet of the air compressor or blower (height of H_2O column), $h_{submergence}$ is the head loss due to submergence, $h_{pipe+fittings}$ is the head loss in the pipe and fittings (height of H_2O column), and $h_{diffuser}$ is the head loss in the diffuser (height of H_2O column).

If head is expressed in terms of water depth, the pressure head required from a blower due to diffuser submergence is equal to the depth of the diffuser. Head losses due to frictional losses in pipes and fittings can also be expressed as the height of a water column. As already discussed in Chapter 2, the frictional losses in pipes with air flow varies with the flow velocity, the pipe inside diameter, the pipe inside surface roughness, and the length of the pipe. Fittings, such as, pipe couplings, elbows, pipe reducer fittings, and valves also must be included in calculations of pipe head losses. Tabulated values of pressure losses and flow velocities for air through PVC Schedule 40 pipe are presented in Table A.71. The pipe fittings can be included in the piping head loss calculations by using tabled values for various fittings that equate the fitting to a length of pipe of the same nominal size as the fitting. The entire piping system is equated to an equivalent total length of straight pipe and the pressure drop can be determined from a tabulated value such as Table A.71. The head loss due to the diffuser can also be determined by using a value of the dimensionless resistance coefficient (also called the loss coefficient), K, which can be determined from the measured value of frictional loss at a given flow through the diffuser (see Equation (2.210)). Tabulated values for loss coefficients are available for most types of fittings; or can be determined from the measured head loss at a given flow.

Manufacturers supply performance curves for their blowers or compressors; and in addition, a design engineer would construct a system head loss requirement curve for the specific diffuser aerator system in question. The intersection of these two curves will predict the operating point for the system. If more than one diffuser is to be supplied from a single compressor or blower, control valves should be installed in the air lines so that pressure drops can be balanced; otherwise, most of the air will flow through the line with the lowest pressure drop. Without flow balancing valves in the branch line, most of the air from a given blower would go to the diffuser at the shallowest depth or with the shortest supply pipe.

Finally, it should be noted that the method of control for the aeration system will also affect the size of aeration equipment required. Accordingly, the simplest control strategy would be to predict the maximum aeration demand, and then install and operate an aeration system to meet the maximum demand on a continuous basis. In addition, a system to utilize less overall power would be a step system capable of two levels of aeration; and the extra aeration capacity would only be operated during peak oxygen demand periods. A proportional control system, on the other hand, requires continuous monitoring of DO with computer control to match the power output of the aerator to the DO requirement. Moreover, the fourth protocol would be to use a management strategy that would even out the peak demand for oxygen to a lower overall average oxygen demand, which can be done by using several feeding intervals spread over the day.

Example 5.40: Calculation of Head Loss for Air Supply

Determine the total pressure head required in the setup shown below if the air supplied to the diffuser is 20 standard cubic feet per minute (scfm).

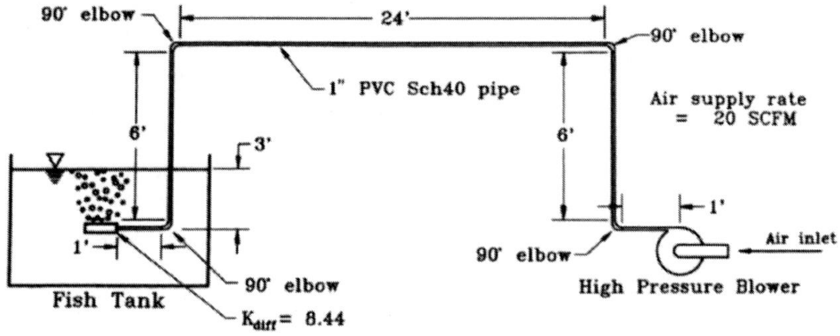

Solution

Let's first determine $h_{submergence}$, the head loss due to submergence in Equation (5.60):

$$h_{submergence} = 3' = 36"H_2O$$

Now, let's determine the head loss due to diffuser, $h_{diffuser}$, using Equation (2.210):

$$h_{f,minor} = K\frac{u^2}{2g} \times \left(\frac{0.0142"H_2O}{foot\ air}\right)$$

Note: The conversion factor from pressure head units in standard feet of air to inches of H_2O is 0.0142 in. of H_2O per foot of standard air.

And if the head loss of air at 20 standard cfm in 1 in. PVC Schedule 40 pipe is 49.2 ft/s (Table A.71), then

$$h_{diffuser} = 8.44\frac{(49.2\,ft/s)^2}{2\times(32.17\,ft/s^2)} \times \left(\frac{0.0142"H_2O}{foot\,air}\right)$$
$$= 4.51"\ H_2O$$

And now, we can calculate the head loss due to pipe and fittings as follows:

$$Length\,of\,Straight\,Pipe = 1' + 6' + 24' + 6' + 1' = 38'$$

$$Equivalent\,length\,of\,pipe\,(L_e)\,to\,replace\,1(one)\,90°\,elbow$$
$$= K_e \times D = 35 \times 1.049" = 36.72"$$

Note: The frictional loss of the 90° elbow expressed as equivalent pipe lengths, K_e, is obtained from Table A.71, and the internal diameter (D) of 90° PVC elbow is 1.049 inches.

$$Equivalent\,length\,of\,pipe\,(L_e)\,to\,replace\,4(four)\,90°\,elbow$$
$$= 4 \times 36.72"/(12"/ft) = 12.24'$$
$$Total\,equivalent\,length\,of\,pipe = 38' + 12.24' = 50.24'$$

$$h_{pipe+fittings} = \left(\frac{0.67\,psi}{100'}\right) \times 50.24' \times \left(\frac{27.67"H_2O}{psi}\right) = 9.31"H_2O$$

Then, using Equation (5.60), the total pressure drop, h_{total} is:

$$h_{total} = h_{submergence} + h_{pipe+fittings} + h_{diffuser}$$
$$h_{total} = 36"H_2O + 9.31"H_2O + 4.51"H_2O = 49.83"H_2O$$
$$h_{total} = 50"\ H_2O\ (\approx 127\ cm\ H_2O) \cong 4.2'\ H_2O$$
$$(\approx 1.3\ m\ H_2O)$$

5.16 AQUAPONICS: INTEGRATED FISH AND CROP PRODUCTION

5.16.1 Introduction

Aquaponics is the combination of the two commercially well-established recirculating systems: aquaculture and hydroponics. Aquaponic systems are essentially recirculating aquaculture systems, incorporating the production of plants, demanding no soil, conserving water and other resources, and producing food not only in the form of fish, but also in the form of vegetables, fruits, and herbs in particular.

Recirculating systems rear quite large quantities of fish in relatively small volumes of water in controlled environments, treating and then reusing the water. In the process of filtering and reusing the water in recirculating systems, non-toxic nutrient-rich fish waste and organic matter accumulate. Rather than wasting these metabolic byproducts, they are channeled into secondary crops and used as a fertilizer with all of the nutrients and minerals needed for plants grown in hydroponic systems. If the secondary crops are aquatic or terrestrial plants grown in conjunction with fish, this integrated system is referred to as an aquaponic system (Timmons and Ebeling, 2010). In brief, combining the aquaculture and hydroponic systems provides nutrient-rich solution required for crop growth while mitigating the waste aquacultural discharge and earn extra income.

In recirculating systems with very little daily water exchange, concentrated organic matter and dissolved nutrients accumulate, and may sometimes pose a threat to the environment, or at least, cause an additional expense in the case they are discharged to a municipal sewer system. As discussed earlier, dissolved nitrogen in particular, occurs at very high levels in recirculating systems, and bacteria convert ammonia to nitrite and then to nitrate, which is relatively harmless and is the preferred form of nitrogen for growth of vegetables and herbs, for instance. In aquaponics, the secondary crops produced recover the major portion of these nutrients, reducing the need to discharge the water to the environment, and therefore minimizing the water exchange rate. Lennard (2006), for instance, demonstrated that nitrate accumulation in aquaculture waters was reduced up to 97% in the aquaponic lettuce production system when compared with the fish-only system. Recirculating systems are not inexpensive to construct and operate, and hence profitability is quite often a major concern. A secondary crop that receives most of the required nutrients with no additional expense, definitely helps the system profitability.

Having all these things said, however, aquaponics is still not being practiced at large-scale commercial applications, rather only at small hobby and backyard levels. Large-scale operations are definitely possible as long as the markets are able to absorb the primary fish and the secondary crop products. No matter how small or large the production scale is, we would be producing food locally, and producing it in a locally sustainable fashion. Moreover, the design principles of aquaponic systems are quite simple as long as the fish are stocked at an adequate rate, which provides a proper feeding rate ratio for the secondary crop production.

5.16.2 System Design and Components

Figure 5.61 shows the essential elements of an aquaponic system, which consists of a fish rearing tank, a settleable and suspended solids removal component, a biofilter, a hydroponic subsystem, and a sump tank (Rakocy and Hargreaves, 1993). In an aquaponic system, water flows from the fish culturing tank through a mechanism that removes solids, then it enters a biofilter where the fish waste is microbially converted into nutrients useful for plant growth before entering the hydroponic subsystem where secondary crop roots receive these nutrients as well as irrigation water. Clean water from the hydroponic subsystem then flows into a sump tank where constant water levels are maintained, and then it is sent back to the fish rearing tank. In the aquaponic system depicted in Figure 5.61, either biofilter and hydroponic subsystem, or the solids removal, biofilter and hydroponic subsystem are combined.

The aquaponic system can be designed in such a way that only a portion of the recirculating water flow is directed into a particular treatment unit (Naegel, 1977; Wren, 1984). For instance, a small portion of water may go to a hydroponic subsystem after the solids removal process, while the remaining portion of water goes through a biofilter first and then returns to the rearing tank. As shown in Figure 5.61, the biofilter and hydroponic subsystem can be combined by using a crop support media, such as gravel (Lewis et al., 1978; Sutton and Lewis, 1982; Rakocy, 1984; Watten and Busch, 1984), or sand (McMurtry et al., 1990), which functions as biofilter media as well. Using a crop support media where plant roots develop can possibly eliminate the need for a separate solids removal system and biofilter. Solids coming the rearing tank are captured within the media and are processed by bacteria developing in the media, whereby acting as a solids removal and biofilter system. Since an overload of suspended solids is always a threat due to variations in fish feeding activities and efficiency of the solids removal component, gravel or sand beds should be avoided for large commercial-scale operations.

If the system does not employ a media, and plant roots are exposed directly to the water, then a separate solids removal system and a biofilter are required to return the water quality to sufficient levels for fish rearing in culture tanks (Rakocy et al., 2006). Floating raft hydroponic systems, which consist of floating sheets of Styrofoam and net pots for plant support can also provide adequate biofiltration if the plant production area is large enough (Rakocy, 1995). Combining the biofiltration process with hydroponics obviously eliminates the expense of a separate biofilter. Another design approach combines solids removal, biofiltration, and hydroponic subsystem as a single unit.

5.16.2.1 Solids Removal

As it has been discussed earlier in depth, recirculating systems generate fecal waste, uneaten fish feed, and organisms

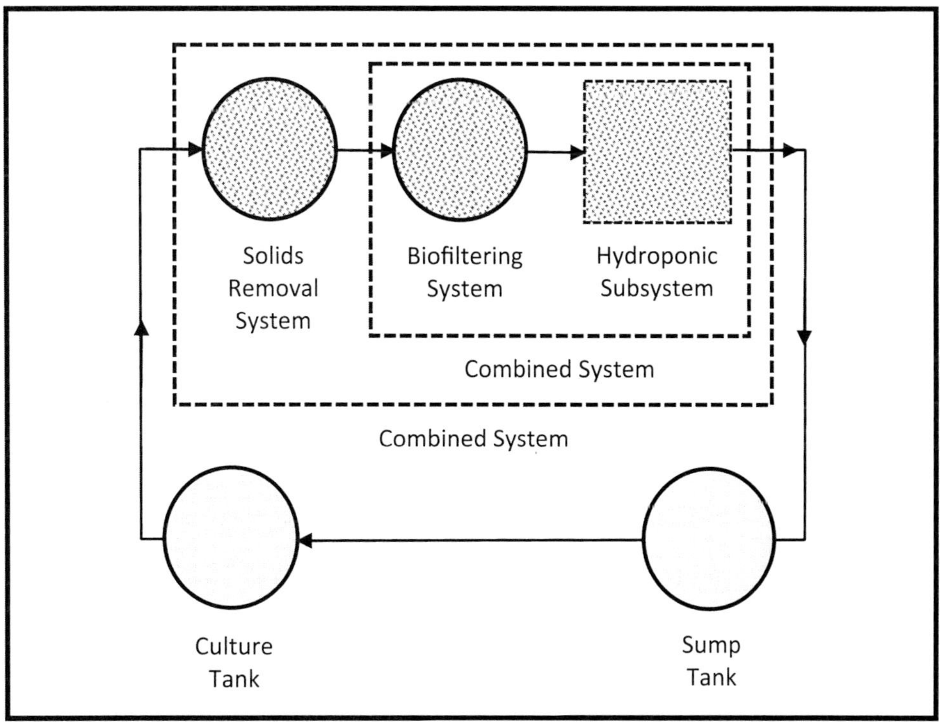

FIGURE 5.61 Schematic view of optimum aquaponic system components.

(e.g., bacteria, fungi, and algae), most of which must be removed from the stream before it enters the hydroponic subsystem. Solids capture and removal mechanisms and some of the devices used for removing solids from recirculating systems have also been discussed earlier in this chapter. These include settling basins, tube or plate separators, microscreen filters, bead filters, etc., many of which may also be used for aquaponic systems.

Moreover, the solids that are discharged from aquaponic systems must be disposed in a sustainable fashion (see Chapter 9). In rural facilities, effluents can be stored in aerated ponds first, and then applied as dilute sludge to land after the organic matter content is stabilized. This method is particularly useful in dry areas where sludge can be used for irrigating and fertilizing field crops. The solid fraction of the sludge can also be separated from the water and used with other waste products from the system (e.g., plant matter) for composting. Urban area facilities, on the other hand, might have to dispose solid waste into city sewer lines to be treated at municipal treatment plants.

5.16.2.2 Biofiltration

In aquaponic systems, ammonia will accumulate and eventually reach toxic levels if it is not removed by nitrification through biofiltration, in which ammonia is first oxidized to nitrite, which is toxic; and then, nitrite is oxidized to nitrate, which is relatively non-toxic. Rakocy and Hargreaves (1993) reported that aquaponic systems have used biofilters with sand, gravel, shells, or various other plastic media as substrate. Biofiltering options, such as trickling tower, rotating biological contactors, floating bead biofilters, and fluidized-bed sand filters are already covered in this chapter earlier. If a separate biofilter is required or if a combined biofiltration and hydroponic subsystem is used, the standard equations used for sizing biofilters may not be applicable to aquaponic systems as additional surface area is provided by plant roots and a considerable amount of ammonia is removed by plants. Since the contribution of various hydroponic subsystem designs and plant species are not very well known, biofilters for aquaponic systems should be sized close to the recommendations for recirculating systems.

5.16.2.3 Fish Production and Crop Selection

The most common fish cultured in aquaponic systems is tilapia. However, other fish species used in aquaponics can be cited as channel catfish, largemouth bass, rainbow trout, common carp, goldfish, Asian sea bass, and Murray cod. Many types of vegetables have, on the other hand, been grown in aquaponic systems. However, the main purpose is to chose a vegetable that can generate the highest level of profitability. Therefore, culinary herbs such as basil, cilantro, chives, parsley, portulaca, and mint seem to be the best choice as they grow fast and find a market with high prices. Lettuce is another good crop candidate for aquaponic systems because it can be produced in a short period with a high proportion of edible biomass. Traditional medicinal plants and plants used for the extraction of modern pharmaceuticals might as well be potential candidates for aquaponic systems.

5.16.2.4 System Design Considerations

Several approaches can be used to design an aquaponic system, and the details of these approaches are provided by Timmons and Ebeling (2010). They suggest that the simplest approach is to duplicate a standard system or scale a standard system down or up, keeping the components proportional. Other approaches include designing for an available space, designing for fish production, and designing for plant production. If a limited amount of space is available, such as an existing greenhouse, then simply the available space will define the size of the aquaponic system. If, on the other hand, the main objective is to produce a certain amount of fish annually, the first step in the design process will then be to determine the number of systems required, the number of rearing tanks required per system and the optimum rearing tank size, and then determine the required plant production area. The solids removal component, water pump, and blowers would then be sized accordingly. Furthermore, if the main objective is to produce a certain amount of crops annually, the first step in the design process will then be to determine the area required for plant production. The area needed will be based on plant spacing, length of the production cycle, number of crops per year or growing season, and the estimated yield per unit area and per crop cycle. Then, after determining a number of fish side requirements, the water volume among multiple culturing tanks per system with the goal of creating a practical system size and tank array, and the fish stocking rate are determined. The solids removal component, water pump, and blowers would then be sized accordingly.

NOMENCLATURE

A	area; surface area of air/water interface area
ACH	air change per hour
BHP	boiler horsepower
c	specific heat
C	coefficient; drag coefficient; concentration; constant
$(C_s)_{20}$	oxygen saturation capacity of tap water at 20 °C used for standard test
$(C_s')_{20}$	oxygen saturation capacity of water being aerated at 20 °C
CFL	compact fluorescent lamp
cfm	cubic feet per minute
CHP	closed heat pump greenhouse system
CON	conventional greenhouse system
COP	coefficient of performance
d	diameter
D	distance; diameter; dosing value
DLI	daily light integral
DO	dissolved oxygen
E	energy; aerator efficiency
F	feeding rate
g	gravitational acceleration
h	enthalpy; head loss
H	height
HID	high intensity discharge
HPS	high-pressure sodium

HRT hydraulic retention time
I content of expansion vessel
INC incandescent
K flow coefficient
K_La volumetric oxygen transfer rate coefficient
$(K_La)_T$ volumetric oxygen transfer coefficient at temperature T
$(K_La)_{20}$ volumetric oxygen transfer rate coefficient for tap water at standard conditions
L length
Le equivalent pipe length
LED light-emitting diode
m mass production rate
M mixing group size
MBBR moving-bed bioreactor
MH metal halide
n excess air
N efficiency
OD outside diameter
OHP open heat pump greenhouse system
OTR_f field oxygen transfer rate
P production
PAR photosynthetically active radiation
PC protein concentration
PPE photosynthetic photon efficacy
PPF photosynthetic photon flux
PPFD photosynthetic photon flux density
PSA pressure swing adsorption
Q heat transfer rate; flow rate; ventilation rate
R return
S supply
SCFM standard cubic feet per minute
SOTR standard oxygen transfer rate
SSA effective biofilm surface area
t time
T temperature
TAN total ammonia nitrogen
TSS total suspended solids
u velocity
U heat transmission rate
UV ultraviolet
V volume; velocity; overflow rate
VSA vacuum swing adsorption
x variable
X percent increase in production
W width; humidity ratio; moisture production
Z diffuser submergence depth

Greek Letters

α oxygen transfer correction factor
β oxygen saturation correction factor
Δ difference
η efficiency
θ temperature correction factor
τ hydraulic retention time
v specific volume
ρ density
ϕ relative humidity
Φ flue gas flow rate

Subscripts

a animals; inlet conditions
adj adjusted
b boiler; buffer; outlet conditions
cg combustion gas
cond conduction
d dimensionless; existing
dgas dosing gas
dm dry matter
e effective; evaporation
evap evaporation
g natural gas
gh greenhouse
i inside; inlet
inf infiltration
max maximum
min minimum
o outside; opening; outlet
p pressur; particles
s saturation conditions; settling particles
ss suspended solids
t total; table
TAN total ammonia nitrogen
TSS total suspended solids
v ventilation
vs volite solids
w water; wet bulb
wq water quality
1,2 position or state numbers

BIBLIOGRAPHY

Plant and Animal Systems

Aldrich, RA and Bartok, JW (1994) *Greenhouse Engineering.* NRAES-33. Northeast Regional Agricultural Engineering Service. Cornell University, Ithaca, NY, USA.

Andrews, JW (1978) Development of a cost effective solar assisted heat pump. Proc. 1978 Ann. Meet. Am. Section Int. Solar Energy Soc. Inc., Vol. 2.1, pp. 281–287.

ASABE (2018) *Standards, Engineering Practices, and Data.* V21(1). American Society of Agricultural and Biological Engineers, St. Joseph, MI, USA.

ASAE (1996) *Standards, Engineering Practices and Data.* 43rd Edition. American Society of Agricultural Engineers, St. Joseph, MI, USA.

ASHRAE (1987) *Handbook, HVAC Syst. & Applic.* American Society of Heating, Refrigeration and Air Conditioning Engineers, Inc., Atlanta, GA, pp. 10.25–10.29.

ASHRAE (1997) *Handbook of Fundamentals.* American Society of Heating, Refrigeration and Air Conditioning Engineers, Inc., Atlanta, GA, pp. 10.25–10.29.

ASHRAE (2021) *Handbook of Fundamentals.* American Society of Heating, Refrigeration and Air Conditioning Engineers, Inc., Atlanta, GA, pp. 26.1–26.23.

Baeza, EJ (2007) Optimización del diseño de los sistemas de ventilación en invernaderos tipo parral. Tesis doctoral. Escuela Politécnica Superior. Departamento de Ingeniería Rural. Universidad de Almería, Almería, Spain.

Boersma, L and Rykbost, KA (1975) Soil warming with power plant waste heat greenhouses. *HortScience*, 10: 28–30.

Bosio, RC and Suryanarayana, NV (1975) Solar assisted heat pump system: A parametric study for space heating of a characteristic house in Madison, WI, USA, ASME Paper 75-WA/Sol-8.

Boulard, T (2006) Greenhouse natural ventilation modelling: A survey of the different approaches. *Acta Horticulturae*, 719: 29–40.

Buchholz, M, Jochum, P and Zaragoza, G (2004) Basic water, heat and food supply from a closed greenhouse—the watergy project. International Symposium GREENSYS 2004—Sustainable Greenhouse Systems, Leuven, Belgium, 12–16 September, published in International Society of Horticultural Science (ISHS) und Wageningen University and Research Center, Wageningen. ISBN: 907601924X, S. 133.

Buchholz, M, Zaragoza, G, Buchholz, R and Pérez-Parra, J (2006) Temperature and humidity control in the watergy greenhouse. *Acta Horticulturae*, 719: 401–408.

Casault, B, Gauthier, L and Boily, R (1989) Comparison of three gas-fired greenhouse heating systems in cold climates. ASAE Paper No: 89-4033, ASAE, St. Joseph, MI 49085, USA.

Castilla, N and Montero, JI (2008) Environmental control and crop production in Mediterranean greenhouses. *Acta Horticulturae*, 797: 25–36.

Christensen, RN and Santoso, M (1990) An evaluation of a Rankine cycle driven heat pump. *Heat Recovery Systems and CHP*, 10(2): 161–175.

Colosimo, DD (1987) Introduction to engine driven heat pump: Concepts, approach and economics. *ASHRAE Transaction*, 93(2): 987–996.

Currey, C and Lopez, R (2013) Comparing LED lighting to high-pressure sodium lamps. *Greenhouse Grower*, October, Meister Media Worldwide, Willoughby, OH, USA., pp. 34–40.

Currey, C and Lopez, R (2014) Managing photoperiodic lighting. In: *GrowerTalks*. March, Ball Publishing, Chicago, IL, USA, pp. 72–73.

DesignLights Consortium (2020) Available on: https://www.designlights.org/horticultural-lighting Retrieved 07-06-2020.

de Zwart, HF and Kempkes, FLK (2008) Characterizing of cooling equipment for closed greenhouses. *Acta Horticulturae*, 801: 409–416.

Duncan, RT and van Zuiden, GJ (1979) Economic viability of the solar-assisted industrial heat pump system. Solar Indust. Process Heat Conf. Proc. Solar Energy Research Institute/TP-333–429, San Fransisco, CA, USA.

Dutta Gupta, S (2017) *Light Emitting Diodes for Agriculture*. Springer, Singapore.

Faust, JE and Logan, J (2018) Daily light integral: A research review and high resolution maps of the United States. *HortScience*, 53(9): 1250–1257.

Freeman, TL, Mitchell, JW and Audit, TE (1979) Performance of combined solar-heat pump systems. *Solar Energy*, 22: 125–135.

Friedman, D (1977) Light commercial Bryton/Rankine space conditioning system. 12th Intersociety Energy Conversion Engineering Conference, Washington, DC, USA, p. 172.

Goldammer, T (2019) *Greenhouse Management: A Guide to Operations and Technology*. Apex Publishers, Centreville, VI, USA.

Gosselin, A and Trudel, MJ (1982) L'utilisation du chauffage au dol dans les cultures de legume de serre. *Revue Horticole*, 232: 21–25.

Grevstad, FS and Coop, LB (2015) The consequences of photoperiodism for organisms in new climates. *Ecological Applications*, 25(6): 1506–1517.

Handarto Hayashi, M and Kozai, T (2006) Air and leaf temperatures and relative humidity in a naturally ventilated single-span greenhouse with a fogging system for cooling. *Acta Horticulturae*, 710: 165–170.

Hellickson, ML and Chen, CF (1987) Simulated and tested performance of a rockbed heat sink for cooling summer ventilation air. *Transactions of the ASAE*, 30(1): 207–214.

Jordan, RC and Preister, GB (1956) *Refrigeration and Air Conditioning*. Prentice-Hall, Inc., New York, NY, USA.

Kacira, M, Sase, S and Okushima, L (2004) Effects of side vents and span numbers on wind-induced natural ventilation of a gothic multi-span greenhouse. *JARQ* 38: 227–233.

Kim, K, Yoon, JY, Hahn, JH, Kwon, HJ, Son, JE, Nam, SW, Giacomelli, GA and Lee, IB (2008) 3-D CFD analysis of relative humidity distribution in greenhouse with fog cooling system and refrigerative dehumidifiers. *Biosystems Engineering*, 100: 245–258.

Knies, P, van de Braak, NJ and Breuer, JJG (1984) Infrared heating in greenhouses. *Acta Horticulturae*, 148: 73–80.

Korczynski, PM, Logan, J and Faust, JE (2002) Mapping monthly distribution of daily light integrals across the contiguous United States. *HortTechnology*, 12: 12–16.

Kozai, T, Fujiwara, K and Runkle, ES (2016) *LED Lighting for Urban Agriculture*. Springer, Singapore.

Kusuma, P, Pattison, PM and Bugbee, B (2020) From physics to fixtures to food: Current and potential LED efficacy. *Horticulturae Research*, 7(56). https://doi.org/10.1038/s41438-020-0283-7

Lee, I, Lee, S, Kim, G, Sung, J and Yoon, Y (2005) PIV verification of greenhouse ventilation air flows to evaluate CFD accuracy. *Transactions of the ASAE*, 48: 2277–2288.

Lopez, R and Currey, C (2014) *Understanding Light*. GrowerTalks, February, Ball Publishing, Chicago, IL, USA., pp. 58–61.

Lopez, R and Runkle, ES (2017) *Light Management in Controlled Environments*. Meister Media Worldwide, Willoughby, OH, USA.

Mistriotis, A, Bot, GPA, Picuno, P and Scarascia-Mugnozza, G (1997) Analysis of the efficiency of greenhouse ventilation using computational fluid dynamics. *Agricultural and Forest Meteorology*, 85: 217–228.

Montero, JI (2006) Evaporative cooling in greenhouses: Effect on microclimate, water use efficiency and plant response. *Acta Horticulturae*, 719: 373–384.

Montero, JI, Short, TH, Curry, RB and Bouerle, WL (1981) Influence of evaporative cooling systems on greenhouse environment. ASAE Paper No: 81-4027, ASAE, St. Joseph, MI 49085, USA.

Montero, JI, van Henten, EJ, Son, JE and Castilla, N (2011) Greenhouse engineering: New technologies and approaches. *Acta Horticulturae*, 893: 51–64.

Morgan, L (2013) Daily Light Integral (DLI) and greenhouse tomato production. *The Tomato Magazine*, Winter, p. 15.

MWPS (1987) *Structures and Environment Handbook*. 11th Edition. Midwest Plan Service, Iowa State University, Ames, IA, USA. ISBN: 0-89373-057-2.

Nam, YI, Yu, IH, Kim, TY, Roh, MY and Cho, MW (2007) Effect of newly developed fan and mist cooling system on greenhouse cooling and growth of cucumber. *Acta Horticulturae*, 761: 49–53.

Nelson, J and Bugbee, B (2014) Economic analysis of greenhouse lighting: Light emitting diodes vs. high intensity discharge fixtures. *PLoS One*, 9: e99010.

Nielsen, OF (2002) Natural ventilation of a greenhouse with top screens. *Biosystems Engineering*, 81: 443–452.

Okusima, L, Sase, S and Nara, M (1989) A support system for natural ventilation design of greenhouses based on computational aerodynamics. *Acta Horticulturae*, 284: 129–136.

OMAFRA (2001) *Growing Greenhouse Vegetables*. Ontario Ministry of Agriculture, Food and Rural Affairs. Publication 371. ISSN: 1492-6601. Toronto, Canada, p. 77.

Runkle, ES (2007) Light quality defined. *Greenhouse Product News Magazine*, June. Sparta, MI, USA, p. 82.

Runkle, ES (2010) The future of greenhouse lighting. *Greenhouse Product News Magazine*, September. Sparta, MI, USA, p. 66.

Runkle, ES (2018a) An update on LED lighting efficacy. *Greenhouse Product News Magazine*, July. Sparta, MI, USA, p. 58.

Runkle, ES (2018b) Horticultural Lighting Applications. *Greenhouse Product News Magazine*, June. Sparta, MI, USA, p. 58.

Runkle, ES (2019) DLI 'Requirements'. *Greenhouse Product News Magazine*, May. Sparta, MI, USA, p. 50.

Runkle, ES and Bugbee, B (2013) Problems with foot-candles, lux and lumens. *Greenhouse Product News Magazine*, September. Sparta, MI, p. 78.

Ryou, YS, Kang, YK, Kang, KC, Kim, YJ and Paek, Y (2008) Cooling performance of horizontal geothermal heat pump system for protected horticulture. *Journal of Bio-Environment Control*, 17: 90–95.

Santoso, M (1989) An alternative configuration of Rankine cycle engine-driven heat pump system. M.S. Thesis. The Ohio State University, Columbus, OH, USA.

Sase, S (1989) The effects of plant arrangement on airflow characteristics in a naturally ventilated glasshouse. *Acta Horticulturae*, 245: 429–435.

Sase, S (2006) Air movement and climate uniformity in ventilated greenhouses. *Acta Horticulturae*, 719: 313–324.

Seager, JCR (1982) Pot plants respond to root zone warming. *Farm and Food Research*, 12: 178–179.

Swensen, PF and Rose, RK (1977) Development of a high seasonal performance factor gas heat pump. Proc. 12th Intersociety Energy Conversion Engineering Conference, Washington, DC, USA, p. 390.

Thimijan, RW and Heins, RD (1983) Photometric, radiometric and quantum light units of measure: A review of procedures for interconversion. *Horticulturae Science*, 18(6): 818–822.

Voogt, J and van Weel, P (2008) Climate control based on stomatal behavior in a semi closed greenhouse system 'AIRCOKAS'. *Acta Horticulturae*, 797: 151–156.

Yildiz, I (1993) A performance study of a Rankine cycle driven heat pump system. M.S. Thesis. The Ohio State University, Columbus, OH, USA.

Yildiz, I (2021) *Greenhouse Engineering: Integrated Energy Management*. 1st Edition. CRC Press, Taylor and Francis Group, FL. ISBN: 978-1-4987-4313-6.

Yildiz, I, Christensen, RN and Drake, MS (1996) Energy consumption of a Rankine cycle heat pump system in a residential building. Süleyman Demirel University, Isparta, Turkiye. *Mechanical Engineering Journal*, 1(9): 57–66.

Yildiz, I, Lin, C and Christensen, RN (1993) A performance study of a Rankine cycle driven heat pump system. AES-Vol. 29, *Heat Pump and Refrigeration Systems Design, Analysis, and Applications. ASME*, 1993: 9–18.

Yildiz, I and Stombaugh, DP (2006a) Dynamic modeling of microclimate and environmental control strategies in a greenhouse coupled with a heat pump system. *Acta Horticulturae*, 718: 331–340.

Yildiz, I and Stombaugh, DP (2006b) Heat pump cooling and greenhouse microclimates in open and confined greenhouse systems. *Acta Horticulturae*, 719: 255–262.

Yildiz, I, Yue, J and Lowrey, J (2014a) Spatiotemporal distributions of energy and water consumption in open, semi-closed and closed loop heat pump greenhouse systems, and associated management strategies in Canada. *Acta Horticulturae*, 1037(1): 119–128.

Yildiz, I, Yue, J and Lowrey, J (2014b) Spatiotemporal distributions of energy and water consumption in open, semi-closed and closed loop heat pump greenhouse systems, and associated management strategies in the Canadian North. *Acta Horticulturae*, 1037(1): 171–177.

Yu, IH (2008) Optimum design of air-circulation fans using CFD model in air-heated greenhouse for chrysanthemum. Ph.D. Dissertation. Seoul National University, Seoul, Korea.

Zeroni, M and Gale, J (1982) Effect of root temperature on the development, growth, and yield of 'Sonia' roses. *Scientia Horticulturae*, 18: 177–184.

AQUATIC SYSTEMS

American Public Health Association (1989) *Standard Methods for the Examination of Water and Wastewater*. 19th Edition. American Public Health Association, Washington, DC.

American Public Health Association (1995) *Standard Methods for the Examination of Water and Wastewater*. 19th Edition. American Public Health Association, Washington, DC, p. 1108.

Arceivala, SJ (1983) Hydraulic modeling for waste stabilization ponds. *Journal of Environmental Engineering*, 109(5): 265–268.

Asato, Y (1976) Ultraviolet light inactivation and photoreactivation of AS-1 cyanophage in Anacystis nidulans. *Journal of Bacteriology*, 126: 550–552.

Bessems, E (1998) The effect of practical conditions on the efficacy of disinfectants. *International Biodeterioration and Biodegradation*, 41: 177–183.

Boyd, CE (1982) *Water Quality Management for Pond Fish Culture*. Elsevier Scientific Publishing Company, New York, NY, Chap. 2.

Braaten, B (1992) Impact of pollution from aquaculture in six Nordic countries. Release of nutrients, effects and waste water treatment. In: DePauw, N and Joyce, J (eds.), *Aquaculture and the Environment*. EAS Special Publications 1992, No. 16. Gent, Belgium, pp. 79–101.

Bullock, GL and Stuckey, HM (1977) Ultraviolet treatment of water for destruction of five gram-negative bacteria pathogenic to fishes. *Journal of Fisheries Research Board of Canada*, 34: 1244–1249.

Chen, S, Ling, J and Blancheton, J-P (2006) Nitrification kinetics of biofilms as affected by water quality factors. *Aquacultural Engineering*, 34: 179–197.

Chen, S and Malone, RF (1991) Suspended solids control in recirculating aquaculture systems. In: Timmons, MB (ed.), *Engineering Aspects of Intensive Aquaculture*. Northeast Regional Agricultural Engineering Service, Cooperative Extension, Ithaca, NY, pp. 170–186.

Chen, S, Timmons, MB, Aneshansley, DJ and Bisogni, JJ (1993) Suspended solids characteristics from recirculating aquacultural systems and design implications. *Aquaculture*, 112: 143–155.

Chen, S, Timmons, MB, Bisogni, JJ and Aneshansley, DJ (1994) Modeling surfactant removal in foam fractionation II: Experimental investigations. *Aquacultural Engineering*, 13: 121–138.

Chesness, JL, Poole, WH and Hill, TK (1975) Settling basin design for raceway effluent. *Transactions of the American Society of Agricultural Engineering*, 19: 192–194.

Chiang, HC and Lee, JC (1986) Study of treatment and reuse of aquacultural wastewater in Taiwan. *Aquacultural Engineering*, 5: 3001–312.

Colt, J and Orwicz, C (1991) Aeration in intensive culture. pp. 198–271. In: Brune, DE and Tomasso, JR (eds.), *Advances in World Aquaculture*. Vol. 3. The World Aquaculture Society, Baton Rouge, LA. pp. 180–319.

Colt, J and Watten, J (1988) Application of pure oxygen in fish culture. *Aquacultural Engineering*, 7: 397–441.

Colt, JE, Orwicz, K and Bouck, G (1991) Water quality considerations and criteria for high-density fish culture with supplemental oxygen. In: Colt, J and White, RJ (eds.), *Fisheries BioEngineering Symposium 10*. American Fisheries Society, Bethesda, MD, pp. 372–385.

Cripps, SJ, Bergheim, A (2000) Solids management and removal for intensive land-based aquaculture production systems. *Aquacultural Engineering*, 22: 22–56.

Crites, R and Tchobanoglous, G (1998) *Small and Decentralized Wastewater Management Systems*. McGraw-Hill, New York.

Davidson, J and Summerfelt, S (2005) Settler evaluation: Radial flow vs. tangential flow. *Global Aquaculture Advocate*, 8: 56–57.

de Veer, I, Moriske, HJ and Ruden, H (1994) Photochemical decomposition of organic compounds in water after U.V.—irradiation: Investigation of positive mutagenic effects. *Toxicology Letters*, 72: 113–119.

Easter, C (1992) System component performance and water quality in a recirculating system produce hybrid striped bass. Paper presented at Aquacultural Expo V. New Orleans, January 11–16.

Ebeling, JM and Wheaton, FW (2006) Batch in-situ determination of nitrification kinetics and performance characteristic curves for a bubble-washed bead filter. *International Journal of Recirculating Aquaculture*, 7: 13–41.

Eding, EH, Kamstra, A, Verreth, JAJ, Huisman, EA and Klapwijk, A (2006) Design and operation of nitrifying trickling filters in recirculating aquaculture: A review. *Aquacultural Engineering*, 34: 234–260.

EIFAC (1980) Symposium on new developments in the utilization of heated effluent and recirculation systems for intensive aquaculture. EIFAC, 11th Session, Stavanger, Norway, May 28–30th.

Eker, APM, Formenoy, L and de Wit, LEA (1991) Photoreactivation in the extreme halophilic archaebacterium Halobacterium cutirubrum. *Photochemistry and Photobiology*, 53: 643–651.

Environmental Protection Agency (EPA) (1975) Process design manual for suspended solids removal. U.S. EPA Technology Transfer Publication.

European Inland Fisheries of Advisory Commission (1973) Water quality criteria for European freshwater fish. *Report on Ammonia and Inland Fisheries*, 7: 1011–1022.

FAO (2014) *The State of World Fisheries and Aquaculture*. Fisheries and Aquaculture Department, VialedelleTerme de Caracalla, 00153 Rome, Italy, 2007. ISBN 978-92-5-108275-1.

Friedberg, EC, Walker, GC and Siede, W (1995) *DNA Repair and Mutagenesis*. ASM Press, Washington, DC.

Gilbert, B (1990) *Aquaculture*, Gilbert, B (ed.), Vol. 1 & 2. Ellis Horwood Limited, Chichester, West Sussex, PO19EB, England. ISBN 0-13-044108-2.

Glass, C and Silverstein, J (1998) Denitrification kinetics of high nitrate concentration water: pH effect on inhibition and nitrite accumulation. *Water Resources*, 32: 831–839.

Grady, CPL and Lim, HC (1980) *Biological Wastewater Treatment: Theory and Applications*. Marcel Dekker, New York.

Gregory, R and Zabel, TF (1990) Sedimentation and flotation. In Pontius, FW (ed.), *Water Quality and Treatment*. 4th Edition, McGraw-Hill, Inc., New York, pp. 367–453.

Harm, W (1968) Dark repair of photorepairable u.v. lesions in Escherichia coli. *Mutation Research*, 6: 25–35.

Haug, RT and McCarty, PL (1972) Nitrification with submerged filters. *Journal of the Water Pollution Control Federation*, 44: 2086.

Hazen, A (1904) On sedimentation. *Transactions ASCE*, 53: 45–88.

Heinzel, M (1998) Phenomena of biocide resistance in microorganisms. *International Biodeterioration and Biodegradation*, 41: 225–234.

Henderson, JP and Bromage, NR (1988) Optimizing the removal of suspended solids from aquaculture effluents in settlement lakes. *Aquacultural Engineering*, 7: 167–188.

Honda, H, Watanaba, Y, Kikuchi, K, Iwata, N, Takeda, S, Uemoto, H, Furata, T and Kiyono, M (1993) High density rearing of Japanese Flounder, Paralichthys olivaceus with a closed seawater recirculation system equipped with a denitrification unit. *Suisanzoshoku*, 41: 19–26.

Hopkins, TA and Manci, WE (1989) Feed conversion, waste and sustainable aquaculture, the fate of the feed. *Aquaculture Magazine ban*, 15(2): 32–36.

Huguenin, JE and Colt, J (1989) *Design and Operating Guide for Aquaculture Seawater Systems*. Elsevier Science Publishing Company, Inc., New York, NY, Chap. 12.

IDEQ (Idaho Division of Environmental Quality) (1988) Idaho Waste Management Guidelines for Aquaculture Operations. Idaho Department of Health and Welfare, Division of Environmental Quality, Twin Falls, ID.

Jacob, B (2015) *A Guide to Recirculation Aquaculture*. The Food and Agriculture Organization of the United Nations (FAO) and EUROFISH International Organization. ISBN 978-92-5-108776-3.

Jagger, J (1967) *Introduction to Research in Ultraviolet Photobiology*. Prentice-Hall Inc., Engelwood.

Jones, RD and Morita, RY (1985) Low temperature growth and whole cell kinetics of marine ammonium oxidizer. *Marine Ecology Progress Series*, 21: 239–243.

Juell, JE (1991) Hydroacoustic detection of food waste—a method to estimate maximum food intake of fish populations in sea cages. *Aquacultural Engineering*, 10: 207–217.

Kelner, A (1961) Historical background of the study of photoreactivation. In: Christensen, BC and Buchmann, B (eds.), *Progress in Photobiology*. Elsevier Publishing Company, Amsterdam, The Netherlands, pp. 276–278.

Kostenbauder, HB (1991) Physical factors influencing the activity of antimicrobial agents. In: Block, SS (ed.), *Disinfection, Sterilization, and Preservation*, pp. 59–71.

Kumar, S (1984) Closed recirculating nitrification systems for soft-shelled crabs. Masters Thesis. Louisiana State University, Baton Rouge, p. 99.

Lennard, WA (2006) Aquaponic integration of Murray Cod (*Macculochella peelii peelii*) aquaculture and Lettuce (*Lactuca sativa*) hydroponics. PhD Thesis. School of Applied Sciences, Department of Biotechnology and Environmental Biology, Royal Melbourne Institute of Technology, Melbourne, Victoria, Australia.

Lewis, WM, Yopp, JH, Schram Jr. HL and Brandenburg, AM (1978) Use of hydroponics to maintain quality of recirculated water in a fish culture system. *Transactions of the American Fisheries Society*, 197: 92–99.

Liao, PB and Mayo, RD (1974) Intensified fish culture combining water reconditioning with pollution abatement. *Aquaculture*, 3: 61–85.

Libey, GS (1993) Evaluation of a drum filter for removal of solids from a recirculating aquaculture system. In: Wang, JK (ed.),

Techniques for Modern Aquaculture. American Society of Agricultural Engineers, St. Joseph, MI, pp. 519–532.

Liltved, H and Cripps, SJ (1999) Removal of particle associated bacteria by prefiltration and ultraviolet irradiation. *Aquaculture Research*, 30: 445–450.

Liltved, H and Landfald, B (1996) Influence of liquid holding recovery and photoreactivation on survival of ultraviolet-irradiated fish pathogenic bacteria. *Water Resources*, 30: 1109–1114.

Loyless, CL and Malone, RF (1997) A sodium bicarbonate dosing methodology for pH management in freshwater-recirculating aquaculture systems. *Progressive Fish-Culturist*, 59: 198–205.

Malone, RF and Beecher, LE (2000) Use of floating bead filters to recondition recirculating waters in warmwater aquaculture production systems. *Aquacultural Engineering*, 22: 57–73.

Malone, RF, Beecher, LE and DeLos Reyes, AA Jr. (1998) Sizing and management of floating bead bioclarifiers. In: Libey, GS (ed.), *Second International Conference on Recirculating Aquaculture (Proc.)*, 16–19 July, Roanoke, VA. Northeast Regional Agricultural Engineering Service, Ithaca, NY, pp. 319–341.

Malone, RF, Chitta, BS and Drennan, DG (1993) Optimizing nitrification in bead filters for warmwater recirculating aquaculture systems. In: Wang J (ed.), *Techniques for Modern Aquaculture*. American Society of Agricultural Engineers, St. Joseph, MI, pp. 315–325.

Manthe, DP, Malone, RF and Kumar, S (1988) Submerged rock filter devaluation using oxygen consumption criterion for closed recirculating systems. *Aquacultural Engineering*, 7: 97–111.

McLaughlin, TW (1981) Hatchery effluent treatment. In: Allen, LJ and Kinney, EC (eds.), *Bio-Engineering Symposium for Fish Culture (FCS Publ. 1)*. American Fisheries Society, Bethesda, MD, pp. 167–173.

McMurtry, MR, Nelson, PV, Sanders, DC and Hughes, L (1990) Sand culture of vegetables using recirculating aquaculture effluents. *Journal of Applied Agricultural Research*, 5(4): 280–284.

Meade, JW (1985) Allowable ammonia for fish culture. *Progressive Fish-Culturist*, 47(3): 135–145.

Miller, RV, Jeffrey, W, Mitchell, D and Elasri, M (1999) Bacterial responses to ultraviolet light. *ASM News*, 65: 535–541.

Montgomery, JM (Consulting Engineers, Inc.) (1985) *Water Treatment Principles and Design*. John Wiley and Sons, p. 696.

Moss, SH and Davies, JG (1974) Interrelationship of repair mechanisms in ultraviolet-irradiated Escherichia coli. *Journal of Bacteriology*, 120: 15–23.

Mudrak, VA (1981) Guidelines for economical commercial fish hatchery wastewater treatment systems. In: Allen, LJ and Kinney, EC (eds.), *Proceedings of the Bio-Engineering Symposium for Fish Culture*. American Fisheries Society, Fish Culture Section, Bethesda, MD, pp. 174–182.

Mudrak, VA and Stark, KR (1981) Guidelines for economical commercial hatchery wastewater treatment systems. U.S. Dept. of Commerce, NOAA.

Muir, JF (1982) Recirculated system in aquaculture. In: Muir, JF and Roberts, RJ (eds.), *Recent Advances in Aquaculture*. Vol. 1, Croom Helm and Westview Press, London, p. 453.

Naegel, LCA (1977) Combined production of fish and plants in recirculating water. *Aquaculture*, 10: 17–24.

Needham, J (1988) Salmon smolt production. In: Laird and Needham (eds.), *Salmon and Trout Farming*. Halsted Press, Div. of John Wiley and Sons.

Ødegaard, H (2006) Innovations in wastewater treatment: The moving bed biofilm process. *Water Science Technology*, 53: 17–33.

Oliver, BG and Carey, JH (1976) Ultraviolet disinfection: An alternative to chlorination. *Journal of the Water Pollution Control Federation*, 48: 2619–2624.

Otte, G and Rosenthal, H (1979) Management of a closed brackish water system for high-density fish culture by biological and chemical water treatment. *Aquaculture*, 18: 169–181.

Park, JK (2000) Biological nutrient removal, theories and design. University of Wisconsin-Madison. Dept. Civil and Environ. Engineering. Available on: http://www.engr.wisc.edu/cee/faculty/park_jae.html

Parker, NC (1981) An air-operated fish culture system with water-reuse and subsurface silos. In: Allen, LJ and Kinney, EC (eds.), *Bio-Engineering Symposium for Fish Culture (FCS Publ. 1)*. American Fisheries Society, Bethesda, MD, pp. 131–137.

Perry, RH and Chilton, CH (1973) *Chemical Engineers Handbook*. 5th Edition. McGraw-Hill Book Company, New York.

Piper, RG, McElwainn, IB, Orme, LE, McCraren, JP, Fowler, LG and Leonard, JR (1982) *Fish Hatchery Management*. United States Department of the Interior, Fish and Wildlife Service, Washington, DC, Chap. 1.

Rakocy, JE (1984) A recirculating system for tilapia culture and vegetable hydroponics. In: Smitherman, RC and Tave, D (eds.), *Proceedings of the Auburn Symposium on Fisheries and Aquaculture*. Auburn University, Auburn, AL, pp. 103–114.

Rakocy, JE (1995) Aquaponics: The integration of fish and vegetable culture in recirculating systems. In: Palada, MC and Clarke, CC (eds.), *Proceedings of the 30th Annual Meeting of the Caribbean Food Crops Society*, St. Thomas USVI, pp. 101–108.

Rakocy, JE and Hargreaves, JA (1993) Integration of vegetable hydroponics with fish culture: A review. In: Wang, J-K (ed.), *Techniques for Modern Aquaculture*. American Society of Agricultural Engineers, St. Joseph, MI, pp. 112–136.

Rakocy, JE, Masser, MP and Losordo, TM (2006) Recirculating aquaculture tank production systems: Aquaponics-integrating fish and plant culture. *Southern Regional Aquaculture Center*, Publication No. 454: 1–16.

Robertson, WD (1992) *Assessment of Filtration and Sedimentation Systems for Total Phosphorus Removal at the Lake Utopia Fish 5 Culture Station*. Canadian Fish Culture Operations, Aquaculture Division, Connors Bros., Limited, Blacks Harbour, New Brunswick, Canada.

Rosenthal, H (1981) Ozonation and sterilization. In: Tiens, K (ed.), *Proceedings from the World Symposium on Aquaculture in Heated Effluents and Recirculation System*. Vol. 1, Heenemann, Berlin, Germany, pp. 219–274.

Russo, RC and Thurston, RV (1991) *Aquaculture and Water Quality*. Brune, DE and Tomasso, JR (eds.). World Aquaculture Society, pp. 58–89.

Rusten, B, Eikebrokk, B, Ulgenes, Y and Lygren, E (2006) Design and operations of the Kaldnes moving bed biofilm reactors. *Aquacultural Engineering*, 35: 322–331.

Shelton, JL and Boyle, CE (1983) Correction factors for calculating oxygen-transfer rates of pond aerators. *Transactions of the American Fisheries Society*, 112: 120–122.

Stechey, M and Trudell, Y (1990) Aquaculture wastewater treatment: Wastewater characterization and development of appropriate treatment technologies for the New Brunswick smolt production industry. Prepared for New Brunswick Department of the Environment and Local Government, PO Box 6000, Fredericton, New Brunswick, Canada, E3B 5H1, p. 101.

Summerfelt, S and Hochheimer, JN (1997) Review of ozone processes and application as an oxidizing agent in aquaculture. *Progressive Fish-Culturist*, 59: 94–105.

Summerfelt, ST and Wade, EM (1997) Recent advances in water treatment processes to intensify fish production in large recirculating systems. Advances in Aquacultural Engineering, Proceedings from Aquacultural Engineering Society Technical Sessions at the 4th International Symposium on Tilapia in Aquaculture, November 9–12, Orlando, FL, NRAES-105, pp. 350–367.

Sutton, RJ and Lewis, WM (1982) Further observations on a fish production system that incorporates hydroponically grown plants. *Progressive Fish-Culturist*, 44: 55–59.

Tchobanoglous, G, Burton, FL (1991) *Wastewater Engineering: Treatment, Disposal, and Reuse.* 3rd Edition. McGraw-Hill, Inc., New York.

Timmons, MB and Ebeling, JM (2010) *Recirculating Aquaculture.* 2nd Edition. NRAC Publication No. 401-2010. ISBN 978-0-9712646-2-5.

Tucker, Jr. JW (1999) *Species Profile Grouper Aquaculture.* SRAC. Pub. 721, p. 10.

van Gorder, SD and Jug-Dujakovic, J (2005) Performance characteristics of rotating biological contactors within two commercial recirculating aquaculture systems. *International Journal of Recirculating Aquaculture*, 6: 23–38.

van Rijn, J, Tal, Y and Schreier, HJ (2006) Denitrification in recirculating systems: Theory and applications. *Aquacultural Engineering*, 34: 364–376.

Vinci, B, Summerfelt, S and Bergheim, A (2001) Solids control. In: Summerfelt, ST (ed.), *Presentation Notebook: Aquacultural Engineering Society Workshop: Intensive Fin-fish Systems and Technologies.* Aquaculture 2001, International Triannual Conference and Exposition of the World Aquaculture Society, January 23, Disney's Coranado Springs Resort, Orlando, FL.

Walker, GC (1984) Mutagenesis and inducible responses to deoxyribonucleic acid damage in Escherichia coli. *Microbiological Reviews*, 48: 60–93.

Warrer-Hansen, I (1982) Evaluation of matter discharged from trout farming in Denmark. In: Alabaster, JE (ed.), *Report of the EIFAC Workshops on Fish Farm Effluents.* Silkegorg, Denmark, Technical Paper No.4.

Watten, BJ and Busch, RL (1984) Tropical production of tilapia (*Sarotherodon aurea*) and tomatoes (*Lycopersicon esculentum*) in a small-scale recirculating water system. *Aquaculture*, 41: 271–283.

Wedemeyer, GA (1996) *Physiology of Fish in Intensive Culture.* International Thompson Publishing, New York, NY.

Wedemeyer, GA, Nelson, NC and Yasutake, WT (1979) Physiological and biochemical aspects of ozone toxicity to rainbow trouts (Salmo gairdneri). *Journal of the Fisheries Research Board of Canada*, 36: 605–614.

Wheaton, FW (1985) *Aquacultural Engineering.* 2nd Edition. Wiley-Interscience, New York, p. 708.

Wren, SW (1984) Comparison of hydroponic crop production techniques in a recirculating fish culture system. M.S. Thesis, Texas A&M University, College Station, TX.

Zhu, S and Chen, S (2002) The impact of temperature on nitrification rate in fixed film biofilters. *Aquacultural Engineering*, 26: 221–237.

6 Instrumentation, Monitoring, and Control Systems

6.1 INTRODUCTION

Measurement, monitoring, and control systems form the foundation of controlled environment systems engineering. In controlled environment systems, such as livestock, poultry, and particularly greenhouse and aquacultural production systems, we are dealing with extremely complex and nonlinear living systems whose behaviors change in time depending on numerous disturbances. Therefore, it is many times difficult to control, for instance, greenhouse environments with classical controllers; hence, complex and smarter approaches are required in order to bring the indoor and outdoor stakeholders, such as weather, the cost of energy, water, CO_2 and others all together, and make proper adjustments in a greenhouse environment instantly for healthy crop development and therefore a profitable operation.

Today, smart digital environmental control systems are available, and they control every single parameter in an integrated fashion in modern commercial greenhouses, poultry, and recirculating aquatic systems providing smart decision making processes for optimum plant, poultry, and fish growth conditions. By and large, this is also true for certain livestock production systems. Further precision controls and therefore savings can be achieved via incorporating smarter strategies with artificial intelligence (AI) algorithms into the environmental control systems. For instance, through reducing ventilation a bit or maintaining little higher indoor relative humidity levels without harming the crop growth and development (and therefore the eventual economic gains), achieving special savings available through power companies, supplying power from co-generation plants and alternate power supply plans, etc., and performing CO_2 enrichment and storing the heat in hot water storage tanks for later use in heating, we can reduce energy losses and save dollars escaping through greenhouse vents or our pockets. The consequence of every management step and strategy need to be evaluated and dynamically optimized in an integrated fashion with continuous environmental monitoring before it is actuated. Today, the remote monitoring and control systems are on duty day and night continuously; and with the help of artificial intelligence, computer vision and enhanced sensor integration, our controlled environment plant, animal, and aquatic systems will soon become more flexible and autonomous operations, and therefore more universal and sustainable.

Before something is controlled, however, its position needs to be sensed first somehow so that proper adjustments can be made in its position. And we call these sensing devices as sensors, which are essential components of any controlled environment monitoring and control systems. Depending upon the nature of controlled environment facility, these sensors measure at least conditions of different environmental variables such as temperature, relative humidity, and other air and water quality parameters such as pH, alkalinity, DO, TAN, and CO_2 concentrations, aeration system pressure, water level, water flow, electric power, light intensity, wind speed, etc. Today, both hardwired and wireless monitoring and control systems are available, and commercial operations have either one of these types or sometimes the combination of both. The latest internet of things (IoT) technology makes it possible to remotely monitor and control devices over the internet. It should be kept in mind that control systems are limited by the accuracy and responsiveness (time constant) of their sensors; that is, they not only need to sense the environment accurately, but also fast enough so that the control system can take proper actions in a timely manner. Recall that fish can die in minutes if we have failure in water flow and DO providing systems. Sensor accuracy and reliability under greenhouse, livestock, poultry, and aquatic conditions are particularly important due to the unfavorable environmental conditions in such facilities.

This chapter briefly covers control theory and classification first, then moves to classical and modern control theories and system classifications, and concludes with future evolution of control systems after covering instrumentation, software, and monitoring systems.

6.2 CONTROL THEORY

A *control system* has four main functions termed as measuring, comparing, computing and correcting functions. These four functions are realized by a detector, a transducer, a transmitter, a controller and a final control element. The detector, transducer and transmitter, all contained in a single unit (e.g., a resistance thermometer) do the measuring task. Comparing and computing tasks are performed by the controller, which may be implemented electronically by a proportional (P) control, integral (I) control, a proportional-integral (PI) controller, a proportional-derivative (PD) controller, a proportional-integral-derivative (PID) controller or programmable logic controller. Older controller units have been mechanical devices, for instance, as in a carburetor. Correcting task is, on the other hand, performed by a final control element, the actuator. The final control element changes an input or output in the control system that influences the adjusted or controlled variable such as temperature, CO_2 concentration, light intensity, etc.

Control theory is an interdisciplinary branch of engineering and mathematics that deals with influencing the

DOI: 10.1201/9781003349297-6

behavior of dynamic systems, and how their behavior is modified by feedback. The usual objective of control theory is to control a system or process so its output tracks a desired control signal (the reference), which may have a fixed or changing value. In order to perform this task, a controller is designed, which monitors the output and compares it with the reference value. The difference between actual and desired outputs (the error) is fed back as the input to the system in order to bring the actual output closer to the set reference.

A system or process to be controlled is represented by a *block diagram* in Figure 6.1. The input-output relationship represents the cause-and-effect relationship of the process, which in turn represents a processing of the input signal to provide an output signal variable, often with power amplification.

There are two main types of control systems. When there is no feedback as shown in Figure 6.2, the control system is called as an *open-loop controller* or a *non-feedback controller* because the system output is not measured, fed back and used to change the position of the control. As a consequence of this configuration, the controller cannot compensate for changes acting on the system or process. If the process transfer function, which theoretically models the output of the process for each possible input, is known and there is very small disturbance, an open-loop controller, as shown in Figure 6.2, would satisfy the most design requirements. A manual control system is an open loop control system. Examples of an open-loop control system can be listed as an electric toaster, automatic coffee maker, light switch, automatic washing machine, timer based clothes drier, etc. we use every day.

In a *closed-loop control system*, as shown in Figure 6.3, on the other hand, data from a sensor monitoring the system output enters a controller, which continuously subtracts the quantity representing the actual output from the reference quantity representing the desired output response. The measure of the output is called the *feedback signal*. An open-loop control system can as well be easily converted into a closed-loop system by providing a feedback.

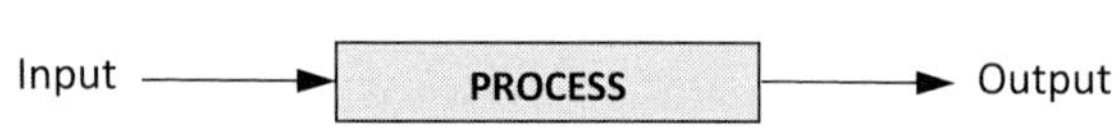

FIGURE 6.1 A process to be controlled.

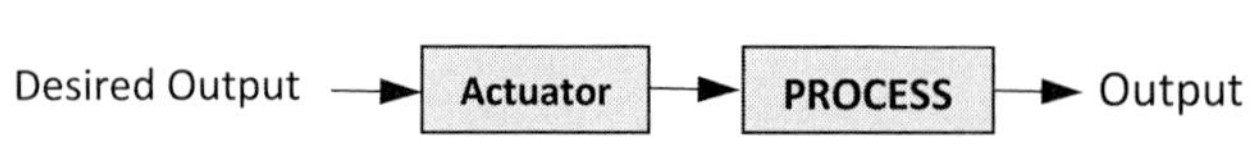

FIGURE 6.2 Open-loop control system (without feedback).

With an accurate sensor, the measured output is equal to the actual output of the system. The difference between the actual (measured) and the desired output (i.e., the error) determines the position of the control. The result is to continuously and dynamically match the actual or measured output to the reference response in order to maintain the desired system output. As the sensed output drops below the reference, obviously the error increases, hence the throttle opens. These control systems using the *feedback loop*, as shown schematically in Figure 6.3, affect the system output, which in turn is measured and fed back to the controller dynamically. Quite often the error is amplified and used to control the process so that the difference is continually reduced. The output of the control device causes the actuator to modulate the process in order to reduce the error. The feedback concept has been the foundation for control system analysis and design. If the feedback signal is positive then system is called a *positive feedback system*, and if the feedback signal is negative then the system is called a *negative feedback system*. With dynamic variations and significant disturbance in the process, feedback control is the only choice, whose simplicity is quite appealing. Significant steady-state error and overshoot are the most common problems with this proportional controller, a control response in proportion to how far away the measured output is from the desired output response.

A common example of a closed-loop control system is a person steering a car by visually inspecting the car's location on the road, and based on this visual "feedback", making proper adjustments and controlling the car's position. The system shown in Figure 6.3 is called a negative feedback control system; because the measured output is subtracted from the desired input and the difference (error) is used as the input signal to the power amplifier. Many other examples can be given for the closed-loop feedback system such as automatic electric iron, water level controller, an air conditioner, etc.

Increasing complexity of systems and interest in achieving optimal performance has forced control systems to grow tremendously in the past several decades. Furthermore, as the systems become more complex, the interrelationship of many controlled variables such as those in greenhouses need to be considered in the control scheme. A block diagram schematically demonstrating a multivariable control system is shown in Figure 6.4.

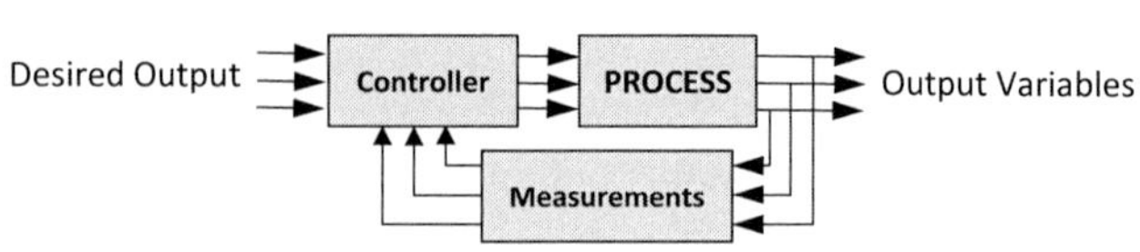

FIGURE 6.4 Multivariable control system.

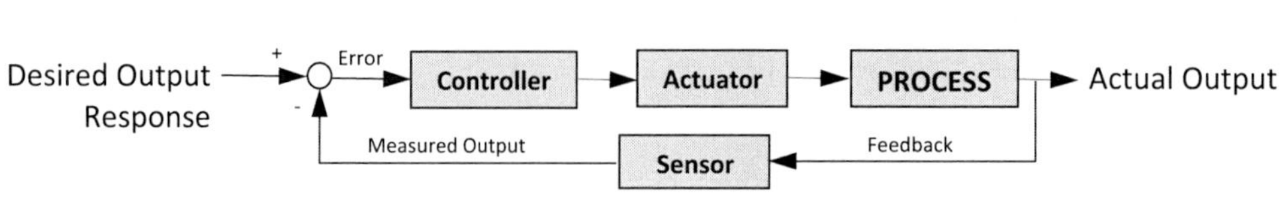

FIGURE 6.3 Closed-loop feedback control system.

6.3 CLASSIFICATION

There are numerous types of control systems, which are broadly categorized and discussed below.

6.3.1 Linear versus Nonlinear Control Theory

The field of control theory can be divided into two branches, linear and nonlinear. In order to comprehend the linear control system, we first need to cover the principle of superposition. The principle of superposition has two important properties termed as homogeneity and additivity. A system is said to be homogeneous if when the input is multiplied by a constant, the output is also multiplied by the same constant. Now, let's suppose there is a system, which is fed by the input a_1 and provides the output b_1 for the input a_1 for the first time; and the next time, when the system is fed by the input a_2, the system provides the output b_2 for the input a_2. When the sum of the previous inputs (i.e., $a_1 + a_2$) is provided as the input to the system and we receive the output as the sum of previous outputs as well (i.e., $b_1 + b_2$), then it is said that the system follows the property of additivity. Now, the linear control systems can be defined as the systems, which obey the superposition principle, i.e., the principle of homogeneity and additivity (only exists in theory). This roughly means that the output is proportional to the input, and these systems are governed by linear differential equations.

A nonlinear control system is, on the other hand, defined as a control system that does not obey the principle of homogeneity. In practice, all control systems are nonlinear systems where we cannot apply the principle of superposition. Sometimes, it is even desirable to introduce a nonlinearity intentionally in order to improve the performance of a system or make its operation safer and more economical. One of the simplest examples of a system with an intentionally introduced nonlinearity is a relay controlled or ON/OFF system. For instance, a furnace in a home heating system is turned ON when the temperature falls below a specified temperature and OFF when the temperature exceeds this time another specified temperature. Analysis of different nonlinear control systems is usually very difficult due to their nonlinear behavior. These systems are often governed by nonlinear differential equations, and such systems are quite often analyzed using numerical methods on computers by simulating their operation using a simulation language, for instance.

6.3.2 Frequency Domain versus Time Domain

Mathematical techniques for analyzing and designing control systems, on the other hand, fall into two different domains called as the *frequency domain* and the *time domain*.

In frequency domain, the state variables, i.e., the input, output and feedback are all represented as functions of frequency. The input signal and the transfer function are converted from time functions to functions of frequency by a transform function such as the Fourier, Laplace, or Z transform. In this technique, the differential equations representing the system are replaced by algebraic equations in the frequency domain, so that the mathematical solutions become much simpler; however, it should be noted that frequency domain techniques cannot be used with nonlinear systems, rather they only be used with linear systems.

As frequency domain techniques can only be used with linear systems, time domain is extensively used for analyzing the real-world nonlinear systems. In time domain type design, the state variables are represented as functions of time using differential equations, which are obviously more difficult to solve compared to algebraic equations; however, they have turned into routine analyses with the help of computer simulation techniques today.

6.3.3 SISO versus MIMO Systems

Based on the number of inputs and outputs, on the other hand, control systems are divided into two different categories: *single-input single-output* (SISO) and *multiple-input multiple-outputs* (MIMO).

SISO is the simplest and most common type, in which one output is controlled by one control signal. Simple example is an audio system where the input audio signal is the control input, and the sound waves from the speaker is the system's output. Other examples include temperature control, position control system, etc.

MIMO are, however, found in more complicated systems. For example, modern greenhouse facilities today are composed of many segments, each of which is controlled by a separate actuator. Complex systems such as nuclear reactors, greenhouses, and many other biological systems are handled by MIMO control systems such as programmable logic controllers (PLCs).

6.3.4 Analog versus Digital Systems

Control systems can be classified into two different categories depending on the input signal: *analog systems* and *digital systems*.

Analog control systems, also known as continuous control systems, have a continuous signal such as 4–20 mA or 0–10 VDC as the input to the system to control different equipment. These signals are continuous input signals such as sinusoidal, square, or triangle, etc., and that's why they are called continuous control systems. This type of control is preferred where a faster equipment response is needed, and is the standard method of control for certain equipment such as proportioning devices and variable speed motors.

Digital control systems, also known as discrete control systems, have a discrete or a pulse input signal to the system. And these signals have a discrete interval of time. Various sources of continuous (analog) input signals such as sinusoidal or square type can be converted into a discrete signal. Simply, digital control is a simple on/off control and used to control on/off devices such as louvers, motors, solenoid valves, etc. Proportional devices such as curtains, motorized valves, vents, and other on/off devices can be controlled with a digital signal as well. Digital control is achieved by activating an open or close relay for a period of time to move the device to the desired position. Then, proportional devices are controlled based on the movement time.

Digital systems handle nonlinear control systems more effectively, and are more reliable and accurate compared to the analog control systems.

6.3.5 Lumped versus Distributed Parameter Systems

Lumped parameter control systems have the various active (current sources, voltage sources, etc.) and passive (resistors, capacitors, etc.) components concentrated at singular points in space, hence the spatial derivatives are neglected and we only have the time derivatives to consider. Therefore, the analyses of such systems include ordinary differential equations and are relatively easy to handle. An electrical circuit with passive elements like resistor, capacitor, etc. is a common example for such systems, and the physical quantities current and voltage are functions of time only.

Distributed parameter control systems, on the other hand, have various active and passive components distributed in space, so that physical quantities depend on both time and space. And therefore, the analyses of such systems include partial differential equations and are slightly difficult compared to lamped parameter control systems. The classical example is the electrical line where inductance, capacity and resistance are not constant but functions of length. This leads to partial derivatives in time and space (length).

6.4 CLASSICAL CONTROL THEORY

In order to overcome the above-discussed limitations of the open-loop controller, feedback is introduced in control theory. As mentioned earlier, a closed-loop controller provides feedback to adjust outputs of a dynamic system. Process inputs such as voltage applied to an electric motor have an effect on the process outputs (e.g., speed or torque of a pump or a fan motor) measured with sensors and processed by the controller; the result (the control signal) is fed back as input to the process and closes the loop; that's why, it is called as a closed-loop controller.

It is possible for some systems to use closed-loop and open-loop controls simultaneously. In such systems, the open-loop control is termed feed-forward and serves to further improve the already improved reference tracking performance of a closed-loop controller. This type of dual loop control provides a higher degree of precision than controls based on closed-loop control only because the response can be predicted ahead of time.

6.4.1 Closed-Loop Transfer Function

In a closed-loop or feedback control system, as discussed earlier, the output of the system is fed back to the desired output response (set-point) via a sensor measurement. Then, the controller takes the difference (error) between the set-point and the output to change the inputs u to the system under control. This is shown in Figure 6.3. This closed-loop control system is called a single-input-single-output (SISO) control system; and as discussed earlier, multi-input multi-output (MIMO) systems with more than one input and output are common as well, especially in modern greenhouse operations.

If we assume that the controller, the process, and the sensor are linear and not time-dependent, then the abovementioned systems can be analyzed using the Laplace transform. Its simplest form can be given as $u = Ke$, where K is a constant. It means that a proportional controller provides a control signal, which is proportional to the error (e), i.e., a proportional controller returns its input multiplied by the proportional gain K. Significant steady-state error and overshooting are the common problems with this type of controller.

6.4.2 PID Controller

In order to correct the abovementioned steady-state error and overshooting problems with the constant gain (proportional) controller, two additional terms, the integral and the derivative terms, are added as shown below in Figure 6.5. All together, these three terms operate on the error signal $e(t)$ to produce a control signal $u(t)$. The end result is the well-known *proportional-integral-derivative* (PID) controller, which is the most-used feedback control system by engineers today.

The proportional term usually plays the key role ensuring the stability, with the integral and derivative terms added to reduce or eliminate the steady-state error by rejecting step disturbance and the overshoot by providing damping, respectively. If an integral action is executed, the integral of the error is added to the control signal as well. In the case the error signal is large, then the control signal increases quickly; but when the error signal is small, then the control signal increases slowly. When a derivative control action is executed, on the other hand, a signal proportional to the signal error change rate is provided only when a change in the error signal is present. The main problems with the PID controller are that the integrator introduces a phase lag that might lead to stability problems and the differentiator makes the controller sort of sensitive to noise. The desired closed loop dynamics is mainly obtained by adjusting the three parameters, Kp, Ki, Kd, and often iteratively and with no specific knowledge of a process model. PID controllers are the most established class of control systems; however, they cannot be used in some complicated cases, especially if MIMO systems are considered. For example, greenhouse systems are quite complex mathematical models with many nonlinearity and MIMO systems; therefore, PID controllers are not the most appropriate systems for solving the pertinent problems in greenhouses. As a consequence, this has led to the development of new control techniques.

6.4.3 Nonlinear PID Controller

Nonlinear PID is an improved PID controller for better performance. Maintaining the simplicity and intuition of PID, nonlinear PID makes the PID stronger with nonlinear gains. In a nonlinear PID controller, when the error is small, the need for the integral control is reduced by making the proportional gain bigger. The differential gain is also reduced for small errors to reduce sensitivities to sensor noise.

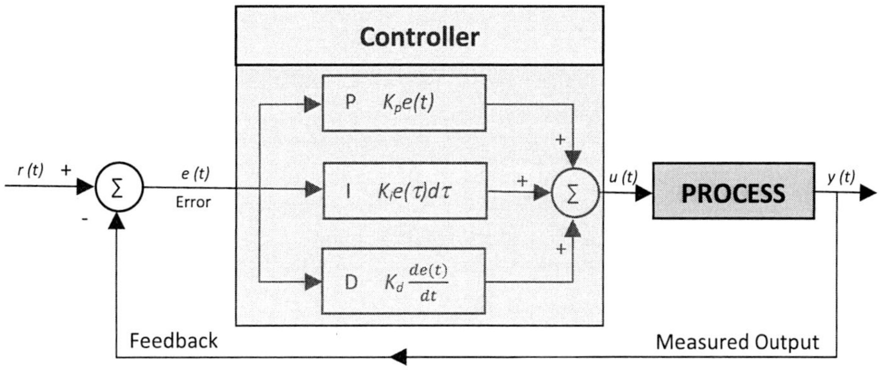

FIGURE 6.5 Block diagram of a PID controller in a feedback system. $r(t)$ is the desired output response or set-point, and $y(t)$ is the actual (measured) output.

6.5 MODERN CONTROL THEORY

In this Chapter, we have covered some fundamental analysis and design methods in classical control theory, primarily the feedback control, the development of which was primarily driven by engineering practice and needs of modern industry and society. With an accurate sensor, the measured output is equal to the actual output; and the difference between the desired output and the actual output is equal to the error, which is then adjusted by the control device as covered earlier. Over the last few decades or so, huge efforts in control research have led to the creation of modern mathematical control theory or advanced control paving the way for intelligent control, which uses a number of artificial intelligence (AI) computing methods such as artificial neural networks (ANN), Bayesian probability, fuzzy logic systems (FLS), machine learning systems (MLS), expert systems (ES), evolutionary computation and genetic algorithms (GA), etc. to control a dynamic system. This development originally started with *optimal control theory* in the late 1950s and early 1960s to study the optimality of control design. Briefly, a cost function is minimized in optimal control theory; and for deriving optimal controllers, analytical or computational techniques are employed. Examples of optimal control include minimized cost, fuel or energy use, water use situations, etc., which are quite applicable in controlled environment operations. As they can guarantee closed-loop stability, the most widely used optimal control design approaches are Model Predictive Control (MPC) and linear-quadratic-Gaussian control (LQG). Along with PID controllers, MPC systems are the most widely used control techniques in process control. In optimal control of greenhouse environments, for instance, it is required to consider aspects like greenhouse energy consumption, water consumption, crop response, etc. as input parameters of the control system, for which ES and MPC are the most commonly used techniques. Other artificial intelligence based approaches are also considered as optimal control when they consider input parameters such as energy consumption, water consumption, crop responses, etc. As an example, Figure 6.6 schematically shows the general classification of greenhouse control systems.

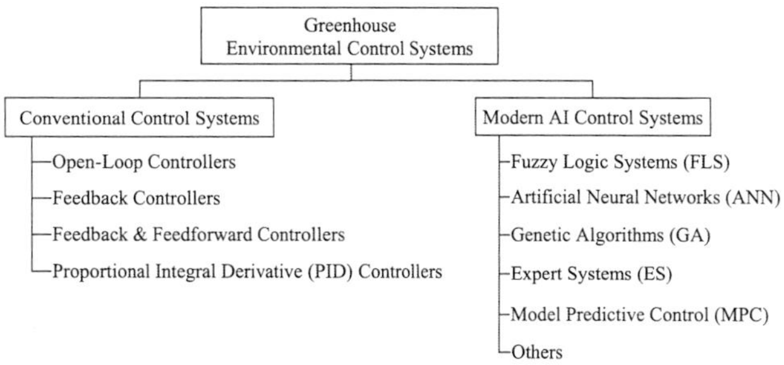

FIGURE 6.6 Classification of greenhouse control systems.

Source: *Greenhouse Engineering*, Edn. 1 by İlhami Yıldız, Copyright © 2021 by Imprint. Reproduced by permission of Taylor & Francis Group (*Greenhouse Engineering*).

Other major achievements in modern control theory include, for instance, multi-input multi-output (MIMO) control systems as discussed earlier, and this achievement extended the well-known SISO concepts to MIMO systems; Adaptive control is another example of modern control approaches, which is primarily designed to extend the operating range of a controller by modulating automatically the controller parameters based on the estimated dynamic changes in the processes; analysis and design of nonlinear control systems, and so on. Modern control theory employs the time-domain state-space representation, a mathematical model of a physical system as a set of input, output and state variables related by first-order differential equations, whereas the frequency domain analysis forms the basis of the classical control theory. In contrast to the frequency domain approach, the utilization of the state-space representation is not limited to systems with linear components and zero initial conditions.

6.6 SYSTEM CLASSIFICATIONS

Control systems can further be classified based on the characteristics of systems being controlled such as linear systems control, nonlinear systems control, decentralized systems control, centralized systems control, etc.

Linear and nonlinear systems have already been discussed, and they can even be elaborated further and we can talk about, for instance, the linearization of control theory, and then the application of linear techniques, etc.; however, such topics are outside the scope of this manuscript.

Furthermore, when the multiple controllers control the system, the control system is called as *decentralized control*. Decentralization is helpful in many ways, for instance, it helps control systems to operate over a larger geographical area. The agents in decentralized control systems can interact with each other using communication channels and coordinate their actions. Computer zone controllers in greenhouses, for instance, are examples of decentralized controllers; and they bring much greater control flexibility and programmability for different zones, and reduce the wiring requirements compared to centralized controls if wireless systems are not in operation or an option.

Centralized or *integrated computer controls* provide coordinated control of distributed systems such as all greenhouse environmental functions in a single, integrated computer control system. Such systems bring together the capability of individual decentralized controllers (e.g., zone controllers) along with the individual control devices such as timers.

6.7 INSTRUMENTATION, SOFTWARE, AND MONITORING SYSTEMS

6.7.1 Sensors, and Parameters to Monitor

Sensors are essential components of any controlled environment monitoring and control systems. Sensors, using different sensing technologies, depending upon the nature of controlled environment facility, measure at least conditions of different environmental variables such as temperature, relative humidity, and other air and water quality parameters such as temperature, pH, alkalinity, DO, TAN, and CO_2 concentrations, aeration system pressure, water level, water flow, electric power, light intensity, wind speed, etc. And today, both hardwired and wireless monitoring and control systems are available, and commercial operations have either one of these types or sometimes the combination of both. The latest internet of things (IoT) technology makes it possible to remotely monitor and control devices over the internet.

6.7.2 Microcontrollers

Needless to say, today, environmental control in greenhouses, livestock, poultry, and aquacultural buildings all rely on microcontrollers. And each microcontroller for these controlled environment applications has a dedicated assignment to measure and control, for instance, air flow to control air temperature, moisture, and odor via ventilation, or internal pressure, water quality and flow rates, electric power, light intensity, or to perform higher-level control of a series of microcontrollers.

A microcontroller is essentially a mini computer, consisting of a central processing unit (CPU), memory units, and input/output (I/O) hardware with a limited memory and processing capacity. The *CPU* (i.e., the microprocessor) is responsible for the computation and internal control, where an arithmetic logic unit (ALU) executes mathematical functions on the data that is organized as binary digits. Moreover, a *memory unit* stores data, addresses, and control instructions, which is used by the CPU during processing. In addition, *digital input and output (I/O) ports* makes the connection between the microcontroller and external devices using digital signals. The high and low voltage in the signal correspond to on and off states, respectively; while the input port is employed to read in the status of the external device (sensor) and the output port is employed to give a control instruction to an external device (actuator). The CPU, memory unit, and I/O ports are connected through electrical signal conductors, and they serve as the central nervous system of the computer (Holden et al., 2020).

6.7.3 Microcontroller Software and Programming

The software of a microcontroller, on the other hand, determines the specific functions of a microcontroller. The programs of the software are stored in the memory unit. It must be remembered that the CPU can only execute binary code, or machine code, and performs low-level operations such as adding a number to a register or moving a register's value to a memory location; and it is very difficult to write a program in machine code. Low-level programming languages (e.g., assembly language) are the ones most similar to machine code. They are not interchangeable among different types of microcontrollers as they are hardware-specific programming languages. High-level programming languages (e.g., BASIC, C, or C++) can, on the other hand, be deployed among different types of microcontrollers with minor modifications. And in order to convert codes from a high-level language to the machine code to be executed by a specific CPU, or from one language to another language, a computer program called a compiler is used (Holden et al., 2020). For example, Arduino is a family of open-source hardware and software, single-board microcontrollers, which are easy to understand and use in real-world applications with sensors and actuators (Arduino, 2019). The programming language of the Arduino microcontrollers is based on a language called Processing, which is similar to C or C++, but it is much simpler, and can be adapted for other microcontrollers, and conveniently used in controlled environment plant, livestock, poultry, and fish production systems.

6.8 FUTURE PERSPECTIVES

In the past, the application of modern control technologies was quite limited due to the limited availability of computational power. Controllers based on artificial intelligence such as ANNs, FLS, GAs, and others alike could not be implemented in the old technological platforms. Modern techniques based on computational intelligence or algorithms on new technological platforms are now becoming more popular than ever in almost every industry.

Control systems and the automatic control of a production process are used to achieve improved performance and high-quality products along with increased productivity in systems in controlled environments. Today, our overall goal by developing and utilizing modern control systems is to provide higher level of flexibility and autonomy. And modern control theory deals with systems that typically have adaptive, robust, self-organizing, learning, and optimum qualities. Therefore, in its modern usage, automation can be defined as a technology that uses algorithms to operate a process with supplied feedback of information to ensure that the programmed commands executed properly. Automated processes operate without any human interference; and indeed, most automated systems are able to perform their functions with much greater accuracy and precision, and in less time.

Two system concepts (control systems and robotics) are approaching the goal of providing higher level of flexibility and autonomy through different evolutionary pathways. When it is programmed properly, today's industrial robot is quite autonomous and normally requires no further human intervention. However, due to the sensory limitations, robotic systems today have sort of limited flexibility in adapting to the changes in work environment, which indeed provides the motivation for computer/machine vision research. The control system is highly adaptable, but it still relies on human supervision. Advanced robotic systems are trying to achieve work adaptability through further enhanced sensory feedback. Sensor integration, particularly crop, animal, and fish physiological sensing via computer/machine vision, and artificial intelligence will soon make our systems, including the commercial controlled environment production facilities to be controlled more precisely and accurately, making them more flexible and autonomous, and therefore more universal and sustainable. Improvements in supervisory control and human–machine interface technologies will make the lives of operators much easier, while big data management via powerful computers will improve the efficiency of operators and intelligent control systems. Involving in improved communication techniques and advanced programming languages, both robotics and control systems research are aimed toward reducing the cost of implementation and eventually reaching the universal application.

NOMENCLATURE

AI	artificial intelligence
ANN	artificial neural networks
ES	expert systems
e	error
FLS	fuzzy logic systems
GA	genetic algorithm
I	integral
LQG	linear-quadratic-Gaussian control
MIMO	multiple-input multiple-output
MLS	machine learning systems
MPC	Model Predictive Control
P	proportional
PD	proportional-derivative
PI	proportional-integral
PID	proportional-integral-derivative
PLC	programmable logic controller
r	desired output response
SISO	single-input single-output
t	time
u	transfer function
y	actual/measured output

Greek Letters

τ	time

Subscripts

d	derivative
i	integral
p	proportional

BIBLIOGRAPHY

Ali, IA and Abdalla, AM (1993) A microcomputer-based system for all-year-round temperature control in greenhouses in dry arid lands. *Computers and Electronics in Agriculture*, 8(3): 195–210.

Ang, KH, Chong, GCY and Li, Y (2005) PID control system analysis, design, and technology. *IEEE Trans Control Systems Tech*, 13(4): 559–576.

Antsaklis, PJ and Michel, AN (1997) *Linear Systems*. McGraw-Hill.

Arduino (2019) Available on: https://www.arduino.cc/ (Retrieved on 15.03.2019).

Arvanitis, KG, Paraskevopoulos, PN and Vernardos, AA (2000) Multirate adaptive temperature control of greenhouses. *Computers and Electronics in Agriculture*, 26(3): 303–320.

Åström, KJ and Murray, RM (2008) *Feedback Systems: An Introduction for Scientists and Engineers*. Princeton University Press. ISBN 0-691-13576-2.

Baptista, FJ, Bailey, BJ, Meneses, JF and Navas, LM (2010) Greenhouses climate modelling. Tests, adaptation and validation of a dynamic climate model. *Spanish Journal of Agricultural Research*, 8(2): 285–298.

Belforte, G, Deboli, R, Gay, P, Piccarolo, P and Ricauda Aimonino, D (2006) Robot design and testing for greenhouse applications. *Biosystems Engineering*, 95: 309–321.

Bellman, R (1964) Control theory. *Scientific American*, 211(3): 186–200.

Bennis, N, Duplaix, J, Enéa, G, Haloua, M and Youlal, H (2008) Greenhouse climate modeling and robust control. *Computers and Electronics in Agriculture*, 61(2): 96–107.

Blasco, X, Martínez, M, Herrero, JM, Ramos, C, Sanchis, J (2007) Model-based predictive control of greenhouse climate for reducing energy and water consumption. *Computers and Electronics in Agriculture*, 55(1): 49–70.

Briat, C (2015) *Linear Parameter-Varying and Time-Delay Systems*. Analysis, Observation, Filtering & Control, Springer Verlag Heidelberg. ISBN 978-3-662-44049-0.

Caponetto, R, Fortuna, L, Nunnari, G, Occhipinti, L, Xibilia, MG (2002) Soft computing for greenhouse climate control. *IEEE Transactions on Fuzzy Systems*, 8(6): 753–760.

Castañeda-Miranda, R, Ventura-Ramos, E, Peniche-Vera, RR and Herrera-Ruiz, G (2006) Fuzzy greenhouse climate control system based on a field programmable gate array. *Biosystems Engineering*, 94(2): 165–177.

Chalabi, ZS, Bailey, BJ and Wilkinson, DJ (1996) A real-time optimal control algorithm for greenhouse heating. *Computers and Electronics in Agriculture*, 15(1): 1–13.

Coelho, JP, De Moura Oliveira, PB and Boaventura Cunha, J (2005) Greenhouse air temperature predictive control using the particle swarm optimisation algorithm. *Computers and Electronics in Agriculture*, 49(3): 330–344.

Contreras-Medina, LM, Torres-Pacheco, I, Guevara-Gonzalez, RG, Romero-Troncoso, RJ, Terol-Villalobos, IR and Osornio-Rios, RA (2009) Mathematical modeling tendencies in plant pathology. *African Journal of Biotechnology*, 8(25): 7391–7400.

Davis, PF and Hooper, AW (2002) Improvement of greenhouse heating control. *Control Theory and Applications, IEE Proceedings D*, 138, IET: 249–255.

De Baerdemaeker, J, Munack, A, Ramon, H and Speckmann, H (2002) Mechatronic systems, communication, and control in precision agriculture. *Control Systems Magazine*, IEEE, 21(5): 48–70.

De Koning, ANM (1990) Long-term temperature integration of tomato. Growth and development under alternating temperature regimes. *Scientia Horticulturae*, 45(1–2): 117–127.

Duarte-Galvan, C, Torres-Pacheco, I, Guevara-Gonzalez, RG, Romero-Troncoso, RJ, Contreras-Medina, LM, Rios-Alcaraz, MA and Millan-Almaraz, JR (2012) Review: Advantages and disadvantages of control theories applied in greenhouse climate control systems. *Spanish Journal of Agricultural Research*, 10(4): 926–938.

El Ghoumari, MY, Tantau, HJ and Serrano, J (2005) Non-linear constrained MPC: Real-time implementation of greenhouse air temperature control. *Computers and Electronics in Agriculture*, 49(3): 345–356.

Fan, X and Zuo-Hua, T (2006) Application of a genetic simulated annealing algorithm in the greenhouse system. *Computer Simulation*, 12: 045.

Fang, L and Zhen-Xiao, L (2008) Research on the control mode of agriculture greenhouse control system in China. *Journal of Agricultural Mechanization Research*, 10: 223–226.

Fitz-Rodríguez, E and Giacomelli, GA (2009) Yield prediction and growth mode characterization of greenhouse tomatoes with neural networks and fuzzy logic. *Transations of ASABE*, 52(6): 2115–2128.

Fourati, F and Chtourou, M (2007) A greenhouse control with feed-forward and recurrent neural networks. *Simulation Modelling Practice and Theory*, 15(8): 1016–1028.

Franklin, GF, Powell, JD and Emami-Naeimi, A (1994) *Feedback Control of Dynamic Systems*. 3rd Edition. Addison-Wesley.

Freeman, JA and Skapura, DM (1991) *Neural Networks: Algorithms, Applications, and Programming Techniques*. Addison Wesley, New York.

Gauthier, L (1992) A smalltalk-based platform for greenhouse environment control. Part I. Modeling and managing the physical system. *Transations of ASABE*, 35(6): 2003–2009.

Gauthier, L, De Halleux, D, Boisvert, A, Trigui, M and Zehrouni, A (1995) A control strategy for the operation of a reversible heat-pump in greenhouses. *Applied Engineering in Agriculture*, 11(6): 873–879.

Gauthier, L and Guay, R (1990) An object-oriented design for a greenhouse climate control system. *Transations of ASABE*, 33(3): 999–1004.

Goggos, V and King, RE (2000) Qualitative-evolutionary design of greenhouse environment control agents. *Computers and Electronics in Agriculture*, 26(3): 271–282.

Goodwin, GC, Graebe, SF and Salgado, ME (2001) *Control System Design*. Prentice Hall.

Gutman, PO, Lindberg, PO, Ioslovich, I and Seginer, I (1993) A non-linear optimal greenhouse control problem solved by linear programming. *Journal of Agricultural Engineering Research*, 55(4): 335–351.

Hashimoto, Y, Murase, H, Morimoto, T and Torii, T (2002) Intelligent systems for agriculture in Japan. *IEEE Control Systems Magazine*, 21(5): 71–85.

Hellerstein, JL, Tilbury, DM and Parekh, S (2004) *Feedback Control of Computing Systems*. John Wiley and Sons. ISBN 0-471-26637-X.

Holden, NM, Wolfe, ML, Ogejo, JA and Commins, EJ (eds.) (2020) *Introduction to Biosystems Engineering*. ASABE in association with Virginia Tech Publishing, Blacksburg, VI. Available on: https://doi.org/10.21061/IntroBiosystemsEngineering

Hooper, AW (1988) Computer control of the environment in greenhouses. *Computers and Electronics in Agriculture*, 3(1): 11–27.

Hooper, AW and Davis, PF (1988) An algorithm for temperature compensation in a heated greenhouse. *Computers and Electronics in Agriculture*, 2(4): 251–262.

Hunt, KJ, Sbarbaro, D, Zbikowski, R and Gawthrop, PJ (1992) Neural networks for control systems—a survey. *Automatica*, 28(1992): 1083–1112.

Ibrahim, IA and Sørensen, CG (2010) A more energy efficient controller for the greenhouses climate control system. *Applied Engineering in Agriculture*, 25(3): 491–498.

Ioslovich, I, Gutman, P and Linker, R (2009) Hamilton-Jacobi-Bellman formalism for optimal climate control of greenhouse crop. *Automatica*, 45(5): 1227–1231.

Ioslovich, I, Gutman, P and Seginer, I (1996) A non-linear optimal greenhouse control problem with heating and ventilation. *Optimal Control Applications and Methods*, 17(3): 157–169.

Jacobson, BK, Jones, PH, Jones, JW and Paramore, JA (1989) Real-time greenhouse monitoring and control with an expert system. *Computers and Electronics in Agriculture*, 3(4): 273–285.

Jewett, TJ and Short, TH (1992) Computer control of a five-stage greenhouse shading system. *Transactions in ASABE*, 35(2): 651–658.

Kilian, C (2005) *Modern Control Technology*. Thompson Delmar Learning. ISBN 1-4018-5806-6.

Körner, O and Challa, H (2003) Process-based humidity control regime for greenhouse crops. *Computers and Electronics in Agriculture*, 39(3): 173–192.

Körner, O and Van Straten, G (2008) Decision support for dynamic greenhouse climate control strategies. *Computers and Electronics in Agriculture*, 60(1): 18–30.

Kuo, BC (1995) *Automatic Control Systems*. 7th Edition. Prentice Hall.

Kurata, K and Eguchi, N (1990) Machine learning of fuzzy rules for crop management in protected cultivation. *Transactions in ASAE*, 33(4): 1360–1368.

Lafont, F and Balmat, JF (2004) Fuzzy logic to the identification and the command of the multidimensional systems. *International Journal of Computational Cognition*, 2: 21–47.

Levine, WS (1996) *The Control Handbook*. CRC Press, New York. ISBN 978-0-8493-8570-4.

Linker, R, Gutman, PO and Seginer, I (2000) Robust model-based failure detection and identification in greenhouses. *Computers and Electronics in Agriculture*, 26(3): 255–270.

Linker, R, Seginer, I and Gutman, P (1998) Optimal CO2 control in a greenhouse modeled with neural networks. *Computers and Electronics in Agriculture*, 19(3): 289–310.

Narendra, KS and Parthasarathy, K (1990) Identification and control of dynamical systems using neural networks, IEEE. *Transactions on Neural Networks*, 1: 4–27.

Nielsen, OF (1995) Climate computer algorithms for peak shaving of greenhouse heating demand. *Computers and Electronics in Agriculture*, 13(4): 315–335.

Ogata, K (1997) *Modern Control Engineering*. 3rd Edition. Prentice Hall.

Pasgianos, GD, Arvanitis, KG, Polycarpou, P and Sigrimis, N (2003) A nonlinear feedback technique for greenhouse environmental control. *Computers and Electronics in Agriculture*, 40(1–3): 153–177.

Pekkeriet, EJ and van Henten, EJ (2011) Current developments of high-tech robotic and mechatronic systems in horticulture and challenges for the future. *Acta Horticulturae*, 893: 85–94.

Pham, DT and Liu, X (1995) *Neural Networks for Identification.* Prediction and Control, Springer, London.

Pinon, S, Camacho, EF, Kuchen, B and Peña, M (2005) Constrained predictive control of a greenhouse. *Computers and Electronics in Agriculture*, 49(3): 317–329.

Rohrs, CE, Melsa, JL and Schultz, DG (1993) *Linear Control Systems.* McGraw-Hill.

Seginer, I (1997) Some artificial neural network applications to greenhouse environmental control. *Computers and Electronics in Agriculture*, 18(2–3): 167–186.

Seginer, I and McClendon, RW (1992) Methods for optimal control of the greenhouse environment. *Transations in ASAE*, 35(4): 1299–1307.

Setiawan, A, Albright, LD and Phelan, RM (2000) Application of pseudo-derivative-feedback algorithm in greenhouse air temperature control. *Computers and Electronics in Agriculture*, 26(3): 283–302.

Sigrimis, N, Antsaklis, P and Groumpos, PP (2002) Advances in control of agriculture and the environment. *IEEE Control Systems Magazine*, 21(5): 8–12.

Sigrimis, N and King, RE (2000) Advances in greenhouse environment control. *Computers and Electronics in Agriculture*, 26(3): 217–219.

Sigrimis, N and Rerras, N (1996) A linear model for greenhouse control. *Transations in ASABE*, 39(1): 253–261.

Soto-Zarazua, GM, Rico-Garcia, E, Ocampo, R, Guevara-Gonzalez, RG and Herrera-Ruiz, G (2010) Fuzzy-logic-based feeder system for intensive tilapia production (*Oreochromis niloticus*). *Aquaculture International*, 18(3): 379–391.

Stengel, RF (1994) *Optimal Control and Estimation.* Dover Publications. ISBN 0-486-68200-5.

Takakura, T, Manning, TO, Giacomelli, GA and Roberts, WJ (1994) Feedforward control for a floor heat greenhouse. *Transactions in ASABE*, 37(3): 939–945.

Trabelsi, A, Lafont, F, Kamoun, M and Enea, G (2007) Fuzzy identification of a greenhouse. *Applied Soft Computing*, 7(3): 1092–1101.

Van Straten, G, Challa, H and Buwalda, F (2000) Towards user accepted optimal control of greenhouse climate. *Computers and Electronics in Agriculture*, 26(3): 221–238.

Van Straten, G, Van Willigenburg, LG, Van Henten, EJ and Van Ooteghem, RJC (2011) *Optimal Control of Greenhouse Cultivation.* CRC Press, New York, USA.

Yildiz, I (1993) Simulation of greenhouse microclimates and environmental control strategies using a Rankine cycle heat pump. Dissertation. The Ohio State University, Columbus, OH, USA.

Yildiz, I (2021) *Greenhouse Engineering: Integrated Energy Management.* 1st Edition. CRC Press, Taylor and Francis Group, FL. ISBN: 978-1-4987-4313-6.

Yildiz, I and Stombaugh, DP (2006a) Dynamic modeling of microclimate and environmental control strategies in a greenhouse coupled with a heat pump system. *Acta Horticulturae*, 718: 331–340.

Yildiz, I and Stombaugh, DP (2006b) Heat pump cooling and greenhouse microclimates in open and confined greenhouse systems. *Acta Horticulturae*, 719: 255–262.

Yildiz, I and Stombaugh, DP (2006c) Simulated performances of a heat pump system for energy and water conservation in open and confined greenhouse systems. *Acta Horticulturae*, 718: 341–350.

Yildiz, I and Stombaugh, DP (2009) Simulated effects of various environmental management practices on energy consumption in open and confined greenhouse systems. *Acta Horticulturae*, 801: 1317–1324.

Yildiz, I, Yue, J and Lowrey, J (2014a) Spatiotemporal distributions of energy and water consumption in open, semi-closed and closed loop heat pump greenhouse systems, and associated management strategies in Canada. *Acta Horticulturae*, 1037 (1): 119–128.

Yildiz, I, Yue, J and Lowrey, J (2014b) Spatiotemporal distributions of energy and water consumption in open, semi-closed and closed loop heat pump greenhouse systems, and associated management strategies in the Canadian North. *Acta Horticulturae*, 1037 (1): 171–177.

7 Energy Estimating

7.1 INTRODUCTION

Energy requirements and fuel consumption of heating, ventilation and air-conditioning (HVAC) systems directly affect the operational cost of a controlled environment building, and indirectly affect the building internal environment. This chapter discusses methods for estimating energy use for two distinct purposes: i) modeling for a controlled environment building, and HVAC system design and pertinent design optimization (also known as *forward modeling*), and ii) modeling energy use of existing controlled environment buildings for calculating retrofit savings (also known as *data-driven modeling*).

A mathematical model of a controlled environment building describes the behavior of a production system, and is basically made up of three main components: i) input (forcing) variables—controllable or uncontrollable (e.g., outside solar radiation, temperature, relative humidity, etc.), ii) system structure and parameters/properties, which provide the necessary physical description of the building system (e.g., thermal mass, etc.), iii) output (response) variables, which describe the reaction of the system to the input variables (e.g., energy use, water use, indoor temperature, leaf temperature, culturing tank water temperature, relative humidity levels, photosynthesis, respiration, overall production, etc.) As an example, Figure 7.1 shows the overall structure of a greenhouse mathematical model.

The objective or purpose of the modeling study dictates what type of approach needs to be taken; that is, the forward (classical) approach or the data-driven (inverse) approach.

And the objective of the forward modeling approach is to predict the output variables of a specified building model with known structure and parameters when subject to specified input variables. With the availability of powerful computers and their easy accessibility, the building models today have become increasingly complex with greater precision and accuracy levels. This approach requires detailed knowledge of the various natural phenomena affecting system behavior along with the magnitude of various interactions (e.g., effective thermal mass, heat and mass transfer coefficients, etc.).

Forward modeling approach starts with a physical description of the building system. For example, geographical location, building orientation and geometry, building envelope, type of equipment and operational strategies, type of HVAC system, etc. are all specified. And then, the energy, water, and other input use profiles of such a production building can for instance be simulated by the forward simulation model. The main advantage of this method is that it is completely based on sound engineering principles; and therefore, the system need not be physically built to predict its behavior. Consequently, the approach is the most commonly used approach by mathematical modelers and

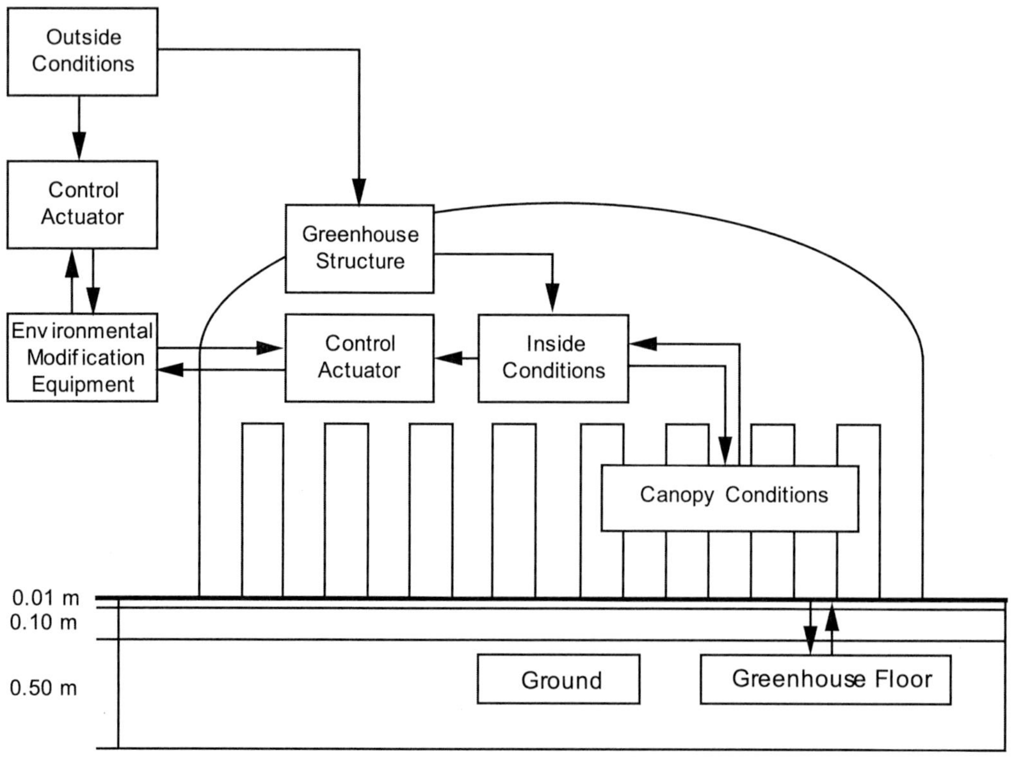

FIGURE 7.1 Overall structure of a greenhouse mathematical model (Yildiz, 1993).

DOI: 10.1201/9781003349297-7

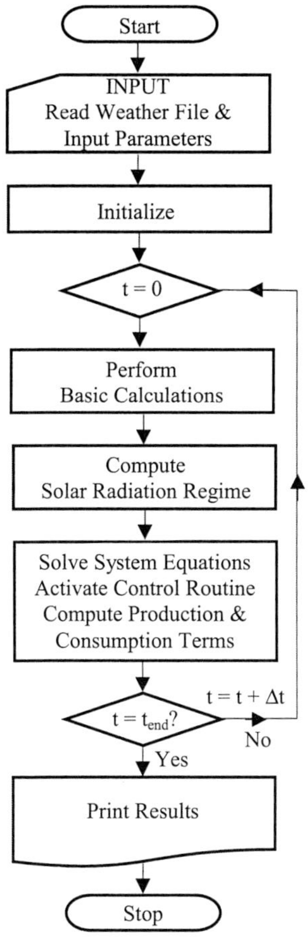

FIGURE 7.2 Typical flowchart of a forward greenhouse simulation model (Yildiz, 1993).

has gained widespread acceptance by the scientific and professional community. Figure 7.2, on the other hand, illustrates the ordering of the analysis typically performed by a greenhouse energy simulation model.

In the case of data-driven approach, on the other hand, input and output variables are measured, and therefore known; and the objective of developing such a model is then to determine a mathematical description of the system and estimate parameters of the system when the system has already been built and actual performance data are available for model development and/or identification. Data-driven models often provide more accurate predictors of future system performance than forward models. Choosing or developing a building energy analysis program, however, depends on its application, the frequency of its use, the experience or knowledge of users, and the available hardware to run the simulation program. Since detailed models are almost all the time computationally intensive, simplified methods such as the degree-day, bin, and correlation methods have also been developed for energy estimating. The most accurate methods for determining building energy consumption are the most complex ones due to their intense computational requirements and the expertise required. Simulation programs that assemble component models into system models and then exercise those models with weather and crop or animal data are preferred by experts for determining energy, water and other input uses in production buildings.

7.2 CHARACTERISTICS OF ENERGY ESTIMATING PROCEDURES

Procedures for energy estimating requirements range quite a bit in their complexity; however, they mostly have the following common elements: i) load calculation for the controlled environment building, ii) load calculation for the secondary equipment, and iii) calculation of primary equipment energy necessities. Secondary equipment means the equipment that distributes the heating, cooling, or ventilating medium to controlled environments, whereas primary equipment are the central plant (e.g., boiler room) equipment that converts fuel or electric energy to thermal (heating or cooling) effect.

Based on the procedure provided above, in order to estimate energy requirements in a controlled environment, we first need to calculate the building space load, which is the amount of energy that must be added to or removed from a controlled environment space for maintaining thermal comfort for plants, animals, fish, and personnel. The simplest procedures typically assume that the energy required for providing and maintaining thermal comfort is a function of the outdoor dry-bulb temperature only. And relatively more detailed and accurate approaches, on the other hand, consider the solar effects, crop, animal or fish effects, internal gains (e.g., artificial lighting), and heat storage in the floor, the inside air, the crop canopy and other interior surfaces, along with the effects of shading, night curtains, insect screens, and wind on the heat loss from the building envelope and infiltration, etc. Calculations for energy estimating are similar to the heating and cooling load calculations used for sizing primary and secondary equipment as performed in Chapters 5; but, they are not the same as we will see shortly. Energy estimating calculations are based on typical weather conditions rather than worst-case weather scenarios. Sophisticated procedures usually use hourly climatic conditions and operational characteristics for a number of typical meteorological days of the year or on 8760 h of operation per year.

After determining the space load, we then take the calculated space load and translate it to the secondary (piping, ducts, pumps, fans, etc.) equipment load. This can only be a simple estimate of duct or piping losses or gains or a complex hour-by-hour, minute-by-minute or even second-by-second simulation of a building air system. This step includes all forms of energy required by all the secondary equipment (e.g., electrical energy to operate fans and/or pumps, including the energy in the form of heated or chilled water, such as culturing tank or irrigation water).

And then finally, we calculate the fuel and energy required by the primary equipment to meet all of the above-mentioned loads and the peak demand considering the efficiencies and partial load characteristics of the equipment. It is quite often necessary to keep track of the form of energy consumptions, such as electrical, natural gas, biomass, oil, etc. so that they can be used in economic analyses or business models for new controlled environment complexes and/or the cost-effectiveness of energy conservation measures.

Before going into further details of the most elaborate detailed simulation or complex modeling approaches, we will spend some time in the following sections on the abovementioned simpler methods for estimating energy consumption in controlled environment buildings. One of these simpler methods is what's called single-measure procedure, which uses only one measure such as annual degree-days (degree-day method). The other procedure is called simplified multiple-measure methods in which more information is used and therefore accuracy is improved. The best known of such kind of methods is the bin method where occurrence hours are used for each temperature interval called as temperature bin.

7.3 BALANCE POINT TEMPERATURE

The *balance point temperature,* T_{bal}, of a building is defined as the outdoor temperature T_o at which, for the specified value of the interior temperature T_i, the total heat losses from a building equal the heat gains in the building (in our case, the building is the controlled environment building).

For a steady-state method, heat gains must be the average for the period in question, not for the peak values. In particular, solar radiation must be based on averages, not peak values. Applying energy balance Equation (2.2) from Chapter 2 to the internal air in a greenhouse, we developed Equation (5.3) for internal heat gains as follows:

$$\dot{Q}_{internal} = \dot{Q}_{equipment} + \dot{Q}_{lighting} + \dot{Q}_{solar} \quad (5.3)$$

And for the internal air in a livestock or poultry building, we developed Equation (5.10) for internal heat gains as follows:

$$\dot{Q}_{internal} = \dot{Q}_{animal\_sensible} + \dot{Q}_{equipment} + \dot{Q}_{lighting} + \dot{Q}_{solar} \quad (5.10)$$

And likewise, for the internal air in an aquatic building, we developed Equation (5.10) for internal heat gains as follows:

$$\dot{Q}_{internal} = \dot{Q}_{fish\_sensible} + \dot{Q}_{equipment} + \dot{Q}_{lighting} + \dot{Q}_{solar} \quad (5.13)$$

Only when the outdoor temperature T_o drops below the balance point temperature T_{bal}, supplemental heat is required to maintain the indoor design temperature T_i. Above the balance point temperature T_{bal}, internal heat gains (solar, equipment and lights, when available) are able to maintain the indoor temperature at or above the indoor design temperature T_i. At the balance point temperature T_{bal}, the total heat loss by conduction and ventilation (or infiltration) equals the internal heat gains $\dot{Q}_{internal}$:

$$\dot{Q}_{internal} = \Sigma UA\left(T_i - T_{bal}\right) + \left(\text{ACH}\right) V \rho c_p \left(T_i - T_{bal}\right) \quad \text{(I-P)} \quad (7.1a)$$

$$\dot{Q}_{internal} = \Sigma UA\left(T_i - T_{bal}\right) + \text{ACH}\left(\frac{1}{3{,}600}\right) V \rho c_p \left(T_i - T_{bal}\right) \quad \text{(SI)} \quad (7.1b)$$

where the first and second terms on the right hand side of Equations (7.1a,b) are the total heat loss terms by conduction and the heat loss terms due to infiltration, respectively.

Then, the balance point temperature T_{bal} in I-P units becomes:

$$T_{bal} = T_i - \left(\frac{\dot{Q}_{internal}}{\left(\Sigma UA\right) + \text{ACH}\, V \rho\, c_p}\right) \quad \text{(I-P)} \quad (7.2a)$$

where T_{bal} is the balance point temperature (°F), $\dot{Q}_{internal}$ is the total internal sensible heat gain (Btu/h) from lights, equipment, etc., $\sum UA$ is the total conduction heat loss term (Btu/h.°F), and ACH is the number of air changes per hour, V is the volume of the controlled environment space (ft³), ρ is the density (lb/ft³) and c_p is the specific heat of greenhouse inside air (Btu/lb.°F).

And the balance point temperature T_{bal} in SI units becomes:

$$T_{bal} = T_i - \left(\frac{\dot{Q}_{internal}}{\left(\Sigma UA\right) + \text{ACH}\left(\frac{1}{3{,}600}\right) V \rho c_p}\right) \quad \text{(SI)} \quad (7.2b)$$

where T_{bal} is the balance point temperature (°C), $\dot{Q}_{internal}$ is the total internal sensible heat gain (W) from lights, equipment, etc., ΣUA is the total conduction heat loss term (W/°C), and ACH is the number of air changes per hour, V is the volume of the controlled environment space (m³), (1/3,600) is for unit conversion (1 h = 3,600 s), ρ is the density (kg/m³) and c_p is the specific heat of greenhouse inside air (J/kg.°C).

7.4 ANNUAL DEGREE-DAY METHOD

Degree-day (single-measure) methods are the simplest methods for energy estimating and are adequate if building use and HVAC equipment efficiency are constant. Controlled environment building heating and cooling are responsible for a large fraction of energy consumption in controlled environment energy production around the world; and heating degree-days (HDDs) and cooling degree-days (CDDs) are proxies for the energy demand needed to heat or cool, respectively, a controlled environment production facility, a home or a business. Both variables are derived from outside air temperature measurements. Heating degree-days or degree-hours for a balance point temperature of 18.3 °C (65 °F) have been widely tabulated (this temperature represents average conditions in typical residential buildings in the past). The 18.3 °C (65 °F) base is assumed whenever T_{bal} is not indicated explicitly. The heating and cooling requirements for a given building such as a greenhouse, livestock, poultry, or aquatic building at a specific location are, to some degree, proportional to the number of HDDs and CDDs at that location. HDDs and CDDs are calculated using daily mean outside temperature ($T_{mean} = \left[T_{max} - T_{min}\right]/2$) and a base temperature—the outside temperature, below or above which a building requires heating or cooling for thermal comfort. The baseline (balance point) temperatures today are quite different from what was developed for residential buildings in the 1940s—still in use—as buildings are much tighter, more insulated and have more internal heat sources today. Therefore, in order to accurately determine energy demands and gauge the impacts of climate change, new baseline temperatures are required for both heating and cooling. Annual and monthly HDD and CDD values have been developed for selected base temperatures in certain geographical jurisdictions; however, further coverage is still required. Outside daily mean temperatures are used in determining HDD and CDD calculations even though cooling a space depends more on the maximum rather than the mean temperature, while the minimum temperature is more relevant for heating. Further, long-term temperature data closer to present time should be used for more accurate calculations of HDD and CDD, and hence, for future energy assessments of buildings including controlled environment buildings. Overall, it can be said that climate change is reducing the demand for space heating and increasing the demand for space cooling globally. In any case, even if a specified base temperature HDDs or CDDs may not be available for the location in question, we can still use the tabulated annual HDD and CDD values determined for classical baseline temperatures for energy estimating, which will be shown numerically later in this section.

We have defined the balance point temperature T_{bal} of a building, and appropriate equations were also developed earlier in this section. It was stated that heating is only needed when outside temperature T_o drops below the balance point temperature T_{bal}. Therefore, the rate of energy consumption for a heating system can be expressed as follows:

$$E = \left(\frac{24 \times (\text{HDD})(\dot{Q}_t)}{\text{HV} \times \eta \times \Delta T} \right) \quad \text{I-P} \qquad (7.3a)$$

where E is the fuel or energy consumption (in trade units), the constant 24 is for unit conversion (i.e., 24 h/d), HDD is the mean annual heating degree days for the location (Table A.72), $\dot{Q}_t$ is the total heat loss from the building (Btu/h), HV is the heating value of fuel (unit consistent with $\dot{Q}_t$ and E), η is the efficiency of heating system (in decimals), ΔT the design temperature difference (°F) for $\dot{Q}_t$.

$$E = \left(\frac{24 \times (3{,}600)(\text{HDD})(\dot{Q}_t)}{\text{HV} \times \eta \times \Delta T} \right) \quad \text{SI} \qquad (7.3b)$$

where E is the fuel or energy consumption (in trade units), the constant 24 is for unit conversion (i.e., 24 h/d), the constant 3,600 is for another unit conversion (i.e., 3,600 s/h) HDD is the mean annual heating degree-days for the location (Table A.72), $\dot{Q}_t$ is the total heat loss from the building (W), HV is the heating value of fuel (unit consistent with $\dot{Q}_t$ and E), η is the efficiency of heating system (in decimals), ΔT the design temperature difference (°C) for $\dot{Q}_t$.

Example 7.1: Heating Energy Estimating in a Greenhouse via Degree–Day Method

Now, let's consider a single glass greenhouse in Columbus, Ohio, with the following dimensions.

GREENHOUSE DIMENSIONS and CHARACTERISTICS

GREENHOUSE COMPONENT	Dimension, ft (m)	
House wall height	8 (2.4)	
House width	24 (7.2)	
House length	192 (57.6)	
Rafter length	13.4 (4.0)	
Lower wall height	1.5 (0.45)	
Upper wall height	6.5 (1.95)	
Gable wall height	6 (1.8)	
Number of bays	4 (1.2)	
	Construction Material	**Reference**
Lower wall	6 in. (0.15 m) poured concrete	See Table A.60
Upper wall	Single glass	See Table A.60
End area	Single glass	See Table A.60
Roof	Single glass	See Table A.60
Perimeter	Uninsulated perimeter	See Table A.60

Estimate the annual heating energy requirement and the costs of heating using fuel oil and natural gas furnaces for this greenhouse. Information on heating values and fuel prices can be found in Chapter 2.

Solution We first need to calculate the UA values as follows:

UA VALUES and AIR CHANGE PER HOUR

GREENHOUSE COMPONENT	A (ft²)	U (Btu/h. ft².°F)	Reference	UA (Btu/h.°F)
Lower Wall Area	864	0.75	Table A.60	648
Upper Wall Area	3,744	1.1		4,118
Gable or Curved Area	576	1.1		634
Roof Area	20,582	1.1		22,640
Perimeter Length	576	0.8		461
ΣUA				28,501
ACH (per hour)	1.0		Table A.61	

Then, using Equation (5.4), we can easily determine the greenhouse envelope and infiltration heat losses as well as the fuel oil and natural gas consumption calculations as shown in the following table:

				Annual Cost of Heating (US$/season)
T_i	65°F			
T_o	6°F (99% level; Table A.52–A55)			
ΔT	59°F			
HDD	5,660 °F.days at 65°F base temp. (Table A.72 or consult your local Weather Bureau)			
Furnace Efficiency	80%			
HV of fuel oil (Btu/gal)	139,000	Price of fuel oil (US$/gal)	2.45	
HV of natural gas (Btu/ft³)	1,036	Price of natural gas (US$/ft³)	0.005	
Q_{cond} (Btu/h)	$UA \times \Delta T$		1,681,559	
Q_{inf} (Btu/h)	$0.02 \times \Delta T \times V \times ACH$		243,302	
Q_t (Btu/h)	$Q_{cond} + Qinf$		1,920,806	
E (gal oil/year)	$24 \times HDD \times Q_t/(HV \times \eta \times \Delta T)$		39,770	97,436
E (ft³ natural gas/year)	$24 \times HDD \times Q_t/(HV \times \eta \times \Delta T)$		5,335,922	26,680

As shown in the above table, the greenhouse heating energy consumption was estimated using the trade units for fuel oil and natural gas, and then they were converted to US$ per season. Based on our calculations, annual natural gas consumption for heating this greenhouse would cost $26,680 while it would cost $97,436 per year if fuel oil is used to heat the same greenhouse facility. For more accurate monetary estimates though, up-to-date local fuel prices should be used.

Please note that the same estimating would be valid for this type of a greenhouse used as an aquatic building under the same operating conditions at the same location.

Example 7.2: Heating Energy Estimating in a Greenhouse with Variable Set-Point Temperatures

Now, using the same greenhouse that was used in Example 7.1, a single glass greenhouse in Columbus, Ohio, estimate the annual heating energy requirement and the costs of heating using fuel oil and natural gas furnaces for different day- and night-time indoor set-point temperatures. Assume that a daytime set-point temperature of 65°F is used, while the night-time set-point temperature is 55°F.

Solution

See Example 8.3 in Chapter 8. Please note that the same estimating would be valid for this type of a greenhouse used as an aquatic building under the same operating conditions at the same location.

Example 7.3: Estimating the Reduction in Heating Energy Requirement When Double-Poly Glazing Used

Let's now consider the same greenhouse facility that was used in Example 7.1; however, this time a double-poly glazing was used instead of single glass, and the lower wall (6 in. poured concrete + 2 in. foamed urethane board) and the perimeter are insulated as shown in the following table. And determine the energy savings.

Solution

See Example 8.1 in Chapter 8. Please note that the same estimating would be valid for this type of a greenhouse used as an aquatic building under the same operating conditions at the same location.

Example 7.4: Estimating the Reduction in Heating Energy Requirement When Thermal Blanket Used

Let's now consider the same greenhouse facility that was used in Example 7.3; however, this time, in addition to using a double-poly glazing instead of single glass, and an insulated lower wall (6 in. poured concrete + 2 in. foamed urethane board) and perimeter, a thermal blanket is used as well. Determine the additional energy savings.

Solution

See Example 8.4 in Chapter 8. Please note that the same estimating would be valid for this type of a greenhouse used as an aquatic building under the same operating conditions at the same location.

7.5 BIN METHOD

The degree-day method provides a simple estimate of annual heating or cooling loads, which is pretty accurate if the indoor temperature and internal gains are relatively constant, and if the HVAC systems operate for a complete heating or cooling season. However, since the HVAC system efficiency, or the balance point temperature T_{bal} may not be sufficiently constant, there is a legitimate concern with the use of degree-day method even with the variable-base method. For instance, if a heat pump is used, the heat pump efficiency changes with outdoor temperature, and the efficiency of other HVAC equipment (boilers, chillers, etc.) may also change as a function of varying load. Furthermore, in controlled environment buildings, free heating energy such as solar and artificial lighting (when used), livestock and poultry heat gains have pronounced patterns, which influence the overall building heat and moisture gains, indoor temperature, and therefore ventilation rate. In such cases, steady-state calculation might yield pretty good estimates for annual energy consumption if different temperature intervals (bins) and time periods are evaluated separately. This approach is called as the *bin* method, a simplified multiple-measure method, because consumption is calculated for several outdoor temperatures and multiplied by the number of occurrence hours (N_{bin}) in the temperature interval (bin) centered around that temperature:

$$E = \left(\frac{(N_{bin})(\dot{Q}_t - \dot{Q}_{internal})}{\text{HV} \times \eta} \right) \quad (7.4)$$

where E is the annual (seasonal) fuel or energy consumption (in trade units), N_{bin} is the number of occurrence hours (see Table A.73), $\dot{Q}_t$ is the total heat loss from the greenhouse, $\dot{Q}_{internal}$ is the free internal heat gains, HV is the heating value of fuel (unit consistent with $\dot{Q}_t$ and E), η is the efficiency of heating system (in decimals), ΔT is the design temperature difference, and the others are as developed in Chapter 5 as follows:

$$\dot{Q}_t = \dot{Q}_{cond} + \dot{Q}_{inf}$$

$$\dot{Q}_{cond} = \Sigma UA(T_i - T_o) \quad \text{(I-P or SI)}$$

$$\dot{Q}_{inf} = (\text{ACH}) V \rho c_p (T_i - T_o) \quad \text{(I-P)}$$

$$\dot{Q}_{inf} = (\text{ACH}) \left(\frac{1}{3{,}600} \right) V \rho c_p (T_i - T_o) \quad \text{(SI)}$$

and $\dot{Q}_{internal}$ is as expressed in Equations (5.3), (5.10), and (5.13) for greenhouses, livestock and poultry, and aquatic buildings, respectively.

No heating is needed when T_o is above T_{bal}. Equation (9.4) is evaluated for each bin, and the total consumption is the sum of the E_{bin} over all bins. In the United States, for instance, the necessary weather data are available in ASHRAE (1986) and USAF (1978). The bins usually have 2.8 °C (5 °F) ranges and are typically collected in three daily 8 h shifts. Hourly weather occurrences can easily be developed for other geographical locations when they are not readily available.

And as discussed earlier, the main advantage of the bin method is that we can analyze the performance of partly loaded heating equipment, weather related variations in capacity and performance of heat pumps, effects of changing stocking density, crop density and canopy in the controlled environment building, *UA* value, minimum winter ventilation, or any other energy saving techniques. Even though easily accessible computers can easily calculate the energy consumption of a controlled environment via complex modeling, the concepts of degree-days and bin approach are nevertheless simple and quite valuable tools.

Example 7.5: Estimating Heating Energy in a Greenhouse via Bin Method

Now, let's consider the same greenhouse, which was used in Example 7.1, a single-glass greenhouse in Columbus, Ohio, and estimate the annual heating energy requirement and the costs of heating using fuel oil and natural gas furnaces for this greenhouse? Assume that internal heat gains are zero.

Solution

We need to follow the steps provided below:

- First, we will calculate the balance point temperature T_{bal} using either Equation (9.2a) or (9.2b) depending on the units used.
- Remember that the internal gains are neglected in this example; therefore, $T_{bal} = T_i = 65°\text{F}$. So, no supplemental heating is required at or above 65 °F outside temperatures.
- See Table A.73 for the number of occurrence hours in each bin.
- Since $T_{bal} = 65°\text{F}$, the highest temperature bin to be considered is the 67 °F bin (64.5 °F and 69.5 °F).
- 67°F bin occurs for 820 h; but, supplemental heating is needed only when $T_o < 65°\text{F}$.
- The outside temperature is between 64.5°F and 67.9°F for about: $820\text{h}\left(65°\text{F}(-)64.5°\text{F}\right)/5°\text{F} = 82\text{h}$
- During this 82-hour period, the average outside (bin) temperature is: $\left(65°\text{F} + 64.5°\text{F}\right)/2 = 64.75°\text{F}$

So, 64.75°F is the highest temperature interval (bin) and the corresponding occurrences of 82 h need to be used first in our calculations. And then, the remaining (lower) temperature bins and the corresponding occurrence hours will be incorporated fully for estimating the seasonal heating energy consumption. Finally, using Equation (9.4) and its components, calculate the full energy estimates as presented in the following table:

BIN METHOD

Bin Temp (°F)	ΔT (°F)	Q_{inf} (Btu/h)	Q_{cond} (Btu/h)	$Q_{internal}$ (Btu/h)	$Q_{inf} + Q_{cond} - Q_{internal}$ (Btu/h)	Occurrence (h)	*E* (Btu/year)
64.75	0.25	1,014	7,125	0	8,139	82	667,399
62	3	12,165	85,503	0	97,668	720	70,321,046
57	8	32,440	228,008	0	260,448	648	168,770,511
52	13	52,716	370,513	0	423,229	622	263,248,139
47	18	72,991	513,018	0	586,009	603	353,363,258
42	23	93,266	655,523	0	748,789	658	492,703,109
37	28	113,541	798,028	0	911,569	730	665,445,458
32	33	133,816	940,533	0	1,074,349	772	829,397,675
27	38	154,092	1,083,038	0	1,237,130	502	621,039,019
22	43	174,367	1,225,543	0	1,399,910	280	391,974,722
17	48	194,642	1,368,048	0	1,562,690	169	264,094,596
12	53	214,917	1,510,553	0	1,725,470	94	162,194,191
7	58	235,192	1,653,058	0	1,888,250	40	75,530,013
2	63	255,468	1,795,563	0	2,051,031	20	41,020,610
–3	68	275,743	1,938,068	0	2,213,811	10	22,138,107
–8	73	296,018	2,080,573	0	2,376,591	1	2,376,591
–13	78	316,293	2,223,078	0	2,539,371	1	2,539,371
–18	83	336,568	2,365,583	0	2,702,151	0	0
							Σ 4,426,823,817

And then, the heating fuel oil and natural gas consumptions were determined as follows:

			Annual Cost of Heating (US$/season)
E (gal oil/year)	Equation (7.4)	39,810	97,533
E (ft³ natural gas/year)	Equation (7.4)	5,341,245	26,706

Our calculations show that both the annual heating degree-day (Example 7.1) and the bin method approaches provided quite similar estimates. We can use the same energy conservation measures that were used in Examples 7.3 and 7.4, and determine similar savings this time using the bin method. One can easily introduce varying heat pump performance (below the balance point temperature) as a function of outside temperature or some sort of internal heat gains and determine new balance point temperature and therefore estimate the heating energy consumption due to the varying heat pump efficiency therefore the heat pump capacity (which must be adjusted for the changing efficiency) or the use of free energy (internal heat gains) available, respectively. We will now introduce a fixed internal heat gain in the following example and see what happens.

Example 7.6: Estimating Heating Energy in a Greenhouse with a Fixed Internal Heat Gain via Bin Method

Let's now consider the greenhouse, which was used in Example 7.5, a single glass greenhouse in Columbus, Ohio, and estimate the annual heating energy requirement and the costs of heating using fuel oil and natural gas furnaces for this greenhouse? Assume an average internal heat gain of 200,000 Btu/h.

Solution

First, using Equation (9.2a) or (9.2b), we have to calculate the balance point temperature T_{bal} for this greenhouse complex:

$$T_{bal} = T_i - \left(\frac{\dot{Q}_{internal}}{\left(\sum UA\right) + \text{ACHV}\,\rho c_p}\right)$$

$$= 65°\text{F} - \left(\frac{200{,}000\ \text{Btu/h}}{\left(28{,}501\,\dfrac{\text{Btu}}{\text{h.°F}}\right) + 1.0 \times \left(202{,}752\ \text{ft}^3\right) \times 0.02}\right)$$

$$T_{bal} = 58.9°\text{F}$$

- Remember that no supplemental heating is required at or above 58.9°F outside temperatures. Even though the inside design temperature $T_i = 65°F$, the difference $T = T_i - T_{bal}$ is handled by free internal heat gains rather than the furnace.
- See Table A.73 for the number of occurrence hours in each bin. Since $T_{bal} = 58.9°F$, the highest temperature bin to be considered is the 57 °F bin (59.5°F and 54.5°F).
- 57°F bin occurs for 648 h, but supplemental heating is needed only when $T_o < 58.0°F$.
- The outside temperature is between 54.5 °F and 58.9 °F for about $648h(58.9°F - 54.5°F)/5°F = 564.6h$
- During this 564.6-hour period, the average outside (bin) temperature is: $(58.9°F + 54.5°F)/2 = 56.7°F$

So, 56.7°F is the highest temperature interval (bin) and the corresponding occurrences of 565 h need to be used first in our calculations. And then, the remaining (lower) temperature bins and the corresponding occurrence hours will be incorporated fully for estimating the seasonal heating energy consumption.

Finally, using Equation (9.4) and its components, we will perform the full energy estimating calculations as presented in the following table:

BIN METHOD

Bin Temp (°F)	ΔT (°F)	Q_{inf} (Btu/h)	Q_{cond} (Btu/h)	$Q_{internal}$ (Btu/h)	$Q_{inf} + Q_{cond} - Q_{internal}$ (Btu/h)	Occurrence (h)	E (Btu/year)
56.7	8.3	33,657	236,558	200,000	70,215	565	39,645,877
52	13	52,716	370,513	200,000	223,229	622	138,848,139
47	18	72,991	513,018	200,000	386,009	603	232,763,258
42	23	93,266	655,523	200,000	548,789	658	361,103,109
37	28	113,541	798,028	200,000	711,569	730	519,445,458
32	33	133,816	940,533	200,000	874,349	772	674,997,675
27	38	154,092	1,083,038	200,000	1,037,130	502	520,639,019
22	43	174,367	1,225,543	200,000	1,199,910	280	335,974,722
17	48	194,642	1,368,048	200,000	1,362,690	169	230,294,596
12	53	214,917	1,510,553	200,000	1,525,470	94	143,394,191
7	58	235,192	1,653,058	200,000	1,688,250	40	67,530,013
2	63	255,468	1,795,563	200,000	1,851,031	20	37,020,610
−3	68	275,743	1,938,068	200,000	2,013,811	10	20,138,107
−8	73	296,018	2,080,573	200,000	2,176,591	1	2,176,591
−13	78	316,293	2,223,078	200,000	2,339,371	1	2,339,371
−18	83	336,568	2,365,583	200,000	2,502,151	0	0
							Σ 3,326,310,737

And then, the heating fuel oil and natural gas consumptions were determined as follows:

			Annual Cost of Heating (US$/season)
E (gal oil/year)	Equation (9.4)	29,913	73,287
E (ft³ natural gas/year)	Equation (9.4)	4,013,406	20,067

The above calculations show that using the available free energy (internal heat gains) reduces the heat provided by the furnace approximately 25% in this example.

Example 7.7: Estimating Heating Energy in a Swine Finishing Building via Bin Method

Let's now consider the swine finishing building in Example 5.13, with 100 pigs weighing 130 lb (60 kg) near Columbus, Ohio. Assume inside design conditions of 60°F (15°C) and heating system efficiency of 82%. Average sensible heat gain from 100 pigs is calculated as 58,000 Btu/h. Estimate the annual heating energy requirement and the cost of heating using natural gas furnaces for this.

HV of natural gas (Btu/ft³)	1,036	Price of natural gas (US$/ft³)	0.005

Solution

First, using Equation (9.2a) or (9.2b), we have to calculate the balance point temperature T_{bal} for this finishing building:

$$T_{bal} = T_i - \left(\frac{\dot{Q}_{internal}}{\left(\sum UA\right) + 1.1Q}\right)$$

$$= 60°F - \left(\frac{58{,}000 \text{ Btu/h}}{\left(246.5 \text{ Btu/h.°F}\right) + 1.1 \times \left(2{,}080 \text{ ft}^3\right)}\right)$$

$$T_{bal} = 37°F$$

- Remember that no supplemental heating is required at or above 37°F outside temperatures. Even though the inside design temperature $T_i = 60°F$, the difference $T = T_i - T_{bal}$ is handled by free internal heat gains rather than the furnace.
- See Table A.73 for the number of occurrence hours in each bin. Since $T_{bal} = 37°F$, the highest temperature bin to be considered is the 37°F bin (39.5°F and 34.5°F).
- 37°F bin occurs for 772 h; but, supplemental heating is needed only when $T_o < 37°F$.
- The outside temperature is between 34.5°F and 37°F for about: $772h(37°F - 34.5°F)/5°F = 386h$
- During this 386-hour period, the average outside (bin) temperature is: $(37°F + 34.5°F)/2 = 35.75°F$

So, 37°F is the highest temperature interval (bin) and the corresponding occurrences of 386 h need to be used first in our calculations. And then, the remaining (lower) temperature bins and the corresponding occurrence hours will be incorporated fully for estimating the seasonal heating energy consumption.

Finally, using Equation (9.4) and its components, we will perform the full energy estimating calculations as presented in the following table:

BIN METHOD

Bin Temp (°F)	ΔT (°F)	Q_{inf} (Btu/h)	Q_{cond} (Btu/h)	$Q_{internal}$ (Btu/h)	$Q_{inf} + Q_{cond} - Q_{internal}$ (Btu/h)	Occurrence (h)	E (Btu/year)
35.75	24.3	55,484	5,978	58,000	3,462	386	1,336,187
32	28	64,064	6,902	58,000	12,966	730	9,465,180
26	34	77,792	8,381	58,000	28,173	502	14,142,846
22	38	86,944	9,367	58,000	38,311	280	10,727,080
17	43	98,384	10,600	58,000	50,984	169	8,616,212

12	48	109,824	11,832	58,000	63,656	94	5,983,664
7	53	121,264	13,065	58,000	76,329	40	3,053,140
2	58	132,704	14,297	58,000	89,001	20	1,780,020
−3	63	144,144	15,530	58,000	101,674	10	1,016,735
−8	68	155,584	16,762	58,000	114,346	4	457,384
−13	73	167,024	17,995	58,000	127,019	1	127,019
−18	78	178,464	19,227	58,000	139,691	0	0
							Σ **56,705,466**

And then, the natural gas consumption and its corresponding cost is determined as follows:

			Annual Cost of Heating (US$/season)
E (ft³ natural gas/year)	Equation (7.4)	66,750	334

Example 7.8: Estimating Heating Energy in a Poultry Building via Bin Method

Let's now consider the poultry confinement building in Example 5.14, with 690, 4 lb laying hens near Columbus, Ohio. Assume inside design conditions of 65°F (18°C) and a heating system efficiency of 82%. Average sensible heat gain from 460 birds is calculated as 11,132 Btu/h. Estimate the annual heating energy requirement and the cost of heating using natural gas furnaces for this.

HV of natural gas (Btu/ft³)	1,036	Price of natural gas (US$/ft³)	0.005

Solution

First, using Equation (9.2a) or (9.2b), we have to calculate T_{bal} for this finishing building:

$$T_{bal} = T_i - \left(\frac{\dot{Q}_{internal}}{\left(\sum UA\right) + 1.1Q}\right)$$

$$= 65^{\circ}\text{F} - \left(\frac{11{,}132 \text{ Btu/h}}{\left(246.5 \text{ Btu/h.}^{\circ}\text{F}\right) + 1.1 \times \left(530 \text{ ft}^3\right)}\right)$$

$$T_{bal} = 51.6^{\circ}\text{F}$$

- Remember that no supplemental heating is required at or above 51.6°F outside temperatures. Even though the inside design temperature $T_i = 65^{\circ}$F, the difference $T = T_i - T_{bal}$ is handled by free internal heat gains rather than the furnace.
- See Table A.73 for the number of occurrence hours in each bin. Since $T_{bal} = 51.6^{\circ}$F, the highest temperature bin to be considered is the 52°F bin (54.5 °F and 49.5°F).
- 52°F bin occurs for 622 h, but supplemental heating is needed only when $T_o < 51.6$°F.
- The outside temperature is between 49.5 °F and 51.6 °F for about $622\text{h}\left(51.6^{\circ}\text{F} - 49.5^{\circ}\text{F}\right)/5^{\circ}\text{F} = 261\text{h}$
- During this 261-hour period, the average outside (bin) temperature is: $\left(51.6^{\circ}\text{F} + 49.5^{\circ}\text{F}\right)/2 = 50.6^{\circ}\text{F}$

So, 52°F is the highest temperature interval (bin) and the corresponding occurrences of 261 h need to be used first in our calculations. And then, the remaining (lower) temperature bins and the corresponding occurrence hours will be incorporated fully for estimating the seasonal heating energy consumption.

Finally, using Equation (9.4) and its components, we will perform the full energy estimating calculations as presented in the following table:

BIN METHOD

Bin Temp (°F)	ΔT (°F)	Q_{inf} (Btu/h)	Q_{cond} (Btu/h)	$Q_{internal}$ (Btu/h)	Q_{inf} + Q_{cond} − $Q_{internal}$ (Btu/h)	Occurrence (h)	E (Btu/year)
50.6	14.4	8,395	3,550	11,132	813	261	212,141
47	18	10,494	4,437	11,132	3,799	603	2,290,797
42	23	13,409	5,670	11,132	7,947	658	5,228,797
37	28	16,324	6,902	11,132	12,094	772	9,336,568
32	33	19,239	8,135	11,132	16,242	730	11,856,295
26	39	22,737	9,614	11,132	21,219	502	10,651,687
22	43	25,069	10,600	11,132	24,537	280	6,870,220
17	48	27,984	11,832	11,132	28,684	169	4,847,596
12	53	30,899	13,065	11,132	32,832	94	3,086,161
7	58	33,814	14,297	11,132	36,979	40	1,479,160
2	63	36,729	15,530	11,132	41,127	20	822,530
−3	68	39,644	16,762	11,132	45,274	10	452,740
−8	73	42,559	17,995	11,132	49,422	4	197,686
−13	78	45,474	19,227	11,132	53,569	1	53,569
−18	83	48,389	20,460	11,132	57,717	0	0
							Σ **57,385,947**

And then, the natural gas consumption, and its corresponding cost is determined as follows:

			Annual Cost of Heating (US$/season)
E (ft³ natural gas/year)	Equation (7.4)	67,551	338

Example 7.9: Estimating Heating Energy in an Aquatic Building via Bin Method

Let's now consider the aquatic building in Example 5.17, with 690 4 lb laying hens near Columbus, Ohio. Assume inside design conditions of 55°F (13°C), no internal sensible heat gain, and a heating system efficiency of 82%. Estimate the annual heating energy requirement and the cost of heating using natural gas furnaces for this.

HV of natural gas (Btu/ft³)	1,036	Price of natural gas (US$/ft³)	0.005

Solution

First, using Equation (9.2a) or (9.2b) we have to calculate the balance point temperature T_{bal} for this finishing building:

$$T_{bal} = T_i - \left(\frac{\dot{Q}_{internal}}{\left(\sum UA\right) + 1.1Q}\right)$$

$$= 60^{\circ}\text{F} - \left(\frac{58{,}000 \text{ Btu/h}}{\left(246.5 \text{ Btu/h.}^{\circ}\text{F}\right) + 1.1 \times \left(2{,}080 \text{ ft}^3\right)}\right)$$

$$T_{bal} = 37^{\circ}\text{F}$$

- Remember that supplemental heating is required below 55°F outside temperatures because there is no internal sensible heat gain in this case (assumed).
- See Table A.73 for the number of occurrence hours in each bin. Since $T_{bal} = 55°F$, the highest temperature bin to be considered is the 57°F bin (59.5°F and 54.5°F).
- 57°F bin occurs for 648 h, but supplemental heating is needed only when $T_o < 55°F$.
- The outside temperature is between 54.5°F and 55°F for about: $648h(55°F - 54.5°F)/5°F = 65h$
- During this 65-hour period, the average outside (bin) temperature is: $(55°F + 54.5°F)/2 = 54.8°F$

So, 57°F is the highest temperature interval (bin) and the corresponding occurrences of 65 h need to be used first in our calculations. And then, the remaining (lower) temperature bins and the corresponding occurrence hours will be incorporated fully for estimating the seasonal heating energy consumption.

Finally, using Equation (9.4) and its components, we will perform the full energy estimating calculations as presented in the following table:

BIN METHOD

Bin Temp (°F)	ΔT (°F)	Q_{inf} (Btu/h)	Q_{cond} (Btu/h)	$Q_{internal}$ (Btu/h)	$Q_{inf} + Q_{cond} - Q_{internal}$ (Btu/h)	Occurrence (h)	E (Btu/year)
54.8	0.2	117	49	0	166	65	10,784
52	3	1,749	740	0	2,489	622	1,547,847
47	8	4,664	1,972	0	6,636	603	4,001,508
42	13	7,579	3,205	0	10,784	658	7,095,543
37	18	10,494	4,437	0	14,931	772	11,526,732
32	23	13,409	5,670	0	19,079	730	13,927,305
26	29	16,907	7,149	0	24,056	502	12,075,861
22	33	19,239	8,135	0	27,374	280	7,664,580
17	38	22,154	9,367	0	31,521	169	5,327,049
12	43	25,069	10,600	0	35,669	94	3,352,839
7	48	27,984	11,832	0	39,816	40	1,592,640
2	53	30,899	13,065	0	43,964	20	879,270
−3	58	33,814	14,297	0	48,111	10	481,110
−8	63	36,729	15,530	0	52,259	4	209,034
−13	68	39,644	16,762	0	56,406	1	56,406
−18	73	42,559	17,995	0	60,554	0	0
							Σ 69,748,508

And then, the natural gas consumption, and its corresponding cost is determined as follows:

			Annual Cost of Heating (US$/season)
E (ft³ natural gas/year)	Equation (7.4)	82,103	411

7.6 ESTIMATING ENERGY VIA COMPLEX MODELING

Mathematical models are needed to accurately predict and aid in the optimization of the controlled environment building thermal and radiant environments. Accurate prediction models for controlled environments and plant, animal, or fish growth performance are also used as a design tool and in economic feasibility analyses. And, simulation of controlled environments requires an accurate prediction of building energy and moisture exchanges as a function of dynamic environmental factors such as solar energy, wind speed, outside temperatures and moisture levels, auxiliary heating, plant moisture and energy exchanges and excess heat removal or storage. Both steady-state and dynamic (transient) analyses techniques have been used in simulating building thermal environments. Since a steady-state analysis does not account for energy storage in the system and does not permit accurate predictions of the instantaneous thermal and biological responses of controlled environment systems, a dynamic analysis is required for more accurate prediction and control of building thermal environments, and therefore energy use estimations.

Steady-state models use one or more variables for energy estimating in controlled environments, and do not account for the effects such as thermal mass or capacitance that cause short-term temperature transients; hence, they are often used for baseline model development. Dynamic models, however, capture the effects of building peak loads, and warm-up or cool-down periods, and are especially useful for controlled environment building heating and cooling load control. A dynamic simulation of a building thermal environment should also incorporate the dynamics of the plant canopy, livestock, poultry, or fish in a reasonably accurate manner. The treatment of the plant canopy, for instance, varies tremendously from model to model. While some simulation models incorporates only a simple plant canopy for energy and moisture exchanges within the greenhouse environment, others do not account for the plant canopy at all. One can easily recognize whether a model is steady-state or dynamic by looking for the existence of time-lagged variables, either in the output or input variables. Steady-state models simply do not have any time-lagged variables.

Several types of steady-state controlled environment building models have been developed for estimating baseline energy consumption: single- or multi-variate, polynomial, etc. Single-variate (i.e., one input variable only) energy estimating models are the easiest of steady-state modeling approaches. They estimate energy use in a controlled environment building based on one forcing function, i.e., the input variable. Many studies have shown that the outdoor air dry-bulb temperature is the most important forcing function. The simplest steady-state data-driven model is the 18.3 °C (65 °F) degree-day model. As mentioned above, the main disadvantage of steady-state single-variate data-driven models is their insensitivity to dynamic effects (e.g., thermal mass) and to variables other than temperature (e.g., humidity, solar gain, and artificial lighting, etc.).

Besides, there also exist two types of steady-state, multi-variate models in general: i) standard multiple-linear model where the data set is treated without having the time-series type data; and ii) Fourier series models having the time-series type data and capturing the diurnal and seasonal cycles, based on which controlled environment buildings are operated. Multi-variate models are naturally an extension of single-variate models, and the goal of such type of energy use modeling is to characterize building energy use with a few readily available and reliable input variables. The environmental variables for modeling heating and cooling energy use include outdoor air dry-bulb temperature, solar radiation, and outdoor specific humidity. In commercial controlled environment buildings, internally generated loads, such as the heat given off by lights and other electrical equipment, also affect heating and cooling energy use. Breaking up energy use data into hourly bins corresponding to each hour and then identifying 24 individual hourly models obviously lead to considerably greater accuracy.

Dynamic data-driven building models, on the other hand, require solving a set of differential equations, and are typically used with hourly or sub-hourly data in controlled environment buildings where the thermal mass is significant enough to delay heat gains or losses. Dynamic models usually require a high level of knowledge of the controlled environment or the overall system being modeled. Dynamic models based on pure statistical approaches have also been developed for the past couple of decades; and such kind of approaches include machine learning and artificial neural networks.

Neural networks have the ability to learn, and they learn by example rather than by programmed rules. The weights of a network are initiated with small random numbers; and then, the weights are adjusted (trained) iteratively with a training data set that consists of many input/output pairs, so that applying a set of inputs eventually produces the desired set of outputs. Neural networks have been useful in modeling energy use in commercial controlled environment buildings for: i) predicting what an adequate operation should be doing compared with the actual operation and correcting any differences and ii) predicting what sort of energy a controlled environment building would have consumed under present conditions, before a retrofit; and when compared to the measured consumption of the retrofitted building, the difference represents quite a reasonable estimate of the energy savings due to the retrofit, for instance.

7.7 MODEL VALIDATION AND TESTING

A developed simulation model's accuracy must be validated and tested before it is used for reliable estimations, controls, or any other strategies. There exist three methods to evaluate the accuracy of a whole-controlled environment building energy simulation model: i) empirical validation, which compares simulated results to monitored data from a real operation; ii) analytical verification, which compares simulated outputs to results from a known analytical solution or a generally accepted numerical method calculation for isolated heat transfer under very simple, highly constrained boundary conditions; and iii) comparative testing, which compares a simulation model to itself or to other simulation models. All of these three validation and testing procedures are fully acceptable and have been used extensively in model development.

NOMENCLATURE

A	area
ACH	air change per hour
c	specific heat
CDD	cooling degree-days
E	energy
HDD	heating degree-days
HV	heating value
HVAC	heating, ventilation and air-conditioning
N	number
Q	ventilation rate
$\dot{Q}$	heat transfer rate
T	temperature
U	conduction heat transmission coefficient
V	volume

Greek Letters

Δ	difference
η	efficiency
ρ	density
Σ	total

Subscripts

a	air
bal	balance point
bin	bin
i	internal
inf	infiltration/ventilation
p	pressure
t	total

BIBLIOGRAPHY

Aldrich, R and Bartok, JW (1992) *Greenhouse Engineering*. Northeast Regional Agricultural Engineering Service, Ithaca, NY, USA.

Arinze, EA, Schoenau, GJ and Besant, RW (1984) A dynamic thermal performance simulation model of an energy conserving greenhouse with thermal storage. *Transactions of the ASAE*, 27: 508–519.

ASHRAE (1986) *Bin and Degree-Hour Weather Data for Simplified Energy Calculations*. American Society of Heating, Refrigeration and Air Conditioning Engineers, Inc., Atlanta, GA 30329.

ASHRAE (1989a) *Handbook of Fundamentals*. American Society of Heating, Refrigeration and Air Conditioning Engineers, Inc., Atlanta, GA 30329, pp. 24.1–24-22.

ASHRAE (1989b) *Handbook of Fundamentals*. American Society of Heating, Refrigeration and Air Conditioning Engineers, Inc., Atlanta, GA 30329, pp. 28.1–28-42.

ASHRAE (1997) *Handbook of Fundamentals*. American Society of Heating, Refrigeration and Air Conditioning Engineers, Inc., Atlanta, GA 30329.

ASHRAE (2005) *Handbook of Fundamentals*. American Society of Heating, Refrigeration and Air Conditioning Engineers, Inc., Atlanta, GA 30329, pp. 32.1–32.39.

Duffie, JA and Beckman, WA (2006) *Solar Engineering of Thermal Processes*. John Wiley and Sons, Inc., Hoboken, NJ, USA.

Engineering ToolBox (2009) US Heating Degree-Days. [online] Available on: https://www.engineeringtoolbox.com/us-degree-days-d_1569.html (Retrieved on 12.08.2020).

Hinrichs, R and Kleinbach, M (2006) *Energy: Its Use and the Environment*. Thompson Higher Education, Belmont, USA.

Incropera, FP and DeWitt, DP (1985) *Introduction to Heat Transfer*. John Wiley & Sons, Inc., New York, USA.

Incropera, F, Dewitt, D, Bergman, T and Lavine, A (2007) *Introduction to Heat Transfer*. John Wiley & Sons, Inc., New York, USA.

Iqbal, M (1983) *An Introduction to Solar Radiation*. Academic Pub., Toronto, Canada.

Kindelan, M (1980) A dynamic modeling of greenhouse environment. *Transactions of the ASAE*, 23(5): 1232–1239.

MWPS (1987) *Structures and Environment Handbook*. 11th Edition. Midwest Plan Service, Iowa State university, Ames, IA, USA. ISBN: 0-89373-057-2.

Navas, LM, de la Plaza, S, Garcia, JL, Duran, JM and Retamal, N (1998) Formulation and sensitivity analysis of a dynamic model of the greenhouse climate validation for a mild Mediterranean climate. *Acta Horticulturae*, 456: 305–312.

NOAA (1973) *Degree-Days to Selected Bases*. U.S. National Climatic Data Center, Asheville, NC, USA.

Rijsdijk, AA and Houter, G (1993) Validation of a model for energy consumption, CO2 consumption and crop production (ECP-Model). *Acta Horticulturae*, 328: 125–131.

Takakura, TK, Jordan, A and Boyd, LL (1971) Dynamic simulation of plant growth and environment in the greenhouse. *Transactions of the ASAE*, 14(5): 964–971.

Thornley, JHM and Johnson, IR (1990) *Plant and Crop Modelling: A Mathematical Approach to Plant and Crop Physiology*. Clarendon Press, Oxford, pp. 48–49 and 128.

USAF (1978) Engineering Weather Data. Air Force Manual AFM 88-29, U.S. Government Printing Office 725-243/511, July.

Yildiz, I (1993) Simulation of greenhouse microclimates and environmental control strategies using a Rankine cycle heat pump. Dissertation. The Ohio State University, Columbus, OH, USA.

Yildiz, I and Stombaugh, DP (2006a) Dynamic modeling of microclimate and environmental control strategies in a greenhouse coupled with a heat pump system. *Acta Horticulturae*, 718: 331–340.

Yildiz, I and Stombaugh, DP (2006b) Simulated performances of a heat pump system for energy and water conservation in open and confined greenhouse systems. *Acta Horticulturae*, 718: 341–350.

Yildiz, I (2021) *Greenhouse Engineering: Integrated Energy Management*. 1st Edition. CRC Press, Taylor and Francis Group, FL, USA. ISBN: 978-1-4987-4313-6.

8 Energy Conservation

8.1 INTRODUCTION

Integrated energy management entails energy conservation, energy mixture and environmental considerations. Since it is already available in hand, we start with energy conservation. Energy conservation is the practice of reducing energy use as a way of extending the lifetime of our fossil-fuel supplies, of being less wasteful, and of reducing our environmental impact. Personal choice and increased efficiency are the two parallel pathways for effective energy conservation. While using more efficient lighting fixtures, cogeneration systems or high efficiency boilers in our controlled environment facilities are some of the examples of energy conservation via personal choice, developing more efficient lighting fixtures, cogeneration systems or high efficiency boilers form the technical fix, i.e., the increased efficiency.

As mentioned earlier, in order to make their operations sustainable, commercial controlled environment operations have to reduce the two largest expenses, labor and energy. Every passing day, the use of mechanization and automation systems is becoming more common in controlled environment production facilities for combating the labor issues. Particularly in colder climates, heating energy is a major expense category especially in commercial controlled environment greenhouses and energy prices have been rising. Consequently, considerable effort is expended to conserve energy and look for alternative energy sources, especially environmentally friendly energy sources and technologies. However, controlled environment operations in certain countries have not been fast enough to adopt energy conservation and environmentally friendly renewable energy technologies.

Controlled environment operations will keep using conventional energy sources for the foreseeable future while incorporating more alternative energy sources in their energy mix and improved energy collection and storage technologies. Paying special attention to energy conservation measures certainly help growers to maximize energy savings and make their operations more sustainable. However, conservation and sustainability require a multi-faceted approach, and there is an additional dimension to energy use and conservation, the environmental dimension. In the years to come, energy investments will increasingly be determined by three factors: using energy more efficiently, improving competitiveness, and protecting the environment. Hence, this book not only covers the conservation techniques and technologies available for controlled environment growers supported with numerical figures, but also dives deeper into the environmental dimension (see Chapter 9) of commercial operations in an effort to convert the global controlled environment industry to a fully sustainable net energy generating industry.

New energy efficient controlled environment facility designs should incorporate features that take maximum advantage of solar energy and are compatible with energy conservation. A number of techniques and technologies have been developed for energy conservation in controlled environments and are currently being used by many commercial operations especially in large operations. Some of these techniques and technologies can be listed as the use of increased insulation on ceiling, foundation, sidewalls and end walls, the use of inflated double or triple plastic alone or over glass, the use of energy curtains at night and different shading devices to control incoming solar radiation, energy efficient lighting systems, and high efficiency condensing boilers or flue gas condensers to extract waste heat that would otherwise be released to the environment and wasted. Other approaches for energy conservation include the use of evaporative cooling, reduced ventilation, and preheating the incoming air. In addition, heat pumps hold great promise for reducing winter heating and warm weather cooling energy requirements, particularly in semi-closed or confined controlled environment operations. This chapter covers all of the abovementioned conservation techniques and technologies available for controlled environment operations supported with a number of numerical examples.

8.2 TRANSPARENT AND OPAQUE INSULATION

As discussed earlier in Section 2.3.5.4, when properly applied in building envelopes or on mechanical systems, insulation is used for conserving energy by reducing heat loss or gains in buildings and HVAC systems, as well as preserving natural and financial resources.

Short and Bauerle (1980) studied energy conservation techniques for commercial greenhouses. They reported that double-layer-air-separated plastic provided a double glazing with low initial costs and a 30 to 35% energy savings over most single glazings (see Example 8.1 below for similar energy saving estimates). Similar findings on the use of double glazing instead of single glazing were published by Roberts and Mears (1969). In inland Australian climatic conditions, Garzoli and Blackwell (1987) reported an average of 32% energy savings by introducing double layer plastic glazing over single layer glazing. Short and Bauerle (1977) as well reported that glass houses can be modified for energy conservation by lap-sealing or using air-inflated-double-plastic-over-glass or rigid-double-wall-acrylic to save up to 20, 60, and 60%, respectively. In another report, Short et al. (1978) indicated that, in Ohio, a polystyrene pellet nighttime insulation system could reduce greenhouse night time energy requirements by 80 to 90%. Miyagawa (1977) reported similar findings in Japan. In addition to

DOI: 10.1201/9781003349297-8

reducing night time heating requirements, the same system can provide variable shading, which can reduce the cooling requirements during hot sunny days.

Polyethylene film is the most common glazing material used in greenhouses today. Two or sometimes three layers of polyethylene films are used in most heated greenhouses to reduce heat loss. In earlier sections, it was discussed that the introduction of each glazing layer impacts light transmission negatively; however, energy losses in an inflated double-poly greenhouse, as shown below, is much less than those of a single layer glass greenhouse. Greenhouse operators need to make sure that outside air is used for inflating the space between the polyethylene film layers to reduce the risk of condensation.

Furthermore, installing an infrared (IR) polyethylene film, which has quite a short payback period (just a few months), can block long-wave (infrared) radiation as well and therefore reduce heat losses at night lowering the heating requirements in winter and help reduce cooling requirements in the summer. IR polyethylene can be used in different ways: i) IR polyethylene as an exterior layer—where outside IR radiation is partly blocked from entering the greenhouse; ii) IR polyethylene as an interior layer—where inside IR radiation is partly blocked from leaving the greenhouse and trapped inside; iii) IR polyethylene as both interior and exterior layer—where both layers of a double-poly glazing are IR polyethylene films preventing both outside infrared radiation entering the greenhouse and inside infrared radiation leaving the greenhouse.

As discussed in Chapter 3 earlier, condensation is quite a common problem in greenhouses, and water droplets on the glazing material reduces light transmission and commonly lead to disease problems particularly when they drip onto plants. AC films prevent early morning and late afternoon fog formation in the greenhouse; hence, fewer water droplets fall on plants and less solar radiation is blocked compared to a condensing plastic film. The anti-condensation agent reduces the surface tension and allows condensate to flow rather than forming droplets. However, it should be kept in mind that the additives are slowly washed out by water in time; and therefore, the AC effect lasts for up to two years or so.

Not only insulating the roof of a greenhouse with transparent materials such as double or triple poly ethylene glazings, but also insulating the sidewalls and end walls with the same type glazing materials is extremely important. Depending on the sidewall and end wall materials, opaque insulation may need to be used as discussed in Chapter 3. Same type of insulation is required for simple and further energy savings on foundations as well.

Special attention must also be given to cracks and other unintended openings, especially where the glazing is attached to the foundation, sidewalls and gable end walls, as well as around doors, fans, fan louvers, and vents. It is the easiest energy conservation measure with great savings and an extremely quick payback period.

In other controlled environment buildings such as livestock, poultry, and aquatic systems as well, we need to make sure that all areas are insulated with enough insulation, and windows are installed only when necessary, and special attention must be given to cracks and other unintended openings particularly around doors, fans, fan louvers, and vents. See Examples 8.1, 8.2, and 8.3 below for relevant numerical examples.

Example 8.1: Reduction in Heating Energy Requirement in a Greenhouse When Double-Poly Glazing Used

Let's now consider the same greenhouse facility that was used in Example 7.1; however, this time a double-poly glazing was used instead of single glass, and the lower wall (6 in. poured concrete + 2 in. foamed urethane board) and the perimeter are insulated as shown in the following table. Determine the energy savings.

Solution

We first need to calculate the UA values as follows:

UA VALUES and AIR CHANGE PER HOUR

GREENHOUSE COMPONENT	A (ft^2)	U (Btu/h.ft^2.°F)	Reference	UA (Btu/h.°F)
Lower Wall Area	864	0.07	Table A.60	60
Upper Wall Area	3,744	0.7	Table A.60	2,621
Gable or Curved Area	576	0.7	Table A.60	403
Roof Area	20,582	0.7	Table A.60	14,407
Perimeter Length	576	0.4	Table A.60	230
ΣUA				17,722
ACH (per hour)		1.0	Table A.61	

And then, using Equations (5.2 and 5.4), we can determine the greenhouse envelop and infiltration heat losses as well as the fuel oil and natural gas consumption calculations as shown in the following table:

			Annual Cost of Heating (US$/season)
Q_{cond} (Btu/h)	$UA \times \Delta T$	1,045,615	
Q_{inf} (Btu/h)	$0.02 \times \Delta T \times V \times ACH$	239,247	
Q_t (Btu/h)	$Q_{cond} + Q_{inf}$	1,284,862	
E (gal oil/year)	$24 \times HDD \times Q_t / (HV \times \eta \times \Delta T)$	26,603	65,177
E (ft^3 natural gas/year)	$24 \times HDD \times Q_t / (HV \times \eta \times \Delta T)$	3,569,294	17,846

As shown in the above table, just by changing the glass glazing to double-poly and insulating the lower wall and the perimeter, we have saved approximately 33% in heating energy.

Example 8.2: Effect of Frame Wall Insulation in a Swine Farrowing Building

Let's now consider the building that was used in Example 5.12; however, this time, assume that the insulation in the frame walls is decreased from R = 11 batt insulation to 25/32 in. insulating sheathing. Determine the change in building heat loss.

Solution

Step 1: We first list the length, width, wall height, and foundation height of the building. Then, we calculate the perimeter, frame wall area (excluding windows and doors), concrete wall, ceiling, window, and door areas.

BUILDING DIMENSIONS & SURFACE AREAS

BUILDING DIMENSIONS	(ft)	(m)
House Width	24	7.2
House Length	48	14.4
Frame Wall Height	6	1.8
Concrete Wall Height	2	0.6
Perimeter	144	43.2
SURFACE AREAS	**(ft^2)**	**(m^2)**
Ceiling Area	1,152	103.7
Window Area	0	0
Door Area	42	3.8
Frame Wall Area (excluding windows and doors)	822	74
Concrete Wall Area	288	25.9

Step 2: We then list the inside and outside design temperatures of the building, and calculate and list the R values of the building sections (see Table 5.6).

Frame Wall Resistance	
Outside surface (15 mph wind)	0.17
Sheet metal	0.00
Air space	0.90
Insulating sheathing (25/32 in.)	2.06
Supplemental vapor barrier	0.00
Plywood (½ in.)	0.62
Inside surface (still air)	0.68
Total resistance $(R_T)_w$	4.43

OTHER PROVIDED or CALCULATED INPUTS

PARAMETER		(I-P)		(SI)
T_i	60	°F	15.5	°C
T_o	6	°F (99%); Table A.52	−14	°C (99%); Table A.53
ΔT	54	°F	29.5	°C
Location		Columbus, Ohio, USA		
R-Values				
Ceiling	13.47	h.ft^2.°F/ Btu; Table 5.6	2.37	m^2.K / W; Table 5.6
Window	–	h.ft^2.°F/ Btu;	–	m^2.K / W;
Door	7.99	h.ft^2.°F/ Btu; Table 5.6	1.41	m^2 K / W; Table 5.6
Frame Wall	4.43	h.ft^2.°F/ Btu; Calculated	0.78	m^2.K / W; Calculated
Concrete Wall	11.58	h.ft^2.°F/ Btu: Calculated	2.04	m^2.K / W; Calculated
Perimeter	2.22	h.ft^2.°F/ Btu; Table 5.6	1.26	m^2.K / W; Table 5.6

Step 3: And then, using Equations (5.5) and (5.6), we calculate the heat loss from the building sections and building.

HEAT LOSS FROM BUILDING SECTIONS

Building Section	$(1/R_T)_i$ (Btu/h.ft^2.°F)	A_i (ft^2 of ft^2)	(T_i/T_o) (°F)	$(Q_{cond})_i$ (Btu/h)	$(Q_{cond})_i$ (W)
Ceiling Area	1/13.47	1,152	54°F	4,618.3	1,354.3
Window Area	0	0		0	0
Door Area	1/7.99	42		283.9	83.3
Frame Wall	1/4.43	822		10,020	2,938
Concrete Wall	1/11.58	288		1,343.0	393.8
Perimeter	1/2.22	144		3,502.7	1,027.2
Total Heat Loss, $\dot{Q}_{cond}$	366 Btu/h.°F			**19,764**	**5,796**

Our calculations show an increase in heat loss through the frame wall of approximately 180%, and the total building heat loss increases almost 50%.

Example 8.3: Effect of Concrete Wall and Perimeter Insulation in a Swine Farrowing Building

Let's now consider the same building in Example 5.12; however, this time, assume that the 2 in. × 48 in. insulation covering the concrete wall and perimeter is neglected. Determine the change in building heat loss.

Solution

Step 1: We first list the length, width, wall height, and foundation height of the building. Then, we calculate the perimeter, frame wall area (excluding windows and doors), concrete wall, ceiling, window, and door areas.

BUILDING DIMENSIONS & SURFACE AREAS

BUILDING DIMENSIONS	(ft)	(m)
House Width	24	7.2
House Length	48	14,4
Frame Wall Height	6	1.8
Concrete Wall Height	2	0.6
Perimeter	144	43.2
SURFACE AREAS	**(ft^2)**	**(m^2)**
Ceiling Area	1,152	103.7
Window Area	0	0
Door Area	42	3.8
Frame Wall Area (excluding windows and doors)	822	74
Concrete Wall Area	288	25.9

Step 2: We then list the inside and outside design temperatures of the building, and calculate and list the R-values (see Table 5.6) of the building sections.

Concrete Wall Resistance	
Outside surface (15 mph)	: 0.17
Concrete (6 in.)	: 0.48
Inside surface (still air)	: 0.68
Total resistance, (R_T)	: 1.33

OTHER PROVIDED or CALCULATED INPUTS

PARAMETER	(I-P)		(SI)	
T_i	60	°F	15.5	°C
T_o	6	°F (99%); Table A.52	−14	°C (99%); Table A.53
ΔT	54	°F	29.5	°C
Location		Columbus, Ohio, USA		
R-Values				
Ceiling	13.47	h.ft².°F/Btu; Table 5.6	2.37	m².K/W; Table 5.6
Window	–	h.ft².°F/Btu;	–	m².K/W;
Door	7.99	h.ft².°F/Btu; Table 5.6	1.41	m².K/W; Table 5.6
Frame Wall	4.43	h.ft².°F/Btu; Calculated	0.78	m².K/W; Calculated
Concrete Wall	1.33	h.ft².°F/Btu: Calculated	0.23	m².K/W; Calculated
Perimeter	1.23	h.ft².°F/Btu; Table 5.6	0.70	m².K/W; Table 5.6

Step 3: And then, using Equations (5.5) and (5.6), we calculate the heat loss from the building sections and building.

HEAT LOSS FROM BUILDING SECTIONS

Building Section	$(1/R_T)_i$ (Btu/h.ft².°F)	A_i (ft² of ft²)	(T_i/T_o) (°F)	$(\dot{Q}_{cond})_i$ (Btu/h)	$(\dot{Q}_{cond})_i$ (W)
Ceiling Area	1/13.47	1,152	54°F	4,618	1,354
Window Area	0	0		0	0
Door Area	1/7.99	42		284	83
Frame Wall	1/12.47	822		3,560	1044
Concrete Wall	1/1.33	288		11,693	3,429
Perimeter	1/1.23	144		6,322	1,854
Total Heat Loss, $\dot{Q}_{cond}$	490.3 Btu/h.°F			**26,477**	**7,764**

Our calculations show that the heat loss through the concrete wall increases about 791%; and the perimeter, approximately 80%. Not insulating the concrete wall and perimeter causes the total heat loss of the building to almost double.

8.3 RETRACTABLE ENERGY CURTAINS, SHADING SYSTEMS, AND REFLECTIVE MULCH

Another energy conservation technique, the use of retractable energy curtains, is a highly effective method in reducing night time heat requirements for greenhouses (see Figure 8.1). Energy curtains are usually used during the winter in higher latitude colder climates for keeping the warm air closer to the plant canopy and reduce heat losses due to the extra insulation layer introduced as well as reduced radiational cooling. The plants under the energy curtains do not have direct interaction with the cold sky temperature any longer; therefore, a couple degrees higher plant temperatures are observed compared to the temperatures of plants directly exposed to the sky in the absence of these curtains. As a result, plants develop faster under an energy curtain, particularly in colder climates.

FIGURE 8.1 Retractable energy curtains reduce night time heat requirements.

Source: ***Greenhouse Engineering*, Edn. 1 by İlhami Yıldız, Copyright © 2021 by Imprint. Reproduced by permission of Taylor & Francis Group (*Greenhouse Engineering*).**

Night curtains are used in the summer months as well in order to provide shading for plants and reduce cooling loads on particularly hot summer days. The savings using energy curtains have been shown to be as much as 50% (Simpkins et al., 1976; Rebuck et al., 1977). However, Albright et al. (1978) reported that a 90% savings could be achieved using a highly insulated, reflective night curtain. Greenhouse operators report energy savings up to 30% with the use of energy curtains depending on the climate and greenhouse location as well as the curtain material and management, and these figures are in line with our energy savings estimates shown in Example 7.3 in Chapter 7. It should also be kept in mind that the effect of the curtain is naturally greater on colder nights and northern climates; however, special attention is required in opening the curtains in the morning for preventing the cold air from rapidly sinking down to the crop level.

Furthermore, as discussed earlier in other chapters, light levels inside greenhouses drop considerably in late fall, winter and early spring months at or beyond certain latitudes. Therefore, providing adequate light for plants might be challenging while trying to take advantage of the energy conserving potential of the energy curtains. It is not that uncommon to see a second energy curtain (with high light transmissivity) installed in greenhouses in an effort to find a solution to this dilemma. This second energy curtain reduces night time energy losses further; and when used during daytime, it transmits more solar radiation compared to the conventional aluminized curtain. It means it serves as a shading cloth optimizing the light during the day when needed. Or sometimes, energy curtains are kept closed until the outside solar radiation exceeds a crop specific level not hindering the plant growth and development.

Example 8.4: Reduction in Heating Energy Requirement in a Greenhouse When Thermal Blanket Used

Let's now consider the same greenhouse facility that was used in Example 7.1; however, this time, in addition to using a double-poly glazing instead of single glass, and an insulated lower wall (6 in. poured concrete + 2 in. foamed urethane board) and perimeter, a thermal blanket is used as well. Determine the additional energy savings.

Solution

Let's first calculate the UA values as follows:

UA VALUES and AIR CHANGE PER HOUR

GREENHOUSE COMPONENT	A (ft²)	U (Btu/h. ft².°F)	Reference	UA (Btu/h.°F)
Lower Wall Area	864	0.07	Table A.60	60
Upper Wall Area	3,744	0.44	Table A.60	1.647
Gable or Curved Area	576	0.44	Table A.60	253
Roof Area	20,582	0.44	Table A.60	9,056
Perimeter Length	576	0.4	Table A.60	230
ΣUA				11,248
ACH (per hour)		1.0	Table A.61	

And then, using Equations (5.2) and (5.4), we determine the greenhouse envelop and infiltration heat losses as well as the fuel oil and natural gas consumption calculations as shown in the following table:

Q_{cond} (Btu/h)	$UA \times \Delta T$	663,618	Annual Cost of Heating (US$/season)
Q_{inf} (Btu/h)	$0.02 \times \Delta T \times V \times ACH$	239,247	
Q_t (Btu/h)	$Q_{cond} + Qinf$	902,865	
E (gal oil/year)	$24 \times HDD \times Q_t / (HV \times \eta \times \Delta T)$	18,694	45,799
E (ft³ natural gas/year)	$24 \times HDD \times Q_t / (HV \times \eta \times \Delta T)$	2,508,123	12,541

As shown in the above table, after changing the glass glazing to double-poly and insulating the lower wall and the perimeter, by introducing an internal thermal blanket, we have saved 30% more in heating energy consumption.

Greenhouse shading is another energy conserving technique, used mostly during warm and bright sunny days, to reduce indoor temperatures, therefore cooling loads. Several shading systems are available and currently used by many commercial greenhouse growers. These include an external spray (whitewash) applied on the outside surface of the glazing (see Figure 5.29) and variable retractable shading (internal or less commonly external) (see Figure 5.30) using different materials. There are advantages and disadvantages of both systems. For instance, whitewash is quite an inexpensive technique to provide shading for plants; however, the proportion of shading it provides is relatively fixed and permanent, and it is washed away slowly with rainfall. Also, it must be washed off when the solar radiation becomes limiting again in the fall. Retractable shading, on the other hand, is a relatively expensive technology, but provides controlled shading when it is needed. Another variable shading system developed by Kuack (1986) is the one with a combination of polystyrene pellets, which provides precise shading in greenhouses; however, it is not commercially adopted yet due to its relatively expensive channelized double acrylic glazing and control system.

Reflective floors (see Figure 8.2) have also been reported to alter crop environmental factors by increasing the light intensity by the reflection of incoming light (Roberts and Anderson, 1994), reducing the soil temperature (Schalk and Robbins, 1987), and increasing the air temperature above the reflective surface (Zalom, 1981). In their study, Mehlitz et al. (2008) focused on reducing energy and water consumption in semi-arid California greenhouses and studied the influences of mulch (having different reflectivities) on energy and water conservation in San Luis Obispo, California, USA (35° 17'N and 120° 39'W). They developed a dynamic computer simulation model to compare different mulch reflectivities, cucumber plant heights, and leaf dimensions to draw a conclusion about energy and water conservation.

The effect of floor surface reflectivity was studied together with four plant heights and three leaf dimension sets (Table 8.1). The study concluded that the greenhouse

FIGURE 8.2 Reflective floor surface in a greenhouse. Source: *Greenhouse Engineering* (2021).

TABLE 8.1
Greenhouse and Crop Characteristics Used in the Simulation Model

Greenhouse length	15.0 m
Greenhouse width	7.50 m
Greenhouse height at eaves	2.50 m
Greenhouse height at ridges	4.50 m
Greenhouse glazing	Double polyethylene
Floor surface	Mulch with reflectivity of 20%, 40%, 60%, or 80%
Crop	Cucumber
Crop row orientation	North—South
Distance between plant rows	0.75 m
Average leaf length x width	Set 1: 0.30 × 0.25 m; Set 2: 0.23 × 0.18 m; Set 3: 015 × 0.10 m
Average plant height	0.5 m, 0.75 m, 1.0 m, 2.0 m

TABLE 8.2
Comparisons of Energy and Water Consumption Means with Respect to Different Mulch Reflectivities Using Tukey's Simultaneous Test at 95% Confidence

	Energy use (all values in MJ/m²/day)					Water use (all values in kg/m²/day)			
	Reflectivity (Mean)	Compared to Reflectivity (Mean)	Difference of Means	Statistically Significant at a 95% level?		Reflectivity (Mean)	Compared to Reflectivity (Mean)	Difference of Means	Statistically Significant at a 95% level?
Winter	0.2 (32.05)	0.4 (32.65)	0.6	YES	Winter	0.2 (2.49)	0.4 (2.45)	0.034	YES
		0.6 (32.91)	0.86	YES			0.6 (2.44)	0.046	YES
		0.8 (33.07)	1.02	YES			0.8 (2.43)	0.057	YES
	0.4 (32.65)	0.6 (32.91)	0.26	YES		0.4 (2.45)	0.6 (2.44)	0.011	YES
		0.8 (33.07)	0.42	YES			0.8 (2.43)	0.023	YES
	0.6 (32.91)	0.8 (33.07)	0.17	YES		0.6 (2.44)	0.8 (2.43)	0.012	YES
Spring	0.2 (25.39)	0.4 (25.95)	0.56	YES	Spring	0.2 (2.509)	0.4 (2.45)	0.055	YES
		0.6 (26.20)	0.8	YES			0.6 (2.44)	0.074	YES
		0.8 (26.51)	1.12	YES			0.8 (2.41)	0.098	YES
	0.4 (25.95)	0.6 (26.20)	0.25	YES		0.4 (2.454)	0.6 (2.44)	0.019	YES
		0.8 (26.51)	0.56	YES			0.8 (2.41)	0.043	YES
	0.6 (26.20)	0.8 (26.51)	0.32	YES		0.6 (2.435)	0.8 (2.41)	0.024	YES
Summer	0.2 (13.17)	0.4 (13.52)	0.35	YES	Summer	0.2 (2.433)	0.4 (2.36)	0.07	NO
		0.6 (13.59)	0.41	YES			0.6 (2.36)	0.072	NO
		0.8 (13.79)	0.62	YES			0.8 (2.22)	0.214	YES
	0.4 (13.52)	0.6 (13.59)	0.07	YES		0.4 (2.363)	0.6 (2.36)	0.002	NO
		0.8 (13.79)	0.27	YES			0.8 (2.22)	0.144	YES
	0.6 (13.59)	0.8 (13.79)	0.21	YES		0.6 (2.361)	0.8 (2.22)	0.142	YES

energy consumption was dependent on the reflectivity of mulch (soil) used in the greenhouse. The higher the reflectivity was, the more energy was consumed. Mulch with a reflectivity of 20% instead of 80% saved as high as 4.2% energy (Table 8.2). The water consumption for transpiration was also dependent on the reflectivity (Mehlitz et al., 2008).

Mehlitz et al. also showed that significant savings in water consumption (as much as 8.6%) could be realized if 80% reflective mulch was used rather than 20% (Table 11.2). However, the effect of mulch on water use faded away as the shade from canopy stand increased. In order to make a conclusion about total energy and water consumption, the greenhouse has to be treated as a complete system. Potential energy savings due to lower reflectivity have to be seen in conjunction with a larger amount of water use. They concluded that total energy and water conservation analysis has to be performed for each case separately, considering the specific infrastructure, local water and energy prices, and the crop type (Mehlitz et al., 2008).

8.4 REDUCED VENTILATION RATES

While lower ventilation rates reduce the heating requirements, they may cause high relative humidity levels in controlled environment buildings, and low CO_2 levels in greenhouses if CO_2 enrichment is not provided. Consequently, low ventilation rates may adversely affect plant and animal responses unless carefully controlled. If conventional ventilation rates are used to maintain reasonable relative humidity levels within greenhouses, livestock, poultry and aquatic buildings, this time, they may result in higher heating energy requirements, and relatively high operating costs.

Yildiz and Stombaugh (2009) studied the effects of minimum ventilation rate on energy and water conservation in a greenhouse heated and cooled with a gas-fired heat pump in Delaware, Ohio (latitude 40° 17'N, longitude 83° 05'W). Table 8.3 shows the greenhouse and crop characteristics used in the dynamic greenhouse simulation model.

In their assessment, a minimum ventilation rate of 0.005 m³/s.m² was used and compared to the conventional rate of 0.01 m³/s.m². Table 8.4 and Figure 8.3 show that energy consumption was significantly affected by the variation in minimum ventilation rate. A 50% reduction in the ventilation rate resulted in 26%, 21%, and 1.5% decreases in total energy consumption in winter, spring, and summer, respectively. In winter, 26% and 50% drops were observed in heating and ventilation energy consumption, respectively.

TABLE 8.3
Greenhouse and Crop Characteristics Used in the Simulation Model

Greenhouse length	7.50 m
Greenhouse width	7.50 m
Greenhouse height at eaves	2.50 m
Greenhouse height at ridges	4.50 m
Greenhouse glazing	Double polyethylene
Floor surface	Mulch with reflectivity of 60%
Crop	Cucumber
Crop row orientation	North—South
Distance between plant rows	0.75 m
Average leaf length and width	0.30 and 0.75 m
Average plant height	2.0 m

TABLE 8.4
Energy Consumption with Respect to Minimum Ventilation Rate in Winter, Spring, and Summer

	WINTER		SPRING		SUMMER	
VENTILATION RATE ($m^3/s.m^2$)	**0.01**	**0.005**	**0.01**	**0.005**	**0.01**	**0.005**
ENERGY ($MJ/d.m^2$)						
Heating	23.45	17.41	8.82	6.95	0.10	0.10
Cooling	–	–	–	–	0.94	0.93
Ventilation	0.11	0.06	0.23	0.19	0.71	0.71
TOTAL	**23.56**	**17.47**	**9.04**	**7.15**	**1.75**	**1.73**

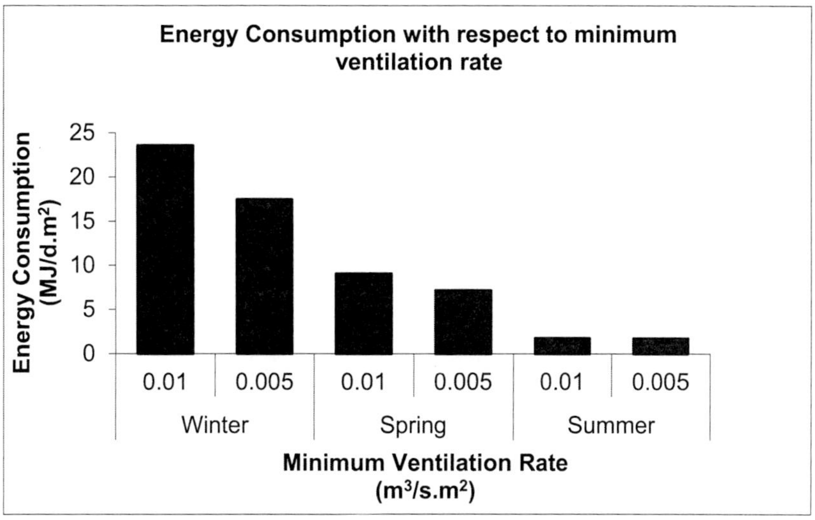

FIGURE 8.3 Energy consumption with respect to minimum ventilation rate in winter, spring, and summer.

This showed that a 50% reduction in minimum ventilation rate resulted in a 50% reduction in energy consumption for ventilation in winter. This was due to the nearly continuous operation of minimum ventilation rates in winter. A 50% reduction in the minimum ventilation rate for spring, however, resulted in only a 15% decrease in energy consumption. This indicated that ventilation rates higher than the minimum were used for some time due to increased relative humidity levels within the greenhouse. The effect of minimum ventilation rate in summer was insignificant for the same reasons that were determined for spring.

8.5 PLANT AND ANIMAL TEMPERATURE MANAGEMENT

As discussed earlier in Chapter 4, temperature is one of the main environmental variables that influences the growth and development of controlled environment crops, livestock, poultry, or fish, which require crop-, livestock-, poultry-, or fish-specific minimum, optimum, and maximum temperatures for their growth. The minimum temperature is the temperature at which a plant, livestock, poultry or fish stops growing. The optimum temperature is the temperature at which they develop most rapidly. Maximum temperature, on the other hand, is the temperature that exerts stress on crops, livestock, poultry, or fish when the day, night, or both temperatures are above some species-specific temperature.

Temperature is an extremely valuable tool for regulating crop timing by changing temperatures during the day and night (DIF) or by controlling the average daily temperature (ADT), the average daily temperature. Plants integrate temperatures over a 24-hour period, and plant growth and development are strongly influenced by this integrated temperature, i.e., the average temperature. DIF is the temperature difference between day- and night-times, which influences plant height, chlorophyll content, etc.; therefore, it is used in greenhouses to control plant height and reduce the use of chemical growth retardants. Plant height can be decreased by decreasing the day-time set-point temperature or increasing the night-time set-point temperature or both by achieving a zero or negative DIF. In order to increase plant height, however, we can increase the day-time temperature or decrease the night-time temperature or both by maintaining a zero or positive DIF. Generally speaking, a night-time temperature beyond 24°C (75°F) is not desirable because many crops are negatively influenced by high temperatures. Another method for manipulating temperature to control plant elongation is to drop the temperature (also known as DIP) by approximately 2.8 to 8.3°C (5 to 15 °F) before sunrise for a period of 2 to 3 hours. Therefore, when greenhouse temperature is managed based on the crop requirements and the crop finish date, further savings would be possible in greenhouse energy consumption.

For example, the estimations performed in the following example show that using variable set-point temperatures in a single glass greenhouse reduces the heating energy supplied by the furnace approximately 29%.

Example 8.5: Heating Energy Savings in a Greenhouse with Variable Set-Point Temperatures

Using the same greenhouse that was used in Example 7.1, a single glass greenhouse in Columbus, Ohio, estimate the annual heating energy requirement and the costs of heating using fuel oil and natural gas furnaces for different day- and night-time indoor set-point temperatures. Assume that a daytime set-point temperature of 65°F is used, while the night-time set-point temperature is 55°F.

Solution

The energy estimation using the degree-day method for the indoor set-point temperature of 65°F was performed in Example 7.1. Now, we need to do the same, but this time for a variable set-point temperature, daytime 65°F and nighttime 55°F. Let's keep in mind that 75%–80% of heating is done during night; so, we can easily calculate the energy savings (by adjusting the HDD) that could be achieved by employing a variable day- and night-time set-point temperatures rather than a continuous set-point temperature as follows:

$$\text{Adjusted HDD} = 0.75(5{,}660)(55°\text{F}/65°\text{F}) + 0.25(5{,}660)(65°\text{F}/65°\text{F})$$

$$\text{Adjusted HDD} = 5{,}007°\text{F.day}$$

In Example 7.1, the energy consumption E for the continuous set-point temperature of 65°F was calculated as 39,770 gal oil/year. Then, the new energy consumption E for the variable set-point temperature would be:

$$E = \left(39{,}770 \text{ gal}\frac{\text{oil}}{\text{year}}\right)(5{,}007°\text{F.day})/(5{,}660°\text{F.day})$$

$$E = 35{,}182 \text{ gal oil/year}$$

And then, the percent energy savings would be calculated as:

$$\left(\frac{(39{,}770 \text{ gal oil/year}) - (35{,}182 \text{ gal oil/year})}{(39{,}770 \text{ gal oil/year})}\right)(100) = 11.5\%$$

The calculations performed in Example 7.1 and above show that using the variable set-point temperature reduces the heating energy provided by the furnace by approximately 11.5% in this particular greenhouse and at this particular location.

Energy consumption for heating indeed decreases as the set-point temperature is lowered. However, the impact of a lower daily average temperature on plant finish time must also be considered when using this strategy. The bottom line is, as mentioned, we have to manage greenhouse temperature based on the crop requirements, providing them with the required daily average temperatures.

It is also a known fact that, when plants are grown at colder temperatures, it takes longer for them to flower. Therefore, the energy consumption per plant basis may sometimes be higher when plants are grown at lower temperatures (due to the heating required for the extended growth period) compared to those grown at higher temperatures in order to make the finish time shorter.

8.6 FLOOR HEATING, INFRARED HEATING, AND HORIZONTAL AIR FLOW

Main goal of greenhouse heating is to provide cost effective uniform heating where and when it is needed. Floor heating in greenhouses is a common practice, and hot-water floor heating systems (see Figure 8.4) today provide the most uniform and quite effective heat distribution for greenhouse crops grown directly on the floor or in the soil such as bedding plants, ornamentals, bag-cultured vegetables, etc.

As discussed earlier, conventional greenhouse heating systems heat the greenhouse air first, and then the heated air heats the plants. A low infrared heater, on the other hand, operates with heated pipes—up to 595°C (1100°F)—and first heat the objects in the greenhouse (including crops) directly through infrared radiation heat exchange, and then the objects exchange this heat with the environment via conduction, convection and again radiation heat transfer. Since the greenhouse air is not heated directly, the temperature of the interior air is typically lower; and this low inside air temperatures result in less heat losses from the greenhouse environment. Furthermore, infrared heaters eliminate the pumping energy requirements and the heat losses through distribution lines typically seen in conventional greenhouse heating systems. As a consequence, when installed properly, infrared heating can provide uniform plant and vertical

FIGURE 8.4 A hot-water in-floor heating system installed in a concrete greenhouse floor. Crops are grown directly on the floor. Source: *Greenhouse Engineering* (2021).

air temperatures and be a cost effective option for greenhouse heating.

Horizontal air flow (HAF) fans, also known as circulation fans, provide a more uniform controlled environment spatial temperature distribution, and reduce the difference between internal air temperature and plant, livestock or poultry temperature (see Figure 5.28). They not only improve uniform temperature distribution, but also humidity and carbon dioxide distributions costing marginal energy expenditure. The flowing air removes moisture and prevents condensation on the leaves, therefore reduce foliar fungal diseases, for instance.

Typically, a total air flow of 0.01 $m^3/s.m^2$ (2 cfm/ft^2) of floor area is provided keeping the greenhouse air mass moving at a velocity of 0.5 m/s (100 ft/min) to 0.7 m/s (140 ft/min) and not exceeding 1.0 m/s (200 ft/min) in the vicinity of crop canopy. HAF systems come with 30 cm (12 in), 40 cm (16 in), 45 cm (18 in), and 50 cm (20 in) diameter blades, and generally move the air approximately 30 times the blade diameter. A 30 cm (12 in) diameter HAF fan would be able to move the air 9 meters (30 ft). In a well designed HAF fan system, the fans are installed in a way that air flow is horizontally directed down the greenhouse bay and aligned with similar fans pushing the greenhouse air in the opposite direction down the next bay. In order to make sure that the air within the canopy is moved rather than the air above the canopy, fans are mounted at between 0.6 m (2 ft) to 0.9 m (3 ft) above the plants. The first and last fans are placed about 4.5 m (15 ft) to 6.0 m (20 ft) from the gable ends to pull the air that is coming from the adjacent bay. Crop canopy and hanging baskets resist horizontal air flow; therefore, the fans should be spaced properly. After the first fan, they are typically placed 9 m (30 ft) to 15 m (50 ft) apart to keep the greenhouse air mass moving smoothly.

8.7 DYNAMIC ROW SPACING

The dynamic row spacing as well might play an important role in energy and water conservation, and have an influence on the plant growth. Papadopoulos and Pararajasingham (1997) reported that greater yields of greenhouse crops could be produced by using narrow spacings (high plant density) compared with wide spacings. Increased fruit yield in narrow spacings result from greater crop biomass generated by the effect of increased light interception and canopy photosynthesis. Increased high plant density also reduces the energy consumption on per plant basis.

Moya et al. (2008) developed and employed a dynamic computer simulation model to compare different row spacings, plant heights, and leaf dimensions of a cucumber crop to draw a conclusion about energy and water conservation. Table 8.5 shows the greenhouse and crop characteristics used in their simulation model. Their findings showed that using smaller spacings between cucumber crop rows (for instance, 0.5 m instead of 0.75 m) reduced energy consumption per unit floor area in average 14.4 per cent. With a decrease in row spacing, the total amount of surface for radiation exchange decreases, and consequently plant canopy shading within the canopy increases. This leads to less radiational and evaporative cooling in smaller row spacings, and hence, results in lower heating requirements during the heating season. By changing the row spacing from 0.75 m to 0.5 m, average water savings (adjusted to the whole greenhouse area) of 27.8% occurred. A complete system analysis, however, is necessary to be able to make a viable conclusion in total energy and water conservation.

TABLE 8.5
Greenhouse and Crop Characteristics Used in the Simulation Model

Greenhouse length	15.0 m
Greenhouse width	7.50 m
Greenhouse height at eaves	2.50 m
Greenhouse height at ridges	4.50 m
Greenhouse glazing	Double polyethylene
Floor surface	Mulch with reflectivity of 60%
Crop	Cucumber
Crop row orientation	North—South
Distance between plant rows	0.50 and 0.75 m
Average leaf length × width	Set 1: 0.30 × 0.25 m; Set 2: 0.23 × 0.18 m; Set 3: 015 × 0.10 m
Average plant height	0.5 m, 0.75 m, 1.0 m, 1.5 m, 2.0 m

The mean energy consumption for heating differed with the season and different row spacings. The colder the outside air, obviously the more heating energy was needed. The wider the row spacing was, the more energy was required in all seasons (i.e., 0.5 m the least, 0.75 m the most). For the row spacing of 0.5 m, the mean heating energy consumptions were 28.3, 21.5, and 11.8 $MJ/m^2/day$ for winter, spring, and summer, respectively. For the row spacing of 0.75 m, however, the mean energy uses for heating were 32.7, 25.9, and 13.5 $MJ/m^2/day$ for winter, spring, and summer, respectively.

Statistical comparisons between all possible pairs were conducted using Tukey's simultaneous test at a 95% confidence level and Table 11.6 shows that all the presented means were statistically different from each other. The radiation exchange surfaces and airflow increased within wider row spacings; as a result, radiational, convective and evaporative cooling increased. Also, canopy shading decreased in wider row spacings; more plant surfaces were exposed to solar radiation inducing transpiration rates and evaporative cooling per unit floor area as a consequence. Therefore, the supplemental heating requirements in 0.75 m row spacing were higher. The row spacing could be kept just enough for the canopy stand, and increased dynamically as the crop grows and canopy stand becomes larger. Their findings showed that such a practice could reduce supplemental heating requirement of a greenhouse, in average of 14.4% (Table 8.6).

Not only the row spacing, but also the plant height and the leaf dimensions were accountable for energy consumption. The most energy was needed when the cucumber plant stands were still short (0.5–1.0 m). As the plants grew, energy needs for heating went down depending on the season and final plant height. The energy consumption also depended on the leaf dimensions, and increased linearly with decreasing leaf area. These results are valuable to estimate different energy consumptions for specific

TABLE 8.6
Comparisons of Energy and Water Consumption Means with Respect to Different Row Spacing Using Tukey's Simultaneous Test at 95% Confidence

	Energy use (all values in MJ/m²/day)				Water use (all values in kg/m²/day)			
	Row Spacing (Mean)	Compared to row spacing (Mean)	Difference of Means	Statistically Significant at a 95% level?	Row Spacing (Mean)	Compared to row spacing (Mean)	Difference of Means	Statistically Significant at a 95% level?
WINTER	0.5 (28.3)	0.75 (32.7)	4.4	YES	0.5(1.72)	0.75 (2.46)	0.74	YES
SPRING	0.5 (21.5)	0.75 (25 9)	4.4	YES	0.5(1.87)	0.75 (2.45)	0.58	YES
SUMMER	0.5(11.8)	0.75(13.5)	1.7	YES	0.5(1.64)	0.75 (2.38)	0.74	YES

plants, but are less important given that a greenhouse operator or designer usually has no choice of which crop to grow with respect to energy consumption. However, this study showed that a greenhouse with tall plants had significantly less heating demand than a greenhouse with small plants. This was because the taller plants overall had much less exposed surface area for radiational heat exchange.

Simulation results of this dynamic model also showed that due to increased plant density within the whole greenhouse space, the water consumption increased an average of 8.5% as row spacing decreased (i.e., 0.50 m the most, 0.75 m the least water consumption). This is in agreement with the findings of Bryla and Strik (2007). In reality, for dynamic row spacing, plant density is kept the same throughout the growing season. The same number of plants exists no matter what the row spacing is.

As crops grow, the row spacing expands covering a larger greenhouse space during the growing season. Considering 0.75 m as the final row spacing covering the whole greenhouse floor area, the 0.5 m spacing transpiration figures can be adjusted for the whole greenhouse space (i.e., to the 0.75 m spacing) reducing the water consumption used in the whole greenhouse space (Table 8.6). Then, for row spacing of 0.5 m, the space-adjusted mean water consumption for transpiration (of three years) was 1.72, 1.87, and 1.64 kg/m²/day for winter, spring, and summer, respectively.

For row spacing of 0.75 m, however, the mean water consumption for transpiration (of three years) was 2.46, 2.45, and 2.38 kg/m²/day for winter, spring, and summer, respectively. Summer means show that less transpiration occurred in summer due to higher air moisture levels maintained inside the greenhouse (due to operational and control strategies employed) compared to those in winter and spring. Comparisons between all possible pairs were also made using Tukey's simultaneous test at a confidence level of 95%. All of the presented means (of 0.5 and 0.75 m row spacings) for winter, spring, and summer were statistically different from each other. From 0.75 m to 0.50 m row spacing, the overall decrease in water consumption was in average of 27.8%. The effect of row spacing definitely changes the dynamics of airflow in the canopy and it affects the canopy resistances, especially, the aerodynamic and stomatal resistances.

The plant heights and leaf dimensions also played an important role with respect to the water consumption for transpiration. In general, the smaller the plants and the larger the leaves were, the less water was consumed. This observation is in agreement with the findings of a previous study by Claypool et al. (1997). In their study, they observed that large-leaved alfalfa plants had 10 to 16% lower transpiration than small (normal) leaves. The yield of large-leaved plants was also 17 to 22% higher than small-leaved. These findings indicate that crop transpiration can be potentially improved by genetically changing leaf size to conserve water.

In conclusion, the findings in this study showed that the energy consumption is influenced by the crop row spacing. The smaller the row spacing, the less energy consumed. If a smaller space requirement for smaller row spacing is considered, then it will be clear that less greenhouse space needs to be heated, increasing the potential to reduce energy consumption further. The growing space could be adjusted dynamically as the row spacing changes throughout the growing season: that is, as the crop grows, the spacing must increase.

A sophisticated design or modification of greenhouses using dynamic row spacing can therefore save both energy and water during the production period of the greenhouse. It seems that using smaller row spacing when the canopy stand is small could save both energy and water. Instead of using fixed row spacing throughout the growing season, one could employ dynamic row spacing using smaller row spacing when the canopy stand is small, and extending dynamically as the canopy stand gets bigger. Also, the smaller the plants and the larger the leaves were, the less water was consumed.

These findings indicate that crop transpiration can potentially be improved by genetically modifying crop leaf sizes and canopy stands. However, to be able to make a conclusion about total energy and water conservation, the greenhouse has to be treated as a complete system. Potential energy savings due to dynamic row spacing have to be seen in conjunction with the energy expenditure, automation, or labor required for providing such a system. Therefore, a total energy and water conservation analysis has to be conducted for each case in hand individually, considering the infrastructure, local water, and energy and labor costs, along with the crop type.

8.8 HIGH-EFFICIENCY FURNACES, COGENERATION, AND HEAT STORAGE SYSTEMS

Recent developments in heating systems technologies help controlled-environment growers to reduce their energy losses and consumptions further. These technologies now include high-efficiency direct-fired unit heaters, high efficiency condensing boilers and boilers coupled with flue gas condensers (see Figures 8.5 and 8.6) that condenses the exhaust gas and captures both waste heat and CO_2 (see Example 2.59). The overall efficiencies of such boilers can go up to 93%–95%, and high-efficiency furnaces can save greenhouse growers up to 40%–45% on energy costs; of course, based on the efficiency of the existing furnace, which is being replaced.

FIGURE 8.5 A greenhouse boiler room equipped with high efficiency boilers.

Source: *Greenhouse Engineering* (2021).

FIGURE 8.6 Flue gas condensers recover heat by allowing condensation of the exhaust gases.

Source: *Greenhouse Engineering* (2021).

There is a clear trend towards local energy production, a trend which is being accelerated by the liberalization of gas and electricity markets. In parallel, technological development also points in the direction of more diversification and small-scale power production. Caring for the environment, on the other hand, requires more sustainable power production. Technological development, financial support and government incentives will in the future all contribute to more sustainable power production. The integration of sustainable power supply and highly efficient use of fossil fuels will be emphasized.

For years, we have been used to thinking of electricity production based on large-scale power stations that are usually remote from electricity consumers, and that require long-distance, high-voltage transmission networks. This is often known as a centralized system. Recent technological developments have introduced a whole range of much-smaller scale generation systems into the energy market, including engines, turbines, fuel cells, and photovoltaics. These are sometimes known as decentralized, distributed, or embedded systems because they can produce electricity near to the consumer, or in the consumer's house, office or production area.

These localized systems are more environmentally friendly, more flexible, more efficient, and more cost-effective than traditional centralized systems—especially when transmission savings are taken into account. Conventional generation of electricity in large central power stations is normally only 30%–40% energy efficient. More recent combined heat and power (CHP) cycle generation can improve this to 55%, excluding the transmission loses of electricity, which can be as great as 20% of the delivered efficiency.

Hence conventional electricity-only stations release large amounts of energy as waste heat, normally via large cooling towers or cooling with sea or river water. They also release large amounts of greenhouse gases (mainly CO_2) to the environment. Through the utilization of the heat, the efficiency of cogeneration unit can reach 90% or more. Cogeneration of heat and power in greenhouse operations not only utilizes the heat and power generated on site, but also, essentially eliminates the CO_2 emissions to the environment (see Figure 8.7). In essence, these systems serve as multi-generation systems in greenhouse operations, generating heat, power, CO_2 and water, all precious resources.

FIGURE 8.7 A micro-cogeneration (CHP) system.

Source: *Greenhouse Engineering* (2021).

It should also be kept in mind that the energy efficiency of greenhouses can be increased further by the addition of heat storage systems to, for instance, a CHP unit or a boiler (during CO_2 dosing in a greenhouse) so that the excess heat can be stored and used later when needed. Typically, these storage systems are the insulated hot water storage tanks (see Figure 5.25) that has already been covered in Chapter 5; however, underground aquifers can also be used as heat storage facilities.

8.9 HEAT PUMPS AND SEMI-CLOSED, AND CLOSED GREENHOUSE SYSTEMS

Heat pumps are reversible refrigerators that can heat or cool the greenhouse environment with a single system depending on the flow direction of the refrigerant. The use of heat pumps for heating and cooling greenhouses makes it possible to use local and renewable energy sources while reducing or totally eliminating CO_2 emissions. Especially, the earth-coupled or geothermal heat pumps are highly efficient and stable systems for heating and cooling greenhouses. Heat pumps also make innovative confined greenhouse operations possible, which would totally eliminate the energy, CO_2 and water losses due to ventilation conserving all of these precious resources, and would make an insect-free greenhouse operation possible.

Yildiz et al. (2014a, 2014b) studied the spatiotemporal distributions and spatial correlations of energy and water consumption in open, semi-closed and completely closed heat pump greenhouse systems in continental Canada in order to help enhance the competitive position of greenhouse and agri-food industries by introducing economically, environmentally and socially sustainable technologies and management strategies. The investigated heat pump technology and operational modes (especially semi-closed and closed) bring increased yield, and energy and water efficient solutions to the doorstep of greenhouse producers. The overall goal of this study was to help reduce the load on power grid, demand for fossil fuels and irrigation water, and also supply CO_2 for the greenhouse production, as well as transform the unfavorable regions in the Canadian North into potential greenhouse regions and open up northern population's limited access to fresh vegetables and locally produced greens for better public health and local sustainability. The greenhouse systems with heat pump units in this study were operated as an open (OHP), semi-closed (SHP) and completely closed (CHP) systems. Simulations were performed starting at the beginning of 1st day of January and ending at the end of 31st day of December for full 365-day simulations except the Canadian North where the seasons varied between seven months (March through September) and nine months (February through October). In order to make reasonable spatial comparisons in the Canadian North, a 7-month growing season was used for all locations, and simulations were performed under natural light conditions starting on the 1st day of March and ending on the 30th day of September for 214-day simulations. The readers of this book are referred to Yildiz et al., 2014a, 2014b for further information on the details of weather file, greenhouse characteristics, operational strategies, and data analyses.

The findings showed that the energy and water consumptions were both spatially and system dependent. Latitude, longitude, and altitude all significantly predicted total energy and water consumptions. Semi-closed (SHP) and closed heat pump (CHP) greenhouse systems were the most energy and water conserving systems, respectively (Table 8.7). Highest energy consuming locations of semi-closed heat pump system were still a lot more efficient than the least energy consuming locations of conventional (CON) systems. Closed systems had the lowest energy consumptions in colder months, but the consumption increased in warmer months due to increased cooling loads; as a result, the overall annual energy consumptions in closed systems were much higher than those in semi-closed systems. Locations close to the Atlantic and Pacific Coasts as well as the Great Lakes region had the lowest energy consumptions in all systems. Semi-closed and closed systems at lower latitudes had more water consumptions (transpiration) due to increased amount of available solar radiation. Open systems at higher latitudes had more water consumptions than those at lower latitudes as the vapor pressure deficits became larger due to decreased outside air temperatures in higher latitudes. Locations close to the Atlantic and Pacific Coasts as well as the Great Lakes region had lower water consumptions due to increased outside relative humidity levels hence reduced vapor pressure deficits. Semi-closed heat pump (SHP) greenhouse system transforms the non-energy-efficient, unfavorable regions under conventional (CON) greenhouse systems into highly energy and water efficient potential greenhouse regions in Canada.

The results in the Canadian North as well showed that the energy and water consumptions were both spatially and system dependent. Both latitude and longitude significantly predicted total energy consumptions. Latitude was significant in semi-closed and open heat pump systems' water consumption as well. Semi-closed (SHP) and open heat pump (OHP) greenhouse systems proved to be the most energy

TABLE 8.7
Comparisons of Means with Respect to Different Greenhouse Systems Using Tukey's Simultaneous Test at a 95% Confidence Interval

SYSTEM	Mean Annual Energy Consumption (GJ/m^2)	Mean Annual Water Consumption (kg/m^2)	Mean Annual Water Collection (kg/m^2)
CON	11.32 a	660.4 p	0.0 w
OHP	5.77 b	700.4 q	6.4 w
SHP	5.27 c	540.7 r	125.3 x
CHP	7.60 d	420.0 s	420.0 y

CON: Conventional greenhouse; OHP, SHP, and CHP: Open, semi-open, and closed heat pump greenhouse systems, respectively. Means in the same column not followed by the same letter are significantly different.

TABLE 8.8
Comparisons of Means with Respect to Different Greenhouse Systems in the Canadian North Using Tukey's Simultaneous Test at a 95% Confidence Interval

SYSTEM	Mean Annual Energy Consumption (MJ/m².d)	Mean Annual Water Consumption (kg/m².d)	Mean Annual Water Collection (kg/m².d)
CON	33.3 a	2.45 p	0.0 w
OHP	16.6 b	2.49 p	0.0 w
SHP	15.1 b	2.08 q	0.34 x
CHP	21.3 c	1.33 r	1.33 y

CON: Conventional greenhouse; OHP, SHP, and CHP: Open, semi-open, and closed heat pump greenhouse systems, respectively. Means in the same column not followed by the same letter are significantly different.

conserving systems, while the closed heat pump (CHP) system was the most water conserving system (Table 8.8).

Even though the heat pump seasonal performance factors dropped at more northern locations due to colder outside temperatures, the highest energy consuming locations of semi-closed (SHP) and open heat pump (OHP) greenhouse systems were almost as efficient as the least energy consuming locations using the conventional (CON) greenhouse system (southwest). Closed heat pump (CHP) systems had the lowest energy consumptions in colder months, but the consumption increased in warmer months especially in the southwest due to increased cooling loads; as a result, the overall annual energy consumptions in closed (CHP) systems were much higher than those in semi-closed (SHP) systems. Locations in the southwest had the lowest energy consumptions in all systems. Water consumptions in all systems increased towards the southwest due to relatively increased available solar radiation. Semi-closed heat pump system transforms the non-energy-efficient, unfavorable regions under conventional systems in the Canadian North into potential greenhouse regions, and opens up the northern population's limited access to fresh vegetables and locally produced greens for better public health and local sustainability.

In hot and arid regions, on the other hand, greenhouses require large quantities of water for irrigation. During the summer months in these regions, inside temperatures rise above the acceptable limits, thence the greenhouses become quite unusable during this time of the year. Using proper technologies and environmental management systems can however significantly change the energy and moisture dynamics of greenhouse production systems. Hardin et al. (2008) focused on reducing natural gas, electricity, and water consumption in semi-arid California greenhouses introducing renewable energy heat pump technologies to both open and confined greenhouses in California. The simulations were performed for three consecutive years 2005, 2006, and 2007. The confined system had no external aeration and therefore need no further water supply. They demonstrated how water and energy consumptions inside the greenhouse changed under conventional (CON), open heat pump (OHP), and closed heat pump (CHP) greenhouse systems (Table 8.9). Outside environment had the most prevailing influence in the conventional system and the least in the closed heat pump system. The open heat pump system depended heavily on the outside weather as well, but it was much better than the conventional greenhouse. The confined system, on the other hand, used only a heat pump to regulate humidity and temperature within the greenhouse and had slightly higher energy consumption than the open heat pump system during summer months. Almost all the transpired water in the closed heat pump system was collected through dehumidification. Water is fully conserved in the confined system, essentially making the water consumption zero. However, additional energy was required to provide cooling in the confined system. Their findings

TABLE 8.9
Comparisons of Energy and Water Consumption Means with Respect to Different Greenhouse Systems Using Tukey's Simultaneous Test at 95% Confidence (CON: Conventional System; OHP: Open Heat Pump System; CHP: Closed Heat Pump System)

Total Energy Consumption (all values in MJ/m²/day)			
System (Mean)	**Compared to System (Mean)**	**Difference of Means**	**Statistically Significant at a 95% level?**
CON (19.29)	OHP (8.25)	10.45	YES
	CHP (11.20)	8.09	YES
OHP (8.25)	CHP (11.20)	−2.35	YES
Water Consumption for Transpiration (all values in kg/m²/day)			
System (Mean)	**Compared to System (Mean)**	**Difference of Means**	**Statistically Significant at a 95% level?**
CON (2.17)	OHP (2.25)	−0.08	NO
	CHP (0.99)	1.18	YES
OHP (2.25)	CHP (0.99)	1.26	YES

suggest that a combination of closed heat pump system and open heat pump system has the greatest benefit. That is, the greenhouse should be operated as a semi-closed system in semi-arid California conditions. During the winter and spring months, the greenhouse system would be operated as a closed system conserving all the water and consuming much less energy for keeping the thermal environment at desirable levels. As mentioned earlier, closed greenhouse operation with a heat pump system would also make it easier to keep high CO_2 levels inside the greenhouse, while reducing the risk of insects and diseases within the confined greenhouse environment. During the summer months, however, the greenhouse would be operated as an open system introducing ventilation into the greenhouse, hence allowing outside air to remove excess humidity within the greenhouse and reducing the cooling load over the heat pump.

Furthermore, for hot and arid regions, Buchholz et al. (2004) developed the Watergy concept—a novel, closed greenhouse system. Watergy follows the development of a technology platform for decentralized basic supply of energy, water and food that can be used within a number of different applications. The system has no external aeration and has almost no need for further water supply. The technological principle is actually proven at two different prototypes, which were built within a European Union (Framework Program Promoting Energy, Environment, and Sustainable Development) funded research project. In 2004, the first prototype was built for the purpose of greenhouse horticulture in Almeria, Spain (see Figure 5.48). The second prototype was built in Berlin, Germany, as a living or office building with an attached façade-greenhouse. It has a great potential to reduce the demand for or to totally displace natural gas, to reduce the load on power grid by electricity savings, and the demand for irrigation water in greenhouse operations. It also allows plant protection without chemical insecticides and the accumulation of carbon dioxide without any aeration losses. In total, the process generates heat, clear water and cooled/dehumidified air with minimum need of primary energy.

8.10 VARIABLE SPEED PUMPS AND FANS

Given that pumping work takes place all around the world not only in controlled environments but also in industrial plants, domestic heating systems, etc., it is probably unsurprising that the pumping energy consumption accounts for approximately 20% of the world's primary energy consumption. This shows that the global pumping industry has quite a large room to make significant contributions towards conserving our energy resources. In pumping applications such as greenhouses and aquatic systems where the required pumping is not constant, installing a variable-speed pump will result in significant energy savings. Variable-speed pumps have a higher initial cost; however, they do not require mixing valves and their operational costs are lower than those of the equivalent fixed-speed pumps; therefore, they have more savings over time. Variable-speed pump systems cost more due to the added complexity of the variable frequency drive; however, this cost is outweighed by the long-term energy cost savings. When greenhouse operations replace their constant-speed pumps with variable-speed pumps, they typically recover the cost of the new pump through energy savings in less than a year.

In a variable speed pump system, when more heat is called for, the pumps run faster to provide this required heat. When less heat is required, on the other hand, the pump slows down. Since variable-speed pumps run below their maximum speeds, there is a reduction in wear and tear as well. Same thing is true for variable speed fans as well. In summary, variable frequency derives save energy when the load varies, and HVAC equipment is typically sized for peak loads, which are a very small portion of the annual operating hours. As we encourage sustainable designs, variable speed control of motors is a central focal point of such designs and discussions.

Example 8.6: Calculating Energy Savings with a Variable-Speed Pump

Now, let's consider a 3 hp single (constant) speed pump, which is used for 5,000 hours annually. Calculate the energy savings if the greenhouse operator switches to a variable-speed pump. Remember that when we pay for electricity we pay for the kWh of energy used. Assume that the average cost of electricity is \$0.12 per kWh.

Solution

<u>3 HP Constant-Speed Pump:</u>

From Chapter 2, one horsepower is equal to 746 Watts, and power is equal to the current times the voltage. Therefore, this constant-speed 3 hp pump uses, 3 hp (746 W/1 hp) = 2,338 W, which is approximately 9.7 Amp at 230 V. In reality, this pump will draw more current than this given amount due to pump and plumbing system inefficiencies; hence, close to approximately 14 Amp at 230 Volt is a more realistic figure. However, for this simplified example, we will use this extremely conservative number of 9.7 Amp.

In order to determine the energy consumption (kWh) of this pump during 5,000 hours of operation, we simply multiply the power (Watts) by the operational hours. Then, the amount of energy this pump would use:

$$(2{,}338\text{W})(5{,}000\text{h}) = 11{,}690{,}000\,\text{Wh or } 11{,}690\text{ kWh}$$

And, the annual cost of electricity would be

$$(11{,}690\text{ kWh})\left(\frac{\$0.12}{\text{kWh}}\right) = \$1{,}403\text{ per year.}$$

Now, let's do this again; but, this time using an equivalent variable-speed pump.

<u>3 HP Variable-Speed Pump:</u>

First of all, the pump affinity laws tell us three things: i) the flow changes proportionally to the speed (RPM; double the RPM, double the flow; ii) the pressure changes by the square of the difference (double the RPM, multiply the pressure by four); iii) the power changes by the cube of the difference (double the RPM, multiply the power by eight).

Hence, if we reduce the speed of the pump's motor, then based on the affinity laws, power is reduced as the cube of the speed. This is important because on a pump curve, the flow and head typically change simultaneously. As the flow increases, pressure capacity drops, and vice versa. In a variable-speed application, however, as the flow decreases and the pump pressure increases, the drive reduces the speed in order to match a constant discharge pressure by slowing the pump speed down. Then again, based on the affinity laws, as the pump slows down, the energy consumed by the pump is reduced by a factor of eight. This means that if we reduce the pump motor RPM by half, then we reduce the electricity consumption by eight times, i.e., power = ([1/2] RPM)^3 = 0.125. Actually, this is the backbone of the efficient technology that variable-speed pumps are based on, and in a constant-speed system, this does not happen because the pump does not change speed.

A pump that uses 2,338 W per hour at 3,450 RPM (single-speed pump RPM) will use only 292 W at 1,725 RPM (half the speed). Then, in order to simplify our calculations, let's assume that the pump operates at the half the speed for performing the same job during the season. Then, the amount of energy this variable-speed pump would use:

$$(292\ \text{W})(5{,}000\ \text{h}) = 1{,}460{,}000\ \text{Wh or} 1{,}460\ \text{kWh}$$

And, the annual cost of electricity would be

$$(1{,}460\ \text{kWh})\left(\frac{\$0.12}{\text{kWh}}\right) = \$175\ \text{per year.}$$

So, this variable-speed pump would save

$$(\$1{,}403 - \$175) = \$1{,}228 \text{per year.}$$

Again, even though the initial costs of variable-speed pump systems are higher, considering the energy savings we have just calculated above, this initial cost can easily be recovered in less than a year.

8.11 ENERGY EFFICIENT LIGHTING SYSTEMS

As discussed earlier in Chapter 5, supplemental lighting is quite often used in commercial controlled environment production facilities in order to enhance crop quality and growth. In addition, photoperiodic lighting is employed to control flowering of certain crops, livestock, poultry and fish. Most greenhouses use the electricity generated at central power plants, while increasing number of greenhouse facilities now generate their own electricity locally in a distributed fashion, that is, via cogeneration on site, conserving the waste heat and eliminating the transmission losses and CO_2 emissions as discussed earlier. Central power plants waste a lot of energy even before the power is transmitted to the consumer's site. And further energy is lost through the transmission lines. Efficiencies of some energy conversion systems and devices are presented in Table 8.10, while a simple case study is provided in Example 8.5 for determining overall "well to wheel" conversion efficiency. Table 8.11, on the other hand, provides a comparison of LED, INC (incandescent light bulb), and CFL bulbs with respect to their energy efficiencies, environmental impacts, some important facts and light outputs.

TABLE 8.10
Efficiencies of Some Energy Conversion Systems and Devices

Energy Conversion System	Efficiency (%)
Gas furnace	70–95
Wind turbine	35–50
Fossil fuel power plant	30–40
LEDs	50
CFLs	20
Incandescent lamp	5

Source: Greenhouse Engineering (2021).

TABLE 8.11
Comparison of LED, INC (incandescent), and CFL Bulbs

	LEDs	INC	CFLs
Life span (hours)	50,000	1,200	8,000
Watts of electricity used (Equivalent to 60 W bulb)	6–8	60	13–15
Contains toxic mercury?	No	No	Yes
CO_2 emissions (30 bulbs per year)	1 unit	10 units	2.3 units
Sensitivity to humidity	No	Some	Yes
On/Off cycling	No	Some	Yes
Durability	Very durable	Not very durable	Not very durable
Heat emitted (Btu/h)	3.4	85	30
Light Output		**Watts**	
400 lumens	4–5	40	9–13
800 lumens	6–8	60	13–15
1,600 lumens	16–20	100	23–30

Source: Compiled from USAI Lighting (2020).

As Tables 8.10 and 8.11 show that quite a significant amount of energy can be saved by simply replacing INC lamps with more efficient lighting fixtures. CFLs, for instance, use approximately a quarter of the energy of incandescent lamps (e.g., a 15-watt CFL replaces a 60-watt incandescent lamp) because they generate less heat. Controlled environment growers can easily replace the INC lamps in work and other areas where lighting is not being used for photoperiodic lighting. Michigan State University's Extension Service recommends installing CFLs for every other INC lamp in the areas where photoperiodic lighting is required due to the fact that flowering of some photoperiodic crops can be delayed if only CFLs are used. Moreover, LED flowering lamps can help greenhouse growers to save further energy and the environment. LEDs consume approximately one tenth of the energy of INC lamps and one half of the energy CFLs; therefore, they can easily replace INC and CFL bulbs for supplemental lighting and provide significant year-round energy savings in greenhouses, livestock and poultry buildings where artificial lighting is used during winter months. In addition, using alternative lighting strategies such as cyclic night-interruption lighting instead of four-hour continuous lighting every night can save further energy and provide one more step towards sustainability.

Example 8.7: Calculating the Overall Efficiency for Chemical Energy to Light Energy Conversion

Calculate the overall efficiency if the power plant efficiency is 35%, transmission efficiency is 90% and the light bulb is an incandescent bulb. What would be the overall efficiency if a CFL bulb replaces the incandescent bulb?

Solution

The overall efficiency is the product of individual efficiencies, and incandescent lamp efficiency is given as 5% in Table 8.10; therefore,

$$\text{Overall Efficiency} = \eta_1 \times \eta_2 \times \eta_3 = 0.35 \times 0.90 \times 0.05 = 0.016$$

And this makes the efficiency 1.6%. If we replace the incandescent bulb with a CFL,

$$\text{Overall Efficiency} = \eta_1 \times \eta_2 \times \eta_3 = 0.35 \times 0.90 \times 0.20 = 0.063$$

And this makes the efficiency 6.3%. Furthermore, if we replace the CFL bulb with an LED, then the overall efficiency moves up to approximately 16%, saving not only our wallets but also the environment further.

Example 8.8: Estimating the Savings in Operational Cost with Different Lighting Fixtures

Determine the seasonal savings when a 100-watt incandescent bulb is replaced with an equivalent CFL or LED bulb if the fixture is used for supplemental heating for an average of 10 hours a day from October through March, and the electricity costs $0.12 per kWh.

Solution

From October through March, we have 6 months; therefore, we have an average of 10 h/day × (180 days) = 1,800 hours of operation. Hence, the cost of electricity consumed per season

With a 100 W incandescent bulb:

$$100\text{ W}\left(\frac{1\text{ kW}}{1{,}000\text{ W}}\right)(1{,}800\text{ h})\left(\frac{\$0.12}{\text{kWh}}\right) = \$21.6$$

With a 25 W CFL (equivalent to 100 W incandescent bulb):

$$25\text{ W}\left(\frac{1\text{ kW}}{1{,}000\text{ W}}\right)(1{,}800\text{ h})\left(\frac{\$0.12}{\text{kWh}}\right) = \$5.40$$

With a 10-W LED (equivalent to 100 W incandescent bulb):

$$10\text{ W}\left(\frac{1\text{ kW}}{1{,}000\text{ W}}\right)(1{,}800\text{ h})\left(\frac{\$0.12}{\text{kWh}}\right) = \$2.16$$

This example shows that significant cost savings can be achieved just replacing a single 100-W incandescent bulb. It should also be kept in mind that the life span of an incandescent light bulb is 1,200 hours, which does not even last for the lighting season and has to be replaced before the season ends.

Examples 8.5 and 8.6 clearly show that, in order to save our funds and the environment, we need to replace our conventional lighting fixtures immediately, including those heavily used in our greenhouses, with more energy-efficient alternatives such as CFL and LED bulbs.

8.12 EFFICIENT ENVIRONMENTAL CONTROL SYSTEMS

Smart digital environmental control systems are available today, and they control every single parameter in mostly an integrated fashion. Further precision controls and therefore savings can be achieved via incorporating smarter strategies with artificial intelligence (AI) algorithms into the environmental control systems. For instance, reducing ventilation a bit or maintaining little higher indoor relative humidity levels without harming the crop growth and development (and therefore the eventual economic gains), heating water when the energy costs less, and storing the heat in hot water storage tanks for later heating, we can reduce the energy losses and save some dollars escaping through the greenhouse vents or our pockets. Particularly greenhouses and aquatic systems are extremely complex living structures; therefore, complex and smart approaches are required in order to bring all the stakeholders (e.g., whether, the cost of energy, water, CO_2, and others) together instantly for healthy crop and fish development and therefore a profitable operation. The cost of every control step and strategy should be evaluated in an integrated fashion before it is actuated; and with today's AI technology, it is perfectly achievable.

8.13 RENEWABLE ENERGY BASED SYSTEMS

Before anything else, greenhouses are passive solar systems, i.e., they are naturally renewable energy-based systems. With the recent improvements in other renewable energy technologies and their dropping expenses, we are getting closer to net energy generating greenhouse complexes every passing year. While natural gas, a fossil fuel, is the predominant fuel used to heat commercial greenhouses in northern climates, renewable biomass, solar, wind and heat pump technologies are increasingly incorporated to commercial greenhouse operations. Soon, we will be able to see greenhouse operations not only selling their green produce, but also selling the net green energy they will be generating in a distributed fashion to the utility companies. In essence, each greenhouse facility is an energy generating station. Dropping prices of renewable energy technologies and various incentives will enable greenhouse growers to easily adopt a number of renewable energy systems in their operations. That's why various renewable energy technologies and a number of numerical examples are intentionally introduced and provided in different sections of this book.

NOMENCLATURE

A	area
ACH	air change per hour
ADT	average daily temperature
CFL	compact fluorescent light
CHP	closed heat pump greenhouse system; combined heat and power
CON	conventional greenhouse system
DIF	day- and night-time temperature difference
E	energy
HDD	heating degree-days
HDPE	high-density polyethylene
HIPS	high-impact polystyrene
HV	heating value
INC	incandescent
LDPE	low-density polyethylene
LED	light-emitting diode
OHP	open heat pump greenhouse system
PP	polypropylene
$\dot{Q}$	heat transfer rate
SHP	semi-closed heat pump greenhouse system
T	temperature
U	conduction heat transmission coefficient

Greek Letters

η	efficiency

Subscripts

cond	conduction
inf	infiltration

BIBLIOGRAPHY

Aldrich, R and Bartok, JW (1992) *Greenhouse Engineering*. Northeast Regional Agricultural Engineering Service, Ithaca, NY, USA.

Argaw, N (2003) Renewable Energy in Water and Wastewater Treatment Applications, National Renewable Energy Laboratory.

ASABE Standards (2018) *Standards, Engineering Practices, and Data*. V21(1). St. Joseph, MI, USA.

Albright, LD, Reines, RG, Anderson, SE, Chandra, P, Price, DR, Langhans, RW and Cerilli, RV (1978) Experimental results of solar heating a Brace Institute style greenhouse. Proceedings, Third Annual Conference on Solar Energy for Heating Greenhouses and Greenhouse-Residence Combinations. Fort Collins, CO, USA.

Albright, LD, Novak, S and Wolfe, DW (1986) A preliminary evaluation of the effect of row cover material properties on the thermal environment under the cover. Project Report. Dept. of Agric. Engng., Cornell University, Ithaca, NY, USA, p. 114.

Alexakhin, R, Anspaugh, L and Balonov, M (2006) Environmental consequences of the Chernobyl accident and their remediation: Twenty years of experience. Report of the Chernobyl Forum Expert group "Environment", International Atomic Energy Agency.

Armienta, MA, Rodriguez, R and Ceniceros, N (2014) Groundwater quality and geothermal energy: The case of Cerro Prieto Geothermal Field, México. *Renewable Energy*, 63: 236–254.

Badger, PC and Poole, HA (1979) Conserving energy in greenhouses. OARDC Special Cir. 102, Extension Bul. 651, p. 32. Ohio Agricultural Research and Development Center, Wooster, OH, USA.

Bryla, DR and Strik, BC (2007) Effects of cultivar and plant spacing on the seasonal water requirements of highbush blueberry. *Journal of the American Society for Horticultural Science*, 132(2): 270–277.

Buchholz, M, Jochum, P and Zaragoza, G (2004) Basic Water, Heat and Food Supply from a Closed Greenhouse—The Watergy Project, International Symposium GREENSYS 2004—Sustainable Greenhouse Systems, Leuven, Belgium, 12–16 September, published in International Society of Horticultural Science (ISHS) und Wageningen University and Research Center, Wageningen. ISBN: 907601924X, S. 133.

Cameron, A (2009) Horticultural plastics recycling—The future is brighter. *Greenhouse Product News Magazine*, June. Sparta, MI, USA, pp. 20–24.

Clark, CE and Veil, JA (2009) Produced Water Volumes and Management Practices in the United States, Argonne National Laboratory ANL/EVS/R-09/1.

Claypool, D, Delaney, R, Ditterline, R and Lockerman, R (1997) Genetic improvement of alfalfa to conserve water. Poster Presentation, Wyoming Water 1997 Conference, April 21–23, Casper, WY.

Deonarine, A, Bartov, G, Johnson, T, Ruhl, L, Vengosh, A and Hsu-Kim, H (2013) Environmental impacts of the Tennessee Valley Authority Kingston coal ash spill. 2. Effect of coal ash on methylmercury in historically contaminated river sediments. *Environmental Science and Technology*, 47(4): 2100–2108.

Ebru, KA and Arif, H (2007) A comparative study on exergetic assessment of two ground-source (geothermal) heat pump systems for residential applications. *Building and Environment*, 42: 2004–2013.

Eurostat (2010) Eurostat database. Brussels, European Commission.

Fontenot, B, Hunt, L, Hildenbrand, Z, Carlton, D and Walton, HO (2013) An evaluation of water quality in private drinking water wells near natural gas extraction sites in the Barnett Shale formation. *Environmental Science and Technology*, 47(17): 10032–10040.

Garzoli, KV and Blackwell, J (1987) An analysis of the nocturnal heat loss from a double skin plastic greenhouse. *Journal of Agricultural Engineering Research*, 36: 75–85.

Hamada, Y, Nakamura, M, Saitoh, H, Kubota, H and Ochifuji, K (2007) Improved underground heat exchanger by using no-dig method for space heating and cooling. *Renewable Energy*, 32: 480–495.

Hardin, C, Mehlitz, T, Yildiz, I and Kelly, SF (2008) Simulated performance of a renewable energy technology—Heat pump systems in semi-arid central California greenhouses. *Acta Horticulturae*, 797: 347–352.

International Commission for the Protection of the Rhine (2010) *Master Plan Migratory Fish Rhine*. Report No. 179. Koblenz, Germany.

IEA (2012) *International Energy Agency, World Energy Outlook 2012*. OECD, Paris.

IEA (2013) *International Energy Agency, World Energy Outlook 2013*. OECD, Paris.

IEA (2015) *International Energy Agency, World Energy Outlook 2015*. OECD, Paris.

Karlsson, F and Fahle'n, P (2007) Capacity-controlled ground source heat pumps in hydronic heating systems. *International Journal of Refrigeration*, 30: 221–229.

Kenny, JF, Barber, NL and Hutson, SS (2009) *Estimated Use of Water in the United States in 2005, USGS Circular 1344*. United States Geological Survey, Reston, VA.

Khademi, F and Yildiz, I (2018) 1.25 energy and solid wastes. In: Dincer, I (ed.), *Comprehensive Energy Systems*. Vol. 1. Oxford: Elsevier, pp. 980–1020.

Khatib, Z and Verbeek, P (2003) Water to value-produced water management for sustainable field development of mature and green fields. *Journal of Petroleum Technology*, 55: 26–28.

Kuack, DL (1986) Shading techniques utilize the unusual. *Greenhouse Grower* (May), pp. 29–30.

Laws, EA (1993) *Aquatic Pollution-An Introductory Text*. Wiley, New York.

Liu, Y and Yildiz, I (2018) 1.24 energy and water pollution. In: Dincer, I (ed.), *Comprehensive Energy Systems*. Vol. 1. Oxford: Elsevier, pp. 950–979.

Macknick, J, Newmark, R, Heath, G and Hallett, KC (2011) A review of operational water consumption and Withdrawal Factors for Electricity Generating Technologies, Golden, CO, National Renewable Energy Laboratory, 303: 275–300.

Mehlitz, T, Yildiz, I, Hardin, C and Rahman, S (2008) Simulated effects of reflective mulch on energy and water conservation in semi-arid central California greenhouses. *Acta Horticulturae*, 797: 353–359.

Mielke, E, Anadon, LD and Narayanamurti, V (2010) Water consumption of energy resource extraction, processing, and conversion, Belfer Center for Science and International Affairs.

Miyagawa I (1977) Studies on the pellet house. *Journal of Agricultural Laboratory*, No. 8. Central Research Institute of Electric Power Industry, Abiko, Chiba Perfecture, Japan.

Moya, A, Mehlitz, T, Yildiz, I, Kelly, SF and Hardin, C (2008) Simulated effects of dynamic row spacing on energy and water conservation in semi-arid central California greenhouses. *Acta Horticulturae*, 797: 307–313.

Olsson, G (2012) *Water and Energy: Threats and Opportunities*. IWA Publishing, London.

Omer, A (2008) Energy, environment and sustainable development. *Renewable and Sustainable Energy Reviews*, 12(9): 2265–2300.

Papadopoulos, AP and Pararajasingham, S (1997) The influence of plant spacing on light interception and use in greenhouse tomato (*Lycopersicon esculentum* Mill.): A review. *Scientia Horticulturae*, 69: 1–29.

Palmer, MA, Bernhardt, ES and Schlesinger, WH (2010) Mountaintop mining consequences. *Science*, 327(5962): 48–149.

Polagye, B, Van Cleve, B, Copping, A and Kirkendall, K (2011) Environmental Effects of Tidal Energy Development: Proceedings of a Scientific Workshop, March 22–25, 2010. U.S. National Oceanographic and Atmospheric Administration.

Rebuck, SM, Aldrich, RA and White, JW (1977) Internal curtains for energy conservation in greenhouses. *Transactions of the ASAE*, 20(4): 732–734.

Reynolds, RR and Kiker, RD (2003) *Produced Water and Associated Issues*. Oklahoma Geological Survey.

Roberts, BW and Anderson, JA (1994) Canopy shade and soil mulch affect yield and solar injury of bell pepper. *HortScience*, 29: 258–260.

Roberts, WJ and Mears, DR (1969) Double covering a film greenhouse using air to separate film layers. *Transactions of the ASAE*, 12(1): 32–33, 38.

Schalk, JM and Robbins, ML (1987) Reflective mulches influences plant survival, production and insect control on fall tomatoes. *HortScience*, 22: 30–32.

Short, TH and Bauerle, WL (1977) A double plastic heat conservation system for glass greenhouses. Proceedings of an International Symposium on Controlled Environment Agriculture, April 7–8, Tucson, AZ, USA.

Short, TH, Badger, PC and Roller, WL (1978) A solar pond-polystyrene bead system for heating and insulating greenhouses. Proc. of International Symposium on Potential Productivity in Protected Cultivation, Kyoto and Tokyo, Japan, August, 1978.

Short, TH and Bauerle, WL (1980) Energy conservation techniques for commercial glass greenhouses. ASAE Paper No: 80-4532. ASAE, St. Joseph, MI 49085, USA.

Simpkins, JC, Mears, DR and Roberts, WJ (1976) Reducing heat losses in polyethylene covered greenhouses. *Transactions of the ASAE*, 19(4): 714–719.

Stephenson, MT (1992) A survey of produced water studies. In: *Produced Water*. Springer US, pp. 1–11.

Teixeira, T, Neves, P and Araújo, L (2012) Thermal impact of a nuclear power plant in a coastal area in Southeastern Brazil: Effects of heating and physical structure on benthic cover and fish communities. *Hydrobiologia*, 684(1): 161–175.

The University of Maine Cooperative Extension (2020) Ag Plastic Recycling in Maine: Current Challenges and New Opportunities. Available on: www.extension.umaine.edu/agriculture/greenhouse-plastic-recycling (Retrieved on 10.07.2020).

UN (2014a) World Water Assessment Program: The United Nations World Water Development Report 2014: Water and Energy, Paris.

UN (2014b) Department of Economic and Social Affairs, World Urbanization Prospects: The 2014 Revision, Highlights (ST/ESA/SER.A/352), New York.

UN (2015) World Water Assessment Program: The United Nation World Water Development Report: Water for a Sustainable World, UNESCO, Paris.

USAI Lighting (2020) Available on: https://www.usailighting.com/ (Retrieved 07.07.2020).

USDOE (2006) Energy Demand on Water Re*Source:* Report to Congress on the Interdependencies of Energy and Water, Washington, DC.

US EPA (2011) *Draft Plan to Study the Potential Impacts of Hydraulic Fracturing on Drinking Water Resources*. Office of Research and Development, Washington, DC.

Wang, Z (2018) 1.23 energy and air pollution. In: Dincer, I (ed.), *Comprehensive Energy Systems*. Vol. 1. Oxford: Elsevier, pp. 909–949.

Wang, T, Yang, Z and Copping, A (2015) A modeling study of the potential water quality impacts from in-stream tidal energy extraction. *Estuaries and Coasts*, 38(1): 173–186.

WEC (2013) *World Energy Council*. Survey of World Energy Resources, London.

Willits, DH and Gurjer, YR (2004) Heat pumps for the heating and night-cooling of greenhouse crops: A simulation study. *Transactions of the ASAE*, 47(2): 575–584.

Wittneben, B (2011) The impact of the Fukushima nuclear accident on European energy policy. *Environmental Science and Policy, Environmental Science and Policy.*

Yildiz, I and Stombaugh, DP (2006a) Dynamic modeling of microclimate and environmental control strategies in a greenhouse coupled with a heat pump system. *Acta Horticulturae*, 718: 331–340.

Yildiz, I and Stombaugh, DP (2006b) Heat pump cooling and greenhouse microclimates in open and confined greenhouse systems. *Acta Horticulturae*, 719: 255–262.

Yildiz, I and Stombaugh, DP (2009) Simulated effects of various environmental management practices on energy consumption in open and confined greenhouse systems. *Acta Horticulturae*, 801: 1317–1324.

Yildiz, I, Yue, J and Lowrey, J (2014a) Spatiotemporal distributions of energy and water consumption in open, semi-closed and closed loop heat pump greenhouse systems, and associated management strategies in Canada. *Acta Horticulturae*, 1037(1): 119–128.

Yildiz, I, Yue, J and Lowrey, J (2014b) Spatiotemporal distributions of energy and water consumption in open, semi-closed and closed loop heat pump greenhouse systems, and associated management strategies in the Canadian North. *Acta Horticulturae*, 1037(1): 171–177.

Yildiz, I (2021) *Greenhouse Engineering: Integrated Energy Management*. 1st Edition. CRC Press, Taylor and Francis Group, FL, USA. ISBN: 978-1-4987-4313-6.

Zalom, FG (1981) Effects of aluminum mulch on fecundity of apterous *Myzus persicae* on heat lettuce in a field planting. *Entomologia Experimentalis and Applicata*, 30: 227–230.

9 Environmental Considerations and Waste Management

9.1 INTRODUCTION

How can we continue to improve human welfare making sure that food, energy, and environmental justices are reasonably realized within the limits of our planet's resources? And then, "What is the maximum number of people our planet can sustain?", that is, what is the sustainable human carrying capacity of our planet while sustaining the natural environment, its species, and its systems in a healthy and functional state? We are using our renewable and non-renewable environmental resources faster than they can be replenished. We are extracting minerals, oil, and ground water without sufficient concern for their limits or the need to recycle them. As a result, there is an existing shortage of some resources and there will be more shortages of other resources in the future. Therefore, some how we have to learn how to sustain our resources so that we can continue to fuel human civilization and other living things locally and therefore globally on our planet within natural limits for many generations.

Sustainability, therefore, means leaving our children and future generations a planet at least as rich as we live in today, and requires maintaining fully functional and healthy natural environment, which basically forms the foundation for human civilization. Due to the population increase, in reality, sustainability is a moving target rather than a fixed one; therefore, more resources will be required to maintain a healthy and functional status. Our choices, up to certain extent, determine the local and global sustainable carrying capacity for the human population of our planet. For instance, more people in the future will mean more food, energy, and fresh water; and in addition, people living in less developed parts of the world today will be consuming more food, energy, and fresh water on per capita basis when they reach higher standards of living or if and when we can actualize food, energy and environmental justices.

The overall goal of sustainable agri-food and energy production in a sustainable environment requires and maintains multiple linkages among agri-food production and justice, energy production and justice, and environmental protection and justice. And exactly at this point, the author would like to emphasize the importance of controlled environments concept. Adopting controlled environment technologies for agri-food production, whenever and where ever it is possible, our producers will truly be able to care for our planet, sustaining the human civilization on it within fully enriching cycles, and provide more diverse, reliable, distributed, and local agri-food and energy productions, encompassing efficiency and conservation measures and protecting our local and hence global environments.

This chapter briefly covers general environmental considerations first; then moves to the special controlled environment systems considerations, waste generation, characteristics, and pertinent planning considerations; and then concludes with waste management functions.

9.2 ENVIRONMENTAL CONSIDERATIONS

The story of mankind's success and his eventual downfall is one that rests largely on the shoulders of our creative exploitation and reimagining of energy and its uses. Throughout history, humankind has been able to utilize energy in ways other species have been unable to grasp. Ingenious use of energy has led to increased brain development, enabled us to travel great distances, allowed for the manufacturing of a variety of products at tremendous speed and helped to power the machines, which influence everything in life from healthcare to communication, and to science, research and development today. The rate at which energy has been exploited however left mankind in a compromising position. Finite resources are rapidly being depleted; carbon emissions and competing water demands and pollution continue to cause large scale environmental issues, increasing the attention to sustainability in all contexts. Renewable energy sources offer one answer to the problem and with increased implementation they may one day power the planet. This is the history of how mankind and society evolved alongside energy and how we have arrived at the current situation, full of challenges and opportunities. From the earliest uses of fire to the steam engine, to massive fossil fuel dependence, and to the beginning of the major shift toward renewable energy sources, humankind has continually innovated and adapted to their situation through desire or necessity. Advancements in energy production and usage improved social and economic aspects of life. Humankind has grown alongside energy and in a world where more and more is demanded, a continued growth and innovation will be required to sustain our societies and their ever-growing needs.

9.2.1 General Considerations

9.2.1.1 Energy Related Air Pollution

Air pollution refers to the pollution, which can harm human health and pollute the atmosphere. When pollutant concentrations in the air reach levels higher than the environmental quality standards, then we call the situation as air pollution. Air pollution can occur due to many different reasons and substances, which may be gaseous, solid or liquid. One of the main sources of air pollution is burning fossil fuels such as oil, gasoline, and coal. During the combustion process, pollutants such as CO_2, particulate matter (PM), SO_2 and NO_x are released. The evaporation or incomplete combustion of fuels releases Volatile Organic Compounds

DOI: 10.1201/9781003349297-9

(VOCs) as well. In addition, when unburned methane (CH_4) is released into the atmosphere, it contributes in the greenhouse effect due to its relatively high global warming potential. Burning fossil fuels also release heavy metals into the environment such as As, Cd, Cr, Ni, Mn, and Pb causing further air pollution. We, as the controlled environment systems energy consumers, need to be conscious about the sources of these pollutants and what they cause, so that we can make proper informed decisions with respect to our preferences in different energy sources.

9.2.1.2 Energy Related Water Pollution

Water resources play influential roles in the energy production process because water is needed in almost all forms of energy production during either extraction or processing. For instance, thermal power and hydropower are the two primary sectors of total electricity production that consume huge amounts of water, accounting for 80% and 15%, respectively (UN, 2014a). The IEA's World Energy Outlook 2012 published that 583 billion cubic meters of water was withdrawn for energy production in 2010 alone, which was up to 15% of global water withdrawals (IEA, 2012). Energy-Water nexus was first introduced in a Department of Energy (DOE) National Laboratory project in the USA in 2004 (Olsson, 2012). From that time, the link between energy and water resources has been more comprehensive and much more profound than initially thought.

The UN (2014a) reported the global water withdrawal per year by sectors in different regions. The total water consumption in different regions demonstrated that more than 60% of water was withdrawn in Asia, more than 20% was withdrawn in the Americas, and around 10% and 5% withdrawals were made in Europe and Africa, respectively. The report also revealed that more water is used for agriculture, especially in developing countries. In these countries, although a small fraction accounted for agricultural, over half of the total water withdrawals were consumed by the cooling system from power plants (Omer, 2008; Olsson, 2012; UN, 2014a). Taking the thermal power sector as an example, it was responsible for approximately 45% of freshwater withdrawals in Europe and 50% of total water withdrawal in the USA, which is much more than the water supply used in the agricultural sector (Kenny et al., 2009; Eurostat, 2010). From the global perspective, agriculture is the largest consumer, responsible for 70% water withdrawals in total for irrigation practice. According to the New Policies Scenario stated by the IEA, water withdrawals are expected to increase by 20% together with an 85% consumption improvement by 2035 (IEA, 2015). However, higher power plant efficiencies could then be achieved with advanced cooling technologies and increased biofuels production.

Energy production and consumption have extraordinary positive impacts on social life, but negative environmental and social consequences of energy production cannot be ignored, including the water issues. Water is an already scarce resource from many aspects, and the expansion of food and energy production is adding further pressures on water resources and exacerbating the water scarcity further (Olsson, 2012). Water shortage will become a serious problem in many areas or the world such as India, Africa, and the Middle East, and New Zealand (WEC, 2013; IEA, 2013; UN, 2015). The United Nations Department of Economic and Social Affairs reported that the available water resources are decreasing, and approximately two-thirds of the population in the world will suffer water scarcity by 2025 (UN, 2014b). Consequently, water scarcity will, in turn, have dramatic impacts on pertinent production sectors and result in regional conflicts.

The footprints of energy projects have significant effects on water quality. Obviously, better management can bring substantial benefits for energy savings and alleviate water pollution. Water requirements for primary energy and related pollution impacts are summarized by the USDOE (2006). According to the USDOE (2006), all forms of energy production pollute water in varying degrees, including clean energy sources. For instance, the extraction of fossil fuels results in a vast volume of produced water discharge. Coal mining activities further destroy the surrounding ecosystems, and produced water that contains toxic substances has critical risk to the aquatic system and human health (Olsson, 2012; UN, 2014a; UN, 2015). The construction of hydropower plants has a serious impact on the physical and chemical parameters of water and brings unexpected impacts on agriculture and nature. In addition, water used for cooling systems in power plants would be returned to the adjacent water bodies at higher temperatures, which may be detrimental to resident flora and fauna (Olsson, 2012; UN, 2014a). Energy can, on the other hand, be used for wastewater treatment and the waste thermal energy can be recovered. Solar energy, for instance, is widely used in wastewater treatment plants; and bioenergy production can as well be integrated with wastewater treatment systems (Argaw, 2003).

9.2.1.2.1 Oil and Water Pollution

Almost each and every step in the oil production process has significant impacts on water resources. For example, steam injection and hydraulic fracturing are the two common technologies used for extracting oil from reservoirs, which contaminate water (Olsson, 2012). Oil fields usually contain a large volume of water and gas, so water or steam are frequently injected into reservoirs to maintain the pressure. After that, more steam is injected and the pressure is utilized to push oil up through the well (US EPA, 2011). Extraction of oil requires a lot of water withdrawal; and meanwhile, it contaminates the water. The polluted water that comes from the oil extraction process is called produced water. The composition of the produced water is quite complex and includes heavy metals, hydrocarbons, hydrogen sulfide, organic acids and other chemicals added through the extraction and separation processes. The amount of produced water that is generated in oil production fields has been estimated about three times as much as the crude oil produced (Khatib and Verbeek, 2003). In the past few decades, most of the produced water is discharged directly into surrounding aquatic systems without any wastewater treatment, resulting in groundwater and even ocean contamination (Stephenson, 1992). Oil spills are the other direct environmental impacts that happen regularly during transportation of oil from place to place. Similarly, a large amount of oil may leak from underground storage tanks and flow into underground water resources. Oil production also causes indirect impacts on water quality. For instance, pollutants are released into the atmosphere during combustion of fuels. However, these particles can return to the water systems through precipitation. Furthermore, unconventional oil mining requires a massive amount of water and energy to refine and separate crude oil (Reynolds and Kiker, 2003). This clearly shows that the generation of produced water is huge, thereby causing water pollution.

9.2.1.2.2 Coal and Water Pollution

Water consumption is tightly linked with coal mining, like oil production, as it is used in extraction, processing, and transportation. Water quality issues frequently accompany mining activities (Palmer et al., 2010). A large volume of water is used to wash the coal to limit the undesirable chemical components such as sulfur, mercury, and others. Then, these elements accumulate in wastewater, causing water pollution as they are discharged or leak into aquatic systems (Mielke et al., 2010). A high load of these detrimental elements, especially heavy metals, persistently accumulate in the environment; and even worse, these heavy metals can eventually incorporate into food chain. Coal spills during transportation can, on the other hand, lead to indirect water contamination. In 2008, a coal ash spill caused terrible impacts on the Emory and Chinch River system in Harriman, Tennessee, USA (Deonarine et al., 2013). Approximately 4.1 million m^3 of fly and bottom coal ash was spilled, and they were responsible for high mercury concentrations in river sediments. The water quality was not significantly impacted by the presence of mercury; however, the total mercury concentration in sediment was 3 to 4 times higher than non-polluted areas (Deonarine et al., 2013). Furthermore, people may have to use coal contaminated water for daily requirements, such as drinking, cooking, and washing. Even worse, it could eventually force local residents to leave their homes and relocate.

9.2.1.2.3 Natural Gas and Water Pollution

A large amount of produced water is generated from the extraction process, usually containing heavy metals, hydrocarbon residues and other chemicals. Clark and Veil (2009) reported that the amount of produced water is about six times gas production. Fontenot et al. (2013) illustrated that water parameters including arsenic, selenium, strontium and TDS (total dissolved solids) exceeded the drinkable water quality standards in some wells near a natural gas extraction site. These wells were located within 3 km of the active natural gas extraction site. Lower concentrations have also been detected in the drinking water wells beyond 3 km radius from the gas extraction site. Moreover, high levels of methanol and ethanol were also observed in about one-third of all drinking water wells.

9.2.1.2.4 Bioenergy and Water Pollution

Agriculture is one of the main polluters of freshwater resources due to the use of fertilizers and pesticides that eventually end up in the aquatic system. Biofuels production expands the agricultural activities and increases the demand for the majority of traditional crops, resulting in water competition and pollution that typically causes what is called the eutrophication phenomenon, which is associated with algal blooms, dissolved oxygen depletion and degraded surface water bodies. Water quality impacts can be tracked throughout the whole bioenergy feedstock production chain. However, the main problem is attributed to the expansion of agricultural activities in order to meet the bioenergy demand. Fertilizer and pesticide usage and other agricultural practices such as tillage and irrigation have damaging effects on water quality.

9.2.1.2.5 Hydropower and Water Pollution

The WEC (2013) indicated that, even though the global installed hydropower capacity has increased by 8% since the last survey, the global electricity generation dropped by 14%, mainly caused by water shortages. Hydropower generation is a large water consumer. However, most of the water used is returned to the river downstream as the reservoir is filled or when the water runs through the turbines. An increased temperature in dams may lead to an increased temperature in the river, which negatively influences the water quality downstream. As water temperature and flow discharges are altered, the life cycle of aquatic animals are threatened further. For instance, migration, spawning and hatching phases of salmon was found to be changed as a result of hydropower operations (International Commission for the Protection of the Rhine, 2010). Furthermore, waterborne diseases and malaria have infected more than 200 million people after the construction of dams and reservoirs, and it is a common health problem in the tropical regions of the world (Olsson, 2012). Detrimental water quality situations can however be mitigated by well-organized and well-managed hydropower plants before implementation.

9.2.1.2.5 Nuclear Power and Water Pollution

The cooling systems in nuclear power plants consume a vast amount of water. The water used for cooling is typically discharged back to the adjacent aquatic system. The water temperature can be elevated by 8°C, depending on the operations and settings (Laws, 1993). The structure of the marine ecosystem and population distribution are tightly linked with water temperature as it affects the physiological processes of fish and others. Therefore, increased temperature in the water bodies can have adverse impacts on aquatic life. Although some resident organisms can tolerate the warm water up to certain extent, most of the organisms are sensitive to environmental changes. As a result, biodiversity and population dynamics eventually change due to thermal pollution in water (Teixeira et al., 2012).

The Chernobyl accident occurred on April 26, 1986 at the Chernobyl nuclear power plant in Ukraine. The bioaccumulation of radiocaesium can still be traced in the aquatic food chain. Some can even be found on lakes in Scandinavia and Germany. People who were living close to the lakes ingested the ^{137}Cs from fish. Besides, the major surface contamination around the accident region is dominated by the groundwater contaminations. A slow declining rate of ^{137}Cs in freshwater and fish has been reported, but it is expected to remain at a high level for decades. Nowadays, the surface contaminations from radioactivity have been dramatically reduced by good countermeasures and natural weather impacts (Alexakhin et al., 2006). Furthermore, on March 11, 2011, the devastating Tohoku earthquake and enormous tsunamis caused serious damages in Fukushima Daiichi Nuclear Power Plant in Japan, and tons of contaminated seawater was dumped back to the Pacific Ocean, which worsened the water pollution.

The international debate on the prospects of nuclear energy development has reignited after the Fukushima nuclear accident. Wittneben (2011) reported that each country has a different response and decision in Europe, mainly due to geographical and political reasons. For instance, the UK maintained its decision to increase nuclear share in the coming years, while the German government decided to shut down the commissioned nuclear reactors and make a comprehensive test to eliminate the potential safety hazard. In reality, the Fukushima accident reduced the public acceptability of nuclear power plants in many developing countries, while increasing the demands for fossil fuels, particularly natural gas (Wittneben, 2011; WEC, 2013).

9.2.1.2.6 Geothermal Energy and Water Pollution

Recent studies reported that geothermal power plants have adverse impacts on water bodies and soils. Geothermal plants require the withdrawal huge amounts of water for cooling and reinjection procedures (Macknick et al., 2011). Geothermal fluids contain various heavy metals and other toxic components such as Cd, Hg, As, Mn, F, and B, due to contact with hot fluids and rocks (Armienta et al., 2014). The geothermal fluid may contaminate the surface water as well if there are spills and if it is released into water ecosystems. They not only damage the aquatic ecosystems but also threaten water security for irrigation and drinking. Heat extraction and cooling by aquifers may, on the other hand, lead to temperature alteration in natural groundwater systems. Similar to the case in nuclear energy, the elevated temperature in water bodies might disturb physical and chemical compositions of water body, causing severe alterations in ecosystem functions. Actually, it is not necessary to use clean freshwater for cooling and reinjection purposes; hence, many geothermal plants recycle the geothermal fluids or wastewater for operating.

9.2.1.2.7 Other Energy Forms and Water Pollution

Tidal energy, for instance, is a renewable and clean ocean energy that is expected to be broadly exploited in the coming years (Wang et al., 2015). A three-dimensional coastal ocean model showed that the water quality impacts were strangely related to the amount of power extracted from the tidal power plant system. The effects on water quality were minor under a lower level tidal energy extraction, while there were adverse impacts, such as low dissolved oxygen content in bottom waters when a larger amount of energy was extracted (Wang et al., 2015). Additionally, antifouling coating in turbines may have potential impacts on water quality (Polagye et al., 2011). It seems that a long-term extraction of tidal energy might alter both physical and biological parameters of water.

9.2.2 Special Considerations

9.2.2.1 Indoor Air Pollution

See Section 4.3.4 and Section 4.4.4 for indoor air pollution in plant controlled environments, and animal controlled environments, respectively; and Section 4.5.3 for water pollutants and their effects on fish. Also, utilizing the correct ozone dosage and monitoring the loading together with proper ventilation in the aquatic building are critically important. See Section 5.14.2 for ozonation in aquatic systems.

9.2.2.2 Light Pollution

See Section 4.3.1.6 for light pollution in and around plant controlled environments.

9.2.2.3 Plastics

Every year, millions of tons of plastic are used around the world for covering greenhouse type structures, mulching greenhouse floors, and making all sorts of containers, silage bags, water pipes and hoses, and plug flats. The uses of plastic are extremely convenient due to its low cost and ease of handling; however, the eventual environmental consequences are not so convenient and friendly due to the vast quantity of plastics generated. Managing plastic waste is a global challenge as landfills and incineration option many times do not provide economically and/or environmentally feasible solutions. Greenhouse glazing and mulching films cannot be reused as they deteriorate under ultraviolet radiation; however, plastic containers can be reused, even though cleaning, sorting, and disease control might still be challenging. In order to be able to maintain the quality of the recycled end-product, greenhouse plastic waste ought to be sorted and handled separately. Contaminants and structural degradation of plastics, particularly soil and organic matter, reduce the recycled plastic quality dramatically. For effective recycling, a lot of waste plastic material must be collected from many greenhouse and other controlled environment operations. Furthermore, waste plastic film cannot be transported very effectively due to its low bulk density. In order to reduce transport costs, bailers are used to compact the plastic film on the facility.

Today, several kinds of plastic are used in controlled environment operations, which includes high-density polyethylene (HDPE), low-density polyethylene (LDPE), polypropylene (PP), and high-impact polystyrene (HIPS) (Table 9.1).

TABLE 9.1
Plastics Commonly Used in Controlled Environment Operations

Type of Plastic	Typical Uses	Reusability	Recyclability
High-Density Polyethylene (HDPE)	Nursery and other containers	Yes	Yes
Low-Density Polyethylene (LDPE)	Greenhouse glazing, plastic mulch, silage bags, cover sheets	No	Yes
Polypropylene (PP)	Pots, shade cloths, and row covers	Yes	Yes
High Impact Polystyrene (HIPS)	Carrier trays, plug and liner trays, and flats	Yes	Yes

Source: Greenhouse Engineering, Edn. 1 by İlhami Yıldız, Copyright © 2021 by Imprint. Reproduced by permission of Taylor & Francis Group.

High-Density Polyethylene (HDPE):

Due to its chemical and thermal resistance, HDPE is commonly used in making nursery containers, particularly for trees, shrubs, pesticides, herbicides, and oil. Since they are quite durable, they can conveniently be reused especially after sterilizing them. Plastic recycling companies recycle waste HDPE, which are typically used for making plastic timber for landscaping, picnic tables, etc. rather than being reformed into containers.

Low-Density Polyethylene (LDPE):

LDPE is extensively used for making greenhouse glazing, mulching material for greenhouse plants, silage bags, and cover sheets for bunker silos. Due to its ultraviolet degradation during its lifecycle, LDPE greenhouse glazings cannot usually be reused. Waste LDPE requires washing (particularly mulching plastic) and baling due to its low bulk density before it is recycled. And most plastic recycling companies do not have the specialized cleaning systems particularly for removing the trapped soil; however, some recycling companies provide the baling equipment for handling the mulching plastic and greenhouse type building glazings.

Polypropylene (PP):

PP is a durable, lightweight and resistant plastic, which is commonly used for making shade cloths, row covers, plant containers for greenhouse production of herbs, potted flowering plants and bedding plants, etc. They can be reused especially after sterilization. And some plastic recycling companies recycle polypropylene by mixing it with HDPE and make plastic timber for landscaping, picnic tables, etc.

High-Impact Polystyrene (HIPS):

HIPS is extremely lightweight plastic, difficult to compress, and used in making flats and plug trays for seedlings and small plants. HIPS plug and liner trays are mostly discarded or recycled, but they are sometimes reused. Plastic recycling companies typically grind, melt and mold HIPS directly back into trays.

Every year, when they are provided with options, greenhouse, livestock, poultry, and aquacultural operations use technologies including environmental technologies to make their operations fully sustainable and also turn to plastic recycling companies for managing their plastic waste rather than dumping the waste into landfills.

Moreover, tremendous amounts of produced greenhouse and aquacultural organic waste should also be considered for co-digesting with livestock and other organic municipal waste to produce both electricity and thermal energy locally using combined heat and power generators. The produced heat, power and CO_2 can be used directly in greenhouse operations and make the operations and our rural communities much more resilient and sustainable.

9.3 WASTE GENERATION AND MANAGEMENT

In this section, we will specifically be dealing with waste generated in controlled environment production facilities and their management other than plastics, which were covered in the above section in this chapter. Waste management and disposal issues are getting more important in all controlled environment production operations as environmental regulations become stricter every passing year depending on the location of operations. Controlled environment operations, particularly greenhouse, swine, poultry, and recirculating aquatic systems have a leading role in the future of intensive food production systems, due to their wise energy and limited water usage, and a relatively easy treated low volume, high concentrated waste streams.

9.3.1 Waste Management

Controlled environment operation wastes are managed employing exactly the same process used for any type of natural resource management system. MWPS (1987) states that the main objective in planning an agricultural waste management system is to help the producer achieve wise use of natural resources, considering the importance of the five resources, i.e., soil, water, air, plants, and animals. In addition to the resources, the social, cultural, and economic effects of alternative management systems on the human environment must be considered. As will be discussed further in depth below, in order make an effective planning, an engineer together with the producer has to analyze the resource data-the six functions: production, collection, storage, treatment, transfer, and utilization (NRCS, 1996). Just to introduce the topic, wastes are typically handled as either solids or liquids, so are the controlled environment operation wastes. Liquid wastes with up to 15% solids, for instance in livestock and poultry operations, are handled with slotted and other floor systems, flushing systems, gravity flow gutters, or scraper systems and blades. Solid wastes (wastes with 20% or more solids) are, on the other hand, handled as a solid with manual or mechanical scrapers or front-end loaders. So, this shows that not only the collection method but also the treatment, transfer, and disposal methods chosen are essentially dependent upon whether the waste is liquid or solid. Solids are usually transferred to and spread on fields with conventional spreaders, while liquids are usually spread on fields with tank wagons, applied with irrigating equipment, or digested in a lagoon before field spreading (MWPS, 1987).

General controlled environment waste management principles include keeping clean water out of the waste produced, i.e., keeping clean water clean, diverting dilute streams to low-cost vegetative treatment, and keeping concentrated manure streams isolated and contained. Overall speaking, controlled environment waste management systems must be developed using the total systems approach, which accounts for all the waste associated with a controlled environment production enterprise throughout the year from production to utilization. Interfacing with other systems, it is the management of all the waste, all the time, and all the way.

9.3.2 Waste Characteristics

As stated by NRCS (1996), different species and ages on different diets and water consumption in different climates produce wastes with different consistencies, i.e. different total solids (TS) concentrations; and the waste of different consistencies require different management techniques and handling equipment. Controlled environment operations may generate waste in the form of a liquid, slurry, semi-solid, or solid. Waste, such as manure, can change consistency throughout the system or throughout the year. The TS concentration can be increased by adding different organic solid material such as bedding, or decreased by adding water, and stabilized by protecting it from additional water. The bottomline is that the total solids concentration of the waste is the main characteristic that dictates how the material can be handled, as either solids or liquids.

Generally speaking, solid waste management systems have a reduced total volume of waste because of the reduction in the amount of water, a lower cost and power requirements, a greater labor requirement for operation and management compared to other methods. On the other hand, liquid waste management systems are easier to automate and require less daily attention than those for solid wastes. Additional water needed in liquid waste management increases the volume of waste requiring management, and the initial cost of the liquid handling equipment may be greater than that for solid waste systems.

Waste generation has already been covered up to certain extent in Chapter 4; in addition, as a quick means of reference, Chen et al. (1993) provides a comparison of the amount of waste generated by different groups of animal species.

Wastewater handling is a major problem in all controlled environment animal systems, but there are great differences between, let's say, manure from hog or dairy systems and aquaculture wastewater. The hog or dairy systems are typically in the range of 5 to 15% suspended solids, while fish wastewater can be anywhere from 0.2 to 4.0% suspended solids (Twarowska et al., 1997; Bergheim et al., 1998; Cripps and Bergheim, 2000).

9.3.3 Waste Management Functions

Here, we will introduce the six waste management functions, namely the production and collection functions first, and then move to the storage, treatment, transfer, and utilization functions specifically in the controlled environment settings.

9.3.3.1 Production

If the quantities of waste produced are sufficient enough to become a resource concern, then a waste management is required. A complete analysis of waste production includes the kind, consistency, volume, location, and timing of the waste produced. The production of unnecessary waste should always be kept to a minimum.

9.3.3.2 Collection

Collection refers to the initial capture and gathering of the waste from the generated point of origin or deposition to a collection point. We must identify the method of collection, location of the collection points, scheduling of the collection, labor requirements, necessary equipment or structural facilities, management and installation costs of the components, and the impact that collection has on the consistency of the waste (NRCS, 1996). Several collection methods are available for controlled environment production systems; and some systems use a combination of collection and storage, such as a built-up manure pack or slotted floors over a liquid pit (MWPS, 1987). However, gases from liquid wastes stored inside a building can be hazardous and create undesirable odors. Accumulating odors and gases are detrimental to animals and operators. Particularly, the most serious gas problems occur when manure is agitated or when ventilation fans fail. Most of the gas produced, approximately 99% or more, is methane, ammonia, hydrogen sulfide, and carbon dioxide.

Liquid wastes, for instance in livestock and poultry operations, are collected with slotted and other floor systems, flushing systems, gravity flow gutters, or scraper systems and blades. Solid wastes are, on the other hand, collected as a solid with manual or mechanical scrapers or front-end loaders. In intensive aquaculture operations, mechanical filters, such as microscreens, sand filters, and floating bead filters generate a separate solids waste effluent stream discharge. These are often combined with waters spilled or flushed from the recirculating system to form the discharge from the controlled environment facility. Drum filter suspended solids concentrations are typically somewhere around 0.5% or so (Ebeling and Summerfelt, 2002), and are captured via straining, settling, or a combination of both. These systems work reasonably effectively, but require periodic flushing of the filters in order to maintain removal efficiency between flushings. For marine recirculating systems, on the other hand, it is critical that most of the waste effluent stream be recycled back into the production system to conserve salts and reduce and/or eliminate the environmental impact of salt discharges (Timmons and Ebeling, 2010).

9.3.3.3 Storage

Storage is the temporary containment place of the waste. The storage facility of a waste management system is the tool that gives the manager control over the scheduling and timing of the system functions. With adequate storage the manager has the flexibility to schedule the land application of the waste when the spreading operations do not interfere with other necessary tasks, when weather and field conditions are suitable, and when the nutrients in the waste can best be used by the crop. While the storage period is determined by the utilization schedule (NRCS, 1996), it is recommended that the plan should have a storage capacity for approximately a period of 10 to 12 months storage (MWPS, 1987). The storage capacity should include storage of precipitation as well as a method for removing solids, and to keep clean water clean, at least a 30 m (100 feet) distance must be allowed between a water supply and the nearest part of a storage (MWPS, 1987).

Sludge storage structures include anaerobic lagoons, earthen ponds, and above and below ground tanks. Sludge thickening can, on the other hand, be accomplished by settling basins, belt filters, lime stabilization, wetlands, or sand beds. Both thickening basins and separate storage structures are utilized for storing sludge, including aquacultural sludge. In many cases, the thickened sludge from thickening basins is transferred to larger sludge storage structures that are able to hold months worth of captured and thickened solids. Earthen ponds are used for holding months worth of captured and thickened sludge that may also be stored in tank structures, above and below ground. There are many commercially available manufactured, modular tanks in reinforced concrete, metal, and wood (NRCS, 1996).

9.3.3.4 Treatment

Waste treatment is designed to reduce waste pollution potential and includes physical, biological, and chemical treatment and pretreatment activities such as separating solids. According to NRCS (1996), the plan should include an analysis of the characteristics of the waste before treatment, a determination of the desired characteristics of the waste following treatment, the selection of the type, estimated size, and location, the installation cost of the treatment facility, and the management cost of the treatment process.

9.3.3.4.1 Physical Treatment: Solids Thickening-Settling Basins

Thickening basins operate based on the particle settling principles discussed earlier in Chapter 5. Off-line settling basins are designed for solids collection, thickening, and storage. They are being successfully used for the treatment of slurries from livestock production, microscreen filter backwash water from recirculation systems and flow from system cleaning activities in aquacultural operations.

9.3.3.4.2 Biological Treatment

A biological treatment system is designed and operated for biodegradation-converting organic matter (feed, bedding, body byproducts) in animal manures to more stable end products. Liquid waste is treated in an anaerobic lagoon, aerobic lagoon, or other facilities. Solids can, on the other hand, be composted.

Anaerobic processes, such as the one takes in anaerobic lagoons, occur without free oxygen and liquefy or degrade high BOD (biochemical oxygen demand) wastes. In these systems, nondegradable wastes settle to the bottom as sludge, whose accumulation depends on management, environment, waste characteristics, and loading rate. Lagoons usually fill with sludge after about 10 to 15 years.

Aerobic processes, on the other hand, require free oxygen, and their odor-free operation can be a valuable advantage. In an aerobic treatment system, biological oxidation converts organic waste to carbon dioxide, water, and microbial cells, and oxygen must be supplied either naturally or artificially to maintain an aerobic system (MWPS, 1987).

Facultative microorganisms can function either anaerobically or aerobically, depending on their environment. Ideal conditions are not required because many varieties of microorganisms are adaptable. Anaerobic and facultative bacteria in animal excreta multiply rapidly. However, extreme environmental changes alter microbial activity. When microorganisms are stressed by the environment, waste treatment processes malfunction and odors can become a problem.

In addition, *composting* is another biological process to degrade organic matter (volatile solids) to a relatively stable humus-like material (MWPS, 1987). Composting can be either anaerobic or aerobic, but is usually aerobic. Composting converts the sludge and other organic waste products from controlled environment operations into a stable organic product by converting nitrogen from the unstable ammonia to a more stable organic form. The end result is a product that is safer to use than the original sludge and when applied to land, will improve the soil fertility, tilth, and water holding capacity. In addition, composting reduces the bulk of material that needs to be spread, improves its handling properties, reduces odor, fly and other vector problems and can destroy pathogens (NRCS, 1996).

NRCS (1996) reports that composting methods include *windrow*, *static pile*, and *in-vessel* methods. The *windrow method* consists of simply piling the compost mix into long, narrow piles or windrows. In order to maintain aerobic conditions, windrows need to be periodically turned and mixed, exposing the decomposing material to air and to keep the temperatures from getting too high. Moreover, the *static pile method* consists of mixing the compost materials and then stacking the mixture on plastic pipe or tubing, through which air can be forced or drawn. Relatively small compost piles may, on the other hand, not need forced ventilation as long as they are highly porous or with a mix that is stacked in layers with highly porous materials. Furthermore, the *in-vessel method* involves in the mixing of the materials in a reactor, building, container, or vessel where forced ventilation might be required. It takes place under a high-level moisture, aeration, and temperature control.

It is also reported that the C:N ratio of the compost mix should be maintained between 25 and 40 to 1. If the C:N ratio is low, a loss of nitrogen generally occurs through rapid decomposition and volatilization of the ammonia. If the C:N ratio is high, this time, the composting time required increases because the nitrogen becomes the limiting factor for microbial growth. Rapid composting of mortalities, on the other hand, occurs when the C:N ratio is maintained at a lower level somewhere between 10 and 20. This is considerably lower than what is normally recommended for more traditional composting, because much of the nitrogen in the dead animal mass is not exposed on the surface. A lower C:N ratio is necessary to ensure rapid composting with elevated temperature to destroy pathogens. The moisture content of the initial compost mixture should, on the other hand, be between 45 and 55%, by weight, to facilitate a rapid decomposition. Also, composting of mortalities should remain aerobic at all times throughout the process, which is easily accomplished by layering the mortalities and amendments in the compost mixture. Layers of high porosity material, such as bark, straw, and wood chips allow lateral movement of air in the compost mix and creates a desirable environment for rapid decomposition (NRCS, 1996).

9.3.3.4.3 Chemical Treatment: Solids Thickening-Coagulation/Flocculation

One of the most commonly used methods for the removal of suspended solids in drinking water is using coagulant and flocculation aids, such as alum, ferric chloride, and long chain polymers (AWWA, 1997). Coagulation/flocculation process is widely used in the drinking water treatment area as a means of attacking fine suspended (< 10 μm) and colloidal particles (< 1 μm). It is normally implemented as a four-step process: 1) coagulation, 2) flocculation (floc formation), 3) sedimentation (gross floc removal), and 4) sand filtration (fine solids removal or polishing).

Coagulation is the process of decreasing or neutralizing the electric charge on suspended particles. As discussed earlier in Chapter 4, similar electric charges on small particles in water cause the particles to naturally repel each other, and force the small, colloidal particles apart and keep them in suspension. This allows the van der Waals force of attraction to encourage initial aggregation of colloidal and fine suspended materials to form microfloc.

Flocculation, on the other hand, is the process of bringing the microfloc particles together to form large agglomerations by physically mixing or through the binding action of flocculants, such as long chain polymers. Polymers consist of simple monomers that are polymerized into high-molecular-weight substances (Metcalf and Eddy, Inc., 1991). With respect to charge, polymers can be cationic, anionic, or nonionic, and they act in two distinct ways: charge neutralization, and bridging between particles. Because wastewater particles are negatively charged, low molecular weight cationic polymers can act as a coagulant that neutralizes or reduces the negative charge on the particles, similar to the effect of alum or ferric chloride. Higher molecular weight polymers are, on the other hand, generally used to promote bridging flocculation. The long chain polymers attach at a relatively few sites on the particles, leaving long loops and tails that stretch out into the surrounding water. Flocculation is usually more effective if the molecular weight of the polymer is higher.

9.3.3.5 Transfer

Transfer function of the waste management means the movement and transportation of the waste throughout the system. It covers the transfer of the waste from the collection point to the storage facility, to the treatment facility, and finally to the utilization site. It also includes waste transfer as a solid, liquid, or slurry. The system plan should include: analysis of the consistency of the waste to be moved, method of transportation, distance between points, frequency and scheduling, necessary equipment, and installation and management costs of the transfer system (NRCS, 1996). In aquacultural systems, on the other hand, before the utilization method(s) discussed below can be used, the sludge and effluent wastewater must first need to be transferred to some form of storage system for both sludge thickening and flow equalization, and then to the utilization site. The transportation of waste from the point of generation to the application site can be a constraining function due the costs of sludge management, because aquacultural sludge is mostly water even after thickened (Timmons and Ebeling, 2010).

9.3.3.6 Utilization

Utilization function of waste management means recycling reusable waste products and reintroducing nonreusable waste products into the environment. Agricultural wastes may be used as a source of energy, bedding, animal feed, mulch, organic matter, or plant nutrients through spreading or irrigation. Once properly treated, they can be marketable products (NRCS, 1996). The most common method used to recycle solid wastes from controlled environment facilities is to use it as a fertilizer for direct land application. Many regions have regulations that govern land application of organic wastes to fertilize agricultural crops by limiting the land application rates and the amount of associated pathogens, heavy metals, and other contaminants. The land application rates are determined based on the nutrient content, soil type, and plant nutrient uptake characteristics to prevent runoff or groundwater contamination. Odor problems can also limit land application in developed, residential, or nearby urban areas (NRCS, 1996).

When the solids content is low, solids and slurries are easily pumped and distributed. Thus, there are a variety of land application systems that can be used, including conventional irrigation via sprinklers and surface flooding. As discussed earlier, crop demands of irrigation and nutrients, i.e., climate constraints on crops and vegetation may require winter storage of the wastewater. Thickened sludge (> 5% solids) can be used as a soil amendment or fertilizer and applied from tanker trucks, by surface spreading with or without incorporation into the soil, and by direct injection into the soil (MWPS, 1987; NRCS, 1996; Timmons and Ebeling, 2010)

NOMENCLATURE

BOD biochemical oxygen demand
HIPS high-impact polystyrene
HDPE high-density polyethylene
LDPE low-density polyethylene
PP polypropylene
TKN total Kjeldahl nitrogen
TS total suspended solids

Subscripts

5 5 day

BIBLIOGRAPHY

Albright, LD, Novak, S and Wolfe, DW (1986) A preliminary evaluation of the effect of row cover material properties on the thermal environment under the cover. Project Report. Dept. of Agric. Engng., Cornell University, Ithaca, NY, USA, p. 114.

Albright, LD, Reines, RG, Anderson, SE, Chandra, P, Price, DR, Langhans, RW and Cerilli, RV (1978) Experimental results of solar heating a Brace Institute style greenhouse. Proceedings, Third Annual Conference on Solar Energy for Heating Greenhouses and Greenhouse-Residence Combinations. Fort Collins, CO.

Alexakhin, R, Anspaugh, L and Balonov, M (2006) Environmental consequences of the Chernobyl accident and their remediation: Twenty years of experience. Report of the Chernobyl Forum Expert Group "Environment", International Atomic Energy Agency.

Argaw, N (2003) *Renewable Energy in Water and Wastewater Treatment Applications*. National Renewable Energy Laboratory.

Armienta, MA, Rodriguez, R and Ceniceros, N (2014) Groundwater quality and geothermal energy. The case of Cerro Prieto Geothermal Field, México, *Renewable Energy*, 63: 236–254.

ASABE Standards (2018) *Standards, Engineering Practices, and Data*. V21(1). St. Joseph, MI.

AWWA (1997) *Water Treatment Plant Design*. American Water Works Association, American Society of Civil Engineers. 3rd Edition. McGraw-Hill, New York.

Badger, PC and Poole, HA (1979) *Conserving Energy in Greenhouses*. OARDC Special Cir. 102, Extension Bul. 651, p. 32. Ohio Agricultural Research and Development Center, Wooster, OH, USA.

Bergheim, A, Cripps, SJ and Liltved, H (1998) A system for the treatment of sludge from land-based fish-farms. *Aquatic Living Resources*, 11: 279–287.

Bryla, DR and Strik, BC (2007) Effects of cultivar and plant spacing on the seasonal water requirements of highbush blueberry. *Journal of the American Society for Horticultural Science*, 132(2): 270–277.

Buchholz, M, Jochum, P and Zaragoza, G (2004) Basic Water, Heat and Food Supply from a Closed Greenhouse—The Watergy Project, International Symposium GREENSYS 2004—Sustainable Greenhouse Systems, Leuven, Belgium, 12–16 September, published in International Society of Horticultural Science (ISHS) und Wageningen University and Research Center, Wageningen. ISBN: 907601924X, S. 133.

Cameron, A (2009) Horticultural plastics recycling—The future is brighter. *Greenhouse Product News Magazine*, June. Sparta, MI, USA, pp. 20–24.

Chen, S, Coffin, DE, and Malone, RF (1993) Production, characteristics, and modeling of aquaculture sludge from a recirculating aquaculture system using a granular media biofilter. In: Wang, JK (ed.) *Techniques for Modern Aquaculture*. American Society of Agricultural Engineers, St. Joseph, MI. pp. 16–25.

Clark, CE and Veil, JA (2009) *Produced Water Volumes and Management Practices in the United States*. Argonne National Laboratory ANL/EVS/R-09/1.

Claypool, D, Delaney, R, Ditterline, R and Lockerman, R (1997) Genetic improvement of alfalfa to conserve water. Poster Presentation, Wyoming Water 1997 Conference, April 21–23, Casper, WY.

Cripps, SJ and Bergheim, A (2000) Solids management and removal for intensive land-based aquaculture production systems. *Aquacultural Engineering*, 22: 22–56.

Deonarine, A, Bartov, G, Johnson, T, Ruhl, L, Vengosh, A and Hsu-Kim, H (2013) Environmental impacts of the Tennessee Valley Authority Kingston coal ash spill. 2. Effect of coal ash on methylmercury in historically contaminated river sediments, *Environmental Science and Technology*, 47(4): 2100–2108.

Ebeling, JM and Summerfelt, ST (2002) Performance Evaluation of a Full-Scale Intensive Recirculating Aquaculture Systems Waste Discharge Treatment System. In: Rakestraw TT, Douglas LS and Flick GF (eds.), *Proceedings of the 4 International Conference on Recirculating Aquaculture*. Northeast Regional Agricultural Engineering Service, Ithaca, NY, pp. 506–515.

Ebru, KA and Arif, H (2007) A comparative study on exergetic assessment of two ground-source (geothermal) heat pump systems for residential applications. *Building and Environment*, 42: 2004–2013.

Eurostat (2010) Eurostat database. Brussels, European Commission.

Fontenot, B, Hunt, L, Hildenbrand, Z, Carlton, D and Walton, HO (2013) An evaluation of water quality in private drinking water wells near natural gas extraction sites in the Barnett Shale formation. *Environmental Science and Technology*, 47(17): 10032–10040.

Garzoli, KV and Blackwell, J (1987) An analysis of the nocturnal heat loss from a double skin plastic greenhouse. *Journal of Agricultural Engineering Research*, 36: 75–85.

Hamada, Y, Nakamura, M, Saitoh, H, Kubota, H and Ochifuji, K (2007) Improved underground heat exchanger by using no-dig method for space heating and cooling. *Renewable Energy*, 32: 480–495.

Hardin, C, Mehlitz, T, Yildiz, I and Kelly SF (2008) Simulated performance of a renewable energy technology—Heat pump systems in semi-arid central California greenhouses. *Acta Horticulturae*, 797: 347–352.

IEA (2012) *International Energy Agency, World Energy Outlook 2012*. OECD, Paris.

IEA (2013) *International Energy Agency, World Energy Outlook 2013*. OECD, Paris.

IEA (2015) *International Energy Agency, World Energy Outlook 2015*. OECD, Paris.

International Commission for the Protection of the Rhine (2010) Master Plan Migratory Fish Rhine. Report No. 179. Koblenz, Germany.

Karlsson, F and Fahle´n, P (2007) Capacity-controlled ground source heat pumps in hydronic heating systems. *International Journal of Refrigeration*, 30: 221–229.

Kenny, JF, Barber, NL and Hutson, SS (2009) *Estimated Use of Water in the United States in 2005, USGS Circular 1344*. United States Geological Survey, Reston, VA.

Khademi, F and Yildiz, I (2018) 1.25 energy and solid wastes. In: Dincer, I (ed.), *Comprehensive Energy Systems*. Vol. 1, Elsevier, Oxford, pp. 980–1020.

Khatib, Z and Verbeek, P (2003) Water to value-produced water management for sustainable field development of mature and green fields. *Journal of Petroleum Technology*, 55: 26–28.

Kuack, DL (1986) Shading techniques utilize the unusual. *Greenhouse Grower* (May), pp. 29–30.

Laws, EA (1993) *Aquatic Pollution-An Introductory Text*. Wiley, New York.

Liu, Y and Yildiz, I (2018) 1.24 energy and water pollution. In: Dincer, I (ed.), *Comprehensive Energy Systems*. Vol. 1, Elsevier, Oxford, pp. 950–979.

Macknick, J, Newmark, R, Heath, G and Hallett, KC (2011) A review of operational water consumption and withdrawal factors for electricity generating technologies. *Golden, CO, National Renewable Energy Laboratory*, 303: 275–300.

Mehlitz, T, Yildiz, I, Hardin, C and Rahman, S (2008) Simulated effects of reflective mulch on energy and water conservation in semi-arid central California greenhouses. *Acta Horticulturae*, 797: 353–359.

Metcalf and Eddy Inc (1991) *Wastewater Engineering: Treatment, Disposal, Reuse*. 3rd Edition. McGraw Hill, New York, p. 929.

Mielke, E, Anadon, LD and Narayanamurti, V (2010) Water consumption of energy resource extraction, processing, and conversion, Belfer Center for Science and International Affairs.

Moya, A, Mehlitz, T, Yildiz, I, Kelly, SF and Hardin, C (2008) Simulated effects of dynamic row spacing on energy and water conservation in semi-arid central California greenhouses. *Acta Horticulturae*, 797: 307–313.

MWPS (1987) *Structures and Environment Handbook*. 11th Edition. Midwest Plan Service, Iowa State university, Ames, IA, USA. ISBN: 0-89373-057-2.

NRCS (1996) *Agricultural Waste Management Field Handbook*. USDA.

Olsson, G (2012) *Water and Energy: Threats and Opportunities.* IWA Publishing, London.

Omer, A (2008) Energy, environment and sustainable development. *Renewable and Sustainable Energy Reviews*, 12(9): 2265–2300.

Palmer, MA, Bernhardt, ES and Schlesinger, WH (2010) Mountaintop mining consequences. *Science*, 327(5962): 48–149.

Papadopoulos, AP and Pararajasingham, S (1997) The influence of plant spacing on light interception and use in greenhouse tomato (*Lycopersicon esculentum* Mill.): A review. *Scientia Horticulturae*, 69: 1–29.

Polagye, B, Van Cleve, B, Copping, A and Kirkendall, K (2011) Environmental Effects of Tidal Energy Development: Proceedings of a Scientific Workshop, March 22–25, 2010. U.S. National Oceanographic and Atmospheric Administration.

Rebuck, SM, Aldrich, RA and White, JW (1977) Internal curtains for energy conservation in greenhouses. *Transactions of the ASAE*, 20(4): 732–734.

Reynolds, RR and Kiker, RD (2003) Produced Water and Associated Issues, Oklahoma Geological Survey.

Roberts, BW and Anderson, JA (1994) Canopy shade and soil mulch affect yield and solar injury of bell pepper. *HortScience*, 29: 258–260.

Roberts, WJ and Mears, DR (1969) Double covering a film greenhouse using air to separate film layers. *Transactions of the ASAE*, 12(1): 32–33, 38.

Schalk, JM and Robbins, ML (1987) Reflective mulches influences plant survival, production and insect control on fall tomatoes. *HortScience*, 22: 30–32.

Short, TH, Badger, PC and Roller, WL (1978) A solar pond-polystyrene bead system for heating and insulating greenhouses. Proc. of International Symposium on Potential Productivity in Protected Cultivation, Kyoto and Tokyo, Japan, August, 1978.

Short, TH and Bauerle, WL (1977) A double plastic heat conservation system for glass greenhouses. Proceedings of an International Symposium on Controlled Environment Agriculture, April 7–8, Tucson, AZ.

Short, TH and Bauerle, WL (1980) Energy conservation techniques for commercial glass greenhouses. ASAE Paper No: 80-4532. ASAE, St. Joseph, MI 49085, USA.

Simpkins, JC, Mears, DR and Roberts, WJ (1976) Reducing heat losses in polyethylene covered greenhouses. *Transactions of the ASAE*, 19(4): 714–719

Stephenson, MT (1992) A survey of produced water studies. In: *Produced Water*, Springer US, pp. 1–11.

Teixeira, T, Neves, P and Araújo, L (2012) Thermal impact of a nuclear power plant in a coastal area in Southeastern Brazil: Effects of heating and physical structure on benthic cover and fish communities. *Hydrobiologia*, 684(1): 161–175.

Timmons, MB and Ebeling, JM (2010) *Recirculating Aquaculture.* 2nd Edition. NRAC Publication No. 401-2010. ISBN 978-0-9712646-2-5.

Twarowska, JG, Westerman, PW and Losordo, TM (1997) Water treatment and waste characterization evaluation of an intensive recirculating fish production system. *Aquacultural Engineering*, 16: 133–147.

UN (2014a) World water assessment program. The United Nations World Water Development Report 2014: Water and Energy, Paris.

UN (2014b) Department of Economic and Social Affairs, World Urbanization Prospects: The 2014 Revision, Highlights (ST/ESA/SER.A/352), New York.

UN (2015) World Water Assessment Program: The United Nation World Water Development Report: Water for a Sustainable World, UNESCO, Paris.

The University of Maine Cooperative Extension (2020) Ag Plastic Recycling in Maine: Current Challenges and New Opportunities. Available on: www.extension.umaine.edu/agriculture/greenhouse-plastic-recycling (Retrieved on 10.07.2020).

USAI Lighting (2020) Available on: https://www.usailighting.com/ (Retrieved on 07.07.2020).

USDOE (2006) *Energy Demand on Water ReSource: Report to Congress on the Interdependencies of Energy and Water.* Washington, DC.

US EPA (2011) *Draft Plan to Study the Potential Impacts of Hydraulic Fracturing on Drinking Water Resources.* Office of Research and Development, Washington, DC.

Wang, T, Yang, Z and Copping, A (2015) A modeling study of the potential water quality impacts from in-stream tidal energy extraction. *Estuaries and Coasts*, 38(1): 173–186.

Wang, Z (2018) 1.23 energy and air pollution. In: Dincer, I (ed.), *Comprehensive Energy Systems.* Vol. 1, Elsevier, Oxford, pp. 909–949.

WEC (2013) World Energy Council, Survey of World Energy Resources, London.

Willits, DH and Gurjer, YR (2004) Heat pumps for the heating and night-cooling of greenhouse crops: A simulation study. *Transactions of the ASAE*, 47(2): 575–584.

Wittneben, B (2011) The impact of the Fukushima nuclear accident on European energy policy. Environmental Science and Policy.

Yildiz, I (2021) *Greenhouse Engineering: Integrated Energy Management.* 1st Edition. CRC Press, Taylor and Francis Group, FL. ISBN: 978-1-4987-4313-6.

Yildiz, I and Stombaugh, DP (2006a) Dynamic modeling of microclimate and environmental control strategies in a greenhouse coupled with a heat pump system. *Acta Horticulturae*, 718: 331–340.

Yildiz, I and Stombaugh, DP (2006b) Heat pump cooling and greenhouse microclimates in open and confined greenhouse systems. *Acta Horticulturae*, 719: 255–262.

Yildiz, I and Stombaugh, DP (2009) Simulated effects of various environmental management practices on energy consumption in open and confined greenhouse systems. *Acta Horticulturae*, 801: 1317–1324.

Yildiz, I, Yue, J and Lowrey, J (2014a) Spatiotemporal distributions of energy and water consumption in open, semi-closed and closed loop heat pump greenhouse systems, and associated management strategies in Canada. *Acta Horticulturae*, 1037(1): 119–128.

Yildiz, I, Yue, J and Lowrey, J (2014b) Spatiotemporal distributions of energy and water consumption in open, semi-closed and closed loop heat pump greenhouse systems, and associated management strategies in the Canadian North. *Acta Horticulturae*, 1037(1): 171–177.

Zalom, FG (1981) Effects of aluminum mulch on fecundity of apterous *Myzus persicae* on heat lettuce in a field planting. *Entomologia Experimentalis and Appllicata*, 30: 227–230.

Index

M

N

O

P

R

S

T

U

V

W

Z